Physics

A marvelous blend of readability and fact, of insight and history, of graphs and pictures, of physics-physics and of physics in the world around. A tremendous piece of work, well done.

John Archibald Wheeler
Princeton University

Physics

With Modern Physics
For Scientists and Engineers

Third Edition

Volume 2

Richard Wolfson

Middlebury College

Jay M. Pasachoff

Williams College

 ADDISON-WESLEY

An imprint of Addison Wesley Longman, Inc.

Reading, Massachusetts · Menlo Park, California · New York · Harlow, England
Don Mills, Ontario · Sydney · Mexico City · Madrid · Amsterdam

On the cover: Optical fibers transmit information in the form of light.

The fiber optics story—told in more detail on page 926—is one of growth of an enabling technology connecting the peoples of the world in a global information system. Fiber optics form the backbone of the global Internet, intercontinental telephone, and, increasingly, local communications. Fiber-optic links are the superhighways that carry the deluge of data, voice, and images that are the currency of the information revolution.

Fiber-optic technology calls on many areas of physics. A solid understanding of electromagnetism and optics has been critical in the field of optic development. Materials science challenges include the fabrication of glass fiber so clear that a kilometer-long piece is as transparent as a window pane. Electrical engineering and solid-state physics are essential in the opto-electronic devices—miniature lasers and high-speed, light-sensitive diodes—that translate fiber-optic signals to and from the world of conventional electronics.

Acquisitions Editor	Sami Iwata
Publisher	Robin J. Heyden
Market Developer	Andy Fisher
Director of Marketing	Stacy Treco
Channel Marketing Manager	Gay Meixel
Publishing Associate	Bridget Biscotti Bradley
Developmental Editor	Catherine Flack
Production Editor	Joan Marsh
Manufacturing Supervisor	Holly Bisso
Photo Researcher	Kathleen Cameron
Design and Production Coordination	Elm Street Publishing Services, Inc.
Illustrators	Precision Graphics
Cover Photograph	© Image Bank
Composition and Film	Interactive Composition Corporation
Cover Printer	Lehigh Press
Printer and Binder	World Color Corporation

Library of Congress Cataloging-in-Publication Data

Wolfson, Richard.
    Physics with modern physics for scientists and engineers / Richard Wolfson,
  Jay M. Pasachoff. — 3rd ed.
       p.   cm.
    Includes index.
    ISBN 0-0321-3574-7 (vol. 2).
    1. Physics.   I. Pasachoff, Jay M.   II. Title.
QC21.2.W655   1999   98-45595
530—dc21   CIP
1 2 3 4 5 6 7 8 9 10 WCV 02 01 00 99 98

Brief Contents

Part 5

Optics 915

Part 6

Modern Physics 1007

Appendixes

Contents

Part 2

Oscillations, Waves, and Fluids 361

Appendixes

Preface

PHYSICS: CHALLENGE AND SIMPLICITY

Physics is fundamental. To understand physics is to understand how the world works, both at the everyday level and on scales of time and space so small and large that they defy intuition.

To the student, physics can be at once fascinating, challenging, subtle, and yet simple. Physics fascinates with its applications throughout science and engineering and in its revelation of unexpected phenomena like superconductivity, black holes, and chaos. It challenges with the need for precise thinking and skillful application of mathematics. Physics can be subtle, especially in describing phenomena at odds with everyday intuition. Most importantly, physics is simple. Its few fundamental laws are succinct, yet they encompass a universe of natural phenomena and technological devices.

GOALS OF THIS BOOK

This text is for science and engineering students. Its main goals are:

- to help students develop an understanding of the physical universe at its most fundamental level
- to explore a wide range of physics applications in science, technology, and everyday life
- to develop students' analytical and quantitative skills as applied to problems in science and engineering

The standard version of this text covers the full sequence of calculus-based university physics, and the extended version adds seven chapters on modern physics.

ORGANIZATION

The book is organized into six parts. Part 1 (Chapters 2 through 14) develops the basics of mechanics. Part 2 (Chapters 15 through 18) applies mechanics to the study of oscillations, waves, and fluids. Part 3 (Chapters 19 through 22) covers thermodynamics. Part 4 (Chapters 23–34) deals with electricity and magnetism. Part 5 (Chapters 35 through 37) treats optics. Part 6 (Chapters 38 and 39) briefly introduces relativity and quantum physics. Part 6 of the extended version (Chapters 38 through 45) expands the coverage of modern physics and its applications to atoms, molecules, and the solid state; nuclear physics and its applications; and particle physics and cosmology. Each part ends with a set of Cumulative Problems that synthesize concepts from the chapters in that part.

Distinguishing Features of the Third Edition

This third edition continues to hone the text and illustrations for clarity and precision, and to ensure that physics is tied closely with the latest in applications that are scientifically exciting and relevant to the student's everyday experience or future professional activities. Below we detail the many changes we've made, as well as describing new and continuing features designed to help students learn.

Improvements in Content This revision includes many substantial changes and improvements, most of which were suggested by instructors who have used the text or reviewed the manuscript. We've also added new pedagogical features to help develop both conceptual and quantitative understanding, and to test students' understanding of subtle concepts. Here are some of the more significant changes we've implemented in this edition.

- We extensively rewrote Chapters 5 (Force and Motion) and 6 (Using Newton's Laws) to account for the results of new research in physics education, especially the Force Concept Inventory.
- We substantially reorganized a number of other chapters to enhance the clarity and logical structure. For example, readers will find the presentations of magnetism and of physical optics much clearer due to chapter reorganization in this new edition. In all chapters, significant rewriting and art revisions improve clarity and provide tie-ins with preceding material.
- We revised the problem sets to include more "real-world" problems and a better balance of easy, multistep, and challenging problems. In each chapter, up to 20 percent of the problems are new.
- A new category of "annotated figures" links text and art, using captions within the figure to guide students through complex concepts.
- A new feature, **Got It!** boxes, provide conceptual and largely non-quantitative reinforcement of students' understanding.
- The new **Math Toolbox** feature presents important mathematical tools separately from the main text.
- We have integrated *ActivPhysics* — the comprehensive simulation and visualization software — throughout the book, to provide a valuable option for both instructors and students. *ActivPhysics* helps students explore the major concepts from all parts of physics. Students can make predictions and test assumptions by changing variables and running the simulations. *ActivPhysics* references in the text, and icons in the margin, are strategically placed at common student trouble spots. Used in the classroom, lab, or for self-study, *ActivPhysics* will help the student develop imagery and intuition, and will reinforce the student's understanding of basic physics.
- We've updated for new scientific discoveries and technological developments, including the finding that neutrinos have mass, advances in superconductivity, and fiber optic technology, and many others.

Other features carry over from the second edition with significant revisions and improvements. These include

Applications and Examples A rich array of practical and up-to-date applications illustrates the use of physics principles in both technology and in natural phenomena. Applications range from the life sciences through the frontiers of technology, and include biomedical technology, antilock brakes, global warming, microelectronics, lasers, and much more. Some new applications and examples in this third edition include hybrid cars with flywheel energy storage; the digital video disc; kayaking; electrical models for the cell membrane; gel electrophoresis for protein separation; capacitive transducers; medical defibrillators, and many others. As in previous editions, we integrate many applications with the text and problem sets, rather than presenting just a few applications as guest essays.

Questions These follow the chapter synopsis and can be used for class discussion, in peer-based learning strategies, or to get students thinking about concepts before they begin work on quantitative problems.

Problems Science and engineering students learn physics best by working physics problems. This third edition contains nearly 3,000 end-of-chapter problems, with significant revision of the problem sets, especially in the early chapters. Problems range from simple confidence builders, to more complex and realistic multi-problems, to difficult **Supplementary Problems** that will challenge the best students. A section of unique **Paired Problems** for each chapter lets students practice problem-solving techniques on a pair of problems whose solutions involve closely related physical concepts or mathematical approaches. **Cumulative Problems** integrate material from the chapters of each major part of the text. Frequent text references to specific problems link text and problems in the common purpose of enhancing the student's understanding of physics.

Worked Examples These reinforce basic concepts, illustrate problem-solving strategies, and often serve to introduce "real-world" applications of physics. Most examples are followed by an exercise that further reinforces concepts and strategies. A **Similar Problems** line after each example points out related end-of-chapter problems.

Tip Boxes Tip boxes appear throughout the text and warn against common pitfalls or give helpful problem-solving hints.

Pedagogical Use of Color The book's design is an essential pedagogical feature. The carefully planned use of color in figures and in highlighting important equations and definitions, as well as the selection of photographs, all reinforce the text and enhance communication of concepts to the student. Specific colors are used consistently for different physical elements in the figures; for example, velocity vectors and graphs of velocity use the same color, while force vectors and graphs consistently use a different color. A list of elements and their colors follows this preface.

Chapter Synopses Chapter summaries emphasize key concepts and remind students of new terms and mathematical symbols. Each summary ends with a reminder of limitations and approximations used in the theories, models, and equations developed in the chapter.

Appendices and Endpapers The book's appendices and endpapers contain a mathematical review and a wealth of up-to-date physical data, conversion factors, and information on measurements systems.

PHYSICS AND MATHEMATICS

For many students, the university physics course is their first contact with practical applications of calculus. We recognize that students have a range of mathematical abilities, from those taking their first calculus course concurrently with physics to those fluent in both differential and integral calculus. The former will find our new Math Toolboxes, along with many of the Tip boxes and figures, helpful in building confidence in their mathematical skills; the latter will find a selection of challenging calculus-based problems.

In this book mathematics is a tool, not an end in itself. Therefore we always relate mathematical solutions and derivations back to the underlying physics. After deriving an equation or solving a sample problem, we frequently ask: "Does this result make sense?" We show that it does by examining easily understood special cases, thus building physical intuition along with mathematical confidence. We explore the meaning of equations verbally and through figures, ensuring that concepts are clear before students begin to use the material quantitatively.

SUPPLEMENTS

Professors and students alike find it useful to have supplements designed to complement their text. For the third edition, we offer a new expanded supplements package.

For the Instructor

Instructor's Solutions Manual (0-321-03578-X), prepared by Edward S. Ginsberg, University of Massachusetts-Boston, includes worked solutions to all problems in the text. The solutions manual is now available on CD-ROM in TEX, techexplorer, and Microsoft Word, for easy editing and posting of homework solutions in documents or on the web (0-321-05149-1).

TestGen-EQ Computerized Test Bank is an expanded testbank that allows instructors to edit and change the order of questions, add entirely new questions in six different formats and print different versions of a test. The testbank is available for Windows (0-321-03579-8) and Macintosh (0-321-03580-1). A print version is also available (0-321-03590-9).

Transparency Acetates (0-321-03577-1) contain over 150 color acetates of figures from the text, useful in lectures and classroom discussions.

For the Student

Student Solutions Manual (0-321-03575-5), by Edward S. Ginsberg, University of Massachusetts-Boston, includes detailed solutions to all the odd-numbered problems in the text.

Study Guide with ActivPhysics, is a book and CD package. The study guide, by Jeffrey J. Braun, University of Evansville, provides an overview of each chapter of *Physics for Scientists and Engineers* and summarizes definitions and key equations. The *ActivPhysics* worksheets and CD-ROM by Alan Van Heuvelen, The Ohio State University, and Paul D'Alessandris help students develop imagery and intuition by changing variables and running the simulations. *ActivPhysics* prompts students with questions that help develop a more thorough understanding, step by step. The *ActivPhysics* CD-ROM is dual platform, for both Macintosh and Windows. *Study Guide with ActivPhysics* is available in two volumes.

- *Volume 1 (covering Wolfson/Pasachoff—Physics for Scientists and Engineers Chapters 1–22)* 0-321-05148-3
- *Volume 2 (covering Wolfson/Pasachoff—Physics for Scientists and Engineers Chapters 23–45)* 0-321-05147-5.

Preparing for General Physics: Math Skill Drills and Other Useful Help, by Arnold D. Pickar of Portland State University, is a self-study workbook that reviews algebra and trigonometry topics for university physics (0-201-53802-4).

ACKNOWLEDGMENTS

A project of this magnitude is not the work of its author alone. Here we acknowledge the many people whose contributions made the book possible. Colleagues using the second edition at universities and colleges throughout the world have volunteered suggestions, and will find most of those incorporated here. Students in Middlebury's Physics 109–110 courses have made significant contributions to the book's accuracy and readability. RW's Middlebury colleagues Ann Broughton, Cris Butler, Jeff Dunham, Bob Prigo, Steve Ratcliff, Susan Watson, and Frank Winkler all provided substantial suggestions or help on this project. We are especially grateful to a group of "on call" reviewers who provided instant feedback to questions from author and editors during the revision: Jerry Hosken, *City College of San Francisco;* Andy Odel, *Northern Arizona University;* James Rohlf, *Boston University;* Gurbax Singh, *University of Maryland;* Robert Weidman, *Michigan Technological University.*

It is frustrating to find numerical errors in a textbook, especially in answers to problems. We have gone to great lengths to make this book as free from error as possible, and we credit the people who helped achieved this goal. Edw. Ginsberg, *University of Massachusetts-Boston,* meticulously checked all numerical results in examples, exercises, and end-of-chapter problems. Mark Coffey of the *University of Colorado at Boulder,* and John Bartelt of the *Stanford Linear Accelerator Center* provided valuable accuracy checks.

Prior to this revision, every chapter of the book was thoroughly reviewed by users of both this and competing texts. Reviewers made general as well as more detailed suggestions, nearly all of which we have been able to incorporate in the third edition. We are grateful to these reviewers, and to reviewers of the second edition:

Edward Adelson, *Ohio State University*
Vijendra Agarwal, *Moorhead State University*
William Anderson, *Phoenix College*
Gordon J. Aubrecht, *Ohio State University—Marion*
Paul Avery, *University of Florida*
Douglas Bennett, *University of California, Los Angeles*
Marvin Blecher, *Virginia Polytechnic Institute and State University*
Gary Bowman, *Northern Arizona University*
John Brient, *University of Texas at El Paso*
James H. Burgess, *Washington University*
Bernard Chasan, *Boston University*
Roger Clapp, *University of South Florida*
Claire D. Dewberry, *Florida Community College at Jacksonville*
Jeffrey Dunne, *Purdue University*
Robert J. Endorf, *University of Cincinnati*
Heidi Fearn, *California State University—Fullerton*
Shechao Feng, *University of California, Los Angeles*
Albert L. Ford, *Texas A & M University*
Ian Gatland, *Georgia Institute of Technology*
Edw. Ginsberg, *University of Massachusetts—Boston*
James Goff, *Pima Community College*
Alan I. Goldman, *Iowa State University*
Philip Goode, *New Jersey Institute of Technology*
Denise S. Graves, *Clark Atlanta University*
Donald Greenberg, *University of Alaska*
Phillip Gutierrez, *University of Oklahoma*
Stephen Hanzely, *Youngstown State University*
Kenneth Hardy, *Florida International University*
Gerald Harmon, *University of Maine at Orono*
Randy Harris, *University of California, Davis*
Warren Hein, *American Association of Physics Teachers*
Roger Herman, *The Pennsylvania State University*
P. G. Hjorth, *Technical University of Denmark*
Francis L. Howell, *University of North Dakota*
J. N. Huffaker, *University of Oklahoma*
Quinton Hurst, *Arizona State University*
Javed Iqbal, *University of British Columbia*
Wayne James, *University of South Dakota*
Karen Johnston, *North Carolina State University*
Evan Jones, *Sierra College*
Randy Knight, *California Polytechnic State University*
Dean Langely, *St. John's University*
Chew-Lean Lee, *Florida Community College*
Brian Logan, *University of Ottawa*

Peter Loly, *University of Manitoba*
Hilliard K. Macomber, *University of Northern Iowa*
David Markowitz, *University of Connecticut*
Daniel Marlow, *Princeton University*
Nolan Massey, *University of Texas at Arlington*
Ralph McGrew, *Broome Community College*
Victor Michalk, *Southwest Texas State University*
Allen Mincer, *New York University*
P. James Moser, *Bloomsburg University*
Vinod Nangia, *University of Minnesota*
Andrew Odell, *Northern Arizona University*
Robert Osborne, *Yakima Valley Community College*
Michael O'Shea, *Kansas State University*
George Parker, *North Carolina State University*
Graham Pearson, *Novia Scotia Agricultural College*
R. J. Peterson, *University of Colorado*
Slawomir Piatek, *New Jersey Institute of Technology*
Gerald Pogatshnik, *Southern Illinois University*
Joseph F. Polen, *Shasta College*
Talat Rahman, *Kansas State University*
Dennis Roark, *The King's College*
James Rohlf, *Boston University*
Alfons Schulte, *University of Central Florida*
Neil Shafer-Ray, *University of Oklahoma*
Gurbax Singh, *University of Maryland*
Roger F. Sipson, *Moorhead State University*
John Sperry, *Sierra College*
Lon D. Spight, *University of Nevada at Las Vegas*

Robert Sprague, *Foothill College*
Julien. C. Sprott, *University of Wisconsin-Madison*
Konrad M. Stein, *Golden West College*
Bryan H. Suits, *Michigan Technological University*
Leo Takahashi, *The Pennsylvania State University—Beaver Campus*
Frank Tanherlini, *College of the Holy Cross*
Karl Trappe, *University of Texas*
Michael Trinkala, *Hudson Valley Community College*
Loren Vian, *Centralia College*
Clarence Wagener, *Creighton University*
Robert Weidman, *Michigan Technological University*
George Williams, *University of Utah*
Robert J. Wilson, *Colorado State University*
Arthur Winston, *Gordon Institute*
David Yee, *City College of San Francisco*
John Yelton, *University of Florida*
Dale Yoder-Short, *Michigan Technological University*

We also thank editors Catherine Flack, Sami Iwata, Joan Marsh, and publisher Robin Heyden at Addison Wesley for their vigorous support of and collaboration on this project, and Ingrid Mount of Elm Street Publishing Services for her skillful efforts in bringing it to fruition. Finally, RW thanks his family, colleagues, and students for their patience during the intense process of revising this book.

Richard Wolfson
Jay M. Pasachoff

A Visual Guide to the Book

Tip

How Does the Battery Know? How does the battery in Fig. 28-5 "know" how much current to supply? How does it even "know" there are two resistors, and what their values are? For the brief instant when the circuit is first connected, the battery doesn't "know," and some rather complicated action occurs as charge moves from the battery, encounters resistance, and accumulates to establish the potential differences across the resistors. During that brief time the current is not the same everywhere. But very soon the circuit reaches a steady state, with the same current throughout. Later in this chapter, when we consider circuits with capacitors, we'll look in more detail at the approach to the steady state; for now, our analyses assume that circuits have already reached that state.

EXAMPLE 28-1 | *Designing a Voltage Divider*

A light bulb with a resistance (when on) of 5.0 Ω is designed to operate at a current of 600 mA. To operate this lamp from a 12-V battery, what resistance should you place in series with it?

Solution

Let R_2 be the lamp and R_1 the unknown series resistor. Since resistors in series add, the current through both resistors is $I = \mathcal{E}/(R_1 + R_2)$, which is supposed to be 600 mA or 0.60 A. Solving for R_1 gives

$$R_1 = \frac{\mathcal{E} - IR_2}{I} = \frac{12\text{ V} - (0.60\text{ A})(5.0\ \Omega)}{0.60\text{ A}} = 15\ \Omega.$$

You can also get this result by noting that the light bulb's proper operating voltage is $V = IR_2 = (0.60$ A$)(5.0\ \Omega) = 3.0$ V. This is one-fourth of the battery voltage, so the light bulb's 5-Ω resistance should be one-fourth of the total. That makes the total 20 Ω, leaving 15 Ω for R_1.

EXERCISE Suppose that in Fig. 28-5 $R_1 = 470\ \Omega$. If the voltage across R_2 is 59% of the battery voltage, find R_2.

Answer: 676 Ω

Some problems similar to Example 28-1: 23–25

Got It!

What is the voltage across the resistor R at the top of each circuit shown Fig. 28-7? In (a) the second resistor has the same resistance R, and in (b) the is an open circuit (infinite resistance). Think about it first, then verify y answers using Equations 28-2.

FIGURE 28-7 What's the voltage across the upper resistor in each circuit?

Tip boxes appear throughout the text to help a student with difficult concepts, warn against common pitfalls, and give problem-solving hints.

Got It! boxes provide conceptual reinforcement of a student's understanding.

Here the student is asked to stop and think about three simple circuits, to probe their understanding before moving on to the next topic. They are asked to apply what they have learned in a qualitative way, and then asked to verify using the equations.

Annotated figures guide a student through complex concepts.

Here we see potential difference in a simple circuit explained step-by-step. Leading the student through the graphical representation of this important, but tricky, concept can lead to a more thorough understanding of the phenomena of voltage.

If I is the current in the circuit of Fig. 28-5, then there must be a voltage $V_1 = IR_1$ across R_1 to drive the current through this resistor. Similarly, the voltage across R_2 is $V_2 = IR_2$. Thus, the voltage across the two resistors together is $V_1 + V_2 = IR_1 + IR_2$. But the battery is connected directly across this series combination, so we have

$$IR_1 + IR_2 = \mathcal{E},$$

or

$$I = \frac{\mathcal{E}}{R_1 + R_2}.$$

Comparison with Ohm's law in the form $I = V/R$ shows that the two resistors in series behave like an equivalent resistance equal to the sum of their resistances. In an obvious generalization to more resistors in series, we have

$$R_{\text{series}} = R_1 + R_2 + R_3 + \cdots \tag{28-1}$$

In other words, resistors in series add.

Given the current, we can use Ohm's law in the form $V = IR$ to solve for the voltage across each resistor:

$$V_1 = \frac{R_1}{R_1 + R_2}\mathcal{E} \tag{28-2a}$$

and

$$V_2 = \frac{R_2}{R_1 + R_2}\mathcal{E}. \tag{28-2b}$$

These expressions show that the battery voltage divides between the two resistors in proportion to their resistance. For this reason a series combination of resistors is called a **voltage divider**. Figure 28-6 depicts the voltages throughout the circuit of Fig. 28-5, and shows explicitly that the resistors divide the battery voltage.

FIGURE 28-6 Voltages in the circuit of Fig. 28-5, with $V = 0$ at the negative battery terminal. Note that there is no potential difference across the wires, since they have negligible resistance, and that potential increases across the battery and decreases across the resistors. Can you tell from the graph which resistance is greater? (Here we've ™unrolled the circuit so we can clearly show the voltage as a function of position around the circuit. But remember that the circuit is a closed loop, so the left and right ends of the graph represent the same physical point, namely the negative battery terminal.)

Voltage—energy per charge—increases by the battery voltage $\mathcal{E}$ as the battery does work on charges, "lifting" them against its internal electric field.

Charge doesn't lose any energy flowing through the wires, since they're assumed to be perfect conductors.

Charge loses energy through collisions within the resistors. The energy loss per unit charge is IR.

In the steady state, charge has lost all the energy it gained from the battery by the time it gets through the entire circuit.

A rich array of **applications** reinforces understanding of underlying physics in both technology and natural phenomena. Applications range from the life sciences through the latest technology. They bring the subject alive by providing relevance of the theory to a student's everyday world.

      Here DVDs are compared to CDs to give the student knowledge of this topical development.

Throughout the book, icons in the margins and references in the text link to the powerful simulation package **ActivPhysics.** The simulations in **ActivPhysics** provide an opportunity for students to make predictions and test those predictions. The student becomes an active participant through manipulation and visualization of hard-to-grasp concepts.

---

998   Chapter 37 Interference and Diffraction

EXAMPLE 37-8   Asteroid Alert!

An asteroid appears on a collision course with Earth, at a distance of 20×10^6 km. What is the minimum size asteroid that the 2.4-m-diameter diffraction-limited Hubble Space Telescope could resolve at this distance, using 550-nm reflected sunlight?

Solution
Resolving the asteroid means being able to distinguish its opposite edges in the telescope image and therefore to assess its size. Suppose the asteroid's long dimension (it might not be spherical) is ℓ. Then at a distance $L \gg \ell$ it subtends an angle given very nearly by $\theta = \ell/L$. Using this result in Equation 37-13b with the mirror diameter and wavelength given, we have

$$\frac{\ell}{L} = \frac{1.22\lambda}{D},$$

or

$$\ell = \frac{1.22\lambda L}{D} = \frac{(1.22)(550 \times 10^{-9}\text{ m})(20 \times 10^9\text{ m})}{2.4\text{ m}} = 5.6\text{ km}.$$

This is a potentially dangerous object, comparable in size to the asteroid that scientists believe caused the extinction of the dinosaurs, and somewhat larger than the comet fragments that slammed into Jupiter in 1994, causing Earth-sized disturbances on the giant planet.

EXERCISE Two ants stand 1 cm apart. Assuming diffraction limited vision, at what distance could a human eye, with an iris aperture of 4 mm, tell that two are indeed separate creatures? Assume a wavelength of 550 nm, near the middle of the visible spectrum. (Other limitations of the human eye prevent its realizing this diffraction limit.)

Answer: 60 m

Some problems similar to Example 37-8: 55–59

APPLICATION   Movies on Disc: From CD to DVD

Standard audio CDs have a maximum playing time of 74 minutes–a value that's set by the optical diffraction limit. In the Application "CD Music, Continued" earlier in this chapter, we showed how a CD encodes information in "pits" 1.6 μm apart and as short as 0.83 μm in length. Standard CDs are read with 780-nm infrared laser light and the pit size and separation are chosen to ensure that diffraction effects at this wavelength don't cause the CD player's optical system to confuse adjacent pits. Ultimately, therefore, the 74-minute playing time is limited by the laser wavelength.

When CDs were first developed, inexpensive semiconductor lasers were available only in the infrared. But today's semiconductor lasers produce light well into the visible–with significantly shorter wavelengths than the infrared light used in CD players and CD-ROM drives. The new DVDs introduced in the late 1990s (read either "digital video disc" or "digital versatile disc") exploit these shorter-wavelength lasers.

Using laser wavelengths of 635 and 650 nm, DVDs can function with significantly lower pit sizes and spacings because of the lower diffraction limit at these shorter wavelengths (Fig. 37-39). Coupled with a two-layer structure (as opposed to the standard CD's single information layer) and more sophisticated data-compression schemes, the smaller pits and pit spacing give DVDs more than 12 times the information-storage capacity of standard CDs. That translates into more than two hours of high quality video and audio, as opposed to the CD's 74 minutes of audio alone. In computer use, it means a DVD can hold some 8.5 gigabytes of information as opposed to a CD's 680 megabytes (by comparison, a 3.5" floppy disc holds only 1.44 MB). And DVDs can be made in two-sided versions, providing another doubling in capacity. Despite their high capacity, DVDs are physically the same size as CDs, and DVD players are compatible with standard CDs as well.

FIGURE 37-39 A comparison of a standard CD and a DVD. The smaller pit size and closer pit spacing of the DVD result in a much higher information storage capacity, and are made possible by the lower diffraction limit associated with the DVD player's shorter wavelength laser light. Both images are microphotographs of the same size region, about 8 μm on a side. The pits lie on a single spiral track that covers most of the disc.

(a)   (b)

---

946   Chapter 36 Image Formation and Optical Instruments

last step in this equality follows from Equation for ℓ' and using the result in the equation

$$f = \frac{R}{2}$$

Thus the focal length of a spherical mirror

We emphasize that the results derived based on the assumption that the spherical curvature radius is large compared with Fig. 36-12c and (2) that all rays reflecting nearly parallel to the mirror axis (the para ings can't accurately reflect these conditions clearly the ray paths.

15.7 ActivPhysics
Spherical Mirror Problems

EXAMPLE 36-2   Sizing Up Hubble: A Concave Mirror

During assembly of the Hubble Space Telescope, a technician stood 3.85 m in front of the telescope's concave mirror and viewed his image (Fig. 36-13). Given the telescope's focal length of 5.52 m, find (a) the location and (b) the magnification of the technician's image. (c) Repeat for the case when the technician stands 15.0 m from the mirror.

Solution
Since the mirror is *concave*, the focal length f is *positive*.
(a) We can solve the mirror equation (Equation 36-2) to get

$$\frac{1}{\ell'} = \frac{1}{f} - \frac{1}{\ell} = \frac{\ell - f}{f\ell},$$

or

$$\ell' = \frac{f\ell}{\ell - f} = \frac{(5.52\text{ m})(3.85\text{ m})}{3.85\text{ m} - 5.52\text{ m}} = -12.7\text{ m}.$$

Why the *negative* answer? Because the technician is closer to the mirror than the focal point, so this is a *virtual* image located *behind* the mirror.
(b) Equation 36-1 gives the magnification:

$$M = -\frac{\ell'}{\ell} = -\frac{(-12.7\text{ m})}{3.85\text{ m}} = 3.30.$$

This *positive* answer indicates an *upright* image, enlarged about three times as shown in Fig. 36-13.
(c) If the technician stands 15.0 m from the mirror, similar calculations give

$$\ell' = \frac{f\ell}{\ell - f} = \frac{(5.52\text{ m})(15.0\text{ m})}{15.0\text{ m} - 5.52\text{ m}} = 8.73\text{ m}.$$

and

$$M = -\frac{\ell'}{\ell} = -\frac{8.73\text{ m}}{15.0\text{ m}} = -0.582.$$

FIGURE 36-13 A technician standing in front of the Hubble Space Telescope mirror. Where and how big is the technician's image?

Now ℓ' is *positive*, so the image is *real* and in *front* of the mirror; the *negative* magnification shows that it's *inverted*, and it's reduced in size.

EXERCISE You're scrutinizing your nose using a handheld concave mirror with curvature radius 2.2 m. How far from your face should you hold the mirror to see your nose doubled in size?

Answer: 55 cm

Some problems similar to Example 36-2: 3–9, 11–14

Worked examples are an integral part of the text and reinforce basic concepts. Real world applications are interwoven to give relevance to a student's experience. Many worked examples are followed by **Exercises. Similar Problems** from the end-of-chapter problems are indicated for students who want to work more problems on the topic.

In principle, we could imagine making the interval Δt as close to zero as we like, and getting ever better approximations to the instantaneous velocity. In practice, given a graph of position versus time, an easy approach is to "eyeball" the tangent line to the graph at the point we're interested in; the slope of the tangent line is the slope of the graph at that point, and is therefore the instantaneous velocity. Figure 2-5 shows this approach. Note that the slope of the tangent line, and therefore the velocity, varies from place to place on the graph. That's the whole point of the instantaneous velocity–it's a quantity that can vary continually.

Got It!

Figure 2-6 shows position-versus-time graphs for several objects. Which object is moving with constant speed? Which reverses direction? Which starts slowly and then speeds up?

Often, we're given the position as a mathematical function of time. We could draw a graph and measure the slopes of tangent lines to find the velocity at different points. But calculus provides a quicker way. In calculus, the result of the limiting process described in Equation 2-2a is called the **derivative** of x with respect to t, and is given the special symbol dx/dt:

$$\frac{dx}{dt} = \lim_{\Delta t \to 0} \frac{\Delta x}{\Delta t}.$$

The quantities dx and dt are called **infinitesimals**; they represent vanishingly small quantities that result from the limiting process. We can then write Equation 2-2a as

$$v = \frac{dx}{dt}, \qquad (2\text{-}2b)$$

When we're given the position x as a function of time t, calculus provid[es a] simple way to calculate the velocity $v = dx/dt$. Consult the Math Toolbox i[f you] haven't yet seen derivatives in your calculus class, or if you need a refres[her.]

FIGURE 2-5 The instantaneous velocity at a given time is the slope of the tangent line at the point on the position-versus-time graph corresponding to that time. If the position-versus-time graph is not a straight line, then the instantaneous velocity varies with time.

Math Toolbox

Derivatives Do we need to go through an elaborate limiting process every [time] we want an instantaneous velocity? No! In calculus you either have or will de[velop] formulas for the derivatives of common functions. For example, any functi[on of] the form

$$x = bt^n,$$

where b and n are constants, has the derivative

$$\frac{dx}{dt} = bnt^{n-1}.$$

Derivatives of other common functions, including trig functions, exponen[tials,] and logarithms, are listed in Appendix A. If a function is made up of a su[m of] terms, its derivative is just the sum of the derivatives of the individual te[rms.]

Math Toolboxes help a student through the math necessary for a full understanding of the physics. They emphasize math as a tool, in context with the physics.

Chapter Synopses emphasize key concepts and remind students of new terms and mathematical symbols. **Limitations to Keep in Mind** remind students of approximations, idealizations, and limited applications of physical concepts and equations. The **Problems You Should Be Able To Solve** section allows a student to evaluate their comprehension of the material in the chapter.

CHAPTER SYNOPSIS

Summary

1. Newton's **first law of motion** states that a body continues in uniform motion unless acted on by a nonzero net force. Uniform motion is a natural state; the study of motion then emphasizes not the cause of motion itself, but the cause of *changes* in motion.
2. Newton's **second law of motion** quantifies the change in motion brought about by a force. The law states that the rate of change of a body's momentum (product of mass and velocity; symbol **p**) is equal to the net force on the body:

$$\mathbf{F}_{net} = \frac{d\mathbf{p}}{dt}.$$

   As long as the mass of the body does not change, Newton's second law can also be written in the form

$$\mathbf{F}_{net} = m\mathbf{a}.$$

   Here $\mathbf{F}_{net}$ is the **net force** on a body—that is, the vector sum of all the individual interaction forces acting on the body. The SI unit of force is the **newton**, defined as the force that gives a 1-kg mass an acceleration of 1 m/s².
3. **Weight** is the force of gravity on an object: $\mathbf{W} = m\mathbf{g}$. In an accelerated frame of reference, the **apparent weight** differs from the gravitational force; in particular, the apparent weight of an object in free fall is zero.
4. Newton's **third law of motion** states that forces always come in pairs. When one object exerts a force on another, the second object exerts an oppositely directed force of equal magnitude back on the first. Newton's second and third laws together permit a consistent description of the motions of interacting objects.
5. Elastic objects such as springs provide a practical way of measuring forces. When the force exerted by a spring is directly proportional to the amount of stretch or compression, the spring is said to obey **Hooke's law**; the force is then given by

$$F = -kx, \quad (\text{Hooke's law})$$

   where k is the **spring constant**, with SI units of N/m.

Terms You Should Understand

(Pairs are closely related terms whose distinction is important; number in parentheses is chapter section where term first appears.)

dynamics (introduction)
Aristotle, Galileo, Newton (5-1)
force, net force, interaction force (5-2)
momentum (5-2)
Newton's laws of motion (5-2, 5-7)
inertia (5-2)
gravitational, electroweak, color forces (5-3)
mass, weight (5-5)
apparent weight (5-5, 5-8), weightlessness (5-5)
third-law pair (5-7)
normal force (5-7)
tension force, compression force (5-8)
Hooke's law (5-8)
spring constant (5-8)

Symbols You Should Recognize

$\mathbf{F}$ (5-2)
$\mathbf{p}$ (5-2)
N (5-4)
$\mathbf{g}$ (5-5)
$\mathbf{W}$ (5-5)
k (5-8)

Problems You Should Be Able to Solve

calculating force, mass, or acceleration from the other two quantities (5-4)
calculating weight or mass given the other (5-5)
solving for forces or accelerations in one-dimensional situations (5-6)
applying Newton's second and third laws together in one-dimensional situations (5-7)
determining spring force, extension/compression, or spring constant given the other two quantities (5-8)
determining apparent weight (5-8)

Limitations to Keep in Mind

Newton's laws apply only when relative velocities are much less than the speed of light; aside from that restriction, the basic material of this chapter is universally applicable.
Newton's laws are valid in inertial reference frames.
Newton's second law in the form $\mathbf{F} = m\mathbf{a}$ is valid only when mass remains constant; the form $\mathbf{F} = d\mathbf{p}/dt$ is more general.
Real springs deviate from ideal Hooke's law behavior when stretched or compressed substantially.

Questions will prompt a student to think about concepts in a chapter before they start on the problems.

652   Chapter 25 Electric Potential

QUESTIONS

1. Why can a bird perch on a high-voltage power line without getting electrocuted?
2. One proton is accelerated from rest by a uniform electric field, the other by a nonuniform electric field. If they move through the same potential difference, how do their final speeds compare?
3. Would a free electron move toward higher or lower potential?
4. The potential difference from A to B in Fig. 25-33 is zero since the two points are equidistant from the charge Q. How can this be, when a charge moving along the path shown clearly experiences an electric force not perpendicular to the path?

DANGER
ELECTROCUTION HAZARD
KEEP CLEAR
DEATH OR SERIOUS INJURY CAN RESULT FROM CONTACT WITH LOAD, MACHINE OR VEHICLE IF THEY BECOME ELECTRICALLY CHARGED

FIGURE 25-34   Question 12.

Paired Problems let a student practice problem-solving on a pair of problems whose solutions involve closely related physical concepts or mathematical approaches.

how this hazard a
to someone on th

Section 25-5 Potentials of Charged Conductors

52. (a) What is the maximum potential (measured from infinity) for the sphere of Example 25-3 before dielectric breakdown of air occurs at the sphere's surface? (Breakdown of air occurs at a field strength of 3 MV/m.) (b) What is the charge on the sphere when it's at this potential?
53. The spark plug in an automobile engine has a center electrode made from wire 2.0 mm in diameter. The electrode is worn to a hemispherical shape, so it behaves approximately like a charged sphere. What is the minimum potential on this electrode that will ensure the plug sparks in air? Neglect the presence of the second electrode.
54. A large metal sphere has three times the diameter of a smaller sphere and carries three times as much charge. Both spheres are isolated, so their surface charge densities

Paired Problems

(Both problems in a pair involve the same principles and techniques. If you can get the first problem, you should be able to solve the second one.)

59. Three 50-pC charges sit at the vertices of an equilateral triangle 1.5 mm on a side. How much work would it take to bring a proton from very far away to the midpoint of one of the triangle's sides?
60. Repeat the preceding problem for the case when one of the charges is −50 pC and the proton is brought to the midpoint of the side between the two positive charges.
61. A pair of equal charges q lies on the x axis at $x = \pm a$. (a) Find expressions for the potential at points on the x axis for which $x > a$ and (b) show that your result reduces to a point-charge potential for $x \gg a$.
62. (a) For the charge distribution of the preceding problem, find an expression for the potential at *all* points on the y axis. (b) Show that your result reduces to a point-charge potential for $y \gg a$.
63. A 2.0-cm-radius metal sphere carries 75 nC and is surrounded by a concentric spherical conducting shell of radius 10 cm carrying −75 nC. (a) Find the potential difference between the shell and the sphere. (b) How would your answer change if the shell charge were changed to +150 nC?
64. A coaxial cable consists of a 2.0-mm-radius central wire carrying 75 nC/m, and a concentric outer conductor of radius 10 mm carrying −75 nC/m. (a) Find the potential

Almost 3000 **end-of-chapter problems,** ranging from confidence builders to realistic multi-problems, will help a student understand concepts and develop problem-solving skills. The problems are thought provoking and encourage critical thinking, and are relevant to a student's everyday life.

PROBLEMS

ActivPhysics can help with these problems:
Activities 11.9, 11.10.

Section 25-2 Potential Difference

1. How much work does it take to move a 50-μC charge against a 12-V potential difference?
2. The potential difference between the two sides of an ordinary electrical outlet is 120 V. How much energy does an electron gain when it moves from one side to the other?
3. It takes 45 J to move a 15-mC charge from point A to point B. What is the potential difference ΔV_{AB}?
4. Show that 1 V/m is the same as 1 N/C.
5. Find the magnitude of the potential difference between two points located 1.4 m apart in a uniform 650 N/C electric field, if a line between the points is parallel to the field.
6. A charge of 3.1 C moves from the positive to the negative terminal of a 9.0-V battery. How much energy does the battery impart to the charge?
7. Two points A and B lie 15 cm apart in a uniform electric field, with the path AB parallel to the field. If the potential difference ΔV_{AB} is 840 V, what is the field strength?
8. Figure 25-37 shows a uniform electric field of magnitude E. Find expressions for (a) the potential difference ΔV_{AB} and (b) ΔV_{BC}. (c) Use your result to determine ΔV_{AC}.

FIGURE 25-37   Problem 8.

9. A proton, an alpha particle (a bare helium nucleus), and a singly ionized helium atom are accelerated through a potential difference of 100 V. Find the energy each gains.
10. The electric field within the membrane separating the inside and outside of a biological cell is approximately 8.0 MV/m, and is essentially uniform. If the membrane is 10 nm thick, what is the potential difference across the membrane?
11. What is the potential difference between the terminals of a battery that can impart 7.2×10^{-19} J to each electron that moves between the terminals?
12. Electrons in a TV tube are accelerated from rest through a 25-kV potential difference. With what speed do they hit the TV screen?
13. A 12-V car battery stores 2.8 MJ of energy. How much charge can move between the battery terminals before it is totally discharged? Assume the potential difference remains at 12 V, an assumption that is not realistic.
14. What is the charge on an ion that gains 1.6×10^{-15} J when it moves through a potential difference of 2500 V?
15. Two large, flat metal plates are a distance d apart, where d is small compared with the plate size. If the plates carry surface charge densities $\pm\sigma$, show that the magnitude of the potential difference between them is $V = \sigma d/\varepsilon_0$.
16. An electron passes point A moving at 6.5 Mm/s. At point B the electron has come to a complete stop. Find the potential difference ΔV_{AB}.
17. A 5.0-g object carries a charge of 3.8 ... It acquires a speed v ... difference ... the same ci

re far apart. One
other −10 nC.
the spheres are
otential on each
uch charge must
eve equilibrium?
s for the isolated,

in diameter and
part. Determine
d strength at the
midway between
ce between the

Most of the problems are **referenced by section** to allow a student to refer easily to the relevant text.

Supplementary Problems drawn from throughout the chapter are more difficult problems that will challenge the best students.

Supplementary Problems

69. A conducting sphere 5.0 cm in radius carries 60 nC. It is surrounded by a concentric spherical conducting shell of radius 15 cm carrying −60 nC. (a) Find the potential at the sphere's surface, taking the zero of potential at infinity. (b) Repeat for the case when the shell also carries +60 nC.
70. Show that the result of Example 25-9 approaches the field of a point charge for $x \gg a$. *Hint:* You will need to apply the binomial theorem to the quantity $1/\sqrt{x^2 + a^2}$.
71. The potential on the axis of a uniformly charged disk at ... V; the potential 10 cm ... he disk radius and its

76. Repeat the preceding problem for the case $\lambda = \lambda_0(x/\ell)$. Why is your answer for $x \gg \ell$ different? *Hint:* What does this charge distribution resemble at large distances?
77. For the situation of Example 25-10, find an equation for the equipotential with $V = 0$ in the x-y plane. Plot the equipotential, and show that it passes through the points described in Example 25-10 and its exercise.
78. A disk of radius a carries a nonuniform surface charge density given by $\sigma = \sigma_0(r/a)$, where σ_0 is a constant. (a) Find the potential at an arbitrary point on the disk axis, a distance x from the disk center. (b) Use the result of (a) to find the electric field on the disk axis, and (c) show that the field reduces to an expected form for $x \gg a$.
79. An open-ended cylinder of radius a and length $2a$ carries charge q spread uniformly over its surface. Find the potential on the cylinder axis at its center. *Hint:* Treat the cylinder as a stack of charged rings, and integrate.

914   Part 4 Cumulative Problems

PART 4   Cumulative Problems

These problems combine material from chapters throughout the entire part or, in addition, from chapters in earlier parts, or they present special challenges.

1. An air-insulated parallel-plate capacitor has plate area 100 cm² and spacing 0.50 cm. The capacitor is charged and then disconnected from the charging battery. A thin-walled, nonconducting box of the same dimensions as the capacitor is filled with water at 20.00°C. The box is released at the edge of the capacitor and moves without friction into the capacitor (Fig. 1). When it reaches equilibrium the water temperature is 21.50°. What was the original voltage on the capacitor?

50 T/s while that in the right-hand solenoid is decreasing at 30 T/s. Find the current in the resistance wire shared by both triangles. Which way does the current flow?

FIGURE 3   Cumulative Problem 3.

4. A long solenoid of length ℓ and radius R has a total of N

rge 92e) decays, emit-
rge 2e) and leaving a
e 90e). At the instant

The student is asked to integrate knowledge from previous chapters to solve the **Cumulative Problems.** They test a student's understanding of a range of topics, and accumulation of knowledge.

The Authors

Richard Wolfson is Professor of Physics and George Adams Ellis Professor of the Liberal Arts at Middlebury College, where he has taught since 1976. He did undergraduate work at the Massachusetts Institute of Technology and Swarthmore College and holds the M.S. degree from the University of Michigan and Ph.D. from Dartmouth. He has published widely in scientific journals, including works ranging from medical physics research to experimental plasma physics, electronic circuit design, solar energy engineering, and theoretical astrophysics. He is also an interpreter of science for the nonspecialist, as a contributor to *Scientific American* and author of the book *Nuclear Choices: A Citizen's Guide to Nuclear Technology*. His videotaped course "Einstein's Relativity and the Quantum Revolution," produced by The Teaching Company, brings the key ideas of modern physics to nonscientists. Professor Wolfson has spent sabbatical years as Visiting Scientist at the National Center for Atmospheric Research in Boulder, Colorado, at St. Andrews University in Scotland, and at Stanford University.

Jay M. Pasachoff is Field Memorial Professor of Astronomy and Director of the Hopkins Observatory at Williams College. He was born and brought up in New York City. After attending the Bronx High School of Science, he received his A.B. degree from Harvard College and his A.M. and Ph.D. from Harvard University. He then held postdoctoral fellowships at Harvard and at the California Institute of Technology before going to Williams in 1972. His research has dealt mainly with solar physics, most recently the solar corona as studied at total eclipses, and with nuclear astrophysics, namely, the abundances of the light elements and their formation in the first minutes of the universe. Professor Pasachoff has spent sabbatical leaves at the University of Hawaii, at l'Institut d'Astrophysique in Paris, at the Institute for Advanced Study in Princeton, and at the Harvard-Smithsonian Center for Astrophysics. He is also author or co-author of major texts in physics, calculus, physical science, and astronomy.

Color Key for Figures

The artwork for this third edition is carefully designed to make effective use of color printing as a learning aid. In particular, we have assigned colors to the vector quantities that are so important in physics, and we have used those colors consistently throughout the book. The table below lists some important physical quantities, along with their text and graphic symbols and color assignments. We also include electric circuit symbols, which, to be consistent with usage in engineering, are printed in black.

Vector	Text Symbol	Graphic Symbol
Displacement	$\mathbf{r}, \boldsymbol{\ell}$	→
Velocity	$\mathbf{v}$	→
Acceleration	$\mathbf{a}$	→
Force	$\mathbf{F}$	→
Linear momentum	$\mathbf{p}$	→
Angular velocity	$\boldsymbol{\omega}$	→
Torque	$\boldsymbol{\tau}$	→
Angular momentum	$\mathbf{L}$	→
Electric field	$\mathbf{E}$	→
Magnetic field	$\mathbf{B}$	→
Electric dipole moment	$\mathbf{p}$	→
Magnetic dipole moment	$\boldsymbol{\mu}$	→
Electric charge		
Positive charge	q, Q	⊕
Negative charge	q, Q	⊖
Circuit symbols		
Battery, emf	$\mathcal{E}$	—⊣⊢—
Resistor	R	—⋀⋀—
Capacitor	C	—⊣⊢—
Inductor	L	—⊸⊸⊸—
Switch	S	—∕·—

Part Four   Electromagnetism

A portion of an integrated circuit, or "chip," magnified 100 times. Integrated circuits are at the heart of computers and many other electronic devices. A single integrated circuit can contain millions of transistors and other electronic components. These miniaturized circuits are but one of a myriad of electromagnetic technologies that we use every day.

Electromagnetism is a fundamental force, and governs the behavior of matter from the atomic scale into the macroscopic world. Electromagnetic technology, from computer microchips to large electric motors and generators, is essential to our technological society. Even our bodies—with their electrochemical nerve impulses, electrically polarized muscle cells, and carefully paced electrical signals controlling heartbeat—rely heavily on electromagnetism.

Four fundamental laws describe electricity and magnetism. Two deal separately with the two phenomena, while the other two reveal profound connections that show electricity and magnetism to be aspects of a single phenomenon—electromagnetism. The twelve chapters in this part will give you a thorough grounding in electromagnetism. You'll come to understand the fundamental laws that govern all electromagnetic phenomena, learn how electromagnetism determines the structure and behavior of nearly all matter, and explore the electromagnetic technologies that play so important a role in your life.

23 Electric Charge, Force, and Field

Lightning strikes at Kitt Peak National Observatory in Arizona relieve the buildup of electric charge in the atmosphere.

What force keeps the molecules in your body together? What force keeps a skyscraper standing or prevents a mountain from spreading into a flat blob? What force holds your car on the road as you round a turn? What force accelerates the electrons that paint the picture on your TV screen? What force underlies the awesome beauty of a thunderstorm? Remarkably, these and all other forces except gravity that we encounter in our everyday lives—and in nearly all scientific work—are manifestations of a single force: the **electromagnetic force**. The friction, tension, and normal forces of mechanics are ultimately electromagnetic. So are interactions as diverse as the focusing of light in the lens of your eye, the extraction of information from a computer disk, or the formation of a water molecule from separate atoms. Electromagnetism is so important in science and engineering that we devote the next twelve chapters to it.

23-1 ELECTROMAGNETISM

Electromagnetism is among the fundamental forces of nature that we introduced in Chapter 5 (see Fig. 5-9) and is the dominant force in a vast range of natural and technological phenomena. Other than gravity, electromagnetism is the only force most of us will ever deal with.

Three distinct themes motivate our study of electromagnetism:

1. The electromagnetic force is solely responsible for the structure of matter from atoms to objects of roughly human size (Fig. 23-1). Much of physics, all of chemistry, and most of biology deal in this realm. Only at much smaller scales do the color and weak forces become important; only at larger scales is gravity significant.

   The wonderful diversity of chemical compounds is testimony to the rich possibilities contained in the electromagnetic interaction. Even life itself, and the DNA replicating mechanism at its heart, are manifestations of electromagnetism (Fig. 23-1). For students of physical and biological sciences, understanding electromagnetism means understanding the most fundamental basis of these disciplines.

2. We live in a technological world increasingly dominated by devices that operate on electromagnetic principles. Electric lights, motors, batteries, and generators have been essential for over 100 years. More recently, electronic technology has led to the proliferation of devices for storing and processing information, for sensing and measuring, and for sophisticated control of industrial, scientific, medical, automotive, and household systems (Fig. 23-2).

3. Studying electromagnetism leads to an understanding of the nature of light and from there to the theory of relativity. Relativity profoundly alters our ideas of space and time—the very basis of physical reality.

We've been speaking of electromagnetism, yet you're probably more familiar with electricity and magnetism separately. Although we begin with separate studies of these seemingly distinct phenomena, the relation between the two will become increasingly central. Eventually you will understand electricity and magnetism as two aspects of a single phenomenon that is basic to the workings of the universe.

FIGURE 23-1 Electric forces govern the transmission of nerve impulses, shape the DNA molecule, and hold atoms together.

FIGURE 23-2 Inside a computer, components as mundane as the cooling fan motor and as sophisticated as a memory chip work on electromagnetic principles. Blowup shows the computer's microprocessor, which includes millions of electronic components on a single silicon chip.

23-2 ELECTRIC CHARGE

Electric charge is a fundamental property of nature. Of the three building blocks of ordinary matter—the electron, the proton, and the neutron—two carry electric charge. What is electric charge? At the most fundamental level, we don't know. We don't know what mass "really" is either, but we're familiar with it because we've spent our lives pushing objects around. By studying electrical interactions we gain a similar familiarity with charge that is as close as we can get to understanding what it "really" is.

Two Kinds of Charge

Electric charge comes in two varieties, which Benjamin Franklin designated **positive** and **negative**. These are convenient labels, but they have no physical significance. There's nothing "missing" about negative charge. Positive and negative charge are complementary properties, not the presence and absence of something. The utility in the names is mathematical, since it's the algebraic sum of charges—described with positive and negative numbers—that has physical significance.

Quantities of Charge

All electrons carry the same charge, and all protons carry the same charge. The proton's charge has *exactly* the same magnitude as the electron's, but with opposite sign. Given that electrons and protons differ substantially in other properties—like mass—this electrical relation is remarkable. Problem 1 shows how dramatically different our world would be if there were even a slight difference between the magnitudes of the electron and proton charges.

The magnitude of the electron or proton charge is the **elementary charge**, e. Electric charge is **quantized**—that is, it comes only in discrete amounts. In a famous experiment in 1909, the American physicist R. A. Millikan used electric forces to suspend small oil drops. From the electric force he computed the charge on each drop and found it was always a multiple of a basic value we now know as the elementary charge (Fig. 23-3).

FIGURE 23-3 Millikan's oil-drop experiment. By balancing the electric force $\mathbf{F}_E$ and gravitational force $\mathbf{F}_g$ on oil drops, Millikan showed that charge is quantized.

X-rays ionize the air, yielding electrons that stick to oil droplets.

Fine mist of oil droplets

X-ray source

$\mathbf{F}_E$

$\mathbf{F}_g$

Metal plates

The charge on the plates is adjusted until a charged droplet remains stationary.

The mass of a droplet is calculated from the rate at which it falls when the charge is turned off.

Adjustable battery

Modern elementary particle theories suggest that the most basic unit of charge is actually $\frac{1}{3}e$. Such "fractional charges" reside on quarks, the basic building blocks of protons, neutrons, and many other particles. Quarks always join to produce particles with integer multiples of the full elementary charge, and it seems impossible to isolate individual quarks.

The SI unit of charge is the **coulomb** (C), named for the French physicist Charles Augustin de Coulomb (1736–1806). Although the coulomb's formal definition is in terms of electric current, it's convenient to describe one coulomb as being about 6.25×10^{18} elementary charges, making the elementary charge approximately 1.60×10^{-19} C.

Charge Conservation

Electric charge is a conserved quantity, meaning that the algebraic sum of the electric charges—i.e., the **net charge**—in a closed region remains constant. Charged particles may be created or annihilated, but always in pairs of equal and opposite charge (Fig. 23-4). The net charge always remains the same.

FIGURE 23-4 The creation of an electron and its antiparticle, a positron. The two particles' paths are the oppositely directed spirals that originate in a common point where the pair was created. The net charge remains zero before and after the particle creation. (The curved paths result from a magnetic force.)

23-3 COULOMB'S LAW

You can transfer charge to a balloon by rubbing it on your clothing; you'll then find that the balloon sticks to you. Charge another balloon and the two balloons will repel each other (Fig. 23-5). You're seeing a manifestation of the fundamental fact that unlike charges attract and like charges repel. Socks clinging to other clothes when they come out of a dryer, dust attracted to the front of a TV screen or computer monitor, and the shocks you get when you cross a carpet and touch a doorknob are other common examples where you're directly aware of this electrical interaction.

But electricity would be rather unimportant if the only significant electrical interactions were these obvious ones. In fact, all interactions of everyday matter—from the motion of a car to the movement of a muscle to the growth of a tree—are dominated by the electric force. It's just that matter on the large scale is nearly perfectly neutral, so electrical effects in bulk matter are not obvious. At the molecular or even cellular level the appearance of individual charged particles makes the electrical nature of matter immediately obvious (Fig. 23-6).

The attraction and repulsion of electric charges implies that a force acts between them. Scientists Joseph Priestley and Charles Augustin de Coulomb investigated this force in the late 1700s. They found that the force between two charges acts along the line joining them, with a magnitude proportional to the product of the charges and inversely proportional to the square of the distance

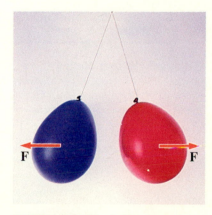

FIGURE 23-5 Two balloons carrying similar electric charge repel each other.

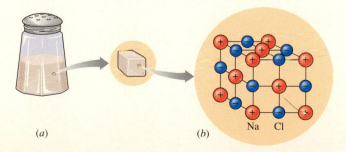

(a)   (b)
Na   Cl

FIGURE 23-6 (a) The salt shaker and even a single salt grain are electrically neutral, so the role of electric forces in these structures is not obvious. (b) At the atomic scale, electric forces bind individual sodium and chloride ions together and are, in fact, responsible for the cubical shape of the salt crystal.

$$\mathbf{F}_{12} = \frac{kq_1q_2}{r^2}\hat{\mathbf{r}}$$

FIGURE 23-7 Quantities in Coulomb's law used in calculating the electric force of charge q_1 on q_2. The unit vector $\hat{\mathbf{r}}$ always points in the direction from q_1 toward q_2, regardless of the signs of the charges. The force $\mathbf{F}_{12}$ points in the same direction as $\hat{\mathbf{r}}$ if the two charges have the same sign, and otherwise in the opposite direction.

11.1

*Activ*Physics

**Electric Force:
Coulomb's Law**

between them. These results are summarized in **Coulomb's law**:

$$\mathbf{F}_{12} = \frac{kq_1q_2}{r^2}\hat{\mathbf{r}} \qquad \text{Coulomb's law .} \qquad (23\text{-}1)$$

Here $\mathbf{F}_{12}$ is the force exerted *by* the charge q_1 *on* the charge q_2 and r is the distance between the two. The quantity k is a proportionality constant whose value in SI units is approximately 9.0×10^9 N·m²/C². The force $\mathbf{F}_{12}$ is a vector, and $\hat{\mathbf{r}}$ is a unit vector giving its direction. Figure 23-7 shows that $\hat{\mathbf{r}}$ lies on the line passing through the two charges and points in the direction *from q_1 toward q_2.*

Coulomb's law is a vector equation that covers all possible combinations of charges and the associated attractive and repulsive forces. If q_1 and q_2 have the same sign—either positive or negative—then the product q_1q_2 is positive, and the force points in the same direction as $\hat{\mathbf{r}}$. The force on q_2 is thus away from q_1, or repulsive, as it should be. But if q_1 and q_2 have opposite signs, then q_1q_2 is negative and the direction of the force is opposite that of $\hat{\mathbf{r}}$—that is, attractive.

What about the force $\mathbf{F}_{21}$ that q_2 exerts on q_1? Equation 23-1 shows it has the same magnitude as $\mathbf{F}_{12}$, but a drawing like Fig. 23-7 gives the opposite direction. Thus the electric force obeys Newton's third law. ActivPhysics Activity 11.1 lets you explore the forces between pairs of electric charges.

EXAMPLE 23-1 | *Two Charges*

A 1.0-μC charge is located at $x = 1.0$ cm, and a -1.5-μC charge at $x = 3.0$ cm. (a) What force does the positive charge exert on the negative one? (b) How would the force change if the distance between the charges were tripled?

Solution

(a) A vector from the positive charge toward the negative charge points in the $+x$ direction, so $\hat{\mathbf{r}}$ is simply the unit vector $\hat{\mathbf{i}}$. Then Equation 23-1 becomes

$$\mathbf{F} = \frac{kq_1q_2}{r^2}\hat{\mathbf{r}}$$

$$= \frac{(9.0 \times 10^9 \text{ N·m}^2/\text{C}^2)(1.0 \times 10^{-6} \text{ C})(-1.5 \times 10^{-6} \text{ C})}{(0.020 \text{ m})^2}\hat{\mathbf{i}}$$

$$= -34\hat{\mathbf{i}} \text{ N .}$$

The minus sign shows that the force is in the negative x direction—toward the positive charge, or attractive.

(b) If the distance were tripled the force would drop by a factor of $1/3^2$, to $-3.8\hat{\mathbf{i}}$ N.

Tip

Let the Algebra Take Care of the Signs Equation 23-1 accounts for all aspects of the force calculation. If you correctly take $\hat{\mathbf{r}}$ to be a unit vector pointing *away* from the charge giving rise to the force—whatever its sign—and *toward* the charge being acted on—whatever its sign—then calculation of $\dfrac{kq_1q_2}{r^2}\hat{\mathbf{r}}$ gives correctly the magnitude and direction of the force. You should, of course, check that your mathematical answer makes sense.

EXERCISE A 10-μC charge is at the origin. Find the force it exerts on a 5-μC charge at the point $x = 4$ m, $y = 3$ m. *Note:* You may need help with this one; see the Math Toolbox following this example.

Answer: $1.44\hat{\mathbf{i}} + 1.08\hat{\mathbf{j}}$ mN

Some problems similar to Example 23-1: 7, 12, 13

Math Toolbox

Finding the Unit Vector In Example 23-1, both charges lay on the x axis and it was obvious that a unit vector from the 1 μC charge toward the -1.5 μC charge is the vector $\hat{\mathbf{i}}$, so in that case $\hat{\mathbf{r}} = \hat{\mathbf{i}}$. But what's $\hat{\mathbf{r}}$ in the exercise with Example 23-1?

Coulomb's law says that $\hat{\mathbf{r}}$ is in the direction from q_1 toward q_2. But that's the same direction as the displacement vector $\mathbf{r}_{12}$ from q_1 to q_2. So $\mathbf{r}_{12}$ has the right direction; it just isn't a *unit* vector. But dividing by its magnitude makes a unit vector, as you can prove in Problem 12. Figure 23-8 shows that $\mathbf{r}_{12} = 4\hat{\mathbf{i}} + 3\hat{\mathbf{j}}$ m, so its magnitude is r = $\sqrt{(4 \text{ m})^2 + (3 \text{ m})^2}$ = 5 m. Then the unit vector from q_1 toward q_2 is

$$\hat{\mathbf{r}} = \frac{4\hat{\mathbf{i}} + 3\hat{\mathbf{j}} \text{ m}}{5 \text{ m}} = 0.80\hat{\mathbf{i}} + 0.60\hat{\mathbf{j}}.$$

Then we're ready to calculate the force, using Coulomb's law (Equation 23-1):

$$\mathbf{F}_{12} = \frac{kq_1q_2}{r^2}\hat{\mathbf{r}} = \frac{(9.0\times10^9 \text{ N·m}^2/\text{C}^2)(10 \text{ } \mu\text{C})(5.0 \text{ } \mu\text{C})}{(5.0 \text{ m})^2}(0.80\hat{\mathbf{i}} + 0.60\hat{\mathbf{j}})$$

$$= 1.44\hat{\mathbf{i}} + 1.08\hat{\mathbf{j}} \text{ mN}.$$

And that's it: a full vector expression for the force. More generally, given the vector $\mathbf{r}_{12}$ from q_1 to q_2, we can write the unit vector as

$$\hat{\mathbf{r}} = \frac{\mathbf{r}_{12}}{r}, \tag{23-2a}$$

where r is the magnitude of $\mathbf{r}_{12}$—i.e., the distance between the charges.

Equivalently, you can find the unit vector using trigonometry. Let θ be the angle that the vector $\mathbf{r}_{12}$ from q_1 to q_2 makes with the positive x axis, as marked in Fig. 23-8. Then the unit vector is

$$\hat{\mathbf{r}} = \cos\theta\,\hat{\mathbf{i}} + \sin\theta\,\hat{\mathbf{j}}. \tag{23-2b}$$

You can see this is a unit vector because it has magnitude $\sin^2\theta + \cos^2\theta = 1$, and the geometry of Fig. 23-8 shows clearly that $\hat{\mathbf{r}}$ as defined in Equation 23-2b is in the direction of $\mathbf{r}_{12}$. Finally, with the hypotenuse $r_{12} = 5$ in Fig. 23-8, the figure shows that $\cos\theta = 4/5$ and $\sin\theta = 3/5$. Thus $\hat{\mathbf{r}} = 0.80\hat{\mathbf{i}} + 0.60\hat{\mathbf{j}}$, as before. The methods of Equations 23-2a and 23-2b are completely equivalent. Equation 23-2a may seem less familiar, but it gives the unit vector directly without the intermediary of trigonometry.

Still another approach is to use Equation 23-1 to find just the magnitude of the force, then figure out whether the force is attractive or repulsive, and then find its components in a chosen coordinate system. This method is completely equivalent to our more elegant and straightforward approach using unit vectors.

FIGURE 23-8 The two charges in the exercise with Example 23-1. You could get from q_1 to q_2 by going 4 m in the x direction and then 3 m in the y direction, so the displacement vector $\mathbf{r}_{12}$ is $\mathbf{r}_{12} = 4\hat{\mathbf{i}} + 3\hat{\mathbf{j}}$ m. The unit vector $\hat{\mathbf{r}}$ in Coulomb's law has the same direction as $\mathbf{r}_{12}$ but magnitude 1, so it's given by $\mathbf{r}_{12}$ divided by $\mathbf{r}_{12}$'s magnitude r, here $\sqrt{(4 \text{ m})^2 + (3 \text{ m})^2}$ = 5 m. More generally, the unit vector in the direction from q_1 toward q_2 is $\hat{\mathbf{r}} = \mathbf{r}_{12}/r$. Equivalently, the unit vector is $\hat{\mathbf{r}} = \cos\theta\,\hat{\mathbf{i}} + \sin\theta\,\hat{\mathbf{j}}$, with θ the angle between $\mathbf{r}_{12}$ and the positive x axis.

Coulomb's law for the electric force is similar to Newton's law for the gravitational force. Both forces show the same inverse-square decrease with distance, and both are proportional to the product of the interacting charges or masses. But there is an important difference: There's only one kind of mass (even antimatter has "positive mass"), and the gravitational force it produces is always attractive. That means large concentrations of mass give rise to large gravitational forces. But charge comes in two varieties that attract each other, so large concentrations of charge tend to be electrically neutral and therefore give rise to weak electric forces. The electric force between individual particles is vastly stronger than the gravitational force between the same particles (see Problem 6), and it's only because of the nearly complete cancellation of positive and negative charge in bulk matter that gravity becomes important on the macroscopic scale.

Got It!

Charge q_1 is located at $x = 1$ m, $y = 0$. What should you use for the unit vector $\hat{\mathbf{r}}$ in Coulomb's law if you're calculating the force q_1 exerts on a charge q_2 located at (a) the origin and (b) the point $x = 0$, $y = 1$ m? Explain why you can answer this question without being told the signs of either charge.

Point Charges and the Superposition Principle

Coulomb's law is strictly true only for **point charges**—charged objects of negligible size. Electrons and protons can usually be treated as point charges; so, approximately, can any two charged objects if their separation is large compared to their size. But often we're interested in the electrical effects of **charge distributions**—arrangements of charge spread over space. The DNA molecule shown in Fig. 23-1 is a very complicated charge distribution. Other charge distributions include water molecules, the innards of a TV tube, a memory cell in a computer memory chip, your heart, and a thundercloud. We need to know how to combine the effects of two or more charges to find the electrical effects of such charge distributions.

Figure 23-9 shows two charges q_1 and q_2 which constitute a simple charge distribution. We want to know the net force on a third charge q_3 from the combination of q_1 and q_2. To find that net force, you might calculate the forces $\mathbf{F}_{12}$ and $\mathbf{F}_{13}$ from Equation 23-1, and then vectorially add them. And you'd be right: the force that q_1 exerts on q_3 is unaffected by the presence of q_2, and vice versa, so you can apply Coulomb's law separately to the pairs q_1q_3 and q_2q_3 and then combine the results. That may seem obvious, but nature needn't have been so simple.

The fact that electric forces add vectorially is called the **superposition principle**. Our confidence in this principle is ultimately based on experiments that show electric and indeed electromagnetic phenomena behave according to the principle. With superposition we can solve relatively complicated problems by breaking them into simpler parts. If the superposition principle did not hold, the mathematical description of electromagnetism would be far more complicated than it is. ActivPhysics Activities 11.2 and 11.3 let you explore the superposition principle.

Although the force that one point charge exerts on another decreases with the inverse square of the distance between them, it's important to recognize that the same is usually not true of the force arising from a charge distribution. The superposition of forces from different charges in a distribution may result in a very different distance dependence of the force due to a charge distribution (see Fig. 23-10). Example 23-2 on the next page provides a case in point.

FIGURE 23-9 The superposition principle allows us to add vectorially the forces from two or more charges.

*Activ*Physics

11.2, 11.3

Electric Force: Superposition Principle

FIGURE 23-10 (*a*) For a charge distribution distant enough to resemble a point charge, the inverse-square law holds approximately. (*b*) Close to charge distribution, the inverse-square law usually does not hold. That's because individual charges in the distribution are located at different distances and directions from the charge experiencing the force.

$\mathbf{F} \propto \dfrac{1}{r^2}$

(*a*)

$\mathbf{F} \propto \dfrac{1}{r^2}$

(*b*)

EXAMPLE 23-2	*Raindrops*

The charging of individual raindrops is ultimately responsible for the electrical activity of a thunderstorm. Suppose two drops with equal charge q are located on the x axis at $\pm a$, as shown in Fig. 23-11. Find the electric force on a third drop with charge Q located at an arbitrary point on the y axis.

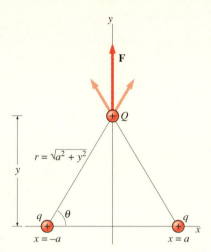

FIGURE 23-11 The force on Q is the vector sum of the forces from the individual charges.

Solution

Charge Q is the same distance $r = \sqrt{a^2 + y^2}$ from the two other drops, so the force from each has the same magnitude. Then it's obvious from Fig. 23-11 that the x components of the two forces cancel, while the y components are the same. Fig. 23-11 and Equation 23-2b show that the y components of the unit vectors from the two charges q toward Q are both given by $\hat{r}_y = \sin\theta = y/\sqrt{a^2 + y^2}$. Then the y component of each force is given by $F_y = (kqQ/r^2)\hat{r}_y$, and the net force on Q becomes

$$\mathbf{F} = F_y\hat{\mathbf{j}} = \left(\frac{2kqQ}{a^2 + y^2}\right)\left(\frac{y}{\sqrt{a^2 + y^2}}\right)\hat{\mathbf{j}} = \frac{2kqQ}{(a^2 + y^2)^{3/2}}\hat{\mathbf{j}},$$

where the factor of two comes from adding the y components of the forces from both charges q.

Does our result make sense? Evaluating $\mathbf{F}$ at $y = 0$ gives zero force. Here, midway between the two charges, Q experi-ences equal but opposite forces from the two q's, so the net force must be zero. At very large distances such that $y \gg a$, on the other hand, we can neglect a^2 compared with y^2, and the force becomes

$$\mathbf{F} = \frac{k(2q)Q}{y^2}\hat{\mathbf{j}}. \qquad (y \gg a)$$

This is just the force we would expect from a single charge of magnitude $2q$ located a distance y from Q—showing that this system of two charges acts like a single charge $2q$ at distances large compared with the charge separation. But this is true only at distances large compared with the charge separation. In closer, where the detailed arrangement of the individual charges is important, the behavior of the field is more compli-cated. In fact, our result shows the field initially increasing as y increases, and only later beginning to decrease with increas-ing y.

In drawing Fig. 23-11 we tacitly assumed that q and Q have the same signs. But our analysis holds even if they don't; in that case the product qQ is negative, and $\mathbf{F}$ therefore points in the opposite direction from $\hat{\mathbf{j}}$.

EXERCISE Find the net force on the charge located at the origin in Fig. 23-12.

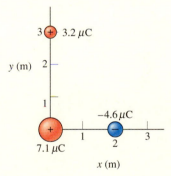

FIGURE 23-12 What is the force on the charge at the origin?

Answer: $73\hat{\mathbf{i}} - 23\hat{\mathbf{j}}$ mN

Some problems similar to Example 23-2: 19–22, 34

EXAMPLE 23-3	*Balancing Forces*

A positive charge $+2q$ lies on the x axis at $x = -a$, and a charge $-q$ lies at $x = +a$, as shown in Fig. 23-13. Find a point where the electric force on a third charge Q would be zero.

Solution

The point must lie on the x axis since off axis the individual force vectors cannot point in opposite directions. But where on

FIGURE 23-13 Where will a third charge experience no net force (Example 23-3)?

the axis? In between the two charges the repulsion of one and attraction of the other would add to give a nonzero net force. To the left of $+2q$, a third charge Q would always be closer to $+2q$ than to $-q$. The force from $+2q$ would always be greater, partly because Q would be closer to it and partly because of its greater charge. So the only place the forces from the two charges might cancel is to the right of $-q$.

So suppose the third charge Q lies at some point x to the right of $-q$, so $x > a$. Then a unit vector from either $+2q$ or $-q$ toward Q points in the $+x$ direction, so is the vector $\hat{\imath}$. Summing the two forces as given by Equation 23-1 and setting the result to zero then gives

$$\mathbf{F} = \mathbf{F}_{2q} + \mathbf{F}_{-q} = \frac{k(2q)Q}{(x+a)^2}\hat{\imath} + \frac{k(-q)Q}{(x-a)^2}\hat{\imath} = \mathbf{0},$$

where the denominators are the distances from any point $x > a$ to the charges located at $x = -a$ and $x = +a$, respectively. Note that Q cancels from the equation, showing that the point we're finding will be a point of zero force for *any* charge, no matter what its sign or magnitude. The quantities k and q also cancel, so the x component of our equation for the net force becomes

$$\frac{2}{(x+a)^2} = \frac{1}{(x-a)^2}.$$

Inverting and taking square roots gives

$$\frac{x+a}{\sqrt{2}} = \pm(x-a).$$

We solve separately for the two possible signs. For the $+$ sign, we have

$$x = a\frac{\sqrt{2}+1}{\sqrt{2}-1} = a\frac{(\sqrt{2}+1)^2}{(\sqrt{2}-1)(\sqrt{2}+1)}$$
$$= (3 + 2\sqrt{2})a = 5.83a.$$

Since this value of x is greater than a, this point does indeed lie to the right of $-q$. So it's the point where the two forces balance. Move a little to the left, and the negative charge dominates, exerting a leftward force on a positive Q. Move a little to the right, and the net positive charge of the whole distribution dominates, exerting a rightward force on a positive Q. In between is the point of zero force.

You can verify that the solution for the minus sign lies to the left of $-q$ and is therefore inconsistent with our choice of direction for the unit vector from $-q$. This second solution is therefore not a meaningful answer.

> **Tip**
>
> *Let the Algebra Take Care of Signs* Once again, Equation 23-1 tells it all—including the direction of the force once you choose the unit vector correctly. You may be tempted in a problem like Example 23-3 simply to set the magnitudes of the two forces equal and solve for x. That may work if you're careful, but it's safer to remember that you're really solving for a point where *the vector sum of two forces is zero*. You can't go wrong if you write those forces carefully with the correct vector directions and then sum them.

EXERCISE Repeat Example 23-3 with the charge $-q$ changed to $+q$.

Answer: $(3 - 2\sqrt{2})a = 0.17a$

Some problems similar to Example 23-3: 16–18, 32, 34

23-4 THE ELECTRIC FIELD

In Chapter 9 we defined the gravitational field at a point as the gravitational force per unit mass that an object at that point would experience. Even without Chapter 9, you're still familiar with this concept. You know that the force on an object of mass m near Earth's surface has magnitude mg and points downward. So we can think of g as a measure of the *force per unit mass* that any object would experience due to Earth's gravity. Full specification of that force includes its direction, which in this case is downward. So we can picture the gravitational field as a set of vectors giving the magnitude and direction of the gravitational force per unit mass at each point (Fig. 23-14a). In so doing, we've generalized to describe the force Earth would exert on *any* mass, rather than just its force on some particular mass.

It's convenient to do the same thing with the electric force. Rather than finding the force that a charge distribution exerts on some particular charge, we define the **electric field** as the force per unit charge:

The electric field at any point is the force per unit charge that would be experienced by a charge at that point. Mathematically,

$$\mathbf{E} = \frac{\mathbf{F}}{q}.$$ (23-3a)

We emphasize that the electric field is a continuous entity, existing at every point in space. When we represent the field by vectors, we can't draw one everywhere, but that doesn't mean there isn't field at all points. Furthermore, we draw vectors as extended arrows, but each vector represents the field at only one point, namely the tail end of the vector. Figure 23-14b illustrates this for the electric field of a point charge.

The field concept implies a shift in our thinking about forces. Instead of the "action-at-a-distance" idea that Earth reaches out across empty space to pull on the moon, the gravitational field concept says that Earth creates a gravitational field and that the moon responds to the field at its location. The same is true for the electric field. Instead of thinking of one charge attracting or repelling another, we view a charge as creating an electric field throughout the space surrounding it. A second charge then responds to the field at its immediate location. Although the field only reveals itself through its effect on a charge, we nevertheless consider that the field exists at all points, whether or not there are charges present to feel a force. Right now you probably find the field concept to be a bit abstract, but as you advance in your study of electromagnetism you'll come to appreciate that fields are an essential feature of our universe, every bit as real as matter itself.

We can use Equation 23-3a as a prescription for measuring electric fields. Place a point charge at some point, measure the electric force it experiences, and divide by the charge to get the field. In practice, we need to be careful because the field generally arises from some distribution of charges, which we call the **source charges**. If the charge we're using to probe the field—the **test charge**—is large, the field it creates may disturb the source charges, altering their configuration and thus the field they create. For that reason, it's important to use a very small test charge.

If we know the electric field **E** at a point, we can rearrange Equation 23-3a to find the force on any point charge q placed at that point:

$$\mathbf{F} = q\mathbf{E}.$$ (23-3b)

If the charge q is positive, then this force is in the same direction as the field, but if q is negative then the force it experiences is opposite to the field direction.

Equations 23-3 show that the units of electric field are newtons per coulomb. Fields of hundreds to thousands of N/C are commonplace, while fields of 3×10^6 N/C will tear electrons from air molecules. Electric fields within atoms can exceed 10^{12} N/C.

FIGURE 23-14 (a) Representing the gravitational field near Earth's surface as a set of vectors. Each has magnitude $g = 9.8$ N/C and points downward. Each field vector gives the force per unit mass at a single point, namely the tail of the vector. Blowup shows that the field exists everywhere; we just can't draw vectors everywhere. (b) The electric field of a positive point charge points radially outward and its magnitude decreases with distance from the point charge (here the arrow lengths don't accurately reflect the exact inverse-square falloff). Each field vector gives the force per unit charge at a single point, namely the tail of the vector. As the blowup suggests, the field is continuous, with a field vector for each point in space. Figure is a 2-dimensional cross section of the field; in three dimensions, field vectors extend radially outward in all directions.

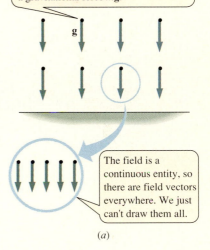

Right at this point the gravitational field is described by the vector **g**. That means a mass m placed here would experience a gravitational force $m\mathbf{g}$.

The field is a continuous entity, so there are field vectors everywhere. We just can't draw them all.

(a)

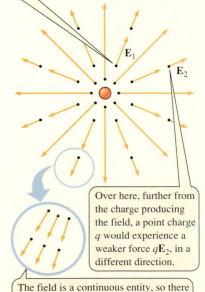

Right at this point the electric field is described by the vector $\mathbf{E}_1$. That means a point charge q placed here would experience an electric force $q\mathbf{E}_1$.

Over here, further from the charge producing the field, a point charge q would experience a weaker force $q\mathbf{E}_2$, in a different direction.

The field is a continuous entity, so there are field vectors everywhere. We just can't draw them all.

(b)

EXAMPLE 23-4	A Thunderstorm: Force and Field

A charged raindrop carrying 10 μC experiences an electric force of 0.30 N in the $+x$ direction. What is the electric field at its location? What would be the force on a -5.0-μC drop at the same location?

Solution

Equation 23-2a defines the electric field; here the force is $0.30\hat{\imath}$ N, so

$$E = \frac{F}{q} = \frac{0.30\hat{\imath}\ N}{10 \times 10^{-6}\ C} = 30\hat{\imath}\ kN/C\ .$$

Acting on a -5.0-μC charge, this field would result in a force

$$F = qE = (-5.0\ \mu C)(30\hat{\imath}\ kN/C) = -0.15\hat{\imath}\ N\ .$$

Tip

The Field Is Independent of the Test Charge You might wonder if the field should point in the $-x$ direction when we talk about putting a negative charge in the field. It doesn't—because the whole point of the field concept is to provide a description that's independent of the particular charge experiencing that force.

The electric field in this example points in the $+x$ direction *no matter what charge* we may choose to put in the field. For a positive charge the force qE points in the *same* direction as the field; for a negative charge $q < 0$, and the force is *opposite* the field direction. As always, the algebra takes care of the signs—although you should check that your algebraic answers make physical sense.

EXERCISE An electric field $E = -450\hat{\imath}$ kN/C is used to accelerate electrons in a portion of a TV picture tube, where the x axis points from the back to the front of the tube. Find the force in this field experienced by (a) an electron and (b) an ion carrying $+2$ elementary charges. (c) Which particle will be accelerated toward the front of the tube?

Answers: (a) $7.2 \times 10^{-14}\hat{\imath}$ N; (b) $-1.4 \times 10^{-13}\hat{\imath}$ N; (c) the electron

Some problems similar to Example 23-4: 26–28

The Field of a Point Charge

Once we know the field of a charge distribution we can calculate its effect on other charges. The simplest charge distribution is a single point charge. Coulomb's law gives the force on a test charge q_1 located a distance r from a point charge q:

$$F = \frac{kqq_1}{r^2}\hat{r}\ ,$$

where $\hat{r}$ is a unit vector pointing *away* from q. The electric field arising from q is the force per unit charge, or

$$E = \frac{F}{q_1} = \frac{kq}{r^2}\hat{r}\ . \qquad \text{(field of a point charge)} \qquad (23\text{-}4)$$

Since it is so closely related to Coulomb's law for the electric force, we also refer to Equation 23-4 as Coulomb's law. Note that the equation contains no reference to the test charge q_1, since the field of q exists independently of any other charge. Since $\hat{r}$ always points *away* from q, the direction of E is radially outward if q is positive and radially inward if q is negative. Figure 23-15 shows some field

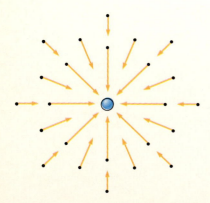

FIGURE 23-15 Field vectors for a negative point charge point radially inward, and their magnitude falls off with increasing distance from the point charge. As in Fig. 23-14b, each vector gives the force per unit charge that a test charge would experience when placed at the vector's tail. In three dimensions the vectors would fill all space in a spherically symmetric fashion.

vectors for a negative point charge, analogous to those of the positive point charge in Fig. 23-14b. ActivPhysics Activity 11.4 explores the point charge electric field.

11.4
Electric Field: Point Charge

23-5 ELECTRIC FIELDS OF CHARGE DISTRIBUTIONS

The electric field is the electric force per unit charge. Since the electric force obeys the superposition principle, so does the electric field. That means the field of a charge distribution is the vector sum of the fields of the individual point charges comprising the distribution:

$$\mathbf{E} = \mathbf{E}_1 + \mathbf{E}_2 + \mathbf{E}_3 + \cdots = \sum_i \mathbf{E}_i = \sum_i \frac{kq_i}{r_i^2}\hat{\mathbf{r}}_i, \qquad (23\text{-}5)$$

where the $\mathbf{E}_i$'s are the fields of the point charges q_i located at distances r_i from the point where we're evaluating the field, and where the $\hat{\mathbf{r}}_i$'s are unit vectors pointing *from* each point charge *toward* where we're evaluating the field. In principle, Equation 23-5 gives the electric field of *any* charge distribution. In practice, the process of summing the individual field vectors is often complicated unless the charge distribution contains relatively few charges arranged in a symmetric way.

EXAMPLE 23-5	*Two Protons*

Two protons are 3.6 nm apart. (a) Find the electric field at the point P shown in Fig. 23-16. (b) Find the force on an electron at point P.

Solution
If we take the x axis along the line joining the protons, then $\hat{\mathbf{r}}_1$ is just $\hat{\mathbf{i}}$ and $\hat{\mathbf{r}}_2$ is $-\hat{\mathbf{i}}$, where the subscripts 1 and 2 refer to the left and right protons, respectively. Then Equation 23-5 gives

$$\mathbf{E} = \mathbf{E}_1 + \mathbf{E}_2 = \frac{ke}{r_1^2}\hat{\mathbf{i}} + \frac{ke}{r_2^2}(-\hat{\mathbf{i}}) = ke\left(\frac{1}{r_1^2} - \frac{1}{r_2^2}\right)\hat{\mathbf{i}}$$

$$= (9.0\times10^9 \text{ N·m}^2/\text{C}^2)(1.6\times10^{-19} \text{ C})$$

$$\times \left[\frac{1}{(1.2\times10^{-9} \text{ m})^2} - \frac{1}{(2.4\times10^{-9} \text{ m})^2}\right]\hat{\mathbf{i}}$$

$$= 750\hat{\mathbf{i}} \text{ MN/C}.$$

FIGURE 23-16 What is the electric field at P (Example 23-5)?

An electron in this field will experience a force

$$\mathbf{F} = q\mathbf{E} = (-1.6\times10^{-19} \text{ C})(750\times10^6 \,\hat{\mathbf{i}} \text{ N/C})$$

$$= -1.2\times10^{-10}\hat{\mathbf{i}} \text{ N}.$$

EXERCISE Find the magnitude of the electric field (a) 4.8 nm and (b) 200 nm to the right of the right-hand proton in Fig. 23-16. (c) Show that your answer to (b) is nearly equal to the field of a single charge 2e located midway between the two protons.

Answers: (a) 82.9 MN/C; (b) 70.7 kN/C

Some problems similar to Example 23-5: 31–34

The Electric Dipole

One of the most important charge distributions is the **electric dipole**, consisting of two point charges of equal magnitude but opposite sign held a fixed distance apart. Many molecules are essentially electric dipoles, so an understanding of

FIGURE 23-17 (left) A water molecule. Electrons spend more time in the vicinity of the single oxygen atom, giving rise to separate regions of negative and positive charge. The molecule therefore approximates the dipole shown at right, and results in an electric field like that of Example 23-6.

the dipole helps explain molecular behavior (Fig. 23-17). During contraction the heart muscle becomes essentially a dipole, and physicians performing electrocardiography are measuring, among other things, the strength and orientation of that dipole. The dipole configuration is also used in a number of technological devices such as radio and TV antennas.

EXAMPLE 23-6	*Modeling a Molecule*

A molecule consists of separated regions of positive and negative charge, modeled approximately as a positive charge q at $x = a$ and a negative charge $-q$ at $x = -a$, as shown in Fig. 23-18. Find a general expression for the electric field at any point on the y axis, and an approximate expression valid at large distances ($y \gg a$).

Solution

Figure 23-18 shows the individual field vectors $\mathbf{E}_-$ and $\mathbf{E}_+$, along with their sum. The y components cancel to give a net field parallel to the x axis, so we can get the net field by summing the two x components. Figure 23-18 and Equation 23-2b show that the x component of the unit vector from the negative charge toward the field point is $\cos\theta$, where $\cos\theta = a/r$; the x component of the unit vector from the positive charge is then $-a/r$. So the net field on the y axis becomes

$$\mathbf{E} = \frac{k(-q)}{r^2}\left(\frac{a}{r}\right)\hat{\mathbf{i}} + \frac{kq}{r^2}\left(-\frac{a}{r}\right)\hat{\mathbf{i}} = -\frac{2kqa}{(y^2 + a^2)^{3/2}}\hat{\mathbf{i}},$$

where the two terms in the sum correspond to the negative and positive charges, and where in the last step we used $r = \sqrt{y^2 + a^2}$. Although the x components of the unit vectors from the two charges have opposite signs, so do the charges—and thus the x components of the fields themselves are actually the same, as is obvious in Fig. 23-18.

Does this result make sense? Midway between the charges the fields from each charge point in the same direction and have the same magnitude, so we expect a resultant field twice that of either charge alone. Setting $y = 0$ at the midpoint gives

$$\mathbf{E}(y=0) = -\frac{2kqa}{(a^2)^{3/2}}\hat{\mathbf{i}} = -\frac{k(2q)}{a^2}\hat{\mathbf{i}},$$

which is indeed twice the field of either charge at the distance a.

We're frequently interested in the field far from a dipole, which is why this example asks for an approximate expression for $y \gg a$. (One needn't go very far from a molecule for its size to be insignificant.) Under this approximation we can neglect a^2 compared with y^2 in our expression for the field, giving

$$\mathbf{E} = -\frac{2kqa}{y^3}\hat{\mathbf{i}}, \qquad y \gg a.$$

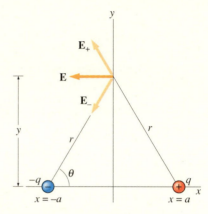

FIGURE 23-18 An electric dipole. The electric field on the dipole's perpendicular bisector (here the y axis) is parallel to the dipole axis.

We'll see later how forces associated with such molecular electric fields contribute to the van der Waals interaction that we considered in Chapter 20.

Tip

Approximations Making approximations requires some care. Here we're basically asking for the field when y is so large that a is negligible *compared with y*. So we neglect a^2 compared with y^2 when the two are summed, but we *don't* neglect a when it appears in the numerator, where it's not summed with y.

EXERCISE Find (a) a general expression for the dipole field at points on the x axis to the right of $x = a$ in Fig. 23-18, and (b) an approximate expression valid for $x \gg a$.

Answers: (a) $\mathbf{E} = \dfrac{4kqax}{(x^2 - a^2)^2}\hat{\mathbf{i}}$; (b) $\mathbf{E} = \dfrac{4kqa}{x^3}\hat{\mathbf{i}}$

Some problems similar to Example 23-6: 31, 37, 38, 41

The dipole fields in Example 23-6 and its exercise both decrease, at large distances, as the inverse *cube* of distance. Physically, this is because the dipole has zero *net* charge. Its field arises entirely from the slight separation of two opposite charges. Because of this separation the dipole field isn't exactly zero, but it is weaker and more localized than the field of a point charge. Many complicated charge distributions exhibit the essential characteristic of a dipole—namely, they're neutral but consist of separated regions of positive and negative charge—and at large distances such distributions all have essentially the same field configuration (see Problems 46 and 82). ActivPhysics Activity 11.5 explores the dipole field. By placing the two charges in simulation 1 close together you can verify approximately the inverse-cube dependence of the dipole field.

11.5
Electric Field:
Due to a Dipole

At large distances the results of Example 23-6 and its exercise show that the dipole's physical characteristics q and a enter the equations for the electric field only through the product qa. We could double q (that is, double both the positive and the negative charge) and halve a, and the dipole's electric field would remain unchanged. In general, a dipole's electrical properties are characterized completely by its **electric dipole moment**, p, defined as the product of the charge q and the separation d between the two charges making up the dipole:

$$p = qd. \quad \text{(dipole moment)} \qquad (23\text{-}6)$$

In Example 23-6 the charge separation was $d = 2a$, so there the dipole moment was $p = 2aq$. In terms of the dipole moment, the fields we derived in Example 23-6 can then be written

$$\mathbf{E} = -\frac{kp}{y^3}\hat{\mathbf{i}} \quad \left(\begin{array}{l}\text{dipole field for } y \gg a,\\ \text{on perpendicular bisector}\end{array}\right) \qquad (23\text{-}7a)$$

and

$$\mathbf{E} = \frac{2kp}{x^3}\hat{\mathbf{i}} \quad \left(\begin{array}{l}\text{dipole field}\\ \text{for } x \gg a, \text{ on axis}\end{array}\right). \qquad (23\text{-}7b)$$

Because the dipole isn't spherically symmetric, its field depends not only on distance but also on orientation; for instance, Equations 23-7a and b show that the field along the dipole axis at a given distance is twice as strong as along the bisector. So it's important to know the orientation of a dipole in space, and therefore we generalize our definition of the dipole moment to make it a vector, of magnitude $p = qd$, and in the direction from the negative toward the positive charge (Fig. 23-19).

FIGURE 23-19 The dipole moment vector has magnitude $p = qd$ given by the product of the charge and separation, and it points from the negative toward the positive charge.

Got It!

Far from a charge distribution, you measure an electric field strength of 800 N/C. What will be the field strength if you double your distance from the charge distribution, if the distribution consists of (a) a point charge or (b) a dipole?

Continuous Charge Distributions

Although any charge distribution ultimately consists of pointlike electrons and protons, it would be impossible in practice to sum all the field vectors from the 10^{23} or so particles in a typical piece of matter. Instead, it's convenient to make the approximation that charge is spread continuously over the distribution. If the charge distribution extends throughout a volume, we describe it in terms of the **volume charge density**, ρ, with units of C/m³. For charge distributions spread over surfaces or lines the corresponding quantities are the **surface charge density**, σ, and **line charge density**, λ. Their units are C/m² and C/m, respectively.

To calculate the field of a continuous charge distribution, we break the charged region into very many small charge elements dq, each small enough that it is essentially a point charge. Each dq then produces an electric field $d\mathbf{E}$ given by Equation 23-3:

$$d\mathbf{E} = \frac{k\,dq}{r^2}\hat{\mathbf{r}}. \tag{23-8a}$$

Charge distribution

FIGURE 23-20 The electric field at point P, given by Equation 23-8b, is the sum of the vectors $d\mathbf{E}$ arising from the individual charge elements dq in the entire distribution, each calculated using the appropriate distance r and unit vector $\hat{\mathbf{r}}$.

We then form the vector sum of all the $d\mathbf{E}$'s. In the limit of infinitely many infinitesimally small dq's and their corresponding $d\mathbf{E}$'s, that sum becomes an integral and we have

$$\mathbf{E} = \int d\mathbf{E} = \int \frac{k\,dq}{r^2}\hat{\mathbf{r}}. \tag{23-8b}$$

The limits of this integral are chosen to include the entire charge distribution. Figure 23-20 shows the meaning of Equation 23-8b; note in particular that both the distance r and the direction specified by $\hat{\mathbf{r}}$ generally vary with position.

EXAMPLE 23-7	A Charged Rod

Wires, antennas, and similar elongated structures can often be treated as thin rods carrying electric charge. Suppose such a rod of length ℓ carries a positive charge Q distributed uniformly over its length. Find the electric field at point P in Fig. 23-21, a distance a from the end of the rod.

FIGURE 23-21 The field at P is the sum—or integral—of the fields arising from all the infinitesimal charge elements dq along the rod.

Solution

Let the y axis lie along the rod, with origin at P. Consider a small length dy of the rod, containing charge dq, and located a distance y from P. A unit vector from dq toward P is $-\hat{\mathbf{j}}$, so the field at P due to dq is

$$d\mathbf{E} = -\frac{k\,dq}{y^2}\hat{\mathbf{j}}.$$

The net field at P is the sum—that is, the integral—of all the fields $d\mathbf{E}$ arising from all the dq's along the rod:

$$\mathbf{E} = \int d\mathbf{E} = -\hat{\mathbf{j}} \int_{y=a}^{y=a+\ell} \frac{k\,dq}{y^2},$$

where we've chosen the limits to cover the entire rod. To evaluate the integral we must relate dq and y. How? The rod carries total charge Q, distributed uniformly over its length ℓ. The line charge density is therefore $\lambda = Q/\ell$. This is the charge per unit length; a length dy therefore carries charge $dq = \lambda\,dy$, or $dq = Q\,dy/\ell$. Then our integral becomes

$$\mathbf{E} = -\hat{\mathbf{j}} \int_a^{a+\ell} \frac{k(Q\,dy/\ell)}{y^2} = -\frac{kQ}{\ell}\hat{\mathbf{j}} \int_a^{a+\ell} \frac{dy}{y^2}$$

$$= -\frac{kQ}{\ell}\hat{\mathbf{j}}\left[-\frac{1}{y}\right]_a^{a+\ell} = -\frac{kQ}{\ell}\hat{\mathbf{j}}\left(\frac{1}{a} - \frac{1}{a+\ell}\right) = -\frac{kQ}{a(a+\ell)}\hat{\mathbf{j}},$$

where we took k, Q, ℓ, and $\hat{\mathbf{j}}$ outside the integral because they are constants.

Does this result make sense? First consider the direction: The negative sign shows that the field is downward for positive Q and upward for negative Q, as we should expect. Now suppose P is very far from the rod, so $a \gg \ell$. Then our result becomes approximately

$$\mathbf{E} = -\frac{kQ}{a^2}\hat{\mathbf{j}}, \qquad (a \gg \ell),$$

which is just what we expect for the field of a point charge Q. In this case we're so far from the rod that its length becomes negligible, and indeed it acts like a point charge. But as we move closer, the field becomes a more complicated superposition of the fields of all the dq's at different locations along the rod, and the field no longer exhibits the inverse-square dependence of the point-charge field.

Knowing the field of the charged rod, we can use superposition to find the fields of charge distributions involving more than one rod or a combination of rods and point charges (see Problems 44 and 46).

Tip

Find a Single Integration Variable Evaluating the integral in Equation 23-8b requires that we relate the charge element dq and the position variable r so we'll have the integral expressed in terms of a single vari-

able. The charge density provides the link needed, since it allows us to write dq as the charge density multiplied by an appropriate element of volume, area, or length:

$$dq = \rho\,dV, \quad dq = \sigma\,dA, \quad \text{or} \quad dq = \lambda\,dx.$$

Which of these we use depends on whether charge is distributed over a volume, area, or length; for the thin rod of Example 23-7, the appropriate charge density was the line charge density λ and the position variable was y, so we had $dq = \lambda\,dy$. Sometimes you'll be given the charge density explicitly, and other times you can compute it from the charge and dimensions of the charge distribution.

EXERCISE A thin rod of length ℓ lies along the y axis with its bottom end at the origin. It carries a line charge density λ that varies with position, being given by $\lambda = Q_0\dfrac{y^3}{\ell^4}$, where Q_0 is a constant and y is the distance from the origin. Find the magnitude of the electric field at $y = 0$.

Answer: $\dfrac{kQ_0}{2\ell^2}$

Some problems similar to Example 23-7: 43–46, 71, 72, 82

Got It!

The point P in Fig. 23-21 is a distance a from the charged rod carrying total charge Q. So why can't you just write kQ/a^2 for the magnitude of the field at P?

EXAMPLE 23-8	*A Charged Ring*

A thin ring of radius a is centered on the origin and carries a total charge Q distributed uniformly around the ring, as shown in Fig. 23-22. Find the electric field at a point P located a distance x along the axis of the ring, and show that the result makes sense when $x \gg a$.

Solution

In Example 23-7 the magnitude but not direction of the individual field vectors $d\mathbf{E}$ from all the charge elements dq varied. Here we have the opposite situation: a point on the ring axis is equidistant from all points on the ring, so the field magnitudes dE are the same but their directions vary. Figure 23-22 shows, however, that components perpendicular to the x axis cancel for any pair of charge elements on opposite sides of the ring,

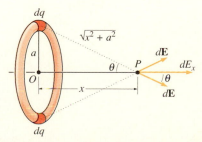

FIGURE 23-22 The electric field of a charged ring points along the ring axis since field components perpendicular to the axis cancel in pairs.

leaving a net field in the x direction. Each charge element contributes an amount dE_x to the field:

$$dE_x = \frac{k\,dq}{r^2}\cos\theta = \frac{k\,dq}{x^2 + a^2}\frac{x}{\sqrt{x^2 + a^2}} = \frac{kx\,dq}{(x^2 + a^2)^{3/2}},$$

where the geometry of Fig. 23-22 gives $r = \sqrt{x^2 + a^2}$ and $\cos\theta = x/\sqrt{x^2 + a^2}$; here $\cos\theta$ is the x component of the unit vector in Equation 23-8a. We now need to integrate this expression over the entire ring. In this integration k, a, and x are all constants—they don't change as we move around the ring. So we have

$$E = E_x = \int_{\text{ring}} dE_x = \int_{\text{ring}} \frac{kx\,dq}{(x^2 + a^2)^{3/2}}$$

$$= \frac{kx}{(x^2 + a^2)^{3/2}}\int_{\text{ring}} dq = \frac{kx\,Q}{(x^2 + a^2)^{3/2}},$$

where the last step follows because $\int_{\text{ring}} dq$ simply means the total charge on the ring. For positive Q this field points away from the ring.

Does this result make sense? At large distances from the ring we can neglect its size a compared with the distance x, and our result reduces to $E = \dfrac{kQ}{x^2}$—just what we would expect for

a point charge Q. As always, a finite-size charge distribution looks essentially like a point charge at distances large compared with its size.

Tip

How Can x Be a Constant? In this exercise x is constant because it's the distance from the center of the ring to the point where we're evaluating the field—that is, it's the distance from the origin to the *field point*, P. In the integration to find the field we're supposed to consider all *source points*—all points where charge is located that contributes to the field at P. Moving around the ring doesn't affect the value of x, so x is a constant for this integration. However, x is arbitrary, so our result holds for *any* value of x.

EXERCISE Find the point on the x axis where the electric field of the ring in Example 23-8 has its greatest magnitude.

Answer: $x = a/\sqrt{2}$

Some problems similar to Example 23-8: 47–50, 76, 83

EXAMPLE 23-9	*A Power Line's Field*

A long, straight electric power line coincides with the x axis and carries a line charge density λ C/m. What is the electric field at a point P on the y axis? Use the approximation that the line is infinitely long.

Solution

Here *both* the direction and magnitude of the field element $d\mathbf{E}$ arising from a charge element on the line vary with the position x of the charge element. But Fig. 23-23 shows that charge

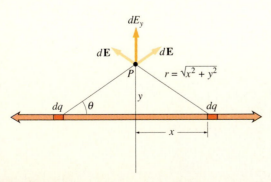

FIGURE 23-23 The field of a charged line is the vector sum of the fields $d\mathbf{E}$ from all the individual charge elements dq along the line. The x components from each pair of charge elements cancel, giving a net field that points directly away from the line.

elements on opposite sides of the y axis give rise to electric fields whose x components cancel. Thus the net field points in the y direction—that is, away from the line if λ is positive. So each charge element contributes an amount dE_y to the net field:

$$dE_y = \frac{k\,dq}{r^2}\sin\theta = \frac{k\lambda\,dx}{x^2 + y^2}\frac{y}{\sqrt{x^2 + y^2}} = \frac{k\lambda y\,dx}{(x^2 + y^2)^{3/2}},$$

where we've written $dq = \lambda\,dx$ and used the geometry of Fig. 23-23 to write $r = \sqrt{x^2 + y^2}$ and $\sin\theta = y/r$. Here, as usual, $\sin\theta$ is the y component of the unit vector $\hat{\mathbf{r}}$. To find the net field, we integrate over the entire line, from $x = -\infty$ to $x = +\infty$. The quantities k, λ, and y don't change as we move along the line, so we have

$$E = E_y = \int_{x=-\infty}^{x=\infty} \frac{k\lambda y\,dx}{(x^2 + y^2)^{3/2}} = k\lambda y \int_{-\infty}^{\infty} \frac{dx}{(x^2 + y^2)^{3/2}}$$

$$= k\lambda y\left[\frac{x}{y^2\sqrt{x^2 + y^2}}\right]_{-\infty}^{\infty} = k\lambda y\left[\frac{1}{y^2} - \left(-\frac{1}{y^2}\right)\right] = \frac{2k\lambda}{y}.$$

$$(23\text{-}9)$$

Here we evaluated the integral using the integral table in Appendix A and applied the limits $\pm\infty$ by noting that as $x \to \pm\infty$ the term y^2 becomes negligible compared with x^2, giving $x/\sqrt{x^2 + y^2} \to x/\sqrt{x^2} = \pm 1$. The integral could also be evaluated by rewriting it in terms of the angle θ (see Problem 87).

Since the line is infinite in both directions and has cylindrical symmetry, Equation 23-9 holds for *any* point a distance *y* from the line. Our result thus shows that the electric field of a long line of positive charge points radially away from the line, with magnitude that drops inversely with distance from the line (Fig. 23-24).

FIGURE 23-24 Field vectors for an infinite line of positive charge point radially outward from the line, with magnitude decreasing inversely with distance.

What about the field at large distances? Shouldn't it resemble that of a point charge, falling with the inverse square of the distance? No: The charged line is infinitely long, so no matter how far away we go it never resembles a point. Its slower dropoff reflects that fact.

Of course our infinite line is an impossibility. But many real charge distributions, including the power line of this example, have long, thin shapes that approximate an infinite line. Equation 23-9 is therefore a good approximation to the field of a *finite* line as long as we're much closer to it than its length, and not too near either end. Very far from a *finite* line, on the other hand, the field does approach that of a point charge.

EXERCISE A thin rod 2.0 m long carries 50 μC distributed uniformly over its length. Find the electric field strength (a) 1.0 cm from the rod axis, not near either end and (b) 500 m from the rod. Make suitable approximations in both cases.

Answers: (a) 45 MN/C; (b) 1.8 N/C

Some problems similar to Example 23-9: 50–54

23-6 MATTER IN ELECTRIC FIELDS

We're ultimately interested in electric fields since they give rise to forces on charged particles. Because matter consists of such particles, much of the behavior of matter is fundamentally determined by electric fields.

Point Charges in Electric Fields

The motion of a single charge in an electric field is governed by the definition of the electric field (Equation 23-3b):

$$\mathbf{F} = q\mathbf{E}$$

and Newton's law:

$$\mathbf{F} = m\mathbf{a} .$$

Combining these equations gives the acceleration of a particle with charge q and mass m in an electric field $\mathbf{E}$:

$$\mathbf{a} = \frac{q}{m}\mathbf{E} . \qquad (23\text{-}10)$$

This equation shows that it's the charge-to-mass ratio, q/m, that determines a particle's response to an electric field. Electrons, nearly 2000 times less massive than protons but carrying the same charge, are readily accelerated by electric fields. Many practical devices, from x-ray machines to TV tubes, make use of electrons accelerated in electric fields.

When the electric field is uniform, problems involving the motion of charged particles reduce to the constant-acceleration problems we considered in Chapter 2. We'll see in the next chapter that uniform fields are easily produced by flat, uniformly charged plates. ActivPhysics Activities 11.7 and 11.8 simulate charged-particle motion in uniform electric fields.

11.7, 11.8
Motion in an Electric Field

EXAMPLE 23-10	*Inside a Heart Monitor*

A heart monitor used in a hospital's intensive care unit includes a cathode-ray tube that gives continuous visual display of a patient's heartbeat (Fig. 23-25). An electron beam "paints" the display on a phosphor screen at the front of the tube, being "steered" to different parts of the screen by electric forces. Figure 23-26 shows a side view of the "steering" region of the tube. The region is 2.0 cm long and contains a uniform electric field of 1.0 kN/C. (The field is produced by charging two flat metal plates with opposite charges, as we'll see in the next chapter.) Electrons enter this region moving horizontally at 4.0 Mm/s, as shown. In what direction are the electrons moving as they leave the field region?

FIGURE 23-25 This heart monitor uses an electron beam deflected by electric fields to display patients' heartbeats.

FIGURE 23-26 A uniform electric field deflects an electron from its straight-line path (Example 23-10).

Solution

As usual in two-dimensional motion, the vertical force due to the electric field does not affect the horizontal component of an electron's velocity. Therefore each electron spends a time $t = \Delta x/v_x$ in the field region. During this time it experiences a vertical acceleration qE/m, and therefore gains a vertical velocity component given by

$$v_y = a_y t = \frac{qE_y}{m}\frac{\Delta x}{v_x}$$

$$= \frac{(-1.6\times10^{-19}\ \text{C})(-1.0\times10^3\ \text{N/C})}{9.11\times10^{-31}\ \text{kg}}\left(\frac{2.0\times10^{-2}\ \text{m}}{4.0\times10^6\ \text{m/s}}\right)$$

$$= 8.78\times10^5\ \text{m/s}\ .$$

Note that the field points downward, as reflected by the minus sign, and combined with the electron's negative charge thus results in an upward velocity. The electron thus leaves the field region moving at an angle of

$$\theta = \tan^{-1}\left(\frac{v_y}{v_x}\right) = \tan^{-1}\left(\frac{8.78\times10^5\ \text{m/s}}{4.0\times10^6\ \text{m/s}}\right) = 12.4°$$

to the horizontal.

As it traverses the field region the electron describes the parabolic trajectory of an object undergoing constant acceleration in two dimensions; once outside the field it continues in a straight line with its new velocity, as indicated in Fig. 23-26. Field configurations like this one are used to "steer" electron beams not only in heart monitors but also in oscilloscopes and other electronic instrumentation.

EXERCISE What electric field strength would be needed to cause the electron of Example 23-10 to leave the field region at a point 4.6 mm above where it entered?

Answer: 2.1 kN/C

Some problems similar to Example 23-10: 57–61, 75, 76

An application of charge motion in uniform electric fields that's important to biochemists is the process of electrophoresis. Here, charged molecules move through a gel substrate under the influence of a uniform electric field. The drag force of the gel (ultimately electric in origin) counters the $q\mathbf{E}$ force of the electric field, resulting in each molecule having a constant velocity that is greater for molecules of lower mass. The field is applied for a given time, and molecules migrate different distances through the gel. The resulting pattern provides an indication of the different molecular species present (Fig. 23-27). A variation on this technique is the most widely used method of determining the molecular weights of proteins. Electrophoresis is also used to separate particles of all sizes

FIGURE 23-27 Gel electrophoresis of DNA produced these characteristic patterns. Each band marks the final position of a different DNA strand.

in studies of such diverse mixtures as blood and geological sediments. It's sometimes coupled with laser Doppler shift measurements (see Chapter 17) to provide the speeds of the migrating charged particles and give high-precision results in very short times.

When the field is not uniform it's generally difficult to calculate particle trajectories. An important exception is the case of a particle moving at right angles to a field that points radially. In that case—the subject of the following example—the electric force changes the particle's direction but not its speed, so the motion is uniform circular.

EXAMPLE 23-11	*An Electrostatic Analyzer*

Two curved metal plates are used to establish an electric field given by

$$E = E_0 \frac{b}{r},$$

where $E_0 = 24$ kN/C and $b = 5.0$ cm. The field points toward the center of curvature, as shown in Fig. 23-28, and r is the distance from that center. A beam of protons with a mix of speeds is incident on the device. Find the speed v for which an incident proton will leave the analyzer moving horizontally in Fig. 23-28.

Solution

To exit with its velocity horizontal, a proton must describe a circular arc while inside the analyzer. The field provides the v^2/r acceleration required for that circular motion:

$$a = \frac{v^2}{r} = \frac{eE}{m} = \frac{e}{m} E_0 \frac{b}{r}.$$

Solving for v gives

$$v = \sqrt{\frac{eE_0 b}{m}} = \sqrt{\frac{(1.6 \times 10^{-19}\ \text{C})(2.4 \times 10^4\ \text{N/C})(0.050\ \text{m})}{1.67 \times 10^{-27}\ \text{kg}}}$$

$$= 3.4 \times 10^5\ \text{m/s}.$$

Note that it doesn't matter where the protons enter the analyzer since the $1/r$ decrease in field strength matches the $1/r$ depen-

FIGURE 23-28 An electrostatic analyzer, showing the trajectories of protons in the radial electric field. Only those entering with the right speed will emerge at the top moving horizontally. Collimators block protons not moving at right angles to the field.

dence of the acceleration. Devices of this sort have been used on spacecraft to analyze charged particles in interplanetary space.

EXERCISE A proton is in circular motion centered on a long charged wire carrying uniform negative line charge density $-\lambda$. Find its speed. *Hint:* Consult Example 23-9 for the field of the charged wire.

Answer: $v = \sqrt{2k\lambda e/m}$, with m the proton mass

Some problems similar to Example 23-11: 60–64

Dipoles in Electric Fields

Earlier in this chapter we calculated the field of an electric dipole, which consists of two opposite charges of equal magnitude. Here we study a dipole's response to electric fields. Since the dipole configuration provides a simple model for many molecules, our results help explain molecular behavior.

Figure 23-29 shows a dipole with charges $\pm q$ separated by a distance d, located in a uniform electric field. The dipole moment vector **p** has magnitude qd and points from the negative to the positive charge (recall Fig. 23-19). Since the field is uniform, it's the same at both ends of the dipole. Since the dipole charges are equal in magnitude but opposite in sign, they experience equal but opposite forces $\pm q\mathbf{E}$—and therefore there's no net force on the dipole.

FIGURE 23-29 A dipole in a uniform electric field experiences no net force, but it does experience a torque.

However, Fig. 23-29 shows that the dipole does experience a torque that tends to align it with the field. In Chapter 13 we described torque as the cross product of the position vector with the force: $\boldsymbol{\tau} = \mathbf{r} \times \mathbf{F}$, where the magnitude of the torque vector is $rF\sin\theta$ and its direction is given by the right-hand rule. Figure 23-29 thus shows that the torque about the center of the dipole due to the force on the positive charge has magnitude

$$\tau_+ = rF\sin\theta = \tfrac{1}{2}d\, qE\sin\theta.$$

The torque associated with the negative charge has the same magnitude and both torques are in the same direction since both tend to rotate the dipole in Fig. 23-29 clockwise. Thus the net torque has magnitude $\tau = qdE\sin\theta$; applying the right-hand rule shows that this torque is into the page. But qd is the magnitude of the dipole moment $\mathbf{p}$, and Fig. 23-29 shows that θ is the angle between the dipole moment vector and the electric field $\mathbf{E}$; therefore we can write the torque vectorially as

$$\boldsymbol{\tau} = \mathbf{p} \times \mathbf{E}. \qquad \text{(torque on a dipole)} \qquad (23\text{-}11)$$

Because of this torque, it takes work to rotate a dipole in an electric field. If we start with the dipole oriented at right angles to the field ($\theta = \pi/2$), then Equation 12-22b gives the work required to rotate it to a new angle θ:

$$W = \int_{\pi/2}^{\theta} \tau\, d\theta = \int_{\pi/2}^{\theta} pE\sin\theta\, d\theta = pE\,[-\cos\theta]_{\pi/2}^{\theta} = -pE\cos\theta,$$

where the last step follows because $\cos(\pi/2) = 0$. This work ends up as stored potential energy U. Since the product of two vector magnitudes with the cosine of the angle between the vectors defines the dot product, we can write the potential energy in compact form as

$$U = -\mathbf{p} \cdot \mathbf{E}, \qquad (23\text{-}12)$$

where the zero of potential energy corresponds to the dipole at right angles to the field.

APPLICATION	*Microwave Cooking and Liquid Crystals*

The torque on dipoles in electric fields forms the basis of two widespread contemporary technologies: the microwave oven and the liquid crystal display (LCD).

A microwave oven works by generating an electric field whose direction changes several billion times per second. Water molecules, whose dipole moment is much greater than that of most other molecules, respond by attempting to align with the field. But the field is constantly changing, so the molecules swing rapidly back and forth. As they jostle against each other, the energy they gain from the field is dissipated as heat that cooks the food (see Fig. 23-30).

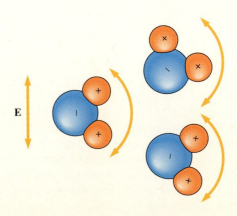

FIGURE 23-30 Water heats in a microwave oven because individual water molecules act as electric dipoles, experiencing torques in the oven's rapidly reversing electric field.

Calculators, laptop computers, gas pumps, watches, and many other devices display numerical and alphabetic information with liquid crystals. This unique state of matter combines the fluidity of a liquid with the order of a solid. The liquid crystal consists of long molecules whose chemical structure gives rise to a dipole-like charge separation. In response to each others' electric fields, the molecules tend to align with each other (see Fig. 23-31). But an external electric field can rotate the liquid crystal dipoles, altering the optical properties of the material. Using optical components we'll study in Chapter 34, different sections of a liquid crystal display can then be made to appear visible or invisible (Fig. 23-32). Liquid crystal displays have the advantage that they consume very little power. On the other hand they generate no light of their own, and therefore an external light source is needed to illuminate the display when it's dark.

Normal
liquid

Liquid
crystal

FIGURE 23-31 Liquid crystals consist of dipole-like molecules that all align in the same direction, in contrast to the random orientation of molecules in an ordinary liquid. Applying an electric field can change the orientation of the liquid-crystal molecules.

FIGURE 23-32 Changing the orientation of the liquid-crystal molecules alters the optical properties of the liquid crystal, making individual segments of this liquid-crystal display either visible or invisible. (More on liquid-crystal displays in Chapter 34.)

When the electric field is not uniform, the charges at opposite ends of the dipole experience forces that differ in magnitude and/or are not exactly opposite in direction. Then the dipole experiences a net force as well as a torque (Fig. 23-33). An important instance of this effect is the force on a dipole in the field of another dipole (Fig. 23-34). Because the dipole field falls off rapidly with distance and because the dipole responding to the field has closely spaced charges of equal magnitude but opposite sign, the dipole-dipole force is quite weak and falls extremely rapidly with distance. This weak force, which Fig. 23-34 shows to be attractive, is partly responsible for the van der Waals interaction between gas molecules that we considered in Chapter 20. Problems 68 and 85 deal with forces on dipoles in nonuniform fields.

FIGURE 23-33 When the electric field differs in magnitude or direction at the two ends of the dipole, then the dipole experiences a nonzero net force as well as a torque.

Conductors, Insulators, and Dielectrics

Bulk matter contains vast numbers of point charges, namely electrons and protons. In some matter—notably metals, ionic solutions, and ionized gases—individual charges are free to move throughout the material. In such materials—called **conductors**—the application of an electric field results in the ordered motion of electric charge that we call **electric current**. We'll consider the behavior of conductors and related materials called semiconductors in subsequent chapters.

Materials in which charge is not free to move are called **insulators**, since they do not support electric current. Insulators, however, still contain charges—it's just that their charges are bound into neutral molecules. Some molecules, such as water, have intrinsic dipole moments and therefore rotate in response to an applied electric field. Even if they don't have dipole moments, molecules may respond to an electric field by stretching and acquiring **induced dipole**

FIGURE 23-34 Dipole B is aligned with the electric field of dipole A. Since the field of A is stronger at the negative end of B, $\mathbf{F}_-$ is greater in magnitude than $\mathbf{F}_+$, so dipole B experiences a net force toward dipole A. This is the origin of the van der Waals force between gas molecules.

FIGURE 23-35 (*a*) A molecule with no dipole moment has negative charge concentric with positive charge. (*b*) In an applied electric field, the molecule stretches and acquires a dipole moment.

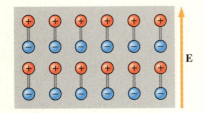

FIGURE 23-36 The alignment of molecular dipoles in a dielectric results in a reduction of the electric field within the dielectric.

moments (Fig. 23-35). In either case, the application of an electric field results in the alignment of molecular dipoles with the field (Fig. 23-36). The fields of the dipoles, pointing from their positive to their negative charges, then reduce the applied electric field within the material. We'll explore the consequences of this effect further in Chapter 26. Materials in which molecules have either intrinsic dipole moments or acquire induced moments are called **dielectrics**.

If the electric field applied to a dielectric becomes too great, individual charges are ripped free, and the material then acts like a conductor. Such **dielectric breakdown** can cause severe damage in electrical equipment (Fig. 23-37). On a larger scale, lightning results from dielectric breakdown in air.

FIGURE 23-37 (*a*) Dielectric breakdown in air results in sparks jumping between two highly charged wires. (*b*) Dielectric breakdown in a solid produced this treelike pattern.

CHAPTER SYNOPSIS

Summary

1. **Electromagnetism** is among the fundamental forces of nature. Electromagnetism comprises electricity and magnetism, seemingly distinct phenomena that are actually intimately related.

2. **Electric charge** is a fundamental property of matter. Charge comes in two types, arbitrarily called positive and negative. Charge is **quantized**, with one elementary charge—the magnitude of the electron or proton charge—equal to 1.60×10^{-19} C, where the coulomb (C) is the SI unit

of charge. Charge is also **conserved**, in that the algebraic sum of the charges in a closed system never changes.

3. Charges interact via the **electric force**. This force is proportional to the product of the charges and inversely proportional to the square of the distance between them. **Coulomb's law** provides a mathematical description of the electric force between point charges:

$$\mathbf{F}_{12} = \frac{kq_1q_2}{r^2}\hat{\mathbf{r}}.$$

4. The **superposition principle** states that the electric force on a charge arising from two or more other charges is the vector sum of the force arising from each according to Coulomb's law. The superposition principle greatly simplifies the calculation of electrical effects of various **charge distributions**.

5. The electric field at a point is a vector giving the electric force per unit charge that would be experienced by a charge at that point:

$$E = \frac{F}{q}.$$

The electric field of a point charge q is therefore

$$E = \frac{kq}{r^2}\hat{r}.$$

6. The superposition principle shows that the electric field of a charge distribution is the sum of the fields of its individual point charges:

$$E = \sum_i E_i = \sum_i \frac{kq_i}{r_i^2}\hat{r}_i.$$

A particularly important charge distribution is the **electric dipole**, consisting of two point charges with equal magnitude but opposite sign, separated by a fixed distance. At large distances from a dipole the field decreases as the inverse cube of the distance.

7. With continuous distributions of charge, the sum over all point charges becomes an integral, giving

$$E = \int dE = \int \frac{k\,dq}{r^2}\hat{r}.$$

For any finite charge distribution with nonzero net charge, the field approaches that of a point charge at large distances.

8. A point charge in an electric field experiences a force $F = qE$; if this is the only force acting, the charge undergoes an acceleration $a = (q/m)E$ in accordance with Newton's law.

9. Because it consists of two equal but opposite point charges, an electric dipole experiences no net force in a uniform electric field. It does, however, experience a torque given by $\tau = p \times E$, where p is the dipole moment vector. A dipole in an electric field has potential energy given by $U = -p \cdot E$, where the zero of potential energy corresponds to the dipole oriented perpendicular to the field. In a nonuniform field, a dipole experiences both a torque and a net force.

Terms You Should Understand

(Pairs are closely related terms whose distinction is important; number in parentheses is chapter section where term first appears.)

electromagnetism (23-1)
electric charge (23-2)
coulomb (23-2)
Coulomb's law (23-3)
superposition principle (23-3)
point charge, charge distribution (23-3)
electric field (23-4)
electric dipole (23-5)
dipole moment (23-5)
volume, surface, and line charge density (23-5)
conductor, insulator, dielectric (23-6)
dielectric breakdown (23-6)

Symbols You Should Recognize

C (23-2)
e (23-2)
k (23-3)
q (23-3)
$\hat{r}$ (23-3)
E (23-4)
p (23-5)
ρ, σ, λ (23-5)

Problems You Should Be Able to Solve

calculating electric forces arising from one or more charges acting on another (23-3)
calculating electric fields of distributions of discrete charges (23-5)
calculating electric fields of continuous charge distributions (23-5)
approximating electric fields at large distances from charge distributions (23-5)
evaluating forces on charges in electric fields (23-4, 23-6)
describing the motion of charged particles in uniform and radial electric fields (23-6)
evaluating torques on dipoles in electric fields (23-6)

Limitations to Keep in Mind

Coulomb's law applies strictly only to one point charge acting on another. The forces and electric fields arising from more than one point charge or from a continuous distribution must be calculated using the superposition principle.

At large distances, the field of a charge distribution approaches that of a point charge only if the distribution is both finite in size and has nonzero net charge.

The dipole field decreases with the inverse cube of the distance only for distances large compared with the dipole's charge separation.

QUESTIONS

1. Discuss this statement: It is precisely because the electric force is so strong that the electrical nature of most everyday interactions is not obvious.
2. The gravitational force between an electron and proton is about 10^{-40} times weaker than the electrical force between the two. Since matter consists largely of electrons and protons, why is the gravitational force ever important?
3. You are given two electric charges. Could you determine whether they had the same or opposite signs? Could you determine the signs of each?
4. In Example 23-3 we found a point where the electric force on a third charge would be zero. Would a charge placed at that point be in stable equilibrium? Why or why not?
5. The gravitational force between an electron and a proton is about 10^{-40} times the electrical force between them. Does this ratio depend on how far apart they are? Explain.
6. In which of the following phenomena does electromagnetism play a dominant role?
   a. Gasoline burns in a car engine.
   b. The moon orbits Earth.
   c. A nerve impulse travels from your brain to a muscle.
   d. Protons and neutrons join to form an atomic nucleus.
   e. A chemist synthesizes a new polymer.
   f. You sit in a chair and the chair doesn't collapse.
7. A free neutron is unstable, and soon decays into other particles. One of the decay products is a proton. Must there be others? If so, what electrical properties must they have?
8. Where in Fig. 23-9 could you put a third charge so it would experience no net force? Would it be in stable or unstable equilibrium?
9. Why should the test charge used to measure an electric field be small?
10. Equation 23-4 gives the electric field of a point charge. Does the direction of $\hat{r}$ depend on whether the charge is positive or negative? Does the direction of **E** depend on the sign of the charge?
11. Is the electric force on a charged particle always in the direction of the field? Explain.
12. Why does a dipole produce an electric field at all? After all, the dipole has no net charge.
13. The ring in Example 23-8 carries total charge Q, and the point P is the same distance $r = \sqrt{x^2 + a^2}$ from all parts of the ring. So why isn't the electric field of the ring just kQ/r^2?
14. The field of a dipole decreases with the inverse cube of the distance. Why doesn't this violate our assertion that the field of a finite size charge distribution with nonzero net charge approaches that of a point charge at large distances?
15. A spherical balloon is initially uncharged. If you spread positive charge uniformly over the balloon's surface, would it expand or contract? What would happen if you spread negative charge instead?
16. Suppose someone argued that the force we call gravity is really an electric force, arising from a net electric charge on Earth. How could you disprove this?
17. Two cubical blocks of wood are each 10 cm on a side and carry electric charge spread over their surfaces. If they're 5 cm apart, would you be justified in writing kq_1q_2/r^2 for the force between them? How about if they're 5 m apart? Explain the difference.
18. A deuteron (heavy hydrogen nucleus) has twice the mass but the same charge as a normal hydrogen nucleus (a proton). Both are released from rest in the same uniform electric field. Compare the distances each goes in the same time.
19. Under what circumstances is the path of a charged particle a parabola? A circle?
20. Explain why a nonuniform field is required for a net force on a dipole.
21. Why should there be a force between two dipoles? After all, each has zero net charge.
22. Dipoles A and B are both located in the field of a point charge Q, as shown in Fig. 23-38. Does either experience a net torque? A net force? If each dipole is released from rest, describe qualitatively its subsequent motion.

FIGURE 23-38 Question 22.

PROBLEMS

ActivPhysics can help with these problems:
Activities 11.1–11.8

Section 23-2 Electric Charge

1. Suppose the electron and proton charges differed by one part in one billion. Estimate the net charge you would carry.
2. A typical lightning flash delivers about 25 C of negativ charge from cloud to ground. How many electrons are involved?
3. Protons and neutrons are made from combinations of the two most common quarks, the u quark and the d quark. The u quark's charge is $+\frac{2}{3}e$ while the d quark carries

$-\frac{1}{3}e$. How could three of these quarks combine to make (a) a proton and (b) a neutron?

4. A 2-g ping-pong ball rubbed against a wool jacket acquires a net positive charge of 1 μC. Estimate the fraction of the ball's electrons that have been removed.

Section 23-3 Coulomb's Law

5. If the charge imbalance of Problem 1 existed, what would be the approximate force between you and another person 10 m away? Treat the people as point charges, and compare the answer with your weight.

6. Find the ratio of the electrical force between a proton and an electron to the gravitational force between the two. Why doesn't it matter that you aren't told the distance between them?

7. The electron and proton in a hydrogen atom are 52.9 pm apart. What is the magnitude of the electric force between them?

8. How far apart should an electron and proton be so the force of Earth's gravity on the electron is equal to the electric force arising from the proton? Your answer shows why gravity is unimportant on the molecular scale!

9. Two charges, one twice as large as the other, are located 15 cm apart and experience a repulsive force of 95 N. What is the magnitude of the larger charge?

10. Earth carries a net charge of -4.3×10^5 C. The force due to this charge is the same as if it were concentrated at Earth's center. How much charge would you have to place on a 1.0-g mass in order for the electrical and gravitational forces on it to balance?

11. A proton is on the x axis at $x = 1.6$ nm. An electron is on the y axis at $y = 0.85$ nm. Find the net force the two exert on a helium nucleus (charge $+2e$) at the origin.

12. Let $\mathbf{V} = V_x\hat{\imath} + V_y\hat{\jmath}$ be an arbitrary vector, with V its magnitude. Show that $\mathbf{V}/V$ is a unit vector—i.e., that its magnitude is 1.

13. A charge q is at the point $x = 1$ m, $y = 0$. Write expressions for the unit vectors you would use in Coulomb's law if you were finding the force that q exerts on other charges located at (a) $x = 1$ m, $y = 1$ m; (b) the origin; (c) $x = 2$ m, $y = 3$ m. Note that you don't know the sign of q. Why doesn't this matter?

14. A proton is at the origin and an electron is at the point $x = 0.41$ nm, $y = 0.36$ nm. Find the electric force on the proton.

15. A 9.5-μC charge is at $x = 16$ cm, $y = 5.0$ cm, and a -3.2-μC charge is at $x = 4.4$ cm, $y = 11$ cm. Find the force on the negative charge.

16. A charge $3q$ is at the origin, and a charge $-2q$ is on the positive x axis at $x = a$. Where would you place a third charge so it would experience no net electric force?

17. A 60-μC charge is at the origin, and a second charge is on the positive x axis at $x = 75$ cm. If a third charge placed at $x = 50$ cm experiences no net force, what is the second charge?

18. You have two charges $+4q$ and one charge $-q$. (a) How would you place them along a line so there's no net force on any of the three? (b) Is this equilibrium stable or unstable?

19. In Fig. 23-39 take $q_1 = 68$ μC, $q_2 = -34$ μC, and $q_3 = 15$ μC. Find the electric force on q_3.

FIGURE 23-39 Problems 19, 20.

20. In Fig. 23-39 take $q_1 = 25$ μC and $q_2 = 20$ μC. If the force on q_1 points in the $-x$ direction, (a) what is q_3 and (b) what is the magnitude of the force on q_1?

21. Four identical charges q form a square of side a. Find the magnitude of the electric force on any of the charges.

22. Three identical charges $+q$ and a fourth charge $-q$ form a square of side a. (a) Find the magnitude of the electric force on a charge Q placed at the center of the square. (b) Describe the direction of this force.

23. Three charges lie in the x-y plane: $q_1 = 55$ μC at $x = 0$, $y = 2.0$ m; q_2 at $x = 3.0$ m, $y = 0$; and q_3 at $x = 4.0$ m, $y = 3.0$ m. If the force on q_3 is $8.0\hat{\imath} + 15\hat{\jmath}$ N, find q_2 and q_3.

24. Two identical small metal spheres initially carry charges q_1 and q_2, respectively. When they're 1.0 m apart they experience a 2.5-N attractive force. Then they're brought together so charge moves from one to the other until they have the same net charge. They're again placed 1.0 m apart, and now they repel with a 2.5-N force. What were the original values of q_1 and q_2?

Section 23-4 The Electric Field

25. An electron placed in an electric field experiences a 6.1×10^{-10} N electric force. What is the field strength?

26. What is the magnitude of the force on a 2.0-μC charge in a 100 N/C electric field?

27. A 68-nC charge experiences a 150-mN force in a certain electric field. Find (a) the field strength and (b) the force that a 35-μC charge would experience in the same field.

28. A -1.0-μC charge experiences a $10\hat{\imath}$-N electric force in a certain electric field. What force would a proton experience in the same field?

29. The electron in a hydrogen atom is 0.0529 nm from the proton. What is the proton's electric field strength at this distance?

30. A 65-μC point charge is at the origin. Find the electric field at the points (a) $x = 50$ cm, $y = 0$; (b) $x = 50$ cm, $y = 50$ cm; (c) $x = -25$ cm, $y = 75$ cm.

Section 23-5 Electric Fields of Charge Distributions

31. In Fig. 23-40, point P is midway between the two charges. Find the electric field in the plane of the page (a) 5.0 cm directly above P, (b) 5.0 cm directly to the right of P, and (c) at P.

FIGURE 23-40 Problem 31.

32. A 1.0-μC charge and a 2.0-μC charge are 10 cm apart, as shown in Fig. 23-41. Find a point where the electric field is zero.

FIGURE 23-41 Problem 32.

33. A proton is at the origin and an ion is at $x = 5.0$ nm. If the electric field is zero at $x = -5$ nm, what is the charge on the ion?

34. For the situation of Example 23-3, (a) write an expression for the electric field as a function of x for points to the right of the charge $-q$ shown in Fig. 23-13. (b) Taking $q = 1.0\ \mu$C and $a = 1.0$ m, plot the field as a function of position for $x = 5$ m to $x = 25$ m.

35. (a) Find an expression for the electric field on the y axis due to the two charges q in Fig. 23-11. (b) At what point is the field on the y axis a maximum?

36. Write an expression for the dipole moment vector of the dipole shown in Fig. 23-18.

37. A dipole lies on the y axis, and consists of an electron at $y = 0.60$ nm and a proton at $y = -0.60$ nm. Find the electric field (a) midway between the two charges, (b) at the point $x = 2.0$ nm, $y = 0$, and (c) at the point $x = -20$ nm, $y = 0$.

38. What is the electric field strength 10 cm from a point dipole with dipole moment 3.8 μC·m (a) on the dipole's perpendicular bisector and (b) on its axis?

39. The dipole moment of the water molecule is 6.2×10^{-30} C·m. What would be the separation distance if the molecule consisted of charges $\pm e$? (The effective charge is actually less because electrons are shared by the oxygen and hydrogen atoms.)

40. You're 1.5 m from a charge distribution whose size is much less than 1 m. You measure an electric field strength of 282 N/C. You move to a distance of 2.0 m and the field strength becomes 119 N/C. What is the net charge of the

distribution? *Hint:* Don't try to calculate the charge. Determine instead how the field decreases with distance, and from that infer the charge.

41. Three charges form an equilateral triangle of side a. At one vertex is a charge $+2q$; at the other two vertices are charges $-q$. The triangle is oriented with the charge $2q$ on the positive x axis and both charges $-q$ on the y axis. (a) Find an expression for the electric field on the x axis, in the approximation $x \gg a$. (b) Compare with Equation 23-7b to show that your result in (a) is a dipole field, and give an expression for the magnitude of the triangle's dipole moment.

42. Three identical charges q form an equilateral triangle of side a, with two charges on the x axis and one on the positive y axis. (a) Find an expression for the electric field at points on the y axis above the uppermost charge. (b) Show that your result reduces to the field of a point charge $3q$ for $y \gg a$.

43. A 30-cm-long rod carries a charge of 80 μC spread uniformly over its length. Find the electric field strength on the rod axis, 45 cm from the end of the rod.

44. A thin rod of length ℓ carries charge Q distributed evenly over its length. A point charge with the same charge Q lies a distance b from the end of the rod, as shown in Fig. 23-42. Find a point where the electric field is zero.

FIGURE 23-42 Problem 44.

45. A thin rod of length ℓ has its left end at the origin and its right end at $x = \ell$. It carries a line charge density given by $\lambda = \lambda_0 \dfrac{x^2}{\ell^2} \sin(\pi x/\ell)$, where λ_0 is a constant. Find the electric field strength at the origin.

46. Two identical rods of length ℓ lie on the x axis and carry uniform charges $\pm Q$, as shown in Fig. 23-43. (a) Find an expression for the electric field strength as a function of position x for points to the right of the right-hand rod. (b) Show that your result has the $1/x^3$ dependence of a dipole field for $x \gg \ell$. (c) What is the dipole moment of this configuration? *Hint:* See Equation 23-7b.

FIGURE 23-43 Problem 46.

47. A uniformly charged ring is 1.0 cm in radius. The electric field on the axis 2.0 cm from the center of the ring has magnitude 2.2 MN/C and points toward the ring center. Find the charge on the ring.

48. Figure 23-44 shows a thin, uniformly charged disk of radius R. Imagine the disk divided into rings of varying radii r, as suggested in the figure. (a) Show that the area of such a ring is very nearly $2\pi r\, dr$. (b) If the surface charge density on the disk is σ C/m², use the result of (a) to write an expression for the charge dq on an infinitesimal ring. (c) Use the result of (b) along with the result of Example 23-8 to write the infinitesimal electric field dE of this ring at a point on the disk axis, taken to be the positive x axis. (d) Integrate over all such rings (that is, from $r = 0$ to $r = R$), to show that the net electric field on the disk axis is

$$E = 2\pi k\sigma\left(1 - \frac{x}{\sqrt{x^2 + R^2}}\right).$$

FIGURE 23-44 Problem 48.

49. Use the result of the preceding problem to show that the field of an *infinite,* uniformly charged flat sheet is $2\pi k\sigma$, where σ is the surface charge density. Note that this result is independent of distance from the sheet.

50. A semicircular loop of radius a carries positive charge Q distributed uniformly over its length. Find the electric field at the center of the loop (point P in Fig. 23-45). *Hint:* Divide the loop into charge elements dq as shown in Fig. 23-45, and write dq in terms of the angle $d\theta$. Then integrate over θ to get the net field at P.

FIGURE 23-45 Problem 50.

51. The electric field 22 cm from a long wire carrying a uniform line charge density is 1.9 kN/C. What will be the field strength 38 cm from the wire?

52. What is the line charge density on a long wire if the electric field 45 cm from the wire has magnitude 260 kN/C and points toward the wire?

53. A straight wire 10 m long carries 25 μC distributed uniformly over its length. (a) What is the line charge density

on the wire? Find the electric field strength (b) 15 cm from the wire axis, not near either end and (c) 350 m from the wire. Make suitable approximations in both cases.

54. Figure 23-46 shows a thin rod of length ℓ carrying charge Q distributed uniformly over its length. (a) What is the line charge density on the rod? (b) What must be the electric field direction on the rod's perpendicular bisector (taken to be the y axis)? (c) Modify the calculation of Example 23-9 to find an expression for the electric field at a point P a distance y along the perpendicular bisector. (d) Show that your result for (c) reduces to the field of a point charge Q for $y \gg \ell$.

FIGURE 23-46 Problem 54.

Section 23-6 Matter in Electric Fields

55. In his famous 1909 experiment that demonstrated quantization of electric charge, R. A. Millikan suspended small oil drops in an electric field. With a field strength of 20 MN/C, what mass drop can be suspended when the drop carries a net charge of 10 elementary charges?

56. How strong an electric field is needed to accelerate electrons in a TV tube from rest to one-tenth the speed of light in a distance of 5.0 cm?

57. A proton moving to the right at 3.8×10^5 m/s enters a region where a 56 kN/C electric field points to the left. (a) How far will the proton get before its speed reaches zero? (b) Describe its subsequent motion.

58. An oscilloscope display requires that a beam of electrons moving at 8.2 Mm/s be deflected through an angle of 22° by a uniform electric field that occupies a region 5.0 cm long. What should be the field strength?

59. An ink-jet printer works by "steering" charged ink drops to the right place on the page by passing moving drops through a uniform electric field that deflects them by the appropriate amount. Figure 23-47 shows an ink drop approaching the field region, which has length ℓ and width d between the charged plates that establish the field. Find an expression for the minimum speed a drop with mass m and

FIGURE 23-47 Problem 59.

charge q must have if it is to get through the region without hitting either plate.

60. An electrostatic analyzer like that of Example 23-11 has $b = 7.5$ cm. What should be the value of E_0 if the device is to select protons moving at 84 km/s?

61. An electron is moving in a circular path around a long, uniformly charged wire carrying 2.5 nC/m. What is the electron's speed?

62. Figure 23-48 shows a device its inventor claims will separate isotopes of a particular element. (Isotopes of the same element have nuclei with the same charge but different masses). Atoms of the element are first stripped completely of their electrons, then accelerated from rest through an electric field chosen to give the desired isotope exactly the right speed to pass through the electrostatic analyzer (see Example 23-11). Prove that the device won't work—that is, that it won't separate different isotopes.

FIGURE 23-48 An isotope separator that won't work (Problem 62).

63. What is the line charge density on a long wire if a 6.8-μg particle carrying 2.1 nC describes a circular orbit about the wire with speed 280 m/s?

64. The electron in a hydrogen atom has kinetic energy 2.18×10^{-18} J. Assuming the electron is in a circular orbit around the central proton, estimate the size of the atom. (Although this problem gives a reasonable answer, the simple model of an electron orbiting a proton is not consonant with the quantum mechanical description of the atom.)

65. A dipole with dipole moment 1.5 nC·m is oriented at 30° to a 4.0-MN/C electric field. (a) What is the magnitude of the torque on the dipole? (b) How much work is required to rotate the dipole until it's antiparallel to the field?

66. A molecule has its dipole moment aligned with a 1.2-kN/C electric field. If it takes 3.1×10^{-27} J to reverse the molecule's orientation, what is its dipole moment?

67. Two identical dipoles, each of charge q and separation a, are a distance x apart as shown in Fig. 23-49. By considering forces between pairs of charges in the different dipoles, calculate the net force between the dipoles. (a) Show that, in the limit $a \ll x$, the force has magnitude $6kp^2/x^4$, where $p = qa$ is the dipole moment. (b) Is the force attractive or repulsive?

FIGURE 23-49 Problem 67.

68. A dipole with charges $\pm q$ and separation $2a$ is located a distance x from a point charge $+Q$, with its dipole moment vector perpendicular to the x axis, as shown in Fig. 23-50. Find expressions for the magnitude of (a) the net torque and (b) the net force on the dipole, both in the limit $x \gg a$. (c) What is the direction of the net force?

FIGURE 23-50 Problem 68.

Paired Problems

(Both problems in a pair involve the same principles and techniques. If you can get the first problem, you should be able to solve the second one.)

69. An electron is at the origin and an ion with charge $+5e$ is at $x = 10$ nm. Find a point where the electric field is zero.

70. A proton is at the origin and an ion is at $x = 5.0$ nm. If the electric field is zero at $x = -6.83$ nm, what is the charge on the ion?

71. A thin rod of length ℓ has its left end at $x = -\ell$ and its right end at the origin. It carries a line charge density given by $\lambda = \lambda_0 \dfrac{x^2}{\ell^2}$, where λ_0 is a constant. Find the electric field at the origin.

72. Repeat the preceding problem for the case when $\lambda = \lambda_0 \dfrac{x^4}{\ell^4}$.

73. A thin, flexible rod carrying charge Q spread uniformly over its length is bent into a quarter circle of radius a, as shown in Fig. 23-51a. Find the electric field strength at the point P, which is the center of the circle. *Hint:* Consult Problem 50.

74. A thin, flexible rod carrying charge Q spread uniformly over its length is bent into a circular arc of radius a, as shown in Fig. 23-51b. Find the electric field strength at the point P, which is the center of the circular arc.

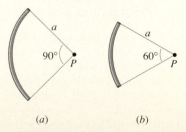

(a)   *(b)*

FIGURE 23-51 Problems 73, 74.

75. Ink-jet printers work by deflecting moving ink droplets with an electric field so they hit the right place on the paper. Droplets in a particular printer have mass 1.1×10^{-10} kg, charge 2.1 pC, speed 12 m/s, and pass through a uniform 97-kN/C electric field in order to be deflected through a $10°$ angle. What is the length of the field region?

76. If the drop speed in the printer of Problem 75 is doubled, what should be done to the electric field to have the drops hit the same point on the paper?

Supplementary Problems

77. A spring of spring constant 100 N/m is stretched 10 cm beyond its 90-cm equilibrium length. If you want to keep it stretched by attaching equal electric charges to the opposite ends, what magnitude of charge should you use?

78. Two small spheres with the same mass m and charge q are suspended from massless strings of length ℓ, as shown in Fig. 23-52. Each string makes an angle θ with the vertical. Show that the charge on each sphere is $q = \pm 2\ell \sin \theta \sqrt{mg \tan \theta / k}$.

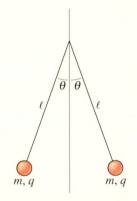

FIGURE 23-52 Problem 78.

79. A charge $-q$ and a charge $\frac{4}{9}q$ are located a distance a apart, as shown in Fig. 23-53. Where would you place a third charge so that all three are in static equilibrium? What should be the sign and magnitude of the third charge?

FIGURE 23-53 Problem 79.

80. Two 34-μC charges are attached to the opposite ends of a spring of spring constant 150 N/m and equilibrium length 50 cm. By how much does the spring stretch?

81. A 3.8-g particle with a 4.0-μC charge experiences a downward force of 0.24 N in a uniform electric field. Find the

electric field, assuming that the gravitational force is *not* negligible.

82. A rod of length 2ℓ lies on the x axis, centered on the origin. It carries a line charge density given by $\lambda = \lambda_0 \frac{x}{\ell}$, where λ_0 is a constant. (a) What is the net charge on the rod? (b) Find an expression for electric field strength at all points $x > \ell$. (c) Show that your result has the $1/x^3$ dependence of a dipole field when $x \gg \ell$. *Hint:* For $\ell \ll x$, $\ln \left(\frac{x - \ell}{x + \ell} \right)$ becomes approximately $-\frac{2\ell}{x} - \frac{2\ell^3}{3x^3}$. (d) By comparing with Equation 23-7b, determine the dipole moment of the rod.

83. The electric field on the axis of a uniformly charged ring has magnitude 380 kN/C at a point 5.0 cm from the ring center. The magnitude 15 cm from the center is 160 kN/C; in both cases the field points away from the ring. Find the radius and charge of the ring.

84. Use the binomial theorem to show that, for $x \gg R$, the result of Problem 48 reduces to the field of a point charge whose total charge is the charge density σ times the disk area.

85. A molecule with dipole moment p is located a distance r from a proton, oriented with its dipole moment vector $\mathbf{p}$ as shown in Fig. 23-54. (a) Use Equation 23-7b to find the force the molecule exerts on the proton. (b) Now find the net force on the molecule in the proton's nonuniform electric field by considering that the molecule consists of two opposite charges $\pm q$, separated by a distance d such that $qd = p$. Take the limit as d becomes very small compared with r, and show that the resulting force has the same magnitude as that of part (a), as required by Newton's third law.

FIGURE 23-54 Problem 85.

86. An *electric quadrupole* consists of two oppositely directed dipoles in close proximity. (a) Calculate the field of the quadrupole shown in Fig. 23-55 for points to the right of $x = a$, and (b) show that for $x \gg a$ the quadrupole field falls off as $1/x^4$.

FIGURE 23-55 Problem 86.

87. Derive Equation 23-9 in Example 23-9 by making θ the integration variable, then evaluating the resulting integral.

24 Gauss's Law

Huge sparks jump to the operator's cage in the Hall of Electricity at the Boston Museum of Science. But the operator is unharmed, for reasons that will become evident in this chapter.

We've seen how it's possible, in principle, to calculate the electric field of any charge distribution by summing the contributions of the many individual charges comprising the distribution. But in practice that process involves a vector integration that becomes difficult for all but the simplest charge distributions. How can we hope to calculate the field of a solid ball of charge, for example, when the individual charge elements are varying distances from the field point and their field vectors point in different directions (Fig. 24-1)?

In this chapter we introduce an elegant way of describing electric fields that makes almost trivial the calculation of fields from certain charge distributions. In the process we will formulate one of the four fundamental laws of electromagnetism—a statement that is equivalent to Coulomb's law but that gives deeper insights into the behavior of the electric field.

24-1 ELECTRIC FIELD LINES

The electric field is a set of vectors defined at all points in space, and we've therefore been representing fields by drawing a number of field vectors. A simpler way to visualize electric fields is with **electric field lines**, continuous lines whose direction is everywhere that of the electric field. To draw a field line, start at some point, and determine the field direction there. Move a small distance in the direction of the field, and evaluate the field direction at the new point. Extending this process in both directions from the starting point traces out an electric field line. The resulting line is a path whose direction at any point is that of the electric field at that point. Drawing many such lines gives a visualization of the overall field structure.

Tracing the field lines of a point charge is particularly simple. Starting at any point near a positive point charge, we find field vectors pointing radially outward from the charge. Move a little way outward, and the field still points in the same direction. So the field lines are straight lines, starting at the point charge and extending radially outward indefinitely (Fig. 24-2).

Field lines show the direction of the field, but what about its magnitude? In Fig. 24-2 the field lines spread apart as they extend farther from the point charge. Coulomb's law tells us that the field weakens farther from the charge. So in Fig. 24-2 the field is stronger where the lines are closer and weaker where they're farther apart. This is generally true, and allows us to infer relative field strength as well as field direction from field line pictures. ActivPhysics Activity 11.4 draws field lines for a point charge.

To trace the field lines of charge distributions, we follow the net field—the vector sum of the field contributions from all the charges in the distribution. Usually the direction of the field varies as we move along a field line, so the line itself is curved. Nevertheless, closeness of the field lines remains a measure of the

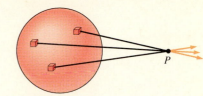

FIGURE 24-1 It would be difficult to find the field of a charged ball by summing vectorially the contributions of all the individual charge elements, three of which are shown here.

11.4
Electric Field: Point Charge

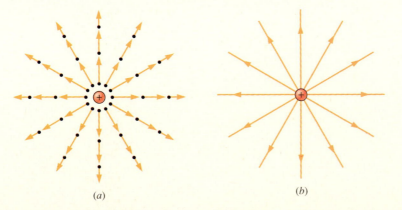

(a)   (b)

FIGURE 24-2 Vectors and field lines provide two ways to visualize the electric field. (*a*) The vector picture, introduced in the preceding chapter, uses individual field vectors drawn at selected points. Each vector gives the magnitude and direction of the field at that point, with the length of the arrow representing the magnitude. (*b*) Field lines are continuous lines that have the same direction as the field. For the point charge shown here, the field lines start at the charge and extend to infinity. The field magnitude is larger where the lines are closer together.

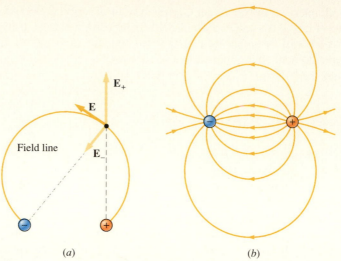

FIGURE 24-3 (*a*) Tracing a field line for an electric dipole. At each point the direction of the field line is that of the *net* electric field $\mathbf{E} = \mathbf{E}_+ + \mathbf{E}_-$. (*b*) Tracing several field lines gives an overall sense of the dipole field. Near each charge the field has the radial structure of a point-charge field, but farther away the influence of both charges becomes important and the field lines curve. Field is stronger where field lines are closer—namely, near the charges and in the region between them.

field strength. Figure 24-3 shows some field lines for a dipole; you can see immediately from Fig. 24-3*b* that the field is strongest near the individual charges and in the region directly between them.

You might argue that "closeness of field lines" is vague because the electric field exists everywhere and therefore we can always draw as many field lines as we want. To make the field-line picture more precise, we associate a fixed number of field lines with a charge of given magnitude. In Fig. 24-4, for example, eight field

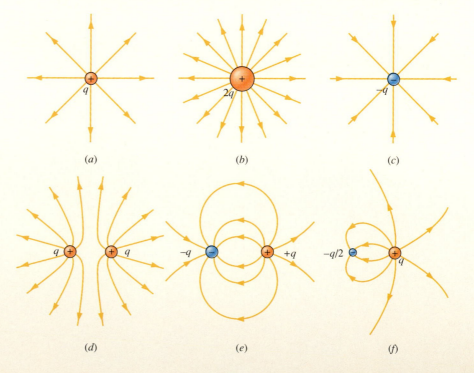

FIGURE 24-4 Field lines for (*a*) a positive charge q, (*b*) $2q$, (*c*) $-q$, (*d*) two identical charges q, (*e*) a dipole, consisting of two equal but opposite charges $\pm q$, and (*f*) opposite and unequal charges q and $-q/2$. In each drawing eight lines are used to represent a charge of magnitude q. Note in (*d*) and (*f*) that the field at large distances begins to resemble that of a single point charge.

lines correspond to a charge of magnitude q. Then eight lines *begin* on the *positive* charge $+q$ (Fig. 24-4a), and 16 on $+2q$ (Fig. 24-4b). Eight lines *end* on the *negative* charge $-q$ in Fig. 24-4c. Figures 24-4d–f show the fields of some two-charge distributions, drawn with the same eight-line convention. Note that field lines always *begin* on *positive* charges and *end* on *negative* charges. Activ-Physics Activity 11.6 draws field lines for other two-charge distributions of your choice.

11.6
Electric Field

24-2 ELECTRIC FLUX

Counting Field Lines

Figure 24-5 shows the charge distributions of Fig. 24-4, each surrounded by several surfaces. (The figure shows only the two-dimensional cross section of each surface.) Each surface is closed, meaning it's impossible to get from inside to outside without crossing the surface. We now ask a simple question: How many field lines emerge from inside each surface?

In Fig. 24-5a the answer for surfaces 1 and 2 is obvious: eight. With surface 3 one field line crosses three times, twice going out and once going in. If we count a field line going inward as negative, then the algebraic sum of field lines is again eight. In fact, any *closed* surface you can draw around $+q$ will have eight field lines emerging from within. That's because eight lines begin on the charge and extend indefinitely outward, so they cross *any* closed surface surrounding the charge.

What about surface 4? Two lines cross going inward and two going outward, so the net number of field lines emerging from this surface is zero. What's different about surface 4 is that it doesn't enclose the charge. By drawing other surfaces you can convince yourself that any surface not enclosing the charge will have as many lines going in as out, and will therefore have zero net field lines emerging.

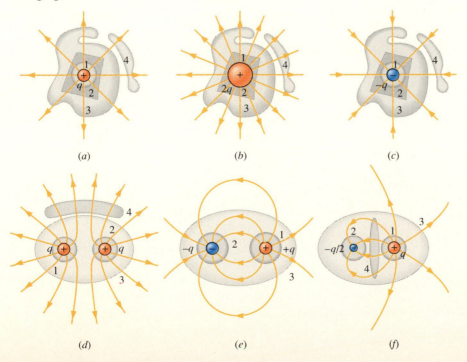

FIGURE 24-5 In all cases, the number of field lines emerging from a closed surface is proportional to the charge enclosed.

Figure 24-5*b* is identical except that now those surfaces enclosing the charge have 16 field lines emerging, reflecting the greater magnitude of the charge. Surfaces that don't enclose the charge still have zero net field lines emerging. Figure 24-5*c* is also like Fig. 24-5*a*, except that now the charge is negative so all field lines point inward. According to our sign convention, −8 field lines emerge from any surface enclosing the charge −*q*.

In Fig. 24-5*d*, surfaces 1 and 2 each enclose one of the charges *q*, and each has eight field lines emerging. Surface 3 encloses *both* charges, for a total enclosed charge +2*q*, and has 16 field lines emerging. Finally, surface 4 encloses no charge and has zero net field lines emerging.

On to Fig. 24-5*e*, the dipole. Surface 1 encloses charge *q* and has eight field lines emerging. Surface 2 encloses −*q* and has −8 field lines emerging. Surface 3 encloses both +*q* and −*q*, giving zero net charge enclosed. And as many field lines enter surface 3 as leave it, giving zero net field lines emerging.

Finally, in Fig. 24-5*f* eight field lines emerge from surface 1—and that surface encloses +*q*. Surface 2 encloses −*q*/2, and has −4 field lines emerging. Surface 3 encloses both charges, for a net enclosed charge +*q*/2—and four field lines emerge from this surface. Surface 4 encloses no charge and has zero net field lines emerging.

Counting the field lines in Fig. 24-5 leads to a simple statement about how electric fields must behave:

> **The number of electric field lines emerging from any closed surface is proportional to the charge enclosed.**

This statement is very general: It doesn't matter what shape the surface is or whether the enclosed charge is a single point charge or a lot of charges adding to the same net charge. Nor does it matter how the charges are arranged, as long as they're *enclosed* by the surface in question. And the presence of charges *outside* the surface doesn't alter the conclusion about the number of field lines emerging—although it may alter the shape of the individual lines. You can explore this statement as applied to the fields of two-charge distributions using ActivPhysics Activity 11.6.

We'll now rephrase our statement in a more mathematically rigorous way, obtaining one of the four fundamental laws of electromagnetism.

Tip

Remember Fig. 24-5 As we define new terms and write equations involving integrals, remember that the mathematics just reflects in a concise way what's clear from Fig. 24-5—that the number of field lines emerging from a closed surface depends only on the net charge enclosed. Go back to that figure any time you begin to lose the physical significance of the mathematics.

Electric Flux

We can make the "number of field lines" more rigorous with Fig. 24-6, which shows several flat surfaces in uniform electric fields. Study the figure and its caption, and you'll see that the number of field lines crossing each surface is

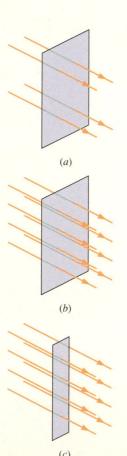

(*a*)

(*b*)

(*c*)

FIGURE 24-6 (*a*) Four field lines cross the surface shown. (*b*) Here the field strength has doubled, and eight lines cross the surface. (*c*) With the same field strength as in (*b*), but with half the area, only four lines cross the surface. In general, the number of field lines crossing a surface is proportional to the surface area and to the field strength.

FIGURE 24-7 The number of field lines crossing a surface also depends on the surface orientation relative to the field **E**. The orientation is specified by a vector **A** normal to the surface. In (a) the surface is perpendicular to the field, so **A** and **E** are parallel; then $\theta = 0$, $\cos\theta = 1$, and the number of field lines crossing the surface is a maximum. In (b) the surface is parallel to the field, so **A** and **E** are perpendicular; here $\theta = 90°$, $\cos\theta = 0$, and no lines cross the surface. (c) In general, the number of field lines varies as $\cos\theta$, where θ is the angle between **E** and **A**.

(a)

(b)

proportional to the surface area A and the field strength E. Figure 24-7 shows that it also depends on the orientation of the surface, specified by a vector normal to the surface. As Fig. 24-7 suggests, the number of field lines crossing the surface is proportional to the cosine of the angle between that normal vector **A** and the field **E**. Putting this all together, we have

$$\text{Number of field lines} \propto E\,A\cos\theta.$$

The quantity on the right-hand side of this equation has a definite value that captures the spirit of the more vague expression "number of field lines crossing a surface." We call this quantity the **electric flux**, ϕ, through the surface. If we make the magnitude of the surface normal vector **A** equal to the surface area A, then we can define the flux compactly using the vector dot product:

$$\phi = \mathbf{E} \cdot \mathbf{A}, \tag{24-1}$$

where the dot product, defined in Chapter 7, is the product of the two vector magnitudes with the cosine of the angle between them. Since the units of **E** are N/C, flux is measured in N·m²/C.

For the open surfaces of Fig. 24-6 and 24-7 there's an ambiguity in the sign of ϕ, since we could have taken **A** in either of the two directions along the perpendicular to the surface. But for *closed* surfaces, we unambiguously define the direction of **A** as that of the outward-pointing normal to the surface.

(c)

Tip

The Flux Is Not the Field The flux ϕ and field **E** are related but distinct quantities. The field is a vector defined at each point in space. Flux is a global property of the field, depending not on a single point but on how the field behaves over an extended surface. Unlike field, flux is a scalar quantity; it's simply a quantification of the "amount of field" or "number of field lines" crossing a surface.

What if a surface is curved and/or the field varies with position? Then we divide the surface into many small patches, each small enough that it's essentially flat and that the field is essentially uniform over each (Fig. 24-8). If a patch has area dA, then Equation 24-1 gives the flux through it:

$$d\phi = \mathbf{E} \cdot d\mathbf{A},$$

where the vector $d\mathbf{A}$ is normal to the patch. The total flux through the surface is then the sum over all the patches. If we make the patches arbitrarily small that sum becomes an integral, and the flux is

$$\phi = \int_{\text{surface}} \mathbf{E} \cdot d\mathbf{A}. \tag{24-2}$$

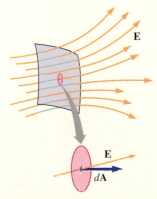

FIGURE 24-8 Even though the surface is curved and the field varies, a small enough patch of surface is essentially flat and the field is uniform over it, so the flux through the patch is $d\phi = \mathbf{E} \cdot d\mathbf{A}$.

FIGURE 24-9 What's the flux through each of the cube faces shown?

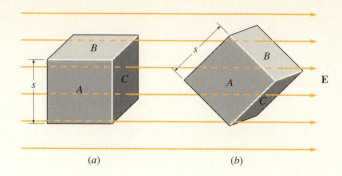

(a) (b)

The limits of the integral range over the entire surface, picking up contributions from all the patches $d\mathbf{A}$. Although the integral can be difficult to evaluate, we'll find it most useful in cases where its evaluation is almost trivial. Again, remember what Equation 24-2 means: The flux ϕ simply serves as a more precise measure of the "number of field lines" crossing a surface.

Got It!

Figure 24-9 shows a cube of side s in a uniform electric field $\mathbf{E}$. What is the flux through each of the three cube faces A, B, C with the cube oriented as in (a)? Repeat for the orientation in (b), where the cube has been rotated through 45°.

24-3 GAUSS'S LAW

We showed in the preceding section that the number of field lines emerging from a closed surface is proportional to the charge enclosed. Now that we've developed electric flux to express more rigorously the notion "number of field lines," we can state the following:

> The electric flux through any closed surface is proportional to the charge enclosed by that surface.

Writing the same thing mathematically gives

$$\phi = \oint \mathbf{E} \cdot d\mathbf{A} \propto q_{\text{enclosed}},$$

where the circle on the integral sign indicates that the integral is over a *closed* surface.

To evaluate the proportionality between flux and enclosed charge, consider a positive point charge q and a spherical surface of radius r centered on the charge (Fig. 24-10a). The flux through this surface is given by Equation 24-2:

$$\phi = \oint \mathbf{E} \cdot d\mathbf{A} = \oint E \, dA \cos\theta.$$

But Fig. 24-10b shows that the surface normal $d\mathbf{A}$ and the electric field $\mathbf{E}$ are parallel at any point on the sphere, so $\cos\theta = 1$. Since the electric field varies as

(a)

(b)

FIGURE 24-10 (a) The electric field of a point charge has the same magnitude everywhere on a spherical surface centered on the charge. (b) At each point on the surface the field and the surface normal $d\mathbf{A}$ are parallel.

$1/r^2$ its magnitude is the same everywhere at the fixed radius r of our sphere. Thus, we can take E outside the integral, giving

$$\phi = \oint_{\text{sphere}} E \, dA = E \oint_{\text{sphere}} dA = E(4\pi r^2),$$

where the last step follows because $\oint dA$ just means the surface area of the sphere. Now, the electric field of a point charge is given by Equation 23-4: $E = kq/r^2$. So we have

$$\phi = E(4\pi r^2) = \left(\frac{kq}{r^2}\right)(4\pi r^2) = 4\pi kq.$$

Since the point charge q is the only charge inside our spherical surface, we see that the proportionality constant between flux and enclosed charge is just $4\pi k$.

Before proceeding, we introduce the so-called permittivity constant, ε_0, defined by the relation

$$\varepsilon_0 = \frac{1}{4\pi k}, \tag{24-3}$$

where k is the Coulomb constant. The value of ε_0 is 8.85×10^{-12} C^2/N·m^2.

Tip

Why Two Constants? There's no physics in Equation 24-3; it's just a definition of a new symbol. The SI unit system requires a proportionality constant in Coulomb's law or, equivalently, in the relation between flux and enclosed charge. The value of that constant conveys information about the strength of the electric force. That same information is contained in the constants k and ε_0; both are different ways of expressing the same thing. That there are two constants in use is purely a historical artifact, and we switch now from k to ε_0 only because doing so makes subsequent formulas simpler. It would have been better to define the units of electric charge so that Coulomb's law didn't need a proportionality constant. There are unit systems where that's done—but they aren't the internationally adopted SI system.

In terms of the new constant ε_0, the proportionality $4\pi k$ between flux and enclosed charge becomes $1/\varepsilon_0$. So our statement that the flux through any closed surface is proportional to the net charge enclosed becomes

$$\oint \mathbf{E} \cdot d\mathbf{A} = \frac{q_{\text{enclosed}}}{\varepsilon_0}. \quad \text{(Gauss's law)} \tag{24-4}$$

Here the integral is taken over *any closed surface,* and q_{enclosed} is the charge enclosed *by that surface.*

Equation 24-4 is **Gauss's law**, after the German mathematician Karl Friedrich Gauss (1777–1855). Gauss's law is one of four fundamental relations that govern the behavior of electromagnetic fields throughout the universe. Whether you journey into a star in some remote galaxy, down among the strands of a DNA molecule, or into the microprocessor chip at the heart of your computer, you will

FIGURE 24-11 The outer sphere has twice the radius of the inner sphere, and thus four times the surface area. Since the same number of field lines crosses both spheres, the density of field lines at the outer sphere has one-fourth its value at the inner sphere. Thus the field-line density accurately reflects the inverse-square falloff of the electric field. If the inverse-square law did not hold, we would have to begin new field lines (or stop them) at different distances from the point charge, and the number of field lines crossing a surface would not depend only on the enclosed charge.

find that the flux of the electric field through any closed surface depends only on the enclosed charge. In over a century of experiments, no electric field has ever been observed to violate Gauss's law.

We stress that Gauss's law, although clothed in the mathematical finery of a surface integral, is just a more rigorous way of saying what's clear in Fig. 24-5: that the number of field lines emerging from a closed surface is proportional to the enclosed charge.

Gauss and Coulomb

Gauss's law and Coulomb's law look completely different, but they're actually equivalent. Figure 24-11 shows that the heart of this equivalence is the inverse-square law. The figure shows two spherical surfaces surrounding a point charge q. Gauss's law tells us that the flux through each surface is the same, and is equal to q/ε_0. But why? Because, as our arguments leading to Gauss's law show, the flux through a spherical surface of radius r centered on a point charge is the product of the surface area $4\pi r^2$ with the electric field strength E at the surface. But Coulomb's law says that the electric field drops off as $1/r^2$. As r increases, the surface area grows as r^2, but the $1/r^2$ decrease in field strength just compensates, giving a constant value for the flux that depends only on the enclosed charge. If the inverse-square law (e.g., Coulomb's law) didn't hold, then the flux would not be constant and Gauss's law wouldn't hold either.

It's also the inverse-square law that makes electric field line pictures useful for visualizing fields. Field lines begin or end only on charges; otherwise they go off to infinity. As the field lines of a point charge spread in three dimensions, the number crossing any spherical surface (or any closed surface, for that matter) remains the same. But larger spheres have larger surface areas, in proportion to r^2—and that means the density of field lines drops as $1/r^2$, accurately reflecting the strength of the field. If the inverse-square law didn't hold, we would have to start additional field lines beyond the point charge (or stop them) to represent the field strength quantitatively, and thus the number of field lines crossing a closed surface would not remain the same. Once again, the inverse-square law (Coulomb) and the relation between flux and enclosed charge (Gauss) are intimately connected. Incidentally, field-line pictures printed in a book or drawn on a blackboard generally cannot be quantitatively correct because they don't show the spreading of field lines in all three dimensions.

We've been talking here only about isolated point charges, but we emphasize that Gauss's law applies to *all* electric fields, no matter how complicated the charge distributions producing them. That's because of the superposition principle. We already know that superposition allows us to add vectorially the electric fields described individually by Coulomb's law for the point-charge field. Since Gauss's and Coulomb's law are equivalent, our argument leading to Gauss's law still applies when the field **E** is a superposition of point-charge fields.

Got It!

A spherical surface surrounds an isolated point charge. If a second point charge is brought to a point just outside the surface, will the flux through the surface change? Will the electric field at points on the surface change?

24-4 USING GAUSS'S LAW

Gauss's law is true for *any* surface enclosing *any* charge distribution; it's a universal statement of how electric fields everywhere must behave. In a few specialized cases Gauss's law can be used to calculate the electric field. In those cases, which we treat in this section, Gauss provides a powerful alternative to Coulomb's law, an alternative that makes electric-field calculations much easier. But despite this section's emphasis on using Gauss's law to calculate the field, remember that the law holds *universally,* even in cases where it's not useful for field calculations.

We can use Gauss to calculate the field when it's possible to evaluate the flux integral without actually knowing **E**. That happens only when the charge distribution has sufficient symmetry. We now illustrate the use of Gauss's law for three important symmetries.

Spherical Symmetry

A charge distribution is spherically symmetric if the charge density depends only on the distance from a central point. A point charge, a uniformly charged solid sphere, and a spherical surface carrying uniform surface charge density are all spherically symmetric charge distributions. Spherical symmetry implies that the magnitude of the electric field depends only on the distance r from the center of symmetry, and that the field direction is radial (Fig. 24-12).

Gauss's law relates the electric flux through a closed surface to the enclosed charge. The law holds for *any* surface whatsoever, provided it's a *closed* surface. The surface over which we're evaluating the flux is called a **gaussian surface**; it's not a real, physical surface but just an imaginary, mathematical surface that we envision for the purpose of applying Gauss's law. Although Gauss holds for *any* surface, it's useful for calculating electric fields only for surfaces that share the symmetry of the charge distribution.

In the case of spherical symmetry, the appropriate surface is a sphere of arbitrary radius r centered on the point of symmetry. Two such surfaces are shown in Fig. 24-13. With spherical symmetry, the magnitude E of the electric field must be the same everywhere on such a gaussian sphere. Furthermore, **E** is everywhere in the same direction as the perpendicular to the surface, so $\cos\theta = 1$. Thus, although the direction of **E** and of the surface vary, the product $E\cos\theta$ remains constant and is simply the field magnitude E. Then the flux through a gaussian sphere becomes

$$\phi = \oint \mathbf{E} \cdot d\mathbf{A} = \oint E\cos\theta \, dA = E \oint dA = 4\pi r^2 E, \qquad (24\text{-}5)$$

where the last step follows because $\oint dA$ is just the surface area of the sphere, $4\pi r^2$. This expression for the flux does not depend on the details of the charge distribution, so long as it is spherically symmetric.

Gauss's law says that the flux through the sphere is given by $q_{\text{enclosed}}/\varepsilon_0$, where q_{enclosed} is the net charge enclosed *by the sphere.* Suppose our spherically symmetric charge distribution carries total charge Q and has radius R. That is, whatever the particular distribution of charge for $r \leq R$, there is no charge at $r > R$. For any gaussian sphere with $r > R$, like the surface 1 in Fig. 24-13,

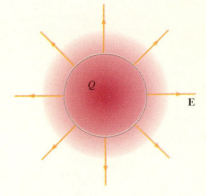

FIGURE 24-12 For a spherically symmetric charge distribution, the field vectors at a given radius all have the same magnitude and point in the radial direction—outward for a positive charge, as shown, or inward for a negative charge.

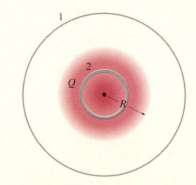

FIGURE 24-13 Two gaussian surfaces surrounding a spherical charge distribution. Surface 1 lies outside the distribution, and encloses all the charge Q. Surface 2 lies inside the distribution, and encloses only part of the charge.

FIGURE 24-14 Hairs on a highly charged person's head trace out the electric field. The essentially spherical head produces a field like that of a point charge.

the enclosed charge is the total charge Q. Equating the flux in Equation 24-5 to Q/ε_0 gives

$$4\pi r^2 E = \frac{Q}{\varepsilon_0},$$

or

$$E = \frac{1}{4\pi\varepsilon_0}\frac{Q}{r^2} = \frac{kQ}{r^2}, \quad (r > R) \qquad (24\text{-}6)$$

where the last step follows because $1/4\pi\varepsilon_0$ is just the Coulomb constant, k. But this is just the field of a point charge! Equation 24-6 says that *outside* any spherically symmetric distribution of charge, the field is identical to that of a point charge located at the center of symmetry (Fig. 24-14). This is *not* an approximation—it is exactly true right up to the surface $r = R$! Imagine how hard it would have been to calculate this field using the superposition principle! Yet somehow all the charge elements throughout the spherically symmetric distribution produce $d\mathbf{E}$'s that add vectorially to give the same field as a single point charge. Like Gauss's law itself, this result is a manifestation of the inverse-square force law.

The field *inside* the charge distribution depends on how charge is distributed. This is because a gaussian sphere with $r < R$, such as surface 2 of Fig. 24-13, does not enclose the entire charge Q. How much it encloses depends on the charge distribution. In the examples below, we consider two important special cases.

| **EXAMPLE 24-1** | *A Uniformly Charged Sphere* |

A total charge Q is spread uniformly throughout a sphere of radius R. What is the electric field at all points in space?

Solution

This charge distribution is spherically symmetric, so the field for $r > R$ is like that of a point charge, given by Equation 24-6.

Inside the charged sphere, our gaussian surface is a sphere like surface 2 in Fig. 24-13. Equation 24-5 for the flux still holds, but now the charge enclosed is some fraction of Q. What fraction? The volume of the sphere is $\frac{4}{3}\pi R^3$, and it contains a total charge Q. Since charge is spread uniformly throughout the sphere, the volume charge density is given by

$$\rho = \frac{Q}{V} = \frac{Q}{\frac{4}{3}\pi R^3}.$$

Then the charge enclosed by our gaussian sphere of radius r is just the volume of that sphere multiplied by the volume charge density:

$$q_{\text{enclosed}} = V\rho = \frac{4}{3}\pi r^3 \frac{Q}{\frac{4}{3}\pi R^3} = Q\frac{r^3}{R^3}.$$

Equating the flux from Equation 24-5 to $q_{\text{enclosed}}/\varepsilon_0$, we have

$$4\pi r^2 E = \frac{Qr^3}{\varepsilon_0 R^3},$$

or

$$E = \frac{1}{4\pi\varepsilon_0}\frac{Qr}{R^3} = \frac{\rho r}{3\varepsilon_0}, \quad (r < R) \qquad (24\text{-}7)$$

where we've written the field in terms of both the total charge Q and the charge density $\rho = Q/\frac{4}{3}\pi R^3$. Equation 24-7 shows that the field *inside* the charge distribution increases linearly with distance from the center. This result is entirely consistent with the inverse-square law for point charges. Although the field of each charge element decreases as $1/r^2$, in this case the amount of charge enclosed increases more rapidly—as r^3—resulting in a field that increases linearly with r. Figure 24-15 shows the combined results for the fields both inside and outside the sphere. The field direction is, of course, radial, pointing outward if Q is positive and inward if Q is negative.

FIGURE 24-15 Field strength versus radial distance for a uniformly charged sphere of radius R. For $r > R$ the field has the inverse-square dependence of a point-charge field.

| **EXAMPLE 24-2** | *A Thin Spherical Shell* |

The thin spherical shell of radius R shown in Fig. 24-16 carries a total charge Q distributed uniformly over its surface. What is the electric field inside and outside the shell?

Solution

Since this distribution is spherically symmetric, we already know that the field outside is just the point-charge field of Equation 24-6.

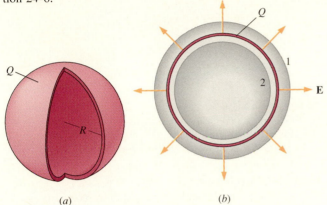

(a)   *(b)*

FIGURE 24-16 (*a*) A thin spherical shell carries charge Q distributed uniformly over its surface of radius R. (*b*) Cross-sectional view, showing two spherical gaussian surfaces. Surface 1 encloses the entire charge Q, while surface 2 encloses zero charge.

For any gaussian sphere inside the shell, the enclosed charge is zero. Equating the flux from Equation 24-5 to this zero enclosed charge gives

$$4\pi r^2 E = 0$$

so the field is zero everywhere inside the shell! How can this be? Again, it's a manifestation of the inverse-square law. At any point inside the shell, the larger fields of nearby portions of the shell are exactly canceled by the weaker fields of more distant, but more extensive, parts of the shell (Fig. 24-17).

FIGURE 24-17 At any point P inside a charged shell, the field arising from the relatively few but nearby charges in region A is exactly canceled by the field arising from the more numerous but more distant charges in region B.

Got It!

A spherical shell carries a charge Q, uniformly distributed over the shell. If the charge on the shell is doubled, what happens to the electric field (a) inside and (b) outside the shell?

| **EXAMPLE 24-3** | *A Point Charge Inside a Shell* |

A point charge $+q$ is at the center of a spherical shell of radius R carrying total charge $-2q$, distributed uniformly over its surface. (a) Draw the electric field lines for this configuration, using eight lines to represent a charge of magnitude q. Find expressions for the field strength for (b) $r < R$ and (c) $r > R$.

Solution

(a) The situation has spherical symmetry and so must the field. Gauss's law tells us that eight field lines must emerge from any surface surrounding $+q$ alone, and that eight field lines must go into any surface surrounding the entire distribution with its net charge $-q$. Fig. 24-18 shows the only way to draw the field that's compatible with Gauss's law. Notice a total of 16 lines end on the charged shell, consistent with its charge of $-2q$.

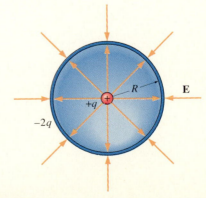

FIGURE 24-18 Field lines for a charge q surrounded by shell carrying $-2q$ (Example 24-3).

(b) For $r < R$ we're inside the shell, so the enclosed charge is just q. Solving Equation 24-4 for E then gives

$$E = \frac{kq}{r^2}.$$

What about the shell? Didn't we forget to take it into account? No! The shell and its charge are *irrelevant* as long as they preserve the spherical symmetry. Example 24-2 showed that the field inside a charged shell *due to the shell itself* is zero. Here we're inside the shell, so the only field we see is that of the point charge.

(c) Outside the shell, a spherical gaussian surface encloses net charge $-q$, so the field we see is that of a point charge $-q$; it has magnitude $E = q/4\pi\varepsilon_0 r^2$, and points radially inward.

What about the field at $r = R$? That's ambiguous; just inside the shell the field points outward, while just outside it points inward. In fact, the field undergoes a discontinuous jump across the infinitesimally thin surface charge layer on the shell.

Tip

Trust Gauss Gauss's law says that the flux—or number of field lines—emerging from any closed surface depends *only* on the *enclosed* charge—not on any other charge that may happen to be outside the surface. (Consult Fig. 24-5 to convince yourself of this.) *When there is enough symmetry* that often means that external charges are totally irrelevant in a field calculation, as is the shell in Example 24-3 for points with $r < R$. *But symmetry matters*: Without enough symmetry, zero net charge inside a gaussian surface is *not* sufficient to ensure that the field on the surface is zero (Fig. 24-19).

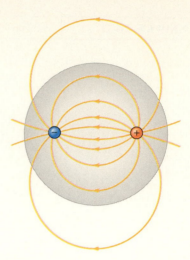

FIGURE 24-19 The net charge enclosed by the gaussian sphere (gray) is zero, but the field on the sphere is not zero. Here the charge distribution—a dipole—is not spherically symmetric, so Equation 24-5 is not a valid expression for the flux.

EXERCISE A solid sphere 10 cm in radius carries a uniform 40-μC charge distributed throughout its volume. It is surrounded by a concentric shell 20 cm in radius, also uniformly charged with $40\,\mu$C. Find the electric field (a) 5.0 cm, (b) 15 cm, and (c) 30 cm from the center.

Answers: (a) 18 MN/C; (b) 16 MN/C; (c) 8.0 MN/C

Some problems similar to Examples 24-1 through 24-3: 17–26, 66

Strategy

Using Gauss's Law This section illustrated the steps needed to calculate the electric field using Gauss's law:

1. Study the symmetry to see if you can construct a gaussian surface on which the field magnitude and its direction relative to the surface are constant. (With spherical symmetry that surface was spherical.) If you can't find such a surface then Gauss's law, while still true, won't help in calculating the field.

2. Evaluate the flux. This should be easy because your choice of gaussian surface makes $E\cos\theta$ constant over the surface. This term then comes outside the integral, leaving the integral equal to the surface area.

3. Evaluate the *enclosed* charge. This is not the same as the total charge if the gaussian surface lies *within* the charge distribution.

4. Equate the flux to $q_{\text{enclosed}}/\varepsilon_0$, and solve for E. The direction of $\mathbf{E}$ should be evident from the symmetry.

Once steps 1 and 2 are done for a particular symmetry, you'll have an equation like Equation 24-5 for the flux, and you can jump right to step 3 to calculate the field in a specific case—just as we did in Examples 24-1 and 24-2.

FIGURE 24-20 A positive charge distribution with line symmetry (red) extends infinitely in both directions, and its charge density depends only on the perpendicular distance from the symmetry axis. A cylindrical gaussian surface (gray) of length ℓ and radius r surrounds the charge distribution. Symmetry requires that the electric field on the surface point radially outward, and that the field magnitude E be constant over the surface. With a negative charge the field would point radially inward.

Line Symmetry

A charge distribution has line symmetry when its charge density depends only on the perpendicular distance r from a line, called the symmetry axis (Fig. 24-20). Symmetry then requires that the field point radially and that the field magnitude depend only on distance from the axis. An appropriate gaussian surface is the cylinder of length ℓ and radius r shown in Fig. 24-20. Being radial, field lines don't cross the circular ends of the cylinder, so there's no flux through these ends. The field is everywhere perpendicular to the curved part of the surface, so $\cos \theta = 1$ in the expression for flux through this part. Since the field magnitude is constant at the fixed radius r of the curved surface, the flux becomes

$$\phi = \int \mathbf{E} \cdot d\mathbf{A} = \int E \, dA \cos \theta = E \int dA = 2\pi r \ell E, \qquad (24\text{-}8)$$

where the last step follows because the cylinder "unrolls" into a rectangular sheet of length ℓ and width $2\pi r$ (Fig. 24-21). Since there is no flux through the cylinder ends, Equation 24-8 gives the total flux through our entire gaussian surface. Solving for the electric field in any situation with line symmetry then amounts to equating the flux given in Equation 24-8 to the enclosed charge divided by ε_0, then solving for E—exactly as we did for spherical symmetry using the analogous flux equation, Equation 24-5. There is one quantity in Equation 24-8 that has no analog in Equation 24-5, namely the length ℓ of the gaussian cylinder. As the following examples show, however, that length also appears in the calculation of the enclosed charge, so it always cancels from the final result.

FIGURE 24-21 (a) A cylindrical gaussian surface. (b) Unrolling the cylinder gives a flat sheet of area $2\pi r \ell$.

EXAMPLE 24-4	*An Infinite Line of Charge*

Use Gauss's law to calculate the electric field of an infinite line carrying line charge density λ.

Solution

This is the same problem we solved in Example 23-9 through a tedious Coulomb's law calculation. With line charge density λ, the charge enclosed by a gaussian cylinder of length ℓ is $q_{\text{enclosed}} = \lambda \ell$. Setting the flux given in Equation 24-8 to $q_{\text{enclosed}}/\varepsilon_0$ and solving for E then gives

$$E = \frac{\lambda \ell}{2\pi \varepsilon_0 r \ell} = \frac{\lambda}{2\pi \varepsilon_0 r}.$$

Since $1/2\pi \varepsilon_0 = 2k$, this is the same result we found in Example 23-9. The Gauss's law calculation is far simpler; symmetry and an intelligent choice of gaussian surface helped us bypass the entire integration of Example 23-9.

Although this problem dealt with an infinitesimally thin charged line, you can see that the same result must hold *outside* any charge distribution with line symmetry. And, as we argued in Example 23-9, the result is a good approximation for long, thin structures of finite length provided we're not too far away nor too close to the ends.

| EXAMPLE 24-5 | *A Hollow Pipe* |

A thin-walled pipe 3.0 m long and 2.0 cm in radius carries a net charge $q = 5.7$ μC, distributed uniformly over its surface. Find the electric field (a) 8.0 mm and (b) 8.0 cm from the pipe axis, not near either end.

Solution

Since the pipe is much longer than its diameter, we can approximate its field as that of an infinitely long charge distribution with line symmetry.

(a) A point 8.0 mm from the axis lies inside the pipe. An 8.0-mm-radius gaussian cylinder therefore encloses zero net charge; equating the flux from Equation 24-8 to zero then shows that the field at this radius—and indeed anywhere deep within the hollow pipe—is zero.

(b) For a point outside the pipe, a gaussian cylinder of length ℓ encloses charge $\lambda\ell$, where the line charge density λ is 5.7 μC/3.0 m = 1.9 μC/m. Setting the flux from Equation 24-8 to this $q_{enclosed}$ and solving for E then gives

$$E = \frac{q_{enclosed}}{2\pi\varepsilon_0 r\ell} = \frac{\lambda}{2\pi\varepsilon_0 r}$$

$$= \frac{1.9\times10^{-6}\ \text{C/m}}{(2\pi)(8.85\times10^{-12}\ \text{C}^2/\text{N·m}^2)(8.0\times10^{-2}\ \text{m})}$$

$$= 4.3\times10^5\ \text{N/C}.$$

In this region the field points radially outward and falls inversely with distance from the axis.

> **Tip**
>
> *That's Distance from the Symmetry Axis* The distance r that arises in applying Gauss's law to situations with spherical or line symmetry is always the distance *from the point or line of symmetry*—as you can see from the derivations of Equations 24-4 and 24-8. Equations for the field are therefore most simply expressed in terms of that distance—*not* in terms of distance from the edge of the charge distribution.

EXERCISE Suppose the pipe in Example 24-5 were surrounded concentrically by a second pipe of the same length and 5.0 cm in diameter. What should be (a) the total charge and (b) the surface charge density (assumed uniform) on this outer pipe in order that there be no electric field outside the entire structure?

Answers: (a) -5.7 μC; (b) -12 μC/m^2

Some problems similar to Examples 24-4 and 24-5: 28, 29, 31, 32

Plane Symmetry

A charge distribution has plane symmetry when its charge density depends only on the distance from a plane. The only electric-field direction consistent with this symmetry is perpendicular to the symmetry plane (Fig. 24-22). We can evaluate the flux integral in Gauss's law using a gaussian surface with sides perpendicular to the symmetry plane, ends of area A parallel to the plane, and which straddles the symmetry plane to extend equal distances on either side of it (Fig. 24-22). Since no field lines cross the sides, the flux through them is zero. Symmetry of the situation implies that the field magnitude E cannot depend on position parallel to the symmetry plane. Therefore E is uniform over each end of the gaussian cylinder so, with the field perpendicular to the ends, the flux through each end is just EA, where A is the end area. Since the ends are the same distance from the symmetry plane, they must have the same field strength E. The total flux through the gaussian surface is therefore

$$\phi = 2EA. \tag{24-9}$$

This equation holds for any charge distribution with plane symmetry; to find the field we evaluate the charge enclosed by the gaussian surface, then apply Gauss's law.

FIGURE 24-22 A charge distribution with plane symmetry. The charge density depends only on the distance from the plane of symmetry (black), and extends infinitely in both directions parallel to that plane. Also shown are the electric field and a gaussian surface with its sides perpendicular and ends parallel to the plane. Dashed lines show that the gaussian surface extends equal distances on both sides of the symmetry plane.

EXAMPLE 24-6	A Sheet of Charge

An infinite sheet of charge carries a uniform surface charge density σ. What is the electric field arising from this sheet?

Solution

Since the surface charge density is uniform, this charge distribution has plane symmetry. Figure 24-23 shows the sheet and an appropriate gaussian surface. The sheet area enclosed by the gaussian surface is clearly equal to the end area A. The surface charge density—charge per unit area—is σ, so the enclosed charge is $q_{enclosed} = \sigma A$. Setting $q_{enclosed}/\varepsilon_0$ to the flux $\phi = 2EA$ given by Equation 24-9 and solving for E then gives

$$E = \frac{\sigma}{2\varepsilon_0}. \qquad (24\text{-}10)$$

This simple result says that the field strength does not depend on distance from the sheet. How can this be? By symmetry, the field must point perpendicular to the sheet. There is no charge anywhere but on the sheet, so that's the only place where field lines can begin or end. Therefore the density of field lines—the measure of field strength—is the same everywhere. Figure 24-24 shows how this result is fully consistent with Coulomb's law.

Although this example treated only an infinitesimally thin sheet of charge, we would find a uniform electric field *outside* any charge distribution with plane symmetry. The field *inside* such a distribution would depend on how the charge density varies in the direction perpendicular to the symmetry plane (see Problem 36).

Perfect plane symmetry requires a charge distribution that is infinite in extent. But close to any large, flat, uniformly charged surface and not near an edge, the assumption of plane symmetry becomes a good approximation, and Equation 24-10 becomes reasonably accurate. We make the same approximation when we treat the acceleration of Earth's gravity as a constant; we're neglecting Earth's curvature and therefore its finite size, and approximating its gravitational field as the uniform field of an infinite sheet of mass.

EXERCISE An electron close to a large, flat sheet of charge is repelled from the sheet with a 1.8×10^{-12} N force. Find the surface charge density on the sheet.

FIGURE 24-23 The area of the charged sheet enclosed by the gaussian surface is the same as the area A of its ends; the enclosed charge is therefore σA.

FIGURE 24-24 As we rise above an infinite sheet, the amount of charge within a given angular region increases just enough to compensate for the inverse-square decrease in field strengths of the individual charges. The result is a field that does not depend on distance from the sheet.

Answer: $-200\ \mu C/m^2$

Some problems similar to Example 24-6: 34–37

24-5 FIELDS OF ARBITRARY CHARGE DISTRIBUTIONS

The examples of Section 24-4 show how easy Gauss's law can sometimes make problems that would be difficult to solve using Coulomb's law. In each case the symmetry allowed us to construct a gaussian surface on which $E\cos\theta$ was constant. Only then could we take E outside the integral and solve for it. But many situations do not possess the symmetry needed to apply Gauss's law in calculating the field. Try, for example, to calculate the field of a dipole using Gauss's law. The attempt fails because it is impossible to draw an appropriate surface.

FIGURE 24-25 The fields of a dipole, a point charge, a charged line, and a charged plane.

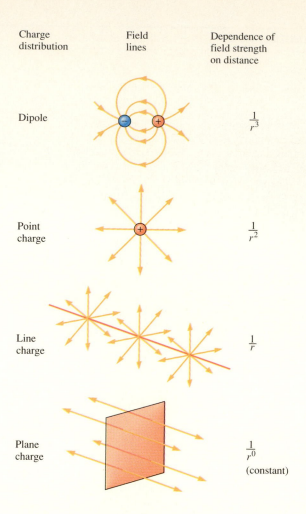

Charge distribution	Field lines	Dependence of field strength on distance
Dipole		$\frac{1}{r^3}$
Point charge		$\frac{1}{r^2}$
Line charge		$\frac{1}{r}$
Plane charge		$\frac{1}{r^0}$ (constant)

We can understand the fields associated with more complicated charge distributions by considering the fields of simpler distributions that we have already calculated using either Coulomb's law or Gauss's law. Figure 24-25 summarizes the fields of a dipole, a point charge, a uniformly charged line, and a uniformly charged plane. For the last three, note the simple relation between the number of dimensions in the charge distribution and the way the field strength depends on distance. The plane has two dimensions and its field strength is independent of distance. The line has one dimension and its field falls as $1/r$. The point has no dimensions and its field falls as $1/r^2$. In a sense, the dipole continues this progression, for it consists of two opposite point charges whose effects very nearly cancel. No wonder its field falls even faster, as $1/r^3$. In fact, one can construct a hierarchy of charge distributions whose fields fall off ever faster as dipoles nearly cancel dipoles, and so on. Such distributions are useful in the mathematical analysis of complicated charge distributions such as complex molecules or radio antennas.

Frequently we have a charge distribution that lacks the symmetry required to make Gauss's law useful and for which a Coulomb's law calculation would be impossibly difficult. Good thinking coupled with knowledge of simpler charge distributions can go a long way toward providing a reasonable approximation to the field. Consider, for example, the uniformly charged disk shown in Fig. 24-26. For points much closer to the disk surface than to the edge, the disk looks almost

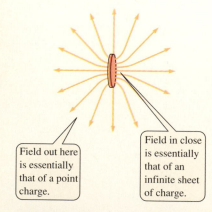

Field out here is essentially that of a point charge.

Field in close is essentially that of an infinite sheet of charge.

FIGURE 24-26 The field of a uniformly charged disk.

like an infinite flat plane of charge. For these points the field is approximately that of an infinite plane—a field that points directly away from the plane and does not fall off with distance. Far from the disk, meanwhile, its exact size and shape are unimportant. Its field closely resembles that of a single charged point: far from the disk the field points radially outward in all directions and falls off as the inverse square of the distance from the disk. The field at intermediate distances is harder to determine. But somehow the infinite-plane field lines close to the disk must connect smoothly to the point-charge field lines far away. If we sketch these in, as in Fig. 24-26, we have a rough picture of the field everywhere. Don't underestimate the value of a simple approximation like this one! It can often tell all we need to know about a situation and may provide a much clearer understanding than would a detailed calculation.

Got It!

A uniformly charged sheet measures 1 m on each side, and you're told the total charge Q. What expressions would you use to get approximate values for the field magnitude (a) 1 cm from the sheet (but not near an edge) and (b) 1 km from the sheet?

24-6 GAUSS'S LAW AND CONDUCTORS

Electrostatic Equilibrium

In the preceding chapter we defined electrical conductors as materials containing free charges—like the free electrons in metals. Figure 24-27 shows what happens when an electric field is applied to a piece of conducting material. Free charges respond to the electric force $q\mathbf{E}$ by moving—in the direction of the field if they are positive, opposite the field if negative. The resulting charge separation gives rise to an electric field within the material that is opposite to the applied field. As more charge moves this internal field becomes stronger until its magnitude eventually equals that of the applied field. At that point free charges within the conductor experience zero net force, and the conductor is in **electrostatic equilibrium**. Although individual charges continue to move about in random thermal motion, there is no longer any net motion of charge. Once equilibrium is reached the internal and applied fields are equal but opposite, and therefore,

| The electric field is zero inside a conductor in electrostatic equilibrium.

It could not be otherwise: Since a conductor contains free charges, the presence of any internal electric field would result in bulk charge motion, and we would not have equilibrium. This result does not depend on the size or shape of the conductor, the magnitude or direction of the applied field, or even the nature of the material as long as it's a conductor. This ability of a conductor to cancel applied fields is the basis of shielding—the use of conductive enclosures to keep out unwanted electric fields.

This discussion of equilibrium is a macroscopic one; it considers only overall average fields within the material. At the atomic and molecular level there are still strong electric fields near individual electrons and protons. But the *average* field, taken over distances many times the separation between individual charges, is zero in electrostatic equilibrium.

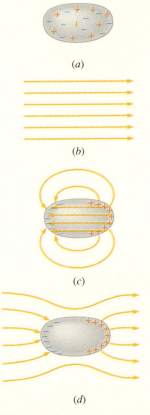

(a)

(b)

(c)

(d)

FIGURE 24-27 (a) A piece of conducting material contains positive and negative charge. (b) A uniform electric field, arising from charges outside the region shown. (c) Charges in the conductor separate in response to the field, resulting in an internal field that cancels the applied field. (d) The net field is the vector sum of the applied field and the field resulting from the redistribution of charge within the conductor. Note that field lines, as always, begin and end on charges.

FIGURE 24-28 Excess charge accumulates at the surface of a charged conductor. In this elongated conductor, mutual repulsion of the excess electrons results in the greatest charge accumulation at the opposite ends.

FIGURE 24-29 Since the electric field inside the conductor is zero, a gaussian surface within the conductor encloses zero net charge. Any excess charge therefore resides on the conductor surface.

Charged Conductors

Although they contain free charges, conductors are normally electrically neutral since they include equal numbers of electrons and protons. But suppose we give a conductor a nonzero net charge, for example, by injecting excess electrons into its interior. There will be a mutual repulsion among the electrons and, because these are *excess* electrons, there is no compensating attraction from protons. We might expect, therefore, that the electrons will move as far apart as possible—namely to the surface of the conductor (Fig. 24-28). (Electrons might even leave the material—but that only occurs with very high charge densities.)

We now use Gauss's law to prove rigorously that excess charge *must* be at the surface of a conductor in electrostatic equilibrium. Figure 24-29 shows a piece of conducting material with a gaussian surface drawn just below the material surface. In equilibrium there is no electric field within the conductor, and thus the field is zero everywhere on the gaussian surface. The flux, $\oint \mathbf{E} \cdot d\mathbf{A}$, through the gaussian surface is therefore also zero. But Gauss's law says that the flux through a closed surface is proportional to the net charge enclosed, and therefore the net charge within our gaussian surface must be zero. This is true no matter where the gaussian surface is as long as it is *inside* the conductor. We can move the gaussian surface arbitrarily close to the conductor surface and it still encloses no net charge. If there is a net charge on the conductor it lies outside the gaussian surface, and therefore we conclude that

> **If a conductor in electrostatic equilibrium carries a net charge, all excess charge resides on the conductor surface.**

EXAMPLE 24-7	*A Hollow Conductor*

An irregularly shaped conductor has a hollow cavity, as shown in Fig. 24-30. The conductor carries a net charge of 1.0 μC. A 2.0-μC point charge is inside the cavity, not touching the conductor. Find the net charge on the cavity wall and on the outer surface of the conductor, assuming electrostatic equilibrium.

+2.0 μC

Gaussian surface

FIGURE 24-30 A conductor with a hollow cavity containing a charge. The gaussian surface shown encloses no net charge, so the charge on the inner wall must be equal but opposite that of the charge within the cavity.

Solution

The electric field is zero everywhere within the conductor, and therefore is zero at all points on the gaussian surface shown in Fig. 24-30. The flux through this surface is thus zero, and Gauss's law tells us that the surface therefore encloses zero net charge. But there is a 2.0-μC charge within the cavity. In order for the gaussian surface to enclose zero net charge, the cavity wall must therefore carry -2.0 μC.

Since the conductor's net charge is 1.0 μC and there is -2.0 μC on its inner wall, the outer surface of the conductor surface must carry 3.0 μC.

EXERCISE A point charge $+q$ is placed inside a hollow conducting shell carrying a net charge $-3q$. What is the total charge on the outside surface of the shell?

Answer: $-2q$

Some problems similar to Example 24-7: 46–49

Got It!

A piece of conductor carries a net charge $+Q$. There's a hollow cavity within the conductor, and in it is a point charge $-Q$. In equilibrium, is there any charge on the outer surface of the conductor?

Experimental Tests of Gauss's Law

The fact that excess charge resides only on a conductor surface provides a very sensitive test of Gauss's law, and thus of the inverse-square law for the electric field. Figure 24-31 shows a charged conducting ball being placed inside a hollow, initially neutral conductor. When the two conductors touch, all the excess charge flows to the outer surface of the hollow conductor, leaving no net charge on the ball. In practice the experiment is often done in reverse, with an uncharged conducting ball placed within a hollow conductor. The outer conductor is then charged, and sensitive instruments used detect any charge moving to the ball. Absence of such charge motion confirms the inverse-square law. Recent experiments of this type show that the exponent 2 appearing in the inverse-square law is indeed 2 to within 3×10^{-16}. Such tests are far more sensitive than direct measurements of how the electric force varies with distance.

The Field at a Conductor Surface

There can be no electric field *within* a conductor in electrostatic equilibrium, but there *may* be a field right *at* the conductor surface. Such a field, though, cannot have a component parallel to the surface; if it did, charge would move along the surface and we would not have equilibrium. So the field at a conductor surface must be perpendicular to the surface (Fig. 24-32a).

We can compute the field strength by considering a small gaussian surface that straddles the conductor surface, as shown in Fig. 24-32b. We make the gaussian surface so small that curvature of the conductor becomes negligible, and we orient the gaussian surface with its sides perpendicular and its top parallel to the conductor surface. Since the field is perpendicular to the conductor, there is then no flux through the sides of the gaussian surface. Since the field is zero inside the conductor, there is also no flux through the inner end of the gaussian surface. The only

FIGURE 24-31 Experimental test of Gauss's law. When the small charged conductor contacts the interior of the hollow conductor, all its charge moves to the outside of the hollow conductor.

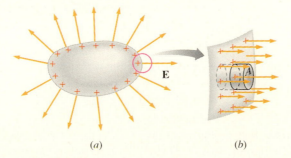

(a)   (b)

FIGURE 24-32 (a) The electric field at the surface of a charged conductor is perpendicular to the conductor surface. (b) A small gaussian surface straddles the conductor surface.

FIGURE 24-33 During thunderstorms, a large charge density builds up at the pointed tip of the Empire State Building. The associated electric field can be strong enough to cause dielectric breakdown in the surrounding air, initiating frequent lightning strikes.

flux is through the outer end, whose area is A. Since the field is essentially uniform and perpendicular to this end, the flux is just EA. The only charge enclosed is right at the conductor surface, where it occupies the same area A. If the surface charge density is σ, then the enclosed charge is σA. Gauss's law equates the flux to $q_{enclosed}/\varepsilon_0$, so we have

$$EA = \frac{\sigma A}{\varepsilon_0}$$

or

$$E = \frac{\sigma}{\varepsilon_0}. \qquad \text{(field at conductor surface)} \qquad (24\text{-}11)$$

This result applies to any conductor in electrostatic equilibrium and shows that large electric fields develop where the charge density is high. Engineers designing electrical devices must avoid high charge densities whose associated fields could lead to sparks, arcing, and breakdown of electrical insulation (Fig. 24-33).

EXAMPLE 24-8	*Earth's Field*

Earth, which is an electrical conductor, carries a net charge of -4.3×10^5 C distributed approximately uniformly over its surface. Find the surface charge density, and use Equation 24-11 to calculate the electric field at Earth's surface.

Solution

Let Q be Earth's charge and R_E its radius (which is given inside the front cover and in Appendix E). Then the surface charge density is

$$\sigma = \frac{Q}{A} = \frac{Q}{4\pi R_E^2} = \frac{-4.3\times10^5\,\text{C}}{(4\pi)(6.37\times10^6\,\text{m})^2}$$
$$= -8.43\times10^{-10}\,\text{C/m}^2\,.$$

Equation 24-11 then gives the electric field at Earth's surface:

$$E = \frac{\sigma}{\varepsilon_0} = \frac{-8.43\times10^{-10}\,\text{C/m}^2}{8.85\times10^{-12}\,\text{C}^2/\text{N·m}^2} = -95\,\text{N/C}\,,$$

where the minus sign indicates that the field direction is downward. This modest field is present near Earth's surface in fair weather; in thunderstorms the local field exceeds this value by many orders of magnitude.

Does our result make sense? We could also treat Earth as a spherical charge distribution, whence its surface field strength would be $E = Q/4\pi\varepsilon_0 R_E^2$. Using the symbolic form of our result for σ in Equation 24-11 gives precisely this expression, showing that these two approaches to the field are indeed consistent.

EXERCISE Dielectric breakdown of air occurs with fields of about 3×10^6 N/C, and results in sparks jumping through the air. What is the maximum surface charge density permissible on a conductor if dielectric breakdown of the surrounding air is to be avoided?

Answer: 27 μC/m^2

Some problems similar to Example 24-8: 43, 44, 51

Equation 24-11 gives a field that depends only on the local charge density. Does that mean the field at any point on a conductor surface arises only from the charge right at that point? No! As always, the field at any point is the vector sum of contributions from all charge elements making up the charge distribution. Remarkably, Gauss's law requires that charges on a conductor arrange themselves in such a way that the field at any point on the conductor surface depends only on

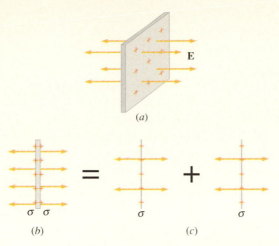

FIGURE 24-34 (*a*) An isolated, charged conducting plate. Its field points outward from both faces. (*b*) Edge-on view of the plate. If the plate is isolated, symmetry requires that there be equal charge densities on both faces. (*c*) The field anywhere is the sum of the fields of the two faces, each treated as a single charged sheet. Within the plate the fields cancel, while outside they sum to give a net field σ/ε_0.

the surface charge density right at that point—even though that field arises from *all* the charges on the surface (as well as from charges elsewhere if there are such)!

Consider a thin, flat, isolated, conducting sheet that has charge density σ on one of its two faces (Fig. 24-34*a*). Equation 24-11 shows immediately that the field at the surface of this plate is σ/ε_0. But if the plate is large and flat we can approximate it as an infinite sheet of charge—for which we found earlier (Equation 24-10 in Example 24-6) that the field should be $\sigma/2\varepsilon_0$. Is there a contradiction here? No! If the plate is isolated from other conductors or charges, then symmetry requires that charge spread itself evenly over *both* faces. If one face has charge density σ, so must the other—so we really have *two* charge sheets, each with density σ (Fig. 24-34*b*). Each gives rise to a field of magnitude $\sigma/2\varepsilon_0$, and *outside* the conductor those fields superpose to give the net field σ/ε_0 (Fig. 24-34*c*). *Inside* the conductor they also superpose, but here their directions are opposite and the result is that there is no field inside the conductor. Application of Equation 24-11 skips all these details. But because Equation 24-11 was derived on the assumption that the field inside the conductor is zero, it "knows" about charges everywhere on the conductor—and in this case that means on the second face of the conductor.

Equation 24-11 also applies to the pair of oppositely charged conducting plates shown in Fig. 24-35; the result, for the field between the plates, is σ/ε_0, where σ is the surface charge density on either plate. Why not $2\sigma/\varepsilon_0$? Again, Equation 24-11 always gives the field at a conductor surface—and it takes into account other charges that may be present. In this case neither plate is isolated, so the symmetry is broken, and charge builds up only on the inner faces of the two plates. Now each plate is a single charge layer, giving rise to a field $\sigma/2\varepsilon_0$, and between the plates the fields sum to Equation 24-11's result, σ/ε_0. Beyond the plates the fields sum to zero—a result that also follows from Equation 24-11 because now there is zero surface charge on the outer faces.

FIGURE 24-35 Edge view of two parallel conducting plates carrying opposite charges. Electrical attraction brings the excess charges to the inner faces. Each face constitutes a charge layer whose surface charge density has magnitude σ; between the plates their fields add to give field strength σ/ε_0.

CHAPTER SYNOPSIS

Summary

1. **Electric field lines** provide a visual representation of the electric field. The direction of a field line is that of the electric field, and the closeness of field lines is a measure of field strength. Electric field lines begin and end only on charges.

2. **Electric flux** quantifies the notion "number of field lines crossing a surface." Flux is the integral of the electric field over the surface:

$$\phi = \int \mathbf{E} \cdot d\mathbf{A},$$

where $d\mathbf{A}$ is an infinitesimal vector perpendicular to the surface at each point.

3. **Gauss's law** is a fundamental relation governing the behavior of electric fields throughout the universe. Gauss's law states that the electric flux emerging from any closed surface is proportional to the charge enclosed:

$$\oint \mathbf{E} \cdot d\mathbf{A} = \frac{q_{enclosed}}{\varepsilon_0},$$

where $\varepsilon_0 = 1/4\pi k$.

4. Gauss's law is true for any surface and any distribution of charge, but it proves useful in calculating the electric field only in cases with sufficient symmetry—in particular, spherical symmetry, line symmetry, and plane symmetry. Table 24-1 lists some fields calculated with Gauss's law. The fields of more realistic charge distributions are often approximated by the fields associated with these symmetries.

5. There is no electric field inside a conductor in **electrostatic equilibrium**. Gauss's law shows that any excess charge on the conductor resides on the conductor surface, and that the field at the conductor surface is perpendicular to the surface and has magnitude σ/ε_0, with σ the surface charge density.

TABLE 24-1 *Some Results from Gauss's Law*

CHARGE DISTRIBUTION	FIELD	WHERE VALID	DERIVED IN . . .
Any spherically symmetric	$E = \dfrac{kQ}{r^2} = \dfrac{Q}{4\pi\varepsilon_0 r^2}$	Outside the charge distribution	Section 24-4
Solid, uniformly charged sphere of radius R	$E = \dfrac{kQr}{r^3}$ $= \dfrac{Qr}{4\pi\varepsilon_0 r^3} = \dfrac{\rho r}{3\varepsilon_0}$	Inside ($r < R$)	Example 24-1
Uniform spherical shell of charge	$E = 0$	Inside	Example 24-2
Any cylindrically symmetric distribution with line charge density λ	$E = \dfrac{2k\lambda}{r} = \dfrac{\lambda}{2\pi e_0 r}$	Outside the charge distribution	Section 24-4
Solid, charged cylinder of radius R, uniform volume charge density ρ	$E = \dfrac{\rho r}{2\varepsilon_0 r}$	Inside ($r < R$)	Problem 24-31
Hollow, uniformly charged pipe	$E = 0$	Inside	Example 24-5
Infinite flat sheet, uniform surface charge density σ	$E = \dfrac{\sigma}{2\varepsilon_0}$	Outside	Example 24-6
Conductor in electrostatic equilibrium	$E = 0$	Inside	Section 24-6
Conductor in electrostatic equilibrium	$E = \dfrac{\sigma}{\varepsilon_0}$	At conductor surface	Section 24-6

Terms You Should Understand

(Pairs are closely related terms whose distinction is important; number in parentheses is chapter section where term first appears.)

electric field lines (24-1)
electric flux (24-2)
Gauss's law (24-3)
electrostatic equilibrium (24-6)

Symbols You Should Recognize

ϕ (24-2)

$\mathbf{A}, d\mathbf{A}$ (24-2)

$\oint$ (24-2)

$\oint \mathbf{E} \cdot d\mathbf{A}$ (24-2)

k, ε_0 (24-3)

σ (24-6)

Problems You Should Be Able to Solve

drawing and interpreting field line patterns for simple charge distributions (24-1)
calculating electric flux (24-2)
using Gauss's law to calculate electric fields with spherical, line, or plane symmetry (24-4)
describing charge distributions in electrostatic equilibrium (24-6)
calculating electric fields at conductor surfaces (24-6)

Limitations to Keep in Mind

Gauss's law is universally true, but it can be used to calculate electric fields only with sufficient symmetry.

QUESTIONS

1. Can electric field lines ever cross? Why or why not?
2. If identical charged particles are placed at points *A* and *B* in Fig. 24-36, which will experience the greater force?

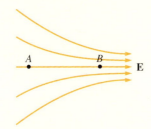

FIGURE 24-36 Question 2.

3. The electric flux through a closed surface is zero. Must the electric field be zero on that surface? If not, give an example.
4. If the flux of the gravitational field through a closed surface is zero, what can you conclude about the region interior to the surface?
5. Under what conditions can the electric flux through a surface be written as *EA*, where *A* is the surface area?
6. Eight field lines emerge from a closed surface surrounding an isolated point charge. Would this fact change if a second identical charge were brought to a point just *outside* the surface? If not, would anything change? Explain.
7. In what sense is Gauss's law equivalent to the inverse-square law?
8. If a charged particle were released from rest on a curved field line, would its subsequent motion follow the field line? Explain.
9. Gauss' law describes the flux of the electric field through a surface that may enclose charge. Must the field in Gauss's law arise only from charges within the closed surface?

10. In a certain region the electric field points to the right and its magnitude increases as you move to the right, as shown in Fig. 24-37. Does the region contain net positive charge, net negative charge, or zero net charge?

FIGURE 24-37 Question 10. (Left ends mark the beginnings of the field lines, which extend indefinitely to the right.)

11. A point charge is located a fixed distance from a uniformly charged sphere, outside the sphere. If the sphere shrinks in size without losing any charge, what happens to the force on the point charge?
12. In applying Equation 24-6 for the field outside a spherically symmetric charge distribution, is *r* the distance from the center or from the edge of the distribution?
13. The field of an infinite line of charge falls as $1/r$. How is this not a violation of the inverse-square law?
14. Why can't you use Gauss's law to determine the field of a uniformly charged cube? Why wouldn't it work to draw a cubical gaussian surface?
15. You're sitting inside an uncharged hollow spherical shell. Suddenly someone dumps a billion coulombs of charge on the shell, distributed uniformly. What happens to the electric field at your location?
16. Why is it that the field inside a uniformly charged sphere actually increases with distance from the center? How is this consistent with Coulomb's law?

17. There is a nonzero flux through each of the surfaces in Fig. 24-6, yet there is no charge in the region shown. Why is this not a violation of Gauss's law?

18. Does Gauss's law apply to a spherical surface not centered on a point charge, as shown in Fig. 24-38? Would this be a useful surface to use in calculating the electric field?

FIGURE 24-38 Question 18.

19. An insulating sphere carries charge spread uniformly throughout its volume. A conducting sphere has the same radius and net charge, but of course the charge is spread over its surface only. How do the electric fields outside these two charge distributions compare?

20. Why must the electric field be zero inside a conductor in electrostatic equilibrium?

21. Why must the electric field at the surface of a conductor in electrostatic equilibrium be perpendicular to the surface?

22. Where in Fig. 24-28 would you find the strongest electric field?

23. The electric field of a flat sheet of charge is $\sigma/2\varepsilon_0$. Yet the field of a flat conducting sheet—even a thin one, like a piece of aluminum foil—is σ/ε_0. Explain this apparent discrepancy.

24. A metal contains free electrons not bound to individual atoms. Does Gauss's law require that all these free electrons be on the metal surface?

PROBLEMS

ActivPhysics can help with these problems:
Activities 11.4–11.6

Section 24-1 Electric Field Lines

1. What is the net charge shown in Fig. 24-39? The magnitude of the middle charge is 3 μC.

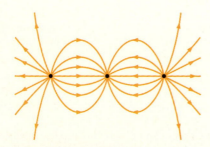

FIGURE 24-39 Problem 1.

2. A charge $+2q$ and a charge $-q$ are near each other. Sketch some field lines for this charge distribution, using the convention of eight lines for a charge of magnitude q.

3. Two charges $+q$ and a charge $-q$ are at the vertices of an equilateral triangle. Sketch some field lines for this charge distribution.

4. The net charge shown in Fig. 24-40 is $+Q$. Identify each of the charges A, B, C shown.

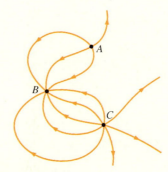

FIGURE 24-40 Problem 4.

Section 24-2 Electric Flux

5. A flat surface with area 2.0 m^2 is in a uniform electric field of 850 N/C. What is the electric flux through the surface when it is (a) at right angles to the field, (b) at 45° to the field, and (c) parallel to the field?

6. What is the electric field strength in a region where the flux through a 1.0 cm $\times$ 1.0 cm flat surface is 65 N·m^2/C, if the field is uniform and the surface is at right angles to the field?

7. A flat surface with area 0.14 m^2 lies in the x-y plane, in a uniform electric field given by $\mathbf{E} = 5.1\hat{\mathbf{i}} + 2.1\hat{\mathbf{j}} + 3.5\hat{\mathbf{k}}$ kN/C. Find the flux through this surface.

8. The electric field on the surface of a 10-cm-diameter sphere is perpendicular to the sphere and has magnitude 47 kN/C. What is the electric flux through the sphere?

9. What is the flux through the hemispherical open surface of radius R shown in Fig. 24-41? The uniform field has magnitude E. *Hint:* Don't do a messy integral! Imagine closing the surface with a flat, circular piece across the open end. What would be the flux through the entire closed surface? And what's the flux through the flat end? So what's the answer?

FIGURE 24-41 Problem 9.

10. The electric field shown in Fig. 24-42 is given by $\mathbf{E} = E_0\dfrac{y}{a}\hat{\mathbf{k}}$, where E_0 and a are constants. Find the flux through the square of side a shown.

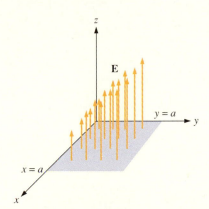

FIGURE 24-42 Problem 10.

Section 24-3 Gauss's Law

11. What is the electric flux through each closed surface shown in Fig. 24-43?

FIGURE 24-43 Problem 11.

12. A 6.8-μC charge and a -4.7 μC charge are inside an uncharged sphere. What is the electric flux through the sphere?

13. A 2.6-μC charge is at the center of a cube 7.5 cm on each side. What is the electric flux through one face of the cube? *Hint:* Think about symmetry, and don't do an integral.

14. If the charge in the preceding problem is still inside the cube but not at the center, (a) what is the flux through the *entire* cube? (b) Could you still calculate the flux through one face without doing an integral?

15. A dipole consists of two charges ± 6.1 μC located 1.2 cm apart. What is the electric flux through each surface shown in Fig. 24-44?

FIGURE 24-44 Problem 15.

16. The electric field in a certain region is given by $\mathbf{E} = 40x\hat{\mathbf{i}}$ N/C, with x in meters. What is the volume charge density in the region? *Hint:* Apply Gauss's law to a cube 1 meter on a side.

Section 24-4 Using Gauss's Law

17. The electric field at the surface of a uniformly charged sphere of radius 5.0 cm is 90 kN/C. What would be the field strength 10 cm from the surface?

18. A solid sphere 25 cm in radius carries 14 μC, distributed uniformly throughout its volume. Find the electric field strength (a) 15 cm, (b) 25 cm, and (c) 50 cm from the sphere's center.

19. A crude model for the hydrogen atom treats it as a point charge $+e$ (the proton) surrounded by a uniform cloud of negative charge with total charge $-e$ and radius 0.0529 nm. What would be the electric field strength inside such an atom, halfway from the proton to the edge of the charge cloud?

20. Positive charge is spread uniformly over the surface of a spherical balloon 70 cm in radius, resulting in an electric field of 26 kN/C at the balloon's surface. Find the field strength (a) 50 cm from the balloon's center and (b) 190 cm from the center. (c) What is the net charge on the balloon?

21. A 10-nC point charge is located at the center of a thin spherical shell of radius 8.0 cm carrying -20 nC distributed uniformly over its surface. What are the magnitude and direction of the electric field (a) 2.0 cm, (b) 6.0 cm, and (c) 15 cm from the point charge?

22. A solid sphere 2.0 cm in radius carries a uniform volume charge density. The electric field 1.0 cm from the sphere's center has magnitude 39 kN/C. (a) At what other distance does the field have this magnitude? (b) What is the net charge on the sphere?

23. A point charge $-2Q$ is at the center of a spherical shell of radius R carrying charge Q spread uniformly over its surface. What is the electric field at (a) $r = \frac{1}{2}R$ and (b) $r = 2R$? (c) How would your answers change if the charge on the shell were doubled?

24. A spherical shell of radius 15 cm carries 4.8 μC, distributed uniformly over its surface. At the center of the shell is a point charge. (a) If the electric field at the surface of the sphere is 750 kN/C and points outward, what is the charge of the point charge? (b) What is the field just inside the shell?

25. A spherical shell 30 cm in diameter carries a total charge 85 μC distributed uniformly over its surface. A 1.0-μC point charge is located at the center of the shell. What is the electric field strength (a) 5.0 cm from the center and (b) 45 cm from the center? (c) How would your answers change if the charge on the shell were doubled?

26. The thick, spherical shell of inner radius a and outer radius b shown in Fig. 24-45 carries a uniform volume charge density ρ. Find an expression for the electric field strength in the region $a < r < b$, and show that your result is consistent with Equation 24-7 when $a = 0$.

FIGURE 24-45 Problems 26, 32.

27. How should the charge density within a solid sphere vary with distance from the center in order that the magnitude of the electric field in the sphere be constant?

28. A long, thin wire carrying 5.6 nC/m runs down the center of a long, thin-walled, hollow pipe with radius 1.0 cm carrying -4.2 nC/m spread uniformly over its surface. Find the electric field (a) 0.50 cm from the wire and (b) 1.5 cm from the wire.

29. A long solid rod 4.5 cm in radius carries a uniform volume charge density. If the electric field strength at the surface of the rod (not near either end) is 16 kN/C, what is the volume charge density?

30. The electric field strength outside a charge distribution and 18 cm from its center has magnitude 55 kN/C. At 23 cm the field strength is 43 kN/C. Does the distribution have spherical or line symmetry?

31. An infinitely long rod of radius R carries a uniform volume charge density ρ. Show that the electric field strengths outside and inside the rod are given, respectively, by $E = \rho R^2/2\varepsilon_0 r$ and $E = \rho r/2\varepsilon_0$, where r is the distance from the rod axis.

32. Repeat Problem 26, assuming that Fig. 24-45 represents the cross section of a long, thick-walled pipe. Now the case $a = 0$ should be consistent with the result of Problem 31 for the interior of the rod.

33. A long, thin wire carries a uniform line charge density $\lambda = -6.8$ μC/m. It is surrounded by a thick concentric cylindrical shell of inner radius 2.5 cm and outer radius 3.5 cm. What uniform volume charge density in the shell will result in zero electric field outside the shell?

34. A square nonconducting plate measures 4.5 m on a side and carries charge spread uniformly over its surface. The electric field 10 cm from the plate and not near an edge has magnitude 430 N/C and points toward the plate. Find (a) the surface charge density on the plate and (b) the total charge on the plate. (c) What is the electric field strength 20 cm from the plate?

35. If you "painted" positive charge on the floor, what surface charge density would be necessary in order to suspend a 15-μC, 5.0-g particle above the floor?

36. A slab of charge extends infinitely in two dimensions and has thickness d in the third dimension, as shown in Fig. 24-46. The slab carries a uniform volume charge density ρ. Find expressions for the electric field strength (a) inside and (b) outside the slab, as functions of the distance x from the center plane.

FIGURE 24-46 Section of an infinite slab of charge (Problem 36).

37. Figure 24-47 shows sections of three infinite flat sheets of charge, each carrying surface charge density with the same magnitude σ. Find the magnitude and direction of the electric field in each of the four regions shown.

FIGURE 24-47 Problem 37.

Section 24-5 Fields of Arbitrary Charge Distributions

38. A rod 50 cm long and 1.0 cm in radius carries a 2.0-μC charge distributed uniformly over its length. What is the approximate magnitude of the electric field (a) 4.0 mm from the rod surface, not near either end, and (b) 23 m from the rod?

39. A nonconducting square plate 75 cm on a side carries a uniform surface charge density. The electric field strength 1 cm from the plate, not near an edge, is 45 kN/C. What is the approximate field strength 15 m from the plate?

40. Two circular plates 10 cm in diameter and 2.0 mm apart carry equal but opposite charges ± 0.50 μC distributed uniformly over their facing surfaces. What is the electric field strength (a) between the plates but not near either edge? (b) 2.5 m from the plates on a plane passing midway between them? *Hint* for (b): See Example 23-6.

41. The electric field strength on the axis of a uniformly charged disk is given by $E = 2\pi k\sigma(1 - x/\sqrt{x^2 + a^2})$, with σ the surface charge density, a the disk radius, and x the distance from the disk center. If $a = 20$ cm, (a) for what range of x values does treating the disk as an infinite sheet give an approximation to the field that is good to within 10%? (b) For what range of x values is the point-charge approximation good to 10%?

42. A nonconducting square 2.0 cm on a side carries a 45-nC charge spread uniformly over its surface. The x axis runs through the plate center, perpendicular to the plate, with $x = 0$ at the plate center. A -45-nC point charge is at $x = 5.0$ cm. Find approximate values for the electric field strength on the x axis at (a) $x = 1.0$ mm; (b) $x = 4.8$ cm; (c) $x = 2.5$ m. *Hint* for (c): Consult Example 23-6.

Section 24-6 Gauss's Law and Conductors

43. What is the electric field strength just outside the surface of a conducting sphere carrying surface charge density 1.4 μC/m^2?

44. Calculate the acceleration of a proton at the surface of a conductor carrying surface charge density 0.60 C/m^2.

45. A net charge of 5.0 μC is applied on one side of a solid metal sphere 2.0 cm in diameter. After electrostatic equilibrium is reached, what are (a) the volume charge density inside the sphere and (b) the surface charge density on the sphere? Assume there are no other charges or conductors nearby. (c) Which of your answers depends on this assumption, and why?

46. A point charge $+q$ lies at the center of a spherical conducting shell carrying a net charge $\frac{3}{2}q$. Sketch the field lines both inside and outside the shell, using 8 field lines to represent a charge of magnitude q.

47. A 250-nC point charge is placed at the center of an uncharged spherical conducting shell 20 cm in radius. (a) What is the surface charge density on the outer surface of the shell? (b) What is the electric field strength at the shell's outer surface?

48. A point charge is placed at the center of an uncharged spherical conducting shell of inner radius 2.5 cm and outer radius 4.0 cm (Fig. 24-48). As a result, the outer surface of the shell acquires a surface charge density $\sigma = 71$ nC/cm^2. Find (a) the value of the point charge and (b) the surface charge density on the inner wall of the shell.

FIGURE 24-48 Problem 48.

49. An irregular conductor containing an irregular, empty cavity carries a net charge Q. (a) Show that the electric field inside the cavity must be zero. (b) If you put a point charge inside the cavity, what value must it have in order to make the surface charge density on the outer surface of the conductor everywhere zero?

50. A neutral dime is placed in a uniform electric field of 6.2×10^5 N/C, with its faces perpendicular to the field. (a) What is the approximate charge density on the faces of the dime? (b) What is the total charge on each face? (Measure a dime!)

51. A total charge of 18 μC is applied to a thin, square metal plate 75 cm on a side. Find the electric field strength near the plate's surface.

52. Two closely spaced parallel metal plates carry surface charge densities ± 95 nC/m^2 on their facing surfaces, with no charge on their outer surfaces. Find the electric field

strength (a) between the plates and (b) outside the plates. Treat the plates as infinite in extent.

53. A conducting sphere 2.0 cm in radius is concentric with a spherical conducting shell with inner radius 8.0 cm and outer radius 10 cm. The small sphere carries 50 nC charge and the shell has no net charge. Find the electric field strength (a) 1.0 cm, (b) 5.0 cm, (c) 9.0 cm, and (d) 15 cm from the center.

54. A coaxial cable consists of an inner wire and a concentric cylindrical outer conductor (Fig. 24-49). If the conductors carry equal but opposite charges, show that there is no net charge on the *outside* of the outer conductor.

FIGURE 24-49 Problem 54.

Paired Problems

(Both problems in a pair involve the same principles and techniques. If you can get the first problem, you should be able to solve the second one.)

55. A point charge $-q$ is at the center of a spherical shell carrying charge $+2q$. That shell, in turn, is concentric with a larger shell carrying charge $-\frac{3}{2}q$. Draw a cross section of this structure, and sketch the electric field lines using the convention that 8 lines correspond to a charge of magnitude q.

56. A point charge $-q$ is at the center of a spherical shell carrying charge $-\frac{3}{2}q$. That shell, in turn, is concentric with a larger shell carrying charge $+2q$. Draw a cross section of this structure, and sketch the electric field lines using the convention that 8 lines correspond to a charge of magnitude q.

57. A point charge q is at the center of a spherical shell of radius R carrying charge $2q$ spread uniformly over its surface. Write expressions for the electric field strength at (a) $\frac{1}{2}R$ and (b) $2R$.

58. A point charge q is at the center of a spherical shell of radius R carrying charge $5q$. At what other distance does the electric field have the same value it does at a point halfway from the center to the shell?

59. A long, thin hollow pipe 4.0 cm in diameter carries charge at a density of $-2.6\ \mu C/m$, uniformly distributed over the pipe. It is concentric with 10-cm diameter pipe carrying $+2.6\ \mu C/m$, also uniformly distributed. Find the magnitude of the electric field at (a) 0.50 cm, (b) 3.5 cm, and (c) 12 cm from the axis of the pipes.

60. Two concentric hollow pipes are 5.0 cm and 12 cm in diameter, respectively. Both carry uniformly distributed electric charges. The electric field 4.0 cm from their common axis is 630 kN/C, radially outward. The field 10 cm from their common axis is 126 kN/C, radially outward. (a) Find the linear charge densities on the two pipes. (b) How would the electric field strengths at 4.0 cm and 10 cm change if the charge density on the outer pipe were doubled?

61. An early (and incorrect) model for the atom pictured its positive charge as spread uniformly throughout the spherical atomic volume. For a hydrogen atom of radius 0.0529 nm, what would be the electric field due to such a distribution of positive charge (a) 0.020 nm from the center and (b) 0.20 nm from the center?

62. A solid sphere of radius R carries a charge spread uniformly throughout its volume. At what point outside the sphere is the electric field strength equal to that at a point halfway from the center to the edge? Express your answer as a distance from the center.

63. A sphere of radius $2a$ has a hole of radius a, as shown in Fig. 24-50. The solid portion carries a uniform volume charge density ρ. Find an expression for the electric field strength within the solid portion, as a function of the distance r from the center.

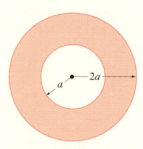

FIGURE 24-50 Problems 63, 64.

64. Repeat the previous problem, now considering that the figure represents the cross section of a thick cylindrical pipe.

Supplementary Problems

65. Repeat Problem 10 for the case $\mathbf{E} = E_0\left(\dfrac{y}{a}\right)^2\hat{\mathbf{k}}$.

66. The volume charge density inside a solid sphere of radius a is given by $\rho = \rho_0 r/a$, where ρ_0 is a constant. Find (a) the total charge and (b) the electric field strength within the sphere, as a function of distance r from the center.

67. A proton is released from rest 1.0 cm from a large sheet carrying a surface charge density of -24 nC/m^2. How much later does it strike the sheet?

68. Fig. 24-51 shows a rectangular box with sides $2a$ and length ℓ surrounding a line of charge with uniform line charge density λ. The line passes directly through the center of the box faces. Using an expression for the field of a line charge, integrate over strips of width dx as shown to find the electric flux through one face of the box. Multiply by 4 to get the total flux through the box, and show that your result is consistent with Gauss's law.

FIGURE 24-51 Problem 68.

69. Repeat Problem 36 for the case when the charge density in the slab is given by $\rho = \rho_0 |x/d|$, where ρ_0 is a constant.

70. The charge density within a uniformly charged sphere of radius R is given by $\rho = \rho_0 - ar^2$, where ρ_0 and a are constants, and r is the distance from the center. Find an expression for a such that the electric field outside the sphere is zero.

71. A small object of mass m and charge q is attached by a thread of length ℓ to a large, flat, nonconducting plate carrying a uniform surface charge density σ with the same sign as q (Fig. 24-52). If the object is displaced slightly sideways from its equilibrium, show that it undergoes simple harmonic motion with period $T = 2\pi\sqrt{2\epsilon_0 m\ell/q\sigma}$. Assume the gravitational force is negligible.

FIGURE 24-52 Problem 71.

72. An infinitely long nonconducting rod of radius R carries a volume charge density given by $\rho = \rho_0(r/R)$, where ρ_0 is a constant. Find the electric field strength (a) inside and (b) outside the rod, as functions of the distance r from the rod axis.

73. A thick spherical shell of inner radius a and outer radius b carries a charge density given by $\rho = \dfrac{ce^{-r/a}}{r^2}$, where a and c are constants. Find expressions for the electric field strength for (a) $r < a$, (b) $a < r < b$, and (c) $r > b$.

74. A solid sphere of radius R carries a nonuniform volume charge density given by $\rho = \pi^2\rho_0\sin(\pi r/R)$, where r is the distance from the center and ρ_0 is a positive constant. Find the magnitude and direction of the electric field at the sphere's surface.

75. A solid sphere of radius R carries a uniform volume charge density ρ. A hole of radius $R/2$ occupies a region from the center to the edge of the sphere, as shown in Fig. 24-53. Show that the electric field everywhere in the hole points horizontally and has magnitude $\rho R/6\epsilon_0$. *Hint:* Treat the hole as a superposition of two charged spheres of opposite charge.

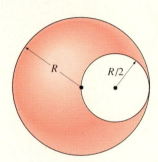

FIGURE 24-53 Problem 75.

76. You're 5.0 m from a charge distribution and you measure an electric field strength of 850 N/C. At 2.5 m the field strength has increased to about 3.4 kN/C. When you're 5.0 mm from the center of the distribution the field strength is 42.5 MN/C, and it increases to 85 MN/C at 2.5 mm. Describe the distribution as fully as you can, including its shape, any dimensions you can find, its total charge, and any appropriate charge density.

77. Two flat, parallel, closely spaced metal plates of area 0.080 m^2 carry total charges of -2.1 μC and $+3.8$ μC. Find the surface charge densities on the inner and outer faces of each plate.

78. Since the gravitational force of a point mass goes as $1/r^2$, the gravitational field **g** also obeys a form of Gauss's law. (a) Formulate this law, and (b) use it to find an expression for the gravitational field strength *within* Earth, as a function of distance r from the center. Treat Earth as a sphere of uniform density.

25 Electric Potential

Lines of constant electric potential on the surface of the human body reflect the electric dipole structure of the heart.

You have to do work to lift your book against Earth's gravitational field. The work you do gets stored as potential energy, which is released if you lower the book. In Chapter 8 we introduced the term **conservative** to describe a force or field that, like gravity, "gives back" all the stored energy. With conservative forces we could associate a potential energy, and then use the powerful conservation of energy principle to solve practical problems.

The electric field of a static charge distribution is also conservative—an experimental fact that's also suggested theoretically by the similarity between Coulomb's law and Newton's law of universal gravitation. Because the electric field is conservative, it is meaningful to talk about potential energy in electrostatic systems. If you do work moving a charge against the electric force, for example, that work ends up stored as electric potential energy.

In this chapter we deal with electric potential, a useful and easily measured quantity that derives from the concept of potential energy. We'll see how the use of electric potential can provide a simpler approach to calculating electric fields, and how electric potential relates to the properties of everyday devices like batteries.

25-1 POTENTIAL ENERGY, WORK, AND THE ELECTRIC FIELD

In Chapter 8 we defined potential energy difference ΔU_{AB} as the negative of the work W_{AB} done by a conservative force $\mathbf{F}$ on an object moved from point A to point B:

$$\Delta U_{AB} = U_B - U_A = -W_{AB} = -\int_A^B \mathbf{F} \cdot d\boldsymbol{\ell}, \qquad (25\text{-}1)$$

where $d\boldsymbol{\ell}$ is a small element of the path from A to B. We also showed that this potential energy difference is independent of the path taken between points A and B (Fig. 25-1).

We'll now look at potential energy changes associated with moving charged objects in an electric field. First, consider a positive charge q being moved between two points A and B a distance ℓ apart in a uniform electric field $\mathbf{E}$, as shown in Fig. 25-2. A conservative electric force $q\mathbf{E}$ acts on the charge, and since the field is uniform that force doesn't vary as we move from A to B. Therefore, we can dispense with the integral and write

$$\Delta U_{AB} = -W_{AB} = -\mathbf{F} \cdot \boldsymbol{\ell} = -q\mathbf{E} \cdot \boldsymbol{\ell} = -qE\ell \cos(180°) = qE\ell,$$

where $\boldsymbol{\ell}$ is a vector from A to B and where the factor $\cos 180° = -1$ appears because the vectors $\mathbf{E}$ and $\boldsymbol{\ell}$ are in opposite directions. Does this result make sense? Pushing a positive charge against the electric field is like pushing a car up a hill: We do positive work, gravity does negative work, and the gravitational potential energy increases. Here we must do positive work to move the charge from A to B, the electric field does negative work, and the electric potential energy change—the *negative* of the work done by the conservative electric force—is positive. Let go of the charge and the field will accelerate it back toward B, changing potential to kinetic energy, just as gravity would accelerate the car back down the hill.

FIGURE 25-1 The gravitational field is conservative, and therefore the net work required to move between two points is independent of the path taken.

FIGURE 25-2 The work required to move a charge q from A to B in a *uniform* electric field is $qE\ell$.

25-2 POTENTIAL DIFFERENCE

We've just found the potential energy change in moving a charge q from point A to point B in Fig. 25-2. If we had moved a charge $2q$ the potential energy change would have been twice as great, and it would have required twice as much work; $\frac{1}{4}q$ would have one-fourth the potential energy change and would require one-fourth the work. Since the potential energy change is directly proportional to the charge it's convenient to consider the *potential energy change per unit charge* involved in moving between two points. We call this quantity the **electric potential difference**, ΔV. Mathematically, we express the potential difference by writing $\mathbf{F} = q\mathbf{E}$ in Equation 25-1 and dividing by q:

$$\Delta V_{AB} = \frac{\Delta U_{AB}}{q} = -\frac{1}{q}\int_A^B q\mathbf{E} \cdot d\boldsymbol{\ell} = -\int_A^B \mathbf{E} \cdot d\boldsymbol{\ell}.$$

Then our definition of potential difference becomes:

The electric potential difference from point A to point B is the potential energy change per unit charge in moving from A to B:

$$\Delta V_{AB} = -\int_A^B \mathbf{E} \cdot d\boldsymbol{\ell}. \qquad \text{(electric potential difference)} \qquad (25\text{-}2a)$$

TABLE 25-1 *Force and Field, Potential Energy and Potential*

GENERAL QUANTITY	ELECTRICAL QUANTITY
Force **F** (Unit: N)	Electric field $\mathbf{E} = \dfrac{\mathbf{F}}{q}$ (Unit: N/C or V/m)
Potential energy difference ΔU $\Delta U = -\int_A^B \mathbf{F} \cdot d\boldsymbol{\ell}$ (Unit: J)	Electric potential difference $\Delta V = \dfrac{\Delta U}{q}$ $\Delta V = -\int_A^B \mathbf{E} \cdot d\boldsymbol{\ell}$ (Unit: J/C or V)

Here we use the Δ and the subscripts AB to show explicitly that we're talking about a *change* or *difference* from one point to another. We'll sometimes use just the symbol V for potential difference, in cases where the starting point A is understood. Note that potential difference, although computed from the dot product of vectors, is itself a scalar quantity.

The change from potential energy to electric potential is analogous to the change we made in Chapter 23 when we introduced the electric field as the electric force *per unit charge;* similarly, electric potential difference is the change in potential energy *per unit charge.* Again, the reason is the same: we want to express the electrical properties of a system in terms that don't involve the specific charges we may put in the vicinity of that system. Table 25-1 summarizes the relations among force and field, potential energy and electric potential.

In the special case of a uniform field, Equation 25-2a reduces to

$$\Delta V_{AB} = -\mathbf{E} \cdot \boldsymbol{\ell}, \quad \text{(uniform field)} \qquad (25\text{-}2b)$$

where $\boldsymbol{\ell}$ is a vector from A to B. Figure 25-2 shows the special case when the field $\mathbf{E}$ and path $\boldsymbol{\ell}$ are in opposite directions; here, Equation 25-2b gives $\Delta V_{AB} = E\ell$.

Equations 25-2 show that potential difference can be positive or negative, depending on whether the path goes against or with the field. Moving a positive charge through a positive potential difference is like going uphill: We must do work on the charge, and its potential energy increases. Moving a positive charge through a negative potential difference is like going downhill: We do negative work or, equivalently, the field does work on the charge, and its potential energy decreases. In both cases the opposite is true for a negative charge; even though the potential difference remains the same, the work and potential energy reverse because of the negative sign on the charge.

We emphasize that potential difference is a property *of two points;* it doesn't depend on the path taken between those points. In Fig. 25-2, considering a straight path from A to B made the calculation of potential difference easy, but we would have found the same result—albeit with much more work—using any path (Fig. 25-3).

FIGURE 25-3 The potential difference ΔV_{AB} depends only on the points A and B, not on the path taken. Application of Equation 25-2a to any of the three paths shown would give the same result, namely $\Delta V_{AB} = E\ell$.

Got It!

What would happen to the potential difference V_{AB} in Fig. 25-2 if (a) the electric field strength were doubled; (b) the length ℓ were doubled; (c) the points were moved so the path lay at right angles to the field; (d) the positions of A and B were interchanged?

The Volt and the Electron Volt

The definition of potential difference shows that its units are joules/coulomb. Potential difference is important enough that 1 J/C has a special name—the **volt** (V). To say that a car has a 12-V battery, for example, means that the battery does 12 J of work on every coulomb that moves between its two terminals. Table 25-2 lists some common potential differences.

We often use the term **voltage** to speak of potential difference, especially in describing electric circuits. Strictly speaking the two terms are not synonymous, since voltage is used even in nonconservative situations that arise when fields change with time. But in common usage this subtle distinction is usually not bothersome.

Tip

Potential Difference Depends on Two Points Specifically, it is the energy per unit charge involved in moving *between those points*. Always think of potential difference in terms of two points. This is ultimately a very practical matter; if you forget it you won't be able to hook up a voltmeter properly, or connect jumper cables safely to your car battery! Figure 25-4 provides a dramatic illustration of this point.

Sometimes we speak of "the potential (or the voltage) at point *P*." This is *always* a shorthand way of talking, and we *must* have in mind some other point. What we mean is the potential difference going from that other point to point *P*.

In molecular, atomic, and nuclear systems it's often convenient to measure energy in **electron volts** (eV), defined as follows:

One electron volt is the energy gained by a particle carrying one elementary charge when it moves through a potential difference of one volt.

Since one elementary charge is 1.6×10^{-19} C, 1 eV is 1.6×10^{-19} J. Energy in eV is particularly easy to calculate when charge is given in elementary charges. However, the eV is *not* an SI unit and should be converted to joules before calculating other quantities.

Got It!

A proton (charge e) and an alpha particle (charge $2e$) each move through the same potential difference. Compare the work done on each.

TABLE 25-2 *Typical Potential Differences*

Between human arm and leg due to heart's electrical activity	1 mV
Across biological cell membrane	80 mV
Between terminals of flashlight battery	1.5 V
Car battery	12 V
Across loudspeaker operating at 50 watts	20 V
Electrical outlet, United States	120 V
Electrical outlet, Europe	230 V
Between accelerating electrodes in TV picture tube	30 kV
Between long-distance electric transmission line and ground	365 kV
Between base of thunderstorm cloud and ground	100 MV

FIGURE 25-4 Potential difference depends on *two* points. This parasailer landed on a 138,000-V power line, but he's not being electrocuted because his body is not contacting *two* points with a potential difference between them.

EXAMPLE 25-1	*A TV Picture Tube: Potential Difference, Work, and Energy*

At the back end of a TV picture tube, a uniform electric field of 600 kN/C extends over a distance of 5.0 cm and points toward the back of the tube (Fig. 25-5). (a) Find the potential difference between the back and the front end of this field region. (b) How much work would it take to move an ion with charge $+2e$ from the back to the front of the field region? (c) What would happen to an electron released at the back of the field region?

Solution

(a) With a uniform field the potential difference is given by Equation 25-2b:

$$V_{AB} = -\mathbf{E} \cdot \boldsymbol{\ell} = E\ell = (600 \times 10^3 \text{ N/C})(0.050 \text{ m})$$
$$= 30 \text{ kV},$$

where the second equality follows because a path from the back toward the front is opposite the field direction, as shown in Fig. 25-2, so $\cos\theta = -1$.

(b) Potential difference is the potential energy change per unit charge, or, equivalently, the work per unit charge needed to move charge between two points. Thus, the work to move the ion of charge $q = 2e$ against the 30-kV potential difference is

$$W_{\text{ion}} = q\Delta V = (2)(1.6 \times 10^{-19} \text{ C})(30 \text{ kV}) = 9.6 \times 10^{-15} \text{ J},$$

where the units work out because 1 V = 1 J/C. Since the ion carries two elementary charges, we could also express this energy as $(2e)(30 \text{ kV}) = 60 \text{ keV}$.

FIGURE 25-5 A potential difference on the order of 30 kV is used to accelerate the electrons that "paint" the picture on the screen of a TV tube, shown here in a cutaway view.

(c) With its negative charge, the electron would *gain* energy given by $e\Delta V$, or 4.8×10^{-15} J (more simply calculated as 30 keV). But energy is conserved, so this gain—an increase in the electron's kinetic energy $\frac{1}{2}mv^2$—comes at the expense of a decrease in potential energy.

EXERCISE A 1.2-μC charge is accelerated through a 3400-V potential difference. How much energy does it gain?

Answer: 4.1 mJ

EXAMPLE 25-2	*Potential of a Charged Sheet*

An isolated, infinite charged sheet carries a uniform surface charge density σ. (a) Find an expression for the potential difference from the sheet to a point a perpendicular distance x from the sheet. (b) What is the potential difference between two points the same distance from the sheet?

Solution

Equation 24-10 gives $E = \dfrac{\sigma}{2\varepsilon_0}$ for the field of a single, isolated sheet of charge. Since the field is uniform we can apply Equation 25-2b. Moving away from the sheet in either direction is going *with* the field (assuming positive σ), so the dot product in Equation 25-2b is positive and therefore the potential difference is negative. Taking $\ell = x$ in Equation 25-2b then gives

$$\Delta V_{0x} = -Ex = -\frac{\sigma x}{2\varepsilon_0},$$

where the subscript $0x$ means we're taking the potential difference from the sheet ($x = 0$) to the point x. Our result makes two important points. First, the minus sign reaffirms that moving in the direction of the electric field entails a decrease in electric potential. Second, the fact that ΔV is directly proportional to x shows that the potential in a *uniform* field varies *linearly* with distance along the field direction. (Here we're beginning to use the term "potential" as a shorthand for, in this case, "the potential difference between the sheet and another point." When we speak of "the potential," we always mean potential difference between two points, one of which is a common reference point—in this case the position of the charged sheet.)

If, on the other hand, we consider moving charge between two points equidistant from the sheet, then the charge moves perpendicular to the field and there is no change in its potential energy; thus, the potential difference between such points is zero.

EXERCISE Two nonconducting charged sheets carry equal but opposite surface charge densities ± 53 nC/m². The negative sheet is located at $x = 0$, the positive sheet at $x = 10$ cm. Find expressions for the potential difference from the negative sheet to the points (a) $x = 2.0$ cm, (b) $x = 5.0$ cm, (c) $x = 25$ cm, and (d) $x = -10$ cm. *Hint:* Think about the different fields *between* and *beyond* the plates.

Answers: (a) 120 V; (b) 300 V; (c) 600 V; (d) 0 V

Some problems similar to Examples 25-1 and 25-2: 5, 7–9, 18

Got It!

Does a proton tend to move in the direction of positive or negative potential difference, as measured from its starting point? How about an electron?

Curved Paths and Nonuniform Fields

Equations 25-2 contain a dot product that accounts for the orientation of the path relative to the field. Figure 25-6, for example, shows several straight paths of the same length ℓ in a uniform electric field. Path AB is the same path we considered in Fig. 25-2; the potential difference between its ends is just $\Delta V_{AB} = E\ell$ because the angle between $\mathbf{E}$ and ℓ is 180°. Path AC is at 135° to the field direction, giving a potential difference $\Delta V_{AC} = -E\ell \cos 135° = E\ell/\sqrt{2}$. Finally, path AD is perpendicular to the field, giving $\Delta V_{AD} = 0$. Quite generally, as Fig. 25-6 suggests, the potential difference depends only on the component of the path *along* the field direction. This is analogous to the situation with gravity, where the work *mgh* needed to lift a mass depends only on the *vertical* distance h and not on any horizontal component of the motion.

If the field is not uniform or the path is not straight, then we must use Equation 25-2a to calculate the potential because the magnitude of $\mathbf{E}$ and/or the angle between $\mathbf{E}$ and the path is changing. Figure 25-7 shows that in such cases we can view the integral in Equation 25-2b as the sum of infinitely many infinitesimal potential differences dV. Each is taken over a path segment $d\ell$ so short that it's essentially straight and the field is essentially uniform over the segment. The limits of the integral are the endpoints of the path, and thus include all the infinitesimal potential differences dVs along the path. Because the electric field is conservative, we can evaluate the integral over any path between the given endpoints. Several examples in the next section illustrate the use of Equation 25-2a.

$$\Delta V_{AB} = \int_A^B dV = -\int_A^B \mathbf{E} \cdot d\boldsymbol{\ell}$$

$$dV = -\mathbf{E} \cdot d\boldsymbol{\ell} = -Ed\ell \cos\theta$$

FIGURE 25-7 The line integral in Equation 25-2a is the sum of infinitely many infinitesimally small potential differences dV. Each dV is taken over a segment $d\ell$ of the overall path short enough that it's essentially straight and that the electric field is essentially uniform over the segment. Then each dV is given by Equation 25-2b for a straight path in a uniform field: $dV = -\mathbf{E} \cdot d\boldsymbol{\ell}$. Summing—that is, integrating—over the whole path gives Equation 25-2a: $\Delta V_{AB} = -\int_A^B \mathbf{E} \cdot d\boldsymbol{\ell}$.

FIGURE 25-6 The potential difference depends only on the component of the path *along* the field. Mathematically, $\Delta V = -\mathbf{E} \cdot \boldsymbol{\ell} = -E\ell \cos\theta$, where the quantity $-\ell \cos\theta$ can be interpreted as the path component along the field.

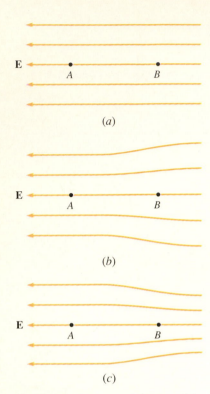

FIGURE 25-8 Which of the potential differences ΔV_{AB} is the lowest? The highest?

Got It!

Figure 25-8 shows three straight paths AB of the same length, each in a different electric field. The field strength at A is the same in each. Rank the potential differences ΔV_{AB} from lowest to highest.

25-3 CALCULATING POTENTIAL DIFFERENCE

Here we use Equation 25-2a to calculate potential differences for a variety of charge distributions. Most important of these is the point charge. Understanding potential differences in the field of a point charge will deepen your understanding of the concept of electric potential, and provides an easy way to compute potential differences in the fields of more complicated charge distributions.

The Potential of a Point Charge

The electric field of a point charge q is given by Equation 23-3:

$$\mathbf{E} = \frac{kq}{r^2}\hat{\mathbf{r}},$$

where $\hat{\mathbf{r}}$ is a unit vector from the charge toward the point where the field is being evaluated. Consider two points A and B at distances r_A and r_B from a positive point charge, as shown in Fig. 25-9. What is the potential difference between these points? The distance between them is $r_B - r_A$, but we can't just multiply this distance by the electric field because the field varies with position. Instead we integrate, using Equation 25-2a:

$$\Delta V_{AB} = -\int_{r_A}^{r_B} \mathbf{E} \cdot d\boldsymbol{\ell} = -\int_{r_A}^{r_B} \frac{kq}{r^2}\hat{\mathbf{r}} \cdot d\boldsymbol{\ell}.$$

As we move from r_A toward r_B, the path element vectors $d\boldsymbol{\ell}$ correspond to small increments dr in the radial direction, so we write $d\boldsymbol{\ell} = \hat{\mathbf{r}}dr$. Then the potential becomes

$$\Delta V_{AB} = -\int_{r_A}^{r_B} \frac{kq}{r^2}\hat{\mathbf{r}} \cdot \hat{\mathbf{r}}dr = -kq\int_{r_A}^{r_B} r^{-2}\,dr,$$

since the dot product of the unit vector $\hat{\mathbf{r}}$ with itself is simply 1. Evaluating the integral gives

$$\Delta V_{AB} = -kq\left[-\frac{1}{r}\right]_{r_A}^{r_B} = kq\left(\frac{1}{r_B} - \frac{1}{r_A}\right). \tag{25-3}$$

Does this result make sense? For $r_B > r_A$ the potential difference is negative, showing that a positive test charge at r_A would "fall down" the potential "hill" toward r_B. Going the other way would require that positive work be done on a positive charge, as it's pushed "up" the potential "hill" against the repulsive force of the charge q. Although we considered q to be positive, our result holds as well for $q < 0$, in which case the sign of the potential difference changes.

Although we derived Equation 25-3 for two points on the same radial line, Fig. 25-10 shows that the result holds for *any* two points in the field of a charge q. It doesn't matter which point is at the greater distance either; if $r_B < r_A$ Equation 25-3 still gives the correct potential difference, which then becomes

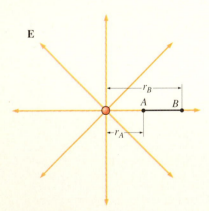

FIGURE 25-9 The potential difference between points A and B is found by integrating between r_A and r_B.

positive to indicate that work must be done moving a positive test charge *toward* a positive q.

The Zero of Potential

So far we've only talked about potential differences, symbolized by the expression ΔV_{AB}. That's because only differences in potential energy—and thus in electric potential, which is potential energy per unit charge—have physical significance. But as we did with potential energy, it's often convenient to define the **electric potential** at some point as zero and then to measure potential differences relative to that point. We then speak of "the potential at point P," designated $V(P)$ or V_P, and meaning the potential difference V_{0P} *from* our reference point *to* point P. Once we've defined a zero of potential, we can then write potential differences as differences in potential between two points; thus, our terminology ΔV_{AB} can equally well be written $V(B) - V(A)$ or $V_B - V_A$. The choice for the zero of potential is arbitrary and is usually made on the basis of mathematical or physical convenience. In electric power systems, Earth, called "ground," is usually taken as the zero. In automobile electric systems, the car's metal structure makes a convenient zero; this is usually connected electrically to the negative battery terminal. (As we'll soon see, every point on a conductor is at the same potential, so it's appropriate to consider an entire conductor like Earth or a car's metal structure as the "point" of zero potential.)

In dealing with isolated point charges, Equation 25-3 shows that it is convenient to choose the zero of potential at infinity. If we let r_A become arbitrarily large, and drop the subscript on r_B because it can be at any radial distance r, Equation 25-3 becomes

$$V_{\infty r} = V(r) = \frac{kq}{r}. \qquad \text{(point-charge potential)} \qquad (25\text{-}4)$$

When we call this expression $V(r)$ "the potential of a point charge," we really mean that $V(r)$ is the potential difference going from a point very far from a charge q to a point a distance r from the charge—an interpretation that is consistent with our definition of potential difference as depending on *two* points. Because the field outside any spherically symmetric charge distribution is that of a point charge, Equation 25-4 also gives the potential outside a spherically symmetric charge distribution.

Does it bother you that potential difference can be finite over an infinite distance? The reason lies in the inverse-square dependence of the field, which drops so rapidly that the work done in moving a charge from infinity to the vicinity of a point charge remains finite. We found an analogous result in Chapter 9, where it took only a finite amount of energy—and therefore a finite "escape speed"—to escape completely from a planet's gravitational attraction. As long as a charge distribution is finite in size—so its field at large distances falls at least as fast as $1/r^2$, then it makes sense to take the zero of potential at infinity.

Got It!

You measure a potential difference of 50 V between two points a distance 10 cm apart in the field of a point charge. If you move closer to the charge and measure the potential difference over another 10 cm interval, will it be more, less, or the same?

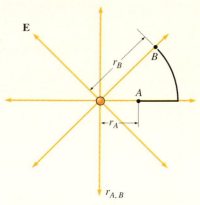

FIGURE 25-10 Here points A and B do not lie on the same radial line. But potential difference is independent of path, and one path between A and B consists of a radial segment and a circular arc. Since **E** is perpendicular to the arc, it takes no work to move a charge along the arc. The potential difference ΔV_{AB} results only from the radial segment, and is therefore given by Equation 25-3. Any other path from A to B would also give the same result.

EXAMPLE 25-3	A Huge Van de Graaff Generator

The Hall of Electricity at the Boston Museum of Science contains the world's largest air-insulated Van de Graaff generator, a device that builds up charge on a metal sphere (Fig. 25-11; you may have seen a smaller version used for demonstrations in your physics course). The Boston museum's generator has a sphere of radius $R = 2.3$ m that develops a charge of 640 μC, spread essentially uniformly over its surface. (a) What is the potential at the sphere's surface, with the zero of potential taken at infinity? (b) How much work would it take to bring a proton from infinity to the sphere's surface? (c) What is the potential difference from the sphere's surface to a point located $2R$ from its center?

Solution

(a) The field outside the sphere is that of a point charge Q located at the sphere's center, so the potential difference between infinity and any point *outside* the sphere is given by Equation 25-4. Therefore the potential at the sphere's surface is

$$V(R) = \frac{kQ}{R} = 2.5 \text{ MV} ,$$

with the numbers given for the museum's device.

(b) This quantity, 2.5 MV, is the work per unit charge needed to move from infinity to the sphere's surface, so the work required for a specific charge q is qV. Since a proton carries one elementary charge e, the work needed to bring the proton to the sphere is 2.5 MeV or 4.0×10^{-13} J.

(c) To get the potential difference from the surface ($r = R$) to $2R$, we could use Equation 25-3 with $r_B = 2R$ and $r_A = R$. But it's perhaps easiest to keep in mind just Equation 25-4 and then find the potential difference by subtracting the potentials at the two points:

$$\Delta V_{R\,2R} = V(2R) - V(R) = \frac{kQ}{2R} - \frac{kQ}{R} = -\frac{kQ}{2R},$$

or -1.25 MV in this example. Does this make sense? Yes: the negative result shows that electric potential decreases as we move in the direction of the electric field—that is, away from the positively charged sphere. Our result also shows that fully half the potential difference between the sphere and infinity

FIGURE 25-11 This Van de Graaff generator at the Boston Museum of Science puts some 640 μC on a 2.3-m-radius sphere, developing a potential difference of several million volts between the sphere and its surroundings. Here, sparks jump to a nearby conductor as the local electric field exceeds the breakdown field in air.

occurs within one radius of the sphere's surface—a consequence of the rapid $1/r^2$ decrease in the field.

You can convince yourself that with a 640-μC charge, the electric field at the sphere's surface is close to the 3 N/C breakdown field in air. Bringing another conductor nearby increases the local field, drawing the sparks shown in Fig. 25-11.

EXERCISE The potential at the surface of a 10-cm-radius sphere is 4.8 kV. (a) What is the charge on the sphere, assuming it's distributed in a spherically symmetric fashion? (b) What is the electric field at its surface? Assume here—and anytime it's not specified—that potential differences are taken from infinity.

Answers: (a) 53 nC; (b) 48 kN/C

Some problems similar to Example 25-3: 19–23, 25, 26, 52, 63, 69

Potentials of Arbitrary Charge Distributions

If we already know the field of a charge distribution, we can calculate potential differences by applying Equation 25-2a, as we did for the point-charge field. Example 25-4 illustrates this approach.

| **EXAMPLE 25-4** | *A Power Line* |

A long, straight power line is made from wire with radius $r_A = 1.0$ cm and carries a line charge density $\lambda = 2.6\ \mu C/m$ (Fig. 25-12). Assuming no other charges are present, what is the potential difference between the surface of the wire and the ground, a distance $r_B = 22$ m below?

Solution

We treat the wire as an infinitely long, charged rod. In the preceding chapter we used Gauss's law to show that the electric field outside any cylindrically symmetric charge distribution is the same as that of a line charge, namely

$$\mathbf{E} = \frac{\lambda}{2\pi\varepsilon_0 r}\hat{\mathbf{r}}.$$

We want to evaluate the potential difference in this field from the wire surface at r_A to the ground at r_B, as suggested in Fig. 25-12b. For a straight path between these points, we can again write $d\boldsymbol{\ell} = \hat{\mathbf{r}}\,dr$ just as we did in evaluating the point-charge potential. Then Equation 25-2a becomes

$$\Delta V_{AB} = -\int_{r_A}^{r_B} \mathbf{E} \cdot d\boldsymbol{\ell} = -\int_{r_A}^{r_B} \frac{\lambda}{2\pi\varepsilon_0 r}\hat{\mathbf{r}} \cdot \hat{\mathbf{r}}\,dr$$

$$= -\frac{\lambda}{2\pi\varepsilon_0}\int_{r_A}^{r_B}\frac{dr}{r} = -\frac{\lambda}{2\pi\varepsilon_0}\ln r \bigg|_{r_A}^{r_B} \quad (25\text{-}5)$$

$$= \frac{\lambda}{2\pi\varepsilon_0}\ln\left(\frac{r_A}{r_B}\right),$$

where the last step follows because $\ln x - \ln y = \ln(x/y)$. For the numbers of this example, this result gives $\Delta V = -360$ kV, a value typical of long-distance electric power transmission lines. Why negative? Because the charge on the wire is positive, so moving away from the wire means moving in the direction of the field—which is always the direction of lower potential.

Note that we cannot let r_B go to infinity in the case of a truly infinite wire, for that would give an infinite potential difference. Physically, this reflects the fact that the charge distribution is itself of infinite extent. Mathematically, it reflects the slow $1/r$ decrease in field strength.

The calculation in this example is not entirely accurate, because the nearby presence of other wires and of charges drawn to the surface of the ground alters the field somewhat.

(a)

(b)

FIGURE 25-12 (a) A long, straight power line can be approximated as an infinite charged rod whose field is that of a line charge. (b) We want the potential difference from the wire surface, a distance r_A from the wire's axis, to the ground a distance r_B below. The wire carries uniform line charge density λ, and extends indefinitely to left and right. Diagram is not to scale, and we're ignoring the presence of other nearby wires.

Nevertheless, the example correctly illustrates the logarithmic dependence of potential on position in the $1/r$ field of a line charge.

EXERCISE An infinitely long rod of radius R carries a uniform volume charge density ρ; Problem 24-31 shows that the electric field *inside* this rod points radially outward and has magnitude $E = \rho r/2\varepsilon_0$. Use Equation 25-2a to find the potential difference from the rod's surface to its axis.

Answer: $\rho R^2/4\varepsilon_0$

Some problems similar to Example 25-4: 28, 29, 64, 73

Finding Potential Differences Using Superposition

When we don't know the field of a charge distribution, or the field is too complicated to integrate easily, we can find the potential using superposition. This often provides an easier approach to calculating the field as well.

Consider a charge q being brought from infinity to a point P in the vicinity of some other charges. We want to know the potential at P—by which we mean the work per unit charge required to move from infinity to P. The superposition principle states that the electric field of a charge distribution is the sum of the fields of the individual charges comprising the distribution. Therefore the work per unit charge—that is, the potential difference—between infinity and P is just the sum of the potential differences associated with the individual point charges. Mathematically, we find $V(P)$ by summing Equation 25-4 over the individual point charges q_i:

$$V(P) = \sum_i \frac{kq_i}{r_i}, \tag{25-6}$$

where the r_i's are the distances from each of the charges to the point P. Equation 25-6 has one enormous advantage over its counterpart for the electric field, Equation 23-4. Electric potential is a *scalar,* so the sum in Equation 25-6 is a scalar sum, and there's no need to consider angles, vector components, or unit vectors.

EXAMPLE 25-5 | *The Dipole Potential*

The dipole of Fig. 25-13 consists of two point charges $\pm q$ separated by a distance $2a$. Find the potential at an arbitrary point P, taking the zero of potential at infinity.

Solution

Note first that this is a more ambitious example than the dipole field calculations in Chapter 23, which for mathematical simplicity were restricted to the dipole axis and perpendicular bisector. Here, since potential is a scalar, we can easily calculate the dipole potential *anywhere.* To do so, we sum the potentials of the individual point charges, as Equation 25-6 suggests:

$$V(P) = \sum_i \frac{kq_i}{r_i} = \frac{kq}{r_1} + \frac{k(-q)}{r_2} = kq\left(\frac{1}{r_1} - \frac{1}{r_2}\right)$$
$$= \frac{kq(r_2 - r_1)}{r_1 r_2},$$

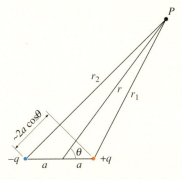

FIGURE 25-13 A dipole and a point P where its potential is to be evaluated. When P is far from the dipole, then r_2 is longer than r_1 by approximately $2a \cos \theta$. (The angle whose vertex is $-q$ is very nearly θ.)

where r_1 and r_2 are the distances from P to the positive and negative charges, respectively.

We're often interested in the electrical effects of a dipole at distances great compared with the separation of its two charges. For example, we saw in Chapter 23 how many molecules can be modeled as dipoles, and how effects such as the van der Waals force involve interactions among distant molecules. If r is the distance from the dipole's center to P, then in the limit $r \gg a$ the quantity $r_1 r_2$ becomes approximately r^2, and, as Fig. 25-13 shows, $r_2 - r_1 \simeq 2a \cos \theta$. Then the dipole potential for $r \gg a$ becomes

$$V(r, \theta) = \frac{k(2aq) \cos \theta}{r^2} = \frac{kp \cos \theta}{r^2}, \tag{25-7}$$

with $p = 2aq$ the dipole moment.

Note that the dipole potential drops more rapidly with distance than the point-charge potential, just as the dipole field drops more rapidly with distance than the point-charge field Note also that Equation 25-7 gives $V = 0$ along the perpendic-

ular bisector of the dipole ($\theta = 90°$). That makes sense because a charge approaching the dipole along its bisector moves at right angles to the dipole field, so no work need be done. Since potential difference is path independent, that means it takes no net work to move on *any* path from infinity to a point on the dipole's bisector (see Question 20).

Figure 25-14 shows a three-dimensional plot of the dipole potential (Equation 25-7) in the plane of Fig. 25-13. The positive charge corresponds to a "hill" of high potential and the negative charge to a deep "hole." Points on the right side of the diagram are closer to the positive charge and are therefore elevated relative to the zero of potential; this is reflected in the positive value of $\cos \theta$ for $\theta < 90°$ in Equation 25-7. Points on the left side are closer to the negative charge and are all depressed relative to zero potential; here $\cos \theta$ in Equation 25-7 is negative. The perpendicular bisector, with $\cos \theta = 0$ and therefore $V = 0$, is a level, straight line, neither elevated nor depressed.

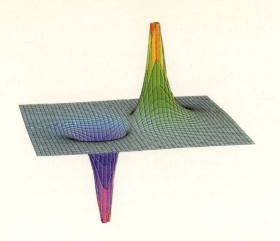

FIGURE 25-14 Three-dimensional plot of potential as a function of position in the plane of Fig. 25-13. Edges of the diagram are at nearly the zero of potential; perpendicular bisector is at exactly $V = 0$.

EXERCISE The dipole moment of a water molecule is 6.2×10^{-30} C·m. (a) Find the potential difference $V_B - V_A$ between two points on the axis of the molecular dipole, where points A and B are 8.2 nm and 5.1 nm, respectively, from the center. Both points are closer to the positive end. (b) How much work would it take to move a proton from A to B?

Answers: (a) 1.32 mV; (b) 1.32 meV $= 2.1 \times 10^{-22}$ J

Some problems similar to Example 25-5: 31–34, 74

Continuous Charge Distributions

We can calculate the potential of a continuous charge distribution by considering it to be made up of infinitely many infinitesimal charge elements dq. Each acts like a point charge and therefore contributes to the potential at some point P an amount dV given by

$$dV = \frac{k \, dq}{r},$$

where the zero of potential is at infinity. The potential at P is the sum—in this case an integral—of the contributions dV from all the charge elements:

$$V = \int dV = \int \frac{k \, dq}{r}, \tag{25-8}$$

where the integration is over the entire charge distribution.

The examples that follow illustrate the use of Equation 25-8 for some charge distributions whose simple shapes make the integration reasonably straightforward. Simple cases such as these can often serve as approximations to more complicated charge distributions.

EXAMPLE 25-6	*A Charged Ring*

A total charge Q is distributed uniformly around a thin ring of radius a, as shown in Fig. 25-15. What is the potential on the axis of this charged ring?

Solution

Let x be the distance from the center of the ring to some arbitrary point P on the axis. The distance from each point on the ring to P is the same, and is given by $r = \sqrt{x^2 + a^2}$. The potential on the axis is the sum of the potentials dV of all the charge elements dq around the ring, as described by

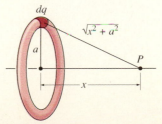

FIGURE 25-15 A charged ring (Example 25-6).

Equation 25-8:

$$V(x) = k \int_{ring} \frac{dq}{r} = \frac{k}{\sqrt{x^2 + a^2}} \int_{ring} dq ,$$

where we've taken $r = \sqrt{x^2 + a^2}$ outside the integral because it's the same for all charge elements. The remaining integral is just the total charge Q, so we have

$$V(x) = \frac{kQ}{\sqrt{x^2 + a^2}} . \qquad (25\text{-}9)$$

Does this result make sense? At great distances from the ring $(x \gg a)$, a^2 in the denominator becomes negligible, and our result becomes

$$V(x) = \frac{kQ}{x} ,$$

which is just the potential of a point charge Q—as we would expect when we're so far from the ring that its size is no longer significant. At the center of the ring, on the other hand, Equation 25-9 gives

$$V(o) = \frac{kQ}{a} .$$

Here we're a distance a from all parts of the ring, and, since potential is a *scalar* the directions to those parts don't matter. The result is therefore the same as being a distance a from a point charge Q.

EXAMPLE 25-7	*A Charged Disk*

A charged disk of radius a carries a total charge Q distributed uniformly over its surface. What is the potential at a point P on the disk axis, a distance x from the disk?

Solution

To use Equation 25-8, we must divide the disk into charge elements dq. In the preceding example we found the potential of a charged ring, so we can take the charge elements of our disk to be thin rings (see Fig. 25-16), and integrate over all the rings that make up the disk. If a ring-shaped charge element has charge dq and radius r, then Equation 25-9 gives its potential dV a distance x from the disk center:

$$dV = \frac{k\,dq}{\sqrt{x^2 + r^2}} .$$

Then the potential of the entire disk is

$$V(x) = \int_{ring} dV = \int_{r=0}^{r=a} \frac{k\,dq}{\sqrt{x^2 + r^2}} .$$

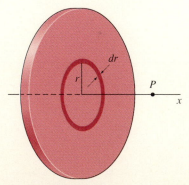

FIGURE 25-16 A charged disk, showing a ring-shaped charge element dq of radius r and width dr.

FIGURE 25-17 Unwinding the thin ring gives a strip of width dr and length $2\pi r$.

To evaluate this integral, we must relate r and dq. "Unwinding" the ring gives a strip of area $2\pi r\,dr$ (Fig. 25-17). The surface charge density σ is the total charge divided by the disk area: $\sigma = Q/\pi a^2$. Then the charge dq on our infinitesimal ring of area $2\pi r\,dr$ is

$$dq = \sigma 2\pi r\,dr = \frac{Q}{\pi a^2} 2\pi r\,dr = \frac{2Q}{a^2} r\,dr .$$

Using this result in the integral for the potential gives

$$V(x) = \int_0^a \frac{2kQ}{a^2} \frac{r\,dr}{\sqrt{x^2 + r^2}} = \frac{kQ}{a^2} \int_0^a \frac{2r\,dr}{\sqrt{x^2 + r^2}} .$$

Note that $2r\,dr = d(r^2) = d(x^2 + r^2)$ since x is a constant with respect to the integration. The integral therefore has the form $u^{-1/2}\,du$, where $u = x^2 + r^2$, and the result is $2u^{1/2}$ or

$$V(x) = \frac{2kQ}{a^2} \sqrt{x^2 + r^2} \Bigg|_{r=0}^{r=a} = \frac{2kQ}{a^2} \left(\sqrt{x^2 + a^2} - |x| \right) .$$

Figure 25-18 shows that this complicated-looking result makes sense: Close to the sheet, the potential resembles that of an infinite sheet, while far from the disk it approaches the potential of a point charge. Example 25-9 and Problem 70 explore these limiting cases further.

FIGURE 25-18 Charged-disk potential of Equation 25-10 approaches the potential of an infinite sheet for points close to the disk, and that of a point charge far from the disk.

EXERCISE Point P in Fig. 25-19 lies a perpendicular distance y from the end of a uniformly charged rod of length ℓ and total charge Q. Find an expression for the potential at P, taking the zero of potential at infinity.

FIGURE 25-19 What is the potential at P? Figure shows a charge element dq, of length dx, to use in the integration for the potential.

Answer: $V(y) = \dfrac{kQ}{\ell}\ln\left(\dfrac{\ell + \sqrt{\ell^2 + y^2}}{y}\right)$

Some problems similar to Examples 25-6 and 25-7: 36–39, 71, 78

25-4 POTENTIAL DIFFERENCE AND THE ELECTRIC FIELD

Equipotential

It takes no work to move a charge at right angles to an electric field. Therefore there can be no potential difference between two points on a surface that is everywhere perpendicular to the electric field. Such surfaces are called **equipotential surfaces**, or simply **equipotentials**. Figure 25-20 shows some equipotential surfaces for both uniform and nonuniform electric fields.

Equipotentials are like contour lines that show land elevation on a map (Fig. 25-21). A contour line is a line of constant elevation, and therefore, it takes

(a)

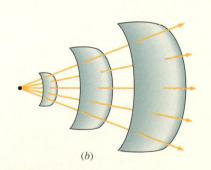

(b)

FIGURE 25-20 (a) Equipotential surfaces in a uniform electric field are planes perpendicular to the field. (b) In a nonuniform field (here that of a point charge) the equipotential surfaces are curved (here they're spherical), but are still perpendicular to the field.

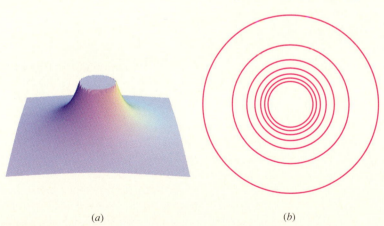

(a)   (b)

FIGURE 25-21 (a) A flat-topped hill and (b) its representation as a contour map. Closely spaced contours indicate steep slopes. Figures can also represent the potential of a uniformly charged spherical shell whose potential is constant inside, where $\mathbf{E} = 0$, and falls as $1/r$ outside. In this case (a) is a 3-dimensional plot of the potential in a plane through the center of the sphere, and (b) shows the intersection of the three-dimensional equipotential surfaces (as shown in Fig. 25-20b) with this plane. Closely spaced contours indicate strong electric field.

11.9

Electric Potential

FIGURE 25-22 Equipotentials and field lines for a dipole, in a plane containing the dipole. The two sets of curves are everywhere perpendicular. Equipotentials on the left side represent negative values of potential; those on the right positive values. The perpendicular bisector is the special equipotential on which $V = 0$. In three dimensions it would be a plane and the other equipotentials would be closed surfaces. You should convince yourself that this figure is essentially a contour map of Fig. 25-14.

no work to move along a contour line. Contour lines are usually spaced at even increments of elevation. Where lines are closely spaced, the elevation changes rapidly. Similarly, closely spaced equipotentials indicate large potential differences between nearby points. That, in turn, means it takes a lot of work to move charge between those points—and therefore there must be a large electric field present. Figure 25-21 might just as well represent electric potential, in which case regions with closely spaced equipotentials—steep slopes on the "potential hill"—indicate large electric fields. Similarly, the equipotentials for a dipole (Fig. 25-22) describe the steep "hill" of the positive charge and a correspondingly deep "hole" of the negative charge that we showed in Fig. 25-14. ActivPhysics Activity 11.9 draws equipotentials for common charge distributions.

Got It!

Figure 25-23 shows cross sections through two sets of equipotential surfaces. In both figures the potential difference between adjacent equipotentials is the same. Which could represent the potential of a point charge? Explain.

(a)

(b)

FIGURE 25-23 Which equipotential diagram is for a point charge?

Calculating the Field from the Potential

Given electric field lines, we can construct equipotentials. Conversely, given equipotentials we can reconstruct the field by sketching field lines at right angles to the equipotentials. Specifying the potential at each point thus conveys all the information needed to determine the field.

We can quantify the relation between potential and field by writing the potential difference dV between two points separated by an infinitesimal displacement $d\ell$:

$$dV = -\mathbf{E} \cdot d\ell = -E_\ell \, d\ell,$$

where E_ℓ designates the field component in the direction of $d\ell$. Rearranging this equation gives

$$E_\ell = -\frac{dV}{d\ell}. \qquad (25\text{-}10)$$

This equation confirms our statement that the electric field is strong where potential changes rapidly with distance. The minus sign in Equation 25-10 is the same one that appears in Equation 25-2; here it tells us that if we move in the direction of *increasing* potential, then we must be moving *against* the electric field. If we want to find the field components in a chosen coordinate system, we simply choose $d\ell$ along one of the coordinate axes and apply Equation 25-10; the x component of the field, for example, is given by $E_x = -dV/dx$. (If V is a function of all three coordinates we should, strictly speaking, write E_x in terms of the *partial* derivative, $\partial V/\partial x$, that we introduced in Chapter 16.) ActivPhysics Activity 11.10 explores graphically the relation between potential and electric field.

11.10

Electric Potential, Field, and Force

Equation 25-10 also shows that the units of electric field—introduced in Chapter 23 as N/C—can be written, equivalently, as V/m. You can see from an equipotential diagram why this unit makes sense: where the equipotentials are close, and the field therefore strong, you traverse a greater potential difference in a given distance—i.e., more "volts per meter." In fact, the electric field is sometimes called the "potential gradient," meaning the rate of change of potential with respect to position.

EXAMPLE 25-8	The Field of a Point Charge

Use the point-charge potential of Equation 25-4 to derive the electric field of a point charge.

Solution

The point-charge potential, $V(r) = kq/r$, depends only on r. Therefore the electric field points in the radial direction, and has the form $\mathbf{E} = E_r\hat{\mathbf{r}}$. Choosing $d\ell = dr$ in Equation 25-10 gives the field component:

$$E_r = -\frac{dV}{dr} = -\frac{d}{dr}\left(\frac{kq}{r}\right) = -kq\frac{d(r^{-1})}{dr} = kqr^{-2} = \frac{kq}{r^2}.$$

Thus $\mathbf{E} = \frac{kq}{r^2}\hat{\mathbf{r}}$, as expected.

EXAMPLE 25-9	The Field of a Charged Disk

Use the result of Example 25-7 to find the electric field on the axis of a charged disk.

Solution

Symmetry shows that the field must point along the disk axis, which is the x axis in Fig. 25-16. So the field has only an x component, given by applying Equation 25-10 to the disk potential:

$$E_x = -\frac{dV}{dx} = -\frac{d}{dx}\left(\frac{2kQ}{a^2}(\sqrt{x^2 + a^2} - |x|)\right)$$

$$= \frac{2kQ}{a^2}\left(1 - \frac{|x|}{\sqrt{x^2 + a^2}}\right).$$

To see that this makes sense, consider the case $x \ll a$, for which the field becomes approximately $2kQ/a^2$. Writing Q as the surface density σ times the area πa^2 gives $E_x = 2\pi k\sigma = \sigma/2\varepsilon_0$, which is the field of an infinite charged sheet. Of course—very close to the disk it looks effectively infinite, and its field should be well approximated by that of an infinite sheet. Problem 70 shows that the field far from the disk approaches that of a point charge Q.

EXERCISE Use Equations 25-7 and 25-10 to calculate the electric field on the axis of a point dipole, and show that your result is equivalent to Equation 23-5b.

Some problems similar to Examples 25-8 and 25-9: 47–49, 78

Examples 25-8 and 25-9 show that it is often much easier to calculate the electric field by first finding the potential and then differentiating, rather than doing a vector integration to get the field of a complicated charge distribution.

It's important to recognize that the *values* of the field and potential are not directly related; rather, as Equation 25-10 indicates, the field measures the *rate of change* of the potential. Field and potential are like acceleration and velocity; the *values* of the two are quite independent, with the former depending on the *rate of change* of the latter. Figure 25-24 and Example 25-10 illustrate the relation between potential and field.

Got It!

Is it possible to have a point where the electric field is zero but the potential is not? What about a point where the potential is zero but the field is not?

FIGURE 25-24 The x component of the electric field is the negative rate of change of the potential with respect to x.

EXAMPLE 25-10 | *Potential and Field*

A positive charge $+2q$ lies at $x = -a$, and a charge $-q$ lies at $x = +a$ as shown in Fig. 25-25. (a) Derive an expression for the potential on the x axis, and find a point on the axis in the region $x > a$ where the potential (with respect to infinity) is zero. (b) Use your expression for potential to find the electric field for $x > a$. Is the electric field zero where the potential is zero? If not, where is the field zero?

Solution

(a) The distance from any point x to the positive charge at $-a$ is $r_+ = |x - (-a)| = |x + a|$; similarly, the distance to the negative charge is $r_- = |x - a|$. Thus, the potential on the axis is

$$V(x) = \frac{k(2q)}{r_+} + \frac{k(-q)}{r_-} = kq\left(\frac{2}{|x + a|} - \frac{1}{|x - a|}\right).$$

We can find where $V = 0$ by setting the quantity in parentheses to zero. For $x > a$ both denominators are positive and we can remove the absolute value signs. For $V = 0$ we then have

$$\frac{2}{x + a} = \frac{1}{x - a}.$$

Solving for x gives $x = 3a$.

(b) The field clearly has only an x component, and Equation 25-10 then gives

$$E_x = -\frac{dV}{dx} = -kq\frac{d}{dx}\left(\frac{2}{x + a} - \frac{1}{x - a}\right)$$

$$= \frac{2kQ}{(x + a)^2} - \frac{kQ}{(x - a)^2}.$$

This result is hardly surprising: It's just the sum of two point-charge fields. We actually solved for the zero-field point in this configuration in Example 23-3; there we found that the electric force would be zero at $x = 5.83a$, which is *not* the same place where the potential is zero.

Figure 25-26 shows the potential of this charge distribution in the region $x > a$, showing clearly the point $x = 3a$ where

FIGURE 25-25 Two opposite but unequal charges.

FIGURE 25-26 Potential on the x axis for a charge distribution consisting of $+2q$ at $x = -a$ and $-q$ at $x = +a$. Shown here is the region $x > a$, including the point $x = 3a$ where the potential is zero and the point $x = 5.83a$ where the electric field—proportional to the *slope* of the potential curve—is zero. This graph shows potential along the x axis only in the region to the right of the negative charge. Leftmost part of curve shows the potential "hole" associated with the negative charge.

$V = 0$. Getting a charge to this point from infinity would take no *net* work, although it would require going up and then down a potential "hill." Also clear in Fig. 25-26 is the point $x = 5.83a$ where the electric field is zero. Note that this is the point where the *slope* of the potential curve is zero, *not* where the potential itself is zero.

There are actually two points on the x axis where $V = 0$; the second lies between the charges. Both lie on an equipotential surface of zero potential that surrounds the negative charge.

Tip

Field and Potential Are Not Proportional In particular, where one is zero, the other need not be zero. You can see that in Fig. 25-26, where it clearly takes work to get from infinity to the point where $E = 0$. Just because a mountaintop is flat doesn't mean it didn't take work to climb it! Potential depends not on the field at a point but on the field over an entire path from infinity to that point. Similarly, the potential can be zero at points where the field is not, as evidenced by the steep slope of the potential curve in Fig. 25-26 at the point where it crosses zero.

EXERCISE Find the second point on the x axis in Example 25-10 where the potential is zero.

Answer: $x = a/3$

Some problems similar to Example 25-10: 50, 77

25-5 POTENTIALS OF CHARGED CONDUCTORS

There is no electric field inside a conductor in electrostatic equilibrium, as we found in Chapter 24, so it takes no work to move a test charge around inside the conductor. We also found that the field at the conductor surface is perpendicular to the surface, so it takes no work to move a test charge along the surface, either. Therefore the potential difference between two points in or on a conductor must be zero, and thus,

> **A conductor in electrostatic equilibrium is an equipotential.**

Consider an isolated, spherical conductor of radius R carrying charge Q. Since the conductor is isolated, charge is distributed uniformly over its surface, so the field and potential outside the sphere are that of a point charge. Then the potential at its surface is

$$V(R) = \frac{kQ}{R}, \tag{25-11}$$

as we found in Example 25-3. The sphere itself is an equipotential, so the potential *difference* between two points *on the sphere* is zero. But that doesn't mean the potential difference between *infinity* and the sphere is zero; since the sphere is charged, it takes work to move charge to its surface from infinity, and that's what Equation 25-11 says.

Now consider two widely separated spheres of different sizes. If we connect them by a thin conducting wire, as shown in Fig. 25-27, then the system constitutes a single conductor, and charge will move through the wire until both spheres are at the same potential. But since the spheres are widely separated, each still has an essentially spherical charge distribution, so Equation 25-11 applies to each. Since the spheres have the same potential, Equation 25-11 implies that

$$\frac{kQ_1}{R_1} = \frac{kQ_2}{R_2},$$

FIGURE 25-27 Two conducting spheres held at the same potential by a conducting wire. The surface charge density is greater on the smaller sphere, in inverse proportion to its radius.

where the subscripts label the two spheres. We can write each charge as the surface area of the sphere multiplied by the surface charge density: $Q = 4\pi R^2 \sigma$. Substituting for the Q's in the above equation and solving for the ratio of surface charge densities then gives

$$\frac{\sigma_1}{\sigma_2} = \frac{R_2}{R_1}.$$

Thus the *smaller* sphere has the *larger* surface charge density. Since the electric field at a conductor surface has magnitude $E = \sigma/\varepsilon_0$, the field must be stronger at the smaller sphere.

This discussion of spherical conductors provides a qualitative description of nonspherical conductors as well. All parts of an irregularly shaped conductor must be at the same potential. Where the conductor surface curves sharply, it is like a small sphere and therefore has a higher surface charge density and a stronger electric field. In general, the field is strongest where the surface curves most sharply.

Because a conductor surface is an equipotential and the electric field is perpendicular to the conductor surface, equipotentials just above the surface must have approximately the same shape as the surface. Because the electric field is stronger where the conductor surface curves sharply, there must be more

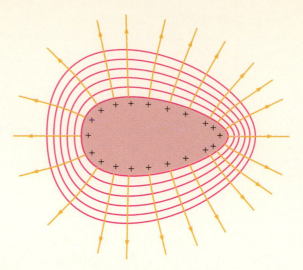

FIGURE 25-28 A charged conductor of irregular shape. Equipotentials near the conductor have approximately its shape, and the field is strongest—and therefore the equipotentials closest—where the conductor curves most sharply.

field lines emerging from such regions. Far from a charged conductor, on the other hand, its field must resemble that of a point charge, with radial field lines and circular equipotentials. With these limiting cases in mind, we can sketch the approximate form for the field of an arbitrarily shaped conductor (Fig. 25-28).

We stress that our conclusion about surface charge density and curvature applies only to *isolated* conductors—those far from any other charges. The field of a nearby charge will modify the charge distribution of a conductor, altering the surface charge distribution (Fig. 25-29).

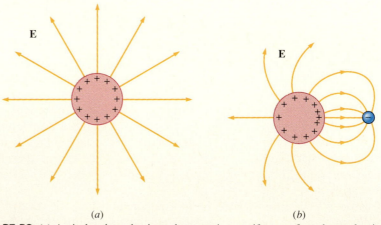

(*a*)   (*b*)

FIGURE 25-29 (*a*) An isolated conducting sphere carries a uniform surface charge density and has a spherically symmetric electric field. (*b*) The presence of a nearby charge distorts the surface charge distribution and therefore the electric field.

APPLICATION	*Corona Discharge, Pollution Control, and Xerography*

The large electric fields that develop where a charged conductor is sharply curved can cause serious problems in electrical equipment; in other applications those fields are put to good use. Fields above 3 MN/C are strong enough to strip electrons from air molecules, making the air a conductor. Breakdown of air is often evidenced by a blue glow around sharply-pointed conductors (Fig. 25-30). Called **corona discharge**, this glow results from the recombination of electrons with atoms. Corona discharge causes loss of power from high-voltage transmission lines, and engineers try to avoid it by eliminating sharp edges on wires and other conducting structures.

Corona discharge is put to good use in the **electrostatic precipitator**, a pollution-control device used especially on coal-burning power plants (Fig. 25-31). A typical precipitator consists of parallel metal plates with thin wires running between them. Application of a high voltage between plates and wires sets up large electric fields near the wires. Exhaust gases flow between the plates and the field ionizes some gas molecules. These charged molecules, in turn, attach themselves to pollutant particles. The charged particles are driven to the collecting plates by the electric field. Every few minutes a mechanical vibrator taps the plates and the particles fall into a hopper, where they can be trucked away to use for fill or in making products such as cinder blocks. A typical power plant produces some 30 pounds of particulates every second, and electrostatic precipitators remove up to 99 percent of this. Since the implementation of the Clean Air Act in the early 1960s, widespread use of electrostatic precipitators has significantly reduced particulate air pollution in the United States. However, precipitators do not remove gases like SO_2 that causes acid rain or CO_2 that contributes to global warming, and they also consume several percent of a power plant's electrical output.

(a)

(b)

FIGURE 25-31 (*a*) Electrostatic precipitators at a 1300-MW coal-burning power plant. (*b*) In the electrostatic precipitator, high voltage applied between the metal chamber and the thin central wire results in a strong, nonuniform electric field near the wire. This field ionizes air molecules, which attach to soot particles with the result that the particles are accelerated to the chamber walls. Mechanical vibration then dislodges the particles into the hopper for eventual collection. Parallel metal plates are often used instead of the cylindrical chamber shown here.

Xerography (literally, "dry writing") used in copiers and laser printers is another widespread application of corona discharge. The process starts with the uniform charging of a special photoconductive plate by corona discharge from a thin wire maintained at about 5 kV relative to the plate (Fig. 25-32*a*). Light, imaged in a copier from the document being copied or from a computer-controlled scanning laser

FIGURE 25-30 Corona discharge on a power-line insulator occurs because of strong electric fields that develop where a highly charged conducting wire is attached to the top of the insulator.

FIGURE 25-32 A step-by-step look at the xerography process, widely used in copiers and laser printers.

Corona discharge from a moving, thin wire lays positive charge over the light-sensitive surface (photoconductive plate).

The photoconductive surface is a good insulator in the dark and becomes a good conductor when exposed to light.

Corona wire

Metal

The potential difference between the wire and the metal below the photo-conductive surface is approximately 5000 V.

Light is projected onto the original to be copied.

Light falls on the plate, except in the areas where the original is dark.

Lens

Photoconductive plate

Original   Light

Where light falls, the photoconductive plate is neutralized.

A positive charge pattern remains, corresponding to the dark areas of the image.

(a)   (b)

Negatively charged toner particles are carried toward the plate on a larger carrier particle, which they cling to due to electrostatic attraction.

As they near the plate, the toner particles are attracted to areas of the plate that are positively charged.

The negatively charged toner is attracted to the positively charged paper.

Heat is used to fuse the toner to the paper.

(c)   (d)

beam in a printer, then strikes the plate. The photoconductive material has the property that it is a good insulator in the dark, but the energy contained in light incident on it can dislodge electrons, creating free charge and making illuminated areas of the material a conductor. These free charges flow to neutralize the surface charge on illuminated areas of the plate, but dark areas retain their charge. At this point, then, the photoconductive plate contains an electrostatic version of the image being reproduced (Fig. 25-32b).

Next, a fine dusting of charged particles called toner is spread over the plate, often delivered on larger so-called carrier particles (Fig. 25-32c). The toner sticks to the charged parts of the photoconductive plate, and is then transferred to a sheet of charged paper (Fig. 25-32d). Finally, heating fuses the toner particles into the paper, making a permanent image.

In practical copiers the photoconductive surface may take the form of a flat plate charged by a moving corona wire, as we've shown in Fig. 25-32, or a moving belt or rotating drum with a fixed wire. The charge spread on the plate may be positive or negative, depending on the particular photoconductive material; inorganic silicon or selenium-based materials are positively charged, as shown in Fig. 25-32, while organic photoconductors, incorporating zinc oxide or cadmium sulfide, are charged negatively. Some copiers, especially smaller ones designed for intermittent use, eliminate the reusable carrier particles that transfer toner to the plate, and instead incorporate magnetic materials within the toner and employ magnets to effect the transfer. Either way, the basic copy process is electrostatic, with toner particles clinging to the unexposed parts of the plate because of the electric force.

CHAPTER SYNOPSIS

Summary

1. The **electric potential difference** between two points is the change in potential energy per unit charge for a charge moved between those points. Its unit is the joule/coulomb, or **volt** (V). Potential difference is a property of two points; because the electric field is conservative, the potential difference between two points is independent of the path taken between them. The potential difference ΔV_{AB} between points A and B is given by

$$\Delta V_{AB} = -\int_A^B \mathbf{E} \cdot d\boldsymbol{\ell}.$$

   When the field is uniform, this expression reduces to

$$\Delta V_{AB} = -\mathbf{E} \cdot \boldsymbol{\ell}.$$

2. Defining the potential to be zero at some point allows us to speak of "the potential at a point," meaning the potential difference from the reference point to the point in question. For isolated point charges, a convenient zero is infinitely far from the charge; then the potential at an arbitrary point a distance r from the point charge q is

$$V(r) = \frac{kq}{r}. \qquad \text{(point-charge potential)}$$

   The potentials of charge distributions may be found by taking the line integral of the field, if the latter is known, or by summing the potentials of the point charges making up the distribution:

$$V = \sum_i \frac{kq_i}{r_i}, \quad \begin{pmatrix} \text{discrete} \\ \text{charges} \end{pmatrix} \quad \text{or}$$

$$V = \int \frac{k\,dq}{r}. \quad \begin{pmatrix} \text{continuous charge} \\ \text{distribution} \end{pmatrix}$$

3. **Equipotentials** are surfaces over which the potential has a constant value. Equipotentials are everywhere perpendicular to the electric field. The field is strong where equipotentials are closely spaced and vice versa. Mathematically, the field component in a given direction is related to the rate of change of potential with position in that direction:

$$E_\ell = -\frac{dV}{d\ell}.$$

4. A conductor in electrostatic equilibrium is an equipotential. The surface charge density and therefore the electric field at the conductor surface are usually greatest where the conductor curves most sharply. Very strong electric fields occur at sharp bends; if strong enough, these fields can result in **corona discharge**, in which the surrounding air becomes a conductor and charge leaks off the charged conductor.

Terms You Should Understand

(Pairs are closely related terms whose distinction is important; number in parentheses is chapter section where term first appears.)

conservative field (introduction)
potential difference (25-2)
volt, electron volt (25-2)
line integral (25-2)
equipotential (25-4)
corona discharge (25-5)

Symbols You Should Recognize

ΔV_{AB} (25-2)	$V(P)$ (25-3)
V, eV (25-2)	$d\ell$ (25-2)

Problems You Should Be Able to Solve

calculating the work needed to move a given charge through a given potential difference (25-2)
calculating potential differences in uniform electric fields (25-2)
calculating potential differences using the line integral of a known electric field (25-3)
evaluating potentials by summing or integrating over point charges (25-3)
finding electric field components given potential as a function of position (25-4)
sketching equipotentials of simple charge distributions (25-4)
sketching equipotentials and fields around conductors (25-5)

Limitations to Keep in Mind

Electric potential difference depends on *two points*. Phrases like "the potential at point P" or "the potential of a point charge" are always shorthand ways of talking about the potential difference between two points. When the second point is not specified, it is often taken to be at infinity.

QUESTIONS

1. Why can a bird perch on a high-voltage power line without getting electrocuted?
2. One proton is accelerated from rest by a uniform electric field, the other by a nonuniform electric field. If they move through the same potential difference, how do their final speeds compare?
3. Would a free electron move toward higher or lower potential?
4. The potential difference from A to B in Fig. 25-33 is zero since the two points are equidistant from the charge Q. How can this be, when a charge moving along the path shown clearly experiences an electric force not perpendicular to the path?

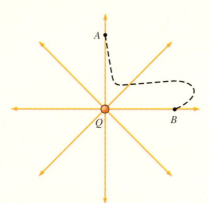

FIGURE 25-33 Question 4.

5. A proton and a positron (a particle with the electron's mass carrying charge $+e$) are accelerated through the same potential difference. How do their final energies compare? Their final speeds?
6. The electric field at the center of a uniformly charged ring is obviously zero, yet Example 25-6 shows that the potential at the center is not zero. How is this possible?
7. Must the potential be zero at any point where the electric field is zero? Explain.
8. Must the electric field be zero at any point where the potential is zero? Explain.
9. The potential is constant throughout an entire volume. What must be true of the electric field within that volume?
10. In considering the potential of an infinite flat sheet, why is it not useful to take the zero of potential at infinity?
11. The potential of a point charge is given by kq/r. Is r the distance between the two points for which this is the potential difference? Explain.
12. "Cherry picker" trucks for working in trees or power lines often carry the warning sign shown in Fig. 25-34. Explain

FIGURE 25-34 Question 12.

how this hazard arises and why it might be more of a danger to someone on the ground than to a worker on the truck.
13. The electric field midway between two equal positive charges is, clearly, zero. Is the potential at this point also zero? (Take the zero of potential at infinity.) Are there any other points (other than infinity) where the potential of this system is zero?
14. Is it possible for equipotential surfaces to intersect? Explain.
15. Is the potential at the center of a hollow, uniformly charged spherical shell higher, lower, or the same as at the surface?
16. A solid sphere contains positive charge uniformly distributed throughout its volume. Is the potential at its center higher, lower, or the same as at the surface?
17. Why do the spheres in Fig. 25-27 need to be far apart for the conclusion that surface charge density is inversely proportional to radius to hold accurately?
18. Figure 25-35 shows a cross section of some equipotential surfaces. Where does the electric field have the greatest magnitude, and in this region what is its approximate direction?

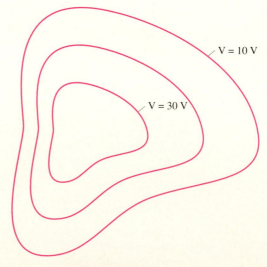

FIGURE 25-35 Question 18.

19. Two equal but opposite charges form a dipole. Describe the equipotential surface on which $V = 0$.
20. Figure 25-36 shows three paths leading from infinity to a point P on the perpendicular bisector of a dipole. For each path, how much work is needed to bring a charge q from infinity to P?
21. The electric potential in a region increases linearly with distance. What can you conclude about the electric field in this region?
22. In Fig. 25-29b the charge density on the conducting sphere is not uniform, yet the field inside must still be zero since the system is in electrostatic equilibrium. How is this possible?

FIGURE 25-36 Question 20.

23. Why is lightning likely to strike an isolated tree?
24. What is the difference between a volt and an electron volt?

PROBLEMS

ActivPhysics can help with these problems:
Activities 11.9, 11.10.

Section 25-2 Potential Difference

1. How much work does it take to move a 50-μC charge against a 12-V potential difference?
2. The potential difference between the two sides of an ordinary electrical outlet is 120 V. How much energy does an electron gain when it moves from one side to the other?
3. It takes 45 J to move a 15-mC charge from point A to point B. What is the potential difference ΔV_{AB}?
4. Show that 1 V/m is the same as 1 N/C.
5. Find the magnitude of the potential difference between two points located 1.4 m apart in a uniform 650 N/C electric field, if a line between the points is parallel to the field.
6. A charge of 3.1 C moves from the positive to the negative terminal of a 9.0-V battery. How much energy does the battery impart to the charge?
7. Two points A and B lie 15 cm apart in a uniform electric field, with the path AB parallel to the field. If the potential difference ΔV_{AB} is 840 V, what is the field strength?
8. Figure 25-37 shows a uniform electric field of magnitude E. Find expressions for (a) the potential difference ΔV_{AB} and (b) ΔV_{BC}. (c) Use your result to determine ΔV_{AC}.

FIGURE 25-37 Problem 8.

9. A proton, an alpha particle (a bare helium nucleus), and a singly ionized helium atom are accelerated through a potential difference of 100 V. Find the energy each gains.
10. The electric field within the membrane separating the inside and outside of a biological cell is approximately 8.0 MV/m, and is essentially uniform. If the membrane is 10 nm thick, what is the potential difference across the membrane?
11. What is the potential difference between the terminals of a battery that can impart 7.2×10^{-19} J to each electron that moves between the terminals?
12. Electrons in a TV tube are accelerated from rest through a 25-kV potential difference. With what speed do they hit the TV screen?
13. A 12-V car battery stores 2.8 MJ of energy. How much charge can move between the battery terminals before it is totally discharged? Assume the potential difference remains at 12 V, an assumption that is not realistic.
14. What is the charge on an ion that gains 1.6×10^{-15} J when it moves through a potential difference of 2500 V?
15. Two large, flat metal plates are a distance d apart, where d is small compared with the plate size. If the plates carry surface charge densities $\pm \sigma$, show that the magnitude of the potential difference between them is $V = \sigma d / \varepsilon_0$.
16. An electron passes point A moving at 6.5 Mm/s. At point B the electron has come to a complete stop. Find the potential difference ΔV_{AB}.
17. A 5.0-g object carries a net charge of 3.8 μC. It acquires a speed v when accelerated from rest through a potential difference V. A 2.0-g object acquires twice the speed under the same circumstances. What is its charge?

Section 25-3 Calculating Potential Difference

Note: In these problems, the zero of potential is taken at infinity unless noted otherwise.

18. An electric field is given by $\mathbf{E} = E_0\hat{\jmath}$, where E_0 is a constant. Find the potential as a function of position, taking $V = 0$ at $y = 0$.

19. The classical picture of the hydrogen atom has a single electron in orbit a distance 0.0529 nm from the proton. Calculate the electric potential associated with the proton's electric field at this distance.

20. Earth carries an electric charge of -4.3×10^5 C, distributed essentially uniformly over its surface. What is the potential difference between Earth's surface and the base of the ionosphere, about 80 km above the surface?

21. Points A and B lie 20 cm apart on a line extending radially from a point charge Q, and the potentials at these points are $V_A = 280$ V, $V_B = 130$ V. Find Q and the distance r between A and the charge.

22. What is the maximum potential allowable on a 5.0-cm-diameter metal sphere if the electric field at the sphere's surface is not to exceed the 3 MV/m breakdown field in air?

23. A 3.5-cm-diameter isolated metal sphere carries a net charge of 0.86 μC. (a) What is the potential at the sphere's surface? (b) If a proton were released from rest at the sphere's surface, what would be its speed far from the sphere?

24. A sphere of radius R carries a negative charge of magnitude Q, distributed in a spherically symmetric way. Find the "escape speed" for a proton at the sphere's surface—that is, the speed that would enable the proton to escape to arbitrarily large distances.

25. A thin spherical shell of charge has radius R and total charge Q distributed uniformly over its surface. What is the potential at its center?

26. A solid sphere of radius R carries a net charge Q distributed uniformly throughout its volume. Find the potential difference from the sphere's surface to its center. *Hint:* Consult Example 24-1.

27. Find the potential as a function of position in an electric field given by $\mathbf{E} = ax\hat{\imath}$, where a is a constant and where $V = 0$ at $x = 0$.

28. A coaxial cable consists of a 2.0-mm-diameter inner conductor and an outer conductor of diameter 1.6 cm and negligible thickness (Fig. 25-38). If the conductors carry line charge densities ± 0.56 nC/m, what is the magnitude of the potential difference between them?

29. The potential difference between the surface of a 3.0-cm-diameter power line and a point 1.0 m distant is 3.9 kV. What is the line charge density on the power line?

FIGURE 25-38 Problem 28.

30. Three equal charges q form an equilateral triangle of side a. Find the potential at the center of the triangle.

31. A charge $+Q$ lies at the origin, and $-3Q$ at $x = a$. Find two points on the x axis where $V = 0$.

32. Two identical charges q lie on the x axis at $\pm a$. (a) Find an expression for the potential at all points in the x-y plane. (b) Show that your result reduces to the potential of a point charge for distances large compared with a.

33. Find the potential 10 cm from a dipole of moment $p = 2.9$ nC·m (a) on the dipole axis, (b) at $45°$ to the axis, and (c) on the perpendicular bisector. The dipole separation is much less than 10 cm.

34. Two points A and B lie 55 cm from a dipole of moment $p = 6.4$ nC·m, whose charge separation is much less than 55 cm. A line from the dipole to A makes a $20°$ angle with the dipole axis, and a line to B makes a $50°$ angle. Find the potential difference $V_B - V_A$.

35. A hollow, spherical conducting shell of inner radius b and outer radius c surrounds is concentric with a solid conducting sphere of radius a, as shown in Fig. 25-39. The sphere carries a net charge $-Q$ and the shell carries a net charge

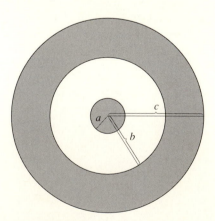

FIGURE 25-39 Problem 35.

+3Q. Both conductors are in electrostatic equilibrium. Find an expression for the potential difference from infinity to the surface of the sphere.

36. A thin plastic rod 20 cm long carries 3.2 nC distributed uniformly over its length. (a) If the rod is bent into a ring, find the potential at its center. (b) If the rod is bent into a semicircle, find the potential at the center (i.e., at the center of the circle of which the semicircle is part).

37. A thin ring of radius R carries a charge $3Q$ distributed uniformly over three-fourths of its circumference, and $-Q$ over the rest. What is the potential at the center of the ring?

38. The potential at the center of a uniformly charged ring is 45 kV, and 15 cm along the ring axis the potential is 33 kV. Find the ring's radius and its total charge.

39. The annulus shown in Fig. 25-40 carries a uniform surface charge density σ. Find an expression for the potential at an arbitrary point P on its axis.

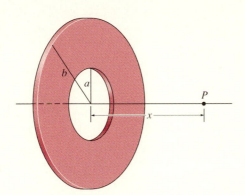

FIGURE 25-40 Problem 39.

40. A thin rod of length ℓ carries a charge Q distributed uniformly over its length. (a) Show that the potential in the plane that perpendicularly bisects the rod is given by

$$V(r) = \frac{2kQ}{\ell} \ln\left[\frac{\ell}{2r} + \sqrt{1 + \frac{\ell^2}{4r^2}} \right],$$

where r is the distance from the rod center. (b) Show that this expression reduces to an expected result when $r \gg \ell$. *Hint:* See Appendix A for a series expansion of the logarithm.

41. (a) Find the potential as a function of position in the electric field $\mathbf{E} = E_0(\hat{\mathbf{i}} + \hat{\mathbf{j}})$, where $E_0 = 150$ V/m. Take the zero of potential at the origin. (b) Find the potential difference from the point $x = 2.0$ m, $y = 1.0$ m to the point $x = 3.5$ m, $y = -1.5$ m.

Section 25-4 Potential Difference and the Electric Field

42. In a uniform electric field, equipotential planes that differ by 1.0 V are 2.5 cm apart. What is the field strength?

43. Figure 25-41 shows a plot of potential versus position along the x axis. Make a plot of the x component of the electric field for this situation.

FIGURE 25-41 Problem 43.

44. Figure 25-42 shows some equipotentials in the x-y plane. (a) In what region is the electric field strongest? What are (b) the direction and (c) the magnitude of the field in this region?

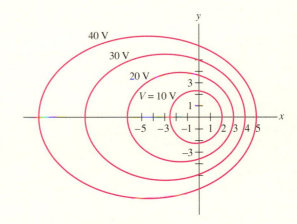

FIGURE 25-42 Problem 44.

45. The potential in a certain region is given by $V = axy$, where a is a constant. (a) Determine the electric field in the region. (b) Sketch some equipotentials and field lines.

46. Sketch some equipotentials and field lines for a distribution consisting of two equal point charges.

47. The electric potential in a region of space is given by $V = 2xy - 3zx + 5y^2$, with V in volts and the coordinates in meters. If point P is at $x = 1$ m, $y = 1$ m, $z = 1$ m, find

(a) the potential at P and (b) the x, y, and z components of the electric field at P.

48. Use Equation 25-7 to calculate the electric field on the perpendicular bisector of a point dipole, and show that your result is equivalent to Equation 23-5a.

49. Use the result of Example 25-6 to determine the on-axis field of a charged ring, and verify that your answer agrees with the result of Example 23-8.

50. A charge $+4q$ is located at the origin and a charge $-q$ is on the x axis at $x = a$. (a) Write an expression for the potential on the x axis for $x > a$. (b) Find a point in this region where $V = 0$. (c) Use the result of (a) to find the electric field on the x axis for $x > a$ and (d) find a point where $\mathbf{E} = \mathbf{0}$.

51. The electric potential in a region is given by $V = -V_0(r/R)$, where V_0 and R are constants, r is the radial distance from the origin, and where the zero of potential is taken at $r = 0$. Find the magnitude and direction of the electric field in this region.

Section 25-5 Potentials of Charged Conductors

52. (a) What is the maximum potential (measured from infinity) for the sphere of Example 25-3 before dielectric breakdown of air occurs at the sphere's surface? (Breakdown of air occurs at a field strength of 3 MV/m.) (b) What is the charge on the sphere when it's at this potential?

53. The spark plug in an automobile engine has a center electrode made from wire 2.0 mm in diameter. The electrode is worn to a hemispherical shape, so it behaves approximately like a charged sphere. What is the minimum potential on this electrode that will ensure the plug sparks in air? Neglect the presence of the second electrode.

54. A large metal sphere has three times the diameter of a smaller sphere and carries three times as much charge. Both spheres are isolated, so their surface charge densities are uniform. Compare (a) the potentials and (b) the electric field strengths at their surfaces.

55. Two metal spheres each 1.0 cm in radius are far apart. One sphere carries 38 nC of charge, the other −10 nC. (a) What is the potential on each? (b) If the spheres are connected by a thin wire, what will be the potential on each once equilibrium is reached? (c) How much charge must move between the spheres in order to achieve equilibrium?

56. Sketch some equipotentials and field lines for the isolated, charged conductor shown in Fig. 25-43.

57. Two conducting spheres are each 5.0 cm in diameter and each carries 0.12 μC. They are 8.0 m apart. Determine (a) the potential on each sphere; (b) the field strength at the surface of each sphere; (c) the potential midway between the spheres; (d) the potential difference between the spheres.

FIGURE 25-43 Problem 56.

58. Two small metal spheres are located 2.0 m apart. One has radius 0.50 cm and carries 0.20 μC. The other has radius 1.0 cm and carries 0.080 μC. (a) What is the potential difference between the spheres? (b) If they were connected by a thin wire, how much charge would move along it, and in which direciton?

Paired Problems

(Both problems in a pair involve the same principles and techniques. If you can get the first problem, you should be able to solve the second one.)

59. Three 50-pC charges sit at the vertices of an equilateral triangle 1.5 mm on a side. How much work would it take to bring a proton from very far away to the midpoint of one of the triangle's sides?

60. Repeat the preceding problem for the case when one of the charges is −50 pC and the proton is brought to the midpoint of the side between the two positive charges.

61. A pair of equal charges q lies on the x axis at $x = \pm a$. (a) Find expressions for the potential at points on the x axis for which $x > a$ and (b) show that your result reduces to a point-charge potential for $x \gg a$.

62. (a) For the charge distribution of the preceding problem, find an expression for the potential at *all* points on the y axis. (b) Show that your result reduces to a point-charge potential for $y \gg a$.

63. A 2.0-cm-radius metal sphere carries 75 nC and is surrounded by a concentric spherical conducting shell of radius 10 cm carrying −75 nC. (a) Find the potential difference between the shell and the sphere. (b) How would your answer change if the shell charge were changed to +150 nC?

64. A coaxial cable consists of a 2.0-mm-radius central wire carrying 75 nC/m, and a concentric outer conductor of radius 10 mm carrying −75 nC/m. (a) Find the potential

difference between the outer and inner conductor. (b) How would your answer change if the outer conductor were charged to +150 nC/m?

65. On the x axis, the electric field of a certain charge distribution is given by $\mathbf{E} = a/x^4\hat{\imath}$, where $a = 55$ V·m^3. Find the potential difference from the point $x = 1.3$ m to the point $x = 2.8$ m.

66. A sphere of radius R carries a nonuniform but spherically symmetric volume charge density that results in an electric field in the sphere given by $\mathbf{E} = E_0(r/R)^2\hat{\mathbf{r}}$, where E_0 is a constant. Find the potential difference from the sphere's surface to its center.

67. The potential as a function of position in a certain region is given by $V(x) = 3x - 2x^2 - x^3$, with x in meters and V in volts. Find (a) all points on the x axis where $V = 0$, (b) an expression for the electric field, and (c) all points on the x axis where $\mathbf{E} = \mathbf{0}$.

68. The potential in a certain region is given by $V(x) = -[2x^2 + (y - 1)^2 - 1]$. Find (a) a point where $V = 0$, (b) an expression for the electric field, and (c) a point in the x-y plane where $\mathbf{E} = \mathbf{0}$.

Supplementary Problems

69. A conducting sphere 5.0 cm in radius carries 60 nC. It is surrounded by a concentric spherical conducting shell of radius 15 cm carrying −60 nC. (a) Find the potential at the sphere's surface, taking the zero of potential at infinity. (b) Repeat for the case when the shell also carries +60 nC.

70. Show that the result of Example 25-9 approaches the field of a point charge for $x \gg a$. *Hint:* You will need to apply the binomial theorem to the quantity $1/\sqrt{x^2 + a^2}$.

71. The potential on the axis of a uniformly charged disk at 5.0 cm from the disk center is 150 V; the potential 10 cm from the disk center is 110 V. Find the disk radius and its total charge.

72. A uranium nucleus (mass 238 u, charge 92e) decays, emitting an alpha particle (mass 4 u, charge 2e) and leaving a thorium nucleus (mass 234 u, charge 90e). At the instant the alpha particle leaves the nucleus, the centers of the two are 7.4 fm apart and are essentially at rest. Find their speeds when they are a great distance apart. Treat each particle as a spherical charge distribution.

73. A power line consists of two parallel wires 3.0 cm in diameter spaced 2.0 m apart. If the potential difference between the wires is 4.0 kV, what is the charge per unit length on each wire? The wires carry equal but opposite charges. *Hint:* The wires are far enough apart that they don't greatly affect each other's fields.

74. For the dipole of Example 25-5, show that the electric field at an arbitrary point far from the dipole can be written

$$\mathbf{E} = \frac{kp}{r^3}[(3\cos^2\theta - 1)\hat{\imath} + 3\sin\theta\cos\theta\,\hat{\jmath}].$$

75. A thin rod of length ℓ lies on the x axis with its center at the origin. It carries a line charge density given by $\lambda = \lambda_0(x/\ell)^2$, where λ_0 is a constant. (a) Find an expression for the potential on the x axis for $x > \ell/2$. (b) Integrate the charge density to find the total charge on the rod. (c) Show that your answer for (a) reduces to the potential of a point charge whose charge is the answer to (b), for $x \gg \ell$.

76. Repeat the preceding problem for the case $\lambda = \lambda_0(x/\ell)$. Why is your answer for $x \gg \ell$ different? *Hint:* What does this charge distribution resemble at large distances?

77. For the situation of Example 25-10, find an equation for the equipotential with $V = 0$ in the x-y plane. Plot the equipotential, and show that it passes through the points described in Example 25-10 and its exercise.

78. A disk of radius a carries a nonuniform surface charge density given by $\sigma = \sigma_0(r/a)$, where σ_0 is a constant. (a) Find the potential at an arbitrary point on the disk axis, a distance x from the disk center. (b) Use the result of (a) to find the electric field on the disk axis, and (c) show that the field reduces to an expected form for $x \gg a$.

79. An open-ended cylinder of radius a and length $2a$ carries charge q spread uniformly over its surface. Find the potential on the cylinder axis at its center. *Hint:* Treat the cylinder as a stack of charged rings, and integrate.

26 Electrostatic Energy and Capacitors

A test firing of the Particle Beam Fusion Accelerator at Sandia National Laboratories involves the sudden release of energy stored in electric fields.

Suppose you hold two positive charges in your outstretched arms (Fig. 26-1). Bringing them closer takes work, as you move each charge against the other's electric field. That work is stored as potential energy associated with the new distribution of charge you create by moving the charges closer together. Because the static electric field is conservative, you could recover the stored energy by releasing the charges and letting them accelerate.

The example of Fig. 26-1 is trivial, but its implications are not. Energy storage in configurations of electric charge is a vital aspect of the natural and technological worlds. The energy produced in chemical reactions—including the metabolizing of food and the burning of coal, oil, and other fuels—is electrical energy released in the rearrangement of molecular charge distributions. Energy storage in systems of charged conductors is essential to the workings of electronic equipment and is important in devices that require large amounts of energy delivered in a short time. In this chapter we explore the energy of charge distributions and their electric fields, and we introduce a practical device—the capacitor—whose function is electrical energy storage.

(a)

26-1 ENERGY OF A CHARGE DISTRIBUTION

In the preceding chapter we defined the electric potential difference between two points as the change in potential energy per unit charge associated with moving charge between those points. To move charge between the two points takes work equal to the change in potential energy—equal, that is, to the electric potential difference multiplied by the charge being moved. In the simple case of two point charges, suppose charge q_1 is initially an infinite distance from a fixed charge q_2. You can assemble a new charge distribution by moving q_1 to a distance r from q_2. Equation 25-4 gives the electric potential difference from infinity to any point a distance r from q_2:

$$V_{\infty r} = \frac{kq_2}{r}.$$

(b)

FIGURE 26-1 (a) Widely separated charges exert little force on each other. (b) Moving them together takes work, which is stored as potential energy. Note how the bending of the field lines suggests the repulsive force.

Multiplying by q_1 gives the work you must do to bring q_1 in from infinity—that is, to assemble the new charge distribution. Since that work, W, gets stored as potential energy, U, we can write

$$W = U = \frac{kq_1q_2}{r}, \tag{26-1}$$

where we've taken the zero of potential energy when the charges are infinitely far apart.

Equation 26-1 gives the **electrostatic potential energy** of two point charges. The equation shows that the potential energy is positive if the charges have the same sign and negative if they have opposite signs. In the latter case, it would take positive work to separate the charges. We would have obtained the same potential energy had we moved q_2 in the field of q_1, showing that the potential energy of a charge distribution is independent of how it is assembled.

These considerations hold for any charge distribution. In general, it takes work to assemble a charge distribution and that work is stored as potential energy. The potential energy can be positive, zero, or negative. Because the electric force obeys the superposition principle, the potential energy is independent of how the charge distribution is assembled: the total energy is simply the sum of the potential energies of every charge pair making up the distribution.

11.11

Electrical Potential Energy and Potential

EXAMPLE 26-1	The Energy of a Charge Distribution

Three point charges each carrying $+q$ and a fourth carrying $-q/2$ are initially infinitely far apart. They are brought together to form the square charge distribution shown in Fig. 26-2. What is the electrostatic potential energy of this charge distribution?

Solution

We can assemble the charge distribution in any order. Assume that the positive charge at the upper left is brought in first. This takes no work because no other charge is present. Next the positive charge q at the upper right is brought to a distance a from the first positive charge. Equation 26-1 tells us that the work required is

$$W_2 = k\frac{q^2}{a}.$$

Now the third positive charge is brought to its place at the lower left. This point is a distance a from the first charge q and $\sqrt{2}a$ from the second, so the work required is

$$W_3 = k\left(\frac{q^2}{a} + \frac{q^2}{\sqrt{2}a}\right).$$

Finally, the negative charge $-q/2$ is brought to a point a distance a from the second and third charges and $\sqrt{2}a$ from the first charge. The work required is again the sum of the potential differences multiplied by the charge $-q/2$:

$$W_4 = k\left(-\frac{q^2}{2a} - \frac{q^2}{2a} - \frac{q^2}{2\sqrt{2}a}\right).$$

This work is negative, indicating that it would take positive work to remove the negative charge. Adding the work required

FIGURE 26-2 How much energy is stored in this square charge distribution? (Example 26-1).

to bring in the second, third, and fourth charges gives the electrostatic potential energy of the charge distribution:

$$W = W_2 + W_3 + W_4$$
$$= k\left(\frac{q^2}{a} + \frac{q^2}{a} + \frac{q^2}{\sqrt{2}a} - \frac{q^2}{2a} - \frac{q^2}{2a} - \frac{q^2}{2\sqrt{2}a}\right)$$
$$= \frac{kq^2(2\sqrt{2} + 1)}{2\sqrt{2}a}.$$

That this is a positive quantity indicates that the work needed to assemble the three positive charges is greater than the energy gained bringing in the negative charge.

EXERCISE Repeat Example 26-1 for the case when the charge in the upper left corner is changed to $-q$.

Answer: $-3kq^2[1 - 1/(2\sqrt{2})]/a$

Some problems similar to Example 26-1: 1-4, 7, 79

26-2 TWO ISOLATED CONDUCTORS

A charge distribution useful for storing electrostatic energy consists of two isolated conductors carrying equal but opposite charges. Figure 26-3 shows two such conductors, each initially uncharged. Imagine moving a small quantity of charge from one conductor to the other, resulting in a net positive charge on one and a net negative charge on the other. This results in an electric field and therefore in a potential difference between the conductors. If we transfer more charge between the conductors, we do work traversing this potential difference. The more charge we move, the harder it gets to transfer additional charge. The work it takes to transfer the charge is stored as potential energy of the charge distribution.

It is generally difficult to calculate the stored potential energy for a pair of irregularly shaped conductors such as those of Fig. 26-3. An important practical case for which the potential energy may be calculated more easily is a pair of identical, flat, parallel conducting plates whose separation is small compared

FIGURE 26-3 A pair of isolated conductors. (*a*) Initially they are uncharged, and there is no potential difference between them. (*b*) When they carry opposite charges, there is an electric field and consequently a potential difference between them.

(a)

(b)

with their width (Fig. 26-4*a*). We start with the plates uncharged, and then transfer charge Q from one plate to the other. (In practice, we would accomplish this by connecting the plates to the terminals of a battery.) Charging the plates results in an electric field between them. For closely spaced plates, this field is essentially uniform except very near the edges (Fig. 26-4*b*), and we may neglect this nonuniform "fringing field."

What is the electric field strength between the plates? In Chapter 24 we used Gauss's law to show that the field near the surface of a conductor carrying surface charge density σ is given by

$$E = \frac{\sigma}{\varepsilon_0}.$$

As we discussed in Section 24-6, charge gathers entirely on the facing surfaces of the two plates, resulting in a charge density of magnitude $\sigma = q/A$, where A is the plate area. So the electric field between the plates is

$$E = \frac{\sigma}{\varepsilon_0} = \frac{q}{\varepsilon_0 A},$$

where q is the magnitude of the charge on either plate. Shouldn't this result be doubled because there are two plates? No! Review the discussion accompanying Figs. 24-34 and 24-35 to convince yourself of this point.

The presence of the electric field means there is a potential difference between the plates. Since the field is uniform, this potential difference is the product of the field strength and the distance d between the plates:

$$V = Ed = \frac{qd}{\varepsilon_0 A},$$

where we're now using V rather than ΔV_{AB} for the potential difference.

Now imagine moving an additional very small positive charge dq from the negative to the positive plate. How much work does this take? That depends on the potential difference between the plates, which, as our expression for V shows, depends on how much charge has already been transferred. Because potential difference is work per unit charge, the work dW required to move the charge dq between the plates is

$$dW = V \, dq = \frac{qd}{\varepsilon_0 A} \, dq.$$

Suppose we start with zero net charge on either plate and gradually transfer a total charge Q from one plate to the other. Each dq that we move requires work dW as given above, so the total work is the sum of all the dW's associated with all the small quantities of charge dq that make up Q. In the limit of infinitely many infinitesimal charges dq, this sum becomes an integral, and we have

$$W = \int_0^Q dW = \int_0^Q \frac{qd}{\varepsilon_0 A} \, dq = \frac{d}{\varepsilon_0 A} \int_0^Q q \, dq.$$

That the variable q remains under the integral sign reflects the physical fact that the continually increasing charge on the plates results in an increasing potential difference and therefore makes it harder to move each additional charge dq. Continuing the integration gives

$$W = \frac{d}{\varepsilon_0 A} \frac{q^2}{2} \Big|_0^Q = \frac{d}{2\varepsilon_0 A} Q^2.$$

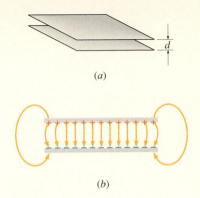

(a)

(b)

FIGURE 26-4 (*a*) A pair of closely spaced conducting plates. (*b*) Edge-on view of the plates when oppositely charged. For closely spaced plates the electric field is essentially uniform, except very near the edges, and the field outside the region between the plates is very small. Neglecting these small "fringing fields" is a good approximation.

Thus the work required to charge the plates increases as the square of the charge Q. The work done in charging the plates ends up as stored potential energy of the final charge distribution, so the stored energy is

$$U = W = \frac{d}{2\varepsilon_0 A} Q^2 .$$

(26-2)

Note that the stored energy in Equation 26-2 depends on the *square* of the charge. Thus a pair of parallel plates on which we can build up lots of charge represents an excellent device for storing electrostatic energy. We'll see later in this chapter how widespread the technological applications of such devices are.

26-3 ENERGY AND THE ELECTRIC FIELD

We've seen that the work required to assemble a distribution of electric charge ends up as electrostatic potential energy. Just where is the energy stored? In Example 26-1, we considered the assembly of four point charges to form a square. Surely the point charges themselves did not change; we only moved them closer together. Similarly, when we charged our pair of metal plates we didn't alter the individual charges; we only moved them from one plate to another. So what has changed in both these cases? The electric field has changed. In the first case we started with four isolated point charges and an electric field that looked like four isolated point-charge fields. We ended with a new charge distribution whose field did not look at all like a point-charge field.

In the case of the parallel plates, we started with uncharged plates and no electric field. As soon as we began transferring charge from one plate to the other, an electric field appeared between the plates, and this field grew in strength as more charge was transferred.

So where is the energy stored? It is stored in the electric field. As we create or alter a charge distribution, we do work and an altered electric field configuration develops. The work done in moving the charges becomes energy stored in the electric field. If that work is positive, the new field configuration contains more energy; if the work is negative, the field has released energy and the new field contains less energy.

Every electric field represents stored energy. Because electric forces are primarily responsible for the behavior of everyday matter, many seemingly different forms of energy storage really involve electric field energy. When you burn gasoline or metabolize food, for example, you are rearranging the charge distributions we call molecules into new configurations whose electric fields contain less energy (Fig. 26-5).

If electric fields store energy, then the amount of stored energy should depend on the field strength. Since the field strength may vary with position, we describe the stored energy in terms of **energy density**, or energy stored per unit volume. We can readily determine the energy density for our parallel plates. There we found that the field strength is given by $E = Q/\varepsilon_0 A$; solving for Q and using the result in Equation 26-2 for the stored energy gives

FIGURE 26-5 Combustion involves the rearrangement of atoms into new molecular structures. Energy released as light and heat comes ultimately from the electric fields associated with the charge distribution in the molecules.

$$U = \frac{d}{2\varepsilon_0 A} Q^2 = \frac{d}{2\varepsilon_0 A} (\varepsilon_0 A E)^2 = \tfrac{1}{2}\varepsilon_0 E^2 A d .$$

Our assumption that the plates are very close together allowed us to conclude that the field is very nearly uniform between the plates and essentially zero outside the plates. Therefore, the energy U is stored in the region between the plates, and is distributed uniformly because the field is uniform. The volume between the plates is just the plate area times the separation, or Ad, and therefore the energy density u_E is given by $u_E = U/Ad$, or

$$u_E = \tfrac{1}{2}\varepsilon_0 E^2. \quad \text{(electric energy density)} \quad (26\text{-}3)$$

Although we derived this expression for the uniform field between two parallel plates, it is in fact a universal expression that holds for *any* electric field. At any point where an electric field exists, there is stored energy whose density, in J/m^3, is given by Equation 26-3.

The deepest significance of Equation 26-3 lies in its statement that every electric field represents stored energy. As we observe a variety of physical phenomena, from everyday happenings on Earth to events in distant galaxies, we can understand that the driving energy for many of these phenomena comes from the release of energy stored in electric fields.

EXAMPLE 26-2	*Electrical Energy of a Thunderstorm*

Electric fields inside a thunderstorm (Fig. 26-6) have typical values of 10^5 V/m and get even higher just before electrical energy is unleashed as lightning (recall that 3×10^6 V/m is the field at which air ionizes and becomes a conductor—which is what happens in a lightning strike). The origin of these fields and hence of the energy stored in them is believed to be associated with charge transfer to rising and falling water droplets or ice crystals in the intense updrafts and downdrafts of the thunderstorm. Consider a typical thundercloud that rises to an altitude of 10 km and has a diameter of 20 km. Assuming an average field strength of 10^5 V/m, estimate the total electrostatic energy stored in the cloud. How many gallons of gasoline would you have to burn to release the same amount of energy?

FIGURE 26-6 Lightning is the sudden release of energy stored in atmospheric electric fields. The most common form of cloud-to-ground lightning, shown here, involves a negative charge layer at the cloud base, and a corresponding positive charge induced in the ground below. Other charge layers exist within the cloud, so electrostatic energy is stored both below and within the cloud.

Solution

The energy density is given by Equation 26-3:

$$u_E = \tfrac{1}{2}\varepsilon_0 E^2 = \tfrac{1}{2}(8.85\times10^{-12}\ \text{C}^2/\text{N}\cdot\text{m}^2)(10^5\ \text{V/m})^2$$
$$= 4.4\times10^{-2}\ \text{J/m}^3.$$

(You should verify that the units work out!) We are assuming that this energy density is the same throughout the storm, so we find the total energy by multiplying the energy density by the volume. The storm is roughly cylindrical in shape, so its volume is

$$V = \pi r^2 h = \pi(10\ \text{km})^2(10\ \text{km}) = 3100\ \text{km}^3$$
$$= 3.1\times10^{12}\ \text{m}^3.$$

Then the total stored energy is

$$U = u_E V = (4.4\times10^{-2}\ \text{J/m}^3)(3.1\times10^{12}\ \text{m}^3) = 1.4\times10^{11}\ \text{J}.$$

A gallon of gasoline contains about 10^8 J (see Appendix C), so the electrical energy stored in a thunderstorm at any given instant is equivalent to about 1000 gallons or 4000 L of gasoline. This comparison is not quite fair to the thunderstorm, though, because its electrical energy is continually dissipated in lightning strikes and at the same time renewed by the violent motion of the air. Problem 77 explores thunderstorm energetics in more detail.

EXERCISE In fair weather, Earth's atmospheric electric field is about 100 V/m. Find the energy stored in each km^3 of the fair-weather atmosphere.

Answer: 44 J

Some problems similar to Example 26-2: 16–18

When the electric field is uniform, as in our thunderstorm example, we can find the stored energy simply by multiplying the energy density by the volume. But when the field changes with position we must resort to calculus. Consider a small volume element dV, so small that the electric field is essentially uniform over this volume. The stored energy dU in the volume element is just the energy density times the volume, or

$$dU = u_E\, dV = \tfrac{1}{2}\varepsilon_0 E^2\, dV.$$

The total energy U is then the sum of all the dUs. In the limit of infinitesimally small volumes dV and energies dU, this sum becomes an integral:

$$U = \tfrac{1}{2}\varepsilon_0 \int E^2\, dV, \tag{26-4}$$

where the limits on the integral are chosen to cover the entire region in which the electric field of interest exists.

We derived Equation 26-4 for the electric field energy using our previously determined expression for the work needed to assemble a simple charge distribution. We can also reverse that process, using the electric field of a charge distribution to calculate the energy density and from it the stored energy and therefore the work needed to assemble the distribution. Example 26-3 illustrates this procedure for a case when the energy density varies with position.

EXAMPLE 26-3	*A Shrinking Sphere*

A sphere of radius R_1 carries a total charge Q distributed evenly over its surface (Fig. 26-7a). How much work does it take to shrink the sphere to a smaller radius R_2? Practical applications in which this question might prove important include the behavior of cell membranes, charged bubbles, and raindrops in thunderclouds.

Solution

Shrinking the sphere moves all the charges on its surface closer together, and therefore requires positive work. That work is equal to the change in stored electric field energy given by Equation 26-4. With all charge distributed evenly over the sphere's surface, Gauss's law tells us that there is no electric field within the sphere. Because of the spherical symmetry, Gauss's law also tells us that the field outside the sphere is identical to that of a point charge Q at the sphere's center. This means that the field at and beyond the original radius R_1 does not change as we shrink the sphere. What does change is the field between R_1 and R_2 (Fig. 26-7b). Originally this field was zero. After the sphere has shrunk, this region, too, is filled with a point-charge field. This newly created field is the site of the additional energy stored in shrinking the sphere.

Because the point-charge field between R_1 and R_2 changes with position, we must use the integral form 26-4 to calculate the stored energy. The electric field outside the sphere is that of a point charge:

$$E = \frac{kQ}{r^2},$$

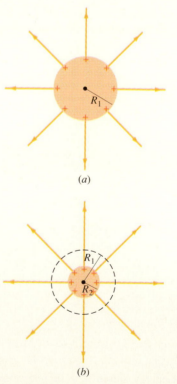

FIGURE 26-7 (*a*) A charged sphere and its electric field. (*b*) Shrinking the sphere creates new field in the region $R_2 < r < R_1$.

so, from Equation 26-3, the energy density as a function of r is

$$u_E(r) = \tfrac{1}{2}\varepsilon_0 E^2 = \tfrac{1}{2}\varepsilon_0 \left(\frac{kQ}{r^2}\right)^2 = \frac{kQ^2}{8\pi r^4},$$

where we used $k = 1/4\pi\varepsilon_0$.

To determine the total stored energy we integrate this energy density over the volume between R_2 and R_1. Because of the spherical symmetry we consider volume elements made of thin spherical shells of thickness dr. Figure 26-8 shows that the volume of the shell is $dV = 4\pi r^2\, dr$, so Equation 26-4 becomes

$$U = \int_{R_2}^{R_1} u_E\, dV = \int_{R_2}^{R_1} \frac{kQ^2}{8\pi r^4}\, 4\pi r^2\, dr$$

$$= \frac{kQ^2}{2}\int_{R_2}^{R_1} r^{-2}\, dr = \frac{kQ^2}{2}\left(-\frac{1}{r}\right)\Big|_{R_2}^{R_1}$$

$$= \frac{kQ^2}{2}\left(\frac{1}{R_2} - \frac{1}{R_1}\right).$$

This is the total energy stored in the new electric field between R_2 and R_1, and is therefore also the work done in shrinking the sphere from R_1 to R_2. If we let R_1 go to infinity, our result becomes the work required to assemble a sphere of radius R_2 carrying surface charge Q, or equivalently, the energy stored in the field of the sphere. Because the stored energy becomes infinite as R_2 approaches zero, our result suggests that the notion of a point charge is an impossible idealization. Problem 90 explores some implications of this result in the theory of elementary particles.

FIGURE 26-8 A thin spherical shell of thickness dr and radius r. Because the shell is very thin, its inner and outer surfaces have essentially the same area, namely $4\pi r^2$. Its volume is therefore $dV = 4\pi r^2\, dr$.

FIGURE 26-9 How much energy is stored in a length ℓ of the cable? (The cable is much longer than the outer conductor radius b.)

EXERCISE A long coaxial cable consists of an inner cylindrical conductor of radius a and an outer cylindrical conducting shell of radius b (Fig. 26-9). The conductors carry equal but opposite line charge densities $\pm\lambda$. Find the electric energy stored in a length ℓ of this cable.

Answer: $k\lambda^2\ell \ln(b/a)$

Some problems similar to Example 26-3: 21–23, 26, 27

26-4 CAPACITORS

In electrical and electronic equipment, electrical energy is often stored using a pair of charged conductors. Such a device is called a **capacitor**. Capacitors are typically used for short-term energy storage in situations where it is necessary to release electrical energy quickly. Most practical electronic devices, including radio, TV, computers, and audio equipment, would be virtually impossible to construct without capacitors. When you tune a radio, you're adjusting a capacitor. Your computer's memory stores information in millions of microscopic capacitors. Failure of a capacitor in your car's ignition system could leave you stranded on the highway. And many high-energy experiments in physics and engineering use so much power that, were it not for capacitors, they could not be done without disrupting the supply of electric power to the entire world!

In an uncharged capacitor both conductors are neutral. A capacitor is charged by transferring charge (usually electrons) from one conductor to the other. The work required is supplied by a source of electrical energy—for example, a battery—connected by wires to the capacitor's two conductors. An electric field develops as a result of the separation of positive and negative charge on the two conductors, and the stored energy resides in this field.

Since there's an electric field in a charged capacitor, there's also a potential difference between its two conductors. The more charge, the greater the electric field, and, proportionately, the greater the potential difference. Thus a capacitor's charge and the potential difference between its conductors are proportional. In practice, we impose a potential difference between the two conductors, and a proportionate charge buildup occurs (that's because common electrical energy sources—such as batteries—produce a fixed potential difference rather than a fixed amount of charge). The ratio of charge to potential difference is characteristic of a given capacitor and is called its **capacitance**:

$$C = \frac{Q}{V}. \tag{26-5}$$

Here Q is the magnitude of the charge on either conductor and V the potential difference (or voltage) between the conductors. Clearly the units of capacitance are coulombs/volt. One coulomb/volt is given the name **farad** (F), in honor of the nineteenth century scientist Michael Faraday. One farad is so large a capacitance that the smaller units microfarad (10^{-6} F; abbreviated μF) and picofarad (10^{-12} F; abbreviated pF and often pronounced "puff") are widely used.

Capacitance depends on the physical construction of a capacitor—the shapes of its two conductors, their separation, and the choice of insulating material between them. Although Q and V enter the defining relation 26-5, capacitance is a constant. If V is increased, Q increases proportionately, maintaining the constant ratio C that characterizes the capacitor.

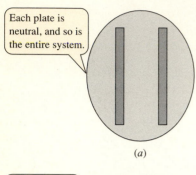

Each plate is neutral, and so is the entire system.

(a)

Each plate now carries a net charge, but the whole system is still neutral.

+Q   −Q

(b)

FIGURE 26-10 (a) In an uncharged capacitor both plates are neutral. (b) In a charged capacitor, the two plates carry opposite charges of equal magnitude. The entire system of two plates remains neutral.

Tip

Charged but Neutral When we speak of a "charged capacitor," we mean a capacitor with one conductor positive and the other negative; the net charge on the entire capacitor—i.e., on *both* plates—remains zero. Similarly, when we say "the charge on a capacitor is Q," we mean that one conductor carries $+Q$ and the other $-Q$. Overall, the capacitor remains neutral even when it's "charged" (Fig. 26-10).

Any arrangement of two insulated conductors constitutes a capacitor. Practical capacitors are manufactured in a variety of configurations (Fig. 26-11). Often they are made from two long strips of aluminum foil separated by thin layers of plastic or paper. This foil "sandwich" is then rolled into a compact cylinder, wires attached, and the whole assembly covered with a protective coating. Another common arrangement is the variable capacitor, whose configuration can be altered to change its capacitance. This change can be accomplished mechanically or, as in many modern electronic devices, electrically. Very large capacitances are achieved with so-called electrolytic capacitors, in which a thin insulating layer develops chemically under the influence of the applied voltage. We'll consider the role of insulating materials further in Section 26-7.

FIGURE 26-11 Typical capacitors. The large blue unit is an 18-mF electrolytic capacitor. At top right is an air-insulated variable capacitor in which a set of metal plates rotates with respect to fixed plates in order to change the capacitance. The remaining smaller capacitors range from 43 pF to 10 μF.

Got It!

One capacitor has $\pm 50\ \mu$C on its plates when the voltage between the plates is 6 V. A second capacitor has $\pm 100\ \mu$C on its plates when the voltage between them is 12 V. Compare the two capacitances.

Calculating Capacitance

Capacitance is defined through Equation 26-5 as the ratio of charge to potential difference. To calculate the capacitance of a particular configuration, we assume there is a charge Q on the capacitor and calculate the corresponding potential difference. When the capacitor design includes sufficient symmetry, this calculation is straightforward.

By far the most important capacitor design is the parallel-plate configuration. In Section 26-2 we examined such a capacitor in some detail, although at that time we did not call the configuration a capacitor. There we found that the potential difference between plates of area A separated by a distance d is

$$V = \frac{Qd}{\varepsilon_0 A}.$$

Solving for the ratio $C = Q/V$ gives

$$C = \frac{\varepsilon_0 A}{d}. \qquad \text{(parallel-plate capacitor)} \qquad (26\text{-}6)$$

Equation 26-6 gives the capacitance of a parallel-plate capacitor in terms of the universal constant ε_0 and factors that describe the physical configuration of the capacitor. (Strictly speaking, this expression holds only for capacitors insulated by vacuum. In Section 26-7 we will modify Equation 26-6 to account for other insulating materials.) Note that neither charge nor potential difference enters the final expression for capacitance, showing that the capacitance is indeed a constant, independent of Q and V. Equation 26-6 suggests that the way to make a capacitor with large capacitance is to use two plates of large area but small separation. Incidentally, Equation 26-6 shows that the units of ε_0 may be expressed as farads/meter (F/m); see Problem 32.

APPLICATION	*Microphones, Keyboards, Elevators, Barometers . . .*

Capacitance depends on the geometrical arrangement of two conductors, and capacitance is an electrical quantity that's relatively easy to measure (you'll see why after you're studied Chapter 33). Those facts make capacitors ideal **transducers**—devices that convert physical quantities into electrical signals for measurement and control purposes.

The mechanical displacement of an object can be measured precisely by linking one plate of a capacitor to the object; capacitance changes as the object moves, and the change in capacitance provides a measure of displacement. This principle is used not only in direct displacement measurement, but also in capacitive microphones, where a metal diaphragm forming one capacitor plate vibrates in response to sound waves, changing the capacitance. Similarly, capacitive pressure transducers use the deflection of a capacitor plate diaphragm to measure changes in air pressure. And, in an application you may use literally thousands of times a day, pressing a key on a computer keyboard varies the spacing of capacitor plates, changing the capacitance and signaling the computer that the key has been pressed (Fig. 26-12).

Contact or even nearby presence of a conducting object can change the charge distribution on capacitor plates, effectively

FIGURE 26-12 The keys of a computer keyboard move capacitor plates, altering the capacitance and signaling which key has been pressed.

altering the capacitance. "Touch sensitive" buttons in elevators and other applications work on this principle; as your finger approaches the button, charge movement in your body alters the charge on a capacitor behind the button, signaling your floor choice to the elevator. Lamps that turn on or off at the touch of a finger work on the same principle.

EXAMPLE 26-4 | *A Parallel-Plate Capacitor*

A capacitor consists of two circular metal plates of radius 10 cm separated by an air gap of 5.0 mm (Fig. 26-13). What is its capacitance? When a 12-volt battery is connected to the capacitor, how much charge appears on the plates?

Solution

Since the plate spacing is much smaller than the plate size, Equation 26-6 holds and the capacitance is

$$C = \frac{\varepsilon_0 A}{d} = \frac{\varepsilon_0 \pi r^2}{d} = \frac{(8.85 \times 10^{-12} \text{ F/m})(\pi)(0.10 \text{ m})^2}{5.0 \times 10^{-3} \text{ m}}$$

$$= 5.6 \times 10^{-11} \text{ F} = 56 \text{ pF}.$$

Equation 26-5 defines capacitance as the ratio of charge to potential difference. We can rewrite this defining relation to solve for the charge:

$$Q = CV = (56 \text{ pF})(12 \text{ V}) = 670 \text{ pC}.$$

What this really means, of course, is that the positive plate carries 670 pC and the negative plate −670 pC. Overall, the capacitor remains neutral. Note that by working with the capacitance in pF, the charge automatically comes out in pC.

EXERCISE A parallel-plate capacitor is to be made from two square pieces of aluminum foil each 8.0 cm on a side. (a) What

FIGURE 26-13 A parallel-plate capacitor connected to a battery. Drawing is not to scale (Example 26-4).

should be the spacing between them if the capacitance is to be 47 pF? (b) What applied voltage will put ±95 nC on the plates?

Answers: (a) 1.2 mm; (b) 2.0 kV

Some problems similar to Example 26-5: 35, 36, 67

EXAMPLE 26-5 | *A Cylindrical Capacitor*

A capacitor consists of two long concentric metal cylinders of length L, as shown in Fig. 26-14. The inner and outer cylinders have radii a and b, respectively. What is the capacitance?

Solution

Equation 26-6 does not apply to this configuration because the field between the cylinders is not uniform. To find the capacitance, we need a relation between charge and potential difference for the cylindrical configuration. In Example 25-4, we found that the potential difference between two points outside a charge distribution with line symmetry can be written

$$V(a) - V(b) = \frac{\lambda}{2\pi \varepsilon_0} \ln\left(\frac{b}{a}\right),$$

where λ is the line charge density. Because our capacitor is long compared with its diameter, this expression is a good approximation to the potential difference due to the field of the inner conductor. What about the outer conductor? Recall (Example 24-5) that the electric field inside an empty, hollow pipe is zero; therefore, the outer conductor contributes nothing to the electric field or the potential difference between the conductors. If the magnitude of the charge on either conductor is Q, then the line charge density is $\lambda = Q/L$, and our expression for

FIGURE 26-14 A cylindrical capacitor (Example 26-5).

potential difference becomes

$$V = V(a) - V(b) = \frac{Q}{2\pi \varepsilon_0 L} \ln\left(\frac{b}{a}\right).$$

Capacitance is the ratio of charge to potential difference, so we have

$$C = \frac{Q}{V} = \frac{Q}{(q/2\pi \varepsilon_0 L)\ln(b/a)} = \frac{2\pi \varepsilon_0 L}{\ln(b/a)}. \quad (26\text{-}7)$$

Does this result make sense? We already found that the capacitance of a parallel-plate capacitor increases with increasing plate area or with decreasing plate separation. With the cylindrical capacitor we can increase the area of both conductors by increasing the length L of the capacitor, and indeed Equation 26-7 shows the capacitance increasing proportionately. We can decrease the spacing of the conductors by making the radii a and b more nearly equal. This makes b/a closer to one, and $\ln(b/a)$ closer to zero, again increasing the capacitance. Although the geometries of the cylindrical and parallel-plate capacitors are quite different, the same physical considerations apply to both: a large capacitance is achieved with large conductor areas and small separation. When the separation is very small, the curvature of the cylindrical capacitor cannot matter, and Equation 26-7 should reduce to Equation 26-6 for the parallel-plate capacitor (see Problem 84).

EXERCISE A conducting sphere of radius R is enclosed in a concentric spherical conducting shell of radius $\frac{3}{2}R$. What is the capacitance of this configuration?

Answer: $3R/k$

Some problems similar to Example 26-5: 35, 36, 67

Working Voltage

The most important consideration in using a capacitor is whether its capacitance is right for the particular application. But it's also important not to exceed a capacitor's **working voltage**, a safety rating set somewhat below the potential difference that would cause dielectric breakdown of the material between the capacitor plates.

Large capacitances are most easily achieved using small plate separations, as Equation 26-6 suggests. But a small plate separation implies a large electric field for a given voltage. Thus in practical capacitors there is a trade-off between capacitance and working voltage. High working voltage and high capacitance together require large plate separation to keep the electric field small and avoid dielectric breakdown, while at the same time requiring large plate area to keep the capacitance up. Thus large, high-voltage capacitors are physically bulky and expensive to build. Often economics as well as physics dictates the final design of a circuit involving capacitors.

26-5 ENERGY STORAGE IN CAPACITORS

In Section 26-3 we found that any electric field represents stored energy. The example that guided us to that conclusion was a parallel-plate capacitor. For that configuration, the stored energy U is given by Equation 26-2:

$$U = \frac{d}{2\varepsilon_0 A} Q^2 .$$

Since $\varepsilon_0 A/d$ is the capacitance of the parallel-plate capacitor, this stored energy may be written

$$U = \frac{Q^2}{2C}. \qquad \text{(energy in a capacitor)} \qquad (26\text{-}8a)$$

It is usually easier to measure voltage than charge. To express the stored energy in terms of voltage, we can solve the equation defining capacitance, $C = Q/V$, for Q and use the result in Equation 26-8a:

$$U = \frac{Q^2}{2C} = \frac{(CV)^2}{2C} = \tfrac{1}{2}CV^2 . \qquad \text{(energy in capacitor)} \qquad (26\text{-}8b)$$

Although Equations 26-8a and b were derived for a parallel-plate capacitor, they hold for any capacitor regardless of its configuration (see Problem 49).

FIGURE 26-15 The large black cylinders are 4700-μF electrolytic capacitors used in the power supply of a stereo amplifier. They store and release electrical energy to smooth out the voltage variations of the power line, and can retain significant charge after the stereo has been turned off.

That the stored energy depends on the *square* of the potential difference implies that more energy can be stored in a small capacitor at high voltage than in a larger one at low voltage. Practically, the difficulties of handling high voltages mitigate this conclusion somewhat, but the fact remains that the stored energy in a capacitor increases rapidly with increasing voltage.

Capacitors can store energy for significant times. In VCRs, for example, capacitors maintain program memory in the event of a power failure lasting up to an hour or so. Large capacitors used to produce steady direct current in TVs, stereos, and other electronic devices may retain their charge after the device is turned off (Fig. 26-15); before beginning work, repair technicians routinely discharge these capacitors by connecting a screwdriver across their terminals (Fig. 26-16). However, capacitors cannot store arbitrarily large amounts of energy indefinitely. One reason is that all capacitors "leak" charge through their imperfect insulation, gradually losing their energy; another is that dielectric breakdown sets a limit on the magnitude of the electric field in a capacitor. In Problem 18, for example, you can show that the maximum energy density for an electric field in air is far less than the energy density stored in gasoline. Thus you'll not see electric cars powered by the energy stored in capacitors. Capacitors do excel as energy storage devices in situations that call for rapid delivery of stored energy. We'll look at several such applications after Example 26-6.

FIGURE 26-16 This screwdriver has been used repeatedly to discharge a large capacitor. Steel vaporized in each of the pitted areas.

EXAMPLE 26-6	*Which Capacitor?*

A 100-μF capacitor has a working voltage of 20 V, while a 1.0-μF capacitor has a 300-V working voltage. Which can store the most energy? The most charge?

Solution

Applying Equation 26-8b with V set to the working voltage gives

$$U_{100\,\mu F} = \tfrac{1}{2}CV^2 = \tfrac{1}{2}(100 \ \mu F)(20 \ V)^2 = 20 \times 10^3 \ \mu J = 20 \ mJ$$

and

$$U_{1.0\,\mu F} = \tfrac{1}{2}CV^2 = \tfrac{1}{2}(1.0 \ \mu F)(300 \ V)^2$$
$$= 45 \times 10^3 \ \mu J = 45 \ mJ .$$

Because of its higher working voltage, the smaller capacitor can store more energy. On the other hand, the larger capacitor

stores more charge, as shown by solving the defining relation $C = Q/V$:

$$Q_{100\,\mu F} = CV = (100 \ \mu F)(20 \ V) = 2000 \ \mu C = 2.0 \ mC$$

and

$$Q_{1.0\,\mu F} = CV = (1.0 \ \mu F)(300 \ V) = 300 \ \mu C = 0.30 \ mC .$$

Again, these numbers refer to the magnitude of the charge on each plate; overall, each capacitor remains neutral.

EXERCISE The "memory" capacitor in a VCR stores 25 J of energy with a potential difference of 3.5 V. (a) What is its capacitance? (b) What is the magnitude of the charge on each plate?

Answers: (a) 4.1 F; (b) 14 C

Some problems similar to Example 26-6: 38–41, 43, 44

Got It!

You need to replace a capacitor with one that can store more energy. Which will allow a greater increase in stored energy, a replacement capacitor with twice the capacitance and the same working voltage, or one with the same capacitance but twice the working voltage?

| **APPLICATION** | *Camera Flashes, Toilet Flushes, Defibrillators, and Fusion* |

You've probably used a camera equipped with a flash. After you snap a picture, you have to wait about 10 seconds before the flash is ready again. Why? Because the power required by the flash is far greater than a small battery could supply. So the battery is used to charge a capacitor at a slow rate. Once the capacitor is charged its stored energy can be dumped suddenly into the flash tube, producing a short burst of intense light. You can't use the flash again until the capacitor is recharged. The capacitor acts as a reservoir, accumulating energy slowly and then releasing it quickly. The average power—averaged over the relatively long capacitor-charging time—is modest and well within the battery's capability. The instantaneous power, during the millisecond or so of the flash, is much larger—and that's why the capacitor is needed. Problem 45 explores the camera flash.

The camera flash and its capacitor are analogous to a situation you encounter every day with household plumbing. Flushing a toilet requires a large amount of water in a short time—far more than typical household plumbing could supply in that time. So the water is accumulated gradually in the toilet tank, then suddenly dumped when needed for flushing. In this analogy the household plumbing corresponds to the small battery of the camera flash. The toilet tank, which gradually accumulates water, corresponds to the capacitor, which gradually accumulates electrical energy. Of course you need to wait between flushes for the toilet tank to fill, just as you need to wait between flash pictures for the capacitor to charge.

Professional photographers needing to take flash pictures in rapid succession often carry around a large, heavy battery pack capable of supplying the flash power directly. Similarly, institutional buildings with large, high-pressure water pipes often have toilets without tanks that can be flushed in rapid succession.

The example of the camera flash, scaled up in size, shows how large amounts of power may be obtained for brief intervals. An analogous medical application is the defibrillator, used to restore rhythmic beating of a heart undergoing the random, ineffectual muscle contractions called fibrillation. Successful defibrillation requires the delivery of several hundred joules of electrical energy in a few milliseconds (Fig. 26-17)—an impossibility for a battery-powered instrument that needs portability for use in medical emergencies. So, just as in the camera, capacitors are used to store the energy and make it available for instant use. Problem 44 explores the energetics of defibrillation.

At a grander scale, experiments involving high-powered lasers for nuclear fusion research require more power than all

FIGURE 26-17 A defibrillator delivers several hundred joules of electrical energy to restore the heart's normal beating. That energy is stored in capacitors which discharge in a few milliseconds.

FIGURE 26-18 This huge capacitor bank stores 60 MJ of energy to drive the Nova laser fusion experiment at Lawrence Livermore National Laboratory. The capacitors can discharge a significant fraction of their energy in 1 ns.

the world's electric generating stations produce (Fig. 26-18). The required energy is accumulated in huge banks of capacitors, which are suddenly discharged to provide energy to the laser. Think here about the difference between energy and power! The pulsed laser is only on for about 10^{-9} seconds, so although it consumes energy at an enormous rate while on, it does not use all that much total energy (see Problem 46). The laser is not fired very often, so there's plenty of time to charge the capacitors. The *average* power consumption of the experiment is modest.

26-6 CONNECTING CAPACITORS

Often a single capacitor with a desired combination of capacitance and working voltage is not available. But we can obtain the combination by connecting two or more capacitors together; that's how the large capacitance needed for laser fusion

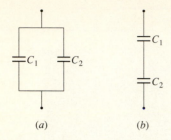

FIGURE 26-19 Connecting capacitors together. (*a*) Parallel; (*b*) series. ⊥ is the standard circuit symbol for a capacitor.

is achieved, for example. There are only two ways to connect two capacitors (and indeed any electrical components with only two wires coming from them). Called **parallel** and **series**, these two possible connections are shown in Fig. 26-19. We would like to know the equivalent capacitance of each combination.

Capacitors in Parallel

Consider first the parallel combination of Fig. 26-19*a*. If we impose a potential difference *V* across the two wires coming from the combination, what will be the potential difference across each capacitor? The key to answering this question is the recognition that the wires connecting the capacitors are conductors and that in electrostatic equilibrium there can be no potential difference between any points connected to the same wire.* Thus all points connected directly together—including the top plates of each capacitor—are at the same potential. Similarly, the bottom plates and the wires connecting them are all at the same potential. Therefore, the potential differences across the two capacitors are equal. This is a very important point in the practical understanding of electric circuits, and applies to any two circuit components that are connected in parallel:

> The potential differences across two electrical components in parallel are equal.

Recognizing this simple fact is essential in developing your understanding of electric circuits!

The equivalent capacitance is the ratio of the total charge on both capacitors to the voltage across the parallel combination. Solving the defining relation $C = Q/V$ for charge, we can write the charges on the two capacitors as

$$Q_1 = C_1 V \qquad \text{and} \qquad Q_2 = C_2 V.$$

The potential difference *V* is the *same* in both cases because the capacitors are connected in *parallel*. Thus the total charge is

$$Q = Q_1 + Q_2 = C_1 V + C_2 V = (C_1 + C_2)V.$$

Taking the ratio of total charge *Q* to the voltage *V* across the parallel combination gives the equivalent capacitance:

$$C = C_1 + C_2. \qquad \text{(parallel capacitors)} \qquad (26\text{-}9a)$$

Equation 26-9a is frequently stated as "capacitors in parallel add." You can understand this result physically by considering two parallel-plate capacitors with equal spacing. Connecting them in parallel amounts to adding their plate areas, giving a larger capacitance. Although we derived Equation 26-9a for two parallel capacitors, the result that parallel capacitances add is easily extended to three or more capacitors (see Problem 56):

$$C = C_1 + C_2 + C_3 + \cdots \qquad \text{(parallel capacitors)} \qquad (26\text{-}9b)$$

What about the working voltage of the parallel combination? Both capacitors experience the full potential difference *V*, so the working voltage of the combination is that of whichever capacitor has the lower working voltage.

Capacitors in Series

Suppose we charge the series capacitor system of Fig. 26-19*b*, putting $+Q$ on the upper plate of C_1 and $-Q$ on the lower plate of C_2. The positive charge on the

---

*Even when we relax the equilibrium assumption, this conclusion will still hold in the approximation that the wires are perfect conductors.

uppermost plate attracts $-Q$ to the lower plate of C_1 and the negative charge $-Q$ on the lowermost plate attracts $+Q$ to the upper plate of C_2 (Fig. 26-20). Note that this leaves the middle two plates together with zero net charge, which must be the case since these two plates are not connected to any external source of charge. With charge of magnitude Q on every plate, we can conclude that

▎ Capacitors in series carry the same charge.

FIGURE 26-20 Capacitors in series carry the same charge.

To find the equivalent capacitance, we first solve the relation $C = Q/V$ for the voltages across the two capacitors:

$$V_1 = \frac{Q}{C_1} \quad \text{and} \quad V_2 = \frac{Q}{C_2}.$$

Here there's no need to label the Q's since series capacitors carry the same charge. But now the voltages need not be the same. Since the electric fields in the two capacitors point the same way (Fig. 26-20), the voltage across the series combination is just

$$V = V_1 + V_2.$$

Inserting our expressions for the individual potential difference gives

$$V = \frac{Q}{C_1} + \frac{Q}{C_2} = Q\left(\frac{1}{C_1} + \frac{1}{C_2}\right).$$

Dividing by Q gives V/Q, which the relation $C = Q/V$ shows is the reciprocal of the equivalent capacitance:

$$\frac{1}{C} = \frac{1}{C_1} + \frac{1}{C_2}. \qquad \text{(series capacitors)} \qquad (26\text{-}10a)$$

This result is frequently described by saying that "capacitors in series add reciprocally." The result is easily extended to three or more capacitors (see Problem 56):

$$\frac{1}{C} = \frac{1}{C_1} + \frac{1}{C_2} + \frac{1}{C_3} + \cdots . \qquad \text{(series capacitors)} \qquad (24\text{-}10b)$$

Equations 26-10 show that the combined capacitance of two series capacitors is less than the capacitance of either. You can make physical sense of this by considering parallel-plate capacitors with equal plate areas. Putting them in series effectively adds the plate separations of the two capacitors, yielding a smaller overall capacitance.

What about the voltage rating of a series combination? The full applied voltage V is the sum of the voltages across each capacitor, so each can be rated for less than the full applied voltage. The fraction of the applied voltage that appears across each capacitor depends on the relative capacitances.

Got It!

You have two identical capacitors, each with capacitance C. How would you connect the two to get equivalent capacitances of (a) $2C$ and (b) $\frac{1}{2}C$? (c) Which equivalent capacitor has the higher working voltage?

EXAMPLE 26-7 | *Connecting Capacitors*

(a) Find the equivalent capacitance of the combination shown in Fig. 26-21a. (b) If the maximum voltage applied between points A and B is 100 V, what should be the working voltage of C_2?

Solution

This is a problem about an electric circuit—an assemblage of interconnected electrical components, in this case four capacitors. The way to handle many such circuit problems is to reduce the circuit to a simpler one by recognizing combinations of series and parallel components, as shown in Fig. 26-21. Here C_3 and C_4 are in parallel, so they add to give an equivalent capacitance of 4.0 μF; the circuit then looks like Fig. 26-21b. This parallel combination C_{34} is in series with C_2; applying

Equation 26-10c gives

$$C_{234} = \frac{C_2 C_{34}}{C_2 + C_{34}} = \frac{(12\ \mu\text{F})(4.0\ \mu\text{F})}{12\ \mu\text{F} + 4.0\ \mu\text{F}} = 3.0\ \mu\text{F}.$$

Now the circuit looks like Fig. 26-21c, with C_1 and C_{234} in parallel. The equivalent capacitance of the entire circuit is their sum, or 7.0 μF (Fig. 26-21d).

To find the working voltage needed for C_2, we need to know the voltage across C_2 with 100 V across the entire combination. Since C_1 and the combination C_{234} are in parallel and connected to points A and B, both these capacitors experience the full 100 V. We can then use the defining relation $C = Q/V$ to find the charge on C_{234}:

$$Q_{234} = C_{234} V = (3.0\ \mu\text{F})(100\ \text{V}) = 300\ \mu\text{C}.$$

But C_{234} is the series combination of C_2 and C_{34} (Fig. 26-21b, c), and the charge on series capacitors is the same. So $Q_2 = 300\ \mu$C as well. We again use the defining relation $C = Q/V$, now to find the voltage across C_2:

$$V_2 = \frac{Q_2}{C_2} = \frac{300\ \mu\text{C}}{12\ \mu\text{F}} = 25\ \text{V}.$$

This is the required working voltage for C_2.

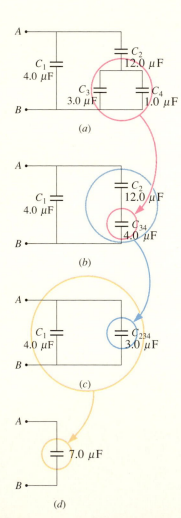

(a)

(b)

(c)

(d)

FIGURE 26-21 What is the equivalent capacitance (Example 26-7)? Analyzing the circuit involves reducing parallel and series combinations using Equations 26-9 and 26-10.

Tip

Analyzing Circuits This example illustrates a very general approach to circuit analysis. You reduce the circuit to its simplest form by recognizing series and parallel combinations, then gradually build back up to find the quantity of interest. The reduction is essential, even when you're only interested in what's happening with one component. Here, for instance, we wanted to know the voltage on C_2 with 100 V across the whole combination. But to find that we had to analyze the entire circuit, and only then could we focus on C_2.

Tip

Recognize Series and Parallel Parallel components have their two ends, respectively, connected *directly* together, as in Fig. 26-19a. Series components are connected in such a way that if you imagine moving through one component, then the *only* place you can go is into the next component (Fig. 26-19b). Many connections are neither; C_2 and C_3 in Fig. 26-21a, for example, are *not* in series because after you go through C_2 the circuit splits and you could go through either C_3 or C_4. And C_1 and C_2 are *not* in parallel; even though their top plates are connected directly together, their bottom plates are not. The series and parallel formulas we've derived apply *only* to true series and parallel combinations.

26-7 CAPACITORS AND DIELECTRICS

The insulating material between the plates of a capacitor serves several purposes. It maintains physical separation of the plates and minimizes charge leakage. Its molecular properties also influence the capacitance. In Chapter 23 we found that electric dipoles tend to align with an applied electric field, and we defined **dielectrics** as materials whose molecules behave as dipoles. The molecular dipole moments may be intrinsic to the molecules, or may be induced by an applied electric field, as we showed in Fig. 23-35. Essentially all insulators are dielectrics.

Suppose we have a parallel-plate capacitor charged to some voltage V_0, with air or vacuum between its plates. What happens if we insert a slab of dielectric material that fills the space between the plates? If the molecules of the dielectric have permanent dipole moments, they'll tend to align with the field, although thermal agitation prevents perfect alignment. If the molecular dipoles are induced by the electric field of the capacitor plates (recall Fig. 23-35), then the induced dipoles form in alignment with the field. Either way, as Fig. 26-22 shows for the case of induced dipoles, the fields of the dipoles themselves *oppose* the capacitor field. The result is a reduction in the net electric field between the plates. How much reduction depends on details of molecular structure and molecular interactions; empirically, though, we find that a given material may be characterized by its **dielectric constant**, κ, which describes the reduction in field. If E_0 is the original field, then the field after insertion of the dielectric will be decreased by a factor $1/\kappa$: $E = E_0/\kappa$.

If the capacitor is not connected to anything, then there's no way for the charge on its plates to change. But the field between the plates has been reduced, and therefore the potential difference $V = Ed$ has decreased by the same factor $1/\kappa$; $V = V_0/\kappa$. But capacitance is the ratio of charge to voltage, so with the dielectric in place we have

FIGURE 26-22 A capacitor with a dielectric between its plates. Molecular dipoles align with the capacitor's electric field, and their own fields point opposite to the original field. The result is a reduction in field strength.

$$C = \frac{Q}{V} = \frac{Q}{(V_0/\kappa)} = \frac{\kappa Q}{V_0} = \kappa C_0, \qquad (26\text{-}11)$$

where $C_0 = Q/V_0$ is the original capacitance. Thus insertion of the dielectric increases the capacitance by a factor κ.

Capacitors are among the most difficult electronic components to miniaturize, so the ongoing revolution in microelectronics has spurred a search for suitable dielectrics with large dielectric constants, good insulating properties, and high breakdown fields. Exotic materials such as tantalum oxide and strontium titanate have become widely used in recent years because of their high dielectric constants. In a few unusual applications water, with its dielectric constant of 78, is used as a dielectric. In high-energy experiments where it is necessary to store a large amount of energy for a short time, water's large dielectric

TABLE 26-1 *Properties of Some Common Dielectrics*

DIELECTRIC MATERIAL	DIELECTRIC CONSTANT	BREAKDOWN FIELD (kV/mm)
Air	1.0006	3
Aluminum oxide	8.4	670
Glass (Pyrex)	5.6	14
Paper	3.5	14
Plexiglass	3.4	40
Polyethylene	2.3	50
Polystyrene	2.6	25
Quartz	3.8	8
Tantalum oxide	26	500
Teflon	2.1	60
Water	78	—

constant outweighs the disadvantage of poor insulating quality. Table 26-1 lists dielectric constants and breakdown fields of selected materials.

In addition to helping build better capacitors, the relation between capacitance and dielectric constant serves as a useful probe of the structure of matter. Introducing a dielectric material between capacitor plates lowers the potential difference and therefore allows us to calculate the dielectric constant. This, in turn, gives information about the density and structure of the individual molecular dipoles. Conversely, we can use the measured dielectric constant to help identify an unknown material.

EXAMPLE 26-8	*Capacitors and Dielectrics*

An air-insulated capacitor is charged by connecting it to a 12-V battery. The battery is then disconnected. When the space between capacitor plates is filled with an unknown plastic, the voltage between the plates drops to 4.6 V. What is the unknown material? If the plate spacing is 0.10 mm, how much voltage can the capacitor withstand with this material between its plates?

Solution

The voltage has dropped by a factor $1/\kappa = 4.6/12 = 1/2.6$. From Table 26-1, we see that a plastic with $\kappa = 2.6$ is polystyrene. With a dielectric breakdown field of 25 kV/mm, the 0.1-mm-thick piece of polystyrene can withstand 2.5 kV.

The rated working voltage would actually be lower, to allow a margin of safety.

EXERCISE An air-insulated capacitor with $C = 25\ \mu F$ is connected to a 10-V battery and the battery is left connected as a quartz slab is inserted to fill the space between the plates. Find the charge on the capacitor (a) before and (b) after the slab is inserted. *Hint:* The battery maintains a fixed 10 V across the plates, but now charge can move from the battery to the plates.

Answers: (a) 250 μC; (b) 950 μC

Some problems similar to Example 26-8: 64–66

What happens to the energy stored in a capacitor when a dielectric is inserted between its plates? The dielectric increases the capacitance by a factor κ, but it also decreases the potential difference by the same factor. If the energy is initially $U_0 = \frac{1}{2} C_0 V_0^2$, then after the dielectric is inserted it becomes

$$U = \frac{1}{2}(\kappa C_0)\left(\frac{V_0}{\kappa}\right)^2 = \frac{1}{2\kappa} C_0 V_0^2 = \frac{U_0}{\kappa}. \tag{26-12}$$

Since $\kappa > 1$, the energy has decreased. Where has it gone? As the dielectric moves into the capacitor, the electric field causes charge separation and rotation of the molecular dipoles. The dipoles thus gain energy from the field. In a solid the molecules interact strongly, and the energy is quickly dissipated as heat.

FIGURE 26-23 The nonuniform fringing field outside the capacitor is stronger nearer the plates and therefore results in a net force on the molecular dipoles in the dielectric slab. As a result, the dielectric is pulled into the capacitor.

That the energy of a capacitor is lower with a dielectric inserted suggests that some attractive force must pull the dielectric slab into the capacitor. Figure 26-23 shows that this force originates in something we have heretofore intentionally ignored—the nonuniform fringing field beyond the plates of the capacitor. This nonuniform field acts on the dipoles in the dielectric to produce a net force toward the interior of the capacitor (see Problem 81).

CHAPTER SYNOPSIS

Summary

1. The work required to assemble a charge distribution is stored as the **electrostatic potential energy** of the distribution. Electrostatic potential energy resides in the electric field. Whenever an electric field is altered, energy is added to or removed from the field.

   a. The **electric field energy density** is given by

   $$u_E = \tfrac{1}{2}\varepsilon_0 E^2,$$

   where the SI units of u_E are J/m^3.

   b. The electrostatic potential energy of a charge distribution may be determined either by computing the work required to assemble the individual charges of the distribution or, knowing the electric field of the distribution, by integrating the energy density over the volume containing the field:

   $$U = \int u_E \, dV = \tfrac{1}{2}\varepsilon_0 \int E^2 \, dV.$$

2. A **capacitor** is an arrangement of two conductors separated by an insulator. Transferring charge from one conductor to the other results in an electric field in the region between the conductors, and energy is stored in the field.

   a. The **capacitance** of a capacitor is the ratio of the charge to the potential difference between its conductors:

   $$C = \frac{Q}{V}.$$

   The capacitance of a parallel-plate capacitor is given by

   $$C = \frac{\varepsilon_0 A}{d},$$

   where A is the plate area and d the spacing. The capacitance of other configurations may be determined by assuming a charge, computing the associated potential difference, and taking the ratio $C = Q/V$.

   b. The energy stored in a capacitor depends on the capacitance and on the square of the potential difference:

   $$U = \tfrac{1}{2}CV^2.$$

   c. Capacitors may be connected in **series** or **parallel**. The capacitances of parallel capacitors add:

   $$C = C_1 + C_2 + C_3 + \cdots. \qquad \text{(parallel capacitors)}$$

   The capacitances of series capacitors add reciprocally:

   $$\frac{1}{C} = \frac{1}{C_1} + \frac{1}{C_2} + \frac{1}{C_3} + \cdots. \qquad \text{(series capacitors)}$$

   d. The **working voltage** of a capacitor is the maximum potential difference that should be applied across the capacitor without risk of dielectric breakdown in the insulating material.

   e. The **dielectric constant**, κ, of the insulating material used in a capacitor affects the capacitance, increasing it by this factor κ.

Terms You Should Understand

(Pairs are closely related terms whose distinction is important; number in parentheses is chapter section where term first appears.)

electrostatic potential energy (26-1)
energy density (26-3)
capacitor (26-4)
capacitance (26-4)
working voltage (26-6)
series, parallel (26-6)
dielectric constant (26-7)

Symbols You Should Recognize

U, u (26-2, 26-3)
C (26-4)
κ (26-7)

Problems You Should Be Able to Solve

calculating the work needed to assemble a distribution of discrete charges (26-1)
calculating electrostatic energy by integrating electric field energy density (26-3)

determining capacitance of simple capacitor configurations (26-4)

evaluating energy stored in capacitors (26-5)

analyzing parallel and series capacitor combinations (26-6)

determining the effect of dielectric materials in capacitors (26-7)

Limitations to Keep in Mind

Equation 26-6 for a parallel-plate capacitor is an approximation that neglects fringing fields at the plate edges; the approximation is good for capacitor plates much larger than their spacing.

QUESTIONS

1. Two positive point charges are initially infinitely far apart. Is it possible, using only a finite amount of work, to move them until they are located a small distance d apart?

2. How does the energy density a certain distance from a negative point charge compare with the energy density the same distance from a positive point charge of equal magnitude?

3. A dipole consists of two equal but opposite charges. Is the total energy stored in the field of the dipole zero? Why or why not?

4. Charge is spread over the surface of a balloon. The balloon is then allowed to expand. What happens to the energy of the electric field? If it is reduced, where does it go? If it is increased, where does the extra energy come from?

5. Why doesn't the superposition principle hold for electric field energy densities? That is, if you double the field strength at some point, why don't you simply double the energy density as well?

6. A student argues that the total energy associated with the electric field of a charged sphere must be infinite because its field extends throughout an infinite volume. Criticize this argument.

7. A capacitor is said to carry a charge Q. What is the net charge on the entire capacitor?

8. Does the capacitance of a capacitor describe the maximum amount of charge it can hold, in the same way that the capacity of a bucket describes the maximum amount of water it can hold? Explain and compare the meanings of capacitance and capacity.

9. A cylinder for storing compressed gas has a fixed volume, yet knowing this volume does not tell how much gas the cylinder can hold. Why not? How is the cylinder like a capacitor? Form analogies between quantities used in describing a capacitor and the amount of gas in the cylinder, the cylinder pressure, and the maximum pressure the cylinder can withstand.

10. A capacitor of capacitance C is charged to a potential difference V and carries charge Q. Why isn't the stored energy given simply by CV^2? After all, the work required to move a charge Q against a potential difference V is QV, and $C = Q/V$, so $Q = CV$.

11. Is a force needed to hold the plates of a charged capacitor in place? Explain.

12. Why do we say that capacitance depends only on the physical configuration of conductors making up a capacitor, not on the charge or potential difference, and yet we define capacitance as $C = Q/V$?

13. Why can't useful capacitors of arbitrarily large capacitance be made by simply reducing the spacing between parallel plates?

14. A solid conducting slab is inserted between the plates of a capacitor, as shown in Fig. 26-24. Does the capacitance increase, decrease, or remain the same?

FIGURE 26-24 Question 14.

15. Why is a capacitor needed for energy storage in a camera flash? After all, the battery is the ultimate source of the flash energy, so it should be capable of supplying the needed energy.

16. Two capacitors are connected in series. Is the equivalent capacitance more or less than that of either one?

17. Two capacitors are connected in series; they have capacitances C_1 and C_2, with $C_1 > C_2$. A potential difference V is imposed across the combination. Is the voltage across the two capacitors the same? If not, which has the greater voltage? Is the charge on each the same? If not, which has the greater charge?

18. Explain why the potential differences across parallel capacitors must be the same.

19. Two capacitors are storing equal amounts of energy, yet one has twice the capacitance of the other. How do their voltages compare?

20. Explain why inserting a dielectric between capacitor plates increases the capacitance.

21. An air-insulated parallel-plate capacitor is connected to a battery that imposes a potential difference V across the capacitor. If a dielectric slab is inserted between the capacitor plates, what happens to (a) the potential difference; (b) the capacitor charge; and (c) the capacitance?

22. A capacitor is charged and left connected to the charging battery. If you insert a dielectric slab between the capacitor plates, do you do work on the slab, or does it do work on you? Explain.

PROBLEMS

Section 26-1 Energy of a Charge Distribution

1. Three point charges, each of $+q$, are moved from infinity to the vertices of an equilateral triangle of side ℓ. How much work is required?

2. Repeat the preceding problem for the case of two charges $+q$ and one $-q$.

3. Four 50-μC charges are brought from far apart onto a line where they are spaced at 2.0-cm intervals. How much work does it take to assemble this charge distribution?

4. Repeat Example 26-1 for the case when the negative charge is $-q$ rather than $-q/2$.

5. Suppose two of the charges in Problem 1 are held in place, while the third is allowed to move freely. If this third charge has mass m, what will be its speed when it's far from the other two charges?

6. To a very crude approximation, a water molecule consists of a negatively charged oxygen atom and two "bare" protons, as shown in Fig. 26-25. Calculate the electrostatic energy of this configuration, which is therefore the magnitude of the energy released in forming this molecule from widely separated atoms. Your answer is an overestimate because electrons are actually "shared" among the three atoms, spending more time near the oxygen.

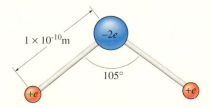

FIGURE 26-25 Problem 6.

7. Four identical charges q, initially widely separated, are brought to the vertices of a tetrahedron of side a (Fig. 26-26). Find the electrostatic energy of this configuration.

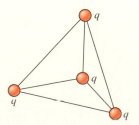

FIGURE 26-26 Problem 7 (All sides have equal length a).

8. A charge Q_0 is at the origin. A second charge, $Q_x = 2Q_0$, is brought to the point $x = a$, $y = 0$. Then a third charge Q_y is brought to the point $x = 0$, $y = a$. If it takes twice as much work to bring in Q_y as it did Q_x, what is Q_y in terms of Q_0?

Section 26-2 Two Isolated Conductors

9. Two square conducting plates 25 cm on a side and 5.0 mm apart carry charges ± 1.1 μC. Find (a) the electric field between the plates, (b) the potential difference between the plates, and (c) the stored energy.

10. Two square conducting plates measure 5.0 cm on a side. The plates are parallel, spaced 1.2 mm apart, and initially uncharged. (a) How much work is required to transfer 7.2 μC from one plate to the other? (b) How much work is required to transfer a second 7.2 μC?

11. (a) How much charge must be transferred between the initially uncharged plates of the preceding problem in order to store 15 mJ of energy? (b) What will be the potential difference between the plates?

12. Two parallel, circular metal plates of 15 cm radius are initially uncharged. It takes 6.3 J to transfer 45 μC from one plate to the other. How far apart are the plates?

13. A conducting sphere of radius a is surrounded by a concentric spherical shell of radius b. Both are initially uncharged. How much work does it take to transfer charge from one to the other until they carry charges $\pm Q$?

14. Show that the energy given by Equation 26-2 can be written as the product of the charge transferred with the *average* value of the potential during the transfer.

15. Two conducting spheres of radius a are separated by a distance $\ell \gg a$; since the distance is large, neither sphere affects the other's electric field significantly, and the fields remain spherically symmetric. (a) If the spheres carry equal but opposite charges $\pm q$, show that the potential difference between them is $2kq/a$. (b) Write an expression for the work dW involved in moving an infinitesimal charge dq from the negative to the positive sphere. (c) Integrate your expression to find the work involved in transferring a charge Q from one sphere to the other, assuming both are initially uncharged.

Section 26-3 Energy and the Electric Field

16. The energy density in a uniform electric field is 3.0 J/m³. What is the field strength?

17. A car battery stores about 4 MJ of energy. If all this energy were used to create a uniform electric field of 30 kV/m, what volume would it occupy?

18. Air undergoes dielectric breakdown at a field strength of 3 MV/m. Could you store energy in a uniform electric field in air with the same energy density as that of liquid gasoline? (See Appendix C.)

19. Find the electric field energy density at the surface of a proton, taken to be a uniformly charged sphere 1 fm in radius.

20. A pair of closely spaced square conducting plates measure 10 cm on a side. The electric field energy density between the plates is 4.5 kJ/m³. What is the charge on the plates?

21. The electric field strength as a function of position x in a certain region is given by $E = E_0(x/x_0)$, where $E_0 = 24$ kV/m and $x_0 = 6.0$ m. Find the total energy stored in a cube 1.0 m on a side, located between $x = 0$ and $x = 1.0$ m. (The field strength is independent of y and z.)

22. A sphere of radius R contains charge Q spread uniformly throughout its volume. Find an expression for the electrostatic energy contained within the sphere itself. *Hint:* Consult Example 24-1.

23. A sphere of radius R carries a total charge Q distributed over its surface. Show that the total energy stored in its electric field is $U = kQ^2/2R$.

24. A uranium-235 nucleus contains 92 protons and 143 neutrons, and has a diameter of 6.6 fm. Assuming that the proton charge is distributed uniformly throughout the nucleus, calculate the total electrostatic energy of this configuration. *Hint:* See the preceding two problems.

25. Two 4.0-mm-diameter water drops each carry 15 nC. They are initially separated by a great distance. Find the change in the electrostatic potential energy if they are brought together to form a single spherical drop. Assume all charge resides on the drops' surfaces.

26. A 2.1-mm-diameter wire carries a uniform line charge density $\lambda = 28$ μC/m. How much energy is contained in a space 1.0 m long within one wire diameter of the wire surface?

27. A long, solid rod of radius a carries uniform volume charge density ρ. Find an expression for the electrostatic energy per unit length contained *within* the rod. *Hint:* See Problem 24-31.

Sections 26-4 Capacitors

28. A capacitor's plates hold 1.3 μC when charged to 60 V. What is its capacitance?

29. The "memory" capacitor in a VCR has a capacitance of 4.0 F and is charged to 3.5 V. What is the charge on its plates?

30. What voltage is needed to put 1.6 mC on a 100-μF capacitor?

31. Figure 26-27 shows data from an experiment in which known amounts of charge are placed on a capacitor and the resulting voltage measured. Fit a line to the data, and use it to determine the capacitance.

FIGURE 26-27 Problem 31 (data plot).

32. Show that the units of ε_0 may be written as F/m.

33. Find the capacitance of a parallel-plate capacitor consisting of circular plates 20 cm in radius separated by 1.5 mm.

34. A parallel-plate capacitor with 1.1-mm plate spacing has ± 2.3 μC on its plates when charged to 150 V. What is the plate area?

35. Find the capacitance of a 1.0-m-long piece of coaxial cable whose inner conductor radius is 0.80 mm and whose outer conductor radius is 2.2 mm, with air in between.

36. A capacitor consists of a conducting sphere of radius a surrounded by a concentric conducting shell of radius b. Show that its capacitance is $C = \dfrac{ab}{k(b - a)}$.

37. Figure 26-28 shows a capacitor consisting of two electrically connected plates with a third plate between them, spaced so its surfaces are a distance d from the other plates. The plates have area A. Neglecting edge effects, show that the capacitance is $2\varepsilon_0 A/d$.

FIGURE 26-28 Problem 37.

Section 26-5 Energy Storage in Capacitors

38. The power supply of a stereo receiver contains a 2500-μF capacitor charged to 35 V. How much energy does it store?

39. Find the capacitance of a capacitor that stores 350 μJ when the potential difference across its plates is 100 V.

40. A certain capacitor stores 40 mJ of energy when charged to 100 V. (a) How much would it store when charged to 25 V? (b) What is its capacitance?

41. Which can store more energy, a 1-μF capacitor rated at 250 V or a 470 pF capacitor rated at 3 kV?

42. A circuit application calls for a 10-μF capacitor that can store 12 mJ. What should be its voltage rating? The capacitors are available with voltage ratings that are multiples of 25 V.

43. A 0.01-μF, 300-V capacitor costs 25¢, a 0.1-μF, 100-V capacitor costs 35¢, and a 30-μF, 5-V capacitor costs 88¢. (a) Which can store the most charge? (b) Which can store the most energy? (c) Which is the most cost-effective energy storage device, as measured by energy stored per unit cost?

44. A medical defibrillator stores 950 J of energy in a 100-μF capacitor. (a) What is the voltage across the capacitor? (b) If the capacitor discharges 300 J of its stored energy in 2.5 ms, what is the power delivered during this time?

45. A camera flashtube requires 5.0 J of energy per flash. The flash duration is 1.0 ms. (a) What is the power used by the flashtube *while it is actually flashing*? (b) If the flashtube operates at 200 V, what size capacitor is needed to supply the flash energy? (c) If the flashtube is fired once every 10 s, what is its *average* power consumption?

46. The NOVA laser fusion experiment at Lawrence Livermore Laboratory in California can deliver 10^{14} W (roughly 100 times the output of all the world's power plants) of light energy when its lasers are on. But the laser pulse lasts only 10^{-9} s. (a) How much energy is delivered in one pulse? (b) The capacitor bank supplying this energy has a total capacitance of 0.26 F. Only about 0.17% (i.e., 0.0017) of the capacitor energy actually appears as light. To what voltage must the capacitor bank be charged?

47. A solid conducting slab is inserted between the plates of a charged capacitor, as shown in Fig. 26-29. The slab thickness is 60% of the plate spacing, and its area is the same as the plates'. (a) What happens to the capacitance? (b) What happens to the stored energy, assuming the capacitor is not connected to anything?

FIGURE 26-29 Problem 47.

48. Consider the two widely separated spheres of Problem 15 as a capacitor. Use energy considerations (i.e., the equation $U = \frac{1}{2}CV^2$ applies to *any* capacitor) and the answers to Problem 15 to find the capacitance.

49. The cylindrical capacitor of Example 26-5 is charged to a voltage V. Obtain an expression for the energy density as a function of radial position in the capacitor, and integrate to show explicitly that the stored energy is $\frac{1}{2}CV^2$.

Section 26-6 Connecting Capacitors

50. You have a 1.0-μF and a 2.0 μF capacitor. What values of capacitance could you get by connecting them in series or parallel?

51. Two capacitors are connected in series and the combination charged to 100 V. If the voltage across each capacitor is 50 V, how do their capacitances compare?

52. (a) What is the equivalent capacitance of the combination shown in Fig. 26-30? (b) If a 100-V battery is connected across the combination, what is the charge on each capacitor? (c) What is the voltage across each?

FIGURE 26-30 Problems 52.

53. You're given three capacitors: 1.0 μF, 2.0 μF, and 3.0 μF. Find (a) the maximum, (b) the minimum, and (c) two intermediate values of capacitance you could achieve with various combinations of all three capacitors.

54. What is the equivalent capacitance of the four identical capacitors in Fig. 26-31, measured between A and B?

FIGURE 26-31 Problem 54.

55. You have an unlimited supply of 2.0-μF, 50-V capacitors. Describe combinations that would be equivalent to (a) a 2.0-μF, 100-V capacitor and (b) a 0.50-μF, 200-V capacitor.

56. Repeat the derivations for parallel and series capacitors, now using combinations of three capacitors.

57. What is the equivalent capacitance in Fig. 26-32?

FIGURE 26-32 Problems 57, 58.

58. In Fig. 26-32, find the energy stored in the 1-μF capacitor when a 50-V battery is connected between points A and B.

59. Two capacitors C_1 and C_2 are in series, with a voltage V across the combination. Show that the voltages across the individual capacitors are

$$V_1 = \frac{C_2 V}{C_1 + C_2} \quad \text{and} \quad V_2 = \frac{C_1 V}{C_1 + C_2}.$$

60. A 0.10-μF capacitor rated at 50 V is in series with a 0.20-μF capacitor rated at 200 V. What is the maximum voltage that should be applied across the series combination? *Hint:* See the preceding problem.

61. A variable "trimmer" capacitor used to make fine adjustments has a capacitance range from 10 to 30 pF. The trimmer is in parallel with a capacitor of about 0.001 μF. Over what percentage range can the capacitance of the combination be varied?

62. Capacitors are often marked with a nominal value for the capacitance and a tolerance range within which the actual capacitance lies. For example, a 1-μF, $\pm$20% capacitor has capacitance between 0.8 μF and 1.2 μF. If you connect a 0.01-μF $\pm$ 20% capacitor in series with a 0.02 μF $\pm$ 30% capacitor, in what range will the resulting capacitance lie? Express as a capacitance and its associated tolerance.

63. A 5.0-μF capacitor is charged to 50 V, and a 2.0-μF capacitor is charged to 100 V. The two are disconnected from their charging batteries and connected in parallel, positive to positive. (a) What is the common voltage across each after they are connected? *Hint:* Charge is conserved. (b) Compare the total electrostatic energy before and after the capacitors are connected. Speculate on the discrepancy.

Section 26-7 Capacitors and Dielectrics

64. A parallel-plate capacitor has plates with 50 cm^2 area separated by a 25-μm layer of polyethylene. Find (a) its capacitance and (b) its working voltage.
65. A 470-pF capacitor consists of two circular plates 15 cm in radius, separated by a sheet of polystyrene. (a) What is the thickness of the sheet? (b) What is the working voltage?
66. An electrolytic capacitor is essentially a parallel-plate configuration in which aluminum plates are separated by a thin layer of aluminum oxide created by chemical action when a voltage is applied. If the effective plate area of a 2000-μF capacitor is 2.5 m^2, what are (a) the oxide layer thickness and (b) the working voltage?
67. Repeat Problem 35 for the more realistic case of a cable insulated with polyethylene.
68. An air-insulated parallel-plate capacitor has plate area 76 cm^2 and spacing 1.2 mm. It is charged to 900 V and then disconnected from the charging battery. A plexiglass sheet is then inserted to fill the space between the plates. What are (a) the capacitance, (b) the potential difference between the plates, and (c) the stored energy both before and after the plexiglass is inserted?
69. The capacitor of the preceding problem is connected to its 900-V charging battery and left connected as the plexiglass sheet is inserted, so the potential difference remains at 900 V. What are (a) the charge on the plates and (b) the stored energy both before and after the plexiglass is inserted?
70. The first accurate estimate of the thickness of biological cell membranes used a capacitive technique, in which the capacitance per unit area of cell membrane was determined through a macroscopic measurement of the electrical properties of a suspension of cells; the result was a value of about 1 μF/cm^2 for a wide range of cells. Assuming a dielectric constant of about 3 for the membrane material, find the membrane thickness. (Your answer is the thickness of the bipolar lipid layer alone, and is lower by a factor of about 3 than values based on x-ray techniques; the full membrane may be thicker still.)

Paired Problems

(Both problems in a pair involve the same principles and techniques. If you can get the first problem, you should be able to solve the second one.)

71. A pair of parallel conducting plates of area 0.025 m^2 carrying equal but opposite charges stores 1.6 J in its electric field. When the magnitude of the charge on both plates is increased by 5.0 μC, the stored energy increases to 2.4 J. Find the plate separation.
72. A capacitor stores 50 mJ of energy at voltage V_0. When the voltage is increased by 150 V, the stored energy increases to 75 mJ. Find the capacitance.
73. A 20-μF air-insulated parallel-plate capacitor is charged to 300 V. The capacitor is then disconnected from the charging battery, and its plate separation is doubled. Find the stored energy (a) before and (b) after the plate separation increases. Where does the extra energy come from?
74. Repeat the preceding problem, except that now the capacitor remains connected to the 300-V battery while the plates are separated.
75. In the capacitor network of Fig. 26-33, take $C = 6.0$ μF. Find (a) the equivalent capacitance between A and B and (b) the charge on C when 30 V is applied between A and B.

FIGURE 26-33 Problems 75, 76.

76. Take C in Fig. 26-33 as an unknown capacitance. If 100 V is applied between A and B, the network stores 5.8 mJ of energy. Find C.

Supplementary Problems

77. A typical lightning flash transfers 30 C across a potential difference of 30 MV. Assuming such flashes occur every 5 s in the thunderstorm of Example 26-2, roughly how long could the storm continue if its electrical energy were not replenished?
78. A capacitor is constructed from a "sandwich" consisting of two long strips of aluminum foil each 2.0 cm wide and 1.6 m long, separated by two strips of 5.0-μm-thick polyethylene (Fig. 26-34). The capacitor is rolled up to make a compact cylinder. Find its capacitance. *Hint:* Because the strips are thin and closely spaced, you can treat this as a parallel-plate capacitor. But note that each foil layer in the rolled-up capacitor "sees" an oppositely charged layer on *both* sides.

FIGURE 26-34 Problem 78.

79. Six charges $\pm q$, initially widely separated, are positioned to form a hexagon of side a, as shown in Fig. 26-35. What is the electrostatic energy of this configuration?

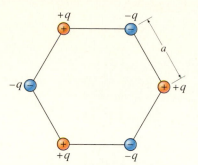

FIGURE 26-35 Problem 79.

80. Show that the result of Problem 36 reduces to that of a parallel-plate capacitor when the separation $b - a$ is much less than the radius a.

81. An air-insulated parallel-plate capacitor of capacitance C_0 is charged to voltage V_0 and then disconnected from the charging battery. A slab of material with dielectric constant κ, whose thickness is essentially equal to the capacitor spacing, is then inserted halfway into the capacitor (Fig. 26-36). Determine (a) the new capacitance, (b) the stored energy, and (c) the force on the slab in terms of C_0, V_0, κ, and the capacitor plate length L.

FIGURE 26-36 Problem 81.

82. Repeat parts (b) and (c) of the preceding problem, now assuming the battery remains connected while the slab is inserted.

83. We live inside a giant capacitor! Its plates are Earth's surface and the ionosphere, a conducting layer of the atmosphere beginning at about 60 km altitude. (a) What is its capacitance? *Hint:* You can treat it as either a spherical or a parallel-plate capacitor. Why? (b) The potential difference between Earth and ionosphere is about 6 MV. Find the total energy stored in this planetary capacitor.

84. Show that the result of Example 26-5 reduces to that of a parallel-plate capacitor when the separation $b - a$ is much less than the radius a. *Hint:* See Appendix A for an approximation to the logarithm.

85. Equation 26-2 gives the potential energy of a pair of oppositely charged plates. (a) Differentiate this expression with respect to the plate spacing to find the magnitude of the attractive force between the plates. (b) Compare with the answer you would get by multiplying one plate's charge by the electric field between the plates. Why do your answers differ? Which is right?

86. A solid sphere contains a uniform volume charge density. What fraction of the total electrostatic energy of this configuration is contained *within* the sphere?

87. A small dipole lies on the x axis, centered at the origin. Find an expression for the total electrostatic energy contained in a thin cylindrical volume of diameter d and length ℓ, with its left end a distance ℓ from the dipole center, as shown in Fig. 26-37. Assume that ℓ is much greater than the dipole spacing. *Hint:* Since the cylinder is very thin, you can use the on-axis dipole field (Equation 23-5b) for the field throughout the cylinder.

FIGURE 26-37 Problem 87.

88. A coaxial cable 15 m long consists of an inner conductor 1.0 mm in radius and an outer conductor 3.0 mm in radius, separated by polyethylene insulation. What is the electrostatic energy contained within this cable when a potential difference of 300 V is applied between its two conductors?

89. A TV antenna cable consists of two 0.50-mm-diameter wires spaced 12 mm apart. Estimate the capacitance per unit length of this cable, neglecting dielectric effects of the insulation.

90. A classical view of the electron pictures it as a purely electrical entity, whose rest mass energy mc^2 (see Section 8-7) is the energy stored in its electric field. If the electron were a sphere with charge distributed uniformly over its surface, what radius would it have to satisfy this condition? (Your answer for the electron's "size" is not consistent with modern quantum mechanics nor with experiments that suggest the electron is a true point particle.)

91. Use the fact that the static electric field is conservative to argue that there *must* be fringing field at the edges of a parallel plate capacitor. *Hint:* Remember that the plates are equipotentials, and consider the potential differences V_{AB} and V_{CD} in Fig. 26-38. What does your argument say about the strength of the fringing field relative to the field between the plates?

FIGURE 26-38 Problem 91.

27 Electric Current

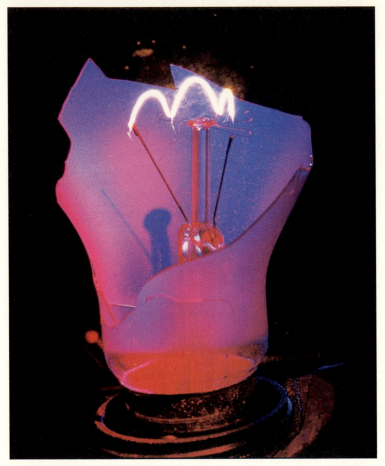

Electrical energy is converted to heat and light as a current of electrons flows through a lightbulb's filament.

So far our discussion of electrical phenomena has been based on the assumption of electrostatic equilibrium. We now relax that assumption, and instead of considering charges at rest, we'll consider situations in which charges are moving. Such motion usually occurs only in materials containing free charges, so our discussion will emphasize electrical conductors.

27-1 ELECTRIC CURRENT

Moving electric charge constitutes **electric current**. Electric currents are essential to many technological and natural processes. Currents flowing in light bulbs, toasters, and electric stoves produce heat and light. Currents in electric motors run our computers' disk drives, our refrigerators and washing machines, subway and railroad locomotives, and, increasingly, even advanced-technology electric cars. Your computer can plot a graph for your physics lab or search for a word in your English paper; fundamentally, these processes involve electric currents within the microchips at the heart of computers and other modern electronic devices. Currents in your body help your brain process information, regulate your heartbeat, and tell your muscles to move. Electric eels sense and kill their prey with electric currents they send through the surrounding water. A lightning strike is a strong but brief flow of current through the atmosphere. Currents in Earth's liquid outer core produce the planetary magnetic field that protects us from cosmic radiation, and currents in the ionized gases making up the stars greatly influence stellar behavior.

The measure of current is the net rate at which charge crosses a given area. Accordingly, its units are coulombs per second (C/s). This unit is given the special name **ampere** (A) after the French physicist André Marie Ampère:

$$1 \text{ A} = 1 \text{ C/s}.$$

In electronics, biomedical applications, and many other practical situations currents are small enough that the milliampere (mA) and microampere (μA) are frequently used. When the current I is steady we write

$$I = \frac{\Delta q}{\Delta t}, \qquad \text{(steady current)} \qquad (27\text{-}1a)$$

where Δq is the net charge crossing the given area in time Δt. If the current is not steady, we consider the ratio of charge to time for arbitrarily small time intervals, giving an instantaneous current that may vary with time:

$$I = \frac{dq}{dt}. \qquad \text{(instantaneous current)} \qquad (27\text{-}1b)$$

The direction of current is the direction in which *positive* charge flows. If the moving charge is negative, as with electrons in a metal, then the current is opposite the charge motion. You can blame Benjamin Franklin for this confusing situation! It was Franklin who assigned the names "positive" and "negative" to the two kinds of electric charge. Had Franklin known that free charges in metals are electrons, he might well have reversed his terminology.

An electric current may consist of only one sign of charge in motion, or it may involve both positive and negative charge. In that case the current is determined by the *net* charge motion—that is, by the algebraic sum of the currents associated with both kinds of charge (Fig. 27-1). That's why the bulk motion of a neutral object—even though it contains many coulombs of positive and of negative charge—does not constitute a current (Fig. 27-1b).

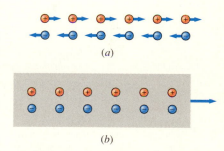

FIGURE 27-1 (*a*) Protons moving to the right constitute a current to the right. Since electrons are negative, *leftward* moving electrons also constitute a current to the *right*. The result in this case is a net current to the right. (*b*) In this bulk motion of neutral matter, the protons constitute a current to the right. But the negative electrons, also moving to the right, constitute a current to the *left*. So the net current is zero. Bulk motion of a *charged* object would, however, constitute a current.

Got It!

In which of the following situations is there a nonzero net current, and in which direction is that current flowing? (a) A beam of electrons moves from left to right; (b) a beam of protons moves upward; (c) in a solution, positive ions move to the left and negative ions to the right; (d) blood, carrying positive and negative ions at the same velocity, flows upward through a vein in your leg; (e) a metal car with no net charge moves westward on a highway.

A Microscopic Look at Current

FIGURE 27-2 A conductor of cross-sectional area A containing n charge carriers per unit volume. Each moves with speed v_d and carries charge q. The total charge in a region of length ℓ is $nA\ell q$, and the current is therefore $nAqv_d$. Although we show positive charge moving, in the most common conductors—metals—charges carrying the current are actually negative electrons.

Current —the rate at which charge crosses a given area—depends on the speed of the charge carriers, their density, and the charge carried by each. Consider a conductor containing n charges per unit volume, each carrying charge q and moving with speed v_d, as shown in Fig. 27-2. The quantity n is called the **number density**, and v_d is the **drift speed**. In some cases—a beam of electrons, for example—v_d is the actual particle speed. More commonly, v_d represents the time-average speed of the charge carriers, averaging out the effects of much faster but random thermal motion and collisions. If A is the conductor's cross-sectional area, then a length ℓ has volume $A\ell$ and therefore contains $nA\ell$ charges. Since each carries charge q, the total charge is $\Delta Q = nA\ell q$. With drift speed v_d, the length ℓ of charge moves past a given point in time $\Delta t = \ell/v_d$, so the current is

$$I = \frac{\Delta Q}{\Delta t} = \frac{nA\ell q}{\ell/v_d} = nAqv_d. \tag{27-2}$$

EXAMPLE 27-1   Current in a Wire

A copper wire with a cross-sectional area of 1.0 mm² carries a current of 5.0 A. The charge carriers in copper are electrons, and each copper atom contributes, on average, 1.3 free electrons. What is the drift speed of the electrons?

Solution
The density of copper (see inside back cover) is 8920 kg/m³, and the periodic table (also inside back cover) lists copper's atomic weight as 63.55—meaning that the mass per atom is 63.55 u. So the number density of copper atoms is

$$n = \frac{8920 \text{ kg/m}^3}{(63.55 \text{ u/atom})(1.66\times10^{-27} \text{ kg/u})}$$

$$= 8.46\times10^{28} \text{ atoms/m}^3.$$

Since each atom contributes 1.3 free electrons, the electron density is $(1.3)(8.46\times10^{28} \text{ m}^{-3}) = 1.10\times10^{29} \text{ m}^{-3}$. Solving Equation 27-2 for v_d then gives

$$v_d = \frac{I}{nAq} = \frac{5.0 \text{ A}}{(1.10\times10^{29} \text{ m}^{-3})(1.0\times10^{-6} \text{ m}^2)(1.6\times10^{-19}\text{C})}$$

$$= 0.284 \text{ mm/s}.$$

This remarkably small value is typical of drift speeds in metallic conductors.

EXERCISE A thin layer of gold, 0.10 mm wide and 3.2 μm thick, is used for connections to an integrated circuit. Find (a) the free electron density in gold and (b) the drift speed of electrons in the gold layer when it carries a current of 140 μA. The density of gold is 1.93×10^4 kg/m³, and each gold atom contributes, on average, 1.5 free electrons.

Answers: (a) 8.85×10^{28} m^{-3}; (b) 31 μm/s

Some problems similar to Example 27-1: 8, 55, 56

How can the drift speed be so small? When you turn on a light switch, the light comes on immediately, not several thousand seconds later as the result of Example 27-1 might imply. Here it's important to distinguish between the speed of the electrons and that of the electrical signal in the wire. As soon as electrons at one end of the wire begin moving, their electric fields affect adjacent electrons, which also begin moving. This effect propagates down the wire at what is in fact nearly the speed of light, so the current begins everywhere almost simultaneously. The same thing happens when you turn on a garden hose full of water: Water comes out the far end even though water at the faucet has not had time to travel down the hose.

Current Density

In many cases electric currents are not neatly confined in a wire. Examples include currents in Earth, in the atmosphere, in chemical solutions, and in the ionized gases that make up the stars and indeed much of the matter in the universe (Fig. 27-3). In these situations the current is spread over a rather ill-defined area, and its magnitude and direction may vary from point to point. It's useful to characterize such diffuse currents in terms of **current density**, J, defined as the current per unit area at a given point. Dividing Equation 27-2 by the area gives

$$J = \frac{I}{A} = nqv_d. \tag{27-3a}$$

The flow of charge can vary in both magnitude and direction, so current density is more generally written as a vector quantity:

$$\mathbf{J} = nq\mathbf{v}_d, \tag{27-3b}$$

where $\mathbf{v}_d$ is the drift velocity.

If current density is uniform, as in a wire, then the total current is simply the product of the current density with the wire's cross-sectional area. When the current density varies with position, then integration is necessary to calculate the total current. Problem 65 explores this case.

FIGURE 27-3 Strong electric currents flow in the bright yellow loops of ionized gas that arch above the Sun's surface in this image made with an orbiting x-ray telescope on the Japanese satellite Yohkoh.

EXAMPLE 27-2	*Through the Cell Membrane*

An ion channel is a narrow pore, formed from a special protein molecule, that allows specific ions to pass through a biological cell membrane (Fig. 27-4). A particular channel carries a current of 1.8 pA, with the charges being singly ionized potassium (K^+). The channel has a circular cross section 0.30 nm in diameter. (a) If the channel is open for 1.0 ms, how many ions pass through it? (b) What is the magnitude of the current density in the channel?

Ion channels

~0.3 nm
Lipid molecules

FIGURE 27-4 Diagram of a cell membrane, showing two ion channels. An ion current is passing through the frontmost channel, shown in cutaway. The current is driven in part by the electric field associated with the charge distribution on the lipid molecules that make up the membrane.

Solution

(a) A current of 1.8 pA means 1.8 pC/s, so the total charge passing through the channel in 1.0 ms is $(1.8 \text{ pC/s})(1.0 \text{ ms}) = 1.8$ fC, where we use the SI prefixes $p = 10^{-12}$, $m = 10^{-3}$, and $f = 10^{-15}$. Each ion carries one elementary charge, so the

number of ions in 1 ms is

$$N = \frac{1.8 \times 10^{-15} \text{ C}}{1.6 \times 10^{-19} \text{ C/ion}} = 1.1 \times 10^4 \text{ ions} .$$

(b) Current density is current per unit area, so

$$J = \frac{I}{A} = \frac{I}{\pi r^2} = \frac{1.8 \times 10^{-12} \text{ A}}{(\pi)(0.15 \times 10^{-9} \text{ m})^2} = 2.5 \times 10^7 \text{ A/m}^2 .$$

This value is about four times the maximum safe current density in copper wire used in household wiring! (See the exercise with this example.) Of course, the ion channel is so narrow that the total current is minuscule.

Tip

Don't Just Plug In! Do you search for an equation as soon as you start a problem? Think first! If you really know what the concepts mean, you may not always need equations. Here we solved the problem entirely by knowing the meanings of the terms *current* and *current density*.

EXERCISE The maximum safe current density in copper wire used in household wiring is about 6 MA/m². (Beyond this level the wire overheats.) What is the minimum safe wire diameter in a circuit that can carry up to 15 A?

Answer: 1.8 mm

Some problems similar to Example 27-2: 2, 5–7, 10, 11

27-2 CONDUCTION MECHANISMS

What causes electric current? Electric charges experience forces in electric fields, so applying a field to a conductor should result in a current. (The simplest way to produce such a field is to connect a battery—a device that provides a fixed potential difference between two points—across the conductor.) You might think that it would suffice to apply the field briefly to get the charges moving; Newton's law suggests they would then keep moving. In most conductors, however, charges do not move unimpeded. They bump into things—usually ions—and quickly lose any energy they've gained from the field. To sustain a current in most materials, it is therefore necessary to maintain an electric field within the material. Having such a field does not violate our conclusion that there can be no electric field in a conductor in electrostatic equilibrium since we're explicitly considering moving charges and are therefore no longer talking about equilibrium.

Although it is generally true that electric fields result in currents, the detailed relation between current and field depends on the type of conductor. In addition to metals, important conductors include ionic solutions, plasmas (ionized gases), semiconductors, and superconductors, all of which we will consider briefly in this section.

In most materials the current density and the electric field are in the same direction, and we can therefore characterize the relation between the two using the equation

$$\mathbf{J} = \sigma \mathbf{E}, \tag{27-4a}$$

where σ is called the **conductivity** of the material. Equation 27-4a shows that conductivity is the ratio of the magnitude of the current density to the magnitude of the electric field.

Ohm's Law: A Microscopic View

For many common conductors, including metals, the conductivity σ is independent of the electric field. Such materials are called **ohmic**, and for these materials Equation 27-4a is one way of stating what is known as **Ohm's law**, namely that the current density at any point in a conductor is proportional to the electric field at that point. In **nonohmic** materials, in contrast, conductivity does depend on field, and therefore $\mathbf{J}$ and $\mathbf{E}$ are not directly proportional.

You may already be familiar with the *macroscopic* version of Ohm's law, which makes the equivalent statement that the total current through a piece of material is proportional to the potential difference, or voltage, across it. In Equation 27-4a we have the *microscopic* version of Ohm's law, which describes what happens at a single point in a material in terms of the electric field and current density at that point. The macroscopic version, in contrast, links the current (a quantity related to current density) and voltage (related, as we saw in Chapter 25, to the electric field) associated with an entire piece of material. We'll derive the macroscopic version in section 27-3, and will find it very useful for analyzing electric circuits. But the microscopic version remains important for many applications in biophysics, geophysics, astrophysics, semiconductor engineering, and other areas where electric fields in conducting materials may vary with position, and where we therefore need to know what's going on at each point.

The conductivity σ tells how large a current density will result from a given electric field; it's thus a measure of how easy it is for an electric field to cause free charges in a material to move. A perfect conductor would have $\sigma = \infty$; a perfect insulator, $\sigma = 0$. A lot of subtle physics goes into determining the conductivity, as we'll soon see for metallic conductors. An equivalent quantity for describing the relation between current density and electric field in a material is the material's **resistivity**, ρ, defined as the reciprocal of the conductivity:

$$\rho = \frac{1}{\sigma},$$

so Equation 27-4a can also be written

$$\mathbf{J} = \frac{\mathbf{E}}{\rho}. \tag{27-4b}$$

Where conductivity tells how *easy* it is for a field to cause current in a material, its reciprocal, resistivity, tells how *hard* it is. Thus the larger the resistivity, the greater the electric field needed to produce a given current density. You may be familiar with electrical *resistance,* which we'll soon see is an electrical property of specific objects that relates to the underlying *resistivity* of the materials they're made from.

TABLE 27-1 *Resistivities*

MATERIAL	RESISTIVITY ($\Omega \cdot m$)
Metallic conductors (at 20°C)	
Aluminum	2.65×10^{-8}
Copper	1.68×10^{-8}
Gold	2.24×10^{-8}
Iron	9.71×10^{-8}
Mercury	9.84×10^{-7}
Silver	1.59×10^{-8}
Ionic solutions (in water at 18°C)	
1-molar copper sulfate ($CuSO_4$)	3.9×10^{-4}
1-molar hydrochloric acid (HCl)	1.7×10^{-2}
1-molar sodium chloride (NaCl)	1.4×10^{-4}
Water, pure (H_2O)	2.6×10^5
Blood, human	0.70
Sea water (typical value)	0.22
Semiconductors (pure, at 20°C)	
Germanium	0.45
Silicon	640
Insulators	
Ceramics	10^{11}–10^{14}
Glass	10^{10}–10^{14}
Polystyrene	10^{15}–10^{17}
Rubber	10^{13}–10^{16}
Wood (dry)	10^8–10^{14}

Equation 27-4b shows that the units of resistivity are V·m/A. One V/A is given the name **ohm** (symbol Ω), after the German physicist Georg Ohm (1789–1854), whose experiments clarified the relation between voltage and current. Thus the SI units of resistivity can be written Ω·m. Reciprocally, the units of conductivity are $(\Omega \cdot m)^{-1}$.

Conductivity and resistivity vary dramatically among different materials; indeed, their measurable range is one of the broadest of any physical quantity, spanning some 24 orders of magnitude. Table 27-1 lists some typical resistivities. In addition to telling us how well materials conduct electricity, measurement of resistivity or conductivity can also provide information on structure or composition. Some devices for routine blood analysis, for example, make use of conductivity measurements, while measurement of blood conductivity in laboratory animals has been used experimentally to assess changes in metabolic activity.

EXAMPLE 27-3	*Household Wiring: The Electric Field*

A 1.8-mm-diameter copper wire carries 15 A to a household appliance. What is the electric field in the wire?

Solution

The current density is $J = I/A$; using this result in Equation 27-4b and solving for E gives

$$E = \frac{I\rho}{A} = \frac{(15 \text{ A})(1.68 \times 10^{-8} \ \Omega \cdot m)}{(\pi)(0.90 \times 10^{-3} \text{ m})^2} = 99 \text{ mV/m} ,$$

where we found the resistivity of copper in Table 27-1. This result is much smaller than the fields we've been discussing in electrostatic situations. Because copper is such a good conductor, only a very small field is needed to drive even a substantial current. In analyzing electric circuits we will often make the approximation that the fields and therefore potential differences in copper wires are essentially zero.

EXERCISE A uniform electric field of 0.76 V/m drives a 10-A current in an iron wire. Find the wire diameter.

Answer: 1.3 mm

Some problems similar to Example 27-3: 17–19

Conduction in Metals

Metals are good conductors because they contain large numbers of free electrons, which respond readily to electric fields. Each atom in the metal contributes, on average, about one electron to this "sea" of free electrons. The resulting positive ions combine in a regular structure, called the crystal lattice (Fig. 27-5). In the absence of an electric field, electrons move randomly through the lattice at high speeds of typically 10^6 m/s. They collide frequently with the ions, bouncing off in random directions. There's no net flow of electrons in any one direction, so there's no electric current. And because the whole system is in thermal equilibrium, there's no net energy transfer from electrons to ions or vice versa.

We'll use this simple description of a metal to develop an understanding of what happens when we apply an electric field to a metal—i.e., the conduction process. Although this will help you see why metals obey Ohm's law, be aware that our explanation is incomplete; a full description of metallic conduction necessarily involves quantum mechanics.

Applying an electric field to a metal gives the negative electrons an acceleration in the direction opposite the field. But the electrons are already whizzing around at high speed, colliding frequently with the ions. So an electron doesn't gain much energy or speed from the electric field before it collides with an ion. When it does, it gives up whatever energy it gained from the field, and rebounds with a random velocity. But the field is still there, and the electron again accelerates (Fig. 27-6). The result is that the electrons acquire an additional, very small average velocity in the direction opposite the field. This is the drift velocity $\mathbf{v}_d$ that we introduced earlier. Unlike their random thermal velocities, all electrons share a common drift velocity, and thus their motion constitutes a net current that is proportional to $\mathbf{v}_d$.

The electrons' behavior in metallic conduction is somewhat like driving through stop-and-go city traffic; you spend a lot of time accelerating, but soon you have to stop again. As a result, you acquire a rather slow average speed through the city. But this analogy is imperfect: your car accelerates from rest at each red light, while the electrons' acceleration introduces only a minuscule velocity change compared with their already high thermal velocity.

What happens to the energy the electrons gain from the field? They give it up in collisions with the ions, ultimately increasing the internal energy of the material. We saw in Chapter 21 that temperature is a measure of internal energy, so the metal's temperature increases when a current flows through it. That's the ultimate cause of heating in electric stoves, hair dryers, water heaters, toasters, and myriad other everyday devices.

So now we have the electrons drifting slowly in a direction opposite the electric field, with velocity $\mathbf{v}_d$. That drift velocity depends on two things: the acceleration of the electrons and the rate at which they undergo collisions. The electric force $-e\mathbf{E}$ gives an acceleration $\mathbf{a} = \mathbf{F}/m = -e\mathbf{E}/m$, with m the electron mass and $-e$ its charge. If we pick a random electron and ask when it last underwent a collision, the answer will be some mean time τ called the **collision time**. During that time the electron has been accelerating with acceleration $\mathbf{a} = -e\mathbf{E}/m$, and therefore its average velocity due to the presence of the electric field—i.e., the drift velocity $\mathbf{v}_d$—will be

$$\mathbf{v}_d = -\frac{e\mathbf{E}}{m}\tau.$$

FIGURE 27-5 Atoms of a metal form a regular structure, or crystal lattice. This image, made with a scanning tunneling microscope that uses the quantum mechanical behavior of electrons to image individual atoms, shows the arrangement of atoms at the surface of a metallic nickel crystal.

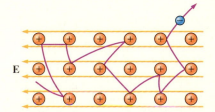

FIGURE 27-6 The path of an electron moving through a metal. The motion is almost completely random, but in the presence of an electric field there is a very slight drift antiparallel to the field.

Of course the electron also has its random thermal motion, but over many collisions this averages to zero while the velocity resulting from the electric field does not. Using our expression for the drift velocity in Equation 27-3b gives

$$\mathbf{J} = nq\mathbf{v}_d = n(-e)\mathbf{v}_d = \frac{ne^2\mathbf{E}}{m}\tau.$$

Comparing this expression with Equation 27-4a shows that the conductivity is

$$\sigma = \frac{ne^2}{m}\tau. \qquad (27\text{-}5)$$

The collision time τ depends on several factors. Obviously, the spacing of ions in the crystal lattice influences how often collisions occur. Less obvious is the fact that the presence of impurities—i.e., atoms of other materials—has a dramatic influence on electron-ion collisions and thus on conductivity. And, also obviously, τ depends on how fast the electrons are moving; the faster they go, the more frequent their collisions, and the lower τ should be. The electrons have two kinds of motion: their random thermal motion and the drift velocity $\mathbf{v}_d$ acquired from the electric field. In Example 27-1 we found that a typical drift speed is on the order 1 mm/s, while thermal speeds are around 10^6 m/s. The drift speed is therefore completely negligible in determining the collision time. That means the collision time and hence the conductivity are essentially independent of the applied electric field—and that makes Equation 27-4a a linear relation for metallic conductors. In other words, metals obey Ohm's law. We stress that this conclusion is only approximate, and that Ohm's law, unlike Gauss's law, is *not* an exact, universal statement that holds everywhere and for all materials.

Although the conductivity of a metal is independent of the applied electric field, it does depend on other factors, especially temperature. From Chapter 20 we know that classical physics gives a thermal speed proportional to the square root of the temperature, so we might expect conductivity to be proportional to $1/\sqrt{T}$ and thus resistivity to $\sqrt{T}$. Experiment, however, shows that resistivity is very nearly proportional to temperature rather than to its square root. We show this relation for copper in Fig. 27-7, and explore it further in Problem 25. Here we've reached the limits of classical physics, which cannot describe fully the behavior of the free electrons in a metal. Readers of the extended version of this text will explore quantum mechanics and how it deals with electrical conduction.

FIGURE 27-7 Resistivity of copper—the inverse of conductivity—shows a nearly linear dependence on temperature, in contrast to the classical prediction of a $\sqrt{T}$ dependence.

EXAMPLE 27-4	*Conductivity and Collision Time*

Copper at room temperature contains 1.1×10^{29} free electrons per cubic meter. What is the collision time for these electrons?

Solution

Writing $\sigma = 1/\rho$ and solving Equation 27-5 for the collision time gives

$$\tau = \frac{m}{ne^2\rho}$$

$$= \frac{9.11\times10^{-31}\text{ kg}}{(1.1\times10^{29}\text{ m}^{-3})(1.6\times10^{-19}\text{ C})^2(1.68\times10^{-8}\text{ }\Omega\cdot\text{m})}$$

$$= 1.9\times10^{-14}\text{ s},$$

where we found the resistivity of copper in Table 27-1. Problem 66 uses this result to estimate the mean thermal speed of the electrons.

EXERCISE Sodium contains 2.5×10^{28} free electrons per cubic meter, and the collision time is 3.4×10^{-14}s. What is the resistivity of sodium?

Answer: $4.2\times10^{-8}\text{ }\Omega\cdot\text{m}$

Some problems similar to Example 27-4: 21, 66

| APPLICATION | *Noise in Electronic Equipment* |

Although the time-average current associated with random thermal motion of charge carriers is zero, at any given instant random fluctuations may result in more charge carriers moving in one direction than in another. The result is a very small current, whose sign and magnitude fluctuate randomly. Called thermal noise, this current may disturb or even overwhelm currents of interest in sensitive electronic instruments. Thermal motion decreases at low temperatures, and sensitive circuits such as the amplifiers in radio telescopes are often cooled to liquid helium temperatures, around 4 K (Fig. 27-8). That way, the very weak signals reaching Earth from cosmic sources are not overwhelmed by thermal noise. Ultimately thermal noise limits our ability to detect and study very weak electrical signals.

FIGURE 27-8 This 12-m radio telescope at Kitt Peak National Observatory uses amplifiers cooled to 4 K with liquid helium in order to reduce electrical noise.

Ionic Solutions

An ionic solution contains positive and negative ions that respond to an electric field by moving in opposite directions, resulting in a net current. Conductivity of the solution is limited by collisions between the ions and neutral atoms. In addition to heating the solution, some of the energy of these collisions may go into chemical reactions that store energy—as in charging a rechargable battery.

Conduction in ionic solutions is essential to life, as the flow of sodium, potassium, and other ions through cell membranes occurs through the process of ionic conduction. Electric eels set up currents in the surrounding water, using ionic conduction to sense and kill their prey (Fig. 27-9). Technological processes that use ionic conduction include electroplating of thin layers of one material onto another, and electrolysis of water to produce hydrogen and oxygen gas. Ionic conduction also plays an important role in the corrosion of metals, for example those exposed to salt solutions. And the presence of an ionic solution—sweat—increases our vulnerability to electric shock. Table 27-1 includes the resistivities of some ionic solutions. Note that these solutions are poorer conductors than metals.

FIGURE 27-9 The electric eel sets up currents in the surrounding water, and can sense the presence of nearby objects by subtle variations in conductivity. It uses larger currents to kill its prey.

Plasmas

Plasma is ionized gas that conducts because it contains free electrons and positive ions. It takes substantial energy to ionize atoms, so plasmas usually occur only in high-temperature environments. The few examples of plasmas on Earth occur in fluorescent lamps, neon signs, devices for fusion research, the ionosphere, flames, and lightning flashes (Fig. 27-10). Yet most of the matter in the universe is probably in the plasma state; the stars, in particular, are almost entirely plasma.

The electrical properties of plasma make it so different from ordinary gas that plasma is often called "the fourth state of matter." Some plasmas, especially in astrophysical situations, are so diffuse that collisions between particles are rare. These "collisionless" plasmas are sometimes far better conductors than metals, and they can sustain large electric currents with very modest electric fields.

FIGURE 27-10 Plasma, or ionized gas, is an excellent electrical conductor. Photo shows glowing plasma in the Tokamak Fusion Test Reactor at the Princeton Plasma Physics Laboratory.

Semiconductors

Even in the best insulators, random thermal motions occasionally dislodge electrons, giving these materials very modest conductivity. This effect is usually insignificant at normal temperatures. But a few materials—notably the element silicon—have more electrons dislodged by thermal motion even at room temperature, and therefore exhibit significant conductivity. Because their conductivity lies between that of insulators and that of good conductors, such materials are called **semiconductors**. The electrical properties of semiconductors make possible the microelectronic technology that plays a pervasive role in modern civilization. Here we give a qualitative description of semiconductors based on classical physics; the extended version of this text revisits semiconductors from a quantum perspective in Chapter 42.

Conduction in semiconductors involves not only electrons dislodged from their places in the material structure, but also the "holes" left behind by those electrons. Nudged by the electric field, an adjacent electron can "fall" into the hole, with the effective result that the hole has moved in the direction of the field (Fig. 27-11). Thus holes act as positive charge carriers, and a pure semiconductor such as silicon contains equal numbers of negative charge carriers (free electrons) and positive charge carriers (holes).

A pure semiconductor has rather low conductivity and is useful only in a few applications. The key to semiconductor technology lies in the control of conductivity by adding very small amounts of impurities—a process called **doping**. Adding an element with five electrons in its outermost shell—as opposed to silicon's four—results in large numbers of free electrons and a much more conductive material whose charge carriers are predominantly negative (Fig. 27-12). Since its charge carriers are negative, such a material is called an **N-type** semiconductor. In contrast, doping with an element containing only three outermost electrons results in a **P-type** semiconductor, whose charge carriers are predominantly positive holes.

The key element in nearly all semiconductor devices is the **PN junction**, where *P*- and *N*-type semiconductors meet. When such a junction forms, free

(a)   (b)

FIGURE 27-11 (*a*) Structure of a silicon crystal, showing each atom bound to each of its neighbors by two shared electrons. Thermal motion has dislodged one of the electrons from the crystal structure, leaving behind a hole. (*b*) In the presence of an electric field, the free electron drifts opposite the field direction. In addition, a nearby bound electron falls into the hole, effectively moving the hole in the direction of the field. Holes thus act as positive charge carriers.

FIGURE 27-12 (*a*) An arsenic atom, showing its five valence electrons. (*b*) Incorporated into a silicon crystal, the arsenic atom does its best to fit into the crystal structure. One of its electrons cannot participate in the bonding, but moves throughout the crystal lattice as a free conduction electron. Because its free charge carriers are predominantly electrons, the arsenic-doped silicon is an *N*-type semiconductor. Doping with aluminum, which has three valence electrons, would result similarly in excess holes, making a *P*-type material. In either case the material remains neutral; *P* and *N* refer to the sign of the free charges, not to a net charge on the material.

(*a*)

(*b*)

electrons, abundant in the *N*-type material but rare in the *P*, diffuse across the junction and combine with ions on the *P* side of the junction (Fig. 27-13*a*). Similarly, holes diffuse the other way and combine with electrons. (This diffusion process is similar to what happens when you open a bottle of perfume or drop food coloring into water; molecules of the substance spread from regions of high concentration to regions of low concentration.) Eventually diffusion stops because buildup of charge on opposite sides of the junction results in an electric field opposing further diffusion. But by now the junction region is depleted of free charges, and is therefore a poor conductor.

A single *PN* junction does conduct, but only in one direction. Applying a potential difference with positive at the *N*-type material only reinforces the charge depletion at the junction, further reducing its conductivity (Fig. 27-13*b*) and preventing significant current from flowing. This situation is called **reverse bias**. In contrast, making the *P*-type material positive reduces the internal electric field, shrinking the depleted region and greatly increasing its conductivity (Fig. 27-13*c*). A substantial current thus flows in this case of **forward bias**. We'll see in Chapter 33 how this "one-way" property of the *PN* junction is used to convert alternating current from household power sources into the direct current needed to power electronic devices.

The wide range of semiconductor devices in use today results largely from carefully engineered combinations of *PN* junctions. The following application presents one of the most important such devices.

(*a*) (*b*) (*c*)

FIGURE 27-13 (*a*) When a *PN* junction is first established, electrons diffuse from the *N*-type material into the *P*, where they combine with holes. The result is a net negative charge to the left of the junction, but few free charges. Similarly, holes diffuse the other way, resulting in net positive charge and few free charges to the right of the junction. (*b*) Applying a potential difference from *N* to *P* reinforces the depletion, and very little current flows. (*c*) Applying a potential difference from *P* to *N* narrows the depletion region and results in substantial current flow.

Few inventions have revolutionized human society as much as the transistor. Transistors are the basis of all modern electronic devices, from stereo amplifiers to automobile ignition systems to VCRs, laboratory instruments, and computers.

Figure 27-14 shows a widely used type of transistor, called the field effect transistor (FET). This particular FET consists of a slab of *P*-type material with two embedded *N*-type regions. Wires connect these *N*-type regions to external circuitry, but normally no current can flow through the transistor in either direction because one of the two *PN* junctions will be reverse biased. Coated on top of the *P* region is a layer of silicon dioxide (SiO_2), an excellent insulator, and over that is a metal layer called the *gate*. Suppose the gate is charged positively, by connecting it to a battery. This charge will repel positive holes in the *P*-type channel below the gate, and attract electrons. As a result, the channel becomes temporarily *N*-type, and the *PN* junctions disappear. The transistor therefore conducts electric current (Fig. 27-14*b*).

Varying the gate charge continuously varies the conductivity and therefore the current; the transistor then functions as an *amplifier*, in which a weak signal applied to the gate controls a stronger signal in the form of the current through the transistor.

Audio and video systems all contain amplifying transistors; ultimately, for example, the weak signal from a microphone or guitar pickup is amplified enough to drive a loudspeaker. In digital circuits like computers, in contrast, the transistor functions as a switch, with its drain–source channel either fully "on" (conducting) or "off" (nonconducting).

(a)

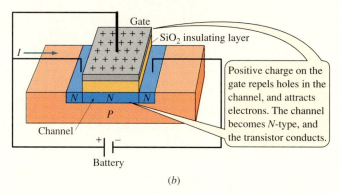

(b)

FIGURE 27-14 A field-effect transistor, or FET. (*a*) Normally current flow between the *N*-type regions is impossible, because one of the two *PN* junctions is necessarily reverse biased. (*b*) Positive charge on the gate makes the channel temporarily *N*-type, allowing current to flow.

FIGURE 27-15 The PowerPC chip is a complex integrated circuit at the heart of Apple's Macintosh computers.

A transistor is fabricated from a single piece of silicon. By exposing various parts of the surface to dopant chemicals, oxygen, and metal atoms, the various *N* and *P* regions as well as the SiO_2 insulator and metallic gate are formed. The same process is used to produce entire circuits containing millions of transistors (Fig. 27-15). These **integrated circuits**—also called **chips**—make possible the complexity, miniaturization, and sophistication of modern electronic devices.

Superconductors

In 1911 the Dutch physicist H. Kamerlingh Onnes, studying the electrical properties of mercury at very low temperatures, found a sudden drop in resistivity at a temperature of 4.2 K. The resistivity below this temperature proved immeasurably low. Subsequent research has identified thousands of substances that become **superconductors** at sufficiently low temperatures. Currents in superconductors persist for years without any measurable decrease, suggesting that the resistivity of a superconductor is truly zero (Fig. 27-16).

FIGURE 27-16 Resistivity versus temperature for a thin film of yttrium-barium-copper-oxide. Below the 93-K transition temperature the resistivity is truly zero.

For decades the known superconductors were largely metals and metal alloys that required cooling in liquid helium to achieve their superconductivity. Then, in 1986, physicists J. Georg Bednorz and K. Alex Müller of IBM's Zurich research laboratory made the stunning discovery that a class of ceramic materials becomes superconducting at temperatures around 100 K—high enough that these materials superconduct when cooled with liquid nitrogen. A flurry of research followed, soon pushing the highest superconducting temperature to well over 100 K. Bednorz and Müller received the 1987 Nobel Prize in physics, only a year after their discovery. The search for higher temperature superconductors continues, with the highest reported temperatures now over 160 K. Sporadic reports of superconductivity at or close to room temperature have appeared in the literature, but so far none have met the test of repeatability. Development of a room-temperature superconductor would revolutionize electromagnetic technology.

Superconductivity offers loss-free flow of electric current. Electromagnets made from metal and metal alloy superconductors cooled with liquid helium have been used for years in such applications as nuclear magnetic resonance spectrometers (your chemistry lab may have one) and, more recently, in magnetic resonance imagers used in medicine. Electric current in these devices, once started, continues forever with no power source—as long as the liquid helium is regularly replenished (Fig. 27-17).

FIGURE 27-17 This superconducting electromagnet at Lawrence Berkeley Laboratory is among the strongest of its kind. It uses 14 miles of niobium-tin superconducting wire, and is cooled with liquid helium.

The liquid nitrogen coolant adequate for high-temperature superconductors is far less expensive than liquid helium, so scientists and engineers are developing a wide array of commercial applications of these materials. Devices based on thin films of high-T superconductors measure weak magnetic fields in biomedical and geophysical applications, sense radiant energy, and produce and process microwaves. Finally, high-T superconducting wires are now commercially available in kilometer lengths.

Although superconductivity was discovered in 1911, a satisfactory explanation of the phenomenon was not given until 1957. In that year John Bardeen, Leon Cooper, and J. Robert Schrieffer showed how superconductivity in the traditional low-temperature materials arises from a quantum-mechanical interaction among the conduction electrons. In a way that has no analog in classical physics, all the electrons move coherently through the crystal lattice, with no energy loss. Their theory earned Bardeen, Cooper, and Schrieffer the 1972 Nobel Prize in physics. A comprehensive theory of the newer, high-temperature superconducting materials has yet to be fully developed. Readers of the extended version of this text will revisit superconductivity in Chapter 42.

27-3 RESISTANCE AND OHM'S LAW

How much current does it take to run this hair dryer? Do I risk a fatal shock if I touch this wire? How long an extension cord can I use with this electric saw? Why are these jumper cables so thick? How long will it take to recharge my dead battery? Is the wiring in my house safe, or could it overheat and cause a fire? All these questions are ultimately about the total electric current flowing in wires, bodies, batteries, and other objects. The answer in each case depends on two things: the voltage (i.e., potential difference), V, applied across the object, and the **resistance**, R, that the object offers to the flow of electric current.

The macroscopic version of Ohm's law—with which you may already be familiar—relates voltage, current, and resistance. Ohm's law states that the current through an object is proportional to the voltage across it and inversely proportional to the object's resistance:

$$I = \frac{V}{R}. \quad \text{(Ohm's law)} \tag{27-6}$$

Ohm's law in this form makes good sense, for it shows that a given voltage can push more current through a lower resistance. It's worth noting two extreme cases: An **open circuit** is a nonconducting gap with infinite resistance. No matter what the voltage across an open circuit, Equation 27-6 shows that no current can flow. A common example of an open circuit is a switch in its "off" position. A **short circuit**, in contrast, has zero resistance. In a short circuit, current of any magnitude is possible without any voltage or electric field. A closed switch approximates a short circuit. An unintentional short circuit is often dangerous; short circuits in household wiring, for example, are a leading cause of fires because they allow large currents to flow, resulting in excessive heating. All real situations, with the exception of superconductors, lie between the extremes of short and open circuits.

We can understand how the macroscopic version of Ohm's law, Equation 27-6, follows from the microscopic version by considering the cylindrical conductor shown in Fig. 27-18. Suppose there is a uniform electric field **E** within the conductor. Then there must be a uniform current density given by Equation 27-4b:

$$\mathbf{J} = \frac{\mathbf{E}}{\rho},$$

where ρ is the resistivity of the material. Then the total current is

$$I = JA = \frac{EA}{\rho},$$

where A is the conductor's cross-sectional area. If the conductor has length ℓ then the potential difference between its ends is

$$V = E\ell,$$

FIGURE 27-18 A cylindrical conductor made from a material with resistivity ρ. The uniform electric field **E** drives a uniform current density **J**, giving a total current $I = JA$ through the conductor's cross-sectional area A. The electric field is associated with a potential difference V across the length ℓ of the conductor.

since the electric field is uniform. Solving to get $E = V/\ell$ and using the result in our expression for I gives

$$I = \frac{VA}{\ell\rho} = \frac{V}{(\rho\ell/A)}.$$

Comparison with the macroscopic Ohm's law, Equation 27-6, lets us identify the resistance with the term $\rho\ell/A$. Thus resistance depends on the resistivity ρ and the geometry—length and area—of the particular piece of material:

$$R = \frac{\rho\ell}{A}. \qquad (27\text{-}7)$$

We derived this expression for a conductor of uniform cross section; although Ohm's law still holds for a nonuniform conductor, integration is required to calculate the resistance (see Problems 71 and 72). Incidentally, Equations 27-6 and 27-7 both show that the units of resistance are ohms (Ω).

We emphasize again that neither the microscopic nor the macroscopic version of Ohm's law is fundamental, but rather that each is an empirical law that provides a good description of electrical conductivity in some materials. Table 27-2 summarizes the relation between microscopic and macroscopic quantities in Ohm's law.

TABLE 27-2 *Microscopic and Macroscopic Quantities and Ohm's Law*

MICROSCOPIC QUANTITY	MACROSCOPIC QUANTITY	RELATION
Electric field, $\mathbf{E}$	Voltage, V	$\mathbf{E}$ is defined at each point in a material; V is the integral of $\mathbf{E}$ over a path containing many points. In a uniform field, $V = E\ell$.
Current density, $\mathbf{J}$	Current, I	$\mathbf{J}$ is defined at each point in a material; I is the integral of $\mathbf{J}$ over an area containing many points. With uniform current density, $I = JA$.
Resistivity, ρ	Resistance, R	ρ is a property of a given material; R is a property of a particular piece of the material. In a piece of uniform cross section, $R = \dfrac{\rho\ell}{A}$.
Ohm's law $\mathbf{J} = \dfrac{\mathbf{E}}{\rho}$	Ohm's law $I = \dfrac{V}{R}$	Microscopic version relates current density to electric field at a point in a material of specified resistivity; macroscopic version relates current through to voltage across a given piece of material with specified resistance.

EXAMPLE 27-5	Resistance and Ohm's Law

A copper wire 0.50 cm in diameter and 70 cm long is used to connect a car battery to the starter motor. What is the wire's resistance? If the starter motor draws a current of 170 A, what is the potential difference across the wire?

Solution
Table 27-1 gives 1.68×10^{-8} $\Omega \cdot$m for the resistivity of copper. Then Equation 27-6 gives

$$R = \frac{\rho \ell}{A} = \frac{(1.68 \times 10^{-8} \ \Omega \cdot \text{m})(0.70 \ \text{m})}{(\pi)(0.25 \times 10^{-2} \ \text{m})^2} = 6.0 \times 10^{-4} \ \Omega .$$

This low resistance is necessary because the starter motor draws such a large current. We now apply Ohm's law to find the voltage across the wire:

$$V = IR = (170 \ \text{A})(6.0 \times 10^{-4} \ \Omega) = 0.10 \ \text{V} .$$

This is small compared with the 12 volts available from the car battery, showing that the wire is well chosen for this application (Fig. 27-19). A larger voltage difference—which would occur with a thinner wire—would mean a significant reduction in current to the motor.

FIGURE 27-19 Thick jumper cables are necessary to carry the large current used by a car's starter motor.

EXERCISE What should be the diameter of an aluminum wire that carries 15 A when the voltage across 1.0 m of the wire is 0.25 V?

Answer: 1.4 mm

Some problems similar to Example 27-5: 30, 35, 36, 38

FIGURE 27-20 Typical resistors. The large unit at the upper left is a wire-wound resistor that can dissipate 30 watts. At upper right is a variable resistor, in which a rotating contact can be positioned at different points on a coil of resistance wire. Colored stripes on the smaller resistors code their resistance values. The smallest resistor shown is a carbon-film resistor that can dissipate at most $\frac{1}{4}$ watt.

A **resistor** is a piece of conductor made to have a specific resistance. Heating elements used in electric stoves, hair dryers, irons, space heaters, and the like are all essentially resistors; so are the filaments of ordinary incandescent light bulbs. In all these cases the resistance—ultimately resulting from collisions between conduction electrons and lattice ions—provides a means of turning electrical energy into heat. Resistors are also used to set appropriate values of current and voltage in electronic circuits; for this purpose, they are made in a wide range of resistances. Resistors are rated not only by their resistance but also by the maximum power they can dissipate without overheating. Figure 27-20 shows some typical resistors.

Got It!

Figure 27-21 shows three pieces of wire. (1) and (2) are made from the same material, while (3) is made from a material with twice the resistivity. (a) Which has the greatest resistance? (b) If the same voltage is applied across each, which will pass the greatest current?

(1)	(2)	(3)

FIGURE 27-21 Which wire has the greatest resistance?

27-4 ELECTRIC POWER

Suppose a potential difference V is imposed across a resistor, driving a current I through it. The quantity V is the energy gained per unit charge as charge "falls" through the potential difference. In a resistor, that energy is dissipated through collisions, heating the material. So V is also the energy per unit charge going into heating. Meanwhile, the current I is the rate at which charge flows through the resistor. Then the energy per unit time—that is, the power dissipated in heating the resistor—is the product of the energy per unit charge times the rate at which charge moves through the conductor:

$$P = IV. \quad \text{(electric power)} \qquad (27\text{-}8)$$

Although we developed Equation 27-8 for power dissipated as heat in a resistance, it holds any time electrical energy is being converted to some other form. If, for example, we measure 5 V across an electric motor and 2 A through the motor, we can conclude that the motor is converting electrical to mechanical energy at the rate of 10 W (actually somewhat less because some of the power goes into heating).

Solving Ohm's law for V and putting the result in Equation 27-8 gives

$$P = I^2 R; \qquad (27\text{-}9\text{a})$$

solving instead for I gives

$$P = \frac{V^2}{R}. \qquad (27\text{-}9\text{b})$$

These are useful forms when we know the resistance and either the voltage or current. Although they may seem contradictory, the forms are equivalent and both are equivalent to the more general Equation 27-8 for the case when current and voltage are related by Ohm's law.

Tip

Think about What's Constant Equation 27-9a seems to imply that power increases with increasing resistance, while Equation 27-9b seems to suggest the opposite. Both implications are correct—*if* I in Equation 27-9a and V in Equation 27-9b are constants. But there's no contradiction because I and V can't both be constant while the resistance R—the ratio of V to I—changes. In most cases we work with sources of constant voltage—for example, the power company promises to maintain 120 V between the two contacts in a household electrical outlet—and in this case the power dissipated is inversely proportional to the resistance we connect across that voltage, as shown by Equation 27-9b. See Question 20 for more on this point.

EXAMPLE 27-6 | *A Light Bulb*

The voltage in typical household wiring is 120 V. How much current does a 100-W light bulb draw? What is the bulb's resistance under these conditions?

Solution
Solving Equation 27-8 for I gives

$$I = \frac{P}{V} = \frac{100 \text{ W}}{120 \text{ V}} = 0.833 \text{ A} .$$

Since we know the current, we can determine the resistance directly from Ohm's law:

$$R = \frac{V}{I} = \frac{120 \text{ V}}{0.833 \text{ A}} = 144 \text{ }\Omega .$$

We could have bypassed the calculation of current and obtained R directly from Equation 27-9b:

$$R = \frac{V^2}{P} = \frac{(120 \text{ V})^2}{100 \text{ W}} = 144 \text{ }\Omega .$$

Finally, had we known the current but not the voltage, we could have used Equation 27-9a:

$$R = \frac{P}{I^2} = \frac{100 \text{ W}}{(0.833 \text{ A})^2} = 144 \text{ }\Omega .$$

The three approaches are equivalent. Use of Equation 27-9a or b merely amounts to solving symbolically for V or I before using Equation 27-8.

Because a light bulb filament undergoes a huge temperature change when turned on, its resistance is not independent of voltage and current. Our value 144 Ω holds when the light is on. When off, it is cool, and its resistance is much lower.

EXERCISE A power line has 0.20 Ω resistance per kilometer of length. If it carries 300 A of current, find (a) the voltage across 1.0 km of the wire and (b) the power dissipated in each km of wire.

Answers: (a) 60 V; (b) 18 kW

Some problems similar to Example 27-6: 44–46

The exercise with Example 27-6 suggests that high currents result in significant power loss even in wires of low resistance—as the quadratic dependence of power on current in Equation 27-9b confirms. That result has important implications for electric power transmission, showing that power companies are better off transmitting a given power $P = IV$ at high voltage and low current rather than vice versa. We'll examine this situation in more detail in Chapter 33, once we've seen how to effect transformations among different voltage levels (see the Application: Electric Power Distribution, on page 873).

The Kilowatt-Hour

The SI unit of power is the watt (W), defined as 1 J/s and thus reflecting the definition of power as energy per time. We could equally well have defined the watt first, then defined the Joule as 1 W·s. The **kilowatt-hour** (kWh), a unit commonly used for electrical energy, is in fact defined in an analogous way. Just as 1 joule is the energy used by a device consuming 1 watt for 1 second, so a kilowatt-hour is the energy used by a device consuming 1 kW for 1 hour. Your household electric bill shows your electrical energy consumption in kWh; your cost for 1 kWh of electrical energy is typically in the range from 5¢ to 15¢. Since there are 3600 s in an hour, 1 kWh is equal to (1000 W)(3600 s) = 3.6 MJ. Burning a 100-W light bulb for 1 hour, for example, uses 100 watt-hours or 0.1 kWh. Although it is usually used only with electrical energy, the kWh is a perfectly good non-SI unit for describing any kind of energy; for example, it's useful to remember that the energy content of a gallon of oil or gasoline is about 40 kWh. If all that energy could be converted to electrical energy, it could power a 100-W light bulb for 400 hours.

CHAPTER SYNOPSIS

Summary

1. **Electric current** is a net flow of electric charge, specified as the charge per unit time crossing a given area:

$$I = \frac{dq}{dt}.$$

   If a material contains n free charges q per unit volume, moving with average speed v_d (called the **drift speed**), then the current through an area A perpendicular to the flow is

$$I = nqAv_d.$$

   Current density is a vector specifying the current per unit area:

$$\mathbf{J} = nq\mathbf{v}_d.$$

2. **Conductivity**, σ, is a property of a given material describing the ratio of current density to electric field in the material:

$$\mathbf{J} = \sigma\mathbf{E}.$$

   For **ohmic** materials conductivity is independent of electric field and this relation constitutes the microscopic version of **Ohm's law**.

   Resistivity, ρ, is the inverse of conductivity.

3. Conduction mechanisms vary with material, and include:
   a. **Metals**, in which the charge carriers are free electrons. Metals are ohmic materials in which resistivity arises from collisions of free electrons with ions.
   b. **Ionic solutions** are conductors because of the presence of negative and positive ions that can move through the solution.
   c. **Plasmas** are ionized gases, often with extremely high conductivity. Plasmas are rare on Earth but comprise much of the matter in the universe.
   d. **Semiconductors** conduct only poorly in their pure state, but their electrical properties can be radically altered by doping with impurities. Charge carriers in semiconductors can be electrons, positive "holes," or both. Semiconductors are the basis of modern electronic technology.
   e. **Superconductors** exhibit zero resistivity at sufficiently low temperatures, and consequently require no electric field or potential difference to drive a current.

4. The macroscopic version of **Ohm's law** relates current I, voltage V, and **resistance** R:

$$I = \frac{V}{R}.$$

   Resistance depends on resistivity and physical dimensions. For an object of resistivity ρ, length ℓ, and uniform cross-sectional area A, the resistance is $R = \rho\ell/A$.

5. The rate at which electrical energy is converted to other forms—i.e., the power, P—is the product of the current I through a device and the potential difference V across it:

$$P = IV.$$

   In a resistance the electrical energy is converted to heat, and the power can be written in the two equivalent forms

$$P = I^2R \quad \text{and} \quad P = \frac{V^2}{R}.$$

Terms You Should Understand

(Pairs are closely related terms whose distinction is important; number in parentheses is chapter section where term first appears.)

electric current, current density (27-1)
ampere (27-1)
drift speed, drift velocity (27-1)
conductivity, resistivity (27-2)
ohm (27-2)
ohmic, nonohmic materials (27-2)
Ohm's law (27-2, 27-3)
plasma (27-2)
semiconductor (27-2)
superconductor (27-2)
resistance (27-3)
open circuit, short circuit (27-3)

Symbols You Should Recognize

I (27-1)	σ, ρ (27-2)
v_d (27-1)	Ω (27-2)
$\mathbf{J}$ (27-1)	R (27-3)

Problems You Should Be Able to Solve

calculating current from drift speed and material properties (27-1)
relating current and current density (27-1)
relating electric field, current density, and conductivity or resistivity (27-2)
calculating resistance from resistivity and dimensions (27-3)
using Ohm's law to relate current, voltage, and resistance (27-3)
calculating electric power (27-4)

Limitations to Keep in Mind

Ohm's law is not a universal statement but an approximate empirical relation that holds with high accuracy for some materials, such as metals.

QUESTIONS

1. If you physically move an electrical conductor, does this constitute a current?
2. In previous chapters we've stressed the absence of electric fields inside conductors in equilibrium. Why now do we allow such fields?
3. A wire carries a steady current. If the wire diameter decreases in the direction of the current, what happens to the current density?
4. When you talk on the telephone, your voice is heard almost immediately at the other end. Yet the drift speed of electrons in the telephone wire is on the order of millimeters per second. Explain the apparent discrepancy.
5. What is the difference between current and current density?
6. A constant electric field generally produces a constant drift velocity. How is this consistent with Newton's assertion that force results in acceleration, not velocity?
7. When caught in the open in a lightning storm, it is better to crouch low with the feet close together rather than lie flat on the ground. Why?
8. Why does the conductivity of a metal depend on the *square* of the electron charge?
9. What are P- and N-type semiconductors? Does either carry a net electric charge?
10. Good conductors of electricity are often good conductors of heat. Why might this be?
11. Why can current persist forever in a superconductor with no applied voltage?
12. A plasma contains equal densities of free electrons and protons. Do you expect each to contribute equally to the net current? Explain.
13. Does an electric stove burner draw more current when it is first turned on or when it's fully hot?
14. A person and a cow are standing in a field when lightning strikes the ground nearby. Why is the cow more likely to be electrocuted?
15. You put a 1.5-V battery across a piece of material and a 100-mA current flows through the material. With a 9-V battery the current increases to 400 mA. Is the material ohmic or not?
16. The resistance of a metal increases with increasing temperature, while the resistance of a semiconductor decreases. Why the difference?
17. How does the fact that the drift speed of electrons in a metal is much less than their thermal speed imply that metals are ohmic conductors?
18. A 50-W and a 100-W light bulb are both designed to operate at 120 V. Which has the lower resistance?
19. A power line with a small but nonzero resistance is used to carry 450 MW of electric power from a nuclear power plant to a city. Is it most efficient to transmit this power at high voltage and low current or vice versa? Explain.
20. Equation 27-9a suggests that no power can be dissipated in a superconductor, since $R = 0$. But Equation 27-9b suggests the power should be infinite. Which is right, and why?
21. A motor made with superconducting wire and frictionless bearings is turning at constant speed and doing no mechanical work. Make an argument showing that the motor cannot be drawing current, even if it's connected to a battery. What would happen if the motor started to do mechanical work, like lifting a weight?
22. The resistivity of a pure semiconductor decreases with increasing temperature. Speculate on what might happen if a fixed voltage were applied across a piece of such material.

PROBLEMS

Section 27-1 Electric Current

1. A wire carries 1.5 A. How many electrons pass through the wire in each second?
2. In an ionic solution, 4.1×10^{15} ions, each carrying charge $+2e$, pass to the right each second; 3.6×10^{15} ions, each carrying $-e$, pass to the left in the same time. What is the net current?
3. A 12-V car battery is rated at 80 ampere-hours, meaning it can supply 80 A of current for 1 hour before it becomes discharged. If you accidentally leave the headlights on until the battery discharges, how much charge moves through the lights?
4. The electron beam that "paints" the image on a computer screen contains 5.0×10^6 electrons per cm of its length. If the electrons move toward the screen at 6.0×10^7 m/s, how much current does the beam carry? What is the direction of this current?
5. Microbiologists measure total current due to potassium ions (K^+) moving through a cell membrane of a rock crab neuron cell to be 30 nA. How many ions pass through the membrane each second?
6. An ion channel in a cell membrane carries 2.4 pA when it's open, but it's only open 20% of the time. (a) What is the average current in the channel? (b) If the channel opens for 1.0 ms, how many singly ionized ions pass through it in this time?
7. The National Electrical Code specifies a maximum current of 10 A in 16-gauge (0.129 cm diameter) copper wire. What is the corresponding current density?
8. Each atom in aluminum contributes about 3.5 conduction electrons. What is the drift speed in a 0.21-cm-diameter aluminum wire carrying 20 A?
9. What is the drift speed in a silver wire carrying a current density of 150 A/mm^2? Each silver atom contributes 1.3 free electrons.

10. The filament of the light bulb in Example 27-6 has a diameter of 0.050 mm. What is the current density in the filament? Compare with the current density in a 12-gauge wire (diameter 0.21 cm) supplying current to the light bulb.

11. A gold film in an integrated circuit measures 2.5 μm thick by 0.18 mm wide. It carries a current density of 6.8×10^5 A/m². What is the total current?

12. A piece of copper wire joins a piece of aluminum wire whose diameter is twice that of the copper. The same current flows in both wires. The density of conduction electrons in copper is 1.1×10^{29} m⁻³; in aluminum it is 2.1×10^{29} m⁻³. Compare (a) the drift speeds and (b) the current densities in each.

13. A plasma used in fusion research contains 5.0×10^{18} electrons and an equal number of protons per cubic meter. Under the influence of an electric field the electrons drift in one direction at 40 m/s, while the protons drift in the opposite direction at 6.5 m/s. (a) What is the current density? (b) What fraction of the current is carried by the electrons?

14. In Fig. 27-22, a 100-mA current flows through a copper wire 0.10 mm in diameter, a 1.0-cm-diameter glass tube containing a salt solution, and a vacuum tube where the current is carried by an electron beam 1.0 mm in diameter. The density of conduction electrons in copper is 1.1×10^{29} m⁻³. The current in the solution is carried equally by positive and negative ions with charges $\pm 2e$; the density of each ion species is 6.1×10^{23} m⁻³. The electron density in the beam is 2.2×10^{16} m⁻³. Find the drift speed in each region.

FIGURE 27-22 Problem 14.

15. In a study of proteins mediating cell membrane transport, microbiologists measure current versus time through the cell membranes of oocytes (nearly mature egg cells) taken from the African clawed frog, *Xenopus*. The measured current versus time is given approximately by $I = 60t + 200t^2 + 4.0t^3$, with t in seconds and I in nA. Find the total charge that flows through the cell membrane in the interval from $t = 0$ to $t = 5.0$ s.

Section 27-2 Conduction Mechanisms

16. The electric field in an aluminum wire is 0.085 V/m. What is the current density in the wire?

17. What electric field is necessary to drive a 7.5-A current through a silver wire 0.95 mm in diameter?

18. A cylindrical tube of sea water carries a total electric current of 350 mA. If the electric field in the water is 21 V/m, what is the diameter of the tube?

19. A 1.0-cm-diameter rod carries a 50-A current when the electric field in the rod is 1.4 V/m. What is the resistivity of the rod material?

20. There is a potential difference of 2.5 V between opposite ends of a 6.0-m-long iron wire. (a) Assuming a uniform electric field in the wire, what is the current density? (b) If the wire diameter is 1.0 mm, what is the total current?

21. Use Table 27-1 to determine the conductivity of (a) copper and (b) sea water.

22. The maximum safe current in 12-gauge (0.21-cm-diameter) copper wire is 20 A. What are (a) the current density and (b) the electric field under these conditions?

23. The free-electron density in aluminum is 2.1×10^{29} m⁻³. What is the collision time in aluminum?

24. A pure silicon crystal contains 4.9×10^{28} atoms/m³. At room temperature, the density of electron-hole pairs is 1×10^{16} m⁻³. In what concentration (aluminum atoms per silicon atom) must aluminum be added to give a conductivity 1000 times that of pure silicon? Assume that each aluminum atom contributes one extra hole, and that the conductivity is proportional to the density of charge carriers.

25. The resistivity of copper as a function of temperature is given approximately by $\rho = \rho_0[1 + \alpha(T - T_0)]$, where ρ_0 is the value listed in Table 27-1 for 20°C, $T_0 = 20°C$, and $\alpha = 4.3 \times 10^{-3}$°C⁻¹. Find the temperature at which copper's resistivity is twice its room-temperature value.

26. Manganin is an alloy whose resistivity hardly changes with temperature; its resistivity is given by the same equation as in the previous problem, but with $\rho_0 = 48.8 \times 10^{-8}$ Ω·m and $\alpha = 2.0 \times 10^{-6}$°C⁻¹. At what temperature does manganin have a resistivity 10 times that of copper?

Section 27-3 Resistance and Ohm's Law

27. What is the resistance of a heating coil that draws 4.8 A when the voltage across it is 120 V?

28. What voltage does it take to drive 300 mA through a 1.2-kΩ resistance?

29. What is the current in a 47-kΩ resistor with 110 V across it?

30. The "third rail" that carries the electric power to a subway train is a rectangular iron bar whose cross section measures 10 cm × 15 cm, as shown in Fig. 27-23. What is the resistance of a 5.0-km piece of this rail?

FIGURE 27-23 A "third rail" (Problem 30).

31. What current flows when a 45-V potential difference is imposed across a 1.8-kΩ resistor?

32. A silver and an iron wire of the same length and diameter carry the same current. How do the voltages across the two compare?

33. The presence of a few ions makes air a conductor, albeit a poor one. If the total resistance between the ionosphere and Earth is 200 Ω, how much current flows as a result of a 300-kV potential difference between Earth and ionosphere?

34. A uniform wire of resistance R is stretched until its length doubles. Assuming its density and resistivity remain constant, what is its new resistance?

35. A cylindrical iron rod measures 88 cm long and 0.25 cm in diameter. (a) Find its resistance. If a 1.5-V potential difference is applied between the ends of the rod, find (b) the current, (c) the current density, and (d) the electric field in the rod.

36. You have a cylindrical piece of material 2.4 cm long and 2.0 mm in diameter. When you attach a 9-V battery to the ends of the piece, a current of 2.6 mA results. Which material from Table 27-1 do you have?

37. How must the diameters of copper and aluminum wire be related if they are to have the same resistance per unit length?

38. Extension cords are often made from 18-gauge copper wire (diameter 1.0 mm). (a) What is the resistance per unit length of this wire? (b) An electric saw that draws 7.0 A is operated at the end of an 8.0-m-long extension cord. What is the potential difference between the wall outlet and the saw?

39. Engineers call for a power line with a resistance per unit length of 50 mΩ/km. What wire diameter is required if the line is made of (a) copper or (b) aluminum? (c) If the costs of copper and aluminum wire are $1.53/kg and $1.34/kg, which material is more economical? The densities of copper and aluminum are 8.9 g/cm³ and 2.7 g/cm³, respectively.

40. A solid, rectangular iron bar measures 0.50 cm by 1.0 cm by 20 cm. Find the resistance between each of the three pairs of opposing faces, assuming that the faces in question are equipotentials.

41. Corrosion at battery terminals results in increased resistance, and is a frequent cause of hard starting in cars. In an effort to diagnose hard starting, a mechanic measures the voltage between the battery terminal and the wire carrying current to the starter motor. While the motor is cranking, this voltage is 4.2 V. If the motor draws 125 A, what is the resistance at the battery terminal?

42. A clear plastic trough 2.5 cm wide, 5.0 cm high, and 15 cm long has the insides of its two long sides coated with metal, as shown in Fig. 27-24. If a 60-V potential difference is applied between these sides, how much current flows when the trough contains (a) pure water and (b) sea water?

FIGURE 27-24 Problem 42.

Section 27-4 Electric Power

43. A car's starter motor draws 125 A with 11 V across its terminals. What is its power consumption?

44. A 4.5-W flashlight bulb draws 750 mA. (a) At what voltage does it operate? (b) What is its resistance?

45. A watch uses energy at the rate of 240 μW. How much current does it draw from its 1.5-V battery?

46. An electric stove burner with 35 Ω resistance consumes 1.5 kW of power. At what voltage does it operate?

47. What is the resistance of a standard 120-V, 60-W light bulb?

48. Use the numbers from Problem 33 to find the electric power dissipation in Earth's atmosphere. If we could harness this power, would it make a dent in humanity's 10^{12}-W electric power consumption?

49. If the electrons of Problem 4 are accelerated through a potential difference of 10 kV, how much power must be supplied to produce the electron beam?

50. The "instant on" feature of all the television sets in the United States requires the continuous power output of a typical large power plant—about 1000 MW. If there are 10^8 TVs in the United States, how much current does the "instant on" circuit of each draw from the 120-V power line?

51. How much total energy could the 12-V battery of Problem 3 supply?

52. During a "brownout," the power line voltage drops from 120 V to 105 V. By how much does the thermal power output of a 1500-W stove burner drop, assuming its resistance remains constant?

53. Two cylindrical resistors are made from the same material and have the same length. When connected across the same battery, one dissipates twice as much power as the other. How do their diameters compare?

54. Your author's house uses approximately 110 kWh of electrical energy each week. If that energy is supplied at 240 V, what average resistance does the house present to the power line?

55. A 2000-horsepower electric railroad locomotive gets its power from an overhead wire with 0.20 Ω/km. The potential difference between wire and track is 10 kV. Current returns through the track, whose resistance is negligible. (a) How much current does the locomotive draw? (b) How far from the power plant can the train go before 1% of the energy is lost in the wire?

56. A 100% efficient electric motor is lifting a 15-N weight at 25 cm/s. If the motor is connected to a 6.0-V battery, how much current does it draw?

Paired Problems

(Both problems in a pair involve the same principles and techniques. If you can get the first problem, you should be able to solve the second one.)

57. Electrons in a fine silver wire 20 μm in diameter drift at 0.14 mm/s. What is the current in the wire? Each silver atom contributes 1.3 free electrons.

58. A potassium slab carries a current density of 470 kA/m², and the drift speed of the electrons is 0.20 mm/s. If the density of potassium is 860 kg/m³, what is the average number of free electrons contributed by each atom?

59. What is the resistance of a column of mercury 0.75 m long and 1.0 mm in diameter?

60. An integrated circuit design calls for a 470-Ω resistor. The resistor is to be 10 μm long, 1.4 μm wide, and 0.85 μm thick, with connections made at the ends 10 μm apart. Its resistivity is to be set by the appropriate amount of doping. What should that resistivity be?

61. A power plant produces 1000 MW to supply a city 40 km away. Current flows from the power plant on a single wire of resistance 0.050 Ω/km, through the city, and returns via the ground, assumed to have negligible resistance. At the power plant the voltage between the wire and ground is 115 kV. (a) What is the current in the wire? (b) What fraction of the power is lost in transmission?

62. What should be the power line voltage in the preceding problem if the transmission loss is not to exceed 2%?

63. A 240-V electric motor is 90% efficient, meaning that 90% of the energy supplied to it ends up as mechanical work. If the motor lifts a 200-N weight at 3.1 m/s, how much current does it draw?

64. An 8.5-kN elevator is powered by a 480-V electric motor that draws 24 A. If the motor is 85% efficient, how long does it take to lift the elevator 18 m?

Supplementary Problems

65. A metal bar has a rectangular cross section 5.0 cm by 10 cm, as shown in Fig. 27-25. The bar has a nonuniform conductivity, ranging from zero at the bottom to a maximum at the top. As a result, the current density

increases linearly from zero at the bottom to 0.10 A/cm² at the top. What is the total current in the bar?

FIGURE 27-25 Problem 65.

66. Metallic copper (atomic weight 64, density 8.9 g/cm²) forms a crystal structure with copper atoms located at the corners of cubes. (a) Use the density and atomic weight to determine the distance between copper atoms. (b) Use your result, and the collision time from Example 27-4, to estimate the mean thermal speed of the electrons in copper. Consider τ to be the mean time between collisions.

67. General Motors' EV1 electric car has a mass of 1500 kg and is powered by 26 12-V batteries connected in series, for a total of 312 V. About 85% of the electrical energy from the batteries ends up as mechanical energy at the drive wheels. How much current do the batteries supply when the car is climbing a 10° slope at 45 km/h? Neglect frictional losses and air resistance.

FIGURE 27-26 General Motors' EV1 electric car is powered by 26 12-V batteries driving a 102 kW (137 horsepower) electric motor. It can accelerate from 0 to 60 mi/h in 9s, and can travel up to 90 miles before recharging (Problem 67).

68. An immersion-type heating coil is connected to a 120-V outlet and immersed in a 250-ml cup of water initially at 10°C. The water comes to a boil in 85 s. What are (a) the power and (b) the resistance of the heater? Assume no heat loss, and neglect the mass of the heater.

69. A 100-Ω resistor of negligible mass is mounted inside a calorimeter. When a 12-V battery is connected for 5.0 min, the temperature inside the calorimeter rises by 26°C. What is the heat capacity of the calorimeter contents?

70. A parallel-plate capacitor has plates of 10 cm² area separated by a 1.0-mm layer of glass insulation with resistivity

$\rho = 1.2 \times 10^{13}$ $\Omega \cdot$m and dielectric constant $\kappa = 5.6$. The capacitor is charged to 100 V and the charging battery disconnected. (a) What is the initial rate of discharge (i.e., the current through the insulation)? (b) At this rate, how long would it take the capacitor to discharge fully? (The rate does not remain constant; more on this in the next chapter.)

71. Figure 27-27 shows a resistor made from a truncated cone of material with uniform resistivity ρ. Consider the cone to be made of thin slices of thickness dx, like the one shown; Equation 27-6 shows that the resistance of each slab is $dR = \rho \, dx/A$. By integrating over all such slices, show that the resistance between the two flat faces is $R = \rho \ell / \pi ab$. (This method assumes the equipotentials are planes, which is only approximately true.)

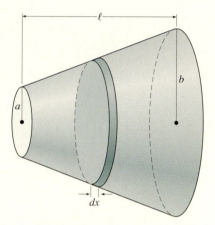

FIGURE 27-27 Problem 71.

72. A circular pan of radius b has a plastic bottom and metallic side wall of height h. It is filled with a solution of resistivity ρ. A metal disk of radius a and height h is placed at the center of the pan, as shown in Fig. 27-28. The side and disk are essentially perfect conductors. Show that the resistance measured from side to disk is $\rho \ln(b/a)/2\pi h$.

FIGURE 27-28 Problem 72.

73. At some point in a material of resistivity ρ the current density is J. Show that the power per unit volume dissipated at that point is $J^2 \rho$.

74. A thermally insulated container of sea water carries a uniform current density of 75 mA/cm^2. How long does it take to raise its temperature from 15°C to 20°C? Use the result of the preceding problem, and assume that both the specific heat (assumed the same as pure water) and the resistivity are constant over this range.

Electric Circuits

This circuit board is a complex interconnection of electronic components, including resistors, capacitors, transistors, light-emitting diodes, and integrated circuits.

28-1 CIRCUITS AND SYMBOLS

An **electric circuit** is a collection of electrical devices, called **circuit elements**, connected by conductors. Most circuits include a source of electrical energy. You're most familiar with human-made electrical circuits, which range from simple flashlights to computers, but important circuits also exist in nature. Examples include nervous systems in living organisms and Earth's global atmospheric circuit, in which thunderstorms are the batteries and the atmosphere a resistor. Your study of electric circuits should prove immensely practical, for it will help you to understand and to use effectively and safely the growing myriad of electrical and electronic devices you encounter. Basic circuit knowledge can even help you design new devices and troubleshoot old ones.

Circuit symbols

Resistor Capacitor Source Voltmeter Switch
of emf

Ammeter Variable Variable Ground Fuse
resistor capacitor

FIGURE 28-1 Common circuit symbols.

FIGURE 28-2 Chemical reactions in this alkaline battery separate electric charges, resulting in a positive charge at the + end (left) and negative charge at the − end (right).

FIGURE 28-3 A gravitational analog of a source of emf. The person lifting the balls does work against the gravitational field **g** to maintain a steady flow of balls. This represents the energy conversion process in the emf, which does work lifting charge against the electric field. The balls roll down the ramp driven by the gravitational field. They bump into the pegs and dissipate their energy. The ramp and pegs are analogous to an external resistance connected across the emf.

It's often helpful to represent circuits symbolically. We do so using standard symbols for circuit elements, with lines to represent the wires connecting them. We usually assume that the wires are perfect conductors, so all points connected by wires alone are at the same potential; such points are electrically equivalent. Realizing this will greatly facilitate your interpretation of circuit diagrams! Figure 28-1 shows some common circuit symbols. With ActivPhysics you can "build" and simulate circuits using such symbols.

28-2 ELECTROMOTIVE FORCE

In the preceding chapter, we found that an electric field is necessary to drive a current in any conductor with nonzero resistance. But if we simply apply an electric field, say by putting excess charge on one end of the conductor, the charge will quickly redistribute itself until electrostatic equilibrium is reached and the electric field disappears. Somehow we must maintain the electric field, and with it the current, despite the tendency toward equilibrium. To keep current flowing through a conductor with nonzero resistance, we need to maintain a potential difference between the ends of the conductor. Otherwise the conductor would quickly come to electrostatic equilibrium, with no electric field to drive the current.

A device that can maintain a fixed potential difference and thus an electric field, despite the flow of current, is called a source of **electromotive force**, or **emf**. (This name, which has historical origins, is not accurate; emf is not a force but a potential difference.) Most sources of emf have two electrical contacts, or **terminals**, for connection to other circuit elements. A source of emf converts some other form of energy into electrical energy by separating charge to maintain the fixed potential difference between its two terminals. The circuit symbol for a source of emf is shown in Fig. 28-1. The most familiar example is a battery, in which chemical energy drives opposite charges to the two terminals (Fig. 28-2). Others include electric generators, which convert mechanical energy to electrical energy; photovoltaic cells, which convert sunlight; fuel cells, which "burn" hydrogen to produce electrical energy and water; and biological cell membranes, which separate charge to control the movement of ions into the cell.

When a source of emf is connected to an external circuit, current flows through the circuit from the emf's positive to the negative terminal. Energy conversion processes within the emf then "lift" charge against the internal electric field, boosting the charge's energy and thus maintaining a fixed voltage across the emf's terminals. The charge then "falls" through the external circuit, dissipating its energy in the circuit resistance. The process repeats as the charge returns to the negative terminal of the emf, and the result is a steady current and a constant voltage across the emf. Figure 28-3 shows a gravitational analogy for an emf with a resistance connected across it.

Electromotive force is quantified by the work per unit charge done as the emf "lifts" charge against the electric field. The units of emf are therefore volts. An **ideal source of emf**—one with no internal energy losses—maintains the same voltage across its terminals under all conditions. Real sources always have internal energy losses, so the terminal voltage may not equal the rated emf. We discuss this situation in the next section.

28-3 SIMPLE CIRCUITS: SERIES AND PARALLEL RESISTORS

In the circuit of Fig. 28-4 an ideal battery of emf $\mathcal{E}$ drives a current through the resistor R. How much current? The voltage across the battery is its emf $\mathcal{E}$. The battery is connected to the resistor by wires, assumed to have zero resistance. Because they have zero resistance, there is no potential difference across either wire. Therefore the voltage across the resistor is the same as the voltage across the battery, and we can immediately apply Ohm's law to find the current:

$$I = \frac{\mathcal{E}}{R}.$$

Energetically, this circuit is analogous to Fig. 28-3. Charge gains energy in the amount $\mathcal{E}$ joules per coulomb as it's "lifted" against the electric field inside the battery, then dissipates that energy in heating the resistor.

(a)

(b)

FIGURE 28-4 A circuit containing a battery and a resistor. (a) Physical circuit. (b) Schematic diagram, using the symbols of Fig. 28-1. Colored stripes on small resistors, like that shown here, code for the value of the resistance.

Tip

Don't Get Hung Up about Wires We assume the wires connecting circuit elements to be essentially perfect conductors—so it takes no potential difference between the ends of a wire to drive current through it. Thus all points on a wire are at the same potential, and are electrically equivalent. That means there are many ways to diagram the same circuit; as long as two components are shown as connected by a wire, that's all that matters. Of course real wires have some resistance, but if it's negligible compared with other resistances in the circuit, then we can approximate the wires as being ideal.

Series Resistors

Figure 28-5 shows a circuit containing two resistors in series. What's the current through these resistors? What's the voltage across each? Note that neither resistor is connected directly across the battery, so we can't argue that the voltage across either is the battery voltage. But we *can* argue that the current through the two resistors is the same. Why? Because the resistors are connected in series, so the only place for current to go after it flows through R_1 is through R_2. In a steady state, with no buildup of charge in the circuit, the current through both resistors—and, for that matter, through the battery as well—must be the same. This situation holds whenever circuit elements are in series:

12.1

DC Series Circuits

| The current through each circuit element connected in series is the same.

(a)   (b)

FIGURE 28-5 A battery and two resistors in series. (a) Physical circuit. (b) Schematic diagram.

If I is the current in the circuit of Fig. 28-5, then there must be a voltage $V_1 = IR_1$ across R_1 to drive the current through this resistor. Similarly, the voltage across R_2 is $V_2 = IR_2$. Thus, the voltage across the two resistors together is $V_1 + V_2 = IR_1 + IR_2$. But the battery is connected directly across this series combination, so we have

$$IR_1 + IR_2 = \mathscr{E},$$

or

$$I = \frac{\mathscr{E}}{R_1 + R_2}.$$

Comparison with Ohm's law in the form $I = V/R$ shows that the two resistors in series behave like an equivalent resistance equal to the sum of their resistances. In an obvious generalization to more resistors in series, we have

$$R_{\text{series}} = R_1 + R_2 + R_3 + \cdots \tag{28-1}$$

In other words, resistors in series add.

Given the current, we can use Ohm's law in the form $V = IR$ to solve for the voltage across each resistor:

$$V_1 = \frac{R_1}{R_1 + R_2} \mathscr{E} \tag{28-2a}$$

and

$$V_2 = \frac{R_2}{R_1 + R_2} \mathscr{E}. \tag{28-2b}$$

These expressions show that the battery voltage divides between the two resistors in proportion to their resistance. For this reason a series combination of resistors is called a **voltage divider**. Figure 28-6 depicts the voltages throughout the circuit of Fig. 28-5, and shows explicitly that the resistors divide the battery voltage.

FIGURE 28-6 Voltages in the circuit of Fig. 28-5, with $V = 0$ at the negative battery terminal. Note that there is no potential difference across the wires, since they have negligible resistance, and that potential increases across the battery and decreases across the resistors. Can you tell from the graph which resistance is greater? (Here we've "unrolled" the circuit so we can clearly show the voltage as a function of position around the circuit. But remember that the circuit is a closed loop, so the left and right ends of the graph represent the same physical point, namely the negative battery terminal.)

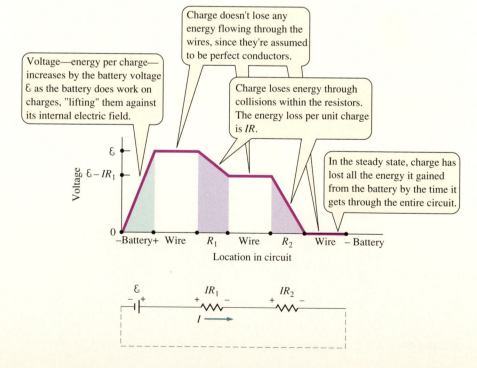

Tip

How Does the Battery Know? How does the battery in Fig. 28-5 "know" how much current to supply? How does it even "know" there are two resistors, and what their values are? For the brief instant when the circuit is first connected, the battery doesn't "know," and some rather complicated action occurs as charge moves from the battery, encounters resistance, and accumulates to establish the potential differences across the resistors. During that brief time the current is not the same everywhere. But very soon the circuit reaches a steady state, with the same current throughout. Later in this chapter, when we consider circuits with capacitors, we'll look in more detail at the approach to the steady state; for now, our analyses assume that circuits have already reached that state.

EXAMPLE 28-1 | *Designing a Voltage Divider*

A light bulb with a resistance (when on) of 5.0 Ω is designed to operate at a current of 600 mA. To operate this lamp from a 12-V battery, what resistance should you place in series with it?

Solution

Let R_2 be the lamp and R_1 the unknown series resistor. Since resistors in series add, the current through both resistors is $I = \mathscr{E}/(R_1 + R_2)$, which is supposed to be 600 mA or 0.60 A. Solving for R_1 gives

$$R_1 = \frac{\mathscr{E} - IR_2}{I} = \frac{12\text{ V} - (0.60\text{ A})(5.0\text{ }\Omega)}{0.60\text{ A}} = 15\text{ }\Omega.$$

You can also get this result by noting that the light bulb's proper operating voltage is $V = IR_2 = (0.60\text{ A})(5.0\text{ }\Omega) = 3.0$ V. This is one-fourth of the battery voltage, so the light bulb's 5-Ω resistance should be one-fourth of the total. That makes the total 20 Ω, leaving 15 Ω for R_1.

EXERCISE Suppose that in Fig. 28-5 $R_1 = 470$ Ω. If the voltage across R_2 is 59% of the battery voltage, find R_2.

Answer: 676 Ω

Some problems similar to Example 28-1: 23–25

Got It!

What is the voltage across the resistor *R* at the top of each circuit shown in Fig. 28-7? In (*a*) the second resistor has the same resistance *R*, and in (*b*) the gap is an open circuit (infinite resistance). Think about it first, then verify your answers using Equations 28-2.

(*a*)   (*b*)   (*c*)

FIGURE 28-7 What's the voltage across the upper resistor in each circuit?

FIGURE 28-8 A 1.5-V calculator battery and a 1.5-V D-cell flashlight battery have the same voltage, but the internal resistance of the calculator battery is higher.

Real Batteries

What's the difference between the two 1.5-V batteries shown in Fig. 28-8? If both were ideal there would be no difference because both would maintain 1.5 V across their terminals no matter how much current was flowing. But these are real batteries. The rate at which internal chemical reactions take place limits the amount of current each can supply. Not surprisingly, the larger battery can supply more current.

We can model a real battery by considering it to be an ideal emf in series with an **internal resistance**, as shown in Fig. 28-9. Of course this is not how batteries are made since no manufacturer can make an ideal emf! The internal resistance is intrinsic to the battery, and there is no way to circumvent it. Some of the internal resistance is actual resistance of the battery materials, but most represents the limited rate of the chemical reactions that power the battery. The more powerful battery is the one with lower internal resistance; it approaches more closely the ideal of zero internal resistance and can therefore supply more current.

We can understand the effect of internal resistance by considering the circuit of Fig. 28-10. This is just the series circuit of Fig. 28-5, with R_1 the internal resistance R_{int} and R_2 the external resistance R_L. R_L is called the *load resistor* because it is the component to which we wish to deliver electric power; it is the electrical load on the battery. From Equation 28-2b we see that if R_{int} is small compared with R_L, then the voltage across the load resistance will be very close to the battery's internal emf. In this case the battery's behavior is nearly ideal, since it has essentially $\mathscr{E}$ volts across its terminals. But if we lower R_L so it becomes comparable with R_{int}, then the voltage across R_L decreases and the battery no longer seems ideal. As we lower R_L we draw more current from the battery. It takes a higher voltage to drive this current through the fixed resistance R_{int}, so more voltage drops across R_{int}, leaving less across R_L. Even if we short-circuit the battery terminals (which is not good for the battery!) we will not get infinite current—in fact, we will simply have

$$I = \frac{\mathscr{E}}{R_{\text{int}}}. \quad \text{(battery short-circuited)}$$

We conclude that a battery or other source of emf behaves more or less ideally depending on the size of its load resistance relative to its internal resistance. A calculator, for example, has a very high resistance and draws little current. It is quite happy with a small battery whose internal resistance, while high, is still small compared with the calculator's resistance. A car starter motor, on the other hand, draws a large current and thus requires a much larger battery with very low internal resistance.

FIGURE 28-9 A real battery modeled as an ideal emf in series with an internal resistance.

FIGURE 28-10 A real battery connected to an external load. Some voltage drops across the internal resistance, making the voltage across the battery terminals less than the battery's rated voltage.

EXAMPLE 28-2	*Starting a Car*

Your car's starter motor draws 125 A. The car has a 12-V battery, but while the starter motor is running the voltage across the battery terminals measures only 9.5 V. What is the internal resistance of the battery?

Solution

This circuit is just that of Fig. 28-10, with the starter being the load. With 9.5 V across the starter, there must be 2.5 V left across the internal resistance to make a total of 12 V. The current is the same throughout this series circuit, so 125 A is the current through R_{int}. Knowing current and voltage, we apply Ohm's law:

$$R = \frac{V}{I} = \frac{2.5 \text{ V}}{125 \text{ A}} = 0.020 \ \Omega \, .$$

A battery voltage between 9 and 11 volts is typical of a car being started. A battery voltage much below 9 volts usually indicates a weak battery, a defective starter motor, or very cold weather!

EXERCISE A 9-V battery has an internal resistance of 13 Ω. What is the maximum current that can be drawn from the battery if its terminal voltage is to remain above 8.0 V?

Answer: 77 mA

Some problems similar to Example 28-2: 12, 13, 15, 17

Parallel Resistors

Figure 28-11 shows two resistors in parallel, connected across an ideal battery. What is the equivalent resistance of this parallel combination? Since the two resistors are connected at top and bottom by ideal wires, the voltage across each must be the same. We made this point in Chapter 26 when we discussed parallel capacitors, and it's worth repeating here:

▎ **The voltage across circuit elements in parallel is the same.**

The parallel resistors are connected directly across the battery, so their common voltage is the battery emf $\mathcal{E}$. Applying Ohm's law then gives the current through each resistor:

$$I_1 = \frac{\mathcal{E}}{R_1}$$

and

$$I_2 = \frac{\mathcal{E}}{R_2} \, .$$

At the point marked A in Fig. 28-11, a current I brings in charge from the battery, while the currents I_1 and I_2 carry charge away. If charge is not to accumulate (see Problem 69), the incoming and outgoing currents must be equal; that is,

$$I = I_1 + I_2 \, .$$

Using our expressions for the two resistor currents gives

$$I = \frac{\mathcal{E}}{R_1} + \frac{\mathcal{E}}{R_2} = \mathcal{E}\left(\frac{1}{R_1} + \frac{1}{R_2}\right) \, .$$

Comparison with Ohm's law in the form $I = V/R$ shows that the equivalent resistance of the parallel combination is given by

$$\frac{1}{R_{parallel}} = \frac{1}{R_1} + \frac{1}{R_2} \, . \tag{28-3a}$$

FIGURE 28-11 Parallel resistors connected across a battery. The resistor currents I_1 and I_2 sum to the battery current I.

12.2
DC Parallel Circuits

This result is readily generalized to more than two parallel resistors:

$$\frac{1}{R_{\text{parallel}}} = \frac{1}{R_1} + \frac{1}{R_2} + \frac{1}{R_3} + \cdots .$$ (28-3b)

In other words, resistors in parallel add reciprocally. Equation 28-3b shows that the resistance of a parallel combination is always lower than that of the lowest resistance in the combination.

An analogy with highway traffic shows why this makes sense: adding a lane to a crowded highway eases congestion (i.e., lowers the overall resistance), allowing a greater traffic flow (e.g., a greater current). Putting one resistor in parallel with another is like adding an extra traffic lane.

When there are only two parallel resistors, we can rewrite Equation 28-3a using a common denominator to obtain

$$R_{\text{parallel}} = \frac{R_1 R_2}{R_1 + R_2} .$$ (28-3c)

Note that *parallel* resistors combine in the same way as *series* capacitors, and vice versa.

Got It!

What's the equivalent resistance if you connect two identical resistors R (a) in series and (b) in parallel?

EXAMPLE 28-3	*Parallel and Series Resistors*

You have available three 2.0-Ω resistors. What different resistances can you make by combining all three resistors?

Solution

Figure 28-12 shows the four possible combinations. Resistors in series add, so combination (*a*) has 6.0 Ω. Resistors in parallel reciprocally add, so combination (*b*) has

$$\frac{1}{R} = \frac{1}{2.0\ \Omega} + \frac{1}{2.0\ \Omega} + \frac{1}{2.0\ \Omega} = 1.5\ \Omega^{-1},$$

for a resistance of 0.67 Ω. Combination (*c*) has two resistors in series, giving 4.0 Ω. This 4.0-Ω combination is in parallel with 2.0 Ω, so Equation 28-3c gives

$$R = \frac{(2.0\ \Omega)(4.0\ \Omega)}{2.0\ \Omega + 4.0\ \Omega} = 1.3\ \Omega .$$

Finally, combination (*d*) has two resistors in parallel, giving 1.0 Ω. This combination is in series with 2.0 Ω, for a total of 3.0 Ω. Thus you can make combinations ranging from 0.67 Ω to 6.0 Ω with these three equal resistors.

EXERCISE A 270-Ω and a 470-Ω resistor are connected in parallel, and the combination is connected in series with a 150-Ω resistor. Find the equivalent resistance of this combination.

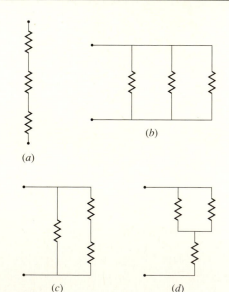

FIGURE 28-12 The four possible combinations of three equal resistors (Example 28-3).

Answer: 321 Ω

Some problems similar to Example 28-3: 16, 18, 19, 59, 60

Analyzing Circuits

Many circuits contain series and parallel combinations of basic circuit elements. We can simplify these circuits by treating each series or parallel combination as a single element, continuing the process until we can determine the voltages and currents throughout the entire circuit. Example 28-4 illustrates this procedure, which is similar to the way we dealt with capacitor combinations in Chapter 26.

12.3

ActivPhysics

DC Circuit Puzzles

EXAMPLE 28-4	*Analyzing a Circuit*

In the circuit of Fig. 28-13, what is the current through the 2-Ω resistor?

FIGURE 28-13 Circuit for Example 28-4.

Solution

We approach this problem by simplifying the circuit until we can solve for something—in this case the total current. Then we reverse the process, analyzing the circuit details until we can solve for the quantity we want. Figure 28-14 shows the steps in simplifying the circuit. We get from the original circuit, Fig. 28-14a, to Fig. 28-14b by calculating the resistance of the parallel combination of 2 Ω and 4 Ω:

$$R_{\parallel} = \frac{(2.0\ \Omega)(4.0\ \Omega)}{2.0\ \Omega + 4.0\ \Omega} = 1.33\ \Omega .$$

Figure 28-14b shows three resistors in series, which add to get the single-resistor circuit of Fig. 28-14c. From here we can calculate the total current:

$$I = \frac{\mathscr{E}}{R} = \frac{12\ V}{5.33\ \Omega} = 2.25\ A .$$

Where does this current flow? It flows from the battery through the 1-Ω resistor, then on through the parallel combination of the 2-Ω and 4-Ω resistors, then through the 3-Ω resistor and back to the battery. It does *not* all flow through the 2-Ω resistor because there are two paths the current can take when it gets to the parallel combination. However, it does all flow through the parallel combination. We already found that this combination has a resistance of 1.33 Ω, and now we know that 2.25 A flows through the combination. So the voltage across the combination is

$$V = IR = (2.25\ A)(1.33\ \Omega) = 2.99\ V .$$

This same voltage appears across each of the two resistors making up the parallel combination so the current through the 2-Ω resistor is

$$I = \frac{V}{R} = \frac{2.99\ V}{2.0\ \Omega} = 1.5\ A .$$

In solving for this current we effectively reversed our original simplification of the circuit, first considering Fig. 28-14b to get the voltage across the parallel combination, and then going to the full circuit to get the answer. At each stage we applied Ohm's law to solve for either a voltage or a current as needed.

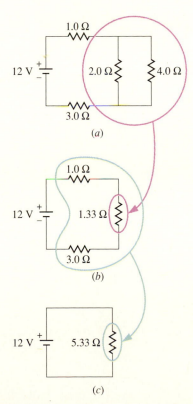

FIGURE 28-14 Simplifying the circuit by forming parallel and series combinations.

Tip

Don't Abuse Ohm's Law Ohm's law relates the voltage *across a resistor* to the current *through that resistor*. It does *not* relate arbitrary voltages and currents anywhere in a circuit. Just because there is a 12-V battery in Fig. 28-13, for example, does *not* mean that 12 V appears across the 2-Ω resistor. Be careful, too, with series and parallel combinations. As it stands, the only individual resistors in Fig. 28-13 in either series or parallel are the parallel pair of 2 Ω and 4 Ω. The 1-Ω and 2-Ω resistors are *not* in series because current flowing through the 1-Ω resistor need not all go through the 2-Ω resistor; some can go through the 4-Ω resistor instead. Equations 28-1 and 28-3 apply *only* to combinations that are strictly series or parallel, respectively.

FIGURE 28-15 What is the voltage across the 180-Ω resistor?

EXERCISE In Fig. 28-15, find (a) the current supplied by the battery and (b) the voltage across the 180-Ω resistor.

Answers: (a) 30 mA; (b) 2.9 V

Some problems similar to Example 28-4: 26–28

FIGURE 28-16 Which bulb will be brightest? What happens if you disconnect bulb C?

Got It!

Figure 28-16 shows a circuit with a 6-V battery and identical 6-V light bulbs—that is, bulbs that light at full brightness when there's a 6-V potential difference between their terminals. You can treat all three bulbs as being resistors. Which, if any, of the bulbs will be brightest? What will happen to the other two bulbs if you disconnect bulb C?

28-4 KIRCHHOFF'S LAWS AND MULTILOOP CIRCUITS

Some circuits cannot be simplified using series and parallel combinations. This often happens when there is more than one source of emf or when circuit elements are connected in complicated ways. In Fig. 28-17, for example, are

FIGURE 28-17 This circuit cannot be analyzed using series and parallel combinations.

resistors R_1 and R_2 in parallel? No, because R_3 separates their lower ends. Are R_1 and R_3 in series? No, because current flowing out the bottom of R_1 can go through either R_3 or R_4. Solving circuits that aren't reducible to series and parallel combinations usually requires more general techniques than we've used so far.

Kirchhoff's Laws

In Fig. 28-6 we looked at changes in electric potential around the loop comprising a simple series circuit. The result was an increase in potential at the battery, followed by decreases at both resistors that summed to the gain in the battery. If we count the increase as a positive change and the decreases as negative changes, we can state the following:

> The sum of the voltage changes across all the circuit elements around any closed loop is zero.

This statement is known as **Kirchhoff's loop law**, and it applies not just to Fig. 28-6 but to *any* closed loop in a circuit. The loop law is ultimately about energy conservation; it says that charge moving around a loop gains as much energy from batteries or other sources as it loses in resistors or other energy-conversion devices.

In analyzing parallel resistors with Fig. 28-11 we noted that the current flowing into point A must equal the total current flowing out through the two parallel resistors. This is really a statement about conservation of charge. It applies to any point in a circuit carrying steady currents since under steady-state conditions charge cannot build up or be depleted. The statement is most useful at a point where three or more wires join; such a point, like A in Fig. 28-11, is called a **node**. If we count currents flowing into a node as positive, and currents flowing out as negative, then the statement of charge conservation becomes the following:

> The sum of the currents at any node is zero.

This is **Kirchhoff's node law**.

Analyzing Multiloop Circuits

Even the most complex circuits can be analyzed using Kirchhoff's laws. Applying the laws amounts to writing equations expressing the loop law and node law for the distinct loops and nodes in the circuit. The number of equations needed is generally one less than the number of loops plus one less than the number of nodes; this is because the quantities in one loop and one node can be expressed entirely in terms of other loops and nodes, making one loop and one node equation each redundant. Here we give just two examples; electrical engineers take entire courses in circuit analysis using these laws.

12.5
Using Kirchhoff's Laws

| EXAMPLE 28-5 | *Rate the Resistor* |

What power dissipation must resistor R_3 of Fig. 28-18a be able to tolerate?

Solution

This is a multiloop circuit in which none of the resistors are in either series or parallel with the others, so we need to use Kirchhoff's laws. In Figure 28-18b we identify three loops and two nodes. It's not always obvious in a multiloop circuit which way the currents flow, so in Fig. 28-18b we've arbitrarily assigned directions to the three currents. If an answer comes out negative, that just means the current is flowing in the opposite direction.

Here we need to know I_3 to find the power dissipation in R_3. Rather than solving symbolically, it will be simpler if we write the equations directly with their numerical values. Let's go counterclockwise around loop 1; starting at node A, the loop law becomes

$$6 - 2I_1 - I_3 = 0, \quad \text{(loop 1)}$$

where we've temporarily dropped the units. Starting from A and going clockwise around loop 2 (remember, the direction is arbitrary), we have

$$9 + 4I_2 - I_3 = 0. \quad \text{(loop 2)}$$

Why did we make the second term positive? Because this quantity is the voltage drop $I_2 R_2$ across the 4-Ω resistor R_2, and we're going through the resistor *against* the direction we've assigned for the current. This implies a voltage *increase* going in this direction. Finally, the node equation at node A reads

$$-I_1 + I_2 + I_3 = 0, \quad \text{(node A)}$$

where the first term is negative because I_1 is indicated as flowing *away* from the node. The equations for node B and loop 3 are redundant, so we're ready to solve the system. The node equation gives $I_1 = I_2 + I_3$; substituting in the loop 1 equation, we have $6 - 2I_2 - 3I_3 = 0$, or

$$I_2 = \tfrac{1}{2}(6 - 3I_3).$$

Finally, we can use this result in the loop 2 equation to get $9 + 2(6 - 3I_3) - I_3 = 0$ or

$$21 - 7I_3 = 0,$$

giving $I_3 = 3$ A. That this answer is positive indicates that I_3 is indeed upward in Fig. 28-18. The power dissipated in R_3 is therefore

$$P_3 = I_3^2 R_3 = (3 \text{ A})^2(1 \ \Omega) = 9 \text{ W},$$

using Equation 27-9a. A 10-W resistor, the next larger size commercially available, would be just adequate.

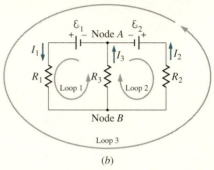

FIGURE 28-18 (a) A circuit that cannot be analyzed using series and parallel combinations. (b) There are 3 loops, 2 nodes, and 3 unknown currents whose direction assignments are, arbitrary (Example 28-5).

Tip

Getting Those Signs Right It really doesn't matter what direction you assign to the currents in a multiloop circuit, so don't spend a lot of time fretting about it; the algebra really will take care of signs. Here we found $I_3 = +3$ A, indicating that our upward direction in Fig. 28-18b was correct. But we found that I_2 was given by $I_2 = \tfrac{1}{2}(6 - 3I_3)$, so, with $I_3 = 3$ A, $I_2 = -1.5$ A. The minus sign says that I_2 is actually flowing downward through R_2, opposite the direction we assigned. Had we assigned a downward direction originally, we would have had a positive answer—but the physical situation would be identical.

The exercise with this example shows that I_2 goes to zero when $\mathcal{E}_2$ drops to 2 V. If $\mathcal{E}_2$ drops lower still, I_2 really will be upward—and then the algebra will give a positive answer.

It also doesn't matter what direction you choose to go around the circuit loops; changing direction will change the sign of *all* terms in the loop equation—leaving the mathematical content of the equation unchanged.

EXERCISE To what value should $\mathcal{E}_2$ in Fig. 28-18 be changed so that the current I_2 becomes zero?

Answer: 2.0 V

EXAMPLE 28-6	*The Cell Membrane: A Microbiological Circuit*

Figure 28-19 shows a simplified version of the so-called Hodgkin-Huxley model, which microbiologists use to describe the flow of ions through cell membranes (Hodgkin and Huxley won the 1963 Nobel Prize for physiology or medicine for development of this model). The batteries labeled $\mathcal{E}_K$, $\mathcal{E}_{Na}$, and $\mathcal{E}_L$ represent, respectively, the electrochemical effects of potassium, sodium, and other ions. (Sodium and potassium dominate, and currents associated with other ions constitute a small "leakage current"; hence the subscript L.) The resistances describe just that—the resistance the cell membrane offers to each ion species passing through it (although the resistances vary with electrical activity of the cell). Find an expression for the voltage across the cell membrane, V_M in Fig. 28-19, in terms of the other quantities shown.

Solution

V_M is clearly the sum of any one battery voltage minus the drop across the corresponding resistor, so we need the current in any one of the resistors. Here we'll go for I_K. We've identified two loops and one node, and marked them on Fig. 28-19, and have assigned directions to all three currents (obviously, they can't all be flowing upward, but once again we'll let the algebra handle that).

All the physics is in writing the loop and node equations. We'll start with the two loop equations, first going clockwise (an arbitrary choice) around loop 1, from the bottom; taking the sum of the voltage differences and setting it to zero gives

$$\mathcal{E}_K - I_K R_K + I_{Na} R_{Na} - \mathcal{E}_{Na} = 0 \, .$$

Here the voltage drop $I_{Na} R_{Na}$ across R_{Na} enters with a positive sign, since we're going *against* the direction we've assigned for the current through this resistor. Similarly, $\mathcal{E}_{Na}$ enters with a negative sign since we're going from the positive to the negative terminal of this battery—that is, in the direction of lower energy per unit charge. Similarly, going clockwise around loop 2 gives

$$\mathcal{E}_{Na} - I_{Na} R_{Na} + I_L R_L - \mathcal{E}_L = 0 \, .$$

Finally, we have an equation for node A. All our currents are assigned as flowing into the node, so

$$I_K + I_{Na} + I_L = 0 \, ,$$

which affirms what should be obvious already: that all three currents can't really flow upward, since all three can't have positive signs and satisfy this equation.

We now have three equations in the three unknown currents, and the rest is algebra. You can either solve and substitute for I_L, then I_{Na}, until you finally have one equation in the single unknown I_K, or you can use the methods of linear algebra if

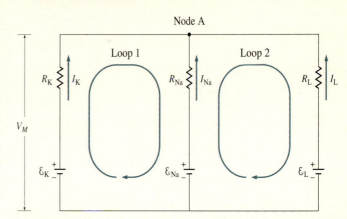

FIGURE 28-19 Simplified version of the Hodgkin-Huxley model for a biological cell membrane.

you've studied them. Either way, we'll skip the algebra and just give the result:

$$I_K = \frac{(\mathcal{E}_K - \mathcal{E}_{Na})R_L + (\mathcal{E}_K - \mathcal{E}_L)R_{Na}}{R_K R_L + R_K R_{Na} + R_L R_{Na}} \, .$$

The membrane voltage itself is then $V_M = \mathcal{E}_K - R_K I_K$. In typical situations this quantity has a value measured in tens of millivolts.

EXERCISE In Fig. 28-20, find an expression for the current through R_3 in terms of the quantities listed.

FIGURE 28-20 What's the current through R_3?

Answer: $I_K = \dfrac{(\mathcal{E}_1 - \mathcal{E}_3)R_2 - (\mathcal{E}_2 + \mathcal{E}_3)R_1}{R_1 R_2 + R_1 R_3 + R_2 R_3}$, with a positive value indicating current from left to right in Fig. 28-20.

Some problems similar to Examples 28-5 and 28-6: 32–38

Strategy

Multiloop Circuits Examples 28-5 and 28-6 illustrate the strategy for solving multiloop circuits, for which simple series-parallel reductions don't work. Here we outline the steps in that strategy.

1. Identify circuit nodes and loops.
2. Assign directions to each current. The directions you assign are completely arbitrary, but once you assign them, use those directions consistently throughout.
3. Go around all but one of the loops you've identified, either clockwise or counterclockwise; the direction doesn't matter (you don't need to do all the loops because one is always redundant). As you go, sum the voltage changes across each circuit element you encounter, following these rules (all of which give the energy change per unit charge in traversing the circuit elements):
   a. For batteries traversed from the negative to the positive terminal, the voltage change is $+\mathscr{E}$, where $\mathscr{E}$ is the battery voltage.
   b. For batteries traversed from positive to negative, the voltage change is $-\mathscr{E}$.
   c. For resistors traversed in the direction you've assigned to the current I through the resistor, the voltage change is $-IR$.
   d. For resistors traversed opposite the direction you've assigned to the current I through the resistor, the voltage change is $+IR$.
   e. For other circuit elements (which we haven't yet encountered) use the characteristics of that element to determine the change in voltage across the element.

   For each loop, set the resulting sum to zero, reflecting the fact that charges going around a circuit gain as much energy as they lose. You then have one equation for each circuit loop but one.
4. For all but one node, sum the currents into the node and set the result to zero, reflecting the fact that there's no charge buildup in the steady state. Use the convention that a current whose assigned direction has it flowing *into* the node is positive, while a current flowing *out* of the node is negative. You now have one equation for each node but one; again, the last node is always redundant.
5. You now have as many equations as there are unknown currents. Solve these as a system of simultaneous equations. Use the algebraic signs of your answers to find the actual current directions. If an answer is positive, your assigned direction is correct; if not, the actual current direction is opposite what you've assigned.

28-5 ELECTRICAL MEASURING INSTRUMENTS

12.4

Using Ammeters and Voltmeters

Voltmeters

A **voltmeter** is a device that indicates the potential difference across its two terminals. The indication is usually by a digital readout, although older meters

FIGURE 28-21 (*a*) Most voltmeters today have digital displays, although (*b*) analog—or moving-needle—meters are still found in older instruments. (*c*) An oscilloscope is essentially a pair of voltmeters in which two different voltages are indicated by the horizontal and vertical deflection of an electron beam. When the horizontal voltage varies linearly with time, the result is a plot of the vertical voltage versus time.

(*a*)

(*b*)

(*c*)

use a moving needle, and oscilloscopes use the deflection of an electron beam (Fig. 28-21). Potential difference—voltage—is a property of two points, and therefore to measure the voltage between two points we connect the two terminals of the voltmeter to those points. So to measure the voltage across resistor R_2 in Fig. 28-22*a*, we connect the voltmeter *across R_2*, as shown. We do *not* break the circuit and insert the meter, as in Fig. 28-22*b*, for then we would not be measuring the voltage *across* the resistor; in fact, we would radically alter the circuit.

How good is our voltage measurement? That depends on two things: first, obviously, is the accuracy of the meter in reading the voltage across it. The second is more subtle: unless the voltmeter in Fig. 28-22 has infinite resistance, its presence will alter the circuit and change the voltage across R_2 from what it was without the meter. Even if the meter is perfectly accurate, it still will not read exactly what we want, namely the voltage that was across R_2 before we connected the meter.

An ideal voltmeter, therefore, should have infinite resistance so it doesn't draw any current and therefore doesn't alter the circuit it's measuring. Practically speaking, a voltmeter's resistance should be large compared with resistances in the circuit where the meter is being used. As a rough rule of thumb, you'll get accuracy within 1% if you use a meter whose resistance is 100 times the circuit resistance.

Older moving-needle meters (Fig. 28-21*b*) fall short of the ideal because they rely on current drawn from the circuit to produce a magnetic force that deflects the needle (more on magnetic force in the next chapter). Modern digital meters (Fig. 28-21*a*) contain amplifiers that greatly reduce current drawn from the circuit; typical meter resistances are around 10 MΩ and can easily be made much higher. Today's digital meters include amplifier, digital converter, and digital display in a single integrated-circuit package, making them the most economical, reliable, and accurate variety.

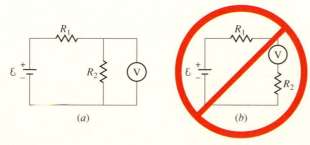

(*a*)   (*b*)

FIGURE 28-22 (*a*) Correct and (*b*) incorrect ways to measure the voltage across R_2.

EXAMPLE 28-7	*Voltmeters*

You wish to measure the voltage across the 40-Ω resistor of Fig. 28-23. What reading would an ideal voltmeter give? A voltmeter with a resistance of 1000 Ω?

Solution

An ideal voltmeter has infinite resistance and therefore would not alter the circuit, which is a simple voltage divider. Applying Equation 28-2b to this divider circuit gives the voltage across the 40-Ω resistor:

$$V_{40} = \frac{(40\ \Omega)(12\ \text{V})}{80\ \Omega + 40\ \Omega} = 4.0\ \text{V}.$$

Because the meter is connected in parallel with the resistor, this is also the voltage read by the meter.

With the nonideal voltmeter in place, the circuit becomes that of Fig. 28-24. The meter and 40-Ω resistor form a parallel combination whose resistance is given by Equation 28-3c:

$$R_{\parallel} = \frac{(40\ \Omega)(1000\ \Omega)}{40\ \Omega + 1000\ \Omega} = 38.5\ \Omega.$$

The circuit now looks like a voltage divider with $R_{\parallel}$ the lower resistor. Applying Equation 28-2b to this circuit gives

FIGURE 28-23 What is the voltage across the 40-Ω resistor?

FIGURE 28-24 A nonideal voltmeter alters the circuit, lowering the overall resistance.

$$V_{\parallel} = \frac{(38.5\ \Omega)(12\ \text{V})}{80\ \Omega + 38.5\ \Omega} = 3.9\ \text{V}.$$

This $V_{\parallel}$ is the voltage across the parallel combination consisting of the meter and 40-Ω resistor. Since the voltage across parallel circuit elements is the same, $V_{\parallel}$ is both the meter reading and the voltage across the 40-Ω resistor. And this value is 0.10 V—about 2.5%—lower than the value indicated by an ideal voltmeter.

EXERCISE A 100-kΩ resistor and a 150-kΩ resistor are in series, with a 250-V potential difference across the combination. A digital meter with a 3-significant-digit display and 1.0-MΩ resistance is used to measure the voltage across the 150-kΩ resistor. (a) What does it read? (b) By what percentage does this reading differ from that of an ideal voltmeter?

Answers: (a) 142 V; (b) 5.3%

Some problems similar to Example 28-7: 39, 40, 41, 61

Ammeters

An **ammeter** measures the current flowing *through* itself. To measure the current through a circuit element it is necessary to break the circuit and insert the ammeter in series with that element (Fig. 28-25); only then will all the current through the element also go through the meter. Connecting the meter across the resistor in Fig. 28-25 would be wrong, for then the current through the resistor would not be going through the meter.

FIGURE 28-25 An ammeter goes in series with the circuit element whose current it is to measure. It doesn't matter whether the meter goes (*a*) before or (*b*) after the circuit element, in this case the resistor *R*. (*c*) Connecting it across the resistor is incorrect, since then the resistor current does not flow through the meter. In fact, this connection would probably destroy the meter!

(*a*)

(*b*)

(*c*)

What electrical properties should the ammeter have so it doesn't alter the circuit in which it is connected? If the meter had any resistance, then the total resistance of the circuit would increase with the meter connected in series. This in turn would decrease the current, resulting in an incorrect reading. We conclude that an ideal ammeter should have zero resistance. In practice, ammeter resistance should be much lower than typical resistances in the circuit being measured.

> ## Tip
>
> *Watch Your Language When Connecting Meters* A voltmeter measures potential difference *between* two points; hence, we connect it *across*—i.e., in parallel with—the circuit element whose voltage we wish to measure. An ammeter measures the current *through* itself; hence, we connect it in *series* with the circuit element whose current we wish to measure. If you get used to voltages appearing *across* things and currents flowing *through* them, you'll have no trouble connecting meters. But if you insist on talking about "the voltage through" something, then you'll be unable to hook up meters accurately or safely. The ways to connect meters, and the words *across* for voltage and *through* for current, go right back to the definitions of potential difference as a property of two points and of current as a flow.

FIGURE 28-26 A simple ohmmeter consists of a known emf and an ammeter. When an unknown resistance is connected across the pair, its resistance may be determined from Ohm's law in the form $R = \mathcal{E}/I$.

Ohmmeters and Multimeters

Often we would like to measure the resistance of a particular circuit element. We can do this by connecting a source of known voltage in series with an ammeter and the unknown resistance, as in Fig. 28-26. Knowing the voltage and measuring the current then allows us to calculate the unknown resistance. A meter used for this purpose can be calibrated directly in ohms even though it is really measuring current; it is then called an **ohmmeter**.

The functions of voltmeter, ammeter, and ohmmeter are often combined in a single instrument called a **multimeter**. Multimeters include switches for selecting the quantity and range to be measured, and may be either analog or digital. Figure 28-27 shows a modern digital multimeter, or DMM.

FIGURE 28-27 This digital multimeter measures voltage, current, and resistance.

Potentiometric Measurement

An elegant way to measure voltage is to compare the unknown voltage with an accurate standard, in much the same way that a pan balance weighs an unknown mass by balancing it against standard masses. Figure 28-28 shows how this scheme works. An accurately known emf $\mathcal{E}_0$ is connected across a resistor along which a sliding contact moves, forming a variable voltage divider in which the position of the sliding contact determines the voltage. The output of this voltage divider—called a **potentiometer**—is connected to the unknown voltage through a meter. The potentiometer is adjusted until the meter reads zero, at which point the potentiometer voltage—which can be read off its calibrated dial—must equal the unknown voltage. The great virtue of this method is that when the system is at null—the condition where source and unknown voltages are equal—then there is no current being drawn from the unknown regardless of the meter resistance, and therefore the method has, in principle, no effect on the circuit being measured.

FIGURE 28-28 Potentiometric voltage measurement compares the unknown voltage with an accurately known voltage.

In the past, potentiometric measurements were made with circuits like that of Fig. 28-28, with special batteries of precisely known emf and accurately calibrated potentiometers. Manual adjustment was used to achieve the null condition. Today, nulling is accomplished electronically or electromechanically through a process known as negative feedback. Circuits using the potentiometric technique are the basis of many powerful measurement and control devices.

28-6 CIRCUITS WITH CAPACITORS

So far we've considered only circuits in which current and voltage are steady in all components. When you turn on a flashlight, for example, current starts to flow almost immediately through the bulb, batteries, and connecting metal parts. The current continues to flow steadily until you turn off the switch.

With a capacitor in a circuit, this picture changes. Circuit quantities change more gradually because of the capacitor. Why is this? Recall that a capacitor is a pair of insulated conductors that stores electrical energy when opposite charges are put on the conductors. A capacitor is characterized by its capacitance

$$C = \frac{Q}{V},$$

where Q is the magnitude of the charge on either conductor and V the voltage between the conductors. (See Section 26-4 for a review of capacitors.) Because charge and voltage are proportional in a capacitor, a change in a capacitor's voltage requires a change in its charge. In a circuit, a capacitor's charge changes only when current flows through the wires connecting the capacitor plates to the rest of the circuit. The magnitude of the current is the rate at which charge enters or leaves the capacitor. Since the current in any real circuit is finite, the charge on the capacitor cannot change instantaneously. But a capacitor's voltage is proportional to its charge, so:

▌ **The voltage across a capacitor cannot change instantaneously.**

This simple statement is the key to understanding circuits containing capacitors, so make sure you're clear on what it means. It says that you'll never see the voltage on a capacitor jump abruptly from one value to a different one; mathematically, capacitor voltage V_C must be a continuous function of time, with its derivative always finite. Just how rapidly the voltage can change depends on capacitance and other circuit quantities, as we'll soon see.

We now consider a simple circuit containing a resistor and a capacitor, and therefore called an *RC* **circuit**. *RC* circuits are ubiquitous, appearing everywhere from microbiological structures to stereo amplifiers to giant energy storage systems. We examine separately the two cases in which the capacitor is (1) charging and (2) discharging.

The *RC* Circuit: Charging

Consider the circuit of Fig. 28-29. The capacitor is initially uncharged, so the voltage across it is zero. What happens when we close the switch?

FIGURE 28-29 An *RC* circuit. The switch is closed at time $t = 0$.

The switch connects the left end of the resistor to the battery's positive terminal, so the left end of the resistor goes to $\mathcal{E}$ volts (here we take the zero of potential at the battery's negative terminal). The right end of the resistor is at the same voltage as the upper capacitor plate. But the voltage across the capacitor cannot change instantaneously, and therefore remains zero just after the switch is closed. With the capacitor plates both at zero volts, the full battery voltage $\mathcal{E}$ appears across the resistor. With $\mathcal{E}$ volts across the resistor, there must be a current $I = \mathcal{E}/R$ through the resistor. This current cannot flow "through" the capacitor but serves instead to pile positive charge on the upper plate, negative charge on the lower. The same current I flows everywhere except in the insulated gap between the capacitor plates.

Now that current is flowing, charge accumulates on the capacitor, and the capacitor voltage increases in proportion to this charge. As the capacitor voltage rises, the resistor voltage falls because the voltage across the series combination of resistor and capacitor is the battery voltage $\mathcal{E}$. But the current through the resistor is proportional to the resistor voltage, so the resistor current falls as well. This in turn decreases the *rate* at which charge accumulates on the capacitor plates, lowering the rate at which the capacitor voltage increases. The voltage across the capacitor continues to increase, and the current through the resistor to decrease, but at an ever slower rate.

What happens if we wait a long time? As the capacitor voltage approaches the battery voltage, the voltage across the resistor, hence the current through the resistor, and therefore the rate of charge buildup on the capacitor, all become very small. The whole system tends more and more slowly toward a final state in which the capacitor is charged to the full battery voltage and the current in the circuit is zero. Figure 28-30 summarizes the interplay among current, charge, and voltage.

We can analyze this circuit quantitatively using the loop law. Going clockwise around the loop, we first encounter a voltage increase $\mathcal{E}$ across the battery, then a drop IR across the resistor, then a drop V_C from the upper to lower capacitor

12.6
Capacitance

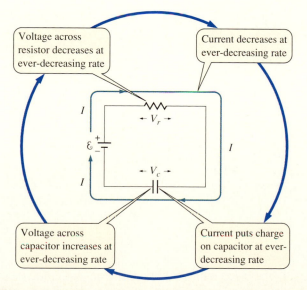

FIGURE 28-30 Interrelationships among circuit quantities in a charging *RC* circuit.

FIGURE 28-31 Voltage changes in a charging RC circuit. Since they sum to zero, the loop equation for this circuit is $\mathcal{E} - (IR) - (Q/C) = 0$.

plate (Fig. 28-31). But the definition of capacitance gives $V_C = Q/C$, so the loop equation becomes

$$\mathcal{E} - IR - \frac{Q}{C} = 0 .$$

This equation contains the two unknown quantities I and Q. Can we relate them? Yes, because the current is the rate at which charge is accumulating on the capacitor. That is:

$$I = \frac{dQ}{dt} .$$

To use this relation, we take the time derivative of the loop equation:

$$-R\frac{dI}{dt} - \frac{1}{C}\frac{dQ}{dt} = 0 .$$

The battery voltage $\mathcal{E}$ does not appear in this differentiated equation because it is constant and thus its derivative is zero. Using $I = dQ/dt$ and rearranging the equation slightly gives

$$\frac{dI}{dt} = -\frac{I}{RC} . \tag{28-4}$$

This equation shows that the rate of change of current is proportional to the current itself, expressing mathematically what Fig. 28-30 shows schematically. Equations like this arise whenever the rate of change of a quantity is proportional to the quantity itself. Population growth, the increase of money in a bank account, and the decay of a radioactive element are all described by similar equations.

Like the equation for simple harmonic motion in Chapter 15, Equation 28-4 is a *differential* equation, so called because the unknown quantity I occurs in a derivative. The solution to a differential equation is not a single number but rather a function expressing the relation between the unknown quantity—in this case current—and the independent variable—in this case time. We can solve this particular differential equation by multiplying both sides by dt/I, in order to collect all terms involving I on one side of the equation. This gives

$$\frac{dI}{I} = -\frac{dt}{RC} .$$

We then integrate both sides, noting that RC is constant:

$$\int_{I_0}^{I} \frac{dI}{I} = -\frac{1}{RC} \int_{0}^{t} dt ,$$

where $I_0 = \mathcal{E}/R$ is the initial current at the time $t = 0$ just after the switch is closed and where the integration runs from $t = 0$ to some arbitrary time t. The integral on the left is just the natural logarithm, and that on the right is just t. Then we have

$$\ln (I/I_0) = -\frac{t}{RC} ,$$

where we used $\ln I - \ln I_0 = \ln(I/I_0)$. To get an equation for I we exponentiate both sides, recalling that $e^{\ln x} = x$. This gives

$$\frac{I}{I_0} = e^{-t/RC},$$

or, since $I_0 = \mathcal{E}/R$,

$$I = \frac{\mathcal{E}}{R}e^{-t/RC}. \tag{28-5}$$

Thus the current in the circuit decreases exponentially with time, in agreement with our qualitative analysis.

What about the capacitor voltage? The capacitor and resistor voltages must add to the battery voltage $\mathcal{E}$, and the resistor voltage is just $V_R = IR$, or

$$V_R = \mathcal{E}e^{-t/RC}.$$

Thus the capacitor voltage is $V_C = \mathcal{E} - V_R$, or

$$V_C = \mathcal{E}(1 - e^{-t/RC}). \qquad \text{(RC circuit, charging)} \tag{28-6}$$

Equation 28-6 shows the capacitor voltage starting at zero, and rising rapidly at first but with its rate of rise ever slowing, as it gradually approaches the battery voltage $\mathcal{E}$—just as we reasoned in our qualitative analysis. Figure 28-32 shows plots of capacitor voltage, charge, and current using the equations we've just derived.

When is the capacitor fully charged? Never, according to our equations! But the rate at which it approaches full charge is determined by the quantity RC that appears in Equations 28-5 and 28-6. (Problem 45 will convince you that this quantity has the units of time.) Called the **time constant**, RC is a characteristic time for changes to occur in a circuit containing a capacitor. Equation 28-6 shows that in one time constant, the voltage rises to $\mathcal{E}(1 - 1/e)$, or to about two-thirds of the battery voltage. A practical rule of thumb says that in five time constants ($t = 5RC$) a capacitor is 99% charged (see Problem 47). The RC time constant clarifies our statement that the voltage across a capacitor cannot change instantaneously. We can now say that in times small compared with the time constant, the voltage across a capacitor cannot change appreciably. On the other hand, if we wait a long time—many time constants—we will find essentially no current flowing to the capacitor. We've shown quantitatively the role of the time constant RC by marking the time in units of RC on Fig. 28-32.

Resistors and capacitors are available in a wide range of values, so practical values for RC span many orders of magnitude. RC circuits with time constants from microseconds to hours are widely used in electronic devices to control the rates at which electrical quantities vary. For example, circuits with RC many times the sixtieth-of-a-second period of standard AC power are used to produce steady, direct current power for audio and video equipment. Bass and treble controls, or equalizers in more sophisticated audio systems, are simply variable resistances in RC circuits; changing the resistance changes the time constant and therefore the way the circuit handles rapidly changing audio signals. Sometimes, though, the time constant can be a nuisance. Capacitance in audio systems can limit high-frequency response, decreasing the quality of music reproduction. As computers reach speeds in the GHz range—meaning basic operations occur a billion times a second—even tiny, unintentional RC time constants associated with the resistance of wires and the capacitance of nearby conductors must be eliminated.

FIGURE 28-32 Time dependence of capacitor voltage, capacitor charge, and current in a charging RC circuit. In one time constant RC the capacitor voltage and charge rise to about 2/3 (actually $1 - 1/e$) of their final value, while the current drops to about 1/3 (actually $1/e$) of its initial value.

FIGURE 28-33 A discharging RC circuit.

The RC Circuit: Discharging

Suppose we connect a charged capacitor across a resistor, as shown in Fig. 28-33. If the capacitor voltage is initially V_0, then when the circuit is connected this voltage will drive a current $I_0 = V_0/R$ through the resistor. This current transfers charge from the positive to the negative capacitor plate, lowering the charge on the capacitor. Since capacitor charge and voltage are proportional, the capacitor voltage drops, too. So, therefore, does the current, and therefore the rate at which the capacitor discharges. We therefore expect both the voltage and current in this circuit to decay toward zero. In terms of energy, that happens because the energy stored in the capacitor's electric field is gradually dissipated as heat in the resistor.

The loop equation for Fig. 28-33 is particularly simple; going clockwise, we have

$$\frac{Q}{C} - IR = 0,$$

where the two terms are the voltage changes across the capacitor and resistor, respectively. Since we've indicated positive current in Fig. 28-33 in the direction that would *reduce* the capacitor charge Q, the rate of change dQ/dt and the current must have opposite signs: $I = -dQ/dt$. Differentiating our loop equation and substituting this expression for I gives

$$\frac{dI}{dt} = -\frac{I}{RC}.$$

This is Equation 28-4; the solution is therefore Equation 28-5, but with $I_0 = V_0/R$ instead of $\mathcal{E}/R$:

$$I = \frac{V_0}{R}e^{-t/RC}. \tag{28-7}$$

12.7
Circuit Time Constant

In this circuit the capacitor and resistor voltage are the same since the two are in parallel. Since the resistor voltage and current are proportional, the voltage across the capacitor and resistor is

$$V = V_0 e^{-t/RC}. \qquad (RC \text{ circuit, discharging}) \tag{28-8}$$

Equations 28-7 and 28-8 show that the capacitor discharges with the same characteristic time constant RC that governs its charging.

EXAMPLE 28-8	*A Camera Flash*

In Chapter 26 we considered an electronic camera flash using a capacitor to store energy. A particular camera flashtube obtains its energy from a 150-μF capacitor and requires 170 V to fire. If the capacitor is charged by a 200-V source through a 30-kΩ resistor, how long must the photographer wait between flashes? What is the peak power drawn from the power source? Assume the capacitor is fully discharged during a flash.

Solution

The time between flashes is the time it takes the capacitor voltage to reach 170 V. To find this time, we solve Equa-

tion 28-6 for the exponential term that contains the time:

$$e^{-t/RC} = 1 - \frac{V_C}{\mathcal{E}}.$$

We then take the natural logarithm of both sides, recalling that $\ln e^x = x$, so

$$-\frac{t}{RC} = \ln\left(1 - \frac{V_C}{\mathcal{E}}\right).$$

Solving for t and setting $V_C = 170$ V, $\mathcal{E} = 200$ V, $R = 30$ kΩ, and $C = 150$ μF gives

$$t = -RC \ln\left(1 - \frac{V_C}{\mathcal{E}}\right)$$

$$= -(30\times10^3 \ \Omega)(150\times10^{-6} \ \text{F}) \ln\left(1 - \frac{170 \ \text{V}}{200 \ \text{V}}\right) = 8.5 \ \text{s}.$$

Problem 74 explores the question of power in this circuit, and shows that energy from the source cannot all end up in the capacitor.

EXERCISE If the flash lamp in Example 28-8 has an effective resistance of 10 Ω, how long does it take the capacitor voltage to drop to 100 V as it discharges through the lamp?

Answer: 0.80 ms

Some problems similar to Example 28-8: 47–49, 51

RC Circuits: Long- and Short-Term Behavior

It's not always necessary to solve exponential equations in analyzing *RC* circuits. If we're concerned only with times short compared with the time constant, it suffices to remember that the voltage across a capacitor cannot change instantaneously. And after many time constants have passed a capacitor has essentially reached its final voltage, and there will be no current flowing to it. These two conditions are sufficient to analyze circuits on short and long time scales.

EXAMPLE 28-9	*Long- and Short-Time Behavior of an RC Circuit*

In Fig. 28-34*a* the capacitor is initially uncharged. What is the current through R_1 the instant after the switch is closed? A long time after the switch has been closed?

Solution

The capacitor voltage cannot change instantaneously, so just after the switch is closed there can be no voltage across the capacitor and therefore none across R_2. Then the full battery voltage is across R_1, so the current in R_1 is $\mathcal{E}/R_1$. After a very long time the capacitor will be fully charged (to what voltage? —see the exercise below), and no current will flow into it. The capacitor then acts like an open circuit, and we simply have two resistors in series. The current in each is $\mathcal{E}/(R_1 + R_2)$.

How simple this example is! The uncharged capacitor has no voltage across it, so it acts instantaneously like a short circuit. The fully charged capacitor has no current into it, so it acts like an open circuit. To solve the problem we could simply redraw the circuit, once with the capacitor replaced by a wire (Fig. 28-34*b*), the second time with the capacitor simply erased from the circuit diagram (Fig. 28-34*c*). Only if we wanted to know what was happening at intermediate times would we have to resort to the solution of an equation describing the circuit.

EXERCISE When the capacitor in Example 28-9 is fully charged, what will be the voltage across it?

Answer: $\mathcal{E}R_2/(R_1 + R_2)$

Some problems similar to Example 28-9: 55–57

(a)

(b)

(c)

FIGURE 28-34 (*a*) A circuit with two resistors and a capacitor. (*b*) Just after the switch is closed the voltage across the capacitor is still zero, so the capacitor acts like a short circuit. (*c*) Long after the switch is closed the capacitor is fully charged. No more current flows into it, so it acts like an open circuit.

FIGURE 28-35 What's the current in the 2-kΩ resistor when a charged capacitor is connected between points A and B?

Got It!

A capacitor is charged to 12 V and then connected between points A and B in Fig. 28-35, with its positive plate to A. What is the current through the 2-kΩ resistor (a) immediately after the capacitor is connected and (b) a long time (i.e., many time constants) after it's connected?

28-7 ELECTRICAL SAFETY

Whether you find yourself in a laboratory hooking up electronic equipment, or in a hospital connecting instrumentation to a patient, or on a job designing electrical devices, or simply at home plugging in appliances and tools, you should be concerned with electrical safety.

Everyone knows enough to be wary of "high voltage." People with a little more electrical sophistication are fond of saying "it isn't the voltage but the current that kills you." In fact, both points of view are partially correct. Current through the body is dangerous, but as with any resistor it takes voltage to drive that current.

Table 28-1 shows typical effects of electric currents introduced into the body through skin contact. A primary danger is disturbance of the biologically generated electrical signals that pace heartbeat; this is reflected in the lethal zone of 100 to 200 mA at which currents the heart is thrown into fibrillation—uncontrolled spasms of the cardiac muscle. With electrical signals applied internally to local regions of the body, much smaller currents can be lethal. Doctors performing cardiac catheterization, for example, must worry about currents at the microampere level.

Above 200 mA, complete cardiac arrest may occur, breathing may stop, and there may be severe burns both internally and at the points of skin contact. Sometimes such high currents are useful: when a heart is fibrillating, doctors or emergency technicians briefly apply a high enough current to stop the heart (Fig. 28-36). The heart often restarts, beating normally. The figures of Table 28-1 are rough averages, and vary widely from person to person as well as with duration of the shock and whether alternating or direct current is involved. In particular, very young children and people with heart conditions are at higher risk.

Under dry conditions, the typical human being has a resistance from one point to another on unbroken skin of about 10^5 Ω. What voltages are dangerous to such a person? At 10^5 Ω it takes

$$V = IR = (0.1 \text{ A})(10^5 \text{ Ω}) = 10,000 \text{ V}$$

to drive the fatal 100 mA. But a person who is wet or sweaty has a much lower resistance and may be electrocuted by 120-V household electricity. People have been electrocuted at voltages as low as 30 V, although such cases are rare.

It takes current to harm a person, but it takes voltage to drive that current. To be dangerous, an electric circuit must have high voltage *and* be capable of

FIGURE 28-36 Defibrillation of a heart attack victim uses a current pulse of some 20 A, delivered over a few milliseconds. This high current shocks the heart out of ineffectual, chaotic muscle spasms called fibrillation. It may then restart with its normal rhythm.

TABLE 28-1 *Effects of Externally Applied Current on the Human Organism*

CURRENT RANGE	EFFECT
0.5–2 mA	Threshold of sensation
10–15 mA	Involuntary muscle contractions; can't let go
15–100 mA	Severe shock; muscle control lost; breathing difficult
100–200 mA	Fibrillation of heart; death within minutes
>200 mA	Cardiac arrest; breathing stops; severe burns

(a)   (b)

FIGURE 28-37 (a) A short circuit in an ungrounded tool could result in a lethal shock. (b) With a grounded tool, the short circuit causes a blown fuse or circuit breaker, thereby protecting the operator.

driving sufficient current. For example, a car battery can deliver 300 A, but it cannot electrocute you because its 12 volts will not drive much current through you (although you could get a bad burn, or be struck by vaporizing metal, if you accidentally short-circuit such a powerful battery). On the other hand the 20,000 V that runs your car's spark plugs will not electrocute you either, since the internal resistance of this high-voltage circuit is so high that it cannot deliver more than a few mA.

Because potential difference is a property of two points, receiving an electric shock requires that two parts of the body be in contact with conductors at different potentials. In typical 120-V wiring used throughout North America, one of the two wires is connected physically to the ground. This ground connection is to prevent the wiring from reaching arbitrarily high potentials with respect to the ground, as might otherwise happen in a thunderstorm or if a short circuit occurred in a power line. At the same time it means that an individual contacting the "hot" side of the circuit and any grounded conductor such as the ground, a water pipe, or a bathtub will receive a shock.

A potentially dangerous situation occurs when power tools, instruments, or appliances are used by an operator who is likely to be in contact with a grounded conductor. Examples include working outdoors with an electric drill, in a kitchen with an electric mixer, or in a laboratory with an oscilloscope. Suppose you're using a power tool that is plugged in through a regular two-wire cord. Normally exposed metal parts of the tool are not connected to either wire. Now suppose something goes wrong in the tool and a wire short-circuits to the metal case. If it happens to be the wire that's plugged into the grounded side of the power line there's no problem, but if it's the other wire the metal is suddenly 120 V above ground. If you're standing on the ground, or in a damp basement, or are leaning on the kitchen sink, you will receive a potentially lethal shock (Fig. 28-37a).

To avoid this danger many electrical devices have three-wire cords (Fig. 28-38). The third wire runs from exposed metal parts to a grounded wire in the outlet, and normally carries no current. If a short circuit occurs the third wire provides a very low resistance path to ground (Fig. 28-37b). A large current will flow and will blow the circuit breaker or fuse, shutting off the current. Held at ground potential by the ground wire, the operator of the device will be safe.

Because many older homes are not wired with grounded outlets, some manufacturers produce two-wire tools and other devices that are "double-insulated"

FIGURE 28-38 Plugs. (Left) grounded; (right) polarized.

FIGURE 28-39 A ground-fault interrupter protects against shock by shutting off the power when it senses small currents leaking to ground. It must then be reset manually by pushing the reset button.

to provide an extra margin of safety. Newer appliances are equipped with "polarized plugs," which can only be plugged in one way (Fig. 28-38), ensuring that exposed metal parts are most likely to end up at ground potential. Finally, electronic devices called ground-fault interrupters (Fig. 28-39) are used in kitchen, bathroom, basement, and other hazardous circuits in new homes. These devices sense a slight imbalance—5 mA or less—in current flowing in the two wires of a circuit, and shut off the circuit in less than a millisecond on the assumption that the excess current is leaking to ground—perhaps through a person. (Do ground-fault interrupters know about the node law?)

CHAPTER SYNOPSIS

Summary

1. A source of **electromotive force (emf)** is a device—like a battery—that converts some form of energy into the electrical energy associated with the buildup of positive and negative charge at its two terminals. An ideal emf maintains a constant potential difference between its terminals, but energy losses in a real emf result in terminal voltage that decreases as more current is drawn from the device.

2. Electric circuits may often be analyzed using series and parallel resistor combinations. Resistors in series add:

$$R_{series} = R_1 + R_2 + R_3 + \cdots,$$

while resistors in parallel add reciprocally:

$$\frac{1}{R_{parallel}} = \frac{1}{R_1} + \frac{1}{R_2} + \frac{1}{R_3} + \cdots.$$

3. More complicated circuits may be analyzed using **Kirchhoff's laws**. The **loop law** follows from conservation of energy and states that the sum of the voltage differences around any circuit loop is zero. The **node law** follows from conservation of charge and states that the sum of the currents at any circuit node is zero.

4. Measuring instruments should not alter the circuit they're measuring. An ideal **voltmeter** has infinite resistance, and should be connected in parallel with the component whose voltage is being measured. An ideal **ammeter** has zero resistance, and should be connected in series with the component whose current is being measured.

5. The voltage across a capacitor cannot change instantaneously. Quantities in a resistor-capacitor (RC) circuit change on a characteristic time scale given by the product RC. When an emf $\mathcal{E}$ charges a capacitor C through a resistor R, the capacitor voltage and circuit current vary with time according to

$$V_C = \mathcal{E}(1 - e^{-t/RC})$$

$$I = \frac{\mathcal{E}}{R}e^{-t/RC}.$$

For a discharging capacitor, both voltage and current decrease exponentially with the same time constant RC.

6. **Electrical safety** is a matter of avoiding currents high enough to cause biological damage. The danger of electric shock depends on the current at which such damage occurs, the resistance of the organism, and the voltage available to drive the current.

Terms You Should Understand

(Pairs are closely related terms whose distinction is important; number in parentheses is chapter section where term first appears.)

circuit, circuit element (28-1)
electromotive force (emf) (28-2)
ideal, real emfs (28-2, 28-3)
voltage divider (28-3)
internal resistance (28-3)
Kirchhoff's loop and node laws (28-4)
voltmeter, ammeter, ohmmeter (28-5)
RC circuit (28-6)
time constant (28-6)

Symbols You Should Recognize

electric circuit symbols (Fig. 28-1)
$\mathcal{E}$ (28-1)

Problems You Should Be Able to Solve

analyzing simple circuits with series and parallel resistors (28-3)
analyzing circuits with the loop and node laws (28-4)
determining the effects of nonideal measuring instruments (28-5)
analyzing charging and discharging RC circuits (28-6)
quickly determining short- and long-term behavior of RC circuits (28-6)
assessing electrical hazards (28-7)

Limitations to Keep in Mind

Real circuit elements deviate from their idealizations. Thus the voltage across a real battery may not be exactly its rated voltage, and voltage drops occur in real wires. Attention to circuit design will minimize these nonideal effects.

Electrical measuring instruments may affect the circuits they are measuring.

Currents may be far more dangerous than Table 28-1 implies if they are introduced beneath the skin.

QUESTIONS

1. In each of the circuits of Fig. 28-40, which, if any, of the resistors are in series? In parallel?

FIGURE 28-40 Question 1.

2. Are the electrical outlets in a home connected in series or parallel? How do you know?
3. In which of the circuits of Fig. 28-41 does the battery supply the same current? All the resistors have the same resistance.

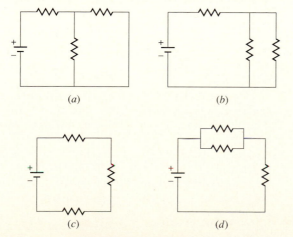

FIGURE 28-41 Question 3.

4. Can the voltage across a battery's terminals differ from the rated voltage of the battery? Explain.
5. Can the voltage across a battery's terminals be higher than the rated voltage of the battery? Explain.
6. In some cities, streetlights are wired in such a way that when one light burns out, they all go out. Are the lights in series or parallel?
7. If you know the battery voltage in Fig. 28-42, can you determine the voltage between points B and C without knowing the resistance R?

FIGURE 28-42 Questions 7, 16.

8. When the switch in Fig 28-43 is open, what is the voltage across the resistor? Across the switch?

FIGURE 28-43 Question 8.

9. Two identical resistors in series dissipate equal power. How can this be, when electric charge loses energy in flowing through the first resistor?
10. What is the current through resistor R_2 in Fig. 28-44? Assume all the wires are ideal.

FIGURE 28-44 Question 10.

11. The resistors in Fig. 28-45 all have the same resistance. If an ideal voltmeter is connected between points A and B, what will it read?

FIGURE 28-45 Question 11.

12. When a large electrical load such as a washing machine, oven, or oil burner comes on, lights throughout a house often dim. Why is this? *Hint:* Think about real wires.

13. If the node law were not obeyed at some node in an electric circuit, what would happen to the voltage at that node?

14. How would you connect a pair of equal resistors across an ideal battery in order to get the most power dissipation in the resistors?

15. You have a battery whose voltage and internal resistance are unknown. Using an ideal voltmeter and an ideal ammeter, how would you determine both these battery characteristics?

16. An ideal voltmeter is used to measure the voltages between points A and C and between B and C in the circuit of Fig. 28-42. How do the measurements compare?

17. You wish to measure the resistance R_2 in Fig. 28-46 with an ohmmeter. Can you do so while R_2 is in the circuit? Why or why not?

FIGURE 28-46 Question 17.

18. A student who is confused about voltage and current hooks a nearly ideal ammeter across a car battery. What happens?

19. A student who is confused about voltage and current tries to measure the voltage across a lighted light bulb by inserting a voltmeter in series with the bulb. What happens to the bulb? Explain.

20. Four identical light bulbs are connected to a battery as shown in Fig. 28-47. How does the brightness of each compare? Explain.

FIGURE 28-47 Questions 20, 21.

21. Suppose bulb A in Fig. 28-47 is unscrewed from its socket. How will the brightness of the three remaining bulbs change? How does the brightness of the three compare?

22. What does it mean for a capacitor to be "fully charged"?

23. Is the current into a charging capacitor in an *RC* circuit greatest when the capacitor voltage is greatest or when it is smallest?

24. If it takes forever to charge a capacitor fully, why is the *RC* time constant of any significance in describing the charging?

25. The two resistors in Fig. 28-48 have equal resistance. If the circuit has been connected for a long time, what is the voltage across the capacitor?

FIGURE 28-48 Question 25.

26. In one time constant, a charging capacitor reaches approximately $\frac{2}{3}$ of full charge. In one time constant, the voltage across a discharging capacitor falls to approximately $\frac{1}{3}$ of its original value. What is the origin of the approximate numerical factors $\frac{2}{3}$ and $\frac{1}{3}$ in these statements?

27. What's wrong with this news report: "A power-line worker was seriously injured when 4000 volts passed through his body"?

PROBLEMS

ActivPhysics can help with these problems:
Section 12, "DC Circuits."

Section 28-1 Circuits and Symbols

1. Sketch a circuit diagram for a circuit that includes a resistor R_1 connected to the positive terminal of a battery, a pair of parallel resistors R_2 and R_3 connected to the lower-voltage end of R_1, then returned to the battery's negative terminal, and a capacitor across R_2.

2. A circuit consists of two batteries, a resistor, and a capacitor, all in series. Sketch this circuit. Does the description allow any flexibility in how you draw the circuit?

3. Resistors R_1 and R_2 are connected in series, and this series combination is in parallel with R_3. This parallel combination is connected across a battery whose internal resistance is $R_{int.}$ Draw a diagram representing this circuit.

Section 28-2 Electromotive Force

4. What is the emf of a battery that delivers 27 J of energy as it moves 3.0 C between its terminals?
5. A 1.5-V battery stores 4.5 kJ of energy. How long can it light a flashlight bulb that draws 0.60 A?
6. If you accidentally leave your car headlights (current drain 5 A) on for an hour, how much of the 12-V battery's chemical energy is used up?
7. A battery stores 50 W·h of chemical energy. If it uses up this energy moving 3.0×10^4 C through a circuit, what is its voltage?

Section 28-3 Simple Circuits: Series and Parallel Resistors

8. A 47-kΩ resistor and a 39-kΩ resistor are in parallel, and the pair is in series with a 22-kΩ resistor. What is the resistance of the combination?
9. What resistance should be placed in parallel with a 56-kΩ resistor to make an equivalent resistance of 45 kΩ?
10. In Fig. 28-49 all resistors have the same value, R. What will be the resistance measured (a) between A and B or (b) between A and C?

FIGURE 28-49 Problems 10, 11.

11. In Fig. 28-49, take all resistors to be 1.0 Ω. If a 6.0-V battery is connected between points A and B, what will be the current in the vertical resistor?
12. A defective starter motor in a car draws 300 A from the car's 12-V battery, dropping the battery terminal voltage to only 6 V. A good starter motor should draw only 100 A. What will the battery terminal voltage be with a good starter?
13. What is the internal resistance of the battery in the preceding problem?
14. Three 1.5-V batteries, with internal resistances of 0.01 Ω, 0.1 Ω, and 1 Ω, each have 1-Ω resistors connected across their terminals. To three significant figures, what is the voltage across each resistor?
15. When a 9-V battery is temporarily short-circuited, a 200-mA current flows. What is the internal resistance of the battery?
16. What possible resistance combinations can you form using three resistors whose values are 1.0 Ω, 2.0 Ω, and 3.0 Ω? (Use all three resistors.)
17. A partially discharged car battery can be modeled as a 9-V emf in series with an internal resistance of 0.08 Ω. Jumper cables are used to connect this battery to a fully charged battery, modeled as a 12-V emf in series with a 0.02-Ω internal resistance. How much current flows through the discharged battery?

18. You have a number of 50-Ω resistors, each capable of dissipating 0.50 W without overheating. How many resistors would you need, and how would you connect them, so as to make a 50-Ω combination that could be connected safely across a 12-V battery?
19. What is the equivalent resistance between A and B in each of the circuits shown in Fig. 28-50? *Hint*: In (c), think about symmetry and the current that would flow through R_2.

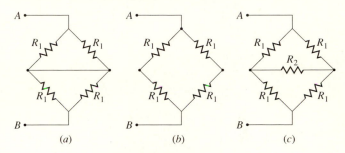

FIGURE 28-50 Problem 19.

20. A 6.0-V battery has an internal resistance of 2.5 Ω. If the battery is short circuited, what is the rate of energy dissipation in its internal resistance?
21. How many 100-W, 120-V light bulbs can be connected in parallel before they blow a 20-A circuit breaker?
22. What is the current through the 3-Ω resistor in the circuit of Fig. 28-51? *Hint*: This is trivial. Can you see why?

FIGURE 28-51 Problem 22.

23. Take $\mathscr{E} = 12$ V and $R_1 = 270$ Ω in the voltage divider of Fig. 28-5. (a) What should be the value of R_2 in order that 4.5 V appear across R_2? (b) What will be the power dissipation in R_2?
24. A voltage divider consists of two 1.0-kΩ resistors connected in series across a 160-V emf. If a 10-kΩ resistor is connected across one of the 1.0-kΩ resistors, what will be the voltage across it?
25. In the circuit of Fig. 28-52, R_1 is a variable resistor, and the other two resistors have equal resistances R. (a) Find an expression for the voltage across R_1, and (b) sketch a graph of this quantity as a function of R_1 as R_1 varies from 0 to 10R. (c) What is the limiting value as $R_1 \rightarrow \infty$?

FIGURE 28-52 Problem 25.

26. In the circuit of Fig. 28-53 find (a) the current supplied by the battery and (b) the current through the 6-Ω resistor.

FIGURE 28-53 Problems 26, 27.

27. In the circuit of Fig. 28-53, how much power is being dissipated in the 4-Ω resistor?

28. A 50-Ω resistor is connected across a battery, and a 26-mA current flows. When the 50-Ω resistor is replaced with a 22-Ω resistor, a 43-mA current flows. What are the battery's voltage and internal resistance?

Section 28-4 Kirchhoff's Laws and Multiloop Circuits

29. In the circuit of Fig. 28-54 it makes no difference whether the switch is open or closed. What is $\mathcal{E}_3$ in terms of the other quantities shown?

FIGURE 28-54 Problem 29.

30. What is the current through the ammeter in Fig. 28-55?

FIGURE 28-55 Problem 30.

31. In Fig. 28-56, what is the equivalent resistance measured between points A and B?

FIGURE 28-56 Problem 31.

32. Find all three currents in the circuit of Fig. 28-18, but now with $\mathcal{E}_2 = 1.0$ V.

33. Find all three currents in the circuit of Fig. 28-18 with the values given, but with battery $\mathcal{E}_2$ reversed.

34. In Fig. 28-57, take $\mathcal{E}_1 = 6.0$ V, $\mathcal{E}_2 = 1.5$ V, $\mathcal{E}_3 = 4.5$ V, $R_1 = 270\ \Omega$, $R_2 = 150\ \Omega$, $R_3 = 560\ \Omega$, and $R_4 = 820\ \Omega$. Find the current in R_3, and give its direction.

FIGURE 28-57 Problems 34, 35, 36.

35. With all the values except $\mathcal{E}_2$ in Fig. 28-57 as given in the preceding problem, find the condition on $\mathcal{E}_2$ that will make the current in R_3 flow upward.

36. Suppose that all resistors in Fig. 28-57 have the same value R and that $\mathcal{E}_1 = \mathcal{E}_3 = \mathcal{E}$ and $\mathcal{E}_2 = 2\mathcal{E}$. Find expressions for the currents in the four resistors, and give their directions.

37. Figure 28-58 shows a portion of a circuit used to model the electrical behavior of long, cylindrical biological cells such as muscle cells or the axons of neurons. Find the current through the emf $\mathcal{E}_3$, given that all resistors have the same value $R = 1.5$ MΩ and that $\mathcal{E}_1 = 75$ mV, $\mathcal{E}_2 = 45$ mV, and $\mathcal{E}_3 = 20$ mV. Be sure to specify the direction of the current.

FIGURE 28-58 Problems 37, 38.

38. An electrochemical impulse traveling along the cell modeled in Fig. 28-58 changes the value of emf $\mathcal{E}_3$ so that now it supplies an upward current of 40 nA. Assuming the rest of the circuit remains as described in the preceding problem, what is the new value of $\mathcal{E}_3$?

Section 28-5 Electrical Measuring Instruments

39. A voltmeter with 200-kΩ resistance is used to measure the voltage across the 10-kΩ resistor in Fig. 28-59. By what percentage is the measurement in error because of the finite meter resistance?

FIGURE 28-59 Problems 39, 40.

40. An ammeter with 100-Ω resistance is inserted in the circuit of Fig. 28-59. By what percentage is the measurement in error because of the nonzero meter resistance?

41. A neophyte mechanic foolishly connects an ammeter with 0.1-Ω resistance directly across a 12-V car battery whose internal resistance is 0.01 Ω. What is the power dissipation in the meter? No wonder it gets destroyed!

42. The voltage across the 30-kΩ resistor in Fig. 28-60 is measured with (a) a 50-kΩ voltmeter, (b) a 250-kΩ voltmeter, and (c) a digital meter with 10-MΩ resistance. To two significant figures, what does each read?

FIGURE 28-60 Problem 42.

43. In Fig. 28-61 what are the meter readings when (a) an ideal voltmeter or (b) an ideal ammeter is connected between points A and B?

FIGURE 28-61 Problem 43.

44. A resistor draws 1.00 A from an ideal 12.0-V battery. (a) If an ammeter with 0.10-Ω resistance is inserted in the circuit, what will it read? (b) If this current is used to calculate the resistance, how will the calculated value compare with the actual value?

Section 28-6 Circuits with Capacitors

45. Show that the quantity RC has the units of time (seconds).

46. If capacitance is given in μF, what will be the units of the RC time constant when resistance is given in (a) Ω, (b) kΩ, (c) MΩ? Your answers eliminate the need for tedious power-of-10 conversions.

47. Show that a capacitor is charged to approximately 99% of the applied voltage in five time constants.

48. An uncharged 10-μF capacitor and a 470-kΩ resistor are connected in series, and 250 V applied across the combination. How long does it take the capacitor voltage to reach 200 V?

49. Figure 28-62 shows the voltage across a capacitor that is charging through a 4700-Ω resistor in the circuit of Fig. 28-29. Use the graph to determine (a) the battery voltage, (b) the time constant, and (c) the capacitance.

FIGURE 28-62 Problem 49.

50. The voltage across a charging capacitor in an RC circuit rises to $1 - 1/e$ of the battery voltage in 5.0 ms. (a) How long will it take to reach $1 - 1/e^3$ of the battery voltage? (b) If the capacitor is charging through a 22-kΩ resistor, what is its capacitance?

51. A 1.0-μF capacitor is charged to 10.0 V. It is then connected across a 500-kΩ resistor. How long does it take (a) for the capacitor voltage to reach 5.0 V and (b) for the energy stored in the capacitor to decrease to half its initial value?

52. A capacitor used to provide steady voltages in the power supply of a stereo amplifier charges rapidly to 35 V every 1/60 of a second. It must then hold that voltage to within 1.0 V for the next 1/60 s while it discharges through the amplifier circuit. If the circuit draws 1.2 A from the 35-V supply (a) what is its effective resistance and (b) what value of capacitance is needed?

53. A capacitor is charged until it holds 5.0 J of energy. It is then connected across a 10-kΩ resistor. In 8.6 ms, the resistor dissipates 2.0 J. What is the capacitance?

54. A 2.0-μF capacitor is charged to 150 V. It is then connected to an uncharged 1.0-μF capacitor through a 2.2-kΩ resistor, by closing switch S in Fig. 28-63. Find the total energy dissipated in the resistor as the circuit comes to equilibrium. *Hint:* Think about charge conservation.

55. For the circuit of Example 28-9, take $\mathscr{E} = 100$ V, $R_1 = 4.0$ kΩ, and $R_2 = 6.0$ kΩ, and assume the capacitor is

FIGURE 28-63 Problem 54.

initially uncharged. What are the currents in both resistors and the voltage across the capacitor (a) just after the switch is closed and (b) a long time after the switch is closed? Long after the switch is closed it is again opened. What are I_1, I_2, and V_C (c) just after this switch opening and (d) a long time later?

56. In the circuit of Fig. 28-64 the switch is initially open and both capacitors initially uncharged. All resistors have the same value R. Find expressions for the current in R_2 (a) just after the switch is closed and (b) a long time after the switch is closed. (c) Describe qualitatively how you expect the current in R_3 to behave after the switch is closed.

FIGURE 28-64 Problem 56.

57. In the circuit of Fig. 28-65 the switch is initially open and the capacitor is uncharged. Find expressions for the current I supplied by the battery (a) just after the switch is closed and (b) a long time after the switch is closed.

FIGURE 28-65 Problem 57.

58. Obtain an expression for the rate (dV/dt) at which the voltage across a charging capacitor increases. Evaluate your result at time $t = 0$, and show that if the capacitor continued charging steadily at this rate it would be fully charged in exactly one time constant.

Paired Problems

(Both problems in a pair involve the same principles and techniques. If you can get the first problem, you should be able to solve the second one.)

59. A 3.3-kΩ resistor and a 4.7-kΩ resistor are connected in parallel, and the pair is in series with a 1.5-kΩ resistor. What is the resistance of the combination?

60. Find the value of R in Fig. 28-66 that will make the resistance between points A and B equal to R.

FIGURE 28-66 Problem 60.

61. A battery's voltage is measured with a voltmeter whose resistance is 1000 Ω; the result is 4.36 V. When the measurement is repeated with a 1500-Ω meter the result is 4.41 V. What are (a) the battery voltage and (b) its internal resistance?

62. An ammeter with a resistance of 1.4 Ω is connected momentarily across a battery (not the way to treat an ammeter!) and it reads 9.78 A. When the measurement is repeated with a meter whose resistance is 2.1 Ω the reading is 7.46 A. What are (a) the battery voltage and (b) its internal resistance?

63. In Fig. 28-67, take $\mathcal{E}_1 = 12$ V, $\mathcal{E}_2 = 6.0$ V, $\mathcal{E}_3 = 3.0$ V, $R_1 = 1.0$ Ω, $R_2 = 2.0$ Ω, and $R_3 = 4.0$ Ω. Find the current in R_2 and give its direction.

FIGURE 28-67 Problems 63, 64.

64. (a) With all values except $\mathcal{E}_2$ as given in the preceding problem, find $\mathcal{E}_2$ such that there is no current in this battery. (b) What are the currents in R_1 and R_2 under these conditions?

65. In Fig. 28-68 what are the meter readings when (a) an ideal voltmeter or (b) an ideal ammeter is connected between points A and B?

FIGURE 28-68 Problems 65, 66.

66. In Fig. 28-68 what are the meter readings when (a) a voltmeter with 50-kΩ resistance or (b) an ammeter with 150-Ω resistance is connected between points *A* and *B*?

67. An initially uncharged capacitor in an *RC* circuit reaches 75% of its full charge in 22.0 ms. What is the time constant?

68. Find the resistance needed in an *RC* circuit to bring a 20-μF capacitor from zero charge to 45% of its full charge in 140 ms.

Supplementary Problems

69. Suppose the currents into and out of a circuit node differed by 1 μA. If the node consists of a small metal sphere with diameter 1 mm, how long would it take for the electric field around the node to reach the breakdown field in air (3 MV/m)?

70. A problem on dairy farms is "stray voltage," caused by corroded wiring, poor wiring practices, or ground currents associated with nearby power lines. These conditions can result in several volts of potential difference between metal objects such as watering bowls, feed troughs, or milking equipment, and the ground. Cows feel slight shocks that make them nervous, resulting in reduced milk output. As a result, farmers can face serious financial losses. Figure 28-69 shows a circuit model for a stray voltage situation; the 1.5-kΩ resistor represents the resistance of corroded connections and poor wiring; you can assume the ground has negligible resistance. (a) The resistance from a cow's mouth to hoof is approximately 350Ω. How much current will flow through the cow in the situation shown? (It takes only about 1 mA to affect milk production.) In an effort to diagnose the problem, the farmer moves the cow aside and connects a multimeter between the watering bowl and ground. What will it read if it's set to measure (b) voltage and (c) current?

FIGURE 28-70 Problem 71.

it's connected with its positive plate at *A*; (h) long after the capacitor in part (g) is connected?

72. A resistance *R* is connected across a battery with internal resistance R_{int}. Show that the maximum power dissipation in *R* occurs when $R = R_{int}$. *Note:* This is not the way to treat a battery! But it is the basis for matching loads in amplifiers and other devices; for example, a stereo amplifier designed to drive 8-Ω speakers has internal resistance close to 8 Ω.

73. A parallel-plate capacitor is insulated with a material of dielectric constant κ and resistivity ρ. Since the resistivity is finite, the capacitor "leaks" charge and can be modeled as an ideal capacitor in parallel with a resistor. (a) Show that the time constant of the capacitor is independent of its dimensions (provided the spacing is small enough that the usual parallel-plate approximation applies) and is given by $\varepsilon_0 \kappa \rho$. (b) If the insulating material is polystyrene ($\kappa = 2.6$, $\rho = 10^{16}$ Ω·m), how long will it take for the stored energy in the capacitor to decrease by a factor of 2?

74. Of the total energy drawn from a battery in charging an *RC* circuit, show that only half ends up as stored energy in the capacitor. *Hint:* What happens to the rest of it? You will need to integrate.

75. Write the loop and node laws for the circuit of Fig. 28-71, and show that the time constant for this circuit is $R_1 R_2 C / (R_1 + R_2)$.

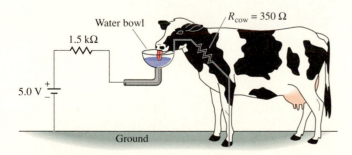

FIGURE 28-69 Stray voltage can bankrupt a dairy farm. (Problem 70)

71. In Fig. 28-70, what is the current in the 4-Ω resistor when each of the following circuit elements is connected between points *A* and *B*: (a) an ideal ammeter; (b) an ideal voltmeter; (c) another 4.0-Ω resistor; (d) an uncharged capacitor, right after it's connected; (e) long after the capacitor of part (d) is connected; (f) an ideal 12-V battery, with its positive terminal at *A*; (g) a capacitor initially charged to 12 V, right after

FIGURE 28-71 Problem 75.

76. The circuit in Fig. 28-72 extends forever to the right, and all the resistors have the same value *R*. Show that the equivalent resistance measured across the two terminals at left is $\frac{1}{2}R(1 + \sqrt{5})$. *Hint:* You don't need to sum an infinite series.

FIGURE 28-72 Problem 76.

29 The Magnetic Field

Magnetic fields shape delicate loops and other structures in the Sun's atmosphere, imaged here in X rays by the Japanese satellite Yohkoh.

People are fascinated with magnets. Magnetism—the seemingly mysterious force you feel when you try to push two magnets together in a way they don't want to go—is always intriguing. Some uses of magnets, like holding notes on refrigerators, are mundane. But others—holding gas at a temperature of one hundred million K in a nuclear fusion reactor, or converting electrical into mechanical energy in the motors of a railroad locomotive—are more impressive. And magnetism, like electricity, is at the heart of many natural phenomena and technological devices. Video and audio cassette recorders, electric motors, TV picture tubes, computer disks, and electric power plants would be impossible without magnetism.

Earth's magnetism helps us find our way around, provides historical evidence for the evolution of our planet, and protects us from harmful radiation. Birds, sea turtles, and some bacteria use Earth's magnetism for navigation. Without magnetism we would not even see, for light itself originates in an interaction between magnetism and electricity. As with electricity, we often do not recognize the magnetic character of everyday phenomena.

In this chapter we introduce magnetism, and will find immediately that it's intimately connected with electricity. In subsequent chapters we explore further the deep relation between electricity and magnetism, and will eventually come to see electricity and magnetism as two inseparable aspects of the same underlying phenomenon.

29-1 WHAT IS MAGNETISM?

You know about magnetism because you've played with magnets or used them in everyday tasks such as sticking notes to your refrigerator. You know that magnets exert forces on each other and on certain materials, such as iron and steel. It's convenient to describe that interaction in terms of a **magnetic field** (symbol **B**). We say that one magnet produces a magnetic field, and that a second magnet responds to the field in its immediate vicinity. That way we avoid the awkward "action at a distance" picture, in which one magnet somehow reaches out across empty space to pull on another. We can picture a magnetic field using field lines, just as we did for electric fields, and can trace out magnetic field lines using small iron filings that align themselves with the field (Fig. 29-1).

FIGURE 29-1 Iron filings align with the magnetic field, tracing out the field from a pair of bar magnets. The field is strongest near the magnetic poles. We'll see in the next chapter why iron is a special magnetic material.

You may think we're now going to give rules for finding the magnetic field produced by a given magnet, and for the force a magnet experiences in a magnetic field. But we're not. That's because the magnetism you're familiar with—involving magnets and their interaction with specific materials such as iron—is not particularly fundamental. Rather, the magnetism that's familiar to you is a specialized and limited manifestation of a more fundamental and universal phenomenon. Here we'll go straight to the basics of magnetism, and later (in the next chapter) will come back to understand how the magnets you're familiar with fit into the larger picture.

In Chapters 23–28 we studied electrical phenomena, all of which ultimately involve electric charge. We saw how a charge produces an electric field, and how other charges respond to that field. We're now dealing with the magnetic field, so you might expect that we're about to introduce magnetic charge and tell how it gives rise to the magnetic field. But again, no. The fact is that no one has ever found a magnetic charge, an entity that would produce a radially outward magnetic field like the electric field of a point charge. Some theories of elementary particles and the origin of the universe suggest that such magnetic charges—called **magnetic monopoles**—should exist. But they've never been observed, and even if they exist, they play no role in the magnetic phenomena with which we're familiar. We'll have more to say about magnetic monopoles in the next chapter.

So if it's not about magnets or magnetic charge, what *is* magnetism about? Simple: it's about *electric* charge—the same electric charge we've been dealing with since Chapter 23. But whereas the phenomena of electricity involve electric charge, period,

> **the phenomena of magnetism involve *moving* electric charge.**

In particular, *moving* electric charge is the source of magnetic fields, and *moving* electric charges experience forces from magnetic fields. In this chapter we'll assume we have a magnetic field and will explore the magnetic force on moving electric charges that results from that field. In the next chapter we'll see just how moving electric charges give rise to magnetic fields. That will lead us to understand technological applications such as electromagnets, the magnetic fields of Earth and other astrophysical bodies, and how the familiar magnetism of materials such as iron arises ultimately from the motions of the electrons in those materials.

29-2 THE MAGNETIC FORCE ON A MOVING CHARGE

In Chapter 23 we defined the electric field $\mathbf{E}$ through the relation $\mathbf{F}_E = q\mathbf{E}$, where $\mathbf{F}_E$ is the electric force on a charge q. We're now going to develop the analogous relation between a *moving* charge q and the force $\mathbf{F}_B$ it experiences in a magnetic field $\mathbf{B}$.

Consider a region with no electric field, but in which a magnetic field is present. You could verify the presence of the magnetic field and determine its direction with a compass, which itself is just a small magnet that's free to pivot into alignment with the magnetic field. More fundamentally, you can take our description of the force experienced by a moving charge as defining what it means for there to be a magnetic field present.

Now let's put a point charge q in the magnetic field. If the charge is stationary, we find that it experiences no magnetic force. But suppose the charge is moving. We won't worry about how it got moving—except that, as we'll soon see, magnetic force can't be responsible. Experiment shows that the moving charge experiences a magnetic force such that

1. The magnetic force is always at right angles both to the velocity $\mathbf{v}$ of the charge and to the magnetic field $\mathbf{B}$ (Fig. 29-2).
2. The strength of the magnetic force is proportional to the product of the charge q, its speed v, and the magnetic field strength B.
3. The force is greatest if the charged particle is moving at right angles to the magnetic field $\mathbf{B}$ and is zero if the charge velocity $\mathbf{v}$ is parallel or antiparallel to the field. In general, the magnetic force is proportional to the sine of the angle between the vectors $\mathbf{v}$ and $\mathbf{B}$ (see Fig. 29-2).

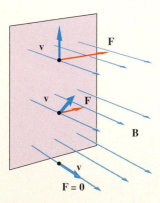

FIGURE 29-2 The magnetic force on a charged particle is perpendicular to the particle's velocity $\mathbf{v}$ and to the magnetic field $\mathbf{B}$. The magnitude of the force depends on the angle between $\mathbf{v}$ and $\mathbf{B}$, and is greatest when $\mathbf{v}$ and $\mathbf{B}$ are perpendicular (top). When $\mathbf{v}$ and $\mathbf{B}$ are parallel there is no magnetic force (bottom).

Putting these facts together allows us to write the magnetic force compactly in terms of the vector cross product introduced in Chapter 13:

$$\mathbf{F}_B = q\mathbf{v} \times \mathbf{B}, \quad \text{(magnetic force)} \qquad (29\text{-}1a)$$

where $\mathbf{F}_B$ is the magnetic force on a particle of charge q moving with velocity $\mathbf{v}$ at a point where the magnetic field is $\mathbf{B}$. Recall from Chapter 13 that the cross product $\mathbf{v} \times \mathbf{B}$ is a vector of magnitude $vB\sin\theta$, where θ is the angle between

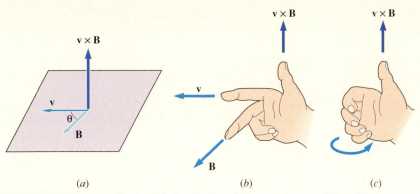

FIGURE 29-3 Ways to apply the right-hand rule. (*a*) Two vectors **v** and **B** and their cross product. The two vectors define a plane, and the cross product is always perpendicular to that plane. (*b*)–(*c*) All the right-hand rule does is distinguish between the two possible directions perpendicular to the plane. In (*b*), the fingers of the right hand point in the directions of the vectors **v** and **B**, and the cross product **v** × **B** in the direction of the thumb; in (*c*) the curved arrow describes a rotation that would take the vector **v** onto the vector **B**.

v and **B**. Thus the magnetic force has magnitude

$$F = qvB\sin\theta. \tag{29-1b}$$

The direction of **v** × **B** is perpendicular to both **v** and **B**, with the choice of the two possible perpendicular directions being made by the right-hand rule (Fig. 29-3). Because the charge q also enters Equation 29-1, the direction of the magnetic force is that of **v** × **B** for a positive charge but opposite **v** × **B** for a negative charge (Fig. 29-4).

We can regard Equation 29-1 as the definition of the magnetic field. If we put a particle of charge q moving with velocity **v** in a region free of other influences (i.e., electric and gravitational fields), then the presence of a force **F** on the particle shows that there is a magnetic field **B** implied by Equation 29-1a. Equations 29-1 show that the SI units of magnetic field are N·s/C·m, a unit given the name **tesla** (T) after the Serbian-American inventor Nikola Tesla (1865–1943). One tesla is a strong magnetic field, and a smaller unit called the gauss (G), equal to 10^{-4} T, is often used. Earth's magnetic field, for example, is a little under 1 G, while the field at the poles of a toy magnet may be 100 G. Strong laboratory magnets (Fig. 29-5) produce fields ranging from several T to

FIGURE 29-5 This superconducting electromagnet is used to steer high-energy beams of heavy atomic nuclei at the Laboratory of High Energies in Dubna, Russia. The Dubna laboratory is one of the world's leading facilities for the production and study of heavy nuclei.

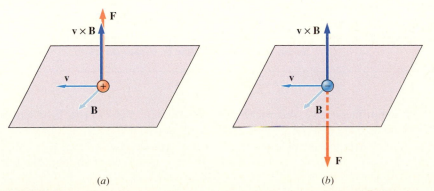

FIGURE 29-4 (*a*) When the charge is positive, the magnetic force is in the same direction as the cross product **v** × **B**. (*b*) When the charge is negative, the force q**v** × **B** is opposite **v** × **B**.

nearly 40 T, while magnetars—incredibly dense, rapidly rotating collapsed stars first identified in 1998—have fields up to 10^{11} T.

Got It!

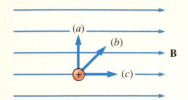

Figure 29-6 shows a proton in a magnetic field. For which of the three possible proton velocities shown will the magnetic force be greatest? What will be the direction of the magnetic force in each case?

FIGURE 29-6 For which velocity is the magnetic force greatest? (The speed is the same in all three cases.)

EXAMPLE 29-1	*Steering Protons*

A magnetic field of 0.10 T is used to steer charged-particle beams in a nuclear physics experiment. The field points vertically upward. Three protons enter the field region, two moving horizontally and one vertically as shown in Fig. 29-7. All three are moving at 2.0×10^6 m/s. What is the magnetic force on each?

Solution

Proton 2 is moving parallel to the field. For it, $\mathbf{v} \times \mathbf{B} = \mathbf{0}$ and it experiences no magnetic force. Protons 1 and 3 are moving at right angles to the field, so $\sin\theta = 1$ and Equation 29-1b gives

$$F = qvB = (1.6\times10^{-19}\text{ C})(2.0\times10^6\text{ m/s})(0.10\text{ T})$$

$$= 3.2\times10^{-14}\text{ N}.$$

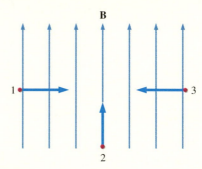

FIGURE 29-7 What is the magnetic force on each proton? (Example 29-1)

Since the protons carry positive charge, the direction of the force is that of the product $\mathbf{v} \times \mathbf{B}$. For proton 1, moving to the right, $\mathbf{v} \times \mathbf{B}$ is out of the page. For proton 3, moving to the left, the force is into the page. This example shows that the magnetic field alone does not determine the magnetic force. Identical particles in the same field experience different forces if their velocities are different. Had the particles been electrons the forces would have been in the opposite directions. (Why?)

EXERCISE A 1.3-T magnetic field points in the x direction. A particle carrying 1.0 μC is moving at 20 m/s in the x-y plane and experiences a magnetic force of 1.4×10^{-5} N. What angle does the particle's velocity make with the x axis?

Answer: 33°

Some problems similar to Example 29-1: 1–11

Although electricity and magnetism are related, the electric field and the magnetic field are distinct. In particular, both may be present simultaneously. In that case, a charged particle experiences both the magnetic force $\mathbf{F}_B$ of Equation 29-1 and also the electric force $q\mathbf{E}$, for a net **electromagnetic force** given by

$$\mathbf{F} = q\mathbf{E} + q\mathbf{v} \times \mathbf{B}. \qquad \text{(electromagnetic force)} \qquad (29\text{-}2)$$

| EXAMPLE 29-2 | *A Velocity Selector* |

A wide range of applications, from manufacturing computer chips to cancer treatment, require charged-particle beams with precise velocities. In other cases—like analyzing charged-particle populations in interplanetary space—scientists need to classify particles by their speeds. Both tasks can be accomplished by a **velocity selector**, a device that uses electric and magnetic fields together to pass only particles with a specific velocity.

Figure 29-8 shows the perpendicular electric and magnetic fields used in a velocity selector; the fields have magnitudes E and B, respectively. A charged particle enters the field region, moving perpendicular to both fields, as shown. What speed must it have if it's to pass through the region undeflected?

Solution

The net force on a particle must be zero if it is to be undeflected. Consider a positively charged particle heading into the page in Fig. 29-8. The electric field points to the right, so the electric force $q\mathbf{E}$ is to the right. Applying the right-hand rule with the velocity vector $\mathbf{v}$ into the page and the magnetic field $\mathbf{B}$ downward gives a magnetic force to the left; since v and B are at right angles, the magnitude of the magnetic force is qvB. So the electric and magnetic forces are in opposite directions; for them to cancel their magnitudes must be equal:

$$qE = qvB,$$

or

$$v = \frac{E}{B}.$$

Particles with any other velocity will be deflected by a nonzero net force, so only those particles emerging undeflected have the precise speed $v = E/B$. Note that this selected speed is independent of the charge—and even of its sign, since the directions of both forces would reverse for a negative charge.

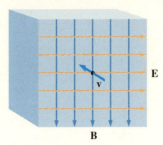

FIGURE 29-8 Crossed electric and magnetic fields give rise to oppositely directed forces on charged particles moving perpendicular to the fields. The magnetic force depends on the particle speed, but the electric force does not; therefore, only those particles with just the right speed can traverse the field region undeflected.

The velocity selector makes use of an important difference between the electric and magnetic forces: the electric force does not depend on a charged particle's velocity, while the magnetic force does. That's what makes possible the velocity selector's cancellation of the two forces at only one specific velocity. You can explore the velocity selector further with ActivPhysics Activity 13.8.

EXERCISE You have a magnet that produces a 0.25-T field. What strength electric field should you apply to make a velocity selector that passes 2.8×10^5 m/s particles undeflected?

Answer: 70 kV/m

Some problems similar to Example 29-2: 12–14

13.8
Velocity Selector

29-3 THE MOTION OF CHARGED PARTICLES IN MAGNETIC FIELDS

Like any force, the magnetic force deflects a particle from the straight-line path that Newton's first law says it would otherwise follow (Fig. 29-9). Use of magnetic forces to "steer" charged particles is the basis of many practical devices, from TV tubes to giant particle accelerators. Magnetic forces shape the trajectories of charged particles in a range of natural situations, especially in astrophysics.

FIGURE 29-9 (*a*) In the absence of a magnetic field, an electron beam moves in a straight line. (*b*) A magnetic field causes the beam to deflect. If the electrons are moving upward in the photo, which way is the field pointing?

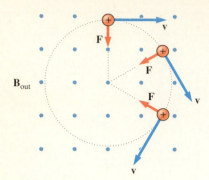

FIGURE 29-10 A positively charged particle moving at right angles to a uniform magnetic field describes a circular path. Note the convention that dots represent magnetic field lines emerging from the page; we designate this field by B_{out}. Crosses, in contrast, would represent field lines going into the page. You can think of the dots as the tips of arrows and crosses as the feathers at the tails.

Unlike most other forces, the magnetic force is necessarily at right angles to the particle velocity. That means force and particle displacement are perpendicular. Since work is the dot product of force with displacement (Equation 7-5; $W = \mathbf{F} \cdot \Delta\mathbf{r}$), and the dot product of perpendicular vectors is zero, we can conclude that

▌ **The magnetic force does no work on a charged particle.**

Both the kinetic energy and thus the speed of a charged particle subject only to a magnetic force are therefore constant. The magnetic force changes only the direction of the particle's velocity, not its magnitude. What sort of motion results from the magnetic force? Suppose a particle with positive charge q is moving at right angles to a uniform magnetic field $\mathbf{B}$, as shown in Fig. 29-10. At the instant the particle's velocity is to the right, the outward-pointing magnetic field gives a force $q\mathbf{v} \times \mathbf{B}$ that is downward. Since this force is at right angles to the velocity, it changes the direction but not the magnitude of the velocity. The magnetic force on the particle with its new velocity is still perpendicular to the velocity, and since the magnitude of the velocity is unchanged, so is the magnitude of the force. What we've just described—a force of constant magnitude always at right angles to a particle's velocity—is the prescription for uniform circular motion. Thus, a charged particle moving in a plane perpendicular to a uniform magnetic field describes a circular path. The magnetic force acts exactly like the tension in a string when you tie a mass to the string and whirl it around in a circle. The tension is perpendicular to the motion and changes its direction but not its speed.

How big is the circle, and how long does it take to get around? We can answer these questions like we do any questions about the effects of forces—namely, by using Newton's second law. In its circular path, the charged particle undergoes an acceleration v^2/r, directed toward the center of the circle. What causes this acceleration? The magnetic force! With the field and velocity at right angles, the magnitude of the magnetic force is just

$$F = qvB,$$

and the force points toward the center of the circle. Writing Newton's law $\mathbf{F} = m\mathbf{a}$ then gives

$$qvB = m\frac{v^2}{r},$$

so
$$r = \frac{mv}{qB}. \tag{29-3}$$

This result makes sense: the larger the particle's momentum mv, the harder it is for the magnetic force to bend it out of a straight line, so the larger the radius of the orbit. On the other hand, if we make the field or charge larger, then the magnetic force increases, giving a tighter orbit.

This variation of orbital radius is used in the mass spectrometer, a device that sorts ionized atoms and molecules according to their charge-to-mass ratio. Problem 27 and ActivPhysics Activity 13.7 show how the mass spectrometer works. Another practical application is the television picture tube, described in the example below.

13.7
Mass Spectrometer

Got It!

A uniform magnetic field points out of this page. Will an electron circle clockwise or counterclockwise as viewed from above the page?

EXAMPLE 29-3	*A TV Picture Tube*

In a TV picture tube, deflection of the electron beam that "paints" the TV picture is accomplished using magnetic fields. The geometry of a certain tube requires that the electron beam be bent in a circular arc with a minimum radius of 4.5 cm (Fig. 29-11). If the electrons are accelerated from rest through a 25-kV potential difference before they enter the magnetic field region, what is the magnetic field strength required? In what direction should the field point to accomplish the deflection shown in Fig. 29-11? Assume that the field is uniform over the deflecting region and zero elsewhere.

Solution

Solving Equation 29-3 for the magnetic field strength B gives

$$B = \frac{mv}{er},$$

with e the elementary charge. "Falling" through a potential difference $V = 25$ kV, the electron of charge e acquires a kinetic energy $\frac{1}{2}mv^2 = Ve$, so its speed is

$$v = \sqrt{\frac{2Ve}{m}}.$$

Our expression for the field then becomes

$$B = \frac{m}{er}\sqrt{\frac{2Ve}{m}} = \frac{1}{r}\sqrt{\frac{2mV}{e}}$$

$$= \frac{1}{0.045 \text{ m}}\left(\frac{(2)(9.1\times10^{-31} \text{ kg})(25\times10^{3} \text{ V})}{1.6\times10^{-19} \text{ C}}\right)^{1/2}$$

$$= 0.012 \text{ T}.$$

To achieve an initially downward deflection, the force $q\mathbf{v} \times \mathbf{B}$ must be initially downward. But the electron charge is negative, so $\mathbf{v} \times \mathbf{B}$ must be upward. Application of the right-hand rule to the rightward-moving electrons shows that the magnetic field $\mathbf{B}$ must be into the page in Fig. 29-11.

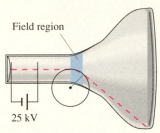

Field region

FIGURE 29-11 A TV picture tube requiring a 4.5-cm bending radius (marked with a diagonal line) for maximum deflection of the electron beam. Which way should the magnetic field point to deflect an electron in the direction indicated?

How does deflection of the electron beam result in a TV picture? In conventional TV, the magnetic field is varied so that three electron beams—one for each of the colors red, green, and blue—sweeps across the front of the screen in horizontal lines, starting at the top of the screen and working downward. A scan of the entire screen occurs thirty times each second. The electron current in each beam varies to match the brightness of its color at each point in the picture to be displayed, controlled by varying electric fields in the TV tube that themselves follow the signal coming into the TV from an antenna, cable, or VCR. The screen itself is coated with materials that glow in red, green, or blue when struck by the electrons, so the result is a reconstruction of the image that entered the TV as an electrical signal.

EXERCISE A beam of protons moving at 2.5×10^{5} m/s is deflected through a 90° turn by a 10-mT magnetic field oriented perpendicular to the beam's velocity. What is the radius of the turn?

Answer: 26 cm

Some problems similar to Example 29-3: 15, 18, 21

The Cyclotron Frequency

How long does it take a charged particle to complete its circular orbit in a uniform magnetic field? The circumference of the orbit is $2\pi r$, so the period T of the circular motion is

$$T = \frac{2\pi r}{v}.$$

Using Equation 29-3 for the radius r gives

$$T = \frac{2\pi r}{v} = \frac{2\pi}{v}\frac{mv}{qB} = \frac{2\pi m}{qB}. \qquad (29\text{-}4)$$

13.4

Magnetic Force on a Particle

FIGURE 29-12 The Crab Nebula, remnant of a supernova explosion observed nearly 1000 years ago. Intense radio emission occurs as electrons undergo circular motion in the Crab's magnetic field. The frequency of the emission is related to the cyclotron frequency of the electrons' motion, and its measurement therefore allows determination of the magnetic field.

This remarkable result shows that the period of the circular motion is independent of the particle's speed and the size of its orbit. You can see why from Equation 29-3: the larger the speed v the larger the radius r, and therefore the circumference, in direct proportion to the speed. So a faster particle describes a larger circle, and ends up taking the same time to go around. Physically, that's because the magnetic force depends on the particle's speed.

Equivalently, we can describe the particle's circular motion in terms of its frequency, in revolutions per second, which is just the inverse of the period:

$$f = \frac{qB}{2\pi m}. \qquad (29\text{-}5)$$

This quantity is called the **cyclotron frequency**, since it's the frequency at which charged particles circulate in a cyclotron particle accelerator. Because the cyclotron frequency depends only on the magnetic field and the particle's charge-to-mass ratio, observing cyclotron motion provides astrophysicists a direct measure of magnetic fields in distant astrophysical objects (Fig. 29-12). Conversely, a fixed magnetic field guarantees a specific cyclotron frequency, regardless of the particles' speeds. Microwave ovens exploit this fact, their microwaves generated by electrons circling 2.4 billion times each second in a special tube called a magnetron (see Problem 21).

APPLICATION	*The Cyclotron*

Physicists studying the basic structure of matter need tools that can probe the atomic nucleus and its constituent particles. The only probes sufficiently small are subatomic particles themselves, accelerated to high enough energies that they can disrupt the strong nuclear force. How is this acceleration accomplished? One way is to accelerate particles of charge q through a large potential difference V, giving each particle energy qV. But there are practical problems in the generation and handling of potential differences much over a million volts. To achieve higher energies, devices are used that circumvent the need for a single large potential difference. One of the earliest and most successful such devices is the **cyclotron** (Fig. 29-13). The device consists of an evacuated chamber between the poles of a magnet. At the center of the chamber is a source of the particles to be accelerated, usually protons or light ions. The ions undergo circular motion in the magnetic field.

Also within the evacuated chamber are two hollow conducting structures each shaped like the letter D (Fig. 29-12). A modest potential difference is applied across these two "dees," and this potential difference is made to alternate in polarity with the same frequency as the circular motion of the ions. Recall that this frequency depends only on the magnetic field strength and the charge-to-mass ratio of the ions, but not on their energy. As the ions circle around in the cyclotron, they are accelerated across the gap between the dees by the strong electric field associated with the potential difference. Because each

dee is a hollow, nearly closed conducting structure, there is no electric field within the dees.

Once inside a dee, the ions simply follow a circle in the magnetic field. Halfway round they again cross the gap between dees. If the electric field were steady, the ions would be *decelerated* at this crossing. But the electric field changes direction in step with the ions' circular motion, so each time the ions cross the dee gap they are accelerated and gain more energy. They move faster and in ever larger circles, but always

FIGURE 29-13 A cyclotron, showing one magnet pole, dees, and a typical ion trajectory (in red). Not shown are the vacuum chamber surrounding the dees, upper pole, and frame supporting pole pieces and electromagnet windings.

with the same orbital period. Eventually the ion orbits become nearly the size of the machine. At this point an electrostatic field provided by a high-voltage electrode deflects the ions out of the magnetic field and toward a target, where their interactions with target nuclei cause nuclear reactions.

In addition to providing experimental data on nuclear structure, cyclotrons are valuable in producing short-lived radioactive isotopes for a variety of purposes, particularly medical research and diagnosis. A number of large hospitals have their own cyclotrons (Fig. 29-14); in particular, the diagnostic procedure known as positron emission tomography (PET) requires cyclotron-produced radioisotopes.

At very high energies, the theory of relativity (Chapter 38) comes into play and alters our conclusion that the cyclotron frequency is independent of particle energy. As a result, the cyclotron cannot be used to achieve these relativistic energies. An alternate accelerator design is the **synchrotron**, in which both the magnetic field and frequency of an alternating electric field are varied to account for increasing particle energy, while the orbital radius is held constant (Fig. 29-15).

FIGURE 29-14 This cyclotron at Massachusetts General Hospital was used to produce radioisotopes for research leading to the development of the medical diagnostic procedure known as positron emission tomography (PET).

FIGURE 29-15 Aerial view of Fermilab, the Fermi National Accelerator Laboratory at Batavia, Illinois. Large circle is the 2-km diameter Tevatron, a synchrotron that accelerates protons to energies approaching 1 TeV (10^{12} eV).

EXAMPLE 29-4	*Designing a Cyclotron*

A cyclotron is to accelerate protons to a kinetic energy of 5.0 MeV. If the magnetic field in the cyclotron is 2.0 T, what must be the radius of the cyclotron and the frequency at which the dee voltage is alternated?

Solution
The cyclotron frequency is given by Equation 29-5:

$$f = \frac{qB}{2\pi m} = \frac{(1.6 \times 10^{-19}\ \text{C})(2.0\ \text{T})}{(2\pi)(1.67 \times 10^{-27}\ \text{kg})} = 3.0 \times 10^7\ \text{Hz}.$$

This is the frequency required to accelerate protons at each crossing of the dee gap; incidentally, it is about the frequency of a citizens band (CB) radio transmitter.

An energy of 5.0 MeV is equal to

$$(5.0 \times 10^6\ \text{eV})(1.6 \times 10^{-19}\ \text{J/eV}) - 8.0 \times 10^{-13}\ \text{J},$$

so the proton kinetic energy is

$$K = \tfrac{1}{2}mv^2 = 8.0 \times 10^{-13}\ \text{J}.$$

Solving for the speed v gives

$$v = \sqrt{\frac{2K}{m}} = \sqrt{\frac{(2)(8.0 \times 10^{-13}\ \text{J})}{1.67 \times 10^{-27}\ \text{kg}}} = 3.1 \times 10^7\ \text{m/s}.$$

Equation 29-3 then gives the radius needed to accommodate 5-MeV protons:

$$r = \frac{mv}{qB} = \frac{(1.67 \times 10^{-27}\ \text{kg})(3.1 \times 10^7\ \text{m/s})}{(1.6 \times 10^{-19}\ \text{C})(2.0\ \text{T})} = 0.16\ \text{m}.$$

To ensure a uniform magnetic field over the particle trajectories, the radii of the magnet pole pieces would have to be somewhat larger than this value.

EXERCISE A microwave oven uses 2.4-GHz microwaves, generated by electrons circling at this frequency in the magnetic field of a magnetron tube. (a) What is the magnetic field strength? (b) If the energy associated with an electron's circular motion is 890 eV, what is the radius of its circular path?

Answers: (a) 86 mT; (b) 1.2 mm

Some problems similar to Example 29-4: 17, 25, 26

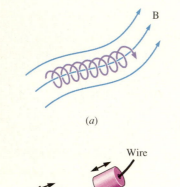

FIGURE 29-16 A particle in a uniform magnetic field describes a spiral path, its motion along the field direction unaffected by the magnetic force.

Particle Trajectories in Three Dimensions

What if a particle's motion is not confined to a plane perpendicular to the magnetic field? Then we can resolve its velocity into two vectors, one perpendicular and the other parallel to the magnetic field. Since the magnetic force is always perpendicular to the field, there is no component of force along the field direction—and therefore the velocity component in this direction is unaffected by the magnetic force. The force acts only on the velocity component perpendicular to the field, and here our previous analysis applies: in the plane perpendicular to the field, the particle's motion is circular with frequency given by Equation 29-5. In the absence of other forces the particle moves uniformly along the field direction while executing circular motion perpendicular to the field. The resulting trajectory is a spiral, as shown in Fig. 29-16.

The absence of magnetic force along the field direction means that particles move readily along the field. But if you try to push them at right angles to the field, they simply move in larger circles about the field direction. As a result, charged particles are often described as being "frozen" to the magnetic field (Fig. 29-17). Nonuniform fields and collisions between particles make this "freezing" of particles and field less than perfect, but in many cases the particle density is low enough that the "frozen" assumption is an excellent approximation. Unlike our relatively cool planet Earth, much of the universe consists of free electrons and protons, not bound into neutral atoms. As a result, magnetic fields are a dominant influence on matter throughout much of the universe (Fig. 29-18).

An important terrestrial application of this "trapping" of particles on magnetic field lines occurs in nuclear fusion reactors. These experimental devices—whose successful development would make 1 gallon of sea water the energy equivalent of more than 300 gallons of gasoline—have to contain ionized gas (plasma) at temperatures around 100 million kelvins. They do so with "magnetic bottles," most commonly toroidal (doughnut-shaped) chambers whose circular magnetic fields never intersect the chamber walls and therefore keep the hot plasma from touching the walls (Fig. 29-19).

FIGURE 29-17 (a) Charged particles undergoing spiral motion about the magnetic field direction are "frozen" to the field like (b) beads sliding along a wire.

FIGURE 29-18 Magnetic fields govern the behavior of matter throughout much of the universe. Here, concentrations of ionized gas trace out magnetic field loops in the Sun's atmosphere.

FIGURE 29-19 (a) Charged particles spiral about the circular field lines in a simplified fusion reactor. With its toroidal shape, the machine has no "ends" from which magnetic field lines emerge. The field therefore keeps charged particles away from the chamber walls. (b) The Tokamak Fusion Test Reactor at Princeton University. This view, taken during construction, shows clearly the machine's toroidal shape.

Earth possesses a magnetic field whose origin we'll consider in the next chapter. This field shields Earth's surface from the direct impact of high-energy particles from the Sun and elsewhere in the cosmos. Particles reaching Earth's vicinity are trapped into spiral motion about the magnetic field, so they can't penetrate directly to the surface. This trapping generally occurs on the field lines that intersect Earth in the polar regions (Fig. 29-20*a*). High-energy particles moving along these field lines slam into the upper atmosphere near the poles, exciting oxygen and nitrogen atoms and thereby producing the spectacular displays we call the aurora (Fig. 29-20*b, c*). Bursts of solar particles also buffet Earth's magnetic field, causing "magnetic storms" that can severely disrupt communications and electric power transmission.

Fig. 29-21 shows a particle spiraling into a region of stronger magnetic field, like the polar regions in Fig. 29-20*a*. Note that there is a component of magnetic force opposite to the general direction of motion. This force eventually reverses the particle's forward motion, with the result that the stronger field region acts like a **magnetic mirror**, reflecting particles that attempt to penetrate. This means that particles trapped in Earth's magnetic field bounce back and forth between northern and southern hemispheres, reflected by the magnetic mirrors of higher field strength near the poles. Auroral displays in opposite hemispheres are often strikingly similar since they are caused by the same particle populations mirroring back and forth. Particles that penetrate the upper atmosphere and cause auroral displays combine with atmospheric atoms and are lost to the mirroring particle population, but they're continually replaced with new particles arriving from the Sun.

(*a*)

(*b*)

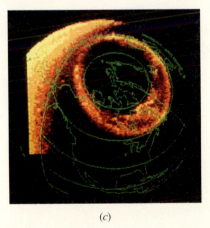
(*c*)

FIGURE 29-20 The aurora arises from the interaction of high-energy particles trapped in Earth's magnetic field. (*a*) Particles trapped in the magnetic field spiral between the northern and southern polar regions. Diagram is highly simplified; Earth's field is actually more complicated because of interactions with the solar wind, a stream of ionized gas flowing outward from the Sun. (*b*) A colorful aurora viewed from Canada's Prince Edward Island. (*c*) The aurora appears as a circle surrounding Earth's north magnetic pole in this satellite image.

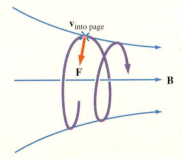
vinto page
F
B

FIGURE 29-21 The magnetic mirror effect. A charged particle is spiraling into a region of stronger magnetic field. At the point shown its velocity is into the page, and the magnetic force—proportional to **v** × **B**—has a component to the left. This force eventually reverses the rightward motion of the particle.

29-4 THE MAGNETIC FORCE ON A CURRENT

So far we've considered the magnetic force on individual charged particles. An electric current is simply a group of charged particles sharing a common motion, so we should expect a current to interact with a magnetic field.

The Force on a Straight Wire

Imagine a long straight wire carrying a current I through a uniform magnetic field $\mathbf{B}$, as shown in Fig. 29-22. If each charge carrier has a drift velocity $\mathbf{v}_d$ along the wire, and if each carries charge q, then the magnetic force on each charge carrier is given by Equation 29-1:

$$\mathbf{F}_q = q\mathbf{v}_d \times \mathbf{B}.$$

(Of course the charge carriers also have random thermal velocities. But these average to zero and therefore make no net contribution to the magnetic force.) If the wire has cross-sectional area A and contains n charge carriers per unit volume, then the net force on all the charge carriers in a length ℓ of the wire is

$$\mathbf{F} = nA\ell q\mathbf{v}_d \times \mathbf{B}.$$

The product $nAqv_d$ is just the current I, as we found in deriving Equation 27-2. If we define a vector $\boldsymbol{\ell}$ whose magnitude is the length ℓ of the wire and whose direction is along the current, then we can write

$$\mathbf{F} = I\boldsymbol{\ell} \times \mathbf{B}. \qquad \text{(magnetic force on a current)} \qquad (29\text{-}6)$$

The direction of this force is at right angles to both the current and the magnetic field, or out of the page in Fig. 29-22. For a given direction of the current, the direction of the force does not depend on the sign of the charge carriers. If the current is to the right, then positive charges move to the right and the force on each is out of the page. If the charges are negative, they move to the left and the force is still out of the page because both the sign of the velocity and the sign of the charge are reversed, given the same sign for the force $q\mathbf{v} \times \mathbf{B}$.

Strictly speaking, Equation 29-6 gives the net magnetic force only on the charge carriers in the wire. But the motion of the charge carriers—typically electrons—under the influence of the magnetic force causes charge separation in the wire, and the resulting electric field exerts a force on the fixed charges—typically ions—in the wire (Fig. 29-23). Thus the entire wire experiences the force. Although its origin is not entirely magnetic, we loosely call the force given by Equation 29-6 the magnetic force on the wire.

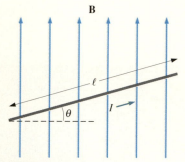

FIGURE 29-22 A straight wire carrying current I through a uniform magnetic field. The charge carriers in the wire experience a magnetic force, in this case out of the page.

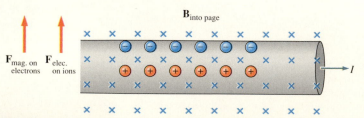

FIGURE 29-23 Electrons moving to the left in a wire are deflected upward by the magnetic force, resulting in charge separation and therefore in an upward electric force on the fixed ions. As a result the entire wire is ultimately influenced by the magnetic force on the moving electrons.

The magnetic force on a current-carrying wire is the basis for many practical devices, including loudspeakers and the electric motors that start cars and run refrigerators, CD players, computer disk drives, subway trains, pumps, food processors, power tools, and myriad other useful instruments of modern society. At the end of this chapter we'll see just how electric motors work.

Got It!

Figure 29-24 shows a flexible, current-carrying wire passing through a region where a magnetic field points out of the page. The wire is deflected to the right, as shown in Fig. 29-24. Is the current flowing upward or downward in the wire?

$\mathbf{B}_{out}$

FIGURE 29-24 A flexible wire passes through a region containing a magnetic field pointing out of the page. Which direction is the current in the wire when it's deflected as shown?

EXAMPLE 29-5	*Magnetic Force on a Power Line*

A power line runs along Earth's equator, where Earth's magnetic field points horizontally from south to north and has a strength of about 0.5 G. The current in the power line is 500 A, flowing from west to east. What are the magnitude and direction of the magnetic force on 1 km of the power line?

Solution

Let eastward be the x direction, northward the y direction, and upward the z direction. After we convert gauss to tesla, Equation 29-6 gives

$$\mathbf{F} = I\boldsymbol{\ell} \times \mathbf{B} = (500 \text{ A})(1000\hat{\mathbf{i}} \text{ m}) \times (0.5 \times 10^{-4}\,\hat{\mathbf{j}} \text{ T})$$
$$= 25\hat{\mathbf{k}} \text{ N}.$$

This 25-N upward force is negligible compared with the weight—on the order of 2×10^4 N—of 1 km of power line.

EXERCISE A 2.0-m-long wire has a mass of 37 g. The wire extends horizontally, at 40° to a horizontal magnetic field of 0.075 T. What current in the wire will result in the magnetic force suspending it against gravity?

Answer: 3.8 A

Some problems similar to Example 29-5: 33, 35, 37

APPLICATION	*Magnetic Levitation and Propulsion*

Engineers throughout the world are at work on transportation systems whose operation results directly from the magnetic force on a straight current-carrying conductor. In so-called maglev vehicles, magnetic forces levitate the vehicle just a few centimeters above a conducting guideway and also provide the horizontal force that propels the vehicle (Fig. 29-25). Such vehicles should be capable of 500-km/h speeds, well above the 300 km/h maximum for high-speed rail systems. Maglev vehicles should prove especially effective for travel between cities in densely populated regions.

A related application is so-called magnetohydrodynamic (MHD) propulsion, a kind of jet propulsion in which a

FIGURE 29-25 A magnetically levitated vehicle under development in Japan.

(a)

Source: Mitsubishi Heavy Industries

(b)

conducting fluid is accelerated by the magnetic force. Engineers are currently experimenting with MHD propulsion for ships. Because sea water is a reasonably good electrical conductor, passage of a current through sea water in a magnetic field results in a magnetic force on the water. Figure 29-26 shows the experimental Japanese ship Yamato-1 and its MHD propulsion system.

FIGURE 29-26 (a) The Japanese ship Yamato-1 uses MHD propulsion. (b) Schematic diagram of the MHD propulsion unit. Superconducting electromagnet produces a magnetic field; current flowing in the sea water between the two electrodes then experiences a magnetic force that ejects a jet of water at right.

APPLICATION	*The Hall Effect*

The force on a current-carrying conductor is independent of the sign of the charge carriers. But there's a subtle difference. Fig. 29-27 shows two conductors, each with the same current I to the right and magnetic field **B** pointing into the page. In Fig. 29-27a the current is carried by electrons moving to the *left*. The product $\mathbf{v} \times \mathbf{B}$ is downward, but since electrons are negative the force $q\mathbf{v} \times \mathbf{B}$ is upward. As a result, the upper edge of the conductor is negative with respect to the lower edge. In Fig. 29-27b the current is carried by protons moving to the *right*. Again the force $q\mathbf{v} \times \mathbf{B}$ is upward, so now the upper edge of the conductor is positive.

This phenomenon of charge separation is the **Hall effect**. The separated charge gives rise to an electric field and therefore to a measurable potential difference—the **Hall potential**—between opposite edges of the conductor. The sign of the Hall potential depends on the sign of the charge carriers. Charge separation will stop once the electric force on the charge carriers cancels the magnetic force. The electric force has magnitude qE and, with **B** perpendicular to the current, the magnetic force has magnitude qv_dB, with v_d the charge-carrier drift speed. Equating these magnitudes gives $E = v_dB$. In a rectangular conductor this field will be essentially uniform, so the Hall potential becomes

$$V_H = Eh = v_dBh,$$

where h is the conductor height shown in Fig. 29-27. But Equation 27-2 gives $I = nAqv_d$, with n the number density of charge carriers, A the conductor's cross-sectional area, and q

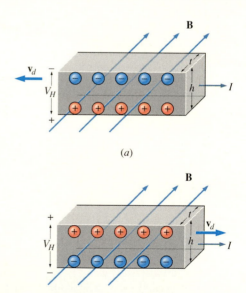

FIGURE 29-27 The Hall potential V_H arises from the deflection of charge carriers by the magnetic force when a current-carrying conductor is in a magnetic field. Although the magnetic force on the conductor is independent of the sign of the charge carriers, the sign of the Hall potential does depend on the sign of the charge carriers. In both (a) and (b) the current is to the right, carried in (a) by negative charge carriers moving to the left and in (b) by positive charge carriers moving to the right.

the charge on each carrier. Solving for v_d and using the result in our expression for V_H gives

$$V_H = \frac{IBh}{nAq},$$

or, since $A = ht$ with t the conductor thickness shown in Fig. 29-27,

$$V_H = \frac{IB}{nqt}. \qquad (29\text{-}7)$$

The quantity $1/nq$ is the **Hall coefficient**. Measurement of V_H in a sample of known thickness carrying known current in a known magnetic field yields the value of this coefficient and, thus, gives information on the nature and density of the charge carriers. Alternatively, measurement of V_H in a material with known Hall coefficient carrying a known current provides a direct measure of the magnetic field strength (see Problem 45).

Nonuniform Fields and Curved Conductors

Equation 29-6 gives the force, $\mathbf{F} = I\boldsymbol{\ell} \times \mathbf{B}$, on a straight conductor of length ℓ carrying current I in a uniform magnetic field $\mathbf{B}$. We can still use this equation if the conductor bends or the field varies with position, if we apply it only to a very small segment of this conductor—so small that the segment is essentially straight and the field essentially uniform over the segment. Designating this segment by $d\boldsymbol{\ell}$, Equation 29-6 becomes

$$d\mathbf{F} = I\,d\boldsymbol{\ell} \times \mathbf{B}$$

for the small force on the segment (Fig. 29-28). To find the total force on the conductor, we sum the forces on all such segments. In the limit of very small segments, this sum becomes an integral, and we have

$$\mathbf{F} = \int d\mathbf{F} = \int I\,d\boldsymbol{\ell} \times \mathbf{B}. \qquad (29\text{-}8)$$

The integration is taken over the entire section of conductor on which we are calculating the force.

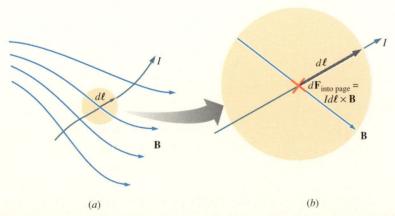

(a) $\qquad\qquad\qquad\qquad\qquad\qquad (b)$

FIGURE 29-28 (a) A curved conductor in a nonuniform magnetic field. (b) A small segment $d\boldsymbol{\ell}$ can be treated as a straight wire in a uniform field; the magnetic force on the segment is $d\mathbf{F} = I\,d\boldsymbol{\ell} \times \mathbf{B}$.

EXAMPLE 29-6   *The Force on a Curved Conductor*

A semicircular wire connects two points C and D a horizontal distance $2R$ apart (Fig. 29-29a). The wire carries a current I from C to D and is in a uniform magnetic field **B** pointing upward. Show that the magnetic force on this semicircular wire is the same as the force that a straight wire from C to D would experience if it carried the same current I.

Solution

Since the orientation of the wire relative to the field varies, we use the integral in Equation 29-8 to calculate the force. Figure 29-29a shows an infinitesimal segment $d\ell$ of the wire, and the angle θ that specifies its position on the semicircular arc. This infinitesimal segment subtends an infinitesimal angle $d\theta$, as shown. A blown-up view of the segment (Fig. 29-29b) shows that the angle between the segment $d\ell$ and the magnetic field **B** is also θ. Then the magnitude of the force on the segment is

$$dF = I\,d\ell\,B\sin\theta.$$

Application of the right-hand rule shows that the direction of the force element is out of the page for all values of θ on the semicircle—that is, for $0 \leq \theta \leq \pi$. Therefore, the integral in Equation 29-8 describes a sum of vectors all of which point in the same direction, so the magnitude of the net force is

$$F = \int dF = \int_0^\pi IB\sin\theta\,d\ell.$$

But $d\ell$ subtends the angle $d\theta$, so $d\ell = R\,d\theta$ (Fig. 29-30). Then, taking the constants I, B, and R outside the integral, we have

$$F = IBR\int_0^\pi \sin\theta\,d\theta = IBR\left.(-\cos\theta)\right|_0^\pi$$

$$= IBR[-(-1-1)] = IB(2R).$$

This is just the force we would get from Equation 29-6 for a straight wire of length $2R$ carrying current I perpendicular to a magnetic field **B**. You can see why this is: Equation 29-6 shows that the magnetic force depends on the quantity $d\ell\sin\theta$, which is the component of the segment $d\ell$ perpendicular to the field. That component is what you get if you take the straight path from C to D instead of the semicircle. Indeed, the force on any piece of wire carrying the same current from C to D would be the same as long as it stayed in the region of uniform field. A corollary is that the net force on any *closed* loop of current in a uniform magnetic field must be zero (see Problem 43).

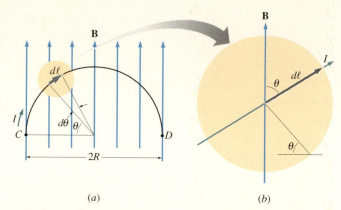

FIGURE 29-29 (a) A semicircular wire carries current I between points C and D. (b) Blowup of an infinitesimal segment $d\ell$ of the wire, showing that the segment makes an angle θ with the magnetic field.

FIGURE 29-30 An angle in radians is defined as the ratio of subtended arc to radius. Here the angle $d\theta$ is so small that the difference between a circular arc and the straight segment $d\ell$ is insignificant. Therefore $d\ell = R\,d\theta$.

EXERCISE A 15-cm-long wire carrying 4.8 A is bent into a quarter circle and oriented like the left half of the semicircular wire in Fig. 29-29a (i.e., from point C to the top of the semicircle). If the magnetic field shown in Fig. 29-29a has magnitude 0.65 T, what is the magnetic force on the wire?

Answer: 0.30 N

Some problems similar to Example 29-6: 40–43

A common application using the force on a current-carrying wire is the loudspeaker, shown in Fig. 29-31. Here an unusually shaped magnet produces a nonuniform magnetic field that results in a net force on a current-carrying coil. The current in the coil alternates direction in accordance with the audio signal

FIGURE 29-31 A loudspeaker uses the magnetic force on a current-carrying coil to convert electrical energy into sound. (*a*) Cutaway view; (*b*) top view of the magnet, showing the nonuniform field and placement of the coil. With the current direction indicated, the net force is into the page. There's a net force on the loop only because the field is nonuniform.

from an amplifier, and this results in an alternating force that moves the coil back and forth. The coil, in turn, is connected to a flexible cone that moves to produce sound waves.

29-5 A CURRENT LOOP IN A MAGNETIC FIELD

Closed loops of current are important in a wide variety of natural and technological systems. Loops of current-carrying wire are at the heart of every electric motor, while miniature current loops associated with charged quarks circulating within the proton are responsible for nuclear magnetic resonance (NMR), widely used to determine the structure of chemical compounds and as the basis of the medical technique called magnetic resonance imaging (MRI). We now show that the magnetic forces on different parts of a current loop result in a torque on the loop; it's that torque that drives electric motors and the motion of protons in NMR.

Consider a rectangular current loop in a uniform magnetic field, with the normal to the loop making an angle θ with the magnetic field, as shown in Fig. 29-32*a*. Since current flows in opposite directions on opposite sides and since the field is uniform, the magnetic forces on opposite sides cancel and there is no net magnetic force on the loop. The right-hand rule shows that the forces on the top and bottom of the loop point directly upward and downward, respectively, and therefore result in no net torque.

The forces on the vertical sections of the loop, however, do produce a torque. Figure 29-32*b* shows a top view of the loop, with the forces resulting from the

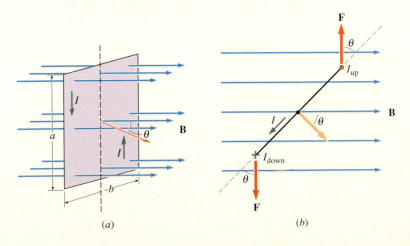

FIGURE 29-32 (*a*) A rectangular current loop in a uniform magnetic field. Shading indicates the plane of the loop. The angle between the loop normal and the magnetic field is θ. (*b*) Top view of the current loop, showing that magnetic forces on the vertical sides result in a net torque. The angles used in the torque calculation are the same as the angle θ between the loop normal and the magnetic field.

upward- and downward-flowing currents in the vertical sides. From the figure, it's clear that these forces give rise to a net torque twisting the loop counterclockwise. Each vertical side is a straight wire of length a carrying current I at right angles to the horizontally directed magnetic field **B**, so the magnitude of the force on each side is simply

$$F_{\text{side}} = IaB.$$

The distance from the loop's central axis to the sides where this force acts is half the loop width, or $b/2$. The geometry of Fig. 29-32b shows that the angle used in calculating the torque is the same as the angle θ between the loop normal and the magnetic field. Therefore, the torque due to the force on each side is

$$\tau_{\text{side}} = F\frac{b}{2}\sin\theta = Ia\frac{b}{2}B\sin\theta.$$

Accounting for the contributions from both sides gives the magnitude of the net torque on the loop:

$$\tau = IabB\sin\theta.$$

We can express the torque in vector notation if we define a vector **A** whose magnitude is the area ab of the loop and which is perpendicular to the loop. We choose the direction of **A** by the right-hand rule: wrap your fingers around the loop in the direction of the current and your thumb points in the direction of **A**. This is the same direction as the loop normal shown in Fig. 29-32. Then we can write

$$\boldsymbol{\tau} = I\mathbf{A} \times \mathbf{B}. \qquad (29\text{-}9)$$

Although we derived this equation for a rectangular loop, it holds in fact for any current loop. The torque on a current loop depends on the current, the loop area, the magnetic field, and the orientation between loop and field.

Equation 29-9 should remind you of Equation 23-12 for the torque on an electric dipole in an electric field. There we had

$$\boldsymbol{\tau} = \mathbf{p} \times \mathbf{E},$$

where **p** is the electric dipole moment and **E** the electric field. Comparison with Equation 29-9 suggests that a current loop in a magnetic field behaves analogously to an electric dipole in an electric field. As far as its response to magnetic fields is concerned, therefore,

▎ **A current loop constitutes a magnetic dipole.**

The quantity $I\mathbf{A}$ in Equation 29-9 plays the same role as the electric dipole moment in the equation $\boldsymbol{\tau} = \mathbf{p} \times \mathbf{E}$. We therefore call this quantity the **magnetic dipole moment, $\boldsymbol{\mu}$**:

$$\boldsymbol{\mu} = I\mathbf{A}. \qquad \text{(single-turn loop)}$$

The direction of the vector $\boldsymbol{\mu}$ is the same as the direction we defined for **A**: curl your right fingers in the direction of the loop current, and your thumb points in the direction of $\boldsymbol{\mu}$ (Fig. 29-33). (With ActivPhysics Activity 13.6 you can

FIGURE 29-33 Using the right-hand rule to find the direction of a current loop's magnetic moment.

graphically resize a current loop to see the effect on magnetic moment and torque.) More generally, a loop may consist of N turns of conducting wire; then each contributes $I\mathbf{A}$ to give the total magnetic moment:

$$\mu = NI\mathbf{A}. \qquad \text{(magnetic dipole moment)} \qquad (29\text{-}10)$$

Clearly the units of magnetic moment are A·m². Using the magnetic moment vector, Equation 29-9 then becomes

$$\tau = \mu \times \mathbf{B}, \qquad \text{(torque on a current loop)} \qquad (29\text{-}11)$$

in analogy with the electric case. The magnetic moment vector of a current loop is perpendicular to the plane of the loop, and the torque given in Equation 29-11 tends to align the magnetic moment with the field (Fig. 29-34). It takes work to twist the loop's magnetic moment vector out of alignment with the field. In exact analogy with the electric-dipole case summarized in Equation 23-10, we express the associated potential energy as

$$U_{\text{magnetic}} = -\mu \cdot \mathbf{B}. \qquad (29\text{-}12)$$

13.6
Torque on a Current Loop

FIGURE 29-34 The torque on a current loop tends to align the loop's magnetic moment vector with the magnetic field.

| **EXAMPLE 29-7** | *Pointing a Satellite* |

Some satellites use the magnetic torque on current loops to point themselves in space. Earth itself provides the magnetic field, while electricity from solar panels powers coils that constitute the current loops; unlike rocket-based pointing systems, there's no fuel to run out. Three mutually perpendicular coils allow the satellite to point itself in any direction. One satellite uses 30-cm-diameter circular coils, each with 1000 turns. (a) If the satellite orbits at an altitude where Earth's magnetic field is 0.24 G, and if specifications call for a maximum torque of 1.3 mN·m from each coil, what should be the coil current? (b) How much work is done on the coil to rotate it from its maximum-torque orientation to an angle of 120° to the field?

Solution

(a) The maximum torque occurs when the coil's magnetic moment is perpendicular to the field; Equation 29-11 shows that this maximum torque has magnitude $\tau_{\text{max}} = \mu B$. The magnetic moment of the coil is the product of the number of turns, the current, and the coil area πR^2, as given by Equation 29-10. Thus the maximum torque is

$$\tau_{\text{max}} = NI\pi R^2 B.$$

Solving for the current gives

$$I = \frac{\tau_{\text{max}}}{N\pi R^2 B} = \frac{1.3 \times 10^{-3} \text{ N·m}}{(1000)(\pi)(0.15 \text{ m})^2(0.24 \times 10^{-4} \text{ T})}$$

$$= 0.766 \text{ A}.$$

(b) The loop's potential energy is given by Equation 29-12: $U = -\mu \cdot \mathbf{B} = -\mu B \cos\theta$. Maximum torque occurs when $\theta = 90°$, so the work in rotating the coil—equal to the difference in its potential energies in the two orientations—is

$$W = U_{120°} - U_{90°} = -\mu B \cos 120° - (-\mu B \cos 90°)$$

$$= (-1.3 \times 10^{-3} \text{ N·m})(\cos 120°)$$

$$= 0.65 \text{ mJ},$$

where we recognized the quantity μB as the maximum torque defined earlier.

EXERCISE An electric motor consists of a 550-turn loop with area 100 cm² in a 0.23-T magnetic field. What loop current is necessary if the motor is to develop a maximum torque of 3.7 N·m?

Answer: 2.9 A

Some problems similar to Example 29-7: 47, 48, 50–52

Electric motors are so much a part of our lives that we hardly think of them. But CD players, car starters, refrigerators, vacuum cleaners, power saws, subway trains, computer disk drives, food processors, washing machines, fans, hair dryers, water pumps, oil burners, and most industrial machinery would be difficult or impossible to build without electric motors (Fig. 29-35).

At the heart of every electric motor is a current loop in a magnetic field. But instead of a steady current, the loop carries a current that reverses periodically. In direct-current (DC) motors, this reversal is achieved through the electrical contacts

FIGURE 29-36 A simple electric motor. Current flows from the battery through the stationary brushes to the rotating commutator and loop. The direction of the loop current reverses as the two halves of the commutator rotate to contact different brushes.

FIGURE 29-35 (*a*) A small electric motor, partially disassembled. At left is the rotating coil (1), surrounded by permanent magnets (2). The commutator (3) is on the forward shaft. At right are the brushes (4), which fit around the commutator to supply current to the rotating coils. (*b*) This electric motor powers General Motors' EV1 electric car.

that provide current to the loop. Figure 29-36 shows a simple motor consisting of a loop in the field of a permanent magnet. Current from an external battery reaches the loop through a set of stationary "brushes," which make contact with a pair of semicircular conductors called the "commutator." The commutator is attached rigidly to the loop, which rotates to align its dipole moment with the field. Just as it reaches alignment, however, the brushes cross the gaps in the commutator. This reverses the connections to the battery, reversing the loop current. This in turn reverses the magnetic moment of the loop, so it's no longer aligned with the field. The loop then rotates another 180° to align its new magnetic moment vector with the field. Just as it reaches alignment, the current again reverses. This process repeats, resulting in continuous rotation of the loop. A rigid shaft through loop and commutator delivers mechanical work to the device powered by the motor. The source of this work is the battery or whatever supplies electrical energy to the motor. The motor itself is a device that converts electrical to mechanical energy; the magnetic field is an intermediary in this conversion.

CHAPTER SYNOPSIS

Summary

1. **Magnetism** is a fundamental interaction described in terms of **magnetic fields**. Magnetic fields interact with **moving electric charges**. A moving charged particle experiences a force that depends on its charge, its velocity, and the magnetic field. The force is at right angles to both velocity and field, and is given by

$$\mathbf{F} = q\mathbf{v} \times \mathbf{B}.$$

Because it's at right angles to the particle's velocity, the magnetic force does no work on a charged particle. In a uniform magnetic field, a particle's trajectory is a circle or spiral, with orbital period independent of the particle's speed.

2. An electric current is made up of moving electric charges, so there is a net magnetic force on an electric current. For a straight wire of length ℓ carrying current I in a magnetic

field **B**, the force is

$$\mathbf{F} = I\boldsymbol{\ell} \times \mathbf{B}.$$

3. A particularly important case of a current in a magnetic field is a closed current loop. Such a loop behaves like a **magnetic dipole** with magnetic dipole moment

$$\boldsymbol{\mu} = NIA,$$

where N is the number of turns in the loop, I the loop current, and **A** a vector perpendicular to the plane of the loop and whose magnitude is the loop area. A current loop in a magnetic field experiences a torque given by $\tau = \boldsymbol{\mu} \times \mathbf{B}$, and the associated potential energy is $U_B = -\boldsymbol{\mu} \cdot \mathbf{B}$.

Terms You Should Understand

(Pairs are closely related terms whose distinction is important; number in parentheses is chapter section where term first appears.)

magnetic field (29-1, 29-2)
magnetic force, electromagnetic force (29-2)
tesla, gauss (29-2)
cyclotron frequency (29-3)
cyclotron, synchrotron (29-3)
magnetic mirror (29-3)
Hall effect, Hall potential (29-4)
current loop (29-5)

magnetic dipole moment (29-5)
electric motor (29-5)

Symbols You Should Recognize

B (29-1, 29-2)
T (29-2)
$\boldsymbol{\mu}$ (29-5)

Problems You Should Be Able to Solve

calculating magnetic force vectors given charge, velocity, and magnetic field (29-2)
analyzing charged-particle trajectories in uniform magnetic fields (29-3)
finding magnetic forces on straight conductors in uniform fields (29-4)
using integration to determine magnetic forces with curved conductors and/or nonuniform fields (29-4)
determining magnetic dipole moments of current loops (29-5)
evaluating torque and potential energy for current loops in magnetic fields (29-5)

Limitations to Keep in Mind

Magnetic forces arise only when charges are *moving*.
The cyclotron frequency is independent of particle energy only for speeds much less than the speed of light.

QUESTIONS

1. A charged particle moves through a region containing only a magnetic field. Under what condition will the particle experience no force?
2. An electron moving with velocity **v** through a magnetic field **B** experiences a magnetic force **F**. Which of the vectors **F**, **v**, and **B** must be at right angles?
3. In Fig. 29-37 a high-energy gamma ray has decayed into an electron and its positively charged antiparticle, a

positron. A magnetic field points out of the plane of the photograph, and the electron and positron spiral in this field. Which path belongs to which particle?

4. A magnetic field points out of this page. Will a positively charged particle moving in the plane of the page circle clockwise or counterclockwise as viewed from above?
5. An electron moves through a region in a straight line at constant speed. Can you conclude that there is no magnetic field in the region? Could you so conclude if you knew that there were no electric or gravitational forces on the electron? Explain.
6. High-resolution color TV monitors sometimes have a built-in circuit that compensates for changes in the orientation of Earth's magnetic field relative to the TV picture tube as the monitor is moved from one place to another. Why is this necessary?
7. What is meant by the statement that charged particles can be "trapped" by magnetic fields?
8. An electron beam comes straight to the center of a TV screen, where it makes a spot of light. If you hold the north pole of a bar magnet on the left side of the picture tube, which way will the spot move? (Magnetic field lines emerge from the north pole of a magnet.)

FIGURE 29-37 Creation of an electron-positron pair in the decay of a gamma ray incident from the top of the figure. Both particles come into existence moving essentially downward in the photo, at the point where the two trajectories diverge. (Question 3)

9. Do particles in a cyclotron gain energy from the electric field, the magnetic field, or both? Explain.

10. A cyclotron is designed to accelerate either hydrogen or deuterium nuclei. If the magnetic field is unchanged, how must the frequency of the alternating dee voltage be changed in order to switch from hydrogen to deuterium?

11. An electron and a proton moving at the same speed enter a region containing a uniform magnetic field. Which is deflected more from its original path?

12. An electron and a proton with the same kinetic energy enter a region containing a uniform magnetic field. Which is deflected more from its original path?

13. For what orientation of electric and magnetic fields could the net force on a particle be zero?

14. Will the velocity selector of Example 29-2 work for particles coming out of the page in Fig. 29-8? Why or why not?

15. What does magnetism have to do with the fact that auroras are seen near Earth's poles?

16. In a certain region uniform electric and magnetic fields are at right angles to one another. A positively charged particle is released from rest in this region. Describe qualitatively its subsequent motion.

17. How do the period and radius of an electron's orbit in a magnetic field depend on its velocity?

18. Current in a certain ionic solution is carried equally by positive and negative ions. Would you expect the Hall effect to occur in this solution?

19. Two identical particles carrying equal charge are moving in opposite directions along a magnetic field, when they collide elastically head-on. Describe their subsequent motion.

20. Repeat the above question for the case when the two particles are moving instantaneously perpendicular to the field when they collide.

21. Under what conditions will a current loop in a magnetic field experience zero torque?

22. What would happen to a motor with no commutator?

PROBLEMS

ActivPhysics can help with these problems:
Activities 13.4, 13.6, 13.7, 13.8

Section 29-2 The Magnetic Force on a Moving Charge

1. (a) What is the minimum magnetic field needed to exert a 5.4×10^{-15}-N force on an electron moving at 2.1×10^7 m/s? (b) What magnetic field strength would be required if the field were at 45° to the electron's velocity?

2. An electron moving at right angles to a 0.10-T magnetic field experiences an acceleration of 6.0×10^{15} m/s². (a) What is the electron's speed? (b) By how much does its *speed* change in 1 ns ($= 10^{-9}$ s)?

3. What is the magnitude of the magnetic force on a proton moving at 2.5×10^5 m/s (a) at right angles; (b) at 30°; (c) parallel to a magnetic field of 0.50 T?

4. A magnetic field of 0.10 T points in the x direction. A charged particle carrying 1.0 μC enters the field region moving at 20 m/s. What are the magnitude and direction of the force on the particle when it first enters the field region if it does so moving (a) along the x axis; (b) along the y axis; (c) along the z axis; (d) at 45° to both x and y axes?

5. A particle carrying a 50-μC charge moves with velocity $\mathbf{v} = 5.0\hat{\mathbf{i}} + 3.2\hat{\mathbf{k}}$ m/s through a uniform magnetic field $\mathbf{B} = 9.4\hat{\mathbf{i}} + 6.7\hat{\mathbf{j}}$ T. (a) What is the force on the particle? (b) Form the dot products $\mathbf{F} \cdot \mathbf{v}$ and $\mathbf{F} \cdot \mathbf{B}$ to show explicitly that the force is perpendicular to both $\mathbf{v}$ and $\mathbf{B}$.

6. Moving in the x direction, a particle carrying 1.0 μC experiences no force. Moving with speed v at 30° to the x axis, the particle experiences a magnetic force of 2.0 N. What magnetic force would it experience if it moved along the y axis with speed v?

7. A proton moving with velocity $\mathbf{v}_1 = 3.6 \times 10^4\,\hat{\mathbf{j}}$ m/s experiences a magnetic force of $7.4 \times 10^{-16}\,\hat{\mathbf{i}}$ N. A second proton moving on the x axis experiences a magnetic force of $2.8 \times 10^{-16}\,\hat{\mathbf{j}}$ N. Find the magnitude and direction of the magnetic field, and the velocity of the second proton.

8. The magnitude of Earth's magnetic field is a little less than 1 G near Earth's surface. What is the maximum possible magnetic force on an electron with kinetic energy of 1 keV? Compare with the gravitational force on the same electron.

9. An alpha particle (2 protons, 2 neutrons) is moving with velocity $\mathbf{v} = 150\hat{\mathbf{i}} + 320\hat{\mathbf{j}} - 190\hat{\mathbf{k}}$ km/s in a magnetic field $\mathbf{B} = 0.66\hat{\mathbf{i}} - 0.41\hat{\mathbf{j}}$ T. Find the magnetic force on the particle.

10. How much charge would you need to put on a 2.7-g ping-pong ball so the magnetic force when it's moving at 6.0 m/s at right angles to Earth's 44-μT magnetic field is equal to its weight? Is this realistic?

11. A 1.4-μC charge moving at 185 m/s experiences a magnetic force $\mathbf{F}_B = 2.5\hat{\mathbf{i}} + 7.0\hat{\mathbf{j}}$ μN in a magnetic field $\mathbf{B} = 42\hat{\mathbf{i}} - 15\hat{\mathbf{j}}$ mT. What is the angle between the particle's velocity and the magnetic field?

12. A velocity selector uses a 60-mT magnetic field and a 24 kN/C electric field. At what speed will charged particles pass through the selector undeflected?

13. A region contains an electric field $\mathbf{E} = 7.4\hat{\mathbf{i}} + 2.8\hat{\mathbf{j}}$ kN/C and a magnetic field $\mathbf{B} = 15\hat{\mathbf{j}} + 36\hat{\mathbf{k}}$ mT. Find the electromagnetic force on (a) a stationary proton, (b) an electron moving with velocity $\mathbf{v} = 6.1\hat{\mathbf{i}}$ Mm/s.

14. A charged particle is moving at right angles to both a 1.1 kN/C electric field and a 0.75-T magnetic field. If the magnitude of the electric force on the particle is twice that of the magnetic force, what is the particle's speed?

Section 29-3 The Motion of Charged Particles in Magnetic Fields

15. What is the radius of the circular path described by a proton moving at 15 km/s in a plane perpendicular to a 400-G magnetic field?

16. How long does it take an electron to complete a circular orbit at right angles to a 1.0-G magnetic field?

17. Radio astronomers detect electromagnetic radiation at a frequency of 42 MHz from an interstellar gas cloud. If this radiation is caused by electrons spiraling in a magnetic field, what is the field strength in the gas cloud?

18. A beam of electrons moving in the x direction at 8.7×10^6 m/s enters a region where a uniform magnetic field of 180 G points in the y direction. How far into the field region does the beam penetrate?

19. Electrons and protons with the same kinetic energy are moving at right angles to a uniform magnetic field. How do their orbital radii compare?

20. The Van Allen belts are regions in space where high-energy charged particles are trapped in Earth's magnetic field. If the field strength at the Van Allen belts is 0.10 G, what are the period and radius of the spiral path described (a) by a proton with a 1.0-MeV kinetic energy? (b) by a 10-MeV proton?

21. Microwaves in a microwave oven are produced by electrons circling in a magnetic field at a frequency of 2.4 GHz. (a) What is the magnetic field strength? (b) The electrons' motion takes place inside a special tube called a magnetron. If the magnetron can accommodate electron orbits with a maximum diameter of 2.5 mm, what is the maximum electron energy?

22. Show that the orbital radius of a charged particle moving at right angles to a magnetic field B can be written

$$r = \frac{\sqrt{2Km}}{qB},$$

where K is the kinetic energy in joules, m the particle mass, and q its charge.

23. Two protons, moving in a plane perpendicular to a uniform magnetic field of 500 G, undergo an elastic head-on collision. How much time elapses before they collide again? *Hint:* Draw a picture.

24. Repeat the preceding problem for the case of a proton and an antiproton colliding head-on (a) if they have the same speed and (b) if they have different speeds. (An antiproton has the same mass as a proton, but carries the opposite charge.)

25. A cyclotron is designed to accelerate deuterium nuclei. (Deuterium has one proton and one neutron in its nucleus.)

(a) If the cyclotron uses a 2.0-T magnetic field, at what frequency should the dee voltage be alternated? (b) If the vacuum chamber has a diameter of 0.90 m, what is the maximum kinetic energy of the deuterons? (c) If the magnitude of the potential difference between the dees is 1500 V, how many orbits do the deuterons complete before achieving the energy of part (b)?

26. Without changing the magnetic field, how could the cyclotron of the preceding problem be modified to accelerate (a) protons and (b) alpha particles (two protons and two neutrons)? What would be the maximum energy achievable with (c) protons and (d) alpha particles?

27. Figure 29-38 shows a simple mass spectrometer, designed to analyze and separate atomic and molecular ions with different charge-to-mass ratios. In the design shown, ions are accelerated through a potential difference V, after which they enter a region containing a uniform magnetic field. They describe semicircular paths in the magnetic field, and land on a detector a lateral distance x from where they entered the field region, as shown. Show that x is given by

$$x = \frac{2}{B}\sqrt{\frac{2V}{(q/m)}},$$

where B is the magnetic field strength, V the accelerating potential, and q/m the charge-to-mass ratio of the ion. By counting the number of ions accumulated at different positions x, one can determine the relative abundances of different atomic or molecular species in a sample.

FIGURE 29-38 A mass spectrometer (Problem 27).

28. A mass spectrometer like that of the preceding problem has $V = 2000$ V and $B = 1000$ G. It is used to analyze a gas sample suspected of containing Ne, O_2, CO, SO_2, and NO_2. Ions are detected at distances of 58 cm, 68 cm, and 87 cm from the entrance to the field region. Which gases are actually present? Assume that all molecules are singly ionized.

29. A mass spectrometer is used to separate the fissionable uranium isotope U-235 from the much more abundant isotope U-238. To within what percentage must the magnetic field be held constant if there is to be no overlap of these two isotopes? Both isotopes appear as constituents of uranium hexafluoride gas (UF_6), and the gas molecules are all singly ionized.

30. An electron is moving in a uniform magnetic field of 0.25 T; its velocity components parallel and perpendicular to the field are both equal to 3.1×10^6 m/s. (a) What is the radius of the electron's spiral path? (b) How far does it move along the field direction in the time it takes to complete a full orbit about the field direction?

31. An electron moving at 3.8×10^6 m/s enters a region containing a uniform magnetic field $\mathbf{B} = 18\hat{\mathbf{k}}$ mT. The electron is moving at 70° to the field direction, as shown in Fig. 29-39. Find the radius r and pitch p of its spiral path, as indicated in the figure.

FIGURE 29-39 Problems 31, 32.

32. A proton in interstellar space describes a spiral path about a 15-mG magnetic field, with velocity component 40 km/s perpendicular to the field. If the pitch of the helix (see Fig. 29-39) is 8.7 km, what is the proton's velocity component parallel to the field?

Section 29-4 The Magnetic Force on a Current

33. What is the magnitude of the force on a 50-cm-long wire carrying 15 A at right angles to a 500-G magnetic field?

34. A wire coincides with the x axis, carrying 2.4 A in the $+x$ direction. The wire passes through a region containing a uniform magnetic field $\mathbf{B} = 0.17\hat{\mathbf{i}} + 0.32\hat{\mathbf{j}} - 0.21\hat{\mathbf{k}}$ T. Find a vector expression for the force per unit length on the wire in the magnetic field region.

35. A wire carrying 15 A makes a 25° angle with a uniform magnetic field. The magnetic force per unit length of wire is 0.31 N/m. (a) What is the magnetic field strength? (b) What is the maximum force per unit length that could be achieved by reorienting the wire in this field?

36. A wire of negligible resistance is bent into a rectangle as shown in Fig. 29-40, and a battery and resistor are connected as shown. The right-hand side of the circuit extends

FIGURE 29-40 Problem 36.

into a region containing a uniform magnetic field of 38 mT pointing into the page. Find the magnitude and direction of the net force on the circuit.

37. In a high-magnetic-field experiment, a conducting bar carrying 7.5 kA passes through a 30-cm-long region containing a 22-T magnetic field. If the bar makes a 60° angle with the field direction, what force is necessary to hold it in place?

38. A 20-cm-long conducting rod with mass 18 g is suspended by wires of negligible mass, as shown in Fig. 29-41. The rod is in a region containing a uniform magnetic field of 0.15 T pointing horizontally into the page, as shown. An external circuit supplies current between the support points A and B. (a) What is the minimum current necessary to move the bar to the upper position shown? (b) Which direction should the current flow?

FIGURE 29-41 Problem 38.

39. A piece of wire with mass per unit length 75 g/m runs horizontally at right angles to a horizontal magnetic field. A 6.2-A current in the wire results in its being suspended against gravity. What is the magnetic field strength?

40. A nonuniform magnetic field points out of the page, as shown in Fig. 29-42. The field strength increases at the rate of 2.0 mT/cm as you move to the right. A square wire loop 15 cm on a side lies in a plane perpendicular to the field, and a 2.5-A current circles the loop in the counterclockwise direction. What are the magnitude and direction of the net magnetic force on the loop?

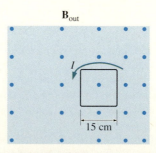

FIGURE 29-42 Problem 40.

41. A wire carrying 1.5 A passes through a region containing a 48-mT magnetic field. The wire is perpendicular to the field and makes a quarter-circle turn of radius 21 cm as it passes through the field region, as shown in Fig. 29-43. Find the magnitude and direction of the magnetic force on this section of wire

1.5 A

$\mathbf{B}_{in}$

21 cm

FIGURE 29-43 Problem 41.

42. A wire coincides with the x axis, and carries a current $I = 2.0$ A in the $+x$ direction. A nonuniform magnetic field points in the y direction, given by $\mathbf{B} = B_0(x/x_0)^2\hat{\mathbf{j}}$, where $B_0 = 0.22$ T, $x_0 = 1.0$ m, and x is the x coordinate. Find the force on the section of wire between $x = 1.0$ m and $x = 3.5$ m.

43. Apply Equation 29-8 to a closed current loop of arbitrary shape in a *uniform* magnetic field, and show that the net force on the loop is zero. *Hint:* Both I and B are constant as you go around the loop, so you can take them out of the integral. What is the remaining vector integral?

44. A rectangular copper strip measures 1.0 mm in the direction of a uniform 2.4-T magnetic field. When the strip carries a 6.8-A current at right angles to the field, the Hall voltage across the strip is 1.2 μV. Find the number density of free electrons in the copper.

45. The probe in a Hall-effect magnetometer uses a semiconductor doped to a charge-carrier density of 7.5×10^{20} m^{-3}. The probe measures 0.35 mm thick in the direction of the magnetic field being measured, and carries a 2.5-mA current perpendicular to the field. If its Hall potential is 4.5 mV, what is the magnetic field strength?

Section 29-5 A Current Loop in a Magnetic Field

46. Earth has a magnetic dipole moment associated with currents flowing in the planet's liquid outer core. Suppose that current flowed in a single loop at the outer edge of the liquid core (radius 3000 km). What current would be needed to give the observed dipole moment of 8.0×10^{22} A·m^2? (The actual current structure is more complex than a single loop.)

47. A single-turn square wire loop 5.0 cm on a side carries a 450-mA current. (a) What is the magnetic moment of the loop? (b) If the loop is in a uniform 1.4-T magnetic field with its dipole moment vector at 40° to the field direction, what is the magnitude of the torque it experiences?

48. An electric motor contains a 250-turn circular coil 6.2 cm in diameter. If it is to develop a maximum torque of 1.2 N·m at a current of 3.3 A, what should be the magnetic field strength?

49. A bar magnet experiences a 12-mN·m torque when it is oriented at 55° to a 100-mT magnetic field. What is the magnitude of its magnetic dipole moment?

50. A single-turn wire loop 10 cm in diameter carries a 12-A current. It experiences a torque of 0.015 N·m when the

normal to the loop plane makes a 25° angle with a uniform magnetic field. What is the magnetic field strength?

51. A simple electric motor like that of Fig. 29-36 consists of a 100-turn coil 3.0 cm in diameter, mounted between the poles of a magnet that produces a 0.12-T field. When a 5.0-A current flows in the coil, what are (a) its magnetic dipole moment and (b) the maximum torque developed by the motor?

52. A satellite with rotational inertia 20 kg·m^2 is in orbit at a height where Earth's magnetic field strength is 0.18 G. It has a magnetic torquing system, as described in Example 29-7, that uses a 1000-turn coil 30 cm in diameter. What should be the current in the coil if the magnetic torque is to give the satellite a maximum angular acceleration of 0.0015 s^{-2}?

53. Nuclear magnetic resonance (NMR) is a technique for analyzing chemical structures and is also the basis of magnetic resonance imaging used for medical diagnosis. The NMR technique relies on sensitive measurements of the energy needed to flip atomic nuclei upside-down in a given magnetic field. In an NMR apparatus with a 7.0-T magnetic field, how much energy is needed to flip a proton ($\mu = 1.41 \times 10^{-26}$ A·m^2) from parallel to antiparallel to the field?

54. A wire of length ℓ carries a current I. (a) Find an expression for the magnetic dipole moment that results when the wire is wound into an N-turn circular coil. (b) For what integer value of N is this moment a maximum?

Paired Problems

(Both problems in a pair involve the same principles and techniques. If you can get the first problem, you should be able to solve the second one.)

55. Find the magnetic force on an electron moving with velocity $\mathbf{v} = 8.6 \times 10^5\hat{\mathbf{i}} - 4.1 \times 10^5\hat{\mathbf{j}}$ m/s in a magnetic field $\mathbf{B} = 0.18\hat{\mathbf{j}} + 0.64\hat{\mathbf{k}}$ T.

56. A proton moving with velocity $\mathbf{v} = 2.0 \times 10^5\hat{\mathbf{i}} + 4.0 \times 10^5\hat{\mathbf{j}}$ m/s experiences a magnetic force $\mathbf{F} = 10\hat{\mathbf{i}} - 5.0\hat{\mathbf{j}} + 21\hat{\mathbf{k}}$ fN. What is the z component of the magnetic field?

57. Proponents of space-based particle-beam weapons have to confront the effect of Earth's magnetic field on their beams. If a beam of protons with kinetic energy 100 MeV is aimed in a straight line perpendicular to Earth's magnetic field in a region where the field strength is 48 μT, what will be the radius of the protons' circular path?

58. Electrons are accelerated through a 30-kV potential difference at the rear of a TV tube. The electron beam is initially headed straight toward the center of the tube. The TV is oriented so the beam is perpendicular to Earth's magnetic field, in a location where the field strength is 62 μT. What will be the radius of the electron beam's curved path?

59. A 170-mT magnetic field points into the page, confined to a square region as shown in Fig. 29-44. A square conducting loop 32 cm on a side carrying a 5.0-A current in the

FIGURE 29-44 Problems 59, 60.

clockwise sense extends partly into the field region, as shown. Find the magnetic force on the loop.

60. Find the force on the circular current loop shown at the right of Fig. 29-44. The loop carries 5.0 A clockwise, has radius 16 cm, and extends 10 cm into the field region.

61. An old-fashioned analog meter uses a wire coil in a magnetic field to deflect the meter needle. If the coil is 2.0 cm in diameter and consists of 500 turns of wire, what should be the magnetic field strength if the maximum torque is to be 1.6 μN·m when the current in the coil is 1.0 mA?

62. A circular wire coil 15 cm in diameter carries a 460-mA current and experiences a 0.020-N·m torque when the normal to the coil makes a 27° angle with a 42-mT magnetic field. How many turns are in the coil?

Supplementary Problems

63. Electrons in a TV picture tube are accelerated through a 30-kV potential difference and head straight for the center of the tube, 40 cm away. If the electrons are moving at right angles to Earth's 0.50-G magnetic field, by how much do they miss the screen's exact center?

64. The coil in the loudspeaker of Fig. 29-31 consists of 100 turns of wire, 3.5 cm in diameter. The magnetic field strength at the coil is 0.64 T. Find the magnitude of the force on the speaker coil when the current in the coil is 2.1 A.

65. A conducting bar with mass 15.0 g and length 22.0 cm is suspended from a spring in a region where a 0.350-T magnetic field points into the page, as shown in Fig. 29-45. With no current in the bar, the spring length is 26.0 cm. The bar is supplied with current from outside the field

FIGURE 29-45 Problem 65.

region, using wires of negligible mass. When a 2.00-A current flows from left to right in the bar, it rises 1.2 cm from its equilibrium position. Find (a) the spring constant and (b) the unstretched length of the spring.

66. In 2.0 μs, an electron moves 15 cm in the direction of a 0.10-T magnetic field. If the electron's velocity components perpendicular and parallel to the field are equal, (a) what is the length of its actual spiral trajectory and (b) how many orbits about the field direction does it complete?

67. A solid disk of mass M and thickness d sits on an incline, as shown in Fig. 29-46. A loop of wire is wrapped around the disk, running along a diameter and oriented so the loop is parallel to the incline. A uniform magnetic field **B** points vertically upward. Find an expression for the current I in the loop that will keep it from rolling down the incline.

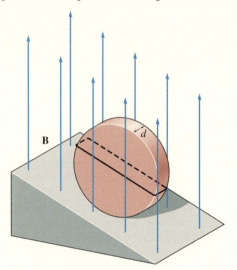

FIGURE 29-46 Problem 67.

68. A disk of radius R carries a uniform surface charge density σ and is rotating with angular frequency ω. Show that its magnetic dipole moment is $\frac{1}{4}\pi\sigma\omega R^4$. *Hint:* Divide the disk into concentric rings. Treat each as a loop carrying an infinitesimal current, and integrate over all the loops.

69. A 10-turn wire loop measuring 8.0 cm by 16 cm carrying 2.0 A lies in a horizontal plane but is free to rotate about the axis shown in Fig. 29-47. A 50-g mass hangs from one

FIGURE 29-47 Problem 69.

side of the loop, and a uniform magnetic field points horizontally, as shown. What magnetic field strength is required to hold the loop in its horizontal position?

70. A closed current loop is made from two semicircular wire arcs of radius R, joined at right angles as shown in Fig. 29-48. The loop carries a current I and is oriented with the plane of one semicircle perpendicular to a uniform magnetic field $\mathbf{B}$, as shown. Find (a) the magnetic moment of this nonplanar loop and (b) the torque on the loop. *Hint:* You can think of the loop as a superposition of two semicircular loops, each closed along the dashed line shown (why?).

FIGURE 29-48 Problem 70.

71. A circular wire loop of mass m and radius R carries a current I. The loop is hanging horizontally below a cylindrical bar magnet, suspended by the magnetic force, as shown in Fig. 29-49. If the field lines crossing the loop make an angle θ with the vertical, show that the strength of the magnet's field at the loop's position is $B = mg/2\pi RI \sin \theta$.

FIGURE 29-49 Problem 71.

72. A square wire loop of mass m carries a current I. It is initially in equilibrium, with its magnetic moment vector aligned with a uniform magnetic field $\mathbf{B}$. The loop is rotated slightly out of equilibrium about an axis through the centers of two opposite sides and then released. Show that it executes simple harmonic motion with period given by $T = 2\pi \sqrt{m/6IB}$.

73. Early models pictured the electron in a hydrogen atom as being in a circular orbit of radius 5.29×10^{-11} m about the stationary proton, held in orbit by the electric force. Find the magnetic dipole moment of such an atom. This quantity is called the *Bohr magneton* and is typical of atomic-sized magnetic moments. *Hint:* The full electron charge passes any given point in the orbit once per orbital period. Use this fact to calculate the average current.

30 Sources of the Magnetic Field

Moving electric charge produces magnetic fields. Here a large electromagnet lifts scrap metal in a recycling operation. When the current through the electromagnet is switched off, it will drop the metal.

The preceding chapter introduced the magnetic field and its effect on matter. Here we consider the opposite question: How does matter produce magnetic fields in the first place? We've seen that magnetic fields exert forces on moving electric charges—a fact that implies a deep relation between electricity and magnetism. That relation in fact goes two ways: Not only do magnetic fields affect moving electric charges, but moving charges themselves produce magnetic fields.

30-1 THE BIOT-SAVART LAW

The first inkling of a relation between electricity and magnetism came in 1820 when the Danish scientist Hans Christian Oersted discovered that a compass needle is deflected by an electric current (Fig. 30-1). A mere month after Oersted's discovery became known in Paris, the French scientists Jean Baptiste Biot and Félix Savart had experimentally determined the form of the magnetic field arising from a steady current. Biot and Savart's result gives the contribution $d\mathbf{B}$ to the magnetic field at some point P due to a small element of current, just as Coulomb's law in the form of Equation 23-9a gives the contribution $d\mathbf{E}$ to the electric field from a small charge element dq:

$$d\mathbf{E} = \frac{k\,dq}{r^2}\hat{\mathbf{r}}$$

(see Fig. 30-2a). In this expression the field $d\mathbf{E}$ depends directly on the charge element dq, inversely on the square of the distance r, and on a universal constant, k (or, equivalently, $1/4\pi\varepsilon_0$). The direction of $d\mathbf{E}$ is determined by the unit vector $\hat{\mathbf{r}}$ from the charge element toward the field point P.

Biot and Savart found similar results for the magnetic field, as described in Fig. 30-2b. Here a steady current I is flowing, and we consider the contribution $d\mathbf{B}$ that a small length $d\ell$ of this current makes to the magnetic field at point P. The product $I\,d\ell$ plays the same role as does dq in Coulomb's law. Both dq and $I\,d\ell$ describe sources of the fields, charge for the electric field, and moving charge for the magnetic field. The quantity $I\,d\ell$ has the units of charge multiplied by velocity—showing that *moving* charge is at the heart of magnetism. Furthermore, the magnetic field due to the current element $I\,d\ell$ falls off inversely as the square of the distance, just like the electric field of a charge element dq.

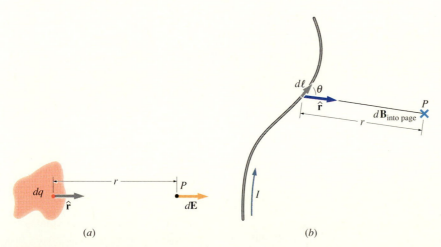

FIGURE 30-2 (a) Coulomb's law gives the electric field $d\mathbf{E}$ at point P due to an infinitesimal charge element dq. (b) The Biot-Savart law gives the magnetic field $d\mathbf{B}$ at point P arising from the current I flowing along the infinitesimal vector $d\ell$. The unit vector $\hat{\mathbf{r}}$ points from the current element $I\,d\ell$ toward the field point P.

But there are also important differences. The source of the electric field—charge—is a scalar quantity, whereas the source of the magnetic field is *moving* charge, and movement has direction. We can account for the direction of the current in Fig. 30-2b by defining a vector $d\boldsymbol{\ell}$ whose magnitude is the length $d\ell$ and which points in the direction of the current. Biot and Savart found that the magnetic field $d\mathbf{B}$ depends on sine of the angle θ between this vector $d\boldsymbol{\ell}$ and the unit vector $\hat{\mathbf{r}}$ from the current element $I\,d\ell$ toward the field point P, and that the direction of $d\mathbf{B}$ is perpendicular to both vectors. Mathematically, the $\sin\theta$ dependence and perpendicular direction together are described by a cross product. Thus our results for the magnetic field $d\mathbf{B}$ can be summarized in a compact vector equation, called the **Biot-Savart law**:

$$d\mathbf{B} = \frac{\mu_0}{4\pi}\frac{I\,d\boldsymbol{\ell} \times \hat{\mathbf{r}}}{r^2}, \quad \text{(Biot-Savart law)} \tag{30-1}$$

where μ_0 is a constant called the permeability constant, which has the exact value $4\pi\times10^{-7}$ N/A^2.

Besides the more complicated directionality evidenced by the cross product in the Biot-Savart law, there's another distinction between the Biot-Savart law and Coulomb's law. Both describe the fields of localized structures—current elements and point charges—that are sources of the fields. It makes sense to talk about the electric field of an isolated point charge. But can we have an isolated current element? Not in a steady-state situation, where the current flowing into a current element must be the same as the current flowing out. Thus any Biot-Savart calculation necessarily involves the fields produced by many small current elements from an entire circuit. Experimentally, we find that the magnetic field obeys the superposition principle, so the net field at any point is the vector sum—or integral—of the fields of individual current elements:

$$\mathbf{B} = \int d\mathbf{B} = \int \frac{\mu_0}{4\pi}\frac{I\,d\boldsymbol{\ell} \times \hat{\mathbf{r}}}{r^2}, \tag{30-2}$$

where the integration is taken over the entire circuit in which the current I flows. The field given by Equation 30-2 depends on the details of the current configuration, but the directionality associated with the cross product means that, quite generally, magnetic field lines encircle the current that is their source (Fig. 30-3).

FIGURE 30-3 Magnetic field lines generally encircle a current. The field direction is given by the right-hand rule: Point your right thumb in the direction of the current, and your fingers curl in the direction of the field.

EXAMPLE 30-1	*The Field of a Current Loop*

Find the magnetic field at an arbitrary point P on the axis of a circular loop of radius a carrying current I.

Solution

Let the loop axis be the x axis, with the origin at the loop center. Figure 30-4a shows the field contribution $d\mathbf{B}$ at P arising from a current element $I\,d\ell$ at the top of the loop, with its direction given by the cross product $d\boldsymbol{\ell} \times \hat{\mathbf{r}}$. The symmetry of the situation shows that all points on the loop produce similar field elements $d\mathbf{B}$, each making the same angle θ with the axis, giving a net field that points along the axis (Fig. 30-4b). To find that net field we therefore need sum only the x components of the field contributions from around the loop.

Figure 30-4a also shows that the vectors $d\boldsymbol{\ell}$ and $\hat{\mathbf{r}}$ are perpendicular; therefore, the product $d\boldsymbol{\ell} \times \hat{\mathbf{r}}$ has magnitude $|d\boldsymbol{\ell}||\hat{\mathbf{r}}|\sin90° = d\ell$ since $\hat{\mathbf{r}}$ is a unit vector. Then the

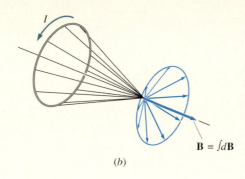

FIGURE 30-4 A current loop whose axis is the x axis. (*a*) The magnetic field $d\mathbf{B}$ arising from a current element of length $d\boldsymbol{\ell}$ on the loop is perpendicular to the vectors $d\boldsymbol{\ell}$ and $\hat{\mathbf{r}}$, and thus makes the angle θ with the loop axis. (*b*) Current elements all around the loop produce similar field elements $d\mathbf{B}$, with the result that the net field $\mathbf{B}$ points along the axis.

magnitude of the field contribution $d\mathbf{B}$, as given by the Biot-Savart law, is

$$dB = \frac{\mu_0 I}{4\pi} \frac{|d\boldsymbol{\ell} \times \hat{\mathbf{r}}|}{r^2} = \frac{\mu_0 I}{4\pi} \frac{d\ell}{x^2 + a^2}.$$

To find the net field, we need to sum the x components $dB_x = dB\cos\theta$, which are the same for all current elements around the loop. Figure 30-4a shows that $\cos\theta = a/r = a/\sqrt{x^2 + a^2}$, so the net field becomes

$$B = \int_{\text{loop}} dB\cos\theta$$

$$= \int_{\text{loop}} \left(\frac{\mu_0 I}{4\pi} \frac{d\ell}{x^2 + a^2}\right)\left(\frac{a}{\sqrt{x^2 + a^2}}\right)$$

$$= \frac{\mu_0 I a}{4\pi(x^2 + a^2)^{3/2}} \int_{\text{loop}} d\ell,$$

where the integral reduces to such a simple form because the distance x is the same for all points on the loop. The remaining integral just means the sum of all the infinitesimal segments $d\ell$

around the loop—or the loop circumference $2\pi a$. Thus, the magnitude of the magnetic field on the loop axis becomes

$$B = \frac{\mu_0 I a^2}{2(x^2 + a^2)^{3/2}}. \tag{30-3}$$

ActivPhysics Activity 13.2 shows you the entire field of this current loop.

EXERCISE Earth's magnetic field at temperate latitudes is about 50 μT. You're experimenting with the effect of magnetic fields on living cells and need a control point where the magnetic field is zero. To get such a point, you orient a 12-cm-diameter wire loop so the field at its center just cancels Earth's field. What loop current is required?

Answer: 4.8 A

Some problems similar to Example 30-1: 1–3, 5, 15, 17

13.2

Magnetic Field of a Loop

We found in the preceding chapter that a current loop behaves like a magnetic dipole when it's in an external magnetic field. Does the loop also produce a dipole-like field? At points far from the loop ($x \gg a$), we can neglect a^2 compared with x^2 in the denominator of Equation 30-3, which then becomes

$$B = \frac{\mu_0 I a^2}{2x^3} = \frac{\mu_0}{2\pi} \frac{\mu}{x^3}, \qquad (x \gg a) \tag{30-4}$$

where the second equality follows by introducing the loop's magnetic dipole moment $\mu = I\pi a^2$ as defined in Equation 29-10. Equation 30-4 shows the inverse-cube behavior we found earlier for the field of an *electric* dipole. Although we derived it for a circular loop, Equation 30-4 in fact holds at large distances from a current loop of *any* shape, suggesting that the field is essentially that of a magnetic dipole. A much more difficult calculation would confirm that the distant field off the axis also shows an inverse-cube dependence and exhibits the angular dependence typical of a dipole. We conclude that, in both its response to magnetic fields and its production of magnetic fields:

▎ **A current loop constitutes a magnetic dipole.**

(a)

(b)

(c)

FIGURE 30-5 (a) The magnetic field of a current loop, as traced by iron filings. Note that field lines encircle the current-carrying wire. (b) The electric field of an electric dipole and (c) the magnetic field of a current loop. Far from their sources, the field patterns are identical, and exhibit the $1/r^3$ dependence of the dipole field.

Figure 30-5a shows the field of a current loop as traced by iron filings, while Figs. 30-5b, c compare the fields of an electric dipole and a magnetic dipole. It's the similarity of the overall field patterns—becoming mathematically identical at large distances, including the $1/r^3$ dependence—that leads us to identify the current loop as a magnetic dipole. Physically, a current-loop dipole doesn't look anything like its electric analog, a pair of opposite electric charges. But its behavior, both in response to magnetic fields and in producing magnetic fields, is completely analogous to that of an electric dipole with respect to electric fields. We'll see later that the familiar bar magnet is nothing but a collection of current-loop magnetic dipoles; what we call the north pole isn't a pole at all but just the end where magnetic field lines emerge from the current loops. The current loop in Fig. 30-5c acts essentially like a very short bar magnet, with its north end upward and south end downward. In fact, the electromagnet shown in the opening photo for this chapter is nothing but a current loop (actually many turns of wire), which produces a strong magnetic field when the current is on. We'll take a deeper look at electric and magnetic dipoles, and magnets, at the end of this chapter.

APPLICATION	*Magnetic Fields of Earth and Sun*

Earth, Sun, and many other astronomical objects possess magnetic fields. Reasonably close to Earth, the field approximates that of a magnetic dipole of dipole moment $\mu = 8.0 \times 10^{22}$ A·m² (Fig. 30-6). The direction of the dipole moment vector differs by about 11° from that of Earth's rotation axis, and this accounts for the difference between magnetic and true north. Locally, the field often deviates significantly from a pure dipole form, and these deviations provide geologists with

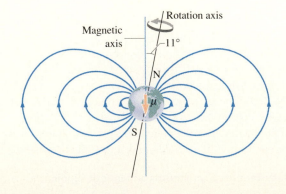

FIGURE 30-6 Earth's magnetic field approximates that of a magnetic dipole located near the center of the Earth and inclined at 11° to the rotation axis. Note from the field direction that Earth's "north" pole is really a magnetic south pole.

FIGURE 30-7 Magma emerges at the midocean ridges, pushing older portions of the ocean floor farther apart. As the magma solidifies, it becomes magnetized in the direction of Earth's magnetic field. (This occurs as atomic-scale current loops in magnetic materials such as iron align with the field and then retain permanently their magnetization, as we'll see later in the chapter.) Analysis of the ocean floor reveals bands of alternating magnetic polarity, showing that Earth's magnetic field reverses on a time scale of 10^5 to 10^6 years. Dark bands represent regions with Earth's present magnetic polarity.

clues to the detailed structure of the planet. The field is not constant; locally, its direction varies significantly over times as short as a few years, and the overall field reverses about every half million years or so (Fig. 30-7). At substantial distances from Earth the field is distorted from its dipole form by the solar wind, a flow of high-speed particles from the Sun (see Fig. 30-57, with Problem 30).

What causes Earth's magnetic field? We know that electric currents produce magnetic fields. Deep inside Earth are a solid inner core and a liquid outer core, both rich in iron. Through an interaction not yet fully understood, the planet's rotation combined with convective motions due to internal heating produces electric currents in the liquid core. For reasons even less well understood, those currents and the resulting magnetic field undergo reversals on a time scale of approximately a million years. Problems 5 and 6 deal with Earth's magnetic field and its origin.

Earth's magnetic field is crucial to our well-being. As we saw in the preceding chapter, high-energy particles from the Sun and elsewhere are trapped in the magnetic field and have difficulty reaching Earth's surface. Thus the field protects us from harmful particulate radiation. During the field reversals, which last about 10,000 years, the magnetic field is significantly reduced and surface exposure to high-energy particulate radiation is accordingly increased. Some scientists speculate that evolutionary changes due to radiation-induced mutations may accelerate at these times.

The Sun's magnetic field probably arises in the same way as does Earth's, although the gaseous nature of the star and the intense energy flow resulting from nuclear fusion in the Sun's core make its magnetic field much more dynamic. The Sun's field reverses every 11 years, giving rise to the solar activity cycle whose best-known indicator is the count of sunspots—regions of intense magnetic field at the solar surface (Fig. 30-8).

The complex behavior of astrophysical magnetic fields is governed entirely by Newton's laws and the laws of electromagnetism. That we do not yet fully understand these magnetic fields is testimony to the rich variety of phenomena subsumed under those laws.

(a)

(b)

FIGURE 30-8 (a) Sunspots are regions of intense magnetic field at the Sun's surface. (b) The solar magnetic activity cycle is clearly evident in this plot of the number of sunspots.

Any steady current ultimately flows in a complete, closed circuit, and therefore at great distances its magnetic field is dipolar. But closer in the field configuration depends on the details of the current. An important case is the field near a straight stretch of current. We can approximate this field as that of an infinite line of current, which we calculate in Example 30-2.

EXAMPLE 30-2	*The Field of a Straight Wire*

What is the magnetic field produced by an infinitely long straight wire carrying a steady current I?

Solution

FIGURE 30-9 Geometry for calculating the field at P due to an infinite straight wire carrying a steady current I along the x axis. All current elements along the wire contribute fields at P that point out of the page.

Let the wire coincide with the x axis, and consider an arbitrary point P on the y axis. Figure 30-9 shows a current element $I\,d\boldsymbol{\ell}$ on the wire. At point P the field $d\mathbf{B}$ of this element is perpendicular to both the wire and the vector $\hat{\mathbf{r}}$—that is, $d\mathbf{B}$ points out of the page. Clearly this is true for all current elements along the wire, so we can find the net field by integrating the field contributions dB without needing a vector integration. Using the Biot-Savart law, we have

$$dB = \frac{\mu_0 I}{4\pi}\frac{|d\boldsymbol{\ell} \times \hat{\mathbf{r}}|}{r^2} = \frac{\mu_0 I}{4\pi}\frac{d\ell \sin\theta}{r^2},$$

where we used the fact that $\hat{\mathbf{r}}$ is a unit vector making an angle θ with $d\boldsymbol{\ell}$. Figure 30-9 shows that $r^2 = x^2 + y^2$ and $\sin\theta = y/r = y/\sqrt{x^2 + y^2}$. And, since the segment $d\ell$ lies along the x axis, $d\ell = dx$. Making these substitutions and integrating over the entire wire (from $x = -\infty$ to $x = +\infty$) gives the net field at point P:

$$B = \int dB = \int_{-\infty}^{\infty} \frac{\mu_0 I}{4\pi} \frac{y\,dx}{(x^2 + y^2)^{3/2}}$$

$$= \frac{\mu_0 I y}{4\pi} \frac{x}{y^2 \sqrt{x^2 + y^2}}\Bigg|_{-\infty}^{\infty},$$

where we found the integral in the table of integrals in Appendix A. (See Problem 71 for another approach.) At the limits $x = \pm\infty$ the expression $x/\sqrt{x^2 + y^2}$ takes on the values ± 1, so we have

$$B = \frac{\mu_0 I}{4\pi y}[1 - (-1)] = \frac{\mu_0 I}{2\pi y}. \tag{30-5}$$

Since the wire has cylindrical symmetry this result must hold anywhere, and the result is circular field lines encircling the wire, as shown in Fig. 30-10 and explored further in Activ-Physics Activity 13.1.

The field given by Equation 30-5 falls as the inverse of the distance from the wire. This should not be surprising: we found

13.1
Magnetic Field of a Wire

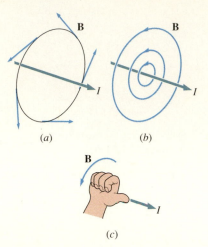

FIGURE 30-10 (*a*) Some magnetic field vectors associated with a straight wire carrying a steady current. (*b*) The corresponding magnetic field lines are circles, concentric with the wire. (*c*) The right-hand rule gives their direction.

the same dependence on distance for the *electric* field of an infinitely long charged line. The field patterns differ, though, in that the electric field points radially from the charged line while the magnetic field encircles the current-carrying wire. Although Equation 30-5 applies strictly only to an infinitely long wire, it is a good approximation for a finite wire at distances small compared with the wire length (Fig. 30-11).

EXERCISE A power line carries 450 A. How close would you have to be to the line for its magnetic field to equal Earth's $50\text{-}\mu\text{T}$ field?

Answer: 1.8 m

Some problems similar to Example 30-2: 4, 11, 57, 58

FIGURE 30-11 Iron filings trace out the circular magnetic field surrounding a current-carrying wire.

Got It!

Figure 30-12 shows two wire loops bent into shapes consisting of straight and semicircular sections. Both loops carry the same steady current in the direction shown. (a) What will be the directions of the magnetic fields at the points P and Q? (b) At which point will the field be stronger?

FIGURE 30-12 Think about the magnetic fields at points P and Q.

30-2 THE MAGNETIC FORCE BETWEEN TWO CONDUCTORS

In the preceding chapter we found the force on a straight wire of length ℓ carrying a current I through a magnetic field **B**:

$$\mathbf{F} = I\boldsymbol{\ell} \times \mathbf{B},$$

where the vector $\boldsymbol{\ell}$ is in the direction of the current flow. Through Equation 30-5 we now know the magnetic field produced by a long wire. If two long, parallel wires carry current in the same direction, as shown in Fig. 30-13, then each will experience a force arising from the other's field. We now determine that force.

If d is the distance between the wires, then at wire 2 the field magnitude B_1 due to the current I_1 is, from Equation 30-5,

$$B_1 = \frac{\mu_0 I_1}{2\pi d}.$$

The field is perpendicular to wire 2, so the magnitude of the force on a length ℓ of wire 2 is

$$F_2 = I_2 \ell B_1 = \frac{\mu_0 I_1 I_2 \ell}{2\pi d}. \tag{30-6}$$

Calculating the force on a length ℓ of wire 1 would amount to interchanging the subscripts 1 and 2, giving a force of the same magnitude.

What is the direction of the force? Evaluating the cross product of $\boldsymbol{\ell}$ and **B** in Fig. 30-13 shows that the force on wire 2 is toward wire 1, and vice versa. By using the right-hand rule, you can convince yourself that reversing one of the currents would reverse the directions of both forces. So we conclude that currents in the same direction experience an attractive magnetic force, while currents in the opposite direction experience a magnetic repulsion.

The force between nearby conductors must be considered in the construction of electrical devices carrying large currents. In electromagnets—coils of wire designed to produce large magnetic fields that can be turned on or off with the current in the coils—nearby conductors must have enough physical support that magnetic forces do not destroy the device (Fig. 30-14).

The hum you often hear around electrical equipment comes from the mechanical vibration of tightly wound conductors in transformers and other electrical components. This vibration results from the changing magnetic force associated with the 60-Hz alternating current.

The magnetic force between conductors is the basis for the definition of the ampere and, consequently, the coulomb. One ampere is defined as the current flowing in two long, parallel conductors 1 m apart when they carry equal currents and experience a magnetic force of 2×10^{-7} N on each meter of their length. It then follows that 1 coulomb is the amount of charge passing in 1 s through a wire carrying 1 A.

FIGURE 30-13 The magnetic force between two parallel wires carrying current in the same direction is attractive.

13.5

Magnetic Force on a Wire

FIGURE 30-14 This superconducting electromagnet was torn apart by the magnetic forces from a current of 50 kA flowing in its coils. The magnetic field reached 55 T—over one million times Earth's field. Coils were made from copper/niobium composite with the strength of steel and were reinforced with Kevlar and a steel casing. The coils originally encircled the channel where a pencil has been placed to show the scale.

FIGURE 30-15 Will the spiral tighten or loosen when current flows as shown?

FIGURE 30-16 Two magnetic field lines surrounding a wire carrying current out of the page. The vector $d\ell$ denotes an infinitesimal portion of a circular path that coincides with the inner field line.

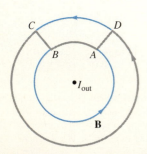

FIGURE 30-17 A closed loop that does not coincide with a single field line. The line integral of the magnetic field around this loop has the same value— $\mu_0 I$—that it has around a circular loop coinciding with a field line.

Got It!

A piece of wire is wound into a flat spiral as shown in Fig. 30-15. If a current is passed through the coil (supplied by flexible wires that aren't shown), will the coil tighten up or become looser? Will your answer change if the current direction reverses?

30-3 AMPÈRE'S LAW

In Chapter 23 we did several relatively cumbersome electric field calculations using Coulomb's law. For example, we found the field of an infinite line charge by integrating over all the charge elements along the line. Then in Chapter 24 we introduced Gauss's law and showed how it made electric field calculations much easier in situations with sufficient symmetry.

Can we make a statement that would enable us to calculate magnetic fields with comparable ease? Figure 30-16 shows two of the circular magnetic field lines surrounding a long wire carrying current I out of the page. Imagine moving around the inner circle. As you move a short way, form the product of the distance $d\ell$ that you travel with the component of the magnetic field in the direction you're going. Here the field is entirely in the direction of your path, so that product is simply $B\,d\ell$, where B is the field magnitude; more generally, it would be the dot product $\mathbf{B} \cdot d\ell$ to account for an arbitrary angle between the field and your path. Now consider adding up all these products as you go around a complete circle. Formally, the result is the line integral $\oint \mathbf{B} \cdot d\ell$, or, in this case where the directions of the path and field coincide, just $\oint B\,d\ell$. Here the circle indicates that the integration is done around a *closed* path. The magnitude of $\mathbf{B}$ is given by Equation 30-5: $B = \dfrac{\mu_0 I}{2\pi r_1}$, where we've replaced y by the radius r_1. Evaluating $\oint \mathbf{B} \cdot d\ell$ then gives

$$\oint \mathbf{B} \cdot d\ell = \frac{\mu_0 I}{2\pi r_1} \oint d\ell = \frac{\mu_0 I}{2\pi r_1}(2\pi r_1) = \mu_0 I,$$

where we took $B = \mu_0 I/2\pi r_1$ outside the integral because r_1 is the constant radius of the circular field line. What remains of the integral, $\oint d\ell$, is just the circumference, $2\pi r_1$. Our final answer, $\mu_0 I$, is thus independent of the radius r_1. We would get the same answer going around the outer circle, or indeed any circular path. On a larger path the distance is greater, but the field—dropping as $1/r$—is correspondingly weaker, making the value of $\oint \mathbf{B} \cdot d\ell$ the same. Thus the line integral $\oint \mathbf{B} \cdot d\ell$ does not depend on the radius of the circular path, but only on the current I encircled by that path.

We'll get the same result even if the path doesn't correspond to a field line. Figure 30-17 shows why for a path consisting of two circular arcs and radial lines joining them. The radial segments DA and BC contribute nothing to the quanity $\oint \mathbf{B} \cdot d\ell$ because the field is perpendicular to the path, and the large arc CD gives the same contribution as would the smaller arc BA because the field is weaker by just the right amount to compensate for arc CD's greater length. So going around the path shown in Fig. 30-17 gives the same value for the line integral—namely, $\mu_0 I$—as would going around a purely circular path. We can generalize this result to an arbitrary path by approximating the path as a series

FIGURE 30-18 Approximating an irregular loop as a series of arcs and radial segments shows that the line integral of the magnetic field is $\mu_0 I$ for *any* closed loop.

of circular arcs and radial segments (Fig. 30-18). Thus the quantity $\oint \mathbf{B} \cdot d\boldsymbol{\ell}$ has the same value, $\mu_0 I$, around *any* path surrounding a line current I. And because the superposition principle holds for magnetic fields, the result must be true for *any* distribution of current, not just the single line current we considered here. This result is in fact a universal statement about the relation between current and magnetic field:

> The value of the line integral $\oint$ B · $d\boldsymbol{\ell}$ around any closed loop is proportional to the current encircled by that loop. Mathematically,
>
> $$\oint \mathbf{B} \cdot d\boldsymbol{\ell} = \mu_0 I_{\text{encircled}} . \quad \text{(Ampère's law, steady currents)} \qquad (30\text{-}7)$$

This statement is a form of **Ampère's law**, one of the four fundamental laws of electromagnetism. Ampère's law says that whatever arrangement of currents we might have, and however complicated the resulting magnetic field, that field is such that the line integral $\oint \mathbf{B} \cdot d\boldsymbol{\ell}$ around any closed loop will have the value $\mu_0 I$, where I is the net current *encircled* by that loop. Compare this with Gauss's law, which says that whatever arrangement of electric charges we might have, the resulting electric field is such that the surface integral $\oint \mathbf{E} \cdot d\mathbf{A}$ has the value q/ε_0, where q is the *enclosed* charge.

There's one important restriction on Ampère's law as expressed in Equation 30-7: it's exactly correct only for *steady* currents—that is, currents that don't change with time. We'll show in Chapter 34 how to relax that restriction—with profound implications. Until then, we restrict ourselves to steady currents for which Equation 30-7 holds, or to currents that change only slowly, for which it's a good approximation.

With the restriction to steady currents, our statement of Ampère's law is true for *any* closed loop whatsoever, no matter what its shape or orientation: for any loop, the line integral of the magnetic field around the loop is proportional to the encircled current. It doesn't matter whether the current is in a single wire or in a number of wires or distributed throughout space; in any case, we simply add the currents to obtain the net current encircled by the loop. If there are currents flowing in opposite directions we add them with appropriate algebraic signs. A current counts as positive if, when you curl the fingers of your right hand in the direction of the loop, your right thumb points in the general direction of the current.

Although Ampère's law describes the relation between the magnetic field on a loop and the current *encircled* by that loop, we emphasize that the field **B** in Ampère's law is the *net field* from all currents, whether inside the loop or not. We found the same thing with Gauss's law, which relates the electric field **E** on a closed surface to the enclosed charge; **E** itself, however, is the *net field* arising from all sources, whether enclosed or not.

Ampère's law, like Gauss's, is a truly universal statement describing the relation between magnetic field and electric current. It holds in the electromagnetic devices we build, in atomic and molecular systems, in the interaction of fluid motion and electric charge that gives rise to Earth's magnetic field, and in distant astrophysical objects (Fig. 30-19). Although it is difficult to show mathematically, the Biot-Savart law follows logically from Ampère's law in the same sense that Coulomb's law follows from Gauss's.

(a)

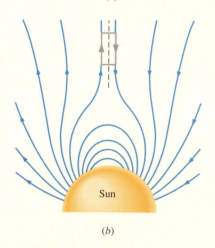

(b)

FIGURE 30-19 (a) A coronal streamer in the Sun's outer atmosphere contains oppositely directed magnetic fields in close proximity. (b) A model calculation of the coronal magnetic field. Since $\oint \mathbf{B} \cdot d\boldsymbol{\ell}$ is clearly nonzero around the loop shown, there must be current in the region encircled by the loop.

EXAMPLE 30-3	*Solar Currents*

The long dimension of the rectangular loop in Fig. 30-19*b* is 4×10^8 m, and the magnetic field strength in the vicinity of the loop is essentially constant at 2 mT. Find the total current encircled by the rectangle. In what direction is this current flowing?

Solution

We can't insert an ammeter into the Sun's atmosphere, but we can determine the current using Ampère's law. The short ends of the rectangle in Fig. 30-19*b* are perpendicular to the field, so here $\mathbf{B} \cdot d\boldsymbol{\ell} = 0$ and there is no contribution to the line integral around the loop. Moving around the loop in the direction shown means going *with* the field along both long dimensions (note that this is because the field direction reverses across the dashed line). Since the field is uniform, the contribution to the line integral from each side is just $B\ell$. Thus $\oint \mathbf{B} \cdot d\boldsymbol{\ell} = 2B\ell$. Ampère's law tells us that the line integral of the magnetic field around a closed loop has the value $\mu_0 I$, so

$$2B\ell = \mu_0 I,$$

or

$$I = \frac{2B\ell}{\mu_0} = \frac{(2)(2 \times 10^{-3}\ \text{T})(4 \times 10^8\ \text{m})}{4\pi \times 10^{-7}\ \text{N/A}^2} = 1 \times 10^{12}\ \text{A}.$$

This is a colossal current by terrestrial standards, but is typical of large-scale currents on the Sun.

Since the line integral is positive, we curl our right fingers around the loop in the direction shown by the arrows and our thumb must point in the direction of the current—that is, into the page in Fig. 30-19. In three dimensions this current flows around the Sun in approximately the equatorial plane.

Tip

Amperian Loops The loop used with Ampère's law is truly arbitrary. It need not coincide with a field line, as Fig. 30-18 showed. Example 30-3 showed this, too; here, the rectangular loop coincided with the straight field lines over its long sides but not along its shorter ends. The loop used with Ampère's law is called an **amperian loop**. Don't confuse amperian loops with field lines!

EXERCISE Figure 30-20 shows a cross-sectional view of three wires carrying current perpendicular to the plane of the page. Wire *A* carries 20 A out of the page. $\oint \mathbf{B} \cdot d\boldsymbol{\ell}$ around loop 1 in the direction shown has the value 1.26×10^{-5} T·m, and the line integral around loop 2 has the value -6.28×10^{-6} T·m. Find the currents in wires *B* and *C*, and the directions of those currents.

Answers: *B*: 10 A into page; *C*: 5 A out of page

Some problems similar to Example 30-3: 28–30

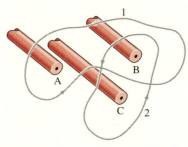

FIGURE 30-20 What are the currents in wires *B* and *C*?

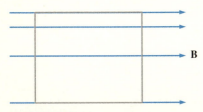

FIGURE 30-21 Which way is the current in this region? Gray rectangle is an amperian loop.

Got It!

Figure 30-21 shows a magnetic field that points to the right but increases in strength as you move up the page. A rectangular amperian loop is also shown. Use Ampère's law to argue that there must be current flowing in this region, and find its direction.

30-4 USING AMPÈRE'S LAW

For charge distributions with sufficient symmetry we used Gauss's law to solve for the electric field in a simple and elegant way. Similarly, for current distributions with sufficient symmetry we can use Ampère's law to solve for the magnetic field. Doing so requires finding suitable amperian loops over which we can evaluate the line integral of the magnetic field.

The Field of a Straight Wire

Here we use Ampère's law to determine the field of an infinite straight wire. This is the same calculation we did in Example 30-2 using the Biot-Savart law.

Figure 30-22 shows the wire, which carries current I. We know that magnetic field lines generally encircle the currents that are their sources. Here cylindrical symmetry requires that those field lines be circles centered on the wire and that the magnetic field strength cannot depend on angular position around the wire. Thus B must be constant on a field line. Applying the right-hand rule—thumb along the current, curling the fingers of the right hand—shows that the magnetic field circles counterclockwise around the wire. And since the wire is infinitely long, Fig. 30-22 represents the field anywhere along the wire.

Given the symmetry, an appropriate amperian loop for evaluating the line integral is itself a circle, also shown in Fig. 30-22. The magnetic field everywhere on this amperian loop is in the same direction as the loop, so the dot product $\mathbf{B} \cdot d\boldsymbol{\ell}$ becomes simply $B\, d\ell$. Then the line integral around the loop is

FIGURE 30-22 Circular field lines surrounding an infinitely long wire. Gray circle is an amperian loop for evaluating the line integral of the magnetic field. Although we've drawn only four field lines, the field exists everywhere, including on the amperian loop.

$$\oint_{\text{loop}} \mathbf{B} \cdot d\boldsymbol{\ell} = \oint_{\text{loop}} B\, d\ell = B \oint_{\text{loop}} d\ell = 2\pi r B\,.$$

Here we can take the field magnitude B outside the integral because it's constant on the circular amperian loop; the remaining integral is then the loop circumference $2\pi r$. Ampère's law equates this line integral to μ_0 times the encircled current, so

$$2\pi r B = \mu_0 I\,,$$

or

$$B = \frac{\mu_0 I}{2\pi r}\,. \tag{30-8}$$

This result is the same as Equation 30-5, but here we derived it in a much simpler way.

> **Tip**
>
> *Symmetry Is Crucial* Our use of Ampère's law to derive the field of a long wire depends crucially on symmetry. We cannot arbitrarily pull B outside the integral unless we know—as we do here from symmetry—that it is constant in magnitude and in direction relative to our amperian loop.

Our calculation made no assumptions about the diameter of the wire. Therefore Equation 30-8 holds for any long cylindrical wire, thick or thin, as long as we restrict ourselves to points *outside* the wire so that our amperian loop encircles the *entire* current. In fact, you can easily convince yourself that Equation 30-8 must hold *outside* any current distribution with cylindrical symmetry. To calculate the field *inside* a current distribution, however, we must be careful to use only the actual current encircled by our amperian loop. Example 30-4 illustrates this point.

EXAMPLE 30-4	Inside a Wire

A long, straight wire of radius R carries a current I uniformly distributed over its cross-sectional area. What is the magnetic field as a function of position within the wire?

Solution

All the symmetry arguments we used to find the field outside a wire still apply here, so the line integral around a circular amperian loop of radius r is still $2\pi r B$. What's different is that an amperian loop within the wire no longer encircles the entire current (Fig. 30-23). How much current is encircled? The current is distributed uniformly, giving a current density (current per unit area) $J = I/A = I/\pi R^2$. The encircled current is then the current density times the loop area, or

$$I_{encircled} = \left(\frac{I}{\pi R^2}\right)(\pi r^2) = I\frac{r^2}{R^2}.$$

Equating the line integral $2\pi r B$ to μ_0 times this encircled current gives

$$2\pi r B = \mu_0 I\frac{r^2}{R^2},$$

or
$$B = \frac{\mu_0 I r}{2\pi R^2}. \qquad (30\text{-}9)$$

FIGURE 30-23 Cross section of a long cylindrical wire carrying current out of the page. Symmetry requires that the magnetic field be circular inside as well as outside the wire. Current is distributed uniformly over the wire; in applying Ampère's law to the amperian loop shown in gray we need the fraction of the total current encircled by the loop.

FIGURE 30-24 The magnetic field inside a wire is zero at the wire's axis and increases linearly with radial distance, while beyond the wire radius R it drops as $1/r$.

Does this result make sense? The field increases linearly with distance from the axis—just as we found for the *electric* field *inside* a uniformly charged cylinder. Here the increase occurs because we encircle more and more current—with $I_{encircled}$ growing as r^2—as long as we're inside the wire. Once we reach the surface, of course, the encircled current remains constant and the field begins to decrease inversely with distance, as described in Equation 30-8. Figure 30-24 plots the field strengths both inside and outside the wire.

This example is not merely academic; in superconducting wires, for example, strong magnetic fields can destroy superconductivity, so it's important to know the field throughout the wire.

EXERCISE A power line 4.0 cm in diameter carries 1.5 kA. Find the magnetic field (a) 1.0 cm from the axis of the wire and (b) 10 cm from the axis.

Answers: (a) 7.5 mT; (b) 3.0 mT

Some problems similar to Example 30-4: 32, 33, 37, 65

Got It!

A coaxial cable consists of an inner wire carrying current I in one direction, and an outer cylindrical conductor carrying the same current in the opposite direction; the current is distributed symmetrically in the outer conductor. (See Fig. 30-60 if the configuration isn't clear). If the current in the outer conductor only is doubled, will the magnetic field change (a) in between the two conductors or (b) outside the outer conductor?

| **EXAMPLE 30-5** | *A Current Sheet* |

An infinite flat sheet carries a current out of the page, as shown in Fig. 30-25. The current is distributed uniformly along the sheet, with the current per unit width along the sheet given by J_s. Find the magnetic field of this sheet.

Solution

What might the field of this sheet look like? Figure 30-26 suggests that we can consider the sheet to be made of many parallel wires. The vector sum of the fields of these wires gives a net field to the left above and to the right below the sheet. We can also argue from symmetry that the fields must point horizontally and with magnitude that is independent of position parallel to the sheet; since the sheet extends infinitely to the right and left, there's nothing to favor an upward or downward deflection of the field lines or any variation in field magnitude

FIGURE 30-25 A current sheet (gray) extends infinitely to the left and right, as well as in and out of the page. Field lines and a rectangular amperian loop are also shown.

parallel to the sheet. We've drawn these horizontal field lines, and an appropriate amperian loop, in Fig. 30-25.

We already evaluated $\oint \mathbf{B} \cdot d\boldsymbol{\ell}$ for a similar geometry in Example 30-3; the result is simply $2B\ell$, with ℓ the width of the amperian rectangle. The current per unit width of the sheet is J_s, so our rectangular loop of width ℓ encircles a current $I = J_s\ell$. Equating μ_0 times this encircled current to the line integral $\oint \mathbf{B} \cdot d\boldsymbol{\ell} = 2B\ell$ gives

$$2B\ell = \mu_0 J_s \ell,$$

or
$$B = \tfrac{1}{2}\mu_0 J_s. \qquad (30\text{-}10)$$

Thus the magnetic field of an infinite current sheet is independent of distance from the sheet. We found the same result for the electric field of an infinite charged plane. Although the infinite sheet is an idealization, Equation 30-10 is a good approximation to the field near long, wide, flat conductors. Thin current sheets also form in conducting plasmas like those in fusion reactors and in the Sun's atmosphere, where they can lead to sudden dissipation of energy.

EXERCISE Current in a printed circuit board is carried in a long copper strip 2.1 mm wide and much thinner than its width. If the current in the strip is 35 mA, spread uniformly over its cross section, find the magnetic field strength near the strip's surface.

Answer: 10 μT

Some problems similar to Example 30-5: 38, 39, 42

FIGURE 30-26 A current sheet approximated by closely spaced parallel wires. The magnetic field of the sheet is the vector superposition of the fields of the individual wires. In the limit of infinitesimally spaced wires, the field lines become straight and strictly horizontal.

Fields of Simple Current Distributions

We've just used Ampère's law to calculate the magnetic fields of two simple, symmetric current distributions. Although those magnetic fields may look quite different from the electric fields of correspondingly symmetric charge distributions, they exhibit the same dependences on distance. Table 30-1 summarizes the electric and magnetic fields of several simple charge and current distributions. Real distributions are usually more complicated, but often they can be approximated by these simple cases. Far from any current loop, for example, the magnetic field is that of a dipole. Very near *any* wire, its magnetic field is essentially that

TABLE 30-1 *Fields of Some Simple Charge and Current Distributions*

DISTRIBUTION	FIELD DEPENDENCE ON DISTANCE*	ELECTRIC OR MAGNETIC FIELD
Electric dipole	$\dfrac{1}{r^3}$	
Magnetic dipole	$\dfrac{1}{r^3}$	
Spherically symmetric charge distribution	$\dfrac{1}{r^2}$	
Spherically symmetric current distribution	Impossible for steady current	
Charge distribution with line symmetry	$\dfrac{1}{r}$	
Current distribution with line symmetry	$\dfrac{1}{r}$	
Infinite flat sheet of change	Uniform field; no variation	
Current sheet	Uniform field; no variation	

*For field *outside* distribution.

of a long straight wire. For situations that lack symmetry and in which approximations are not adequately accurate, we can always calculate the magnetic field of a steady current distribution from the Biot-Savart law, just as we can always calculate the electric field of an arbitrary charge distribution from Coulomb's law.

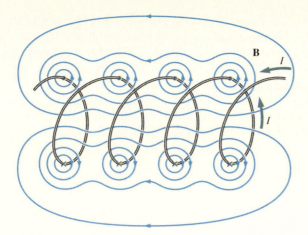

FIGURE 30-27 The magnetic field arising from the current in a loosely wound coil of wire is strongest within the coil and weaker outside. Field is shown only in the plane of the page; dots show current emerging from the page and crosses mark current going into the page.

30-5 SOLENOIDS AND TOROIDS

We found in Chapter 26 that we can produce a uniform electric field between the two closely spaced, charged conducting plates of a capacitor. Is there an analogous device that will produce a uniform magnetic field?

Figure 30-27 shows a coil of wire carrying current *I*. Close to the wire are magnetic field lines encircling the wire. We show these field lines in Fig. 30-27 at the top and bottom of the coil, where they cross the plane of the page. The net field anywhere is, of course, just the vector sum of the field contributions from all parts of the coil. You can see that inside the coil, the fields from current elements at the top and bottom all have a component to the right and so tend to reinforce. Outside the coil, though, field components from top and bottom are in opposite directions and thus reduce the net field, which points to the left as shown in the figure.*

Imagine making the coil longer and winding it more tightly. Such a long, tightly wound coil is a **solenoid**. Figure 30-28 shows what happens as the solenoid gets longer: the interior field stays essentially the same, but exterior field lines spread into more distant regions before closing back on themselves. Therefore, the field strength outside the solenoid—as evidenced by the density of field lines— decreases. In the limit of an infinitely long solenoid, the interior field becomes perfectly straight and the exterior field goes to zero. A real solenoid approaches this ideal limit when its length is much greater than its diameter. Figure 30-29 shows the field of a solenoid as traced by iron filings. Incidentally, the solenoid fields of Figs. 30-28 and 30-29 might remind you of the field of a bar magnet. They should: we'll soon see that a bar magnet and a solenoid, although superficially very different things, are magnetically very similar.

---

* Because the coil carries a net current from right to left, there is also a weak field component outside that encircles the coil. This component plays no role in the discussion that follows, and we ignore it.

FIGURE 30-28 As a solenoid gets longer, the interior field stays nearly constant but the exterior field weakens as the field lines spread ever farther apart.

13.3
Magnetic Field of a Solenoid

FIGURE 30-29 Iron filings trace the magnetic field of a solenoid. Note that the field is strong inside the solenoid, causing the filings to align strongly, and relatively weak outside, except near the ends. (With ActivPhysics Activity 13.3 you can simulate the pattern of iron filings around a solenoid.)

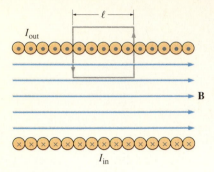

FIGURE 30-30 Cross section of a long solenoid, showing a rectangular amperian loop straddling the region where the solenoid coils emerge from the plane of the page.

In the long-solenoid limit, we can use Ampère's law to find the field in the solenoid. Figure 30-30 shows a cross section through a long solenoid, with a rectangular amperian loop. Since the exterior field is zero, there's no contribution to the line integral $\oint \mathbf{B} \cdot d\boldsymbol{\ell}$ from the top of the amperian loop. The loop's vertical sides are at right angles to the field, so they, too, contribute nothing to the line integral. The only contribution is from the bottom of the loop. Going counterclockwise around the loop, we move with the field on this segment. Since the field cannot vary with position along the infinite solenoid, the contribution to the line integral is just $B\ell$; since the other three sides contribute nothing, this is the value of $\oint \mathbf{B} \cdot d\boldsymbol{\ell}$.

To apply Ampère's law we need the encircled current. If there are n turns per unit length of solenoid, the amperian loop encircles $n\ell$ turns. The same current I flows in each turn, so the encircled current is $n\ell I$. Applying Ampère's law then gives

$$B\ell = \mu_0 n\ell I,$$

or

$$B = \mu_0 nI. \quad \text{(solenoid)} \tag{30-11}$$

Since the vertical dimension of the amperian rectangle never entered our calculation, this field magnitude is the same anywhere inside the solenoid. The magnetic field in the solenoid is therefore uniform. Although it looks very different, the long solenoid with its uniform magnetic field is the magnetic analog of a closely spaced parallel-plate capacitor, which produces a uniform electric field.

Although Figs. 30-27 and 30-28 depict circular coils, the derivation of Equation 30-11 is based only on Fig. 30-30, which could represent a solenoid whose coils are circular, square, or any other shape.

APPLICATION	*Everyday Solenoids, Medical Solenoids*

Solenoids of all sizes are used in a wide variety of experimental and practical devices. Because a solenoid is hollow, magnetic materials such as iron will be pulled into the solenoid by the nonuniform magnetic field near its ends, and solenoids are therefore used to produce straight-line motion in mechanical devices. For example, turning the key to start your car sends current through a solenoid in the car's starter motor. A magnetic field develops in the solenoid, pulling in a steel plunger. Movement of the plunger joins a set of electrical contacts allowing current to flow to the starter motor. At the same time, the rod moves a small gear to the end of the motor shaft, engaging a gear on the engine's flywheel so that the starter motor can turn the engine (Fig. 30-31).

Running a dishwasher or washing machine also involves solenoids. The valves that control the flow of water in these machines are solenoid valves, opened and closed as a steel rod moves in or out of a solenoid.

Finally, huge solenoids with human-size interiors are used to produce the strong, uniform magnetic fields needed for magnetic resonance imaging (MRI). This high-tech medical

FIGURE 30-31 Cross section of a car starter motor, showing the workings of the solenoid. Turning the key to start the car sends current through the solenoid coils, creating a magnetic field that pulls the steel plunger into the solenoid. The lever attached to the plunger then pushes the gear outward on the shaft, engaging a larger gear on the gasoline engine. The electric starter motor turns to start the engine. Armature of the motor is the coil that forms the heart of the motor (recall Figs. 29-35 and 29-36).

technique images the body's interior without the need for X rays or other radiation (Fig. 30-32). The MRI solenoids use superconducting wire cooled with liquid helium to achieve high currents and correspondingly high magnetic fields.

FIGURE 30-32 Patient being inserted into a superconducting solenoid used for magnetic resonance imaging (MRI), a powerful and relatively new medical diagnostic technique.

EXAMPLE 30-6	*An MRI Solenoid*

A solenoid used in magnetic resonance imaging is 2.4 m long and 95 cm in diameter. It is wound from a niobium-titanium superconducting wire 2.0 mm in diameter, with adjacent turns separated by an insulating layer of negligible thickness. What current is necessary to produce a 1.5-T magnetic field inside the solenoid?

Solution

The magnetic field is given by Equation 30-12, $B = \mu_0 n I$, so if we know n—the number of turns per unit length—we can solve for I. Since the wire has diameter $d = 2.0$ mm, 5 turns will occupy 1 cm and therefore $n = 500$ turns/meter. Solving

Equation 30-11 for I then gives

$$I = \frac{B}{\mu_0 n} = \frac{(1.5 \text{ T})}{(4\pi \times 10^{-7})(500 \text{ m}^{-1})} = 2.4 \text{ kA}.$$

This is a large current, but readily handled by the niobium-titanium superconductor.

EXERCISE Copper wire 0.40 mm in diameter is tightly wound into a long solenoid, and a 3.6-A current passed through it. What is the magnetic field in the solenoid?

Answer: 11 mT

Some problems similar to Example 30-6: 43–45, 48

Toroids

A **toroid** is a solenoid bent into a doughnut shape (Fig. 30-33). Toroidal geometry is widely used in fusion reactors, as discussed in the preceding chapter, and toroidal coils help produce the magnetic field that confines the fusion plasma.

Symmetry requires that the toroid's field lines be circular, with constant magnitude on any line. We can readily calculate the line integral of this magnetic field around a circular amperian loop, like that shown in Fig. 30-34. This loop coincides with a field line, and symmetry ensures that the magnetic field has the same magnitude over the loop. Therefore the line integral becomes

$$\oint_{\text{loop}} \mathbf{B} \cdot d\boldsymbol{\ell} = \oint_{\text{loop}} B\, d\ell = B \oint_{\text{loop}} d\ell = 2\pi r B.$$

FIGURE 30-33 A toroidal coil.

To apply Ampère's law we equate this quantity to μ_0 times the encircled current. If the toroid consists of N turns and carries a current I, then an amperian loop inside the toroid encircles a total current NI. Then Ampère's law becomes

$$2\pi r B = \mu_0 N I,$$

or

$$B = \frac{\mu_0 N I}{2\pi r}. \quad \text{(toroid)} \quad (30\text{-}12)$$

FIGURE 30-34 Cross section of a toroid showing current emerging at inner edge of coils and descending into plane of page at outer edge. Also shown are circular field lines and an amperian loop of radius r (gray) for use in calculating the field. You can confirm the field direction using the right-hand rule.

This result holds when the amperian loop is within the toroid itself. But if it's inside the inner edge (inside the "doughnut's hole"), there's no encircled current and, therefore, no magnetic field. And beyond the outer edge, the amperian loop encircles equal but opposite currents, giving zero net current and again no magnetic field. As with a solenoid, the individual turns of a toroid need not be circular.

The toroidal field of Equation 30-12 is not uniform but exhibits a $1/r$ decrease. This nonuniformity causes problems with plasma confinement in fusion reactors, where it causes a drift of particles perpendicular to the field and therefore toward the walls of the machine.

30-6 MAGNETIC MATTER

We began our study of magnetism in Chapter 29 with a discussion of magnets. But we quickly moved on to concentrate on the behavior of electric charges in magnetic fields. In this chapter we've further developed the relation between magnetism and electric charge, as we've seen that moving electric charge is the source of magnetic fields. What does electric charge have to do with the familiar magnets we used to introduce magnetism?

In fact the two are one and the same phenomenon. When an electron orbits an atomic nucleus, its circular motion and charge make a miniature current loop, and it therefore constitutes a magnetic dipole (Fig. 30-35); if you worked Problem 73 in the preceding chapter then you've calculated the dipole moment for a simple atomic model. In addition the electron possesses an intrinsic magnetic dipole moment associated with an intrinsic angular momentum called "spin." Interactions among these magnetic dipole moments determine the magnetic properties of individual atoms and of bulk matter. Although an accurate description of magnetism in matter necessarily involves quantum mechanics, we can nevertheless use our knowledge of magnetic dipoles to gain a qualitative understanding of magnetic matter.

FIGURE 30-35 In the classical model of the atom, the circling electron constitutes a miniature current loop. The atom is therefore a magnetic dipole. Here the current—carried by the *negative* electron and therefore opposite the electron's motion—is counterclockwise when viewed from above, so the magnetic moment points upward.

Ferromagnetism

Experiment reveals three types of magnetic behavior in bulk matter. **Ferromagnetism**, the most familiar, is actually limited to a few substances, notably the elements iron, nickel, and cobalt, and some compounds. In a ferromagnetic material, a quantum-mechanical interaction among nearby atomic magnetic moments results in regions—called **magnetic domains**—in which all the atomic magnetic moments point in the same direction. A typical domain contains 10^{17} to 10^{21} atoms and occupies a volume on the order of 10^{-12} to 10^{-8} m^3. The magnetic moment of a single domain can be large since it's the sum of many atomic magnetic moment vectors all pointing in the same direction. A typical piece of ferromagnetic material, however, contains many domains with their moments in random directions and, therefore, exhibits no net magnetic moment. But when an external magnetic field is applied, a net magnetic moment develops. This occurs because domains that already happen to be aligned with the field can grow by realignment of individual atomic moments in adjacent domains (Fig. 30-36a). In addition, the magnetic moments of entire domains can rotate (Fig. 30-36b).

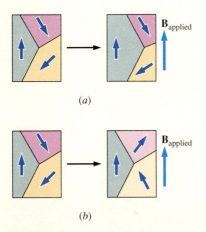

FIGURE 30-36 Domain changes in a ferromagnetic material occur through (a) domain growth and (b) domain realignment. These changes occur when a magnetic field is applied, and result in the material acquiring a net magnetic moment.

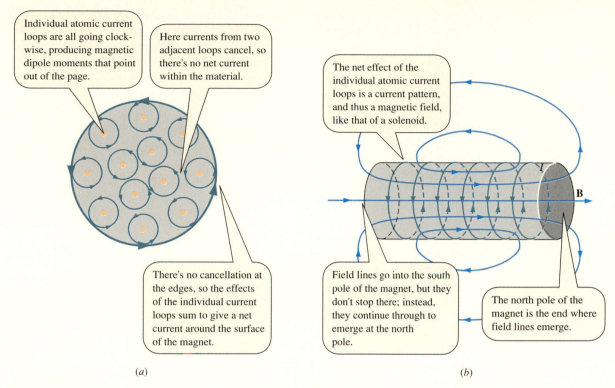

FIGURE 30-37 (*a*) Cross section of a bar magnet, showing individual atomic current loops all aligned the same way (dots represent the magnetic moment vectors). *Within* the material, adjacent atomic currents cancel. But they don't cancel at the edges, leaving a net current encircling the magnet. (*b*) Side view shows that this magnetization current is similar to the current in the coils of a solenoid. The bar magnet's field is therefore similar to that of a solenoid.

Since any dipole experiences a force in a nonuniform field, a piece of ferromagnetic material in such a field experiences a net force; this is why iron and other ferromagnetic materials are attracted to magnets even though they themselves are not magnets.

Some ferromagnetic materials retain significant magnetization even after the applied field has been removed. These so-called *hard* ferromagnetic materials make permanent magnets. A common bar magnet is a piece of hard ferromagnetic material that's been magnetized with its internal magnetic dipoles aligned along its long dimension. We suggested earlier that a bar magnet is magnetically similar to a solenoid, and Fig. 30-37 shows why: the internal atomic current loops give a net current flowing around the magnet, just like the current in a solenoid. This so-called magnetization current, however, arises not from the steady flow of charge in the material, but as the sum of the individual atomic current loops.

Permanent magnetization persists unless a strong magnetic field is applied that changes its direction or strength. Computer disks, credit cards, and audio- and videotapes (Fig. 30-38) make use of hard ferromagnetic materials that

FIGURE 30-38 Videotapes, credit cards, and computer disks all store information using hard ferromagnetic materials.

retain information as patterns of permanent magnetization; these magnetic media can be "rewritten" by applying a magnetic field strong enough to alter the magnetization and thus the stored information. We'll take a closer look at magnetic recording in the next chapter.

Permanent magnetization in so-called "soft" ferromagnetic materials, in contrast, is negligible. Such materials are used in applications that require magnetization to be turned on and off rapidly; these include the "heads" that read and write information on computer disks and audio- and videotapes, as well as electromagnets such as the one shown at the beginning of this chapter.

Random thermal motions tend to disrupt the alignment of individual magnetic moments. Thus ferromagnetic effects weaken with increasing temperature. Above the so-called **Curie temperature**, ferromagnetism ceases altogether. Curie temperatures for the common ferromagnetic elements nickel, iron, and cobalt are 631, 1043, and 1395 K, respectively. The rarer ferromagnetic elements dysprosium and gadolinium have much lower Curie temperatures of 85 and 289 K, respectively. The disappearance of ferromagnetism at the Curie temperature is an example of a phase transition, analogous to the solid/liquid/gas transitions we studied in Chapter 20.

Paramagnetism

Many substances that are not ferromagnetic nevertheless consist of atoms or molecules with permanent magnetic dipole moments. What distinguishes these **paramagnetic** materials from ferromagnetic substances is the absence of a strong interaction that tends to align nearby moments. As a result, individual atomic moments are not organized into domains, but point in random directions. An applied magnetic field still brings the atomic magnetic moments into some degree of alignment, but at all but the coldest temperatures this alignment is far less complete than in a ferromagnetic material. Therefore paramagnetic materials are attracted only weakly to magnets (Fig. 30-39), and paramagnetism decreases with increasing temperature. Unlike ferromagnetism, paramagnetism is not a strong enough effect to have many technological applications, but it does provide an important way of probing the structure of materials and can also be exploited to help achieve very low temperatures.

Diamagnetism

Even materials with no intrinsic magnetic moments can have magnetic moments induced when a magnet approaches or when an applied magnetic field otherwise changes. Such materials are termed **diamagnetic**. In contrast to paramagnetic and ferromagnetic materials, diamagnetic materials are repelled by magnets. We will explore the origins of diamagnetism in the next chapter.

Magnetic Susceptibility

We can characterize the effect of atomic magnetic dipoles on bulk matter just as we did in Chapter 23 for electric dipoles in dielectrics. There we found that atomic electric dipoles align with an electric field to reduce the field within the material, and we introduced the dielectric constant κ as the factor quantifying that reduction.

FIGURE 30-39 Liquid oxygen (O_2) is one of the more strongly paramagnetic materials. Here liquid oxygen is suspended between the poles of a magnet.

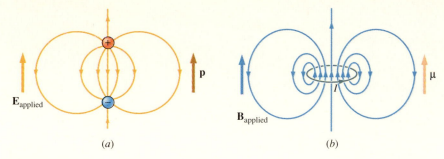

FIGURE 30-40 Although the electric fields far from electric and magnetic dipoles have the same form, the fields *within* the atomic structure giving rise to the different dipole moments have opposite directions. This is because electric dipoles consist of separated point charges, while magnetic dipoles are current loops. When external electric and magnetic fields are applied, respectively, to these two structures, the result is a reduction in the net electric field but an increase in the net magnetic field.

In a magnetic material, alignment of atomic magnetic dipoles has the opposite effect: the field in the material is increased. Figure 30-40 shows why. An electric dipole consists of two separated point charges. The strongest field associated with this dipole is the internal field pointing straight from the positive to the negative charge. When the dipole aligns with an applied electric field, this internal field is *opposite* the applied field and, therefore, *reduces* the net electric field (Fig. 30-40a). But, as we've seen, magnetic field lines always *encircle* a current. As Fig. 30-40b shows, the field inside the current loop therefore points in the *same* direction as the applied field and, therefore, *increases* the net magnetic field.

Because of the very strong alignment of atomic dipoles that occurs in ferromagnetic materials, these materials greatly increase an applied magnetic field. For that reason the coils of electromagnets, including the heads of computer disk drives and VCRs, are usually wound on a ferromagnetic core to give a much stronger field than would be provided by the coil current alone.

In analogy with the dielectric constant, we introduce the quantity κ_M, called the **relative permeability**, as the factor by which the magnetic field within a material increases as a result of the alignment of atomic magnetic dipoles. For paramagnetic materials, κ_M is slightly greater than 1; for ferromagnetic materials it is much greater than 1. In diamagnetic materials the dipoles align antiparallel to the field; this makes $\kappa_M < 1$ for these materials. Because the relative permeabilities of paramagnetic and diamagnetic materials are very close to 1, it's more convenient to work with the **magnetic susceptibility** χ_M, defined by

$$\chi_M = \kappa_M - 1.$$

The internal field in the material is then given by

$$B_{int} = \kappa_M B_{applied} = (\chi_M + 1)B_{applied}. \qquad (30\text{-}13)$$

Equation 30-13 is most useful for paramagnetic and diamagnetic materials. In a ferromagnetic material, the internal and external fields are not proportional, and the internal field depends on the past history of the material's exposure to magnetic fields. This phenomenon, called **hysteresis**, is what makes possible

(a)   (b)

FIGURE 30-41 Internal versus applied magnetic field for different types of magnetic materials. (a) In diamagnetic and paramagnetic materials the relationship is linear; paramagnetic materials strengthen the applied field slightly, while diamagnetic materials weaken it slightly. (b) In ferromagnetic materials, the relationship depends on the strength of the applied field and on the past history of the material. As the applied field is increased, an initially unmagnetized ferromagnetic material responds by developing an increasing internal field (curve a). Eventually the field saturates (point 1) at its maximum possible value. Decreasing the applied field to zero leaves the material with a permanent magnetization (point 2); only a strong field in the opposite direction can eliminate (point 3) and eventually reverse the magnetization (point 4). Curve is for a "hard" ferromagnetic material that retains strong permanent magnetization; curve for a "soft" material would be much narrower, with a much smaller internal field at point 2. Axes are not to scale; field strengths on the vertical axis are much larger, reflecting the high magnetic susceptibility of a ferromagnetic material. Arrows indicate that the applied field is increased (curve a), decreased (curve b), and increased again (curve c).

permanent magnets. Figure 30-41 shows a ferromagnetic material's response to an externally applied magnetic field. If the material is initially unmagnetized, its internal field first increases with increasing applied field (curve a in Fig. 30-41). Eventually, though, increasing the applied field leads to little or no increase in the internal field; the material is then **saturated**, and no higher internal field is possible. Saturation occurs when essentially all the magnetic domains in the material are aligned, so there's no way to increase the internal field further. If the applied field is then reduced, the internal field also falls, but not as quickly. Even at zero applied field, there can remain a strong internal field due to the permanent magnetization we discussed earlier (point 2 in Fig. 30-41). Finally, a sufficiently strong field in the opposite direction results in saturation with the opposite magnetization direction.

Table 30-2 lists some magnetic susceptibilities for all three types of magnetic materials. That the ferromagnetic materials listed have susceptibilities many orders of magnitude higher than others shows why we normally think of magnetism only in the context of these special materials. The highest susceptibilities are achieved with special alloys such as permalloy and supermalloy; these are used in making high-field permanent magnets for a variety of applications. The material called "mu metal" is used for magnetic shielding—for example, to prevent distortion by external fields in TV picture tubes and computer monitors, or to shield the magnets used in loudspeakers (recall Fig. 29-31) so they can be placed near videotapes or TV sets without causing damage or distortion. Of the paramagnetic materials listed, liquid oxygen has by far the highest susceptibility, showing again the effect of temperature on paramagnetism. Finally, note the entry for superconductors, which are perfectly diamagnetic. Their susceptibility of -1 implies that $\kappa_M = 0$, showing that they completely exclude magnetic fields. We will explore the reasons for this in the next chapter.

TABLE 30-2 *Magnetic Susceptibilities**

MATERIAL	MAGNETIC SUSCEPTIBILITY, χ_M
Diamagnetic materials	
Copper	-9.6×10^{-6}
Lead	-1.6×10^{-5}
Mercury	-2.8×10^{-5}
Nitrogen (gas, 293 K)	-6.7×10^{-9}
Sodium chloride	-1.4×10^{-5}
Any superconductor	-1
Water	-9.1×10^{-6}
Paramagnetic materials	
Aluminum	2.1×10^{-5}
Chromium	3.1×10^{-4}
Oxygen (gas, 293 K)	1.9×10^{-6}
Oxygen (liquid, 90 K)	3.5×10^{-3}
Sodium	8.5×10^{-6}
Ferromagnetic materials (field and history dependent; maximum value listed)	
Iron (annealed)	5.5×10^{3}
Permalloy (55% Fe, 45% Ni)	2.5×10^{4}
Supermalloy (15.7% Fe, 79% Ni, 5.0% Mo; 0.30% Mn)	8.0×10^{5}
μ-metal (77% Ni, 16% Fe, 5% Cu, 2% Cr)	1.0×10^{5}

*At 300 K unless noted.

30-7 MAGNETIC MONOPOLES AND GAUSS'S LAW

Electric and magnetic fields, at least as we've studied them so far, have very different configurations. Electric fields begin and end on their sources—namely electric charges. Magnetic fields, in contrast, encircle their sources—namely *moving* electric charges. Magnetic fields generally form closed loops, while the electric fields we've encountered so far do not (Fig. 30-42).

Are there particles analogous to electric charges, from which magnetic field lines might originate? There might be. Symmetry arguments based on the existence of electric charge and the many similarities between electricity and magnetism have long suggested to physicists that such **magnetic monopoles**—isolated magnetic north or south poles—might exist. Furthermore, theories of elementary particles suggest that magnetic monopoles should have been created in the Big Bang event that began our universe. However, recent versions of the

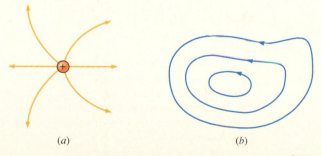

(a)   (b)

FIGURE 30-42 At this point in our study of electromagnetism, electric and magnetic field configurations are clearly distinguished. Electric fields (a) begin and end on electric charges, while magnetic fields (b) generally form closed loops.

FIGURE 30-43 You can't separate the north and south poles of a bar magnet by cutting it in half; that only produces smaller bar magnets, each with a north and a south pole and thus each a magnetic dipole. That's because the field of a bar magnet arises from atomic current loops (recall Fig. 30-37), not from separate magnetic north and south poles, or monopoles.

Big Bang theory suggest that monopoles should be spread so thinly as to make the chance of detecting one completely negligible.

If they existed, magnetic monopoles would be to the magnetic field what electric charges are to the electric field. Isolated monopoles would give rise to radial magnetic fields like the electric fields of point charges. Monopoles would experience forces in magnetic fields, forces given simply by the product of the monopole's "magnetic charge" and the magnetic field. And *moving* magnetic monopoles would produce electric fields—just as moving electric charges produce magnetic fields.

No one has yet found a magnetic monopole, however, although serious experimental searches continue. Every magnetic field we've ever seen has its origin in moving electric charge, not in separate magnetic north and south poles. That's why you can't make an isolated magnetic north pole by chopping a bar magnet in half (Fig. 30-43). The observed absence of magnetic monopoles means that the most fundamental magnetic field configuration is that of the dipole—in contrast to the simpler electric case, where the spherically symmetric point-charge field is possible. That's why Table 30-1 contains no magnetic entry under spherical geometry.

In the absence of magnetic monopoles, there is no place where magnetic field lines begin or end. That means there is no closed surface through which a nonzero net number of magnetic field lines emerges. In complete analogy with our discussion of Gauss's law in Chapter 24, we can say that the **magnetic flux**—mathematically, the surface integral $\oint \mathbf{B} \cdot d\mathbf{A}$—is zero for every closed surface. This statement is **Gauss's Law for magnetism**:

$$\oint \mathbf{B} \cdot d\mathbf{A} = 0. \tag{30-14}$$

Equation 30-14 is another of the four fundamental laws of electromagnetism. We've now seen three of them: Gauss's law for electricity, Ampère's law (still incomplete because of its restriction to *steady* currents), and Gauss's law for magnetism. In the next chapter we'll encounter the fourth of these fundamental laws.

Although Gauss's law for magnetism has zero on its right-hand side, the law is far from being devoid of content. It says what we've already found to be the case: that magnetic fields must configure themselves so they have no beginnings or endings.

CHAPTER SYNOPSIS

Summary

1. Magnetic fields arise from **moving electric charges**. The field at some point P arising from a steady electric current is described by the **Biot-Savart law**:

$$d\mathbf{B} = \frac{\mu_0}{4\pi} \frac{I d\boldsymbol{\ell} \times \hat{\mathbf{r}}}{r^2},$$

where $d\mathbf{B}$ is the contribution to the field from a current element flowing along the infinitesimal vector $d\boldsymbol{\ell}$ a distance r from the point P, and $\hat{\mathbf{r}}$ is a unit vector from the current element toward P. The field at P of an entire current distribution is found by integrating all its contributions $d\mathbf{B}$.

Important special cases include a current loop, which at large distances produces a dipole field, and an infinite straight current, whose field encircles the current and decreases as $1/r$.

2. The field arising from a current exerts a magnetic force on nearby currents; as a result, parallel currents in the same direction attract and those in opposite directions repel.

3. **Ampère's law** relates the line integral of the magnetic field around an arbitrary closed loop to the current encircled by that loop:

$$\oint_{\text{loop}} \mathbf{B} \cdot d\boldsymbol{\ell} = \mu_0 I_{\text{encircled}}.$$

This form holds for all steady currents. It may be used to calculate magnetic fields in cases with sufficient symmetry, including line symmetry (straight wires), plane symmetry (current sheets), solenoids, and toroids.

4. Individual elementary particles and orbiting atomic electrons constitute miniature current loops, which are responsible for magnetic effects in bulk matter. The **relative permeability**, κ_M, gives the ratio of the internal magnetic field in a material to the applied field; this relation is also described by the **magnetic susceptibility**, $\chi_M = \kappa_M - 1$.

    a. In **ferromagnetic** materials, atomic magnetic moments group into domains with net magnetic moments. These align with an applied field, greatly increasing the field within the material. Ferromagnetic materials retain magnetization even after the applied field is removed; this phenomenon of **hysteresis** accounts for permanent magnets. $\kappa_M \gg 1$ for ferromagnetic materials, although the internal field depends on both the applied field and the material's history.

    b. In **paramagnetic** materials, individual atomic magnetic moments become partially aligned with an applied field, but there is no cooperative interaction among the individual moments. $\kappa_M > 1$ for paramagnetic materials.

    c. Diamagnetic materials have no intrinsic magnetic moments. They have $\kappa_M < 1$ and are weakly repelled from magnets.

5. No **magnetic monopoles**, or magnetic analogs of electric charge, have ever been found. Magnetic fields—originating from moving electric charge—therefore form closed loops without beginnings or ends. In the absence of magnetic monopoles, **Gauss's law for magnetism** says that the flux of the magnetic field through any closed surface is zero:

$$\oint \mathbf{B} \cdot d\mathbf{A} = 0.$$

Terms You Should Understand

(Pairs are closely related terms whose distinction is important; number in parentheses is chapter section where term first appears.)

Biot-Savart law (30-1)
Ampère's law (30-3)
solenoid, toroid (30-5)
ferromagnetism, paramagnetism, diamagnetism (30-6)
permeability, susceptibility (30-6)
magnetic monopole (30-7)
Gauss's law for magnetism (30-7)

Symbols You Should Recognize

μ_0 (30-1)

$\oint \mathbf{B} \cdot d\boldsymbol{\ell}$ (30-3)

κ_M, χ_M (30-6)

Problems You Should Be Able to Solve

calculating magnetic fields of simple current distributions by integration using the Biot-Savart law (30-1)
evaluating forces between adjacent conductors (30-2)
evaluating magnetic fields using Ampère's law in situations with sufficient symmetry (30-4)
applying results derived in the text for the fields of straight wires, current loops, solenoids, and toroids (30-1–30-5)
relating internal and applied magnetic fields given magnetic susceptibility (30-6)

Limitations to Keep in Mind

The Biot-Savart law, and Ampère's law as expressed in this chapter, are exactly valid only for *steady* currents—those that never change with time.

QUESTIONS

1. In what two senses does a current loop behave like a magnetic dipole?

2. The electric field far from a pair of equal but opposite charges has the same configuration as the magnetic field far from a current loop. Yet inside the structure giving rise to the fields, the two point in opposite directions. Why?

3. The Biot-Savart law shows that the magnetic field of a current element decreases as $1/r^2$. Could you put together a complete circuit whose field exhibits this $1/r^2$ decrease? Why or why not?

4. Do currents in the same direction attract or repel? Explain.

5. If a current is passed through an unstretched spring, will the spring contract or expand? Explain.

6. Why is it advantageous to define the ampere in terms of magnetic force rather than a standard ammeter?

7. An intense beam of electrons or other charged particles tends to "pinch" or compress to a smaller diameter. Explain the origin of this "pinch effect" in terms of magnetic forces.

8. The field of a long, straight wire consists of circular field lines. Does Ampère's law hold for a square loop surrounding the wire? For a circular loop not concentric with the wire? See Fig. 30-44.

9. If the line integral around a closed loop is zero, does that mean the magnetic field on the loop is zero?

10. Must the integration path in Ampère's law coincide with a field line?

FIGURE 30-44 Question 8.

11. What must be going on inside Earth in order that it have a magnetic field?

12. Figure 30-45 shows some magnetic field lines associated with two parallel wires carrying equal currents perpendicular to the page. Are the currents in the same or opposite directions? How can you tell?

FIGURE 30-45 Question 12.

13. A solid cylinder and a hollow pipe of the same outer diameter carry the same current along their long dimensions, with the current distributed symmetrically about their axes. How do the magnetic fields at their surfaces compare?

14. Why is a piece of iron attracted into a solenoid?

15. Would there be a net magnetic force on a piece of iron deep inside a long solenoid? Explain.

16. In what sense is a long solenoid the magnetic analog of a parallel-plate capacitor?

17. What would happen to the magnetic field inside a solenoid if you (a) doubled the solenoid length without changing the number of turns per unit length or (b) doubled the length without changing the total number of turns?

18. Identify *three* regions in the magnetic field of Fig. 30-46 where there *must* be a current. Which way is the current flowing in each region?

FIGURE 30-46 Question 18.

19. An unmagnetized piece of iron has no net magnetic dipole moment. Yet it is attracted to either pole of a bar magnet. Why?

20. How would you determine experimentally whether a substance was paramagnetic or diamagnetic?

21. Would permanent magnets be possible if the relative permeability of ferromagnetic materials were strictly constant? Explain.

22. Why do paramagnetic and ferromagnetic effects weaken with increasing temperature?

23. If magnetic monopoles existed, then a net motion of monopoles would constitute a "magnetic current." What sort of field should arise from such a "current?"

24. In the absence of magnetic monopoles, why do we even bother to write Gauss's law for magnetism? Does the law have any physical significance?

PROBLEMS

ActivPhysics can help with these problems:
Activities 13.1–13.3, 13.5

Section 30-1 The Biot-Savart Law

1. A wire carries 15 A. You form the wire into a single-turn circular loop with magnetic field 80 μT at the loop center. What is the loop radius?

2. A single-turn wire loop is 2.0 cm in diameter and carries a 650-mA current. Find the magnetic field strength (a) at the loop center and (b) on the loop axis, 20 cm from the center.

3. A 2.2-m-long wire carrying 3.5 A is wound into a tight, loop-shaped coil 5.0 cm in diameter. What is the magnetic field at its center?

4. What is the current in a long wire if the magnetic field strength 1.2 cm from the wire's axis is 67 μT?

5. Suppose Earth's magnetic field arose from a single loop of current at the outer edge of the planet's liquid core (core radius 3000 km), concentric with Earth's center. What current would be necessary to give the observed field strength of 62 μT at the north pole? (The currents responsible for Earth's field are more complicated than this problem suggests.)

6. Earth's magnetic dipole moment is 8.0×10^{22} A·m². What is the magnetic field strength on Earth's surface at either pole?

7. A single-turn current loop carrying 25 A produces a magnetic field of 3.5 nT at a point on its axis 50 cm from the loop

center. What is the loop area, assuming the loop diameter is much less than 50 cm?

8. Two identical current loops are 10 cm in diameter and carry 20-A currents. They are placed 1.0 cm apart, as shown in Fig. 30-47. Find the magnetic field strength at the center of either loop when their currents are in (a) the same and (b) opposite directions.

10 cm

1.0 cm

FIGURE 30-47 Problem 8.

9. You have a spool of thin wire that can handle a maximum current of 0.50 A. If you wind the wire into a looplike coil 20 cm in diameter, how many turns should the coil have if the magnetic field at its center is to be 2.3 mT at this maximum current?

10. A single piece of wire is bent so that it includes a circular loop of radius a, as shown in Fig. 30-48. A current I flows in the direction shown. Find an expression for the magnetic field at the center of the loop.

FIGURE 30-48 Problem 10.

11. Two long, parallel wires are 6.0 cm apart. One carries 5.0 A and the other 10 A, with both currents in the same direction. Where on a line perpendicular to both wires is the magnetic field zero?

12. Four long, parallel wires are located at the corners of a square 15 cm on a side. Each carries a current of 2.5 A, with the top two currents into the page in Fig. 30-49 and the bottom two out of the page. Find the magnetic field at the center of the square.

13. A power line carries a 500-A current toward magnetic north and is suspended 10 m above the ground. The horizontal component of Earth's magnetic field at the power line's latitude is 0.24 G. If a magnetic compass is placed on the ground directly below the power line, in what direction will it point?

I_{in}

15 cm

I_{out}

15 cm

FIGURE 30-49 Problems 12, 23.

14. An electron is moving at 3.1×10^6 m/s parallel to a 1.0-mm-diameter wire carrying 20 A. If the electron is 2.0 mm from the center of the wire, with its velocity in the same direction as the current, what are the magnitude and direction of the force it experiences?

15. Part of a long wire is bent into a semicircle of radius a, as shown in Fig. 30-50. A current I flows in the direction shown. Use the Biot-Savart law to find the magnetic field at the center of the semicircle (point P).

I

a

P

FIGURE 30-50 Problem 15.

16. Use the result given in Problem 57 to find the magnetic field strength at the center of a square loop of side a carrying current I.

17. Figure 30-51 shows a conducting loop formed from concentric semicircles of radii a and b. If the loop carries a current I as shown, find the magnetic field at point P, the common center.

I

P

a b

FIGURE 30-51 Problem 17.

Section 30-2 The Magnetic Force between Two Conductors

18. In standard household wiring, parallel wires about 1 cm apart carry currents around 15 A. What is the magnitude of the force per unit length between such wires?

19. It would take a rather large apparatus to implement the definition of the ampere given at the end of Section 30-2. Suppose you wanted to use a smaller apparatus, with wires 50 cm long separated by 2.0 cm. What force would correspond to a current of 1 A?

20. Two parallel copper rods supply power to a high-energy experiment, carrying the same current in opposite directions. The rods are held 8.0 cm apart by insulating blocks mounted every 1.5 m. If each block can tolerate a maximum tension force of 200 N, what is the maximum allowable current?

21. The structure shown in Fig. 30-52 is made from conducting rods. The upper horizontal rod is free to slide vertically on the uprights, while maintaining electrical contact with them. The upper rod has mass 22 g and length 95 cm. A battery connected across the insulating gap at the bottom of the left-hand upright drives a 66-A current through the structure. At what height h will the upper wire be in equilibrium?

FIGURE 30-52 Problem 21.

22. Three parallel wires 4.6 m long each carry 20 A in the same direction. They are spaced at the vertices of an equilateral triangle 3.5 cm on a side. Find the magnitude of the force on each wire.

23. The wires in Fig. 30-49 carry 2.5-A currents in the directions indicated. Find the net force per unit length on the wire at lower left.

24. A long, straight wire carries 20 A. A 5.0-cm by 10-cm rectangular wire loop carrying 500 mA is located 2.0 cm from the wire, as shown in Fig. 30-53. Find the net magnetic force on the loop.

25. A solenoid 10 cm in diameter is made with 2.1-mm-diameter copper wire wound so tightly that adjacent turns touch, separated only by enamel insulation of negligible thickness. The solenoid carries a 28-A current. In the long, straight wire approximation, what is the net force between two adjacent turns of the solenoid?

FIGURE 30-53 Problem 24.

26. A long, flat conducting ribbon of width w is parallel to a long straight wire, with its near edge a distance a from the wire (Fig. 30-54). Wire and ribbon carry the same current I; this current is distributed uniformly over the ribbon. Use integration to show that the force per unit length between the two has magnitude $\dfrac{\mu_0 I^2}{2\pi w} \ln\left(\dfrac{a+w}{a}\right)$.

FIGURE 30-54 Problem 26.

Section 30-3 Ampère's Law

27. The line integral of the magnetic field on a closed path surrounding a wire has the value $8.8\ \mu\text{T}\cdot\text{m}$. What is the current in the wire?

28. In Fig. 30-55, $I_1 = 2$ A flowing out of the page; $I_2 = 1$ A, also out of the page, and $I_3 = 2$ A, into the page. What is the line integral of the magnetic field taken counterclockwise around each loop shown?

29. The magnetic field shown in Fig. 30-56 has uniform magnitude 75 μT, but its direction reverses abruptly. How much current is encircled by the rectangular loop shown?

30. The solar wind drags Earth's magnetic field out into a long "magnetotail," where the field reverses direction abruptly

FIGURE 30-55 Problem 28.

FIGURE 30-56 Problem 29.

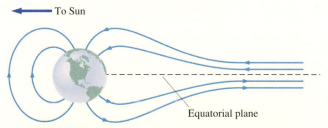

FIGURE 30-57 A simplified view of Earth's magnetic field, showing the long "magnetotail" on the nighttime side, caused by the solar wind dragging Earth's field lines with it. What's the current in the region to the right, where the field lines are essentially straight? (Problem 30)

across the equatorial plane (Fig. 30-57). (a) If the field strength in a section of the magnetotail is 20 nT, what is the current per unit length flowing in the magnetotail? (b) What is the direction of this current in Fig. 30-57?

31. Figure 30-58 shows a magnetic field pointing in the x direction. Its strength, however, varies with position in the y direction. At the top and bottom of the rectangular loop shown the field strengths are 3.4 μT and 1.2 μT, respectively. How much current flows through the area encircled by the loop?

FIGURE 30-58 Problems 31, 66.

Section 30-4 Using Ampère's Law

32. (a) What is the magnetic field strength 0.10 mm from the axis of a 1.0-mm-diameter wire carrying 5.0 A distributed uniformly over its cross section? (b) What is the field strength at the surface of the wire?

33. A solid wire 2.1 mm in diameter carries a 10-A current with uniform current density. What is the magnetic field strength (a) at the axis of the wire, (b) 0.20 mm from the axis, (c) at the surface of the wire, and (d) 4.0 mm from the wire axis?

34. Show that Equations 30-8 and 30-9 give the same results when evaluated at the surface of the wire.

35. A long conducting rod of radius R carries a nonuniform current density given by $J = J_0 r/R$, where J_0 is a constant and r is the radial distance from the rod's axis. Find expressions for the magnetic field strength (a) inside and (b) outside the rod.

36. A long, hollow conducting pipe of radius R carries a uniform current I along the pipe, as shown in Fig. 30-59. Use Ampère's law to find the magnetic field strength (a) inside and (b) outside the pipe.

FIGURE 30-59 Problem 36.

37. Typically, cylindrical wires made from yttrium-barium-copper-oxide superconductor can carry a maximum current density of 6.0 MA/m^2 at a temperature of 77 K, as long as magnetic field at the conductor surface does not exceed 10 mT. Suppose such a wire is to carry the maximum current density. (a) At what wire diameter would the surface magnetic field equal the 10-mT limit? (b) Is this a maximum or minimum value for the diameter if the field is not to exceed the limit? (c) What current would a wire with this diameter carry?

38. A copper ribbon 1.0 cm wide and 0.15 mm thick is rated for a maximum safe current density of 8.8×10^6 A/m^2.

What is the maximum magnetic field strength achievable at the surface of this ribbon?

39. Two large, flat conducting plates lie parallel to the x-y plane. They carry equal currents, one in the $+x$ and the other in the $-x$ direction. In each plate the current per meter of width in the y direction is J_s. Find the magnetic field strength (a) between and (b) outside the plates.

40. The coaxial cable shown in Fig. 30-60 consists of a solid inner conductor of radius a and a hollow outer conductor of inner radius b and thickness c. The two carry equal but opposite currents I, uniformly distributed. Find expressions for the magnetic field strength as a function of radial position r (a) within the inner conductor, (b) between the inner and outer conductors, and (c) beyond the outer conductor.

FIGURE 30-60 Problems 40, 61.

41. A hollow conducting pipe of inner radius a and outer radius b carries a current I parallel to its axis and distributed uniformly through the pipe material (Fig. 30-61). Find expressions for the magnetic field for (a) $r < a$, (b) $a < r < b$, and (c) $r > b$, where r is the radial distance from the pipe axis.

FIGURE 30-61 Problem 41.

42. A conducting slab extends infinitely in the x and y directions and has thickness h in the z direction. It carries a uniform current density $\mathbf{J} = J\hat{\imath}$. Find the magnetic field strength (a) inside and (b) outside the slab, as functions of the distance z from the center plane of the slab.

Section 30-5 Solenoids and Toroids

43. A superconducting solenoid has 3300 turns per meter and can carry a maximum current of 4.1 kA. What is the magnetic field strength in the solenoid?

44. A solenoid used in a plasma physics experiment is 10 cm in diameter, 1.0 m long, and carries a 35-A current to produce a 100-mT magnetic field. (a) How many turns are in the solenoid? (b) If the solenoid resistance is 2.7 Ω, how much power does it dissipate?

45. You have 10 m of 0.50-mm-diameter copper wire and a battery capable of passing 15 A through the wire. What magnetic field strengths could you obtain (a) inside a 2.0-cm-diameter solenoid wound with the wire as closely spaced as possible and (b) at the center of a single circular loop made from the wire?

46. A toroidal coil of inner radius 15 cm and outer radius 17 cm is wound from 1200 turns of wire. What are (a) the minimum and (b) the maximum magnetic field strengths within the toroid when it carries a 10-A current?

47. A toroidal fusion reactor requires a magnetic field that varies by no more than 10% from its central value of 1.5 T. If the minor radius of the toroidal coil producing this field is 30 cm, what is the minimum value for the major radius of the device?

48. A long solenoid with n turns per unit length carries a current I. The current returns to its driving battery along a wire of radius R that passes through the solenoid, along its axis. Find expressions for (a) the magnetic field strength at the surface of the wire and (b) the angle the field at the wire surface makes with the solenoid axis.

49. We noted that there is a nonzero magnetic field component outside a solenoid, encircling the device, associated with the component of current flow parallel to the solenoid axis. For a long solenoid of radius R, find an expression for the ratio of this external encircling field just outside the solenoid to the field inside, and show explicitly that this ratio tends to zero as the number of turns per unit length becomes large.

50. Derive Equation 30-11 for the field of a solenoid by considering the solenoid to be made of a large number of adjacent current loops. Use Equation 30-3 for the field of a current loop, and integrate over all loops.

Section 30-6 Magnetic Matter

51. When a sample of a certain substance is placed in a 250.0-mT magnetic field, the field inside the sample is 249.6 mT. Find the magnetic susceptibility of the substance. Is it ferromagnetic, paramagnetic, or diamagnetic?

52. A container of liquid oxygen at 90 K is placed in a 500.0-G magnetic field. What is the field strength within the liquid oxygen?

53. A ferromagnetic material is placed in a 2.5-G magnetic field and the field within the material is determined to be 1.8 T. What is the magnetic susceptibility of this material?

54. Figure 30-62 shows the hysteresis curve for the ferromagnetic alloy Alnico V, commonly used in permanent magnets. Use the graph to find approximate values for the maximum field strength obtainable inside this material, (a) in the absence and (b) in the presence of an externally applied field.

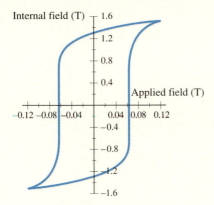

FIGURE 30-62 Problem 54.

Paired Problems

(Both problems in a pair involve the same principles and techniques. If you can get the first problem, you should be able to solve the second one.)

55. Two concentric, coplanar circular current loops have radii a and $2a$. If the magnetic field is zero at their common center, how does the current in the outer loop compare with that in the inner loop?

56. A thin conducting washer of inner radius a and outer radius $2a$ carries a current I distributed uniformly with radial position, as suggested in Fig. 30-63. Find an expression for the magnetic field strength at its center.

FIGURE 30-63 Problem 56.

57. Figure 30-64 shows a wire of length ℓ carrying current fed by other wires that are not shown. Point A lies on the perpendicular bisector, a distance y from the wire. Adapt the calculation of Example 30-2 to show that the magnetic field at A due to the straight wire alone has magnitude $\dfrac{\mu_0 I \ell}{2\pi y \sqrt{\ell^2 + 4y^2}}$. What is the field direction?

FIGURE 30-64 Problems 57, 58.

58. Point B in Fig. 30-64 lies a distance y perpendicular to the end of the wire. Show that the magnetic field at B has magnitude $\dfrac{\mu_0 I \ell}{4\pi y \sqrt{\ell^2 + y^2}}$. What is the field direction?

59. The largest lightning strikes have peak currents around 250 kA, flowing in essentially cylindrical channels of ionized air. How far from such a flash would the resulting magnetic field be equal to Earth's magnetic field strength, about 50 μT?

60. A particle-beam weapon being tested for ballistic missile defense delivers a 2.5-cm-diameter electron beam carrying a current of 10 kA at an altitude where Earth's magnetic field is 30 μT. How close to the beam would an adversary's surveillance apparatus need to be in order to detect the beam by observing a 1% change in the local magnetic field strength? Assume the beam's magnetic field is parallel to Earth's at the point of detection.

61. A coaxial cable like that shown in Fig. 30-60 consists of a 1.0-mm-diameter inner conductor and an outer conductor of inner diameter 1.0 cm and 0.20 mm thickness. A 100-mA current flows down the center conductor and back along the outer conductor. Find the magnetic field strength (a) 0.10 mm, (b) 5.0 mm, and (c) 2.0 cm from the cable axis.

62. Repeat the preceding problem if the current in the outer conductor only is increased to 200 mA.

Supplementary Problems

63. A circular wire loop of radius 15 cm and negligible thickness carries a 2.0-A current. Use suitable approximations to find the magnetic field of this loop (a) in the loop plane, 1.0 mm outside the loop, and (b) on the loop axis, 3.0 m from the loop center.

64. A long, flat conducting bar of width w carries a total current I distributed uniformly, as shown in Fig. 30-65. Use suitable approximations to write expressions for the

magnetic field strength (a) near the conductor surface ($r \ll w$) and (b) very far from the conductor ($r \gg w$).

FIGURE 30-65 Problem 64.

65. A long, hollow conducting pipe of radius R and length ℓ carries a uniform current I flowing around the pipe, as shown in Fig. 30-66. Find expressions for the magnetic field (a) inside and (b) outside the pipe. *Hint:* What configuration does this pipe resemble?

FIGURE 30-66 Problem 65.

66. The magnetic field in Fig. 30-58 is given by $\mathbf{B} = by\hat{\mathbf{i}}$, where $b = 55$ μT/m and y is the vertical coordinate in Fig. 30-58, measured in meters. Find the current density in the region.

67. A wide, flat conducting spring of spring constant $k = 20$ N/m and negligible mass consists of two 6.0-cm-diameter turns, as shown in Fig. 30-67. In its unstretched configuration the coils are nearly touching. A 10-g mass is hung from the spring, and at the same time a current I is passed through it. The spring stretches 2.0 mm. Find I, assuming the coils remain close enough to be treated as parallel wires.

FIGURE 30-67 Problem 67.

68. A solid conducting wire of radius R runs parallel to the z axis and carries a current density given by $\mathbf{J} = J_0(1 - r/R)\hat{\mathbf{k}}$, where J_0 is a constant and r the radial distance from the wire axis. Find expressions for (a) the total current in the wire, (b) the magnetic field strength for $r > R$, and (c) the magnetic field strength for $r < R$.

69. A disk of radius a carries a uniform surface charge density σ, and is rotating with angular speed ω about the central axis perpendicular to the disk. Show that the magnetic field at the disk's center is $\frac{1}{2}\mu_0\sigma\omega a$.

70. Calculate the magnetic field of an infinite current sheet by considering the sheet to be made up of infinitesimal line currents, as suggested in Fig. 30-26, and integrating the fields of these line currents.

71. Work Example 30-2 by expressing all variables in terms of the angle θ and integrating over the appropriate range in θ.

Electromagnetic Induction

31

These wind-driven electric generators use electromagnetic induction to convert the mechanical energy of wind into electrical energy. Underground cables combine the electric power produced by all the generators and the power is then fed into the utility company's power grid. Most of the world's electrical energy is produced by generators driven by steam from burning fossil fuels or from nuclear fission—and all such generators use electromagnetic induction. So do videotapes, computer disks, credit cards, and other magnetic information storage systems.

All the electric and magnetic fields we encountered in the previous chapters had their ultimate origins in electric charge, either stationary or moving. We stressed a relation between electricity and magnetism, whereby electric charge gives rise to and interacts with both the electric field and the magnetic field. The remainder of our study of electromagnetism is devoted to a much more intimate relation between electricity and magnetism, a relation in which the fields themselves interact directly. This interaction forms the basis of new electromagnetic technologies, leads toward an understanding of the nature of light, and points the way to the theory of relativity.

31-1 INDUCED CURRENTS

In 1831, the English scientist Michael Faraday and the American Joseph Henry independently carried out experiments in which electric currents arose in circuits subjected to changing magnetic fields. We now describe four simple experiments that illustrate this phenomenon:

Experiment 1 (Fig. 31-1): A bar magnet is moved in the presence of a circuit consisting of a coil of wire connected to an ammeter. There's no battery or other obvious source of emf in the circuit. As long as the magnet is stationary, there's no current. But when the magnet moves, the ammeter registers a current—which we call an **induced current**, because it's been somehow induced by the movement of the magnet. If we move the magnet faster, the induced current increases. If we reverse the direction of the magnet's motion, the current reverses.

Experiment 2 (Fig. 31-2): A coil is moved in the presence of a stationary magnet. An induced current results that behaves exactly like the current in Experiment 1. So we get the same results whether we move the magnet or the circuit; all that matters is the relative motion between magnet and circuit.

Experiment 3 (Fig. 31-3): Replace the bar magnet with an electromagnet made from a coil of wire carrying a steady current driven by a battery. The new coil creates a magnetic field like that of the bar magnet, and, not surprisingly given the results of Experiments 1 and 2, we get an induced current that depends on the relative motion of the two coils.

Experiment 4 (Fig. 31-4): Hold both coils stationary; then there's no induced current, just as there was none in Experiments 1 and 2 when the magnet and coil were not in relative motion. But now open the switch connecting the battery to the left-hand coil. The current drops quickly to zero, and during the brief interval while it's doing so, the ammeter registers a current in the right-hand coil. Then the induced current ceases as the current in the left-hand coil remains at zero. Now close the switch again: as the current briefly builds up in the left-hand coil, the ammeter registers an induced current in the right-hand coil—and its direction is opposite what it was when the switch was opened. Once the current reaches a steady value, the induced current again ceases.

The common feature in these experiments is a *changing magnetic field*. It does not matter whether the field changes because a magnet is moved, or because a circuit is moved near a magnet, or because the current giving rise to the

FIGURE 31-1 When a magnet is moved near a closed circuit, current flows in the circuit.

FIGURE 31-2 We get the same results if we move the coil instead of the magnet.

FIGURE 31-3 An induced current also arises when the magnet is replaced by a current-carrying circuit.

FIGURE 31-4 A current is also induced—even in the absence of any motion—when the current in an adjacent circuit changes.

field changes. In each case, an induced current appears in a circuit subjected to a changing magnetic field. We are observing here a new phenomenon— **electromagnetic induction**—whereby electrical effects arise from *changing* magnetic fields.

31-2 FARADAY'S LAW

We know from Chapter 27 that a source of electromotive force—such as a battery that supplies energy to electric charges—is needed to establish a current in a circuit. When an induced current flows in a circuit, even though there's no battery present, there must still be an emf. This so-called **induced emf** is usually not localized at one point in the circuit, as in a battery, but may be spread throughout the conductors making up the circuit.

Experimentally, we find that the induced emf in a circuit depends on the rate of change of **magnetic flux** through the circuit. Before quantifying this relationship, we show how to calculate magnetic flux.

Magnetic Flux

We introduced magnetic flux at the end of the preceding chapter, and showed that the magnetic flux through any *closed* surface is zero. Here we're interested in the flux through *open* surfaces, which need not be zero (Fig. 31-5). Like the electric flux defined in Chapter 24, magnetic flux is the integral of the magnetic field over a surface:

$$\phi_B = \int \mathbf{B} \cdot d\mathbf{A} . \tag{31-1}$$

As with electric flux, magnetic flux is proportional to the number of field lines passing through the surface. With electromagnetic induction, we're interested in the flux through a surface bounded by the circuit in question. For a circular loop such as the one in Fig. 31-5, that surface can be the circular disk whose circumference is the loop. More generally, the surface can be *any* surface bounded by the loop.

For a flat surface perpendicular to a uniform magnetic field, as in the example below, Equation 31-1 reduces to the simple expression $\phi_B = BA$. Note that the units of magnetic flux are T·m^2.

FIGURE 31-5 A circular wire loop in the magnetic field of a bar magnet. The flux ϕ_B is proportional to the number of field lines crossing the circular surface bounded by the loop. If the magnet moves closer to the loop, the number of field lines—and hence the flux— increases.

EXAMPLE 31-1	*The Magnetic Flux through a Solenoid*

A solenoid of circular cross section has radius R, consists of n turns per unit length, and carries a current I. What is the magnetic flux through each turn of the solenoid?

Solution
The field of a solenoid is uniform and parallel to the solenoid axis, as we found in deriving Equation 30-11:

$$B = \mu_0 nI .$$

A flat surface bounded by one turn of the solenoid lies at right angles to this uniform field, so the flux is simply the product of the magnetic field and the area:

$$\phi_B = \int \mathbf{B} \cdot d\mathbf{A} = BA = \mu_0 nI\pi R^2 .$$

We're being a little loose here in thinking of a single turn of the solenoid as a closed loop, but if the solenoid is tightly wound this is an excellent approximation. We will be concerned frequently with the flux through a multiturn coil, and in calculating this flux it's convenient to view each turn as an individual loop. The flux through an N-turn coil in a uniform magnetic field is just N times the flux through each turn.

EXERCISE A rectangular wire loop measuring 10 cm by 15 cm is oriented so the normal to the loop makes a 30° angle with a uniform 50-mT magnetic field. Find the magnetic flux through the loop.

Answer: 6.5×10^{-4} T·m^2

Some problems similar to Example 31-1: 3, 4

EXAMPLE 31-2 | *Magnetic Flux in a Nonuniform Field*

A long, straight wire carries a current I. A rectangular wire loop of dimensions ℓ by w lies with its closest edge a distance a from the wire, as shown in Fig. 31-6. What is the magnetic flux through the loop?

Solution

The magnetic field of the wire is given by Equation 30-8:

$$B = \frac{\mu_0 I}{2\pi r},$$

where r is the distance from the wire. At the site of the loop, this field points straight into the page, perpendicular to the plane of the loop. However, the field varies with distance from the straight wire, so we can't simply multiply the field by the loop area to get the flux. Instead, we divide the loop into thin strips of width dr and area $dA = \ell\, dr$, as shown in Fig. 31-6. With the field at right angles to each strip, $\mathbf{B} \cdot d\mathbf{A} = B\, dA$, and the flux through any strip is

$$d\phi_B = B\, dA = B\ell\, dr = \frac{\mu_0 I}{2\pi r}\, \ell\, dr.$$

Then the total flux through the loop is the integral over all such strips contained within the loop, that is, over all strips between $r = a$ and $r = a + w$:

$$\phi_B = \int_a^{a+w} \frac{\mu_0 I}{2\pi r}\, \ell\, dr = \frac{\mu_0 I \ell}{2\pi} \int_a^{a+w} \frac{dr}{r} = \frac{\mu_0 I \ell}{2\pi} \ln\, r \Big|_a^{a+w}$$

$$= \frac{\mu_0 I \ell}{2\pi} \ln\!\left(\frac{a+w}{a}\right).$$

FIGURE 31-6 A rectangular loop in the magnetic field of a long wire. The field to the right of the wire points into the plane of the page, and its magnitude drops inversely with distance from the wire. Area elements for the flux calculation are strips of length ℓ and infinitesimal width dr.

EXERCISE A square wire loop 45 cm on a side lies in the plane of this page, taken to be the x-y plane. The left edge of the loop coincides with the y axis. A magnetic field points into the page; its magnitude varies with position, being given by $B = ax$, where $a = 2.0$ T/m and x is position in meters. Find the flux through this loop.

Answer: 91 mT·m²

Some problems similar to Example 31-2: 43, 44

Flux and Induced EMF

Having quantified the notion of magnetic flux, we are now ready to state rigorously the experimental fact that changing magnetic flux induces an emf in a circuit. Our statement is a special case of **Faraday's law of induction**, which constitutes another of the four basic laws of electromagnetism:

> **The induced emf in a circuit is proportional to the rate of change of magnetic flux through any surface bounded by that circuit.**

This statement is a special case of Faraday's law that describes electromagnetic induction specifically in circuits; later we'll present a more general form that applies even when no circuit is present. The induced emf tends to oppose the change in flux—a crucial point to which we'll devote all of Section 31-3—and so in SI units the proportionality between emf and rate of change of flux is just -1. Thus Faraday's law is

$$\mathcal{E} = -\frac{d\phi_B}{dt}, \tag{31-2}$$

where $\mathcal{E}$ is the emf induced in a circuit and ϕ_B the magnetic flux through any surface bounded by that circuit. Problem 1 will help convince you that the units of the rate of change of magnetic flux are indeed volts, the same as the units of emf.

Faraday's law relates the induced emf to the *change* in flux. It isn't magnetic flux or magnetic field that causes an induced emf—it's the *change* in flux. The relation between emf and flux is like the relation between acceleration and velocity, where the acceleration depends not on the actual value of the velocity, but on its *rate of change*. In Fig. 31-1 we changed the flux by moving a magnet near a coil, increasing the flux as more field lines went through the area bounded by the coil. In Fig. 31-4 we changed the current in an adjacent coil, changing the magnetic field and thus the flux. More generally, since the flux $\mathbf{B} \cdot \mathbf{A} = BA \cos \theta$, we can change flux by changing the field strength B, the area A, or the orientation of the surface to the field (i.e., the angle θ between the field and the normal to the surface). The following examples explore these different ways of changing magnetic flux.

13.9
Electromagnetic Induction

EXAMPLE 31-3	*Changing Magnetic Field*

A wire loop of radius 10 cm has a resistance of 2.0 Ω. The loop is at right angles to a uniform magnetic field **B**, as shown in Fig. 31-7. The field strength is increasing at 0.10 tesla/second. What is the magnitude of the induced current in the loop?

Solution

To find the induced current, we need to know the induced emf. Faraday's law tells us that the induced emf is related to the rate of change of magnetic flux through the circuit. Here the field is uniform and at right angles to the loop, so the flux is just the product of the field with the loop area:

$$\phi_B = \int \mathbf{B} \cdot d\mathbf{A} = \pi r^2 B .$$

Why could we take B outside the integral when the field is changing? Because it's changing with *time,* and the integral is over *space.* At any given instant the field is *uniform* in space, and that's all we need to reduce the flux integral to a simple product.

We don't know B, but this doesn't matter because we are really interested in the *rate of change* of the flux, not in the flux itself. With the loop area constant, the rate of change of flux is

$$\frac{d\phi_B}{dt} = \pi r^2 \frac{dB}{dt} = (\pi)(0.10 \text{ m})^2(0.10 \text{ T/s}) = 3.14 \times 10^{-3} \text{ V} .$$

By Faraday's law, this is the magnitude of the induced emf. We then calculate the current using Ohm's law, as we would for any emf:

$$I = \frac{\mathcal{E}}{R} = \frac{3.14 \times 10^{-3} \text{ V}}{2.0 \text{ Ω}} = 1.6 \times 10^{-3} \text{ A} = 1.6 \text{ mA} .$$

FIGURE 31-7 A circular conducting loop at right angles to a uniform magnetic field.

Tip

It's the Change That Counts Faraday's law relates induced emf to the *rate of change* of magnetic flux. The actual value of the flux—or of the magnetic field if that is what's changing—doesn't matter. You may be troubled by induction problems where the value of the field is not given. You don't need it; what you do need or may be asked to find is the *rate of change* of the field.

EXERCISE A square conducting loop 25 cm on a side lies at right angles to a uniform magnetic field. The loop's resistance is 8.0 Ω, and it carries a current of 14 mA. At what rate is the magnetic field changing?

Answer: 1.8 T/s

Some problems similar to Example 31-3: 4–8

EXAMPLE 31-4	*Changing Area*

A circuit consists of two parallel conducting rails a distance ℓ apart connected at one end by a resistance R. A conducting bar completes the circuit, joining the two rails electrically but being free to slide along them. The whole circuit is in a constant, uniform magnetic field **B** at right angles to the plane of the circuit, as shown in Fig. 31-8. The bar is pulled to the right with constant speed v, increasing the area of the circuit as it moves. What is the current in the circuit? Assume the bar and rails are ideal conductors, so the total circuit resistance is R.

FIGURE 31-8 Pulling the bar to the right increases the circuit area, thereby increasing the magnetic flux and inducing an emf that drives a current.

Solution

Here the current is driven by an induced emf arising from the change in magnetic flux that occurs as the circuit area increases. We determine the emf using Faraday's law, $\mathcal{E} = -d\phi_B/dt$. The circuit area is the rail spacing ℓ times the distance x from resistor to bar. With a uniform field perpendicular to the circuit, the flux integral $\int \mathbf{B} \cdot d\mathbf{A}$ reduces to the product of field strength with the area:

$$\phi_B = \int \mathbf{B} \cdot d\mathbf{A} = BA = B\ell x.$$

Never mind that we don't know x—we do know its rate of change and that's all we need. With the field and the rail spacing constant, the rate of change of flux is

$$\frac{d\phi_B}{dt} = B\ell\frac{dx}{dt} = B\ell v,$$

since dx/dt is just the bar velocity v. Faraday's law tells us that the magnitude of the induced emf is equal to the rate of change of flux, so

$$|\mathcal{E}| = B\ell v.$$

This emf drives a current I around the circuit:

$$|I| = \frac{|\mathcal{E}|}{R} = \frac{B\ell v}{R}.$$

ActivPhysics Activity 13.10 provides a simulation of this example.

EXERCISE A square wire loop 25 cm on a side is moving into a region containing a uniform 1.2-T magnetic field oriented at right angles to the loop, as shown in Fig. 31-9. The loop wire is essentially resistanceless, but inserted in the loop is a 3-V flashlight bulb. How fast must the loop move to keep the bulb lit at its normal brightness?

Answer: 10 m/s

Some problems similar to Example 31-4: 9, 10, 16, 27, 35

FIGURE 31-9 How fast must the loop move to light the 3-V bulb?

13.10
Motional EMF

EXAMPLE 31-5	*Changing Orientation: An Electric Generator*

A circular wire loop of radius a and resistance R is initially perpendicular to a constant, uniform magnetic field B. The loop rotates with angular velocity ω about an axis through a diameter, as shown in Fig. 31-10. What is the current in the loop?

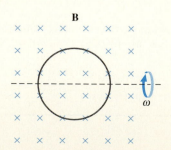

FIGURE 31-10 A wire loop rotating in a uniform magnetic field. The flux changes because of the changing orientation of the loop relative to the field, thereby inducing an emf that drives a current around the loop.

Solution

Again, we must find the rate of change of magnetic flux, from which we can get the induced emf and then the current. Here the field is uniform over the area, but the orientation of the field relative to the area is changing. The definition of magnetic flux contains a dot product to account for this orientation. Evaluating the flux, we have

$$\phi_B = \int \mathbf{B} \cdot d\mathbf{A} = \int B \, dA \cos\theta = B\cos\theta \int dA$$
$$= \pi a^2 B \cos\theta \, ,$$

where θ is the angle between the field and a perpendicular to the loop area. Here we could take the field magnitude B outside the integral because the field is uniform over the loop. And even though the orientation between loop and field changes with *time,* it does not change with *position* at a given instant, so we could also take $\cos\theta$ outside the integral.

The changing orientation is described by giving θ as a function of time. Since the loop rotates with constant angular velocity ω, we can write simply $\theta = \omega t$, where we take the zero of time when $\theta = 0$. Then the flux is

$$\phi_B = \pi a^2 B \cos\omega t \, ,$$

so

$$\frac{d\phi_B}{dt} = \pi a^2 B \frac{d}{dt}(\cos\omega t) = -\pi a^2 B\omega \sin\omega t \, .$$

By Faraday's law, the emf is then

$$\mathcal{E} = -\frac{d\phi_B}{dt} = \pi a^2 B\omega \sin\omega t \, ,$$

giving a current

$$I = \frac{\mathcal{E}}{R} = \frac{\pi a^2 B\omega}{R} \sin\omega t \, .$$

(Check the units!)

Unlike the current in the previous two examples, this one changes with time. Its sinusoidal time dependence is in fact just like that of standard alternating current used for electric power—and with good reason: our rotating loop constitutes a simple alternating-current generator. Sinusoidally varying emf's and currents occur whenever conducting loops rotate in uniform magnetic fields.

Tip

Peak Values You'll often be asked for the peak value of voltage or current when either quantity is changing with time (more on this in Chapter 33). You've derived a formula like $\mathcal{E} = \pi a^2 B\omega \sin\omega t$ in Example 31-5, but you don't have a value for the time t. It doesn't matter! The peak value occurs when $\sin\omega t = 1$, so just replace $\sin\omega t$ by 1. In Example 31-5, for example, the peak emf is just $\mathcal{E}_p = \pi a^2 B\omega$.

EXERCISE Take the radius $a = 11$ cm in Example 31-5 and the magnetic field strength $B = 0.63$ T. With what angular speed ω must the loop rotate in order to produce a peak emf of 6.0 V?

Answer: 250 s^{-1}

Some problems similar to Example 31-5: 12, 24, 58, 63

You might wonder about the direction of the induced emf and current in these examples since we were concerned only with magnitudes. Formally, that direction follows mathematically from the minus sign in Faraday's law. But in the next section we'll provide a physical justification for the minus sign that makes it much clearer how to determine the direction of induced emf's and currents.

31-3 INDUCTION AND THE CONSERVATION OF ENERGY

Move a bar magnet toward a wire loop, as shown in Fig. 31-11. Without the loop present, it would take no work to move the magnet horizontally at constant

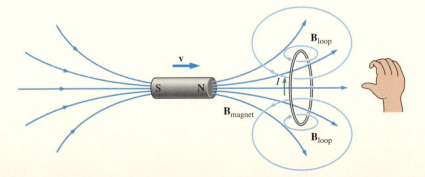

FIGURE 31-11 As the bar magnet moves toward the conducting loop, the changing magnetic flux through the loop induces an emf that drives a current around the loop. Conservation of energy requires that the magnetic field produced by this current *oppose* the motion of the magnet, so the agent moving the magnet does work that ends up heating the loop.

velocity. But with the loop present, moving the magnet induces a current in the loop. Electrical energy is dissipated in the loop's resistance, heating the loop. Where does the energy come from?

There's only one source for the energy—the agent moving the magnet. As you move the magnet, you must be doing positive work. Otherwise you would be getting energy—ultimately heating the loop—for nothing. So you must have to push the magnet against some force. What force? A magnetic force, caused by the interaction of the magnet with the magnetic field produced by the induced current. Think about this! The induced current, like any current, produces a magnetic field. The magnet experiences a force in this field, and you must do work to move the magnet through that field. With the magnetic field as intermediary, the work you do ultimately ends up heating the loop.

As you push the magnet toward the loop, the force the loop's field exerts on the magnet must be repulsive in order that you do positive work on the magnet. In Fig. 31-11 the bar magnet's north pole is toward the loop. To provide a repulsive force the loop must therefore present a north pole to the approaching magnet; that is, field lines must emerge from the loop's interior pointing to the left. What current direction will provide such a field? Application of the right-hand rule gives the answer: Wrap the fingers of your right hand around the loop in the direction of the current, and your right thumb points in the direction of the field in the loop's interior. Thus the current direction is as shown in Fig. 31-11.

This analysis leading to the direction of the induced current was based on conservation of energy, a fundamental principle that we believe applies universally. For electromagnetic induction, we've just seen that conservation of energy requires the following:

The direction of the induced emf and current is such that the magnetic field created by the induced current opposes the change in magnetic flux.

This statement, called **Lenz's law**, is represented mathematically in the minus sign on the right-hand side of Faraday's law (Equation 31-2).

What happens, for example, if you pull the bar magnet away from the loop in Fig. 31-11? Now the loop must present a south pole to the receding magnet, creating an attractive force opposing the magnet's withdrawal (Fig. 31-12). The loop current must be opposite its direction in Fig. 31-11, as again you do work to overcome the magnetic force.

FIGURE 31-12 When the bar magnet is pulled away from the loop, the direction of the induced current is such that the magnetic force opposes the magnet's withdrawal.

Got It!

You push a bar magnet toward a loop, with the north pole toward the loop as in Fig. 31-11. If you keep pushing the magnet straight through the loop, what will be the direction of the current as you pull it out the other side? Will you need to do work, or will work be done on you?

Motional EMF and Lenz's Law

When a conductor moves through a magnetic field, we can understand the origin of the induced emf in terms of the magnetic force on charge carriers in the wire; this emf is called **motional emf**. In the case of motional emf we can show explicitly that Lenz's law requires energy conservation.

Consider a square conducting loop of side ℓ and resistance R being pulled with constant speed v out of a uniform magnetic field $\mathbf{B}$, as shown in Fig. 31-13. Since the magnetic flux through the loop is changing, there's an induced emf, and that emf drives a current. Energy is dissipated as heat because of the loop's resistance, and so, as we've just argued, the agent pulling the loop must do work. We'll now demonstrate quantitatively that energy is conserved, by showing that rate of heating in the loop is exactly equal to the rate at which the agent pulling the loop does work.

In Chapter 29 we found that the magnetic force on a charge is given by $\mathbf{F} = q\mathbf{v} \times \mathbf{B}$. Pulling the loop to the right moves its free electrons through the magnetic field; the magnetic force $q\mathbf{v} \times \mathbf{B}$ on these electrons is downward in Fig. 31-13 (opposite $\mathbf{v} \times \mathbf{B}$ since electrons are negative). The resulting downward motion of the negative electrons in the left-hand side of the loop constitutes an upward current. This current continues clockwise around the loop, driven by an electric field associated with the separation of charge in the loop's left side.

Now we have a current I in the loop. We found in Chapter 29 that the magnetic force on a current-carrying conductor of length ℓ is $\mathbf{F} = I\boldsymbol{\ell} \times \mathbf{B}$. Applying this expression to the conducting loop in Fig. 31-13 shows that there is no magnetic force on the right-hand side (since $\mathbf{B} = \mathbf{0}$ there) and that the forces on top and bottom cancel (Fig. 31-14). So the total magnetic force on the loop is that on the left side alone; since the current is upward while the magnetic field is into the page, the magnitude of this force is $I\ell B$, and the right-hand rule shows that it points to the left. This leftward magnetic force has just the direction needed to cancel the rightward applied force, giving the zero net force that Newton's law requires for the loop to move with constant velocity.

We could equally well determine the current direction from magnetic flux considerations. As the loop leaves the field region, the flux through it decreases. The direction of the induced current is such as to oppose this decrease in flux. Therefore the magnetic field of the induced current points into the page, as the induced current tries to maintain the flux. By the right-hand rule, a field within the loop and into the page requires that the induced current flow clockwise.

To calculate the current, we must find the induced emf, which in turn is related to the rate of change of magnetic flux through the loop. With the field and

FIGURE 31-13 A conducting loop being withdrawn from a magnetic field. The magnetic force $q\mathbf{v} \times \mathbf{B}$ on charge carriers in the left side of the loop drives a current clockwise around the loop. The current direction also follows from Lenz's law: The magnetic field produced by the loop current acts to oppose the *decrease* in flux through the loop, and therefore points in the *same* direction as the original field. This requires a clockwise loop current.

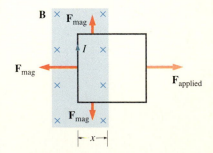

FIGURE 31-14 Forces on the loop. The total magnetic force is that on the left side alone, and the agent pulling the loop must exert an equal but opposite force to maintain constant velocity.

loop perpendicular, and with the field uniform in the region where it is nonzero, the magnetic flux is the product of the magnetic field strength and the loop area that lies within the field:

$$\phi_B = B\ell x,$$

where x is the distance between the left edge of the loop and the right edge of the magnetic field region. The magnetic field remains constant, but as the loop moves the distance x decreases at the rate $dx/dt = -v$ (the minus sign indicates a decrease). Then the rate of change of flux is

$$\frac{d\phi_B}{dt} = \frac{d(B\ell x)}{dt} = B\ell \frac{dx}{dt} = -B\ell v,$$

so Faraday's law gives
$$\mathcal{E} = -\frac{d\phi_B}{dt} = B\ell v.$$

This induced emf drives a current I around the loop, where

$$I = \frac{\mathcal{E}}{R} = \frac{B\ell v}{R}.$$

The rate of energy dissipation in the loop is the product of the emf and the current (Equation 27-8):

$$P = I\mathcal{E} = \frac{B\ell v}{R} B\ell v = \frac{B^2 \ell^2 v^2}{R}. \qquad \left(\begin{array}{c}\text{power dissipated}\\ \text{in loop}\end{array}\right)$$

We've found that the magnetic force on the loop has magnitude $I\ell B$; since the loop is moving with constant velocity, this is also the magnitude of the applied force. Equation 7-21 shows that the rate at which work is done by a force **F** acting on an object moving with velocity **v** is $P = \mathbf{F} \cdot \mathbf{v}$; here, with the applied force and velocity in the same direction, we have

$$P = Fv = I\ell Bv = \frac{B\ell v}{R} \ell Bv = \frac{B^2 \ell^2 v^2}{R}, \qquad \left(\begin{array}{c}\text{power supplied}\\ \text{to pull loop}\end{array}\right)$$

in agreement with our expression for the electric power dissipated in the loop. Thus, all the work done by the agent pulling the loop ends up heating the resistor, showing explicitly that energy is indeed conserved.

Got It!

What will be the direction of the current when the loop in Fig. 31-13 first enters the field, coming in from the left side?

Electromagnetic induction is the principle behind many important technologies, from information storage on computer disks and credit cards to electric power generation. Induction also gives us the flexibility to transform voltage levels in electric power systems, and to provide wireless charging systems for devices ranging from electric cars to toothbrushes. The applications that follow explore some uses of induction.

APPLICATION	*Electric Generators*

Probably the most important technological application of induction is the electric generator. The world currently uses electrical energy at the phenomenal rate of about 10^{13} watts—roughly equal to the power output of 100 billion human bodies—and virtually all this power comes from generators. A generator is just a system of conductors in a magnetic field (Fig. 31-15). Mechanical energy is supplied to rotate the conductors, resulting in a changing magnetic flux. An emf is induced and current flows through the generator and on to whatever electrical loads are connected to it. Because the changing flux results from a change in the orientation of the loop relative to the field—i.e., a change in θ in the expression $\phi_B = BA\cos\theta$—a generator such as that of Fig. 31-15 produces an alternating emf that varies sinusoidally with time (Fig. 31-16), as we saw in Example 31-5.

Any source of mechanical energy can power the generator, but the most common is steam from burning fossil fuels or from nuclear fission (Fig. 31-17). Electrical energy may be generated from the kinetic energy of water or wind—as shown in this chapter's opening photo. A small electric generator, driven by the car's gasoline engine, is used to recharge a car's battery.

Lenz's law, the conservation of energy in electromagnetic induction, is very much applicable to electric generators. Were it not for Lenz's law, which requires that induced currents *oppose* the changes giving rise to them, generators would turn on their own and happily supply electricity without the need for coal, oil, or uranium! The voluminous quantities of fuel (Fig. 31-18) consumed by power plants are dramatic testimony to the minus sign appearing on the right-hand side of Equation 31-2!

An instructive introduction to Lenz's law comes about if you have access to a hand-cranked electric generator. Without any electrical load across the generator, it is easy to turn. But as you switch on increasingly heavy loads the generator gets

FIGURE 31-15 Simplified diagram of an electric generator. As the loop rotates in the magnetic field, the changing flux induces an emf that drives a current through the rotating contacts and stationary brushes and on through the electrical load.

FIGURE 31-17 This large power-plant generator produces over 1000 MW of electric power. It's basically a large version of the generator shown in Fig. 31-15.

FIGURE 31-16 A generator such as that shown in Fig. 31-15 produces an alternating emf whose period is the rotation period of the conducting loop.

FIGURE 31-18 A 110-car trainload of coal arriving at a power plant in Texas. Some fourteen such trains arrive at the plant each week—a testimony to the minus sign in Equation 31-2!

(a)   (b)

FIGURE 31-19 (a) With no electrical load, a hand-cranked generator is easy to turn. (b) With 200 watts of light bulbs connected, turning the generator becomes much more difficult. The generator emf is the same in both cases, because the generator is being turned at the same rate. But only in (b) is current flowing, giving nonzero power $P = IV$ that must be supplied by turning the generator.

harder to turn (Fig. 31-19). Most people find they can just sustain a 100-W light bulb with a hand generator. Think about this next time you leave a light on! You also experience Lenz's law when you turn on the headlights of a car that's idling slowly. You can hear the engine speed drop, and the car may even stall, as the car's generator gets harder for the engine to turn.

Got It!

If you lower the resistance across an electric generator, while turning the generator at a constant speed, will the generator get easier or harder to turn?

EXAMPLE 31-6	*Designing a Generator*

An electric generator consists of a 10-turn square wire loop 50 cm on a side. The loop is turned at 60 revolutions per second, to produce standard 60-Hz alternating current like that used throughout the United States and Canada. How strong must the magnetic field be for the peak output voltage of the generator to be 170 V? (This is actually the peak voltage of standard 120-V household wiring; 120 V is an appropriate average value.)

Solution

We need to evaluate the induced emf as a function of magnetic field strength. With a uniform magnetic field, the flux through one turn of the loop is $\int \mathbf{B} \cdot d\mathbf{A} = BA\cos\theta$, where θ is the angle between the field and the normal to the loop. But the loop rotates with angular frequency $\omega = 2\pi f$, so $\theta = 2\pi ft$. The loop area A is s^2, with s the length of the loop side, so the flux through N turns of the loop is

$$\phi_B = NBs^2\cos(2\pi ft).$$

To find the induced emf, we take the time rate of change of this flux:

$$\mathcal{E} = -\frac{d\phi_B}{dt} = -NBs^2\left[-2\pi f\sin(2\pi ft)\right]$$

$$= 2\pi NfBs^2\sin(2\pi ft).$$

The peak emf is the quantity multiplying the sine; we want this to be 170 V. Solving for the unknown magnetic field B then gives

$$B = \frac{\mathcal{E}_{\text{peak}}}{2\pi Nfs^2} = \frac{170\text{ V}}{(2\pi)(10)(60\text{ Hz})(0.50\text{ m})^2} = 0.18\text{ T}.$$

This is a typical field strength near the poles of a strong permanent magnet.

EXERCISE A generator includes a circular coil 30 cm in diameter, spinning at 3600 rpm in a uniform 0.50-T magnetic field. How many turns should it have if the peak output voltage is to be 2400 V?

Answer: 180

Some problems similar to Example 31-6: 24, 26, 49, 50

APPLICATION	*Magnetic Recording*

Magnetic materials are widely used as information storage media. Examples include audio- and videocassette tapes, computer disks, and the magnetic strips on credit cards. Retrieving the stored information involves electromagnetic induction.

In a typical magnetic recording system the magnetic medium—usually a plastic tape or a circular disk coated with ferromagnetic oxides—moves past a small coil called the **head**. While recording, current passing through the coil creates a magnetic field that impresses a magnetization pattern on the tape or disk (Fig. 31-20). With analog systems, like most audio- and videotapes, the current varies continuously and produces a smoothly varying magnetization pattern. With digital systems, including computer data storage and the newer digital audio and video formats, coil currents of the same magnitude but opposite polarity produce regions of oppositely directed magnetization; these represent the 1's and 0's in which digital information is coded.

To retrieve the stored information, the magnetic medium is again moved past a coil, which may be the same one used for recording. The changing magnetization on the medium results in a varying magnetic flux through the coil, which induces an emf that's amplified and processed to produce images, sound, or digital data.

The rate at which information can be stored and extracted depends on how densely the regions of varying magnetization can be packed without interfering and on how fast the medium moves in relation to the heads. Audiocassette tapes move past the heads at a mere 4.8 cm/s, a speed that limits these tapes' ability to record high-frequency sound faithfully. The much higher information content of video images requires higher tape-to-head speeds; in standard videocassette recorders a speed of 39.52 m/s is achieved by moving the tape past rapidly spinning heads. High-speed computer disks boast even greater speeds, allowing rapid data storage and retrieval (Fig. 31-21).

FIGURE 31-20 Recording on magnetic tape. The iron body of the head confines the magnetic field of the coil except at the gap, where a fringing field emerges to impose a magnetization pattern on the moving tape. When the tape is played back, its changing magnetization pattern imposes a changing magnetic flux at the head, inducing an emf in the head coil.

FIGURE 31-21 Mechanism of a computer disk drive includes a head (arrow) that literally flies as close as 0.25 μm above a spinning aluminum disk coated with ferromagnetic material.

APPLICATION	*Eddy Currents*

Our discussion of induced currents has centered on conducting loops. But induced currents also appear in solid conductors through which magnetic flux is changing. The resistance of a solid piece of conductor is low, so the induced currents are large, resulting in substantial energy dissipation. The presence of these **eddy currents** can make it difficult to move a conduc-

tor rapidly through a magnetic field. For example, if you try to push a piece of metal—it need not be a ferrous metal like iron—between the poles of a magnet, you'll find yourself working against a magnetic force.

A common demonstration of eddy currents consists of a pendulum with a metal bob that swings between the poles of a

(a) (b)

FIGURE 31-22 (a) An eddy-current pendulum consists of a conductor swinging between magnet poles. As it enters the field region, the conductor experiences a changing magnetic flux that induces a current. You should convince yourself that the current is in the direction shown as the bob begins to enter the field region. (b) The current dissipates energy in the resistance of the conductor, at the expense of its kinetic energy. Strobe photo shows rapid deceleration as the bob passes between the magnet poles.

magnet (Fig. 31-22). As it swings toward the magnet, the pendulum experiences an increasing magnetic flux that induces eddy currents. The energy dissipated by these currents comes ultimately from the kinetic energy of the pendulum, which therefore slows abruptly between the magnet poles.

Eddy currents provide an alternative to friction brakes for stopping moving machinery. Rapidly rotating saw blades, for example, can be stopped abruptly by an electromagnet activated next to the blade. Similarly, eddy-current brakes are sometimes used on trains and in other applications involving rotating conductors.

In some instances eddy currents are a nuisance, acting just like friction in reducing the efficiency of machinery. To solve this problem, slots are often cut into moving conductors to make the current paths longer, thus increasing the electrical resistance and reducing the eddy currents. For example, if the solid pendulum bob in Fig. 31-22 is replaced by a slotted piece, it then swings more freely through the magnet.

Lenz's Law and Changing Magnetic Fields

Lenz's law—conservation of energy applied to electromagnetic induction—determines the direction of the induced current even when no motion is involved. Figure 31-23 shows a conducting loop in a magnetic field that points into the page. Suppose the field strength—and, therefore, the magnetic flux through the loop—is *decreasing*. Then the direction of the induced current must be such that the magnetic field it creates *opposes* this decrease. Therefore, the loop's field must reinforce the existing field, which means that the loop's field points into the page. Applying the right-hand rule then shows that the loop current is clockwise in Fig. 31-23.

What if the magnetic field in Fig. 31-23 had been increasing in strength? Then the loop current would have been in the opposite direction, to create a magnetic field opposite the existing field in order to oppose its increase.

FIGURE 31-23 A uniform magnetic field is decreasing in strength, causing a decrease in magnetic flux through the conducting loop. The induced current in the loop must *oppose* this *decrease,* and therefore loop current goes clockwise in order to produce a magnetic field within the loop that reinforces the original field.

Tip

Induction Opposes Change Faraday's law relates the induced emf to the *rate of change* of magnetic flux. Lenz's law says that the induced emf and current are in such a direction as to oppose that *change* in flux—not the flux or the field itself. Figure 31-23 is a case in point; here the induced current is in a direction that actually reinforces the original field and flux—precisely because they are both *decreasing*. The same thing happened in Fig. 31-13, where movement of the loop decreased the flux, giving an induced current that reinforced the existing field.

EXAMPLE 31-7	*Lenz's Law*

Two coils are arranged as shown in Fig. 31-24. If the resistance of the variable resistor is being increased, what is the direction of the induced current in the fixed resistor R?

Solution

Applying the right-hand rule, we find that the magnetic field of coil A emerges from the right side of the coil, pointing toward coil B. As the resistance increases, the current in coil A decreases, and with it the strength of coil A's magnetic field. This results in a decrease in the magnetic flux through coil B. The induced current in coil B acts to oppose this decrease in flux, so the magnetic field resulting from the induced current reinforces the field from coil A. Thus the field of coil B emerges from the right end of the coil and enters on the left end. By the right-hand rule, this requires a current from right to left in the fixed resistor.

FIGURE 31-24 As the variable resistance is increased, the current in coil A decreases. This decreases the magnetic flux through coil B, resulting in an induced current whose direction is such that the magnetic field it creates in the interior of coil B reinforces the field from coil A, thereby opposing the decrease in flux through B.

FIGURE 31-25 Which way is the current in the circular loop?

EXERCISE A circular conducting loop lies next to a long, straight wire carrying current in the direction indicated in Fig. 31-25. In which direction is the induced current in the loop when the current in the wire is (a) increasing and (b) decreasing?

Answers: (a) counterclockwise; (b) clockwise

Some problems similar to Example 31-7: 4, 10, 19, 27, 32

Got It!

A wire loop lies in the plane of this page, and a uniform magnetic field points directly out of the page. What is the direction of the current in the loop if the magnetic field strength is (a) increasing or (b) decreasing?

FIGURE 31-26 An open loop. In a changing magnetic field, there is an induced emf that results in charge buildup at the gap. The polarity shown results when the magnetic field strength is decreasing.

Induction in Open Circuits

An induced emf also arises in an open circuit subject to a changing magnetic field, but here the emf cannot drive a steady current. Figure 31-26 shows the loop and magnetic field of Fig. 31-23, but now with a small gap in the loop. Although the loop is not quite closed, there is still an induced emf. In response to this emf, charge piles up on either side of the gap, creating a voltage across it. Once the gap voltage equals the induced emf, the electric field associated with the separated charge opposes the emf's tendency to move charge, and a steady state is reached. In the open-circuit case, the entire emf implied by Equation 31-2 is available at the gap. The polarity can be determined by considering what would happen if current did flow. Since current flowed clockwise in the closed loop of Fig. 31-23, positive charge will accumulate at the bottom of the gap, as shown in Fig. 31-26.

Another example of an induced emf in an open circuit is the motional emf arising as a single conductor moves through a magnetic field. Example 31-8 treats this case quantitatively.

EXAMPLE 31-8	*The Tethered Satellite*

In 1996 astronauts on the space shuttle Columbia performed a novel experiment to explore a new method of generating electric power in space. Once in orbit, Columbia deployed a 520-kg satellite attached to the shuttle through a 20-km long conducting tether (Fig. 31-27). With the tethered satellite flying vertically above Columbia, the two moved at approximately right angles to Earth's magnetic field, whose strength was about 30 μT at the orbital position. Find the motional emf that developed between the tethered satellite and the shuttle with the tether extended its full 20-km length. (Unfortunately the tether broke just short of its full length, but five hours of data collected during deployment clearly demonstrated the power-generating potential of the tethered satellite.)

Solution

Figure 31-28 shows the pair flying at right angles to the magnetic field. We can imagine forming a closed circuit by letting the satellite and shuttle slide along a system of conducting rails, just like that of Example 31-4. The system is then identical to that example, and we have

$$\mathcal{E} = B\ell v = (30 \times 10^{-6} \text{ T})(20 \times 10^3 \text{ m})(7.8 \times 10^3 \text{ m/s})$$

$$= 4.7 \text{ kV},$$

where we used the orbital speed calculated in Example 9-3.

FIGURE 31-27 Tethered satellite being deployed from the space shuttle.

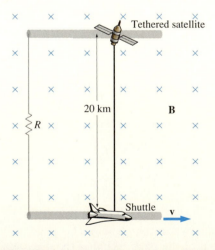

FIGURE 31-28 The tethered satellite generates electric power from the motional emf developed between the spacecraft and a satellite connected by a conducting tether. The system is analogous to the bar and rails shown in Fig. 31-8.

Since the magnetic flux through the rail arrangement would be increasing, the current would flow counterclockwise so that the magnetic field of the induced current would oppose this increase. That means positive charge will accumulate on the satellite and negative charge on the shuttle—a result that can also be obtained by considering the magnetic force on free charges in the tether.

At the shuttle's orbital altitude there are enough free electrons from ionization of the upper atmosphere for the tenuous gas surrounding the shuttle to carry a current, completing a circuit and allowing exploitation of the induced emf for electric power generation.

EXERCISE An airplane with a wingspan of 44 m is in level flight at 300 m/s over the north pole, where Earth's magnetic field has magnitude 0.62 G and points vertically downward. Find the induced emf between the wing tips.

Answer: 0.82 V

Some problems similar to Example 31-8: 10, 13, 56

31-4 INDUCED ELECTRIC FIELDS

What do we really mean by an induced emf? In a circuit containing a battery, the notion of emf is clear—the emf arises in a specific device where chemical energy is converted to electrical energy associated with charge separation. This charge separation sets up an electric field that drives current in an external circuit, as we discussed in Chapter 28. In the case of motional induced emf we also have a clear picture: the emf arises from the separation of charge associated with magnetic forces on the free charges in a moving conductor; again, the electric field associated with this charge separation may drive a current.

Now consider the current induced in a conducting loop by a magnetic field that changes with time. No motion is involved, yet there must be a force on the free charges in the conductor. The one force we know that acts on stationary charges is the electric force. Electric forces arise from electric fields, so there must be an electric field in a conducting loop in a changing magnetic field. This field is called an **induced electric field**. It has the same effect on charges—exerting a force $q\mathbf{E}$—as do the electric fields we considered earlier. The field itself, however, originates not in electric charges but in changing magnetic fields.

An induced electric field results whenever a magnetic field changes with time—whether or not an electric circuit is present. When a circuit is present, then the induced field may drive a current. But the induced field, not the current, is fundamental. A single, stationary electron placed in a changing magnetic field will experience an *electric* force—clear evidence for the existence of an electric field.

Until now we've been thinking of Faraday's law as a relation between the emf induced in a circuit and the rate of change of magnetic flux through that circuit. But we've just seen that induced electric fields are the fundamental manifestation of changing magnetic flux, and that these fields arise whether or not circuits are present. So we need to reformulate Faraday's law to describe induced electric fields without reference to circuits. The induced emf $\mathcal{E}$ that we've been writing on the left-hand side of Faraday's law means simply the work per unit charge gained by a test charge moved around a circuit. Since work is the line integral of force over distance, and electric field is the force per unit charge, we can write the emf as the line integral of the electric field. Then Faraday's law becomes

$$\oint \mathbf{E} \cdot d\boldsymbol{\ell} = -\frac{d\phi_B}{dt} \quad \text{(Faraday's law)} \qquad (31\text{-}3)$$

(a)

(b)

FIGURE 31-29 (*a*) Electric field lines arising from electric charges begin and end on the charges, or extend to infinity. (*b*) In contrast, the lines of induced electric fields have no beginning or ending points, but generally form closed loops encircling regions of changing magnetic field.

Here the line integral on the left-hand side is taken over *any* closed loop, which need not coincide with a circuit. The flux on the right-hand side is the surface integral of the magnetic field over *any* open surface bounded by the loop on the left-hand side.

Faraday's law in the form 31-3 makes no reference to wires or other circuits. It simply describes induced electric fields, which occur whenever there are changing magnetic fields. If electric circuits are present, then induced currents occur as well—but it is the induced electric fields that are fundamental. We can state Faraday's law loosely but powerfully by saying the following:

▌ A changing magnetic field creates an electric field.

This direct interaction between the fields is the basis for numerous practical devices and, as we'll see in Chapter 34, is essential to the existence of light.

Note the similarity between Faraday's law and Ampère's law (Equation 30-7). Faraday's law gives the line integral of the electric field around a closed loop in terms of the rate of change of magnetic flux through the loop. Ampère's law gives the line integral of the magnetic field around a closed loop in terms of the current through the loop. Both give fields that *encircle* their sources—current for the source of magnetic field and changing magnetic field for the induced electric field. That means the configuration of an induced electric field is very different from that of an electric field originating in electric charge. Induced fields have no beginning or end; their field lines generally form closed loops encircling regions of changing magnetic field (see Fig. 31-29).

When a changing magnetic field has sufficient symmetry, we can evaluate the induced electric field in the same way we did the magnetic field of a symmetric current distribution. Example 31-9 illustrates this procedure.

EXAMPLE 31-9	*An Induced Electric Field*

A long solenoid has circular cross section of radius R. The current in the solenoid is increasing and, as a result, the uniform magnetic field within the solenoid increases with time; the field magnitude is given by $B = bt$, with b a constant and t the time. Find the induced electric field outside the solenoid, a distance r from the solenoid axis.

Solution

The induced electric field has no beginning or end, so the field lines must encircle the solenoid. The only field consistent with the symmetry consists of circular field lines centered on the solenoid axis, as suggested in Fig. 31-30. Since the solenoid field points into the page and is increasing in strength, the direction of the induced electric field must be such that any current it might drive would produce a magnetic field *opposing* the increase in the solenoid field. Applying the right-hand rule then shows that the induced electric field runs counterclockwise.

Faraday's law relates the line integral of the electric field to the rate of change of the encircled magnetic flux. Here a suit-

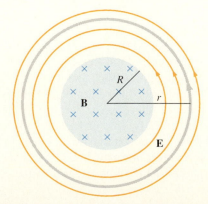

FIGURE 31-30 Cross section of a solenoid of radius R whose magnetic field is increasing with time. Field lines of the induced electric field are circles concentric with the solenoid axis. Also shown is a circular loop (gray) of radius r for evaluating the line integral in Faraday's law.

able integration loop is itself a circle centered on the solenoid axis. Symmetry shows that the field magnitude is constant over this loop, and if we circle the loop in the direction of the field then $\cos\theta$ in the dot product $\mathbf{E} \cdot d\boldsymbol{\ell}$ is 1. Thus

$$\oint \mathbf{E} \cdot d\boldsymbol{\ell} = \oint E\, d\ell = E \oint d\ell = 2\pi r E,$$

since $\oint d\ell$ is just the loop circumference. Faraday's law relates this quantity to the rate of change of the encircled magnetic flux. Here the flux is due to the uniform magnetic field inside the solenoid, perpendicular to our loop and confined to a circular area of radius R. (We use the long-solenoid approximation, neglecting the small magnetic field that must exist outside a solenoid of finite length.) So the encircled flux is just $\phi_B = \pi R^2 B = \pi R^2 bt$, and its rate of change is

$$\frac{d\phi_B}{dt} = \frac{d}{dt}(\pi R^2 bt) = \pi R^2 b.$$

Then Faraday's law gives

$$2\pi r E = -\pi R^2 b,$$

or

$$E = -\frac{R^2 b}{2r}.$$

We've already accounted for the minus sign in arguing that the field circles counterclockwise. The $1/r$ dependence of the field strength on distance should come as no surprise; points with $r > R$ are outside a cylindrically symmetric distribution, in this case a distribution of changing magnetic flux. We found the same $1/r$ dependence for the electric and magnetic fields outside, respectively, a cylindrically symmetric charge distribution and a cylindrically symmetric current distribution.

EXERCISE Show that the electric field at points *inside* the solenoid has magnitude $br/2$.

Some problems similar to Example 31-9: 36–38, 41, 42

In Chapter 30 we derived the magnetic field of a solenoid using Ampère's law—which, as we've formulated it so far, applies only to steady currents. But a solenoid whose field is changing—as in Example 31-9—must have a changing current, so how can we talk about it? In fact, Ampère's law gives a good approximation to the field produced by changing current, provided that change is sufficiently slow. In the examples and problems of this chapter we assume that to be the case. In Chapter 34 we'll see just what "sufficiently slow" means, and will explore what happens if we relax the assumption of slowly changing currents.

Conservative and Nonconservative Electric Fields

We've seen that static electric fields—those beginning and ending on stationary charge distributions—are conservative, meaning that the work required to move a charge between two points is independent of the path taken. A consequence is that it takes no work to move around a closed path in an electrostatic field; mathematically, we express this by writing

$$\oint \mathbf{E} \cdot d\boldsymbol{\ell} = 0. \qquad \text{(electrostatic field)}$$

In contrast, induced electric fields generally form closed loops, and Faraday's law shows that the line integral of the electric field around a closed path in such a field is decidedly not zero. That means the induced electric field does work on a charge moved around a *closed* path and that the work done in moving between two points cannot be independent of the path taken (Fig. 31-31). The induced electric field is therefore not conservative.

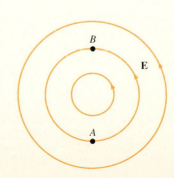

FIGURE 31-31 Two points in an induced electric field. If a positively charged particle moves counterclockwise along the field line from A to B the field does work on it. But if it moves from A to B in the clockwise direction the field does negative work; that is, an external force must do work on the charge. The work done is not path independent, and the induced electric field is not conservative.

FIGURE 31-32 This induction cooktop has no burners. Instead, coils beneath the cooktop produce time-varying magnetic fields that induce currents in the conducting cookware. Resistance in the cookware then results in heating. The cooktop itself remains cool because it's not a conductor.

The work done on charges moving around closed paths in nonconservative electric fields can be useful. A simple example is the induction cooktop (Fig. 31-32), where food cooks in a special conducting pan heated by currents in the pan itself. Those currents are driven by an induced electric field originating from a changing magnetic field in coils just below the cooktop's surface. Much more generally, any electrical device that derives its power from a generator—and that includes almost every electrical thing you use—is part of a circuit in which a nonconservative electric field drives the current.

APPLICATION	*The Tokamak*

In Chapter 29 we described fusion reactors, devices that may someday yield copious amounts of energy through nuclear fusion of hydrogen nuclei, using fuel derived from seawater. We saw how magnetic fields confine very hot plasma in fusion reactors and showed why most reactors have a toroidal shape. The most promising of the toroidal reactor designs is the **tokamak**, in which the confining magnetic field arises in part from current within the plasma itself. Discharging a large capacitor bank through a coil produces a rapidly changing magnetic field in the "hole" of the toroidal "doughnut," resulting in an induced electric field within the torus. This field drives the plasma current that produces a magnetic field component around the minor radius of the device. Other coils around the torus provide a magnetic field component around the long dimension of the torus, giving a net magnetic field that spirals around the plasma (Fig. 31-33).

As in any conductor, particle collisions dissipate energy gained from the electric field, in this case heating the plasma. Unfortunately the resistance of a plasma drops at high temperature, making it difficult to achieve the temperature needed for fusion with resistive heating alone. Nevertheless, the induced electric field helps bring the plasma close to the fusion temperature, as well as driving currents that produce the confining magnetic field.

Since the tokamak uses an induced electric field, its operation requires a changing magnetic field. Therefore, the tokamak

FIGURE 31-33 A tokamak fusion reactor. Current in the plasma is driven by an electric field induced by the magnetic field change associated with a rapid change in current through the central coil. Additional coils wound around the toroid give a steady magnetic field in the same direction as the plasma current. Combining with the field from the plasma current gives a net field that spirals around the plasma.

cannot run continuously but operates instead in a pulsed mode as current builds up rapidly and then decays in preparation for another pulse. Today's experimental tokamaks can achieve pulse durations of 10 s or more, during which they produce significant fusion energy.

| **FOR FURTHER THOUGHT** | *Nonconservative Electric Fields* |

The nonconservative nature of the induced electric field is strikingly demonstrated if you attempt to measure potential differences in nonconservative fields. In Chapter 25, we defined potential difference as the work required to move a unit charge between two points, and stressed that this work is independent of path for a conservative field. But when the field is nonconservative, the work is not independent of path, and the concept of potential becomes ambiguous.

Figure 31-34 shows an end view of a long solenoid surrounded by three identical resistors bent into circular arcs. If the solenoid current is increasing, an induced electric field appears in the resistors, and drives a current I in the counterclockwise direction.

Because they have the same resistance and carry the same current, the potential difference across each resistor should be the same. We could try to measure the potential difference across one resistor, for example, by connecting a voltmeter as with the right-hand meter in Fig. 31-34. With current I flowing through the resis-tance R, this meter reads IR. Since the current flows counterclockwise, we must connect the positive voltmeter terminal to point B.

But now try to measure the potential difference across the other two resistors together, as with the left-hand meter in Fig. 31-34. With the current I flowing through the total resistance $2R$, the meter now reads $2IR$. We have two voltmeters with their terminals connected to the same points, and yet they indicate different voltages. Not only are the magnitudes of the voltages different, but even their polarities differ. How can this be? It's because of the nonconservative nature of the induced

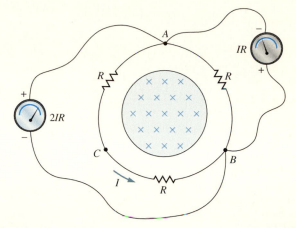

FIGURE 31-34 End view of a long solenoid surrounded by three resistors in series. A changing magnetic field in the solenoid induces an emf in the resistors, and the same current I flows through each. A voltmeter connected between points A and B indicates the voltage IR across one resistor. A second voltmeter connected to the *same points* indicates voltage $2IR$ with the opposite sign.

electric field. The two voltmeters are positioned differently with respect to the changing magnetic flux, so they sample different regions of the induced electric field. Even though the meters are connected to the same points, they measure the line integral of the induced electric field over *different* paths, and so they don't read the same voltage.

31-5 DIAMAGNETISM

In Section 30-6 we discussed paramagnetic and ferromagnetic materials, in which atomic magnetic dipoles align with an applied magnetic field, causing an attractive interaction between the material and a magnet. We also mentioned diamagnetic materials, in which induced magnetic dipoles align antiparallel to the applied field, causing a repulsive force. We are now ready to understand diamagnetism as a manifestation of Faraday's law at the microscopic level.

In a purely diamagnetic material, current loops associated with pairs of atomic electrons exactly cancel, leaving atoms with no intrinsic magnetic moments. Figure 31-35*a* shows a simplified model to describe such an atom.

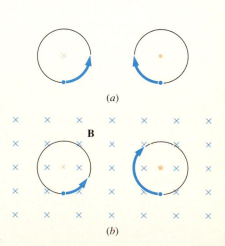

FIGURE 31-35 Simplified, classical model for a diamagnetic atom. (*a*) In the absence of an applied field, magnetic moments (cross and dot) associated with a pair of electrons exactly cancel. (*b*) The changing flux associated with an increasing magnetic field induces an electric field that speeds up one electron and slows down the other, giving rise to a net magnetic moment. (In applying the right-hand rule, remember that the electrons are negative.)

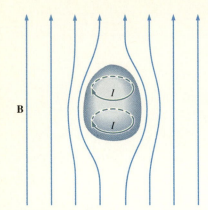

FIGURE 31-36 Induced currents in a superconductor completely cancel an applied magnetic field. Net result is that a magnetic field cannot penetrate the superconductor.

FIGURE 31-37 A small magnet is levitated above a wafer of high-temperature superconductor in a bath of liquid nitrogen at 77 K.

The picture should not be taken too literally, for it uses classical physics to describe a phenomenon properly within the domain of quantum mechanics. Nevertheless, it shows qualitatively how diamagnetism is an electromagnetic induction effect.

The electron orbiting clockwise in Fig. 31-35a has a magnetic moment pointing out of the page (apply the right-hand rule and remember that the electron carries a negative charge). In the absence of an applied magnetic field, this moment is exactly cancelled by the equal but oppositely directed moment of the other electron. But now suppose a magnetic field is applied, for example by moving the north pole of a magnet down toward the plane of Fig. 31-35. This produces a changing magnetic field in the plane of the figure, resulting in an induced electric field that alters the speeds of the orbiting electrons. The direction of this effect must be such as to oppose the increase in field. Thus the electron whose magnetic moment points out of the page has its moment increase, while the moment pointing into the page decreases. Figure 31-35b shows how these changes result from the alteration of electron speeds by the induced electric field.

Once the field is applied and the orbital speeds of the two electrons are no longer identical, the atom now has a net magnetic moment that points out of the page in Fig. 31-35, opposing the incoming magnet and resulting in a repulsive force. This repulsion is the distinguishing characteristic of diamagnetism. We listed a number of diamagnetic materials in Table 30-2.

A material that is entirely superconducting is perfectly diamagnetic, with magnetic susceptibility -1. This means that the magnetic field resulting from induced currents within the material completely cancels any applied field. Since these induced currents persist in the zero-resistance superconductor, the material completely excludes magnetic fields from its interior, a phenomenon known as the Meissner effect (Fig. 31-36). The repulsive force associated with the magnetic moments of a permanent magnet and a nearby superconductor results in the widely publicized phenomenon of magnetic levitation (Fig. 31-37). Readers of the extended version of this text will explore superconductivity further in Chapter 42.

CHAPTER SYNOPSIS

Summary

1. **Electromagnetic induction** is a fundamental phenomenon linking magnetism and electricity. Induction is described by **Faraday's law**, which states that a **changing magnetic field** produces an **induced electric field**. Unlike the conservative electrostatic field of an electric charge, this induced field is **nonconservative**, meaning it can do work on charges as they move around a closed loop. Faraday's law relates the line integral of this nonconservative electric field around an arbitrary loop to the rate of change of magnetic flux through a surface bounded by that same loop:

$$\oint \mathbf{E} \cdot d\boldsymbol{\ell} = -\frac{d\phi_B}{dt}.$$

2. In order for energy to be conserved, the induced electric field is in such a direction as to oppose the change in flux that gives rise to it. This energy-conserving aspect of Faraday's law is called **Lenz's law** and is reflected mathematically in the minus sign on the right-hand side of Faraday's law.

3. When a conductor is present, the nonconservative electric field manifests itself as an **induced emf**:

$$\mathcal{E} = -\frac{d\phi_B}{dt}.$$

This emf drives an **induced current** in any circuit with finite resistance. It does not matter whether the magnetic flux is changed by moving a conductor in a magnetic field, or by moving a magnetic field near a conductor, or by altering the shape or orientation of the conductor. The generation of electric power by moving conducting loops in magnetic fields is an important technological example of induced currents.

4. **Diamagnetism** is a manifestation of electromagnetic induction on the atomic scale. Application of a magnetic field to a diamagnetic material results in induced atomic magnetic moments that cause the material to be repelled from a magnet.

Terms You Should Understand

(Pairs are closely related terms whose distinction is important; number in parentheses is chapter section where term first appears.)

electromagnetic induction (31-1)
induced current, induced emf (31-1, 31-2)
magnetic flux (31-2)
Faraday's law, Lenz's law (31-2, 31-3)
motional emf (31-3)

generator (31-3)
eddy currents (31-3)
induced electric field (31-4)

Symbols You Should Recognize

ϕ_B (31-2)
$\oint \mathbf{E} \cdot d\boldsymbol{\ell}$ (31-4)

Problems You Should Be Able to Solve

calculating magnetic flux (31-2)
calculating induced emf's (31-2)
finding directions of induced emf's and currents (31-3)
calculating induced electric fields (31-4)

Limitations to Keep in Mind

Electromagnetic induction requires a *change* in magnetic flux. Values of magnetic field or flux alone don't matter; only the rate of change of those quantities is important.

QUESTIONS

1. A copper penny falls on a vertical path that takes it between the poles of a magnet. Does it hit the ground faster or slower than if no magnet were present?

2. A bar magnet is moved toward a conducting ring, as shown in Fig. 31-38. What is the direction of the induced current in the ring?

FIGURE 31-38 Question 2.

3. Figure 31-39 shows two concentric conducting loops, the outer connected to a battery and a switch. The switch is initially open. It is then closed, left closed for a while, then

FIGURE 31-39 Question 3.

opened again. Describe the currents in the inner loop during the entire procedure.

4. An electric generator is being turned at constant speed. A load resistor R is connected across the generator terminals. If the electrical resistance of the load is lowered, does the generator get easier or harder to turn?

5. Service manuals for cars often tell you to set the idle speed of the engine with the headlights on. Why? What does this have to do with electromagnetic induction?

6. Figure 31-40 shows two square wire loops, the first containing a battery and variable resistor. The resistor is initially at the midpoint of its resistance range. Should its resistance be lowered or raised in order to induce a clockwise current in the right-hand loop?

FIGURE 31-40 Question 6.

7. Consider the simple motor shown in Fig. 29-36. What happens if you connect a resistor across the motor terminals, and turn the motor by hand? What is the difference between a motor and a generator?

8. Figure 31-41 shows an open wire loop in a magnetic field. The field is changing, and charge has piled up at the loop gap with the polarity indicated. Is the magnetic field strength increasing or decreasing?

FIGURE 31-41 Question 8.

9. A constant, uniform magnetic field points into the page. A flexible, circular conducting ring lies in the plane of the page. If the ring is stretched, maintaining its circular shape, what is the direction of the induced current?

10. When a magnet is moved near a superconductor, the magnetic field lines never enter the superconductor. Why not?

11. Is it possible to produce an induced current that never changes? How or why not? Could you produce an induced current that was steady for some finite time? How or why not?

12. When you push a bar magnet into a conducting loop, you do work. What happens when you pull it out the other side?

13. You are turning a generator in such a way that the current it delivers remains constant. As you lower the load resistance across the generator, does the generator get easier or harder to turn?

14. Devise a way of measuring a magnetic field using Faraday's law.

15. In Fig. 31-42, a copper ring was originally resting on the wooden structure, surrounding the coil. When a rapidly

changing current was applied to the coil, the ring was ejected into the air. Explain this phenomenon.

16. Fluctuations in Earth's magnetic field due to changing solar activity can wreak havoc with communications, even those using underground cables. How is this possible?

17. Why is it not possible to run a tokamak on a continuous basis?

18. Conventional brakes on a car need large surface areas to dissipate the heat of friction when the brakes are applied. Would eddy-current brakes have the same problem?

19. Which way would the eddy currents flow in Fig. 31-22a as the bob continues its swing and begins to emerge from the field?

20. In Chapter 29, we pointed out that a static magnetic field cannot change the energy of a charged particle. Is this true of a changing magnetic field? Discuss.

21. A long solenoid of circular cross section is oriented so that its magnetic field points out of the page, as shown in Fig. 31-43. The solenoid current is increasing. (a) What is the direction of the induced electric field at points A and B in the figure? (b) What is the magnitude of the induced electric field in the center of the solenoid? (Don't calculate! Argue from symmetry.)

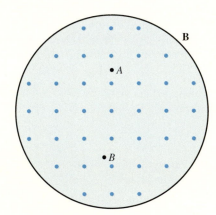

FIGURE 31-43 Question 21.

22. Is the concept of electric potential (Chapter 25) useful in a nonconservative electric field? Give an example to substantiate your answer.

23. Can an induced electric field exist in the absence of a conductor?

24. Could you tell whether a given electric field arises from electric charge or from a changing magnetic field? How or why not?

25. Does a diamagnetic material experience a force in a uniform magnetic field?

FIGURE 31-42 Question 15.

PROBLEMS

ActivPhysics can help with these problems:
Activities 13.9, 13.10

Sections 31-2 and 31-3 Faraday's Law and Induction and the Conservation of Energy

1. Show that the volt is the correct SI unit for the rate of change of magnetic flux, making Faraday's law dimensionally correct.
2. A bar magnet is moved steadily through a conducting ring, as shown in Fig. 31-44. Sketch qualitatively the current and power dissipation in the ring as functions of time. Take as positive a current flowing out of the plane of the page at the top of the ring, and indicate the position of the magnet on your time axis.

(a)   (b)   (c)

FIGURE 31-44 Problem 2.

3. Find the magnetic flux through a circular loop 5.0 cm in diameter oriented with the loop normal at 30° to a uniform 80-mT magnetic field.
4. A circular wire loop 40 cm in diameter has 100-Ω resistance and lies in a horizontal plane. A uniform magnetic field points vertically downward, and in 25 ms it increases linearly from 5.0 mT to 55 mT. Find the magnetic flux through the loop at (a) the beginning and (b) the end of the 25 ms period. (c) What is the loop current during this time? (d) Which way does this current flow?
5. A conducting loop of area 240 cm^2 and resistance 12 Ω lies at right angles to a spatially uniform magnetic field. The loop carries an induced current of 320 mA. At what rate is the magnetic field changing?
6. A conducting loop of area A and resistance R lies at right angles to a spatially uniform magnetic field. At time $t = 0$ the magnetic field and loop current are both zero. Subsequently, the current increases according to $I = bt^2$, where b is a constant with the units A/s^2. Find an expression for the magnetic field strength as a function of time.
7. A conducting loop with area 0.15 m^2 and resistance 6.0 Ω lies in the x-y plane. A spatially uniform magnetic field points in the z direction. The field varies with time according to $B_z = at^2 - b$, where $a = 2.0$ T/s^2 and $b = 8.0$ T.

Find the loop current (a) when $t = 3.0$ s and (b) when $B_z = 0$.

8. The magnetic field inside a 20-cm-diameter solenoid is increasing at the rate of 2.4 T/s. How many turns should a coil wrapped around the outside of the solenoid have in order that the emf induced in the coil be 15 V?
9. A square wire loop of side ℓ and resistance R is pulled with constant speed v from a region of no magnetic field until it is fully inside a region of constant, uniform magnetic field **B** perpendicular to the loop plane. The boundary of the field region is parallel to one side of the loop. Find an expression for the total work done by the agent pulling the loop.
10. A 1.8-m high runner sprints eastward at 9.5 m/s along the equator, where Earth's magnetic field points horizontally with a strength of 31 μT. (a) What is the magnitude of the emf induced between the runner's head and feet? (b) Which end is positive?
11. In Fig. 31-26 the loop radius is 15 cm, and the magnetic field is decreasing at the rate of 550 T/s. If the gap width is small compared with the loop circumference, what is the voltage across the gap?
12. A 5-turn coil 1.0 cm in diameter is rotated at 10 rev/s about an axis perpendicular to a uniform magnetic field. A voltmeter connected to the coil through rotating contacts reads a peak value of 360 μV. What is the magnetic field strength?
13. The wingspan of a 747 jetliner is 60 m. If the plane is flying at 960 km/h in a region where the vertical component of Earth's magnetic field is 0.20 G, what emf develops between the plane's wingtips?
14. A square wire loop 3.0 m on a side is perpendicular to a uniform magnetic field of 2.0 T. A 6-V light bulb is in series with the loop, as shown in Fig. 31-45. The magnetic field is reduced steadily to zero over a time Δt. (a) Find Δt such that the light will shine at full brightness during this time. (b) Which way will the loop current flow?

FIGURE 31-45 Problem 14.

15. In Example 31-2 take $a = 1.0$ cm, $w = 3.5$ cm, and $\ell = 6.0$ cm. Suppose the rectangular loop is a conductor with resistance 50 mΩ and that the current I in the long wire is increasing at the rate of 25 A/s. Find the induced current in the loop. In what direction does it flow?

16. A windmill with conducting blades of length ℓ is rotating with angular speed ω about a horizontal axis; the horizontal component of Earth's magnetic field at its location is B_x. Find an expression for the emf developed between the blade tips and the central axis.

17. A square conducting loop of side $s = 0.50$ m and resistance $R = 5.0$ Ω moves to the right with speed $v = 0.25$ m/s. At time $t = 0$ its rightmost edge enters a uniform magnetic field $B = 1.0$ T pointing into the page, as shown in Fig. 31-46. The magnetic field covers a region of width $w = 0.75$ m. Plot (a) the current and (b) the power dissipation in the loop as functions of time, taking a clockwise current as positive and covering the time until the entire loop has exited the field region.

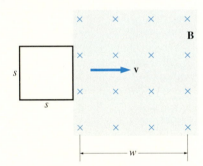

FIGURE 31-46 Problem 17.

18. A 2.0-m-long solenoid is 15 cm in diameter and consists of 2000 turns of wire. The current in the solenoid is increasing at the rate of 1.0 kA/s. (a) Find the current in a wire loop with diameter 10 cm and resistance 5.0 Ω, lying inside the solenoid in a plane perpendicular to the loop axis. (b) Repeat for a similarly oriented loop with diameter 25 cm, lying entirely outside the solenoid.

19. A solenoid 2.0 m long and 30 cm in diameter consists of 5000 turns of wire. A 5-turn coil with negligible resistance is wrapped around the solenoid and connected to a 180-Ω resistor, as shown in Fig. 31-47. The direction of the current in the solenoid is such that the solenoid's magnetic field points to the right. At time $t = 0$ the solenoid current begins to decay exponentially, being given by $I = I_0 e^{-t/\tau}$, where $I_0 = 85$ A, $\tau = 2.5$ s, and t is the time in seconds. (a) What is the direction of the current in the resistor as the solenoid current decays? What is the value of the resistor current at (b) $t = 1.0$ s and (c) $t = 5.0$ s?

20. Make a qualitative plot of the resistor current in the preceding problem as a function of time if the solenoid current

FIGURE 31-47 Problems 19–21.

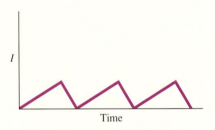

FIGURE 31-48 Problem 20.

has the form shown in Fig. 31-48. Take a left-to-right resistor current as positive.

21. (a) Find an expression for the resistor current in Problem 19 if the solenoid current is given by $I = I_0 \sin \omega t$, where $I_0 = 85$ A and $\omega = 210$ s^{-1}. (b) What is the peak current in the resistor? (c) What is the resistor current when the solenoid current is a maximum?

22. A magnetic field is described by $\mathbf{B} = B_0 \sin \omega t \, \hat{\mathbf{k}}$, where $B_0 = 2.0$ T and $\omega = 10$ s^{-1}. A conducting loop with area 150 cm^2 and resistance 5.0 Ω lies in the x-y plane. Find the induced current in the loop (a) at $t = 0$ and (b) at $t = 0.10$ s.

23. In the preceding problem, what is the first time after $t = 0$ when the loop current will be zero?

24. A car alternator consists of a 250-turn coil 10 cm in diameter in a magnetic field of 0.10 T. If the alternator is turning at 1000 revolutions per minute, what is its peak output voltage?

25. A credit-card reader extracts information from the card's magnetic stripe as it is pulled past the reader's head. At some instant the card motion results in a magnetic field at the head that is changing at the rate of 450 μT/ms. If this field passes perpendicularly through a 5000-turn head coil 2.0 mm in diameter, what will be the induced emf?

26. A generator consists of a rectangular coil 75 cm by 1.3 m, spinning in a 0.14-T magnetic field. If it is to produce a 60-Hz alternating emf (i.e., $\mathcal{E} = \mathcal{E}_0 \sin 2\pi f t$, where $f = 60$ Hz) with peak value 6.7 kV, how many turns must it have?

27. Figure 31-49 shows a pair of parallel conducting rails a distance ℓ apart in a uniform magnetic field **B**. A resistance R is connected across the rails, and a conducting bar of negligible resistance is being pulled along the rails with velocity **v** to the right. (a) What is the direction of the current in the resistor? (b) At what rate must work be done by the agent pulling the bar?

FIGURE 31-49 Problems 27–29, 33, 60.

28. The resistor in the preceding problem is replaced by an ideal voltmeter. (a) To which rail should the positive meter terminal be connected if the meter is to indicate a positive voltage? (b) At what rate must work be done by the agent pulling the bar?

29. A battery of emf $\mathcal{E}$ is inserted in series with the resistor in Fig. 31-49, with its positive terminal toward the top rail. The bar is initially at rest, and now no agent pulls it. (a) Describe the bar's subsequent motion. (b) The bar eventually reaches a constant speed. Why? (c) What is that constant speed? Express in terms of the magnetic field, the battery emf, and the rail spacing ℓ. Does the resistance R affect the final speed? If not, what role does it play?

30. A toroidal coil of square cross section has inner radius a and outer radius b. It consists of N turns of wire and carries a time-varying current $I = I_0 \sin \omega t$. A single-turn wire loop encircles the toroid, passing through its center hole as shown in Fig. 31-50. Find an expression for the peak emf induced in the loop.

FIGURE 31-50 Problem 30.

31. A pair of parallel conducting rails 10 cm apart lie at right angles to a uniform magnetic field **B** of magnitude 2.0 T, as

shown in Fig. 31-51. A 5.0-Ω and a 10-Ω resistor lie across the rails and are free to slide along them. (a) The 5-Ω resistor is held fixed, and the 10-Ω resistor is pulled to the right at 50 cm/s. What are the direction and magnitude of the induced current? (b) Now the 10-Ω resistor is held fixed, and the 5-Ω resistor is pulled to the left at 50 cm/s. What are the direction and magnitude of the induced current? (c) What is the power dissipation in the 10-Ω resistor in both cases?

FIGURE 31-51 Problems 31, 32, and 64.

32. In Fig. 31-51 the 10-Ω resistor is being moved to the right at a constant 50 cm/s. The 5-Ω resistor, initially at rest, is placed across the conducting rails. Describe qualitatively its subsequent motion, and determine its final speed.

33. In Fig. 31-49, take $\ell = 10$ cm, $B = 0.50$ T, $R = 4.0$ Ω, and $v = 2.0$ m/s. Find (a) the current in the resistor, (b) the magnetic force on the bar, (c) the power dissipation in the resistor, and (d) the mechanical work done by the agent pulling the bar. Compare your answers to (c) and (d).

34. A rectangular conducting loop of resistance R, mass m, and width w falls into a uniform magnetic field **B**, as shown in Fig. 31-52. If the loop is long enough and the field region has a great enough vertical extent, the loop will reach a terminal speed. (a) Why? (b) Find an expression for the terminal speed. (c) What will be the direction of the loop current as the loop enters the field?

FIGURE 31-52 Problem 34.

35. A circular loop 40 cm in diameter is made from a flexible conductor and lies at right angles to a uniform 12-T magnetic field. At time $t = 0$ the loop starts to expand, its radius increasing at the rate of 5.0 mm/s. Find the induced emf in the loop (a) at $t = 1.0$ s and (b) at $t = 10$ s.

Section 31-4 Induced Electric Fields

36. The induced electric field 12 cm from the axis of a solenoid with 10 cm radius is 45 V/m. Find the rate of change of the solenoid's magnetic field.

37. Find the electric force on a 50-μC charge inside the solenoid of Problem 18, if the charge is 5.0 cm from the solenoid axis.

38. Figure 31-53 shows a top view of a tokamak. The magnetic field in the center is confined to a circular area of radius 50 cm, and during a pulse it increases at the rate of 5.1 T/ms. (a) What is the magnitude of the induced electric field in the tokamak, 1.2 m from the center of the field region in Fig. 31-53? (b) What is the field direction? (c) If a proton circles the tokamak once at this radius, going with the electric field, how much energy does it gain?

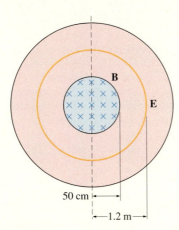

FIGURE 31-53 Top view of a tokamak. (Problem 38)

39. A uniform magnetic field points into the page in Fig. 31-54. In the same region an electric field points straight up, but increases with position at the rate of 10 V/m^2 as you move to the right. Apply Faraday's law to a rectangular loop to show that the magnetic field must be changing with time, and calculate the rate of change.

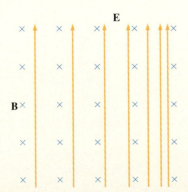

FIGURE 31-54 The magnetic field must be changing. Why? (Problem 39)

40. Use Faraday's law to show that the electric field produced by charges on the plates of a parallel-plate capacitor cannot end abruptly at the edges of the plates. *Hint:* Apply Faraday's law to the loop shown in Fig. 31-55.

FIGURE 31-55 Problem 40.

41. Figure 31-56 shows a magnetic field pointing into the page; the field is confined to a layer of thickness h in the vertical direction but extends infinitely to the left and right. The field strength is increasing with time: $B = bt$, where b is a constant. Find an expression for the electric field at all points outside the field region. *Hint:* Consult Example 30-5.

FIGURE 31-56 Problem 41.

42. The magnetic field inside a solenoid of circular cross section is given by $\mathbf{B} = bt\,\hat{\mathbf{k}}$, where $b = 2.1$ T/ms. At time $t = 0.40$ μs a proton is inside the solenoid at the point $x = 5.0$ cm, $y = 0$, $z = 0$ and is moving with velocity $\mathbf{v} = 4.8 \times 10^6\,\hat{\mathbf{j}}$ m/s. Find the net electromagnetic force on the proton.

Paired Problems

(Both problems in a pair involve the same principles and techniques. If you can get the first problem, you should be able to solve the second one.)

43. A magnetic field is given by $\mathbf{B} = B_0(x/x_0)^2\,\hat{\mathbf{k}}$, where B_0 and x_0 are constants. Find an expression for the magnetic flux through a square of side $2x_0$ that lies in the x-y plane with one corner at the origin and two sides coinciding with the positive x and y axes.

44. A circular region of radius R lies in the x-y plane and contains a magnetic field given by $\mathbf{B} = B_0\dfrac{r}{R}\hat{\mathbf{k}}$, where r is the radial distance from the central axis of the field region. Find an expression for the magnetic flux through this region.

45. A uniform magnetic field is given by $\mathbf{B} = bt\,\hat{\mathbf{k}}$, where $b = 0.35$ T/s. Find the current in a conducting loop with area 240 cm^2 and resistance 0.20 Ω that lies in the x-y plane. In what direction is the current, as viewed from the positive z axis?

46. A uniform magnetic field is given by $\mathbf{B} = bt^3\,\hat{\mathbf{k}}$. A square conducting loop 15 cm on a side has 0.32-Ω resistance and lies in the x-y plane. At time $t = 2.5$ s, the current in the loop is 4.1 A. Find b.

47. A pair of vertical conducting rods are a distance ℓ apart and are connected at the bottom by a resistance R. A conducting bar of mass m runs horizontally between the rods and can slide freely down them while maintaining electrical contact. The whole apparatus is in a uniform magnetic field **B** pointing horizontally and perpendicular to the bar. When the bar is released from rest it soon reaches a constant speed. Find this speed.

48. A conducting bar of mass m slides down the conducting wedges shown in Fig. 31-57. The wedges are separated by a distance ℓ, connected at the top by a resistance R, and make an angle θ with the vertical. A uniform magnetic field **B** points horizontally, as shown. When released from rest the bar soon reaches a constant speed. Find an expression for this speed.

FIGURE 31-57 Problem 48.

49. Figure 31-58 shows an unusual design for a generator, consisting of a conducting bar that rotates about a central axis while making contact with a conducting ring of radius R. A uniform magnetic field is perpendicular to the ring. Wires from the axis and ring carry power to a load. Find an expression for the emf induced in this generator when the bar rotates with angular speed ω.

FIGURE 31-58 Problem 49.

50. A copper disk 90 cm in diameter is spinning at 3600 rpm about a conducting axle through its center, as shown in Fig. 31-59. A uniform 1.5-T magnetic field is perpendicular to the disk, as shown. A stationary conducting brush maintains contact with the disk's rim, and a voltmeter is

FIGURE 31-59 Problem 50.

connected between the brush and the axle. (a) What does the voltmeter read? (b) Which voltmeter lead is positive?

51. An electron is inside a solenoid, 28 cm from the solenoid axis. It experiences an electric force of magnitude 1.3 fN. At what rate is the solenoid's magnetic field changing?

52. It takes 0.43 J to push a 84-μC charge around a closed path surrounding a 1.5-m-diameter solenoid. At what rate is the solenoid's magnetic field changing?

Supplementary Problems

53. At times prior to $t = 0$, there is no current in either the solenoid or the small coil of Problem 19. Subsequently, the current in the small coil is observed to increase at 10 μA/s. What is the solenoid current as a function of time?

54. A conducting loop of area A and resistance R lies perpendicular to a uniform magnetic field B. The loop is then rotated at a uniform rate until it is upside down; this takes time Δt. Find an expression for the work done in flipping the loop.

55. So-called magnetohydrodynamic generators have been proposed as a means of extracting electrical energy from charged particles released in fusion reactions; they've also been suggested as a way to generate electricity from flowing water. An MHD generator consists of two metal plates on either side of a channel carrying conducting fluid in a magnetic field, as shown in Fig. 31-60. The magnetic force on free charges in the fluid drives positive charge to one plate, negative to the other. If there's no electrical load connected across the plates, the electric field that develops eventually halts any further charge motion. (a) Show in this case that the voltage between the plates is $V = vBd$, where v is the fluid velocity and d the plate spacing. (b) Now suppose a resistance R is connected between the plates. Show that the current through R is $I = \dfrac{vABd}{\rho d + AR}$, where A is the plate area and ρ is the resistivity of the fluid.

FIGURE 31-60 Problem 55.

56. The tethered satellite system of Example 31-8 is flown in a low circular orbit about a planet of mass M and radius R. If the tether has length ℓ and generates an emf $\mathcal{E}$, find an expression for the planet's magnetic field strength.

57. Clever farmers whose lands are crossed by large power lines have been known to steal power by stringing wire near the power line and making use of the induced current—a practice that has been ruled legally to be theft. The scene of a particular crime is shown in Fig. 31-61. The power line carries 60-Hz alternating current with a peak current of 10 kA (that is, the current is given by $I = I_0 \sin \omega t$, where $I_0 = 10$ kA and $\omega = 2\pi f$, with $f = 60$ Hz). (a) If the farmer wants a peak voltage of 170 V, what should be the length ℓ of the loop shown in Fig. 31-61? (170 V is the peak of standard 120-V AC power.) (b) If all the equipment the farmer connects to the loop has an equivalent resistance of 5.0 Ω, what is the farmer's average power consumption? *Note:* The *average* power consumption is half the product of the peak voltage and peak current. (c) If the power company charges 10¢ per kWh, what is the monetary value of the energy stolen each day? (d) Without examining the farmer's lands, how, in principle, could the power company know that a crime is being committed?

$I = 10^4$ A
5 m
ℓ
0.5 m

To farmer's equipment

FIGURE 31-61 Problem 57.

58. A circular wire loop of resistance R and radius a lies with its plane perpendicular to a uniform magnetic field. The field strength changes from an initial value B_1 to a final value B_2. Show, by integrating the loop current over time, that the total charge that moves around the ring is

$$q = \frac{\pi a^2}{R}(B_2 - B_1).$$

Note that this result is independent of how the field changes with time.

59. A conducting disk with radius a, thickness h, and resistivity ρ is inside a solenoid of circular cross section. The disk axis coincides with the solenoid axis. The magnetic field in the solenoid is given by $B = bt$, with b a constant. Find expressions for (a) the current density in the disk as a function of the distance r from the disk center and (b) the rate of power dissipation in the entire disk. *Hint:* Consider the disk to be made up of infinitesimal conducting loops.

60. The bar in Problem 27 has mass m and is initially at rest. A constant force F is applied to the bar, pulling it to the right. (a) Formulate Newton's second law for the bar as an equation involving both v and $a = dv/dt$. (b) Use your equation to show that the bar's acceleration becomes zero when its speed reaches the value $FR/B^2\ell^2$. (c) Show by direct substitution that your equation is satisfied if v as a function of time is given by $v(t) = \dfrac{FR}{B^2\ell^2}(1 - e^{-B^2\ell^2 t/mR})$.

61. Find an expression for the speed of the left-hand resistor in Problem 32 as a function of time, in terms of its mass m, the field strength B, the speed v of the right-hand bar, the time t, and the resistances R_{left} and R_{right}.

62. A pendulum consists of a mass m suspended from two identical copper wires of negligible mass. At equilibrium the mass is a vertical distance ℓ below its supports, and the wires make 45° angles with the vertical, as shown in Fig. 31-62. A uniform magnetic field **B** points into the page. The pendulum is displaced from the plane of the page by a small angle θ_0, and at time $t = 0$ it is released. Find an expression for the voltmeter reading as a function of time.

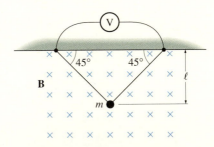

FIGURE 31-62 Problem 62.

63. A *flip coil* consists of a small coil used to measure magnetic fields. The flip coil is placed in a magnetic field with its plane perpendicular to the field, and then rotated abruptly through 180° about an axis in the plane of the coil. The coil is connected to instrumentation to measure the total charge Q that flows during this process. If the coil has N turns of area A and if its rotation axis is perpendicular to the magnetic field, show that the field strength is given by $B = QR/2NA$, where R is the coil resistance.

Inductance and Magnetic Energy

This transformer at an electrical substation uses mutual inductance to change the voltage at which power is transmitted.

Faraday's law implies that a changing magnetic flux through a circuit produces an induced emf in the circuit. In this chapter we consider the special case when that changing flux is itself caused by a changing current in an electric circuit. We then speak of the **inductance** of the circuit.

32-1 MUTUAL INDUCTANCE

FIGURE 32-1 The mutual inductance between two coils is the ratio of magnetic flux through one coil to the current in the other coil. A changing current in one coil induces an emf in the other.

Consider two coils arranged so that some of the magnetic flux associated with current in one coil also passes through the second coil, as in Fig. 32-1. If we change the current I_1 in the first coil, an induced emf $\mathcal{E}_2$ appears in the second. As we discussed in the preceding chapter, $\mathcal{E}_2$ depends on the rate of change of magnetic flux through the second coil. The magnetic flux depends, in turn, on the current in the first coil and on the geometrical arrangement of the two coils which determines how much flux from the first coil actually links the second. We characterize this geometrical arrangement by the ratio of the total magnetic flux through the second coil to the current in the first coil. This ratio defines the **mutual inductance**, M, of the two coils:

$$M = \frac{\phi_2}{I_1}. \tag{32-1}$$

Solving Equation 32-1 for ϕ_2 and differentiating, we obtain

$$\frac{d\phi_2}{dt} = M\frac{dI_1}{dt}.$$

Faraday's law says that $d\phi_2/dt$ is $-\mathcal{E}_2$, the induced emf in coil 2. So we have

$$\mathcal{E}_2 = -M\frac{dI_1}{dt}, \tag{32-2}$$

where the minus sign describes the polarity of the induced emf.

In the context of Fig. 32-1, our discussion of mutual inductance assumed that a changing current in the left-hand coil caused an induced emf in the right-hand coil. But we could have equally well considered the effect a changing current in the right-hand coil would have on the left-hand coil. Although it's not at all obvious, the mutual inductance M is the same either way; it's a constant that characterizes the physical arrangement of two coils regardless of which is the source of the changing magnetic flux.

From Equation 32-2, we see that the unit of mutual inductance is the volt-second/ampere. This unit is given the name henry (H) in honor of the American scientist Joseph Henry (1797–1878), who was also the first secretary of the Smithsonian Institution. Mutual inductances found in common electronic circuits usually range from microhenrys (μH) on up to several henrys.

EXAMPLE 32-1	*Mutual Inductance*

A 2-turn coil is wrapped around a long solenoid with cross-sectional area $A = 26$ cm^2, wound with $n = 3500$ turns per meter of length (Fig. 32-2). Find the mutual inductance of this arrangement.

Solution
Since the solenoid field is uniform and confined to the solenoid interior, the flux through each turn of the small coil is just BA. Accounting for the two turns and using Equation 30-11 for the solenoid's magnetic field, the total magnetic flux through the small coil becomes

$$\phi_2 = 2BA = 2\mu_0 nIA,$$

FIGURE 32-2 What is the mutual inductance of the two coils? (Example 32-1)

where I is the solenoid current. Equation 32-1 shows that mutual inductance is the ratio of the flux in one coil to the current in the other, or

$$M = \frac{\phi_2}{I} = \frac{2\mu_0 nIA}{I} = 2\mu_0 nA$$

$$= (2)(4\pi \times 10^{-7} \text{ N/A}^2)(3500)(26 \times 10^{-4} \text{ m}^2)$$

$$= 23 \ \mu\text{H} .$$

EXERCISE Suppose the solenoid of Example 32-1 has circular cross section, and that the smaller coil is *inside* the solenoid, with its diameter half that of the solenoid. Find the mutual inductance.

Answer: 5.7 μH

Some problems similar to Example 32-1: 6, 7, 10

Mutual inductance is the basis of **transformers** and similar devices that use the mutual inductance of two adjacent coils to transfer electric power from one circuit to another without direct electrical contact. Transformers also shift voltage levels, allowing great flexibility in the handling and use of electric power. But because they're based fundamentally on Faraday's law, these devices work only with alternating current or in situations where current is interrupted to make it change with time. We'll discuss transformers more in Chapter 33; for now, Example 32-2 and its exercise illustrate two transformer-like devices using mutual induction.

EXAMPLE 32-2	*Spark Plugs and Toothbrushes*

Electric sparks ignite the gasoline that powers your car's engine. High voltage to fire the spark plugs is provided by the ignition coil, an arrangement of two different coils with mutual inductance. Current from the car's 12-V battery flows through the coil with fewer turns, and is interrupted periodically by a switch in the car's distributor. The sudden change in current induces a large emf in the coil with more turns, and this emf drives the spark.

A typical ignition coil draws 3.0 A and supplies 20 kV to the spark plugs. If the current decays in 0.10 ms when the switch opens, what is the mutual inductance of the ignition coil?

Solution

The rate of change of current is

$$\frac{dI}{dt} = \frac{3.0 \text{ A}}{0.10 \times 10^{-3} \text{ s}} = 3.0 \times 10^4 \text{ A/s} .$$

Solving Equation 32-2 for M then gives

$$M = \frac{|\mathcal{E}|}{|dI/dt|} = \frac{20 \times 10^3 \text{ V}}{3.0 \times 10^4 \text{ A}} = 0.67 \text{ H} .$$

EXERCISE An electric toothbrush has no electrical connection to the power line (Fig. 32-3). But when the toothbrush is in its stand, a coil inside the toothbrush itself rests inside another coil in the stand, and alternating current from the power line flows in the stand coil. The mutual inductance of the two coils results in an induced current in the toothbrush

FIGURE 32-3 The batteries in this electric toothbrush are charged with energy transferred via mutual induction of coils located in the base unit, which is connected to the AC power line, and in the bottom of the brush unit. There is no direct electrical connection to the brush unit.

coil, and this current charges the batteries that power the toothbrush. Suppose the mutual inductance of this arrangement is 100 mH. At an instant when the current in the stand coil is changing at the rate of 40 A/s, what is the emf in the toothbrush coil?

Answer: 4.0 V

Some problems similar to Example 32-2: 1–4

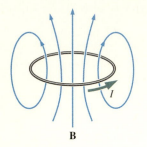

FIGURE 32-4 Magnetic flux from a circular current loop passes through the loop itself.

32-2 SELF-INDUCTANCE

So far we've considered emf and current induced in a circuit by changes—like moving a magnet or varying the current in another circuit—that were external to the circuit in question. But the changing magnetic field associated with the changing current in a circuit also affects that same circuit.

Consider a circular loop carrying current I, as shown in Fig. 32-4. Magnetic field lines arising from this current loop pass through the loop, making a magnetic flux through the loop. If the current is steady, this flux is constant, and there is no induced electric field. But if we change the loop current, then the flux changes, and an induced electric field arises. In order to conserve energy this field opposes the change that causes it—in this case the change in loop current. If the current is counterclockwise and we increase its strength, an induced electric field will appear in the clockwise direction to oppose the current increase. If we decrease the current, the induced electric field will have the opposite sense, now trying to maintain the current. The induced electric field therefore makes it difficult to change the current in the circuit.

This property of a circuit whereby its own magnetic field opposes changes in current is termed **self-inductance**. All circuits possess self-inductance, but this inductance is important only when the circuit encircles a great many of its own magnetic field lines or when current changes very rapidly. A simple piece of wire exhibits very little opposition to current changes in the 60-Hz alternating current used in electric power systems. But in a TV or computer, where currents change on time scales of billionths of a second, self-inductance of the wires themselves must be taken into account.

An **inductor** is a device designed specifically to exhibit self-inductance. Inductors have many uses in electric circuits, including establishing the frequencies of radio transmitters or the channel settings of TV sets, and helping "steer" high and low frequency signals to the tweeters and woofers of loudspeaker systems. We'll explore some of these uses in the next chapter. A typical inductor consists of a coil of wire, constructed so that a great deal of its own magnetic flux is encircled. Some inductors are wound on iron cores to promote flux concentration (Fig. 32-5). Ideally, the only electrical property of an inductor is its inductance. But real inductors are made from wire, so they have resistance as well.

As long as the current in an inductor is steady, the magnetic flux is constant, so there's no induced emf and the inductor acts just like a piece of wire. But when the current changes, the changing magnetic flux induces an emf that opposes the change in current. The more rapidly the current changes, the greater the rate of change of flux and so the greater the emf. The induced emf depends also on how much of its own magnetic flux the inductor encircles; consequently, we define self-inductance, L, as the ratio of magnetic flux through the inductor to current in the inductor:

$$L = \frac{\phi_B}{I}. \tag{32-3}$$

FIGURE 32-5 Typical inductors. At left is a 20-H unit, wound on an iron core for increased inductance. The others have inductances in the mH range.

The unit of self-inductance, like that of mutual inductance, is the henry. Inductance is a constant determined by the physical design of an inductor. In principle, we can calculate the inductance of any inductor, but in practice this is difficult unless the geometry is particularly simple. Inductors for use in electronic circuits are available commercially in a wide range of inductance values.

| **EXAMPLE 32-3** | *Inductance of a Solenoid* |

A long solenoid of cross-sectional area A and length ℓ has n turns per unit length. What is its self-inductance?

Solution

Equation 32-3 shows that self-inductance is the ratio of magnetic flux to current. In Chapter 30 we used Ampère's law to find the magnetic field of a long solenoid: $B = \mu_0 n I$. Since the field is uniform and perpendicular to the solenoid coils, the flux through each turn is just BA. With n turns per unit length, our solenoid has $N = n\ell$ turns. The total flux through the solenoid is then

$$\phi_B = NBA = (n\ell)(\mu_0 n I)(A) = \mu_0 n^2 A \ell I.$$

Then Equation 32-3 gives

$$L = \frac{\phi_B}{I} = \mu_0 n^2 A \ell . \qquad (32\text{-}4)$$

Does this result make sense? As the area increases, so does the flux and therefore the inductance. As the length increases so does the number of turns, and therefore the flux and the inductance increase. Equation 32-4 reflects these trends. Can you see why the inductance should be proportional to the *square* of the number of turns per unit length?

EXERCISE What is the self-inductance of the MRI solenoid in Example 30-6 (page 787)?

Answer: 0.53 H

Some problems similar to Example 32-3: 11, 15, 20

The induced emf in an inductor is determined by Faraday's law, which relates the emf to the rate of change of magnetic flux:

$$\mathcal{E} = -\frac{d\phi_B}{dt}.$$

Differentiating Equation 32-3, the definition of inductance, gives

$$\frac{d\phi_B}{dt} = L\frac{dI}{dt}.$$

Then Faraday's law becomes

$$\mathcal{E} = -L\frac{dI}{dt}. \qquad (32\text{-}5)$$

This equation gives the emf $\mathcal{E}$ induced in an inductor L when the current in the inductor is changing at the rate dI/dt. The minus sign again tells us that the emf *opposes* the change in current. For this reason the inductor emf is often called a **back emf**; it works *against* changes brought about by an externally applied emf. Figure 32-6 shows how to interpret the signs in Equation 32-5.

When the current in an inductor is steady, then $dI/dt = 0$ and there is no emf in the inductor. In this case, the inductor acts just like a piece of wire. But when the current changes the inductor responds by producing a back emf that opposes the change in current. Now the inductor acts very much like a battery, with the magnitude of its emf dependent on how fast the current changes. If we try to start or stop current suddenly, dI/dt is very large and a very large back emf appears. This is not just mathematics! Rapid switching of inductive devices such as solenoids, solenoid valves, or motors can result in destruction of delicate electronic devices by induced currents. And people have been killed opening switches in circuits containing large inductors. In the next section we'll take a closer look at the effect the rapid opening and closing of switches has in circuits containing inductors.

FIGURE 32-6 The inductor emf of Equation 32-5 is positive if, when going in the direction of the inductor current, you encounter the negative end of the inductor first. Because of the minus sign in Equation 32-5, this happens when the current is *decreasing* ($dI/dt < 0$).

FIGURE 32-7 Is the current increasing, decreasing, or steady?

Got It!

Current flows from left to right through the inductor shown in Fig. 32-7. A voltmeter connected across the inductor gives a constant reading, and shows that the left end of the inductor is positive. Is the current in the inductor increasing, decreasing, or steady?

EXAMPLE 32-4	*A Dangerous Inductor*

A current of 5.0 A is flowing in a 2.0-H inductor. The current is reduced steadily to zero in 1.0 ms. What is the magnitude of the emf in the inductor while the current is being turned off?

Solution
Because the current changes steadily, its time rate of change has magnitude

$$\frac{dI}{dt} = \frac{5.0 \text{ A}}{1.0 \text{ ms}} = 5000 \text{ A/s} \,,$$

so

$$|\mathcal{E}| = L\frac{dI}{dt} = (2.0 \text{ H})(5000 \text{ A/s}) = 10{,}000 \text{ V} \,,$$

enough to produce a lethal shock. Note that this voltage is quite unrelated to the voltage of the battery or whatever else was supplying the inductor current. We could have a 6-volt battery and still be electrocuted trying to open the circuit rapidly when a large inductance is present.

EXERCISE A neon lamp that glows only when the voltage across it exceeds 90 V is connected across a 1.2-H inductor carrying 750 mA. When the current is interrupted the lamp flashes. Find the maximum time over which the current could have dropped to zero, assuming a steady decrease.

Answer: 10 ms

Some problems similar to Example 32-4: 12–14

32-3 INDUCTORS IN CIRCUITS

Here we examine circuits containing batteries, resistors, and inductors, analogous to the *RC* circuits of Chapter 28. In the qualitative analysis of *RC* circuits we found a useful guiding principle: The voltage across a capacitor cannot change instantaneously. We can make an analogous statement for inductors. Because the inductor emf is proportional to the rate of change of current in the inductor and because an infinite emf is physically impossible, we conclude that

The current through an inductor cannot change instantaneously.

Thus, the effect an inductor has on current is analogous to the effect a capacitor has on voltage. Much of our understanding of capacitors can be applied to inductors if we interchange the words *voltage* and *current*.

Building Up the Current

Figure 32-8*a* shows a circuit containing a battery, switch, resistor, and inductor (symbol ⁓⁓⁓). What happens when we close the switch? Initially the inductor current is zero; since it can't change instantaneously, it must remain zero immediately after the switch is closed. But this is a series circuit, so the inductor

FIGURE 32-8 (*a*) An *RL* circuit. With the switch open no current flows. (*b*) Immediately after the switch is closed there is still no current. The inductor produces a back emf equal but opposite the battery emf, and therefore the rate of change of current is not zero. (*c*) After a long time, the rate of change of current and therefore the inductor emf approach zero. The inductor then acts like a piece of wire.

and resistor currents are equal. With zero current immediately after the switch is closed, there must be no voltage across the resistor. Therefore the inductor must produce a back emf equal to that of the battery, with the polarity shown in Fig. 32-8*b*. Even though there is, at this instant, no current in the inductor, the presence of an emf indicates that the *rate of change* of the current is not zero. Going around the circuit in the direction that the battery would drive a current, the inductor polarity indicated in Fig. 32-8*b* shows that the inductor emf is *negative*—and Equation 32-5 therefore shows that dI/dt is *positive*—i.e., the current is *increasing*.

As the current rises, so does the voltage across the resistor (since $V_R = IR$). Since the battery emf is constant, that means the inductor emf goes *down*—and that means the rate of change of current goes down. Thus, the current in the circuit builds up, but at an ever-decreasing rate. Concurrently, the inductor emf goes down. Eventually the current reaches a steady value, at which point dI/dt and, therefore, the inductor emf are zero. In this ultimate steady state the inductor acts like a piece of wire, and the circuit looks like Fig. 32-8*c*. The steady-state current is just $\mathcal{E}_0/R$, where $\mathcal{E}_0$ is the battery emf. Figure 32-9 summarizes this qualitative analysis of the *RL* circuit, while Fig. 32-10 shows the behavior of the current and inductor emf over time.

ActivPhysics

14.1

The *RL* Circuit

As current increases, so does resistor voltage *IR*, at an ever decreasing rate.

Current *I* increases at an ever decreasing rate.

Battery emf stays constant. Resistor and inductor voltages sum to the battery voltage.

Change in current results in an emf in the inductor. Emf decreases with decreasing rate of current change.

FIGURE 32-9 Interrelationships among circuit quantities as current builds up in an *RL* circuit. Compare with Fig. 28-30, which is a related diagram for a charging capacitor.

FIGURE 32-10 Inductor current and magnitude of the inductor emf as functions of time. Vertical axis represents either current or voltage.

To analyze the *RL* circuit quantitatively, we apply the loop law. Going clockwise from the negative terminal of the battery, we first encounter a voltage increase $\mathcal{E}_0$ due to the battery. Then the voltage decreases by IR in the resistor. Finally, there is a voltage change $\mathcal{E}_L$ in the inductor. This change is actually a decrease, because the inductor emf opposes the increase in current. However, we'll simply call the inductor emf $\mathcal{E}_L$ and let Equation 32-5 take care of the signs. Then the loop law becomes

$$\mathcal{E}_0 - IR + \mathcal{E}_L = 0 . \qquad (32\text{-}6a)$$

If we differentiate this equation with respect to time, the battery emf $\mathcal{E}_0$ drops out because it's constant, giving

$$\frac{d\mathcal{E}_L}{dt} = R\frac{dI}{dt} .$$

Equation 32-5 shows that $dI/dt = -\mathcal{E}_L/L$, so our differentiated loop equation becomes

$$\frac{d\mathcal{E}_L}{dt} = -R\frac{\mathcal{E}_L}{L} . \qquad (32\text{-}6b)$$

This differential equation describes a quantity—$\mathcal{E}_L$—whose rate of change is proportional to itself. We discussed such equations in Chapter 28 when we considered the *RC* circuit. Equation 32-6*b* is similar to Equation 28-4, but with current I replaced by the inductor emf $\mathcal{E}_L$, capacitance C by L, and R by $1/R$. The solution to Equation 32-6*b* is that of Equation 28-4, provided we make the appropriate substitutions for I, C, and R:

$$\mathcal{E}_L = -\mathcal{E}_0 e^{-Rt/L} . \qquad (32\text{-}7)$$

This equation shows that the inductor emf decays exponentially to zero, starting from an initial value of $-\mathcal{E}_0$ (negative because the inductor emf *opposes* the battery emf). We can now solve for the current using Equation 32-6*a*:

$$I = \frac{\mathcal{E}_0 + \mathcal{E}_L}{R} = \frac{\mathcal{E}_0 + (-\mathcal{E}_0 e^{-Rt/L})}{R} = \frac{\mathcal{E}_0}{R}(1 - e^{-Rt/L}) . \qquad (32\text{-}8)$$

With a capacitor, we characterized the exponentially changing quantities in terms of the capacitive time constant RC. With an inductor, we have an **inductive time constant** L/R. Significant changes in current cannot occur on time scales much shorter than L/R. On the other hand, an *RL* circuit will approach a steady state, with zero $\mathcal{E}_L$, only after many inductive time constants.

We include inductors in circuits when we want to limit the rate at which current can change, and we select L in conjunction with the circuit resistance R to give the desired time constant. But self-inductance occurs in *any* circuit, whether or not there's an explicit inductor present, and in high-speed circuits such as computers and TVs it's important to keep the time constant L/R small to allow rapid current changes.

Why is the inductive time constant a quotient of L and R rather than a product, as in the capacitor case? In Problem 23 you can convince yourself that L/R has the units of seconds. But you can also understand this physically. The instant the switch is closed in Fig. 32-8, the inductor emf must equal the battery emf to keep the current instantaneously at zero. The *larger* the inductance L, the smaller the rate of current change, dI/dt, needed to produce that emf—and so the longer it will take to build up the current. And the *smaller* the resistance, the larger the final current and therefore, again, the longer it will take to build up that current.

EXAMPLE 32-5	*Firing Up an Electromagnet*

A large electromagnet used for lifting scrap metal has a self-inductance of 56 H. It is connected through a switch to a constant 440-V power source; the total resistance of the circuit is 2.8 Ω. When the switch is closed, how long does it take to bring the magnet current to 75% of its final value?

Solution

Letting $t \rightarrow \infty$ in Equation 32-8 shows that the final current is $I_{final} = \mathcal{E}_0/R$, as we argued in our qualitative analysis. Setting the current I in Equation 32-8 to $0.75\mathcal{E}_0/R$ gives

$$0.75 = 1 - e^{-Rt/L},$$

or

$$e^{-Rt/L} = 0.25.$$

Taking the natural logarithm of both sides (recall that $\ln e^x = x$) gives

$$-Rt/L = \ln(0.25),$$

or

$$t = -\frac{L}{R}\ln(0.25) = -\frac{56 \text{ H}}{2.8 \text{ }\Omega}\ln(0.25) = 28 \text{ s}.$$

(The minus sign canceled since the logarithm of a number less than 1 is negative.) Our answer is approximately one time constant ($L/R = 20$ s). This should not be surprising since we found with RC circuits that quantities following equations such as Equation 28-6 (and, therefore, its analog, Equation 32-8) reach $1 - 1/e$, or about two-thirds, of their final value in one time constant.

EXERCISE A 1.0-kΩ resistor is in series with an inductor, and a 12-V battery is connected across the pair. The current rises to 8.5 mA in 21 μs. Find the inductance.

Answer: 17 mH

Some problems similar to Example 32-5: 27–29, 32

The Current Decays

Figure 32-11a shows an *RL* circuit with a two-way switch. Throwing the switch to position *A* allows current to build up in the inductor as we've already described. Then, at time $t = 0$, we throw it to position *B*. This disconnects the battery, leaving a circuit electrically equivalent to Fig. 32-11b. Just prior to $t = 0$ there was some current I_0 flowing downward in the inductor. Since the inductor current cannot change instantaneously, that current must continue just after the switch is closed, as shown in Fig. 32-11b. To drive this current, the inductor must develop an emf in the direction shown. Now the inductor emf is positive, so Equation 32-5 shows that the current is *decreasing,* as we might well expect since the battery has been disconnected. As the current decreases, so does the voltage across the resistor. So, therefore, does the inductor emf and, therefore, the rate of change of current. We thus expect both the current and the inductor emf to decrease, but at an ever decreasing rate.

Note that the inductor emf here is like any other induced emf: it *opposes* the change giving rise to it. In this case that change is the decrease in current caused by disconnecting the battery. The inductor responds with an emf in such a direction as to keep that current flowing.

(a)   (b)

FIGURE 32-11 (a) Throwing the switch to position *A* allows current to build up in the inductor. (b) Throwing it to position *B* gives a circuit containing only the inductor and resistor. The current then decays exponentially.

Applying the loop law to Fig. 32-11b gives

$$\mathcal{E}_L - IR = 0 \,.$$

Using $\mathcal{E}_L = -L\, dI/dt$ from Equation 32-5, the loop equation becomes

$$\frac{dI}{dt} = -\frac{R}{L}I \,.$$

This is just like Equation 32-6b, but with I replacing $\mathcal{E}_L$. The solution follows by analogy with Equation 32-7:

$$I = I_0 e^{-Rt/L}, \tag{32-9}$$

where I_0 is the inductor current when the switch is thrown from A to B. Equation 32-9 shows that the current decays with the same exponential time constant L/R that described its buildup (Fig. 32-12). The resistor voltage IR and therefore the inductor emf also decay in the same way.

It is not always necessary to use Equations 32-8 and 32-9 in describing RL circuits. For times very short compared with the time constant L/R, it suffices to use the fact that inductor currents cannot change instantaneously. And after many time constants, inductors in a circuit containing only steady sources will act like wires. Example 32-6 explores this situation.

FIGURE 32-12 Exponential decay of the current in the circuit of Fig. 32-11b.

EXAMPLE 32-6	*Short Times, Long Times*

In the circuit of Fig. 32-13a, the switch is initially open. What is the current in resistor R_2 immediately after the switch is closed? A long time after the switch is closed? Long after the switch is closed, it is again opened. What is the current in R_2 just after it is opened? A long time after?

Solution

Just before we close the switch, the current in the inductor is zero. The current cannot change instantaneously, so it remains zero just after the switch is closed. At this instant the inductor might as well be an open circuit, giving the circuit shown in Fig. 32-13b. Then all the current from R_1 flows through R_2, so

$$I = \frac{\mathcal{E}_0}{R_1 + R_2} \,.$$

If we wait long enough, the circuit will reach a steady state in which $dI/dt = 0$. So then there is no inductor emf, and the inductor acts like a wire. We can redraw the circuit as Fig. 32-13c, in which all the current from R_1 goes through L, and none through R_2. The resulting current in R_1 and L is just $\mathcal{E}_0/R_1$.

Now the switch is opened again. Current in R_1 stops abruptly since there is no way charge can get through the open switch. But the current through the inductor, which was $\mathcal{E}_0/R_1$ just before the switch was opened, remains $\mathcal{E}_0/R_1$ the instant after the switch is opened. There is only one place this current can

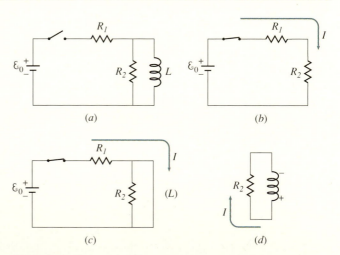

FIGURE 32-13 (*a*) Circuit for Example 32-6. (*b*) Just after the switch is closed the inductor acts like an open circuit. (*c*) A long time later the inductor emf is zero, so it acts like a wire. (*d*) When the switch opens again, current continues to flow in the inductor and on through R_2.

go—through R_2, from bottom to top (Fig. 32-13d). So just after the switch is opened, the current in R_2 is $\mathcal{E}_0/R_1$. Notice that the value of R_2 has no effect on this current, which is determined entirely by the battery emf and the resistance R_1.

What about the voltage across R_2? This is given by Ohm's law:

$$V_2 = I_2 R_2 = \frac{\mathscr{E}_0 R_2}{R_1}.$$

The larger R_2, the larger the voltage that appears when the switch is opened. If there's no resistor connected across the inductor (i.e., $R_2 = \infty$), then the voltage will be arbitrarily large as the inductor seeks at all cost to keep the current flowing. This dangerous situation can result in arcing and vaporization of circuit conductors, and even in electric shock. In circuits with large inductance, resistors are often placed in parallel with inductors to alleviate these dangers.

Finally, the current in Fig. 32-13d decays exponentially to zero. Plots of the currents in R_2 and L as functions of time are shown in Fig. 32-14.

EXERCISE In Fig. 32-13 take $\mathscr{E}_0 = 12$ V, $R_1 = 56$ Ω, $L = 48$ mH, and suppose the switch has been closed for a long time. What is the maximum value of R_2 for which the inductor emf will not exceed 100 V when the switch is opened?

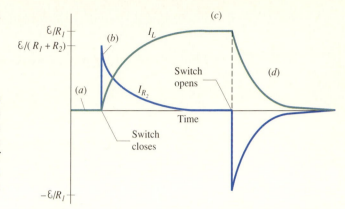

FIGURE 32-14 Currents in R_2 and L for Example 32-6. Note that the time constants before and after the switch opens are different; this reflects the fact that R_1 is out of the circuit after the switch opens. (a)–(d) indicate portions of the graph that correspond to the diagrams in Fig. 32-13.

Answer: 467 Ω

Some problems similar to Example 32-6: 36, 37, 61, 62

Got It!

In a circuit such as that of Fig. 32-13a, can the inductor emf ever exceed the battery emf? If so, can that happen (a) when the switch is closed or (b) when it's opened?

32-4 MAGNETIC ENERGY

In the situations we considered in Figs. 32-11b and 32-13d, current flows in circuits containing only a resistor and an inductor. Energy is dissipated, heating the resistor. Where does this energy come from?

Because there is a current in the inductor, there is also a magnetic field. The change in that magnetic field is what produces the emf that drives current around the circuit. As the current decreases, so does the inductor's magnetic field. Eventually the circuit reaches a state where there is no current, no magnetic field—and a hot resistor. So where did the resistor's thermal energy come from? It came from the magnetic field.

Like the electric field, the magnetic field contains stored energy. Our decaying RL circuit is analogous to a discharging RC circuit, in which the electric field between the capacitor plates disappears as thermal energy appears in the resistor. As in the electric case, magnetic energy is not limited to circuits. *Any* magnetic field contains energy. Release of magnetic energy drives a number of practical devices and also powers violent events throughout the universe (Fig. 32-15).

We can reinterpret the RL circuit of Fig. 32-11 in terms of energy. With the switch in position A, the battery supplies energy to the resistor and inductor. In the resistor the energy is dissipated as heat, but in the inductor it gets stored as magnetic field energy. When the switch is thrown to position B the battery—the

FIGURE 32-15 This eruption of a huge prominence from the Sun's surface involves the release of energy stored in magnetic fields. The intense magnetic fields responsible for such eruptive solar events are often associated with sunspots.

FIGURE 32-16 Energy transfers in the circuit of Fig. 32-11.

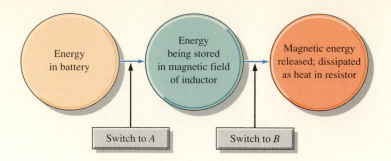

ultimate energy source for the circuit—is disconnected. Energy dissipated in the resistor now comes from the decaying magnetic field of the inductor. That energy came originally from the battery but was stored temporarily in the magnetic field. Figure 32-16 outlines these energy transfers.

Magnetic Energy in an Inductor

How much energy is stored in an inductor's magnetic field? We can answer this by considering first the *rate* of energy storage. If we multiply Equation 32-6a—the loop law for an *RL* circuit—by the current *I*, the result is

$$I\mathscr{E}_0 - I^2R + I\mathscr{E}_L = 0,$$

or, using Equation 32-5 for $\mathscr{E}_L$,

$$I\mathscr{E}_0 - I^2R - LI\frac{dI}{dt} = 0.$$

What do the terms in this equation mean? The first is the product of the battery's current and emf—a product we know gives electrical power. That this term is positive means the battery supplies energy *to* the circuit at the rate $I\mathscr{E}_0$. The second term, I^2R, is the rate of energy dissipation in the resistor (recall Equation 27-9). The negative sign means that this energy is taken *from* the circuit. The third term is also negative (since the current is increasing, dI/dt is positive) and represents the rate at which the inductor takes energy from the circuit. But the inductor does not dissipate this energy; instead, it stores the energy in its growing magnetic field. The rate at which the inductor stores energy is thus

$$P = LI\frac{dI}{dt}.$$

Suppose we increase the current in an inductor by some small amount dI over a small time interval dt. Since the power is the rate of energy storage, the energy dU stored during this time is thus

$$dU = P\,dt = LI\,dI$$

We find the total energy stored in bringing the inductor current from zero to some final value *I* by summing—that is, integrating—all the dU's:

$$U = \int dU = \int P\,dt = \int_0^I LI\,dI = \tfrac{1}{2}LI^2\big|_0^I$$

Evaluating at the two limits then gives the stored energy:

$$U = \tfrac{1}{2}LI^2. \tag{32-10}$$

This much energy is therefore released when the magnetic field decays.

EXAMPLE 32-7	*Quenching a Superconducting Magnet*

Loss of coolant is a danger in superconducting electromagnets. The current is suddenly left without its zero-resistance path, and energy stored in the magnetic field is rapidly released. To prevent explosive energy release, copper or silver is incorporated into the conducting system to lengthen the time constant L/R in the event of such a "quench" (Fig. 32-17). A superconducting MRI solenoid carries a 2.4-kA current and has 0.53-H inductance. In its nonsuperconducting state, the total resistance is 31 mΩ. (a) How much energy is stored in the solenoid's magnetic field? (b) If the coils suddenly lose their superconductivity, what is the initial rate of energy release?

Solution

(a) Equation 32-10 gives the magnetic energy:

$$U = \tfrac{1}{2}LI^2 = (\tfrac{1}{2})(0.53 \text{ H})(2.4 \times 10^3 \text{ A})^2 = 1.53 \text{ MJ} .$$

(b) When superconducting ceases, the current decays according to Equation 32-9. At the instant the decay starts the current still has its original value. Since the magnetic energy released is dissipated in the resistor, the rate of energy release is just I^2R, or

$$P = I^2R = (2.4 \times 10^3 \text{ A})^2 (0.031 \text{ Ω}) = 180 \text{ kW} .$$

This is a substantial power; equivalent to 1,800 100-W light bulbs burning in the space of this roughly human-size device.

FIGURE 32-17 Superconducting cables in various stages of manufacture. The hollow tubes are made of silver, and are packed with superconducting material. In the event of a coolant loss, the silver—the best metallic conductor—will carry the current and thus prevent explosive release of magnetic energy.

The following exercise explores the duration of this power surge.

EXERCISE How long is it before 90% of the magnetic energy in Example 32-7 has been dissipated?

Answer: 20 s

Some problems similar to Example 32-7: 42, 44, 47

Magnetic Energy Density

A long solenoid is a particularly simple inductor in which the magnetic field is essentially uniform. We can readily evaluate the energy density using this magnetic field, just as we found the electric field energy density using a parallel-plate capacitor.

In Example 32-3 we found that the inductance of a long solenoid of length ℓ and cross-sectional area A is $L = \mu_0 n^2 A\ell$, with n the number of turns per unit length. Equation 32-10 then gives the magnetic energy stored in the solenoid when it carries current I:

$$U = \tfrac{1}{2}LI^2 = \tfrac{1}{2}\mu_0 n^2 A\ell I^2 = \frac{1}{2\mu_0}(\mu_0 nI)^2 A\ell = \frac{B^2}{2\mu_0}A\ell ,$$

where we recognized the quantity $\mu_0 nI$ as B, the magnetic field in the solenoid (Equation 30-11). The quantity $A\ell$ is the volume containing this field, so the energy per unit volume—the **magnetic energy density**—is

$$u_B = \frac{B^2}{2\mu_0} . \qquad \text{(magnetic energy density)} \qquad (32\text{-}11)$$

Although we derived this expression for the field of a solenoid, it is, in fact, a universal expression for the local magnetic energy density. Wherever there is a magnetic field, there is stored energy.

Compare Equation 32-11 with Equation 26-3 for the energy density in an electric field:

$$u_E = \tfrac{1}{2}\varepsilon_0 E^2 .$$

The expressions for electric and magnetic energy densities are similar. Each is proportional to the *square* of the field strength, and each contains the appropriate constant. That the constant appears in the numerator in one case and the denominator in the other has no deep significance; it is merely a consequence of the way SI units are defined.

EXAMPLE 32-8	*Energy in Earth's Magnetic Field*

The magnetic field strength near Earth's surface is about 50 μT. (a) How much energy is contained in 1 cubic kilometer of this field? (b) How does this compare with the electrical energy in the same volume, given a fair-weather electric field of 100 V/m?

Solution

(a) The magnetic field is essentially constant over this volume, so the total energy is the product of the energy density with the volume V:

$$U = u_B V = \frac{B^2}{2\mu_0} V = \frac{(50\times10^{-6}\text{ T})^2}{(2)(4\pi\times10^{-7}\text{ N/A}^2)}(1\times10^3\text{ m})^3$$

$$= 1\text{ MJ}.$$

This is not a particularly large energy; a mere gallon of gasoline, for example, stores about 100 times as much.

(b) Since we're considering equal volumes, it suffices to compare energy densities. Then

$$\frac{u_B}{u_E} = \frac{B^2/2\mu_0}{\tfrac{1}{2}\varepsilon_0 E^2} = \frac{B^2}{\mu_0\varepsilon_0 E^2}$$

$$= \frac{(50\times10^{-6}\text{ T})^2}{(4\pi\times10^{-7}\text{ N/A}^2)(8.85\times10^{-12}\text{ F/m})(100\text{ V/m})^2}$$

$$= 2\times10^4$$

Thus the electrical energy is even smaller, by a factor of 20,000. The energy stored in both of Earth's electromagnetic fields is very small compared, for example, with the gravitational energy or thermal energy of an equivalent volume of air. In other systems, though, including the atmospheres of stars or lab experiments in plasma physics, magnetic energy may dominate.

EXERCISE A typical sunspot is about 50,000 km in diameter, and extends into the Sun as a cylinder about 30,000 km long. Its magnetic field is about 0.2 T. What is the magnetic energy in such a spot?

Answer: 9×10^{26} J

Some problems similar to Example 32-8: 51, 52

CHAPTER SYNOPSIS

Summary

1. The **mutual inductance** of a pair of coils is defined as the ratio of the total flux in the second coil to the current in the first:

$$M = \frac{\phi_2}{I_1}.$$

By Faraday's law, the emf in the second coil is proportional to the rate of change of current in the first:

$$\mathcal{E}_2 = -M\frac{dI_1}{dt}.$$

The same mutual inductance M describes the emf developed in the first coil as a result of changing current in the second coil.

2. A changing current in a coil or circuit results in a changing magnetic flux that gives rise to an induced electric field opposing the original change in current. A device constructed to exploit this property of **self-inductance** is called an **inductor**. The self-inductance L of an inductor is the ratio of magnetic flux to current:

$$L = \frac{\phi}{I}.$$

Faraday's law relates the emf in an inductor to the rate of change of current:

$$\mathcal{E} = -L\frac{dI}{dt}.$$

The direction of the emf is such as to oppose changes in the inductor current. Self-inductance prevents the inductor current from changing instantaneously.

3. In a circuit containing a resistor R and inductor L, changes occur exponentially with an **inductive time constant** L/R. The rising current in a series RL circuit is given by

$$I = \frac{\mathcal{E}_0}{R}(1 - e^{-Rt/L}),$$

where $\mathcal{E}_0$ is the battery emf and t the time. If the current is subsequently allowed to decay, it goes exponentially to zero:

$$I = I_0 e^{-Rt/L},$$

where t is measured from the start of the decay.

4. Electrical energy supplied to an inductor ends up stored in the inductor's magnetic field. When an inductance L carries current I, the stored magnetic energy is

$$U = \tfrac{1}{2}LI^2.$$

5. All magnetic fields, not only those of inductors, contain stored energy, with the **magnetic energy density** given by

$$u_B = \frac{B^2}{2\mu_0}.$$

Terms You Should Understand

(Pairs are closely related terms whose distinction is important; number in parentheses is chapter section where term first appears.)

mutual inductance, self-inductance (32-1, 32-2)
inductor (32-2)
back emf (32-2)
inductive time constant (32-3)
magnetic energy, magnetic energy density (32-4)

Symbols You Should Recognize

M, L (32-1, 32-2)
u_B (32-4)

Problems You Should Be Able to Solve

calculating mutual inductance for simple coil configurations (32-1)
finding induced emf in a second coil given mutual inductance and rate of change of current in the first coil (32-1)
calculating self-inductance for simple configurations (32-2)
calculating back emf given self-inductance and rate of change of current (32-2)
analyzing time-dependent behavior of RL circuits (32-3)
calculating magnetic energy (32-4)

Limitations to Keep in Mind

Real inductors, other than those made from superconductors, have resistance as well as inductance.

QUESTIONS

1. Figure 32-18 shows two pairs of identical coils in different geometrical arrangements. For which arrangement is the mutual inductance greatest? Why?

(a)   (b)

FIGURE 32-18 Question 1.

2. A car battery has an emf of 12 V, yet energy from the battery provides the 20,000-V spark that ignites the gasoline. How is this possible?

3. When two coils are connected in series but are physically far apart, they behave as a single inductor whose inductance is the sum of the individual inductances. Why might this not be true if they are close together?

4. You have a fixed length of wire to wind into an inductor. Will you get more inductance if you wind a short coil with large diameter, or a long coil with small diameter?

5. You have a fixed length of wire of resistance R. You want to wind the wire into a small space and use it as a resistor. How would you wind it so as to minimize its self-inductance?

6. In wiring circuits that operate at high frequencies, like TV sets or computers, it is important to avoid extraneous loops in wires. Why?

7. In a popular demonstration of induced emf, a light bulb is connected across a large inductor in an LR circuit, as shown in Fig. 32-19. When the switch is opened, the bulb flashes brightly and burns out. Why?

FIGURE 32-19 Question 7.

8. In the RL circuit of Fig. 32-8a, can the inductor emf exceed the battery emf (a) when the switch is first closed?

(b) When the switch is opened after being closed for a long time?

9. Does it take more or less than one time constant for current in an *RL* circuit to build up to half its steady-state value?

10. If you increase the resistance in an *RL* circuit, what effect does this have on the inductive time constant?

11. Speculate on what would happen if you connected an ideal battery directly across an ideal inductor, with no resistance anywhere in the circuit.

12. How could you modify the simple *RL* circuit of Fig. 32-8*a* to prevent dangerous voltages from developing when the switch is opened?

13. List some similarities and differences between inductors and capacitors.

14. A 1-H inductor carries 10 A, and a 10-H inductor carries 1 A. Which inductor contains more stored energy?

15. Does the energy density in a magnetic field depend on the direction of the field?

16. The field of a magnetic dipole extends to infinity. Is there an infinite amount of energy stored in the dipole field? Why or why not?

17. It takes work to push two bar magnets together with like poles facing each other. Where does this energy go?

PROBLEMS

ActivPhysics can help with these problems:
Activity 14.1

Section 32-1 Mutual Inductance

1. Two coils have a mutual inductance of 2.0 H. If current in the first coil is changing at the rate of 60 A/s, what is the emf in the second coil?

2. A 500-V emf appears in a coil when the current in an adjacent coil changes at the rate of 3.5 A/ms. What is the mutual inductance of the coils?

3. The current in one coil is given by $I = I_p \sin 2\pi ft$, where $I_p = 75$ mA, $f = 60$ Hz, and $t = $ time. Find the peak emf in a second coil if the mutual inductance between the coils is 440 mH.

4. Two coils have a mutual inductance of 580 mH. One coil is supplied with a current given by $I = 3t^2 - 2t + 4$, where I is in amperes and t in seconds. What is the induced emf in the other coil at time $t = 2.5$ s?

5. An alternating current given by $I_p \sin 2\pi ft$ is supplied to one of two coils whose mutual inductance is M. (a) Find an expression for the emf in the second coil. (b) When $I_p = 1.0$ A and $f = 60$ Hz, the peak emf in the second coil is measured at 50 V. What is the mutual inductance?

6. Find the mutual inductance of the two-coil system described in Problem 19 of Chapter 31.

7. Two long solenoids of length ℓ both have n turns per unit length. They have circular cross sections with radii R and $2R$, respectively. The smaller solenoid is mounted inside the larger one, with their axes coinciding. Find the mutual inductance of this arrangement, neglecting any nonuniformity in the magnetic field near the ends.

8. Coils A and B have mutual inductance 25 mH. At time $t = 0$ the current in coil A is zero. Subsequently a time-varying current is supplied to A, and the induced emf in coil B is given by $\mathscr{E} = 50 + 0.2t$, with $\mathscr{E}$ in V and t in ms. Find an expression for the time-varying current in coil A.

9. A rectangular loop of length ℓ and width w is located a distance a from a long, straight wire, as shown in Fig. 32-20. What is the mutual inductance of this arrangement?

FIGURE 32-20 Problem 9.

10. Two wire loops of radii a and b lie in the same plane and have a common center. Find the mutual inductance of this arrangement, assuming $b \gg a$. *Hint:* With $b \gg a$ the magnetic field will be essentially uniform over the smaller loop. See Example 30-1.

Section 32-2 Self-Inductance

11. What is the self-inductance of a solenoid 50 cm long and 4.0 cm in diameter that contains 1,000 turns of wire?

12. The current in an inductor is changing at the rate of 100 A/s, and the inductor emf is 40 V. What is its self-inductance?

13. A 2.0-A current is flowing in a 20-H inductor. A switch is opened, interrupting the current in 1.0 ms. What emf is induced in the inductor?

14. A 60-mA current is flowing in a 100-mH inductor. Over a period of 1.0 ms the current is reversed, going steadily to 60 mA in the opposite direction. What is the inductor emf during this time?

15. A cardboard tube measures 15 cm long by 2.2 cm in diameter. How many turns of wire must be wound on the full length of the tube to make a 5.8-mH inductor?

16. The current in a 2.0-H inductor is given by $I = 3t^2 + 15t + 8$, where t is in seconds and I in amperes. Find an expression for the magnitude of the inductor emf.

17. The emf in a 50-mH inductor has magnitude $|\mathscr{E}| = 0.020t$, with t in seconds and $\mathscr{E}$ in volts. At $t = 0$ the inductor current is 300 mA. (a) If the current is increasing, what will be its value at $t = 3.0$ s? (b) Repeat for the case when the current is decreasing.

18. The current in a 40-mH inductor is given by $I = I_0 e^{-bt}$, where $I_0 = 10$ A and $b = 20$ s^{-1}. What is the magnitude of the inductor emf at (a) $t = 0$, (b) $t = 25$ ms, and (c) $t = 50$ ms?

19. A 2,000-turn solenoid is 65 cm long and has cross-sectional area 30 cm^2. What rate of change of current will produce a 600-V emf in this solenoid?

20. You have a plastic rod 20 cm long and 1.5 cm in diameter. What inductance will you get if you wind the entire rod with a single layer of (a) 22-gauge (0.64-mm-diameter) and (b) 34-gauge (0.16-mm-diameter) wire? Assume adjacent turns are touching, separated only by a negligible thickness of enamel insulation.

21. The emf in a 50-mH inductor is given by $\mathcal{E} = \mathcal{E}_p \sin \omega t$, where $\mathcal{E}_p = 75$ V and $\omega = 140$ s^{-1}. What is the peak current in the inductor? (Assume the current swings symmetrically about zero.)

22. A coaxial cable consists of an inner conductor of radius a and outer conductor of radius b, as shown in Fig. 32-21. Current flows along one conductor and back along the other. Show that the inductance per unit length of the cable is

$$\frac{\mu_0}{2\pi} \ln(b/a).$$

FIGURE 32-21 Problems 22, 68.

Section 32-3 Inductors in Circuits

23. Show that the inductive time constant has the units of seconds.

24. What inductance should you put in series with a 100-Ω resistor to give a time constant of 2.2 ms?

25. The current in a series RL circuit rises to 20% of its final value in 3.1 μs. If $L = 1.8$ mH, what is the resistance R?

26. The current in a series RL circuit rises to half its final value in 7.6 s. What is the time constant?

27. A 10-H inductor is wound of wire with resistance 2.0 Ω. If the inductor is connected across an ideal 12-V battery, how long will it take the current to reach 95% of its final value?

28. In a series RL circuit like Fig. 32-8a, $\mathcal{E}_0 = 45$ V, $R = 3.3$ Ω, and $L = 2.1$ H. If the current is 9.5 A, how long has the switch been closed?

29. In Fig. 32-8a, take $R = 2.5$ kΩ and $\mathcal{E}_0 = 50$ V. When the switch is closed, the current through the inductor rises to 10 mA in 30 μs. (a) What is the inductance? (b) What will be the current in the circuit after many time constants?

30. A series RL circuit like Fig. 32-8a has $\mathcal{E}_0 = 60$ V, $R = 22$ Ω, and $L = 1.5$ H. Find the rate of change of the current (a) immediately after the switch is closed and (b) 0.10 s later.

31. In Fig. 32-8a, take $R = 100$ Ω, $L = 2.0$ H, and $\mathcal{E}_0 = 12$ V. At 20 ms after the switch is closed, what are (a) the circuit current, (b) the inductor emf, (c) the resistor voltage, (d) the rate of change of the circuit current, and (e) the power dissipation in the resistor?

32. Show that a series RL circuit reaches 99% of its final current in approximately 5 time constants.

33. Resistor R_2 in Fig. 32-22 is to limit the emf that develops when the switch is opened. What should be its value in order that the inductor emf not exceed 100 V?

FIGURE 32-22 Problem 33.

34. In Fig. 32-11a take $\mathcal{E}_0 = 12$ V, $R = 2.7$ Ω, and $L = 20$ H. Initially the switch is in position B and there is no current anywhere. At $t = 0$ the switch is thrown to position A, and at $t = 10$ s it is thrown back to position B. Find the inductor current at (a) $t = 5.0$ s and (b) $t = 15$ s.

35. A 5.0-A current is flowing through a nonideal inductor with $L = 500$ mH. If the inductor is suddenly short circuited, the inductor current drops to 2.5 A in 6.9 ms. What is the resistance of the inductor?

36. In Fig. 32-23, take $\mathcal{E}_0 = 12$ V, $R_1 = 4.0$ Ω, $R_2 = 8.0$ Ω, $R_3 = 2.0$ Ω, and $L = 2.0$ H. What is the current I_2 (a) immediately after the switch is first closed and (b) a long time after the switch is closed? (c) After a long time the switch is again opened. Now what is I_2?

FIGURE 32-23 Problem 36.

37. In Fig. 32-24, take $\mathcal{E}_0 = 20$ V, $R_1 = 10$ Ω, $R_2 = 5.0$ Ω, and assume the switch has been open for a long time. (a) What is the inductor current immediately after the switch is closed? (b) What is the inductor current a long time after the switch is closed? (c) If after a long time the switch is again opened, what will be the voltage across R_1 immediately afterward?

FIGURE 32-24 Problem 37.

Section 32-4 Magnetic Energy

38. How much energy is stored in a 5.0-H inductor carrying 35 A?

39. What is the current in a 10-mH inductor when the stored energy is 50 μJ?

40. A 220-mH inductor carries 350 mA. How much energy must be supplied to the inductor in raising the current to 800 mA?

41. A 12-V battery, 5.0-Ω resistor, and 18-H inductor are connected in series and allowed to reach a steady state. (a) What is the energy stored in the inductor? (b) Once in the steady state, over what time interval is the energy dissipated in the resistor equal to that stored in the inductor?

42. A battery, switch, resistor, and inductor are connected in series. When the switch is closed the current rises to half its steady-state value in 1.0 ms. How long does it take for the magnetic energy in the inductor to rise to half its steady-state value?

43. The current in a 2.0-H inductor is decreased linearly from 5.0 A to zero over 10 ms. (a) What is the average rate at which energy is being extracted from the inductor during this time? (b) Is the instantaneous rate constant?

44. When a nonideal 1.0-H inductor is short-circuited, its magnetic energy drops to one-fourth of its original value in 3.6 s. What is its resistance?

45. The current in a 2.0-H inductor is increasing. At some instant, the current is 3.0 A and the inductor emf is 5.0 V. At what rate is the inductor's magnetic energy increasing at this instant?

46. A 500-turn solenoid is 23 cm long, 1.5 cm in diameter, and carries 65 mA. How much magnetic energy does it contain?

47. A superconducting solenoid with inductance $L = 3.5$ H carries 1.8 kA. Copper is embedded in the coils to carry the current in the event of a quench (see Example 32-7). (a) What is the magnetic energy in the solenoid? (b) What is the maximum resistance of the copper that will limit the power dissipation to 100 kW immediately after a loss of superconductivity? (c) With this resistance, how long will it take the power to drop to 50 kW?

48. Show that the quantity $B^2/2\mu_0$ has the units of energy density (J/m³).

49. The Alcator fusion experiment at MIT has a 50-T magnetic field. What is the magnetic energy density in Alcator?

50. What is the magnetic field strength in a region where the magnetic energy density is 7.8 J/cm³?

51. The magnetic field of a neutron star is about 10^8 T. How does the energy density in this field compare with the energy density stored in (a) gasoline and (b) pure uranium-235 (mass density 19×10^3 kg/m³)? Consult Appendix C.

52. A loop of magnetic field arches above the Sun's surface, forming a tube approximately 10^5 km long and 10^4 km in diameter. If the magnetic field strength in the tube is 50 G, what is its magnetic energy content?

53. A single-turn loop of radius R carries current I. How does the magnetic energy density at the loop center compare with that of a long solenoid of the same radius, carrying the same current, and consisting of n turns per unit length?

54. A magnetic field is given by $\mathbf{B} = B_0(x/a)^2\hat{\mathbf{j}}$, where B_0 and a are constants. Find an expression for the magnetic energy in a cube of side a with one corner at the origin and sides extending along the coordinate axes.

55. A toroidal coil has inner radius R and a square cross section of side ℓ (Fig. 32-25). It is wound with N turns of wire, and carries a current I. Show that the magnetic energy in the toroid is given by

$$U = \frac{\mu_0 N^2 I^2 \ell}{4\pi} \ln\left(\frac{R + \ell}{R}\right).$$

FIGURE 32-25 Cross section of a toroidal coil. Also shown is a part of a volume element, in the shape of a ring of thickness dr.

56. A toroid of inner radius 1.5 m and square cross section is wound with 2,500 turns. What must be the length ℓ of its cross-sectional square if the toroid contains 80 J of magnetic energy at a current of 63 A?

Paired Problems

(Both problems in a pair involve the same principles and techniques. If you can get the first problem, you should be able to solve the second one.)

57. Two coils have mutual inductance M. The current supplied to coil A is given by $I = bt^2$. Find an expression for the magnitude of the induced emf in coil B.

58. Two coils have mutual inductance M, and a time-varying current is supplied to coil A; at time $t = 0$ that current is zero. The magnitude of the induced emf in coil B is given by $\mathcal{E} = b\sqrt{t}$. Find an expression for the current in coil A.

59. In the circuit of Fig. 32-8a, take $\mathcal{E}_0 = 5.0$ V and $R = 1.8\ \Omega$. At 2.5 s after the switch is closed, the circuit current is 250 mA. Find the inductance.

60. In the circuit of Fig. 32-8a, take $\mathcal{E}_0 = 5.0$ V and $R = 1.8\ \Omega$. At 2.5 s after the switch is closed, the inductor emf is 2.1 V. Find the inductance.

61. In Fig. 32-13a, take $\mathcal{E}_0 = 25$ V, $R_1 = 1.5\ \Omega$, and $R_2 = 4.2\ \Omega$. What is the voltage across R_2 (a) immediately after the switch is first closed and (b) a long time after the switch is closed? (c) Long after the switch is closed it is again opened. Now what is the voltage across R_2?

62. In Fig. 32-13, take $R_2 = 5R_1$. If the maximum possible value for the inductor emf in this circuit is 300 V, what is the battery emf $\mathcal{E}_0$?

63. A wire of radius R carries a current I distributed uniformly over its cross section. Find an expression for the magnetic energy per unit length in the region from R to $100R$.

64. A wire of radius R carries a current I distributed uniformly over its cross section. Find an expression for the total magnetic energy per unit length *within* the wire.

Supplementary Problems

65. (a) Use the result of Problem 55 to determine the inductance of a toroid. (b) Show that your result reduces to the inductance of a long solenoid when $R \gg \ell$.

66. Two long, flat parallel bars of width w and spacing d carry equal but opposite currents I, as shown in Fig. 32-26. (a) Use Ampère's law to find the magnetic field between the bars. Take $d \ll w$ so you can neglect fringing fields. (b) Use your result to find the magnetic energy per unit length stored between the bars. (c) Compare your result in (b) with the expression $U = \frac{1}{2}LI^2$ to find an expression for the inductance per unit length of the bars.

FIGURE 32-26 Problem 66.

67. (a) Use Equation 32-9 to write an expression for the power dissipation in the resistor as a function of time, and (b) integrate from $t = 0$ to $t = \infty$ to show that the total energy dissipated is equal to the energy initially stored in the inductor, namely, $\frac{1}{2}LI_0^2$.

68. (a) Find the magnetic energy density as a function of radial distance for the coaxial cable of Problem 22. (b) Integrate over the volume between the cable conductors to show that the total energy per unit length of the cable is

$$U = \frac{\mu_0}{4\pi}I^2 \ln(b/a).$$

Hint: Your volume element should be a cylindrical shell of radius r, thickness dr, and length ℓ. What is its volume dV? (c) Use the expression $U = \frac{1}{2}LI^2$ to find the inductance per unit length, and show that your result agrees with that of Problem 22.

69. An electric field and a magnetic field have the same energy density. Obtain an expression for the ratio E/B, and evaluate this ratio numerically. What are its units? Is your answer close to any of the fundamental constants listed inside the front cover?

70. Two long, straight, parallel wires are a distance d apart. The wires have radius a, where $a \ll d$. Current flows down one wire and back along the other. Find the inductance per unit length of the parallel wires. Assume the wire radius is so small that you can neglect the magnetic flux within the wires themselves.

71. The switch in the circuit of Fig. 32-27 is closed at time $t = 0$, at which instant the inductor current is zero. Write the loop and node laws for this circuit, and show that they are satisfied if the inductor current is given by $I = (\mathcal{E}_0/R_1)(1 - e^{-R_\parallel t/L})$, where $R_\parallel$ is the resistance of R_1 and R_2 were they connected in parallel.

FIGURE 32-27 Problem 71.

72. Earth's magnetic field ends abruptly on the sunward side at approximately the point where the magnetic energy density has dropped to the same value as the kinetic energy density in the solar wind. Near Earth, the solar wind contains about 5 protons and 5 electrons per cubic centimeter, and flows at 400 km/s. Treating Earth's field as that of a dipole with dipole moment 8×10^{22} J/T, estimate the distance to the point above the equator where the field ends.

33 Alternating-Current Circuits

Nearly all electric power systems use alternating current (AC) at a frequency of 50–60 Hz. Higher frequency AC is essential to radio, television, and computers.

Chapter 28 considered direct-current (DC) circuits, in which the source of electrical energy is a battery or other device whose emf does not change with time. When we turn on a circuit containing only resistors and a DC emf, current starts to flow immediately and remains steady until the circuit is turned off. Even when we add capacitance, as in Section 28-6, or inductance, as in Section 32-3, all currents and voltages eventually reach steady values.

We now turn our attention to alternating-current (AC) circuits, in which sources of electrical energy vary with time. Alternating current is used almost universally for electric power, for reasons we'll discuss in Section 33-6, so you make use of AC circuits every time you switch on a light, a toaster, a stereo, or another device plugged into your household wiring. More rapidly varying alternating currents include the audio and video signals inside stereos and TVs, and the "clock" signal that sets the basic speed of your computer.

In this chapter we'll show first how to characterize AC current and voltage, and then will look at the behavior of resistors, capacitors, and inductors in AC circuits.

33-1 ALTERNATING CURRENT

We saw in Chapter 31 how the rotation of conducting loops in electric power generators naturally leads to voltage and current that vary sinusoidally with time. In audio and video equipment, computers, and other electronic devices, AC signals may exhibit more complicated time variation. But, as we described in Section 16-6, these more complicated variations can be analyzed as superpositions of sinusoidal functions. For that reason we can limit our description of AC circuits to those where current and voltage vary sinusoidally with time.

A sinusoidal AC voltage or current is characterized by its amplitude, frequency, and phase constant—the same quantities we developed in Chapter 15 to describe simple harmonic motion. Amplitude is specified by giving the peak value (V_p, I_p) or the **root-mean-square** value (V_{rms}, I_{rms}). The rms value is an average obtained by squaring the signal, taking the time average, and then taking the square root. This procedure is used because the direct average of an AC signal is zero since it spends as much time below zero as above. Use of rms values also facilitates the calculation of power in AC circuits. For a sine wave, rms and peak values are related by

$$V_{\text{rms}} = \frac{V_p}{\sqrt{2}} \quad \text{and} \quad I_{\text{rms}} = \frac{I_p}{\sqrt{2}}, \qquad (33\text{-}1)$$

as you can show in Problem 7. When we speak of 120-V household wiring, for example, we are giving the rms voltage. Figure 33-1 shows graphically the relation between rms and peak values.

In practical and engineering situations we usually describe frequency f in cycles per second, or hertz (Hz). In mathematical analysis of alternating current, it's usually more convenient to use the angular frequency ω, measured in radians per second or, equivalently, inverse seconds (s^{-1}). The relation between the two,

$$\omega = 2\pi f, \qquad (33\text{-}2)$$

is the same as for rotational and simple harmonic motion, and for the same reason: a full cycle contains 2π radians.

Sometimes we are interested in the phase constant ϕ of an AC signal, which describes when the sine curve crosses zero (see Fig. 33-1). A full mathematical description of an AC voltage or current then includes its amplitude (V_p, I_p), frequency (ω), and phase constant (ϕ):

$$V = V_p \sin(\omega t + \phi_V) \quad \text{and} \quad I = I_p \sin(\omega t + \phi_I). \qquad (33\text{-}3)$$

FIGURE 33-1 A sinusoidally varying AC voltage, showing peak, rms, and peak-to-peak amplitudes; the latter is just the difference between the extreme values, here equal to $2V_p$. The waveform completes a full cycle as ωt increases by 2π. The phase constant ϕ is $\pi/6$ or 30°; note that a positive phase constant shifts the curve to the left.

Here we've labeled the phase constants with subscripts V and I to indicate that voltage and current—even in the same circuit element—need not have the same phase constant. We will often take one phase constant to be zero; then the other describes the *phase difference* between the voltage and current. Phase difference, rather than the actual value of the phase constant, is the important quantity in AC circuits.

EXAMPLE 33-1	*Household Voltage*

Standard household wiring in North America supplies 120 V rms at 60 Hz. Express this voltage mathematically in the form of Equation 33-3, assuming that the voltage is rising through zero at time $t = 0$.

Solution

The rms and peak voltages are related by Equation 33-1, so

$$V_p = \sqrt{2}V_{rms} = (\sqrt{2})(120 \text{ V}) = 170 \text{ V} .$$

The angular frequency is 2π times the frequency in Hz, so

$$\omega = 2\pi f = (2\pi)(60 \text{ Hz}) = 377 \text{ s}^{-1} .$$

When $t = 0$ we want Equation 33-3 to give $V = 0$; that requires $\phi = 0$ or $\phi = \pi$. But only with $\phi = 0$ will the curve be *rising* at $t = 0$, so Equation 33-3 becomes

$$V = 170 \sin(377t) ,$$

with V in volts and t in seconds. We will frequently take the phase constant to be zero as in this example; only when we're

comparing signals with different phase does the value of ϕ become significant.

Although household outlets in North America provide 120 V rms (or 170 V peak, as we just calculated), wires entering the house actually carry a 240-V potential difference. The full 240 V is available for major appliances such as electric dryers, stoves, and water heaters; it's split into separate 120-V circuits that supply groups of outlets and lights. In Europe, in contrast, the standard outlet voltage is about 230 V, and the frequency is 50 Hz.

EXERCISE A 1.0-kHz sinusoidal current with rms amplitude 1.5 A drives a loudspeaker in a test of a stereo system. Express this current in the form of Equation 33-3, assuming zero phase constant.

Answer: $I = 2.12 \sin(6.28 \times 10^3 t)$, with I in A and t in s.

Some problems similar to Example 33-1: 1, 4, 5

33-2 CIRCUIT ELEMENTS IN AC CIRCUITS

Here we examine separately the AC behavior of resistors, capacitors, and inductors so we can subsequently understand what happens when we combine these elements in AC circuits.

Resistors

An ideal resistor is a device whose current and voltage are always proportional:

$$I = \frac{V}{R} .$$

$V_p \sin \omega t$ R

FIGURE 33-2 A resistor connected across an AC generator (symbol ⊝).

Figure 33-2 shows a resistor R connected across an AC generator, making the voltage across the resistor equal to the generator voltage. The generator voltage is described by Equation 33-3, where we take the phase constant $\phi_V = 0$. Then the current is

$$I = \frac{V}{R} = \frac{V_p \sin \omega t}{R} = \frac{V_p}{R} \sin \omega t .$$

The current has the same frequency as the voltage, and, since its phase constant is also zero, the voltage and current are *in phase*—that is, they peak at the same time. The peak current is simply the peak voltage divided by the resistance: $I_p = V_p/R$. Because both voltage and current are sinusoidal, their rms values are in the same ratio as their peak values; thus $I_{rms} = V_{rms}/R$.

Capacitors

Figure 33-3 shows a capacitor connected across an AC generator. In Chapter 26, we defined a capacitor as a device in which voltage and charge are directly proportional:

$$q = CV.$$

Differentiating this relation gives

$$\frac{dq}{dt} = C\frac{dV}{dt}.$$

FIGURE 33-3 A capacitor across an AC generator.

But dq/dt is the current flowing to the capacitor plates (which we'll call the "capacitor current" even though charge doesn't actually flow through the space between the capacitor plates). So we have

$$I = C\frac{dV}{dt}.$$

The generator voltage $V_p \sin \omega t$ appears directly across the capacitor, so the capacitor voltage V is the generator voltage, and therefore

$$I = C\frac{d}{dt}(V_p \sin \omega t)$$

$$= \omega C V_p \cos \omega t = \omega C V_p \sin(\omega t + \pi/2). \tag{33-4}$$

This equation shows clearly the phase and amplitude relations between current and voltage in a capacitor. Because the cosine curve is just a sine curve shifted left by $\pi/2$ or 90°, Equation 33-4 tells us that:

> The current in a capacitor leads the voltage by 90°.

Figure 33-4 shows graphically this relation between current and voltage in a capacitor.

The term $\omega C V_p$ multiplying the cosine in Equation 33-4 is the amplitude of the current, so we can write

$$I_p = \omega C V_p,$$

or, in a form resembling Ohm's law,

$$I_p = \frac{V_p}{1/(\omega C)} = \frac{V_p}{X_C}, \tag{33-5}$$

FIGURE 33-4 Current in a capacitor leads the voltage by $\pi/2$ or 90°, meaning that the current peaks *before* the voltage by one-fourth of a full cycle.

where we have defined $X_C = 1/\omega C$.

Equation 33-5 shows that the capacitor acts somewhat like a resistor of resistance $X_C = 1/\omega C$. But not quite! This "resistance" does give the relation between the peak voltage and peak current, but it doesn't tell the whole story. The capacitor also introduces a phase difference between voltage and current. This phase difference reflects a fundamental physical difference between resistors and capacitors. A resistor dissipates electrical energy as heat. A capacitor stores and releases electrical energy. Over a complete cycle, the agent turning the generator in Fig. 33-3 does no net work, while the agent turning the generator with the resistive load of Fig. 33-2 continually does work that gets dissipated as heat in the resistor. Because the quantity X_C in Equation 33-5 does not act quite like a resistance, we give it the special name **capacitive reactance**. Like resistance, reactance is measured in ohms (Ω).

Does it make sense that X_C depends on frequency? Yes—as frequency goes to zero, X_C goes to infinity. At zero frequency nothing is changing; there is no need to move charge on or off the plates, so no current flows, and the capacitor might as well be an open circuit. As frequency increases, larger currents flow to move charge on and off the capacitor plates in ever shorter times, so the capacitor looks increasingly like a short circuit. We often summarize this behavior qualitatively by saying that a capacitor at low frequencies acts like an open circuit, while at high frequencies it acts like a short circuit.

Why does the capacitor current *lead* the voltage? Because the capacitor voltage is proportional to its charge, and it takes current to move charge onto the capacitor plates. Therefore current must flow *before* the voltage can change significantly. We found the same thing with the *RC* circuit of Chapter 28; there current flowed as soon as the switch was closed, but it took time to build up the capacitor voltage.

Inductors

FIGURE 33-5 An inductor across an AC generator.

Figure 33-5 shows an inductor connected across an AC generator. The loop law for this circuit is

$$V_p \sin \omega t + \mathscr{E}_L = 0 .$$

From the preceding chapter, we know that the inductor emf is given by $\mathscr{E}_L = -L\dfrac{dI}{dt}$, so the loop law becomes

$$V_p \sin \omega t = L\frac{dI}{dt} .$$

To obtain a relation involving the current I rather than its derivative, we integrate this equation:

$$\int V_p \sin \omega t\, dt = \int L\frac{dI}{dt} dt = \int L\, dI ,$$

or

$$-\frac{V_p}{\omega} \cos \omega t = LI .$$

Here we have set the integration constants to zero because nonzero values would represent a DC emf and current that are absent in this circuit. Solving for I then gives

$$I = -\frac{V_p}{\omega L} \cos \omega t = \frac{V_p}{\omega L} \sin(\omega t - \pi/2) , \qquad (33\text{-}6)$$

where the last step follows because $\sin(\alpha - \pi/2) = -\cos \alpha$ for any α.

Equation 33-6 shows that the current in the inductor lags the applied voltage by $\pi/2$ or 90° (i.e., Equation 33-6 is Equation 33-3 with $\phi = -\pi/2$). Equivalently:

The voltage across an inductor leads the inductor current by 90°.

Figure 33-6 plots this phase relation.

Equation 33-6 also shows that the peak current is

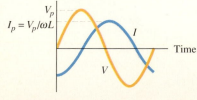

FIGURE 33-6 Voltage across an inductor leads the current by $\pi/2$ or 90°.

$$I_p = \frac{V_p}{\omega L} = \frac{V_p}{X_L} . \qquad (33\text{-}7)$$

TABLE 33-1 *Phase and Amplitude Relations in Circuit Elements*

CIRCUIT ELEMENT	PEAK CURRENT/VOLTAGE	PHASE RELATION
Resistor	$I_p = \dfrac{V_p}{R}$	V, I in phase
Capacitor	$I_p = \dfrac{V_p}{X_C} = \dfrac{V_p}{1/\omega C}$	I leads V by $90°$
Inductor	$I_p = \dfrac{V_p}{X_L} = \dfrac{V_p}{\omega L}$	V leads I by $90°$

Again, this equation resembles Ohm's law, with a "resistance" $X_L = \omega L$. But as with the capacitor, no power is dissipated in the inductor. Instead, energy is alternately stored and released as the inductor's magnetic field builds up, then decays. To distinguish it from dissipative resistance, we call X_L the **inductive reactance**. Inductive reactance, too, is measured in ohms.

Does it make sense that the inductive reactance increases with increasing ω and increasing L? An inductor is a device that, through its induced back emf, opposes changes in current. The greater the inductance, the greater the opposition to changing current. And the more rapidly the current is changing, the more vigorously the inductor opposes the change, so the inductive reactance increases at high frequencies. In the extreme case of very high frequencies, an inductor looks like an open circuit. But at very low frequencies it looks more and more like a short circuit, until with direct current (zero frequency), an inductor exhibits zero reactance because there is no change in current.

Why does the inductor voltage *lead* the current? Because a changing current in an inductor induces an emf. *Before* the current can build up significantly there must first, therefore, be voltage across the inductor.

Table 33-1 summarizes the phase and amplitude relations in resistors, capacitors, and inductors.

Got It!

A capacitor and an inductor are connected across separate but identical electric generators, and the same current flows in each. If the frequency of the generators is doubled, which component will now carry the most current?

EXAMPLE 33-2	*Inductors and Capacitors*

A capacitor is connected across the 60-Hz, 120-V rms power line, and an rms current of 200 mA flows. What is the capacitance? What inductance would have to be connected across the power line for the same current to flow? Would there be anything different about the circuit containing the inductor?

Solution

The peak current and voltage are related through Equation 33-5: so

$$I_p = \frac{V_p}{1/\omega C},$$

$$C = \frac{I_p}{\omega V_p}.$$

We are given the rms voltage and current, but since only the ratio of these quantities appears in our equation, it doesn't matter whether we use rms or peak values. With $f = 60$ Hz, $\omega = 2\pi f$ or 377 s^{-1}, so

$$C = \frac{I}{\omega V} = \frac{0.20 \text{ A}}{(377 \text{ s}^{-1})(120 \text{ V})} = 4.4 \ \mu\text{F}.$$

An inductor that passes the same current must have the same reactance, so

$$\omega L = \frac{1}{\omega C},$$

or

$$L = \frac{1}{\omega^2 C} = \frac{1}{(377 \text{ s}^{-1})^2(4.4 \times 10^{-6} \text{ F})} = 1.6 \text{ H}.$$

Although the currents are the same, the two situations differ in that current leads voltage by 90° in the capacitor and lags by 90° in the inductor.

EXERCISE Inductors (called "ballast") are often used to limit the current in fluorescent lamps. In a particular lamp operating at 60 Hz, 80 V (rms) appears across a 0.53-H ballast inductor.

(a) What is the rms inductor current? (b) What capacitance could be used in place of the inductor to provide the same current?

Answers: (a) 400 mA; (b) 13 μF

Some problems similar to Example 33-2: 13, 14, 17, 20

Phasor Diagrams

Phase and amplitude relations in AC circuits may be summarized graphically in **phasor diagrams**. A phasor is an arrow whose fixed length represents the amplitude of an AC voltage or current. The phasor rotates counterclockwise about the origin with the angular frequency ω of the AC quantity. The component of the phasor on the vertical axis then represents the sinusoidally varying AC signal. Figure 33-7a shows phasors for the current and voltage in a resistor.

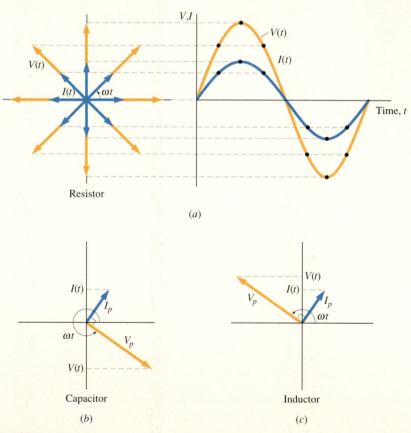

FIGURE 33-7 Phasor diagrams showing voltage and current in (a) a resistor, (b) a capacitor, and (c) an inductor. Lengths of phasors correspond to amplitudes V_p and I_p, while projection on the vertical axis gives the instantaneous value. Angle between voltage and current phasors gives their relative phase. In (a), the phasors are shown at several times; projection of the vertical components onto a plot of current and voltage versus time then shows that these quantities vary sinusoidally with time.

The lengths of the phasors are related by Ohm's law, $V_p = I_p R$. The current and voltage phasors always point in the same direction, showing that current and voltage in the resistor are in phase. For this case we show the phasors at several times in the cycle, to illustrate how, through their rotation, the phasors contain information about the time variation of the quantities they represent. Normally, however, it suffices to show phasors at one arbitrary time. We've done this in Figs. 33-7b and c for a capacitor and an inductor. The phasors are at right angles in each case, indicating the phase relations between current and voltage. Their lengths are related by the appropriate reactance, so $V_p = I_p X$. As phasors rotate, a diagram such as Fig. 33-7a would show them tracing out current and voltage graphs like those of Figs. 33-4 and 33-6. You should convince yourself that all the relationships of Table 33-1 are correctly described by the phasor diagrams of Fig. 33-7. Although phasor diagrams do not add much to our understanding of AC circuits containing only one circuit element, they will greatly simplify the analysis of more complicated circuits.

Capacitors and Inductors: A Comparison

Here and in previous chapters, we have considered separately the behavior of capacitors and inductors. Many of the properties of these devices are analogous. A capacitor opposes instantaneous changes in voltage, while an inductor opposes instantaneous changes in current. In an RC circuit with a DC emf, voltage builds up exponentially across the capacitor, with time constant RC. In the analogous RL circuit, current builds up exponentially in the inductor, with time constant L/R. A capacitor stores electrical energy given by $\frac{1}{2}CV^2$. An inductor stores magnetic energy given by $\frac{1}{2}LI^2$. A capacitor acts like an open circuit at low frequencies, an inductor like a short circuit at low frequencies. Each exhibits the opposite behavior at high frequencies.

Capacitors and inductors are complementary devices, reflecting a deeper complementarity between electric and magnetic fields. Any verbal description of a capacitor applies to an inductor if we replace the words "capacitor" with "inductor," "electric" with "magnetic," and "voltage" with "current." Table 33-2 summarizes the complementary aspects of capacitors and inductors.

TABLE 33-2 *Capacitors and Inductors*

	CAPACITOR	INDUCTOR
Defining relation	$C = \dfrac{q}{V}$	$L = \dfrac{\phi_B}{I}$
Defining relation, differential form	$I = C\dfrac{dV}{dt}$	$\mathcal{E} = -L\dfrac{dI}{dt}$
Opposes changes in	Voltage	Current
Energy storage	In electric field $U = \frac{1}{2}CV^2$	In magnetic field $U = \frac{1}{2}LI^2$
Behavior in low-frequency limit	Open circuit	Short circuit
Behavior in high-frequency limit	Short circuit	Open circuit
Reactance	$X_C = 1/\omega C$	$X_L = \omega L$
Phase	Current leads by 90°	Voltage leads by 90°

APPLICATION	*Loudspeaker Systems*

Loudspeakers in high-quality audio systems invariably contain two or more individual *drivers*—devices for converting electrical energy to sound—within the same enclosure (Fig. 33-8a). The most common form of driver includes a wire coil attached to a diaphragm and suspended in the field of a permanent magnet (Fig. 33-8b). Varying current in the coil then leads to a varying force that drives the diaphragm back and forth to produce sound.

Faithful reproduction of low-frequency sound requires a large driver, called a *woofer*. The woofer is large both because much of the sound power in music lies in the low-frequency range (think of a drum versus a flute!) and because the human ear is much less sensitive to low frequencies (recall Fig. 17-3). But the woofer's large size gives it a large mechanical inertia, so it can't undergo the rapid movements needed to reproduce high-frequency sound. Consequently, a much smaller driver, the *tweeter,* is used for high frequencies.

Power from an amplifier doesn't "know" about the mechanical properties of the drivers, so connecting both drivers directly to the same amplifier would result in low-frequency power being dissipated ineffectually in the tweeter and high-

frequency power in the woofer. To prevent this inefficiency, a speaker system contains a *crossover network* to "steer" power to the appropriate drivers. Figure 33-8c shows a simple crossover network that exploits the frequency-dependent behavior of capacitors and inductors. At low frequencies the inductor reactance ωL is low, and current flows readily through the inductor to the woofer coil. But at high frequencies the inductor reactance is high, and little high-frequency power reaches the woofer. The capacitor's behavior is the opposite: It blocks low-frequency power from reaching the tweeter, while passing high-frequency power. Many speaker systems also employ a *midrange* driver, with a capacitor and inductor in series to block power at both high- and low-frequency extremes. Example 33-4, later in this chapter, explores quantitatively the behavior of the midrange circuitry.

The circuits of Fig. 33-8c are examples of *filters,* widely used in electronic systems to pass preferentially a range of frequencies. For instance, bass and treble controls on audio systems adjust filter properties to boost low or high frequencies as desired. Problem 74 explores a simple filter.

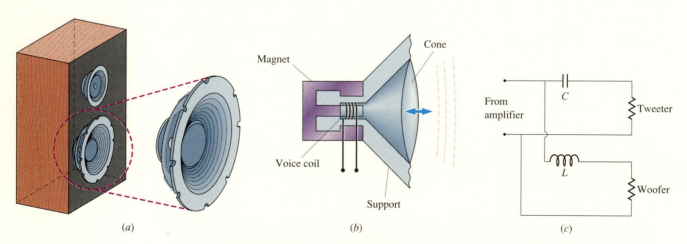

FIGURE 33-8 (*a*) A loudspeaker system, showing the woofer and tweeter drivers for low- and high-frequency sound production. (*b*) Cutaway view of a moving-coil driver. (*c*) The crossover network "steers" power at the appropriate frequencies to the woofer and tweeter. Resistances represent the coils in the two drivers.

33-3 *LC* CIRCUITS

In this section we consider circuits containing both inductors and capacitors. The properties of these circuits reflect directly the complementary nature of the two devices.

LC Oscillations

Figure 33-9 shows a circuit with a capacitor C and inductor L. Suppose the capacitor is initially charged to some voltage V_p and corresponding charge q_p, and is then connected to the inductor. What happens?

Initially, the capacitor is fully charged, while the inductor current is zero. There is electrical energy stored in the capacitor, but none in the inductor. This initial state is shown in Fig. 33-10a. Then the capacitor begins to discharge through the inductor. It cannot do so all at once, because the inductor opposes changes in current. So current in the inductor rises gradually, and with it the magnetic energy stored in the inductor. At the same time the capacitor voltage, charge, and stored electrical energy decrease. Some time later, the initial energy is divided equally between the capacitor and inductor, as in Fig. 33-10b. But the capacitor keeps discharging, eventually reaching zero charge, as in Fig. 33-10c. Now there is no voltage across the capacitor and no stored electrical energy. All the energy that was initially in the electric field of the capacitor is in the magnetic field of the inductor.

Does everything stop at this point? No, because current is flowing in the inductor. Current in an inductor cannot change abruptly, so the current keeps flowing and piles positive charge on the bottom plate of the capacitor (Fig. 33-10d). Stored electrical energy increases as the capacitor charges,

FIGURE 33-9 An *LC* circuit.

FIGURE 33-10 Oscillation in an *LC* circuit, showing energy transfer between electric and magnetic fields.

FIGURE 33-11 An *LC* circuit is the electrical analog of a mass-spring system. The capacitor corresponds to the spring, the inductor to the mass, and current to velocity.

TABLE 33-3 *LC Circuits and Mass-Spring Systems*

LC CIRCUIT	MASS-SPRING
Charge q	Displacement x
Current $I = dq/dt$	Velocity $v = dx/dt$
Inductance L	Mass m
Capacitance C	$1/k$ (k = spring constant)
Magnetic energy $U_B = \frac{1}{2}LI^2$	Kinetic energy $U_K = \frac{1}{2}mv^2$
Electric energy $U_E = \frac{1}{2}(1/C)q^2$	Potential energy $U = \frac{1}{2}kx^2$
Frequency $\omega = 1/\sqrt{LC}$	Frequency $\omega = \sqrt{k/m}$
Resistance	Friction

and the inductor current and stored magnetic energy decrease. Eventually (Fig. 33-10*e*), the capacitor is fully charged in the opposite direction from its initial state. Again all the energy is in the capacitor. Again the capacitor begins to discharge, and the process repeats, now with a counterclockwise current (Fig. 33-10*f*). All the energy is transferred to the inductor (Fig. 33-10*g*), and then back to the capacitor, which again attains its initial state (Fig. 33-10*a*). Provided there is no energy loss, the oscillation repeats indefinitely.

This *LC* oscillation should remind you of the mass-spring system we studied in Chapter 15. There, energy was transferred back and forth between kinetic energy of the mass and potential energy of the spring. Here, energy is transferred back and forth between magnetic energy of the inductor and electrical energy of the capacitor. The mass-spring system oscillates with frequency determined by the mass m and spring constant k. Similarly, the *LC* circuit oscillates with frequency determined by the inductance L and capacitance C. Figure 33-11 shows the corresponding elements of each system and Table 33-3 shows some analogies between mass-spring systems and *LC* circuits. We will develop these analogies more rigorously in the next section.

Analogies with *LC* circuits are so useful that engineers sometimes simulate complicated systems, such as bridges, automobile suspensions, or world energy usage, with networks of *LC* circuits. Such a network is called an analog computer because its behavior is analogous to that of the system under study.

Analyzing the *LC* Circuit

We described the *LC* circuit qualitatively in terms of transfer between electric and magnetic energy. This description suggests a way to analyze the circuit quantitatively. The total energy in the circuit is the sum of the magnetic and electric energy:

$$U = U_B + U_E = \tfrac{1}{2}LI^2 + \tfrac{1}{2}\frac{q^2}{C}.$$

The time derivative of this equation is

$$\frac{dU}{dt} = \frac{d}{dt}\left(\tfrac{1}{2}LI^2 + \tfrac{1}{2}\frac{q^2}{C}\right).$$

But since the total energy in our ideal circuit does not change, $dU/dt = 0$. Carrying out the differentiations in our expression for dU/dt, we then have

$$LI\frac{dI}{dt} + \frac{q}{C}\frac{dq}{dt} = 0. \tag{33-8}$$

Substituting $I = dq/dt$ and $dI/dt = d^2q/dt^2$ gives

$$L\frac{d^2q}{dt^2} + \frac{1}{C}q = 0. \tag{33-9}$$

Equation 33-9 is a differential equation describing the capacitor charge as a function of time. We encountered a similar equation in Chapter 15 when we studied the mass-spring system:

$$m\frac{d^2x}{dt^2} + kx = 0. \tag{15-4}$$

We found that Equation 15-4 could be satisfied by a sinusoidal function of time, with frequency given by

$$\omega = \sqrt{k/m}. \tag{15-11}$$

Equation 33-9 is identical to Equation 15-4 except that q replaces x, L replaces m, and $1/C$ replaces k. Therefore the solution of Equation 33-9 is a sinusoidal oscillation,

$$q(t) = q_p \cos \omega t, \tag{33-10}$$

whose frequency is given by Equation 15-8a with L replacing m and $1/C$ replacing k:

where
$$\omega = \frac{1}{\sqrt{LC}}. \tag{33-11}$$

Here we chose cosine rather than sine since our qualitative description (Fig. 33-10) started with the capacitor charge at its peak value. Differentiating the charge (Equation 33-10) gives the current:

$$I(t) = \frac{dq}{dt} = \frac{d}{dt}(q_p \cos \omega t) = -\omega q_p \sin \omega t. \tag{33-12}$$

All other circuit quantities follow from Equations 33-10 and 33-12. The capacitor voltage, obtained from the relation $q = CV$, is

$$V_C(t) = \frac{q}{C} = \frac{q_p}{C}\cos \omega t.$$

The electrical energy in the capacitor is, therefore,

$$U_E = \tfrac{1}{2}CV^2 = (\tfrac{1}{2}C)\left(\frac{q_p}{C}\cos \omega t\right)^2 = \frac{q_p^2}{2C}\cos^2 \omega t,$$

while the magnetic energy in the inductor is

$$U_B = \tfrac{1}{2}LI^2 = \tfrac{1}{2}L(-\omega q_p \sin \omega t)^2 = \tfrac{1}{2}L\omega^2 q_p^2 \sin^2 \omega t.$$

We can verify that our solution conserves energy by adding the electric and magnetic energies:

$$U_{total} = U_E + U_B = \frac{q_p^2}{2C}\cos^2 \omega t + \tfrac{1}{2}L\omega^2 q_p^2 \sin^2 \omega t.$$

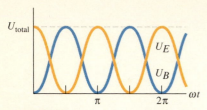

FIGURE 33-12 Electric and magnetic energies in an *LC* circuit. Their sum is constant.

14.2

The *RLC* Oscillator

But Equation 33-11 shows that $\omega^2 = 1/LC$, so we have

$$U_{\text{total}} = \frac{q_p^2}{2C}\cos^2\omega t + \tfrac{1}{2}\frac{L}{LC}q_p^2\sin^2\omega t$$

$$= \frac{q_p^2}{2C}(\cos^2\omega t + \sin^2\omega t) = \frac{q_p^2}{2C},$$

since $\cos^2\omega t + \sin^2\omega t = 1$. Thus the total energy is independent of time and is equal to the initial energy stored in the capacitor. Figure 33-12 is a plot of the electric and magnetic energies as functions of time, showing that the two always sum to a constant.

You can simulate the *LC* circuit we've just analyzed with ActivPhysics Activity 14.2 if you set $R = 0$ in that simulation.

Got It!

You have an *LC* circuit that oscillates at a typical AM radio frequency, namely 1 MHz. If you want to change the capacitor so that the circuit will now oscillate at a typical FM radio frequency of 100 MHz, should you make the capacitance larger or smaller? By what factor?

EXAMPLE 33-3	A Piano Tuner

You wish to make an *LC* circuit oscillate at 440 Hz (the note A above middle C); connected to an amplifier and loudspeaker it will produce a pure tone to be used in tuning pianos. You have available a 2.0-H inductor. What value of capacitance should you use? If you initially charge the capacitor to 5.0 V, what will be the peak charge on the capacitor and the peak current in the circuit?

Solution

The oscillation frequency is given by Equation 33-11. Solving for *C* gives

$$C = \frac{1}{\omega^2 L} = \frac{1}{(2\pi f)^2 L} = \frac{1}{[(2\pi)(440\ \text{Hz})]^2(2.0\ \text{H})} = 65.4\ \text{nF}.$$

The capacitor charge and voltage are related through $C = q/V$, so

$$q_p = CV_p = (65.4\ \text{nF})(5.0\ \text{V}) = 327\ \text{nC}.$$

Equation 33-12 then shows that the peak current is

$$I_p = \omega q_p = 2\pi f q_p = (2\pi)(440\ \text{Hz})(327\ \text{nC}) = 0.90\ \text{mA}.$$

Problem 38 shows how the current also follows from energy considerations.

EXERCISE An FM radio transmitter requires an *LC* circuit that oscillates at the 89.5-MHz transmitter frequency. What should be the inductance if the circuit capacitance is 47 pF?

Answer: 67 nH

Some problems similar to Example 33-3: 27, 28, 31, 32

Resistance in *LC* Circuits—Damping

Real inductors, capacitors, and wires have resistance. Both this intrinsic resistance and any external resistance are represented by the resistor *R* in Fig. 33-13. What happens in such a resistive *LC* circuit?

Provided the resistance is small—small enough that only a small fraction of the energy is lost in one cycle—then our analysis of the preceding section applies. The circuit oscillates at a frequency given very nearly by Equation 33-11. But as current flows back and forth through the resistor, energy is dissipated as heat. On each successive cycle, the total energy decreases. Consequently, the amplitude of the oscillations decreases with time.

FIGURE 33-13 An *RLC* circuit.

We can analyze this *RLC* circuit by starting with Equation 33-8, except that now energy is being lost at the rate I^2R (the power dissipation in a resistor; recall Equation 27-9a). So instead of zero, the right side of Equation 33-8 becomes $-I^2R$, with the minus sign indicating energy loss from the circuit:

$$LI\frac{dI}{dt} + \frac{q}{C}\frac{dq}{dt} = -I^2R.$$

Writing $I = dq/dt$ as we did in the preceding section leads to

$$L\frac{d^2q}{dt^2} + R\frac{dq}{dt} + \frac{q}{C} = 0.$$

This equation is mathematically identical to Equation 15-19 for damped simple harmonic motion, showing that our analogies of Table 33-3 continue to hold when resistance is present. Using Equation 15-20, which is the solution to Equation 15-19, and the appropriate analogies from Table 33-3, we can construct the solution for our decaying *RLC* circuit:

$$q(t) = q_p e^{-Rt/2L}\cos\omega t. \tag{33-13}$$

Other quantities show similar behavior, with oscillation amplitude decaying exponentially with time constant $2L/R$. Figure 33-14 shows an oscilloscope trace of the capacitor voltage in a circuit undergoing damped oscillations.

As the electrical resistance increases, the oscillations decay more rapidly and the frequency of oscillation decreases. Finally, when the exponential time constant $2L/R$ equals the inverse of the natural frequency given by Equation 33-11, much of the energy is lost in the time of one undamped oscillation period. This situation is termed **critical damping**, and at this value of R circuit quantities decay exponentially to zero, in analogy with a critically damped mechanical system (Section 15-6). For greater values of R, the circuit is **overdamped** and also exhibits no oscillation. In circuits that are designed to oscillate, such as radio transmitters or TV tuners, engineers obviously want to minimize damping. But in other situations, circuit oscillations would be a nuisance, and then it's important to introduce enough resistance to suppress oscillation. You can simulate the *RLC* circuit with ActivPhysics Activity 14.2.

FIGURE 33-14 An oscilloscope displays the capacitor voltage in an *RLC* circuit. Note the exponential decline in amplitude of the oscillations.

33-4 DRIVEN *RLC* CIRCUITS AND RESONANCE

Figure 33-15 shows an *RLC* circuit connected across an AC generator. The *RLC* circuit itself is analogous to a mass-spring system, so adding the generator is like adding the external driving force on the mechanical oscillator that we considered in Section 15-7. We'll call the generator frequency ω_d, the driving frequency, just as we did in Chapter 15. Pursuing the mechanical analogy, we expect the driven *RLC* circuit to exhibit resonant behavior as we discussed in Section 15-7. Such electrical resonance is crucial to the operation of radio, television, and other frequency-specific devices.

FIGURE 33-15 A series *RLC* circuit driven by an AC generator.

Resonance in the *RLC* Circuit

Suppose we vary the generator frequency ω_d in Fig. 33-15 while keeping the generator's peak voltage constant. How much current flows in the *RLC* circuit? At low frequencies the capacitor acts almost like an open circuit (its reactance $X_C = 1/\omega C$ is large), so little current flows. At high frequencies the inductor acts almost like an open circuit (its reactance $X_L = \omega L$ is large), so little current

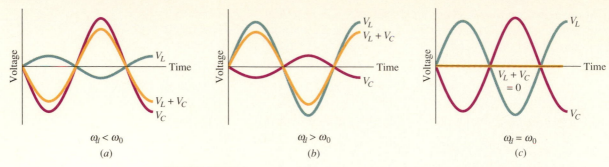

FIGURE 33-16 In a series RLC circuit, capacitor and inductor voltages are 180° out of phase at all frequencies. (*a*) At low frequencies, the voltage is greatest across the capacitor. (*b*) At high frequencies, the voltage is greatest across the inductor. (*c*) At resonance, capacitor and inductor voltages have equal amplitude, and therefore they cancel completely.

flows. At some intermediate frequency the current must be a maximum. We now show that this **resonant frequency** is in fact the undamped natural frequency $\omega_0 = 1/\sqrt{LC}$.

Figure 33-15 is a series circuit, so the *same* current flows through all the components. We know that the voltage in a capacitor lags the current by 90°, while the voltage in an inductor leads by 90°. Since the same current flows through the inductor and capacitor, the inductor and capacitor voltages are therefore 180° out of phase. This phase relation holds at *any* frequency. At low frequencies, however, the capacitor's reactance is greatest and the capacitor voltage is therefore greater than the inductor voltage (Fig. 33-16*a*). At high frequencies the opposite is true, with the inductor voltage being greatest (Fig. 33-16*b*). Since the two voltages are 180° out of phase, they tend to cancel, but at high or low frequencies Figs. 33-16*a* and *b* show that their different amplitudes mean this cancellation is not complete.

Is there some frequency, ω_0, at which the capacitor and inductor voltages exactly cancel? The peak current and voltage in these two components are related by Equations 33-5 and 33-7, respectively. Since the current is the same, comparison of these equations shows that the voltages will be the same when the capacitive reactance $1/\omega C$ is equal to the inductive reactance ωL, which gives

$$\omega_0^2 = \frac{1}{LC}. \quad \text{(resonant frequency)}$$

This is precisely Equation 33-11's condition for the undamped natural frequency.

At resonance, then, the capacitor and inductor voltages completely cancel. The voltage across the pair together is zero, and—at the resonant frequency only—the pair behave just like a wire. The circuit current at resonance is determined entirely by the resistance. At any other frequency the effects of capacitance and inductance do not cancel completely, and the current is lower.

14.3
The Driven Oscillator

Got It!

You measure the capacitor and inductor voltages in a driven RLC circuit, and find 10 V for the rms capacitor voltage and 15 V for the rms inductor voltage. Is the driving frequency above or below resonance?

Frequency Response of the *RLC* Circuit

Here we derive a general expression for the current as a function of frequency in the series *RLC* circuit, using the phasor diagrams introduced in Section 33-2. Since the same current flows through all components of the series circuit, we represent this current by a single phasor of length I_p in Fig. 33-17. Also shown are phasors for the resistor, capacitor, and inductor voltages. The resistor voltage phasor is in the same direction as the current because these two are in phase, but the capacitor and inductor voltages are at 90° relative to the current.

In analyzing circuits, we've often used the fact that the sum of the voltage differences around a circuit is zero. This statement is true *at each instant*, meaning that it applies to *instantaneous* values of the voltages. Because they have different phases, the voltages in the circuit of Fig. 33-15 peak at different times—so we can't simply sum the peak voltages around the circuit and set the result to zero. Instead, we have to sum the instantaneous voltages. We can do that by adding the voltage phasors *vectorially*—a process that accounts for both magnitude and phase. Figure 33-17*b* shows that this vector addition gives the following relation between the peak generator voltage, V_p, and the voltages across the individual components:

$$V_p^2 = V_{Rp}^2 + (V_{Lp} - V_{Cp})^2 .$$

Expressing this in terms of the current and the resistances and reactances gives

$$V_p^2 = I_p^2 R^2 + (I_p X_L - I_p X_C)^2 ,$$

or

$$I_p = \frac{V_p}{\sqrt{R^2 + (X_L - X_C)^2}} = \frac{V_p}{Z}, \qquad (33\text{-}14)$$

where we have defined

$$Z = \sqrt{R^2 + (X_L - X_C)^2} .$$

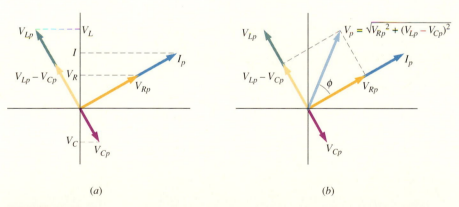

(a)   (b)

FIGURE 33-17 Phasor diagrams for the driven *RLC* circuit. As usual, the length of each phasor indicates the peak value of voltage or current, while the projection on the vertical axis is the instantaneous value. As the phasors rotate with angular frequency ω_d, the instantaneous values vary sinusoidally with time. (*a*) The current in the three circuit elements is the same, and is in phase with the resistor voltage. The voltages differ, and are out of phase with each other. (*b*) Voltage phasors for the inductor, capacitor, and resistor sum vectorially to give the generator voltage *V*. In this case the inductor voltage exceeds the capacitor voltage, so the driving voltage *V* leads the current. The phase difference ϕ between *V* and *I* is given by Equation 33-16.

Equation 33-14 has the form of Ohm's law, with Z playing the role of resistance. We call Z the **impedance** of the circuit. Impedance is a generalization of resistance to include the frequency-dependent effects of capacitance and inductance. Putting in our expressions for the reactances gives

$$Z = \sqrt{R^2 + \left(\omega L - \frac{1}{\omega C}\right)^2}. \tag{33-15}$$

In agreement with our earlier analysis, this equation shows that the circuit impedance becomes very large at high and low frequencies, and has its lowest value, R, at resonance.

Figure 33-18 is a plot of Equation 33-14, showing peak current versus frequency for several values of resistance. As we lower the resistance, the peak current at resonance rises. Although the current at other frequencies rises, too, it does so to a much lower extent than at resonance. This is because the impedance at resonance depends only on the resistance, but includes reactive effects at other frequencies. As a result, the resonance curve becomes more sharply peaked as the resistance drops. For a circuit with very low resistance, the current at resonance is dramatically different from that at even a slightly different frequency. Such a circuit, called a **high-Q** (for high-quality) circuit, does a good job of distinguishing its resonant frequency from nearby frequencies. A rigorous definition of Q can be given in terms of the width of the resonance curve (see Problem 81). High-Q circuits are important in radio and TV transmitters, where they help establish the precise broadcast frequency. At the receiving end, when you tune a simple radio, you're adjusting the variable capacitor in a high-Q LC circuit that selects the desired station from among the hodgepodge of radio signals reaching the antenna (Fig. 33-19).

Equation 33-14 relates the peak current in the RLC circuit and the voltage across it, but that equation doesn't tell the whole story. In general, current and voltage are not in phase. Figure 33-17b shows that the phase difference ϕ between the driving voltage and the circuit current is

$$\tan \phi = \frac{V_{Lp} - V_{Cp}}{V_{Rp}}.$$

Because the voltages are proportional to the reactances and resistance, this expression may be written

$$\tan \phi = \frac{X_L - X_C}{R} = \frac{\omega L - 1/\omega C}{R}. \tag{33-16}$$

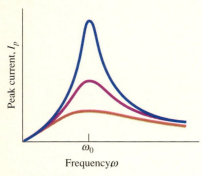

FIGURE 33-18 Resonance curves for an RLC circuit with three different resistance values. Upper curve is for half the resistance, and lower curve for twice the resistance, of the middle curve.

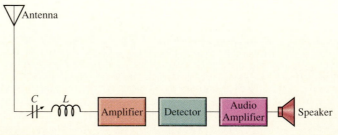

FIGURE 33-19 Diagram of a simple radio, using an LC circuit with a variable capacitor to select the desired station frequency. The signal is subsequently amplified, the audio information is extracted in the detector, and the amplified audio signal drives the loudspeaker. More sophisticated receivers often use digital circuitry in place of the LC frequency selector.

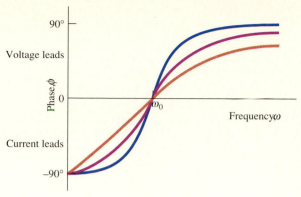

FIGURE 33-20 Phase relations for the *RLC* circuits whose resonance curves are shown in Fig. 33-18.

Does this equation make sense? At resonance, $X_L = X_C$ and the phase difference is zero. Here the effects of capacitance and inductance cancel, and the circuit behaves like a pure resistance. At frequencies below resonance, capacitive reactance is greatest and the phase difference is negative, indicating that current leads voltage—as we expect in a circuit dominated by capacitance. Above resonance $X_L > X_C$ and the phase difference is positive, indicating that voltage leads current—as we expect in a circuit dominated by inductance. At high- and low-frequency extremes, $\tan\phi$ becomes arbitrarily large and the phase differences approach 90°, as shown in Fig. 33-20.

We've defined the phase difference ϕ in Equation 33-16 as negative when the current leads. This means that if we describe voltage and current with Equation 33-3, then the arbitrary choice $\phi_V = 0$ requires $\phi_I = -\phi$. Conversely, if we choose $\phi_I = 0$, then $\phi_V = +\phi$.

Tip

Phase Matters You can't analyze an AC circuit by treating resistors, capacitors, and inductors all as "resistors" with resistances R, X_C, and X_L. Why not? Because associated with each component is a different phase relation between current and voltage. Only a phasor diagram—or mathematical analysis that includes phase relations—correctly characterizes the circuit. The phase relations in a series *RLC* circuit show up in the minus sign joining the capacitive and inductive reactance, and in the Pythagorean-like addition of the resistance and reactances in Equations 33-14 and 33-15.

EXAMPLE 33-4	*Designing a Loudspeaker System*

Current flows to the midrange speaker in a loudspeaker system through a 2.2-mH inductor and a capacitor, in series. (See *Application: Loudspeaker Systems,* earlier in this chapter.) (a) What should be the capacitance in order that a given applied voltage produces the maximum current at a frequency of 1.0 kHz? (b) What should be the speaker's resistance in order that the same voltage produces a current with half the maximum value at 600 Hz? (c) If the peak output voltage of the amplifier is 20 V, what should be the capacitor's peak voltage rating?

Solution

(a) Inductor, capacitor, and resistor comprise a series *RLC* circuit, with resonant frequency $f_0 = \omega_0/2\pi = 1/2\pi\sqrt{LC}$. Solving for C gives

$$C = \frac{1}{4\pi^2 f_0^2 L} = \frac{1}{(4\pi^2)(1.0\times10^3 \text{ Hz})^2(2.2\times10^{-3} \text{ H})} = 11.5 \ \mu\text{F}.$$

(b) Equation 33-14 gives the current in an *RLC* circuit. The denominator in this equation—the impedance $Z = \sqrt{R^2 + (X_L - X_C)^2}$—has the value R at resonance, when

$X_L = X_C$. Thus the current will have half its maximum value where the $Z = 2R$. Using $X_L = \omega L$ and $X_C = 1/\omega C$, we want R such that

$$Z = \sqrt{R^2 + \left(\omega L - \frac{1}{\omega C}\right)^2} = 2R$$

when $f = \omega/2\pi = 600$ Hz, or $\omega = 3.77 \times 10^3$ s^{-1}. Squaring and solving for R gives

$$R = \frac{1}{\sqrt{3}}\left|\omega L - \frac{1}{\omega C}\right| = 8.53\ \Omega,$$

where we used the 11.5-μF answer to part (a) for the capacitance C. This resistance value is typical for a loudspeaker. The exercise below shows that the current also has half its maximum value at 1.7 kHz, so the midrange speaker gets at least half its maximum current in the range from about 600 to 1700 Hz.

(c) The maximum current flows at the resonant frequency, so the Ohm's-law-like Equation 33-5 shows that the maximum capacitor voltage also occurs at this frequency and is given by

$$V_{Cp} = I_p X_C = \left(\frac{V_p}{R}\right)\left(\frac{1}{\omega C}\right)$$

$$= \left(\frac{20\ \text{V}}{8.53\ \Omega}\right)\left(\frac{1}{(2\pi)(1000\ \text{Hz})(11.5 \times 10^{-6}\ \text{F})}\right)$$

$$= 32\ \text{V}.$$

How can this value be *greater* than the peak applied voltage? Remember that there's another source of emf in the circuit—the inductor, whose emf depends on the rate of change of current and may therefore exceed the applied voltage. In this relatively low-Q circuit the peak capacitor voltage is not too much greater than the peak applied voltage, but in high-Q circuits like those used in radio transmitters, capacitors may have to withstand voltages hundreds of times those applied to the circuit.

EXERCISE Find a second frequency where the speaker current in Example 33-4 has half its maximum value. *Hint:* Square Equation 33-15, isolate the frequency-dependent terms, take a square root, and solve a quadratic equation. Or use graphical techniques.

Answer: 1.7 kHz

Some problems similar to Example 33-4: 43, 44, 46, 69, 70, 76

33-5 POWER IN AC CIRCUITS

In Section 33-2, we noted that average power dissipation is zero in a circuit containing only a capacitor or an inductor. We can understand this physically because the reactive element alternately stores and releases energy rather than dissipating it as heat. Mathematically, we can see this from Fig. 33-21a, which

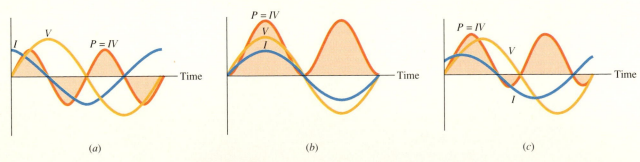

(a)   (b)   (c)

FIGURE 33-21 Current I, voltage V, and power consumption IV. (a) Current and voltage in a capacitor are out of phase by 90°. The power is alternately positive (absorbing energy from the source) and negative (returning energy to the source), and the net energy consumption (shaded area) over one cycle is zero. (b) Current and voltage are in phase in a resistor and therefore the power is always positive, meaning that the resistor is a net energy consumer. (c) Average power consumption decreases as current and voltage go out of phase.

shows the current, voltage, and instantaneous power in a capacitor. The power is the product of the current and voltage. Because these two are out of phase, the power is positive half the time, and negative half the time. When the power is positive, the capacitor absorbs energy from the source that drives the current. When the power is negative, the capacitor returns energy to the driving source. The net energy transferred to the capacitor over one cycle is $\int P\,dt$, or the area under the power-versus-time curve, and is zero in this case. Figure 33-21b, in contrast, shows current, voltage, and instantaneous power in a resistor. Since current and voltage are always in phase, the power is always positive, and the resistor always takes energy from the source. Comparison of Figs. 33-21a and b suggests that the phase difference between current and voltage is important in determining the average power consumption of an AC circuit. We can see this more clearly if we imagine slipping the current and voltage just slightly out of phase, as in Fig. 33-21c. Then there are narrow regions where the power is negative, so the average power over one cycle is slightly less than in the resistor case. As the phase difference increases, so does the time that the power is negative, until at 90° phase difference, the time-average power is zero.

We can develop a general expression for power in AC circuits by considering the time-average product of voltage and current with arbitrary phase difference ϕ:

$$\langle P \rangle = \langle [I_p \sin(\omega t + \phi)][V_p \sin \omega t] \rangle,$$

where $\langle \, \rangle$ indicates a time average over one cycle. Expanding the current term using a trig identity (see Appendix A) gives

$$\langle P \rangle = I_p V_p \langle (\sin^2 \omega t)(\cos \phi) + (\sin \omega t)(\cos \omega t)(\sin \phi) \rangle.$$

The average of $(\sin \omega t)(\cos \omega t)$ is zero, as we've just shown for two signals 90° out of phase. The quantity $\sin^2 \omega t$ swings from 0 to 1, and is symmetric about $\frac{1}{2}$, so its average value is $\frac{1}{2}$ recall (recall Fig. 16-16). Then we have

$$\langle P \rangle = \tfrac{1}{2} I_p V_p \cos \phi .$$

Writing the peak values as $\sqrt{2}$ times the rms values gives

$$\langle P \rangle = \tfrac{1}{2}\sqrt{2} I_{rms} \sqrt{2} V_{rms} \cos \phi = I_{rms} V_{rms} \cos \phi . \qquad (33\text{-}17)$$

This equation confirms our earlier graphical arguments. When the voltage and current are in phase, the average power is just the product $I_{rms} V_{rms}$. (This, in fact, is a principal reason for using rms values: with them, the expression for average power is the same as in the DC case.) But with current and voltage out of phase, the average power is smaller; at 90° phase difference it is zero.

The factor $\cos \phi$ is called the **power factor**. A purely resistive circuit has a power factor of 1, while a circuit with only inductance and capacitance has a power factor of zero. In circuits containing resistance along with capacitance and/or inductance, the power factor generally depends on frequency; in the series RLC circuit, for example, it is 1 at resonance but lower at other frequencies.

Equation 33-17 shows that delivering a given average power at a fixed voltage requires greater current when the power factor is less than 1. Because energy is dissipated at the rate I^2R in the resistance of transmission lines, motor coils, and the like, electrical loads with low power factors result in wasted energy. For that reason, electric power engineers often add capacitance to inductive devices such as motors, bringing the power factor closer to 1 and thus lowering the current and therefore the resistive energy losses.

33-6 TRANSFORMERS AND POWER SUPPLIES

A **transformer** is a pair of wire coils in close proximity. We analyzed such a coil pair in Section 32-1, and showed how a change in current in one coil— which we'll call the **primary**—results in a changing magnetic flux and therefore, by Faraday's law, an emf in the other coil, or **secondary**. This emf can drive a current in circuitry connected to the secondary coil. Thus, the transformer transfers electrical energy from the primary circuit to the secondary circuit, even though there is no direct electrical connection between the two.

Practical transformers are often wound on iron cores that concentrate flux, ensuring that essentially all the magnetic flux produced by the primary coil goes through the secondary. Figure 33-22 shows a simplified diagram of a transformer, along with its circuit symbol.

The transformer in Fig. 33-22a has two turns in its primary and four in its secondary. Since the same changing flux passes through each turn of each coil, the total emf induced in the secondary must be twice that of the primary. This transformer is therefore a **step-up transformer**; it steps up the voltage by a factor of 2. Interchanging primary and secondary would give a **step-down** transformer, with the secondary voltage half that of the primary. In general, the ratio of the peak (or rms) primary voltage V_1 to the peak (or rms) secondary voltage V_2 is the same as the ratio of the numbers of turns in the two coils:

FIGURE 33-22 (a) A transformer, consisting of two coils wound on an iron core. (b) Transformer circuit symbol.

$$V_2 = \frac{N_2}{N_1}V_1 . \tag{33-18}$$

Aren't we getting something for nothing with a step-up transformer? No— a step-up transformer increases voltage but not power. In an ideal transformer, all the power supplied to the primary is transferred to the secondary; therefore,

$$I_1 V_1 = I_2 V_2 . \tag{33-19}$$

If the voltage goes up, the current goes down, and vice versa. Real transformers have losses associated with resistance in their windings and heating of their iron cores, but good engineering holds these losses to a few per cent of the total power.

Transformers are widely used to provide voltages different from those of available sources. Most electronic circuits, for example, require a few tens of volts or less; step-down transformers bring the 120-V AC power line down to this level. Television tubes and computer monitors require tens of kV, calling for step-up transformers.

Transformers work only with alternating current, since they make use of electromagnetic induction and therefore require *changing* current for their operation. With the sinusoidally varying currents in AC power systems, the rate of change of current is also a sinusoidal function (the derivative of sine is cosine), so both the primary and secondary currents are sinusoidally varying AC. One of the main reasons for the near-universal use of AC power is the ease with which it can be transformed from one voltage level to another. To use transformers with DC—from a battery, for example—requires that the current be interrupted periodically. The car ignition coil of Example 32-2 is, in fact, a step-up transformer that works only because the car's distributor regularly interrupts the current to the coil's primary.

APPLICATION	*Electric Power Distribution*

The power loss in a resistance R carrying current I is given by I^2R. That means it's most efficient to transmit electric power at high voltage and low current. Lower voltages, on the other hand, are safer and easier to handle. How can we satisfy both these considerations in practical power systems? With transformers!

Power-plant generators operate at about 20 kV. Transformers at the power plant then step this up to several hundred kilovolts for long-distance transmission. At a city or town the voltage is dropped to several kilovolts for distribution within the municipality. Transformers reduce the voltage further for use in individual homes and businesses. The typical household voltage levels of 120 V and 240 V represent a compromise between the desire to minimize shock hazards—best met with low voltages—and the need to keep currents low. Low current allows the use of thinner wiring and also minimizes the danger of fire—but requires higher voltage to deliver a given power. Individual electrical devices within a building use transformers to meet their particular voltage requirements. Increasingly, small electrical devices such as modems, answering machines, portable stereos, and the like use external transformers that plug right into the wall; the power cord to the device then carries only low-voltage power. Figure 33-23 outlines the voltage transformations in power transmission.

FIGURE 33-23 Transformers are used throughout the power distribution network.

FIGURE 33-24 Circuit symbol for a diode, with preferred direction of current indicated.

(a)   (b)

FIGURE 33-25 (a) A simple DC power supply consists of a transformer and diode, which supply current to the load resistance. (b) The diode passes current in only one direction, cutting off the negative half of each cycle (dashed gray) and giving the load current shown in blue.

DC Power Supplies

Devices such as light bulbs and electric heaters work equally well on AC or DC. But others, especially electronic circuits, require DC power. How do we get DC from the AC power line?

In Chapter 27's discussion of semiconductors, we showed how a junction between *P*-type and *N*-type semiconductors passes current in one direction but not the other. A **diode** is a *PN* junction designed specifically to be such a "one-way valve" for electric current. An ideal diode acts like a short circuit to current flowing in the preferred direction and like an open circuit in the opposite direction (Fig. 33-24).

Figure 33-25*a* shows a diode and resistor connected to the secondary of a transformer. The resistor represents the load—that is, whatever circuit we wish to supply with DC power. The transformer brings the 120-V AC power line to a level appropriate to the load. The diode passes current in only one direction, resulting in the load current shown in Fig. 33-25*b*.

Although the load current in Fig. 33-25 flows in one direction, it still varies drastically with time. If the load were a stereo amplifier, for example, its speakers would emit a loud hum. So we smooth, or **filter**, the output of the diode. The simplest filter is a capacitor, as shown in Fig. 33-26*a*. As the voltage on the left side of the diode rises, the capacitor voltage rises rapidly because of the short time constant associated with the low resistance of the diode in its preferred direction. But then the AC voltage begins to fall, and the diode "turns off." The capacitor cannot discharge from right to left through the diode but only through

(a)   (b)

FIGURE 33-26 (a) Adding a capacitor reduces variations in the load voltage and therefore in the load current. This circuit is a complete DC power supply. (b) Load voltage as a function of time.

the load resistance. If the time constant RC is long enough, the capacitor voltage will drop only slightly before the AC voltage again rises and sends a new surge of current through the diode to bring the capacitor to its maximum charge (Fig. 33-26b). By making the capacitance large enough—so large that the time constant is much longer than the period of the AC power—we can make load current and voltage arbitrarily smooth. Large capacitors are expensive, so in practice additional filters—often using transistors or integrated circuits—are added to the simple capacitive filter in critical applications. High-quality power supplies for audio equipment or electronic instrumentation achieve so-called ripple factors—the ratio of the fluctuation amplitude to the DC level—of 10^{-5} or better. More complicated diode circuits can also be used to produce signals requiring less filtering (see Problem 80).

CHAPTER SYNOPSIS

Summary

1. **Alternating current** varies with time. A sinusoidal AC signal is characterized by its peak amplitude, frequency, and phase constant; for example:

$$I = I_p \sin(\omega t + \phi).$$

The root-mean-square (rms) amplitude of a sinusoidal AC signal is the peak value divided by $\sqrt{2}$; here $I_{rms} = I_p/\sqrt{2}$.

2. In a resistor, the ratio of voltage to current is always constant, so

$$I_p = \frac{V_p}{R},$$

and the current and voltage are in phase.

3. In a capacitor the ratio of peak voltage and current is determined by the **capacitive reactance**:

$$I_p = \frac{V_p}{X_C}, \quad \text{where} \quad X_C = \frac{1}{\omega C}.$$

The current in the capacitor **leads** the voltage by 90°.

4. In an inductor the ratio of peak voltage and current is determined by the **inductive reactance**:

$$I_p = \frac{V_p}{X_L}, \quad \text{where} \quad X_L = \omega L.$$

The current in the inductor **lags** the voltage by 90°.

5. **Phasors** are vector-like arrows showing the amplitude and phase of AC signals, and are useful in analyzing AC circuits.

6. In an undriven LC circuit, energy oscillates between electric and magnetic forms at the **resonant frequency** $\omega_0 = 1/\sqrt{LC}$. The amplitude of the oscillation decays exponentially as energy is dissipated in the circuit resistance.

7. In a series RLC circuit, the effects of inductance and capacitance exactly cancel at the resonant frequency. At this frequency the circuit exhibits the minimum **impedance**, and therefore passes the maximum current. At resonance the current and voltage are in phase. At lower frequencies the capacitor dominates and current leads voltage, while at higher frequencies the inductor dominates and voltage leads current. The impedance of a series RLC circuit is

$$Z = \sqrt{R^2 + (X_L - X_C)^2},$$

while the phase difference between current and voltage is given by

$$\tan \phi = \frac{X_L - X_C}{R}.$$

8. The power dissipated in an AC circuit depends on the relative effects of resistance and reactance. In a purely reactive circuit, current and voltage are 90° out of phase, and no power is dissipated. In a purely resistive circuit, the average power dissipation is $I_{rms}V_{rms}$. When both resistance and reactance are present, the power dissipation depends on the **power factor**, $\cos \phi$, where ϕ is the phase difference between current and voltage. In general, the time-average power consumed in an AC circuit is

$$\langle P \rangle = I_{rms} V_{rms} \cos \phi.$$

9. A **transformer** uses electromagnetic induction to transfer electric power between two circuits. The ratio of the peak or rms voltages in the transformer's two windings is the same as the ratio of the numbers of turns in the windings:

$$V_2 = \frac{N_2}{N_1} V_1,$$

while the power VI is the same in each winding of an ideal transformer.

10. DC **power supplies** use **diodes** to change alternating to direct power. Capacitors then smooth out the remaining time variation to produce the steady voltages and currents needed to power electronic equipment.

Terms You Should Understand

(Pairs are closely related terms whose distinction is important; number in parentheses is chapter section where term first appears.)

AC, DC (introduction)
peak, rms amplitudes (33-1)
frequency, angular frequency (33-1)
capacitive reactance, inductive reactance (33-2)
phasor (33-2)
resonant frequency (33-4)
impedance (33-4)
power factor (33-5)
transformer (33-6)
diode (33-6)

Symbols You Should Recognize

I_p, V_p; I_{rms}, V_{rms} (33-1) f, ω (33-1)
ϕ (33-1) X_C, X_L (33-2)
Z (33-4)

Problems You Should Be Able to Solve

calculating current and voltage in AC circuits involving individual resistors, capacitors, and inductors (33-2)
analyzing oscillating LC circuits (33-3)
analyzing driven RLC circuits (33-4)
calculating power in AC circuits (33-5)
designing simple power supplies using transformers, diodes, and capacitive filters (33-6)

Limitations to Keep in Mind

This chapter considers only sinusoidally varying AC signals. More complex time variations must first be analyzed as sums of sinusoidal terms, using techniques mentioned in Chapter 16.

Care must be taken in adding AC signals with different phases. Phasor diagrams provide a convenient way of doing this.

QUESTIONS

1. Two AC signals have the same amplitude but different frequencies. Are their rms amplitudes the same?
2. Does it make sense to talk about the phase difference between two AC signals of different frequencies? Sketch a diagram to confirm your answer.
3. What is meant by the statement "a capacitor is a DC open circuit"?
4. How can current keep flowing in an AC circuit containing a capacitor? After all, a capacitor contains a gap between two conductors, and no charge can cross this gap.
5. Why does it make sense that inductive reactance increases with frequency?
6. The same AC voltage appears across a capacitor and a resistor, and the same rms current flows in each. Is the power dissipation the same in each?
7. When a particular inductor and capacitor are connected across the same AC voltage, the current in the inductor is larger than in the capacitor. Will this be true at all frequencies?
8. An inductor and capacitor are connected in series across an AC generator, and the rms voltage across the inductor is found to be larger than that across the capacitor. Is the generator frequency above or below resonance?
9. When the capacitor voltage in an undriven LC circuit reaches zero, why don't the oscillations stop?
10. Why is the quantity ωL not called the resistance of an inductor?
11. Why is Equation 33-5 not a full description of the relation between voltage and current in a capacitor?
12. If you double both the capacitance and inductance in an LC circuit, what effect does this have on the resonant frequency?
13. In a series RLC circuit, the applied voltage lags the current. Is the frequency above or below resonance?
14. In a series RLC circuit, the applied voltage leads the current. Is the peak voltage greater across the capacitor or the inductor?
15. What does it mean to say that the capacitor in an RLC circuit dominates at low frequencies?
16. The voltage across two circuit elements in series is zero. Is it possible that the voltages across the individual elements are nevertheless not zero? Give an example.
17. If you measure the rms voltages across the resistor, capacitor, and inductor in a series RLC circuit, will they add to the rms value of the generator voltage? Reconcile your answer with the loop law. (See also Problem 47.)
18. In a fluorescent light fixture an inductor, called the ballast, is used to limit current to the lamp. Why is an inductor preferable to a resistor?
19. What is the power factor in a circuit containing only a resistor? Does this power factor change with frequency?
20. To save electrical energy, should you strive for a large or small power factor?
21. When an AC motor runs with no significant mechanical load, its power factor is very nearly zero. What must happen when the motor begins doing mechanical work?
22. A step-up transformer increases voltage, or energy per unit charge. How is this possible without violating conservation of energy?
23. A battery charger runs off the 120-V AC power line. It supplies up to 30 A to recharge a 12-V car battery, yet it can be plugged into a 15-A circuit without blowing the circuit breaker. How is this possible?

PROBLEMS

ActivPhysics can help with these problems:
Activities 14.2, 14.3

Section 33-1 Alternating Current

1. Much of Europe uses AC power at 230 V rms and 50 Hz. Express this AC voltage in the form of Equation 33-3, taking $\phi = 0$.

2. An rms voltmeter connected across the filament of a TV picture tube reads 6.3 V. What is the peak voltage across the filament?

3. An oscilloscope displays a sinusoidal signal whose peak-to-peak voltage (see Fig. 33-1) is 28 V. What is the rms voltage?

4. An industrial electric motor runs at 208 V rms and 400 Hz. What are (a) the peak voltage and (b) the angular frequency?

5. An AC current is given by $I = 495\sin(9.43t)$, with I in milliamperes and t in milliseconds. Find (a) the rms current and (b) the frequency in Hz.

6. What are the phase constants for each of the signals shown in Fig. 33-27?

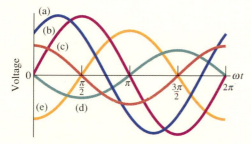

FIGURE 33-27 Problem 6.

7. The rms amplitude is defined as the square root of the average of the square of the signal. For a periodic function the time average is the integral over one period, divided by the period. For a sinusoidal voltage given by $V = V_p \sin \omega t$, show explicitly that $V_{rms} = V_p/\sqrt{2}$.

8. How are the rms and peak voltages related for the square wave shown in Fig. 33-28? See Problem 7.

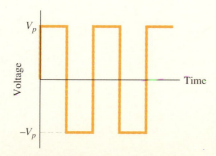

FIGURE 33-28 A square wave (Problem 8).

9. How are the rms and peak voltages related for the triangle wave in Fig. 33-29? See Problem 7.

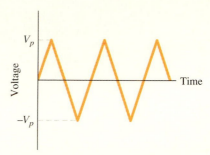

FIGURE 33-29 A triangle wave (Problem 9).

10. A 10-V rms AC signal with frequency 1.0 kHz is 90° ahead of another signal of the same frequency but with peak voltage 7.1 V. On the same graph, plot voltage versus time for both these signals. Choose scales appropriate to display one full cycle.

11. The most general expression for a sinusoidal AC current may be written either $I = I_1 \sin \omega t + I_2 \cos \omega t$ or $I = I_p \sin(\omega t + \phi)$. Find relations between I_1, I_2, I_p, and ϕ that make these expressions equivalent. (See Appendix A for trig identities.)

Section 33-2 Circuit Elements in AC Circuits

12. Show that the unit of both capacitive and inductive reactance is the ohm.

13. What is the rms current in a $1.0\text{-}\mu\text{F}$ capacitor connected across the 120-V rms, 60-Hz AC line?

14. A 470-Ω resistor, 10-μF capacitor, and 750-mH inductor are each connected across 6.3-V rms, 60-Hz AC power sources. Find the rms current in each.

15. Find the reactance of a 3.3-μF capacitor at (a) 60 Hz, (b) 1.0 kHz, and (c) 20 kHz.

16. A 15-μF capacitor carries an rms current of 1.4 A. What is the minimum safe voltage rating for the capacitor if the frequency is (a) 60 Hz or (b) 1.0 kHz?

17. A capacitor and a 1.8-kΩ resistor pass the same current when each is separately connected across a 60-Hz power line. What is the capacitance?

18. (a) A 2.2-H inductor is connected across the 120-V rms, 60-Hz power line. What is the rms inductor current? (b) Repeat if the same inductor is connected across the 230-V rms, 50-Hz power line commonly used in Europe.

19. A 50-mH inductor is connected across a 10-V rms AC generator, and an rms current of 2.0 mA flows. What is the generator frequency?

20. A 2.0-μF capacitor has a capacitive reactance of 1.0 kΩ. (a) What is the frequency of the applied voltage? (b) What inductance would give the same value for inductive reactance at this frequency? (c) How would the two reactances compare if the frequency were doubled?

21. A 1.2-μF capacitor is connected across a generator whose output is given by $V = V_p \sin 2\pi ft$, where $V_p = 22$ V, $f = 60$ Hz, and t is in seconds. (a) What is the peak current? (b) What are the magnitudes of the voltage and (c) the current at $t = 6.5$ ms?

22. A voltage $V = V_p \sin \omega t$ is applied across a capacitor. What is the minimum frequency for which the current will be zero at time $t = 20$ μs?

23. What is the maximum charge on the plates of a 16-μF capacitor connected across the 120-V rms, 60-Hz AC power line?

24. At 10 kHz an inductor has 10 times the reactance of a capacitor. At what frequency will their reactances be equal?

25. A 0.75-H inductor is in series with a fluorescent lamp, and the series combination is across the 120-V rms, 60-Hz power line. If the rms inductor voltage is 90 V, what is the rms lamp current?

26. A 2.2-nF capacitor and a capacitor of unknown capacitance are connected in parallel across a 10-V rms sine-wave generator. At 1.0 kHz, the generator supplies a total current of 3.4 mA rms. The generator frequency is then decreased until the rms current has dropped to 1.2 mA. Find (a) the unknown capacitance and (b) the lower frequency.

Section 33-3 *LC* Circuits

27. Find the resonant frequency of an *LC* circuit consisting of a 0.22-μF capacitor and a 1.7-mH inductor.

28. An *LC* circuit with $C = 18$ mF undergoes *LC* oscillations with period 2.4 s. What is the inductance?

29. You have a 2.0-mH inductor and wish to make an *LC* circuit whose resonant frequency spans the AM radio band (550 kHz to 1600 kHz). What range of capacitance should your variable capacitor cover?

30. The FM radio band covers the frequency range from 88 MHz to 108 MHz. If the variable capacitor in an FM receiver ranges from 10.9 pF to 16.4 pF, what inductor should be used to make an *LC* circuit whose resonant frequency spans the FM band?

31. You want to use an *LC* circuit in a timing application. The circuit is to start with the capacitor fully charged, and the voltage should drop to zero in 15 s. You have available a 25-H inductor. What capacitance should you use?

32. An *LC* circuit includes a 20-μF capacitor and has a period of 5.0 ms. The peak current is 25 mA. Find (a) the inductance and (b) the peak voltage.

33. An *LC* circuit includes a 0.025-μF capacitor and a 340-μH inductor. (a) If the peak voltage on the capacitor is 190 V, what is the peak current in the inductor? (b) How long after the voltage peak does the current peak occur?

34. If the capacitor in the preceding problem is rated at 600 V, what is the maximum inductor current that will keep the capacitor within its rating?

35. At the instant when the electric and magnetic energies are equal in the *LC* circuit of Problem 33, the current is 540 mA. (a) What is the instantaneous voltage? Find (b) the peak voltage, (c) the peak current, and (d) the total energy.

36. In an *LC* circuit, what fraction of a cycle passes before the energy in the capacitor falls to one-fourth of its peak value?

37. One-eighth of a cycle after the capacitor in an *LC* circuit is fully charged, what are each of the following as fractions of their peak values: (a) capacitor charge, (b) energy in the capacitor, (c) inductor current, (d) energy in the inductor?

38. Show from conservation of energy that the peak voltage and current in an *LC* circuit are related by $I_p = V_p \sqrt{\dfrac{C}{L}}$.

39. The 2000-μF capacitor in Fig. 33-30 is initially charged to 200 V. (a) Describe how you would manipulate switches A and B to transfer all the energy from the 2000-μF capacitor to the 500-μF capacitor. Include the times you would throw the switches. (b) What will be the voltage across the 500-μF capacitor once you've finished?

FIGURE 33-30 Problem 39.

40. A damped *LC* circuit consists of a 0.15-μF capacitor and a 20-mH inductor with resistance 1.6 Ω. How many cycles will the circuit oscillate before the peak voltage on the capacitor drops to half its initial value?

41. A damped *RLC* circuit includes a 5.0-Ω resistor and a 100-mH inductor. If half the initial energy is lost after 15 cycles, what is the capacitance?

Section 33-4 Driven *RLC* Circuits and Resonance

42. A series *RLC* circuit has $R = 75$ Ω, $L = 20$ mH, and a resonant frequency of 4.0 kHz. (a) What is the capacitance? (b) What is the impedance of the circuit at resonance? (c) What is the impedance at 3.0 kHz?

43. If the speaker system of Example 33-4 is driven by a 10-V peak, 1.0-kHz sine wave, what will be the peak voltage across the capacitor?

44. An *RLC* circuit includes a 1.5-H inductor and a 250-μF capacitor rated at 400 V. The circuit is connected across a sine-wave generator whose peak voltage is 32 V. What minimum resistance must the circuit have to ensure that the capacitor voltage does not exceed its rated value when the generator is at the resonant frequency?

45. TV channel 2 occupies the frequency range from 54 MHz to 60 MHz. A series *RLC* tuning circuit in a TV receiver includes an 18-pF capacitor and resonates in the middle of the channel 2 band. (a) What is the inductance? (b) To let the whole signal in, the resonance curve must be broad enough that the current throughout the band be no less than 70% of the current at the resonant frequency. What constraint does this place on the circuit resistance?

46. An *RLC* circuit includes a 10-Ω resistor, 1.5-μF capacitor, and 50-mH inductor. The capacitor is rated at 1200 V. The circuit is driven by an AC source whose peak voltage is 100 V. (a) What would be the peak capacitor voltage at resonance? (b) Make a graph of the peak capacitor voltage as a function of frequency, and from it determine the frequency range that should be avoided for the capacitor to stay within its voltage rating.

47. A 2.0-H inductor and a 3.5-μF capacitor are connected in series with a 50-Ω resistor, and the combination is connected to an AC generator supplying 24 V peak at 60 Hz. (a) At the instant the generator voltage is at its peak, what is the instantaneous voltage across each circuit element? Show explicitly that these sum to the generator voltage. (b) If rms voltmeters are connected across each of the three components, what will they read? Do their readings sum to the rms generator voltage? Does this contradict the loop law?

48. Show that the impedance of an *RLC* circuit driven at frequency ω_d can be written

$$Z = \sqrt{R^2 + \omega_d^2 L^2 (1 - \omega_0^2/\omega_d^2)^2},$$

where ω_0 is the resonant frequency.

49. Figure 33-31 shows the phasor diagram for an *RLC* circuit. (a) Is the driving frequency above or below resonance? (b) Complete the diagram by adding the applied voltage phasor, and from your diagram determine the phase difference between applied voltage and current.

FIGURE 33-31 Problem 49.

50. An AC voltage of fixed amplitude is applied across a series *RLC* circuit. The component values are such that the cur-

rent at half the resonant frequency is half the current at resonance. (a) Show that the current at twice the resonant frequency is also half that at resonance. (b) Sketch phasor diagrams for both off-resonance cases.

51. For the circuit of Problem 46, find the phase relation between applied voltage and current at frequencies of (a) 550 Hz and (b) 700 Hz.

52. For the circuit of Problem 46, find two frequencies at which the voltage and current will be 60° out of phase. Which leads in each case?

Section 33-5 Power in AC Circuits

53. An electric drill draws 4.6 A rms at 120 V rms. If the current lags the voltage by 25°, what is the drill's power consumption?

54. A series *RLC* circuit has resistance 100 Ω and impedance 300 Ω. (a) What is the power factor? (b) If the rms current is 200 mA, what is the power dissipation?

55. A series *RLC* circuit has power factor 0.80 and impedance 100 Ω at 60 Hz. (a) What is the circuit resistance? (b) If the inductance is 0.10 H, what is the resonant frequency?

56. An *RLC* circuit with $R = 10~\Omega$, $C = 2.0~\mu$F, and $L = 500$ mH is connected to an AC generator supplying 80 V rms. (a) Find the power factor when the generator frequency is half the resonant frequency. Find the power dissipation (b) at half the resonant frequency and (c) at the resonant frequency.

57. A power plant produces 60-Hz power at 365 kV rms and 200 A rms. The plant is connected to a small city by a transmission line with total resistance 100 Ω. What fraction of the power is lost in transmission if the city's power factor is (a) 1.0 or (b) 0.60? (c) Is it more economical for the power company if the load has a large power factor or a small one? Explain.

Section 33-6 Transformers and Power Supplies

58. A rural power line carries 2.3 A rms at 4000 V. A step-down transformer reduces this to 235 V rms to supply a house. Find (a) the turns ratio of the transformer and (b) the current in the 235-V line to the house.

59. A transformer steps up the 120-V rms AC power line voltage to 23 kV rms for a TV picture tube. If the rms current in the primary is 1.0 A, and the transformer is 95% efficient, what is the secondary current?

60. A car battery charger runs off the 120-V rms AC power line, and supplies 10 A DC at 14 V. (a) If the charger is 80% efficient in converting the line power to the DC power it supplies to the battery, how much current does it draw from the AC line? (b) If electricity costs 9.5¢/kWh, how much does it cost to run the charger for 10 hours? Assume the power factor is 1.

61. The transformer in the power supply of Fig. 33-26a has an output voltage of 6.3 V rms at 60 Hz, and the capacitance is 1200 μF. (a) With an infinite load resistance, what would be the output voltage of the power supply? (b) What is the minimum load resistance for which the output would not drop more than 1% from this value? Assume that the discharge time in Fig. 33-26b is essentially a full cycle.

62. A power supply like that in Fig. 33-26a is supposed to deliver 22 V DC at a maximum current of 150 mA. The transformer's peak output voltage is appropriate to charge the capacitor to a full 22 V, and the primary is supplied with 60 Hz AC. What value of capacitance will ensure that the output voltage stays within 3% of the rated 22 V?

Paired Problems

(Both problems in a pair involve the same principles and techniques. If you can get the first problem, you should be able to solve the second one.)

63. A sine-wave generator delivers a signal whose peak voltage is independent of frequency. Two identical capacitors are connected in parallel across the generator, and the generator supplies a peak current I_p at frequency f_1. The capacitors are then connected in series across the generator. To what frequency should the generator be tuned to bring the current back to I_p?

64. A 1.0-μF capacitor and a 2.0-μF capacitor are connected in parallel across a sine-wave generator, and the generator supplies a total rms current of 25 mA at 1.0 kHz. Assuming the generator voltage is independent of frequency, at what frequency will it supply the same current when the capacitors are in series?

65. The peak current in an oscillating LC circuit is 850 mA. If $L = 1.2$ mH and $C = 5.0$ μF, what is the peak voltage?

66. An LC circuit with $C = 470$ pF is oscillating at 7.3 MHz. If the peak voltage is 95 V, what is the peak current?

67. An RLC circuit includes a 3.3-μF capacitor and a 27-mH inductor. The capacitor is charged to 35 V, and the circuit begins oscillating. Ten full cycles later the capacitor voltage peaks at 28 V. What is the resistance?

68. An RLC circuit with $R = 1.2$ Ω and $C = 10$ μF loses 2% of its initial energy in one oscillation cycle. What is its inductance?

69. A series RLC circuit with $R = 5.5$ Ω, $L = 180$ mH, and $C = 0.12$ μF is connected across a sine-wave generator. If the inductor can handle a maximum current of 1.5 A, what is the maximum safe value for the generator's peak output voltage when it is tuned to resonance?

70. A series RLC circuit with $R = 1.3$ Ω, $L = 27$ mH, and $C = 0.33$ μF is connected across a sine-wave generator. If the capacitor's peak voltage rating is 600 V, what is the maximum safe value for the generator's peak output voltage when it is tuned to resonance?

Supplementary Problems

71. Two capacitors are connected in parallel across a 10-V rms, 10-kHz sine-wave generator, and the generator supplies a total rms current of 30 mA. When the capacitors are rewired in series, the rms generator current drops to 5.5 mA. Find the values of the two capacitances.

72. An LC circuit starts at $t = 0$ with its 2000-μF capacitor at its peak voltage of 14 V. At $t = 35$ ms the voltage has dropped to 8.5 V. (a) What will be the peak current? (b) When will the peak current occur?

73. An undriven RLC circuit with inductance L and resistance R starts oscillating with total energy U_0. After N cycles the energy is U_1. Find an expression for the capacitance, assuming the circuit is not heavily damped.

74. Figure 33-32 shows a **low-pass filter**. When an alternating voltage is applied at the V_{in} terminals, the output voltage V_{out} depends on frequency. (a) Show that $V_{out} = V_{in}/\sqrt{1 + (RC\omega)^2}$, where the voltages are either peak or rms values (there is also a phase difference). (b) At what frequency is the output voltage down from the input voltage by a factor of $1/\sqrt{2}$? (This is called the *half-power point* since the power—proportional to V^2—is down by a factor of $\frac{1}{2}$.) *Hint:* You can repeat the phasor analysis of Section 33-4, but without the inductor. Or you can start from Equation 33-14, with X_L set to zero.

FIGURE 33-32 Problem 74.

75. You wish to make a "black box" with two input connections and two output connections, as shown in Fig. 33-33. When you put a 12-V rms, 60-Hz sine wave across the input, a 6.0-V, 60-Hz signal should appear at the output, with the output voltage leading the input voltage by 45°. Design a circuit that could be used in the "black box."

FIGURE 33-33 Problem 75.

76. A series RLC circuit with $R = 47$ Ω, $L = 250$ mH, and $C = 4.0$ μF is connected across a sine-wave generator whose peak output voltage is independent of frequency.

Find the frequency range over which the peak current will exceed half its value at resonance. *Hint:* You can solve this problem graphically or, with appropriate algebraic manipulations, using quadratic equations.

77. A sine-wave generator with peak output voltage of 20 V is applied across a series *RLC* circuit. At the resonant frequency of 2.0 kHz the peak current is 50 mA, while at 1.0 kHz it is 15 mA. Find R, L, and C.

78. Use phasor analysis to show that the parallel *RLC* circuit of Fig. 33-34 has impedance $Z = \left[\dfrac{1}{R^2} + \left(\dfrac{1}{X_L} - \dfrac{1}{X_C}\right)^2\right]^{-1/2}$.

FIGURE 33-34 Problem 78.

79. A 2.5-H inductor is connected across a 1500-μF capacitor. A 5.0-kg mass is connected to a spring. What should be the spring constant if the mechanical and electrical systems have the same resonant frequency?

80. Figure 33-35 shows a diode circuit called a *full-wave bridge,* whose output requires less filtering than the single-diode circuit of Fig. 33-25. (a) Sketch a graph of the resis-

FIGURE 33-35 A full-wave diode bridge (Problem 80).

tor current as a function of time, covering two full cycles of the AC generator, and (b) explain why less filtering is needed.

81. For *RLC* circuits in which the resistance is not too large, the Q factor may be defined as the ratio of the resonant frequency to the difference between the two frequencies where the power dissipated in the circuit is half that dissipated at resonance. Show, using suitable approximations, that this definition leads to the expression $Q = \omega_0 L/R$, with ω_0 the resonant frequency.

82. Consider a series circuit containing an AC generator, a resistor, and a capacitor. Construct a phasor diagram, and derive expressions for the circuit impedance and the phase angle between the applied voltage and the current. Show that the current always leads the voltage.

83. Consider a series circuit containing an AC generator, a resistor, and an inductor. Construct a phasor diagram, and derive expressions for the circuit impedance and the phase angle between the applied voltage and the current. Show that the voltage always leads the current.

34 Maxwell's Equations and Electromagnetic Waves

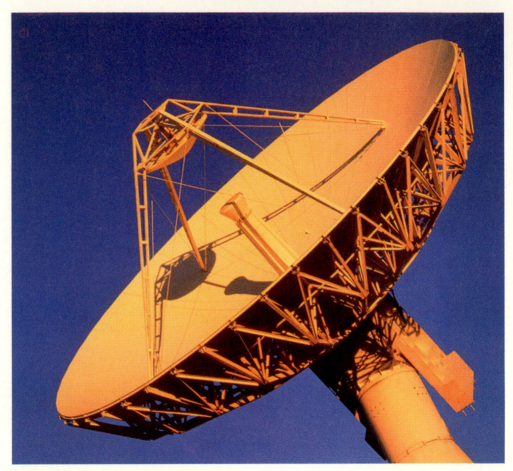

Electromagnetic waves include light and the radio waves essential to modern communications. This antenna communicates with orbiting satellites.

At this point we have introduced the four fundamental laws of electromagnetism—Gauss's law for electricity, Gauss's law for magnetism, Ampère's law, and Faraday's law—that govern the behavior of electric and magnetic fields throughout the universe. We have seen how these laws describe the electric and magnetic interactions that make matter act as it does and have explored many practical devices that exploit the laws of electromagnetism. Here we extend the fundamental laws to their most general form and show how they predict the existence of electromagnetic waves. Those waves include the light, radio, TV, microwaves, X-rays, ultraviolet, and infrared with which we see, communicate, cook our food, diagnose our diseases, learn about the universe beyond Earth, and perform a myriad of other tasks from the mundane to the profound.

Electromagnetic waves are certainly among the most important manifestations of electromagnetism. After showing how electromagnetic waves arise from fundamental principles, we'll take a brief look at some of their properties. The next three chapters, on optics, then examine in detail the behavior of the electromagnetic waves we call visible light.

34-1 THE FOUR LAWS OF ELECTROMAGNETISM

Table 34-1 summarizes the four laws as we introduced them in earlier chapters. As you look at these four laws together, you can't help noticing some strong similarities. On the left-hand sides of the equations, the two laws of Gauss are identical but for the interchanging of **E** and **B**. Similarly, the laws of Ampère and Faraday have left-hand sides that differ only in the interchange of **E** and **B**.

On the right-hand sides, things are more different. Gauss's law for electricity involves the charge enclosed by the surface of integration, while Gauss's law for magnetism has zero on the right-hand side. Actually, though, these laws are similar. Since we have no experimental evidence for the existence of isolated magnetic charge, the enclosed magnetic charge on the right-hand side of Gauss's law for magnetism is zero. If and when magnetic monopoles are discovered, then the right-hand side of Gauss's law for magnetism would be nonzero for any surface enclosing net magnetic charge.

The right-hand sides of Ampère's and Faraday's laws are distinctly different. In Ampère's law we find the current—the flow of electric charge—as a source of magnetic field. We can understand the absence of a similar term in Faraday's law because we have never observed a flow of magnetic monopoles. If we had such a flow, then we would expect this magnetic current to produce an electric field.

Two of the differences among the laws of electromagnetism would be resolved if we knew for sure that magnetic monopoles exist. That current theories of elementary particles suggest the existence of monopoles is a tantalizing hint that there may be a fuller symmetry between electric and magnetic phenomena. The search for symmetry, based not on logic or experimental evidence but on an intuitive sense that nature should be simple, has motivated some of the most important discoveries in physics. We'll see next how symmetry considerations help lead us, finally, to the complete set of fundamental laws of electromagnetism.

TABLE 34-1 *Four Laws of Electromagnetism (still incomplete)*

LAW	MATHEMATICAL STATEMENT	WHAT IT SAYS
Gauss for **E**	$\oint \mathbf{E} \cdot d\mathbf{A} = \dfrac{q}{\varepsilon_0}$	How charges produce electric field; field lines begin and end on charges
Gauss for **B**	$\oint \mathbf{B} \cdot d\mathbf{A} = 0$	No magnetic charge; magnetic field lines do not begin or end
Faraday	$\oint \mathbf{E} \cdot d\boldsymbol{\ell} = -\dfrac{d\phi_B}{dt}$	Changing magnetic flux produces electric field
Ampère (steady currents only)	$\oint \mathbf{B} \cdot d\boldsymbol{\ell} = \mu_0 I$	Electric current produces magnetic field

34-2 AMBIGUITY IN AMPÈRE'S LAW

There remains one difference between the equations of electricity and magnetism that would not be resolved by the discovery of magnetic monopoles. On the right-hand side of Faraday's law we find the term $d\phi_B/dt$ that describes changing magnetic flux as a source of electric field. We find no comparable term in Ampère's law. Are we missing something? Is it possible that a changing electric flux produces a magnetic field? So far, we've given no experimental evidence for such a conjecture. It's suggested only by our sense that the near symmetry between electricity and magnetism is not a coincidence. If a changing electric flux did produce a magnetic field, just as a changing magnetic flux produces an electric field, then we would expect a term $d\phi_E/dt$ on the right-hand side of Ampère's law.

When we first stated Ampère's law in Chapter 30, we emphasized that it applied only to *steady* currents. Why that restriction? Figure 34-1 shows a situation in which current is *not* steady, namely the *RC* circuit we discussed in Chapter 28. Initially current flows in this circuit to carry charge onto the plates of the capacitor. But the current gradually decreases to zero as the capacitor becomes fully charged. While it's flowing, the current should produce a magnetic field. Let's try to use Ampère's law to calculate that field.

Ampère's law says that the line integral of the magnetic field around any closed loop is proportional to the encircled current:

$$\oint \mathbf{B} \cdot d\boldsymbol{\ell} = \mu_0 I.$$

By the encircled current, we mean the current through *any open surface* bounded by the loop. Figure 34-2 shows four such surfaces. The same current flows through surfaces 1, 2, and 4, because each of them is pierced by a current-carrying wire. But no current pierces surface 3, because the right end of that surface lies in the insulating gap between the capacitor plates. Charge flows onto the plates of the capacitor, but it doesn't flow through that gap. So for surfaces 1, 2, and 4, the right-hand side of Ampère's law is $\mu_0 I$, but for surface 3 it's zero. Thus Ampère's law is ambiguous in this case of a changing current.

This ambiguity does not arise with steady currents. In an *RC* circuit the steady-state current is everywhere zero, and thus the right-hand side of Ampère's law is zero for *any* surface. It's only when currents are changing with time that there may be situations like that of Fig. 34-2 where Ampère's law becomes ambiguous. That's why the form of Ampère's law we have used until now is strictly valid only for steady currents.

FIGURE 34-1 A charging *RC* circuit, showing some magnetic field lines surrounding the current-carrying wire.

FIGURE 34-2 Ampère's law relates the line integral around an amperian loop to the current through *any open surface* bounded by that loop. Figure 34-2 shows four such surfaces, all bounded by the same circular loop. Surface 1 is the flat, circular disk bounded by the loop. The others are all like soap bubbles in the process of being blown after dipping the loop into a soap solution; they're open at the left end, so if current does pass through a surface, it does so at the right end only. Current in the wires passes through surfaces 1, 2, and 4, but not 3, so Ampère's law is ambiguous.

Can we salvage Ampère's law, extending it to cover unsteady currents without affecting its validity in the steady case? Symmetry between Ampère's and Faraday's laws has already suggested that a changing electric flux might produce a magnetic field. Between the plates of a charging capacitor there is an electric field whose magnitude is increasing (Fig. 34-3). That means there is a changing electric flux through surface 3 of Fig. 34-2.

It was the Scottish physicist James Clerk Maxwell who, about 1860, suggested that a changing electric flux should give rise to a magnetic field. Since that time many experiments, including direct measurement of the magnetic field inside a charging capacitor, have confirmed Maxwell's remarkable insight. Maxwell quantified his idea by introducing a new term into Ampère's law:

FIGURE 34-3 There is a changing electric field in the charging capacitor and, therefore, a changing electric flux through surface 3 of Fig. 34-2.

$$\oint \mathbf{B} \cdot d\boldsymbol{\ell} = \mu_0 I + \mu_0 \varepsilon_0 \frac{d\phi_E}{dt}. \tag{34-1}$$

Now there's no ambiguity. The integral is taken around any loop, I is the current through *any* surface bounded by the loop, and ϕ_E is the electric flux through that surface. With our charging capacitor, Equation 34-1 gives the same magnetic field no matter which surface we choose. For surfaces 1, 2, and 4 of Fig. 34-2, the current I makes all the contribution to the right-hand side of the equation. For surface 3, through which no current flows, the right-hand side of Equation 34-1 comes entirely from the changing electric flux. You can readily verify that the term $\varepsilon_0 d\phi_E/dt$ has the units of current, and that, for the charging capacitor, this term is numerically equal to the current I (see Problem 3).

Although changing electric flux is not the same thing as electric current, it has the same effect as a current in producing a magnetic field. For this reason Maxwell called the term $\varepsilon_0 d\phi_E/dt$ the **displacement current**. The word "displacement" has historical roots that don't provide much physical insight. But the word "current" is meaningful in that the effect of displacement current is indistinguishable from that of real current in producing magnetic fields. Although we developed the idea of displacement current using the specific example of a charging capacitor, we emphasize that Ampère's law in its now complete form (Equation 34-1) is truly universal: *any* changing electric flux results in a magnetic field. That fact will prove crucial in establishing the existence of electromagnetic waves.

EXAMPLE 34-1	*Displacement Current Produces Magnetic Field*

A parallel-plate capacitor with circular plates a distance d apart is charged through long, straight wires as shown in Fig. 34-4. The potential difference between the plates is increasing at the rate dV/dt. Find an expression for the magnetic field as a function of position between the plates.

Solution

With long, straight feed wires, the situation has cylindrical symmetry. The only magnetic field with this symmetry has

FIGURE 34-4 A circular capacitor (Example 34-1).

FIGURE 34-5 Electric and magnetic fields between the circular capacitor plates. The electric field strength is increasing, so the displacement current is in the same direction as the electric field. Pointing the right thumb in this direction then shows that the magnetic field circles clockwise.

circular field lines and a magnitude that depends only on the radial distance r from the symmetry axis, as shown in Fig. 34-5. A magnetic field line within the capacitor encircles no conduction current—no flow of charge—but it does encircle a changing electric field and therefore a displacement current. If the field line has radius r, the encircled electric flux is

$$\phi_E = \int \mathbf{E} \cdot d\mathbf{A} = \pi r^2 E = \pi r^2 \frac{V}{d},$$

where the uniformity of the field allows us to calculate the field as the ratio of potential difference to plate spacing and the flux as a simple product of field and area. Then the displacement current is

$$I_D = \varepsilon_0 \frac{d\phi_E}{dt} = \frac{\varepsilon_0 \pi r^2}{d} \frac{dV}{dt}.$$

With cylindrical symmetry, the line integral on the left-hand side of Ampère's law becomes

$$\oint \mathbf{B} \cdot d\boldsymbol{\ell} = 2\pi r B.$$

Equating this quantity to μ_0 times the encircled displacement current gives

$$2\pi r B = \frac{\mu_0 \varepsilon_0 \pi r^2}{d} \frac{dV}{dt},$$

so

$$B = \frac{\mu_0 \varepsilon_0 r}{2d} \frac{dV}{dt}.$$

This field, with its magnitude increasing linearly with r, should remind you of the magnetic field inside a cylindrical wire (see Example 30-4). Problem 4 extends this calculation to the field outside the capacitor.

We can find the direction of the induced magnetic field just as we did for the fields of ordinary conduction currents: Point your right thumb in the direction of the current and your right fingers curl in the direction of the magnetic field. But which way does the displacement current go? In this example the electric field strength is increasing, so $d\phi_E/dt$ is positive, and the displacement current is in the direction of the electric field (see Fig. 34-5). If the electric field strength were decreasing, the displacement current would be opposite the field.

The induced magnetic field in a practical capacitor is minuscule, as the following exercise illustrates. We'll soon see, however, that the significance of displacement-current-induced magnetic fields is vastly greater than in this simple example.

EXERCISE In 1984 D. F. Bartlett and T. R. Corle of the University of Colorado first measured the magnetic field inside a charging capacitor using a sensitive magnetometer called a superconducting quantum interference detector (SQUID). They used a capacitor with circular plates spaced 1.22 cm, connected across a 340-V peak sine-wave generator operating at 1.25 kHz. What was the peak magnetic field strength 3.0 cm from the capacitor axis?

Answer: 3.65×10^{-11} T, less than one millionth of Earth's magnetic field

Some problems similar to Example 34-1: 4–6

34-3 MAXWELL'S EQUATIONS

It was Maxwell's genius to recognize that Ampère's law should be modified to reflect the symmetry suggested by Faraday's law. The consequences of Maxwell's discovery go far beyond anything he could have imagined. To honor Maxwell, the four complete laws of electromagnetism are given the collective name **Maxwell's equations**. This full and complete set of equations, first published in 1864, governs the behavior of electric and magnetic fields everywhere. Table 34-2 summarizes Maxwell's equations.

These four simple, compact statements are all it takes to describe classical electromagnetic phenomena. Everything electric or magnetic that we have considered and will consider—from polar molecules to electric current; resistors, capacitors, inductors, and transistors; solar flares and cell membranes; electric generators and thunderstorms; computers and TV sets; the northern lights and fusion reactors—all these can be described using Maxwell's

TABLE 34-2 *Maxwell's Equations*

LAW	MATHEMATICAL STATEMENT	WHAT IT SAYS	EQUATION NUMBER
Gauss for **E**	$\oint \mathbf{E} \cdot d\mathbf{A} = \dfrac{q}{\varepsilon_0}$	How charges produce electric field; field lines begin and end on charges	(34-2)
Gauss for **B**	$\oint \mathbf{B} \cdot d\mathbf{A} = 0$	No magnetic charge; magnetic field lines do not begin or end	(34-3)
Faraday	$\oint \mathbf{E} \cdot d\boldsymbol{\ell} = -\dfrac{d\phi_B}{dt}$	Changing magnetic flux produces electric field	(34-4)
Ampère	$\oint \mathbf{B} \cdot d\boldsymbol{\ell} = \mu_0 I + \mu_0\varepsilon_0\dfrac{d\phi_E}{dt}$	Electric current and changing electric flux produce magnetic field	(34-5)

equations. And despite this wealth of phenomena, we have yet to discuss a most important manifestation of electromagnetism, namely electromagnetic waves. We've put off waves until now because they, unlike electromagnetic phenomena we've already introduced, depend crucially on Maxwell's extension of Ampère's law. It's easiest to understand electromagnetic waves when they propagate through empty space, so before beginning our study of waves we'll first simplify Maxwell's equations for the case of a vacuum.

Maxwell's Equations in Vacuum

Consider Maxwell's equations in a region free of any matter—in vacuum. We've learned enough about electromagnetism to anticipate that the fields themselves will still be able to interact, change, and carry energy even in the absence of matter. To express Maxwell's equations in vacuum, we simply remove all reference to matter—that is, to electric charge:

$$\oint \mathbf{E} \cdot d\mathbf{A} = 0 \qquad \text{(Gauss, }\mathbf{E}) \qquad\qquad (34\text{-}6)$$

$$\oint \mathbf{B} \cdot d\mathbf{A} = 0 \qquad \text{(Gauss, }\mathbf{B}) \qquad\qquad (34\text{-}7)$$

$$\oint \mathbf{E} \cdot d\boldsymbol{\ell} = -\frac{d\phi_B}{dt} \qquad \text{(Faraday)} \qquad\qquad (34\text{-}8)$$

$$\oint \mathbf{B} \cdot d\boldsymbol{\ell} = \mu_0\varepsilon_0\frac{d\phi_E}{dt}. \qquad \text{(Ampère)} \qquad\qquad (34\text{-}9)$$

In vacuum the symmetry is complete, with electric and magnetic fields appearing in the equations on an equal footing. With charge and current absent in a vacuum, the only source of either field is a change in the other field—as shown by the time derivatives on the right-hand sides of Faraday's and Ampère's laws.

34-4 ELECTROMAGNETIC WAVES

Faraday's law shows that a changing magnetic field induces an electric field. Ampère's law shows that a changing electric field induces a magnetic field. Together, the two laws suggest the possibility of **electromagnetic waves**, in which each type of field continually induces the other, resulting in an electromagnetic disturbance that propagates through space as a wave. We'll now confirm this qualitative suggestion with a rigorous demonstration, directly from Maxwell's equations, that electromagnetic waves are indeed possible. In the process we'll discover the properties of electromagnetic waves and will come to a deep understanding of the nature of light.

A Plane Electromagnetic Wave

Here we describe the simplest type of electromagnetic wave—a plane wave in vacuum. Recall from Chapter 16 that a plane wave is one whose properties don't vary in directions perpendicular to the wave propagation, so its wavefronts are planes (see Fig. 16-18). A plane wave is an approximation to the more realistic case of a spherical wave expanding from a localized source, but it's a good approximation at distances from the wave source that are large compared with the wavelength. Light waves reaching Earth from the Sun, for example, or radio waves miles from the transmitter, are essentially plane waves.

In vacuum, it turns out that the electric and magnetic fields of an electromagnetic wave are perpendicular to each other. They're both also perpendicular to the direction of wave propagation—making the electromagnetic wave a transverse wave, as defined in Chapter 16. To be concrete, we'll take the x direction to be the direction of propagation, the y direction that of the electric field, and the z direction that of the magnetic field (Fig. 34-6). We won't prove that this configuration of three mutually perpendicular directions is the only

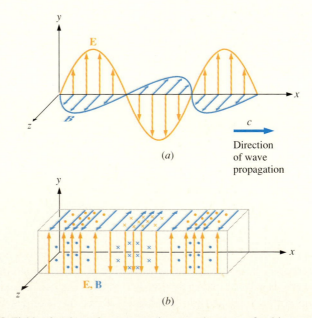

FIGURE 34-6 Fields of a plane electromagnetic wave, shown at a fixed instant of time. (a) Field vectors for points on the x axis show the sinusoidal variation in the fields. The electric and magnetic fields are perpendicular but in phase; they're also perpendicular to the propagation direction. (b) A partial representation of the field lines, which actually extend forever in both directions. Here the lines are pictured only in a finite rectangular slab. Lines on the facing surfaces of the slab are shown as arrows; lines going through the slab appear as dots or crosses depending on whether they're emerging from or going into the slab. Spacing of the field lines reflects the sinusoidal variation shown in (a).

one possible for an electromagnetic wave (although in vacuum it is; see Problem 11). What we will do, though, is prove that this configuration satisfies Maxwell's equations—thus showing that such electromagnetic waves are indeed possible. Before we can do that, we need a mathematical description of our plane electromagnetic wave.

In Chapter 16 we found that a sinusoidal wave propagating in the x direction is represented by a function of the form $A \sin(kx - \omega t)$, where A is the wave amplitude, k the wave number, and ω the angular frequency. For the mechanical waves of Chapters 16 and 17, the function $A \sin(kx - \omega t)$ described some physical quantity such as the height of a water wave or pressure variation in a sound wave. In an electromagnetic wave, the corresponding physical quantities are the electric and magnetic fields—which are vector quantities. It turns out that these two wave fields, although perpendicular, are in phase—meaning that their peaks and troughs coincide, as shown in Fig. 34-6. Having chosen the y direction for the electric field and the z direction for the magnetic field, we can then write the fields of our plane electromagnetic wave as

$$\mathbf{E}(x, t) = E_p \sin(kx - \omega t)\hat{\mathbf{j}} \tag{34-10}$$

and

$$\mathbf{B}(x, t) = B_p \sin(kx - \omega t)\hat{\mathbf{k}}, \tag{34-11}$$

where the peak amplitudes E_p and B_p are constants and where $\hat{\mathbf{j}}$ and $\hat{\mathbf{k}}$ are unit vectors in the y and z directions. Figure 34-6a is a "snapshot" of some field vectors of this wave at points on the x axis, shown at a fixed instant of time. That $\mathbf{E}$ and $\mathbf{B}$ are perpendicular is obvious from the figure, as is the fact that they're perpendicular to the propagation direction, in this case the x direction. You can also see the sinusoidal variation, as the field vectors get alternately longer, then shorter, then reverse direction, and so on. And you can see that $\mathbf{E}$ and $\mathbf{B}$ are in phase: they peak at the same points, and are zero at the same points. We emphasize that Fig. 34-6a shows field *vectors* for points on the x axis only; the fields extend forever throughout space, and because this is a plane wave, a picture of field vectors along any line parallel to the x axis would look just the same.

We can also draw field *lines* for our wave, in contrast to the field vectors of Fig. 34-6a. We can't draw complete field lines, because they extend forever in both directions. So in Fig. 34-6b we've shown the field lines only in a rectangular slab; that's enough to give a picture of what the fields look like everywhere. You should convince yourself that Figs. 34-6a and 34-6b show exactly the same thing, namely a plane electromagnetic wave described by Equations 34-10 and 34-11. In one case we use field *vectors,* whose lengths are proportional to the field magnitudes, and in the other we use field *lines,* which extend forever and whose spacing indicates the field magnitudes.

We'll now use Maxwell's equations to show that the electric and magnetic fields pictured in Fig. 34-6 and described by Equations 34-10 and 34-11 really do satisfy Maxwell's equations. We've chosen a sinusoidal waveform for our wave fields because of its mathematical simplicity. But the superposition principle holds for electric and magnetic fields, and we know from Section 16-6 that by superposition we can represent *any* waveform in terms of sinusoidal waveforms. So our proof that electromagnetic waves can exist actually holds for any wave shape (see Problem 74 here and Problem 54 in Chapter 16). That means we can use electromagnetic waves to communicate the complex waveforms of music, TV images, or computer data encoded as pulses of light.

We have four Maxwell equations to satisfy: the two Gauss's laws, and the laws of Faraday and Ampère.

Gauss's Laws

In vacuum, Gauss's laws for the electric and magnetic fields both have zero on the right-hand side (Equations 34-6 and 34-7), reflecting the absence of charge. That means, as we've seen many times before, that the field lines can't begin or end; they either form closed loops or extend to infinity. Our electromagnetic wave is a plane wave, meaning that no wave property varies in directions perpendicular to the wave propagation. Therefore the field lines shown partially in Fig. 34-6*b* must extend, straight, forever in both directions. So they don't begin or end, and thus the two Gauss laws are satisfied.

Faraday's Law

To see that Faraday's law is satisfied, consider an observer looking directly toward the *x-y* plane in Fig. 34-6. Such an observer would see electric field lines going up and down and magnetic field lines coming straight in and out, as shown in Fig. 34-7. Consider the small rectangular loop of height *h* and infinitesimal width *dx* shown in the figure. Evaluating the line integral of the electric field **E** around this loop, we get no contribution from the short ends because they are at right angles to the field. Going around counterclockwise, we get a contribution $-Eh$ as we go down the left side against the field direction. Then we get a positive contribution going up the right side. Because of the variation in field strength with position, the field strength on the right side of the loop is different from that on the left. Let the change in field be *dE*, so the field on the right side of the loop is $E + dE$, giving a contribution of $(E + dE)h$ to the line integral. Then the line integral of **E** around the loop is

$$\oint \mathbf{E} \cdot d\boldsymbol{\ell} = -Eh + (E + dE)h = h\,dE.$$

Physically, this nonzero line integral implies an induced electric field. Induced by what? By a changing magnetic flux through the loop. The electric field of the wave arises because of the changing magnetic field of the wave. The area of the loop is $h\,dx$, and the magnetic field **B** is at right angles to this area, so the magnetic flux through the loop is just

$$\phi_B = Bh\,dx.$$

The rate of change of flux through the loop is then

$$\frac{d\phi_B}{dt} = h\,dx\,\frac{dB}{dt}.$$

Faraday's law relates the line integral of the electric field to the rate of change of flux:

$$\oint \mathbf{E} \cdot d\boldsymbol{\ell} = -\frac{d\phi_B}{dt},$$

or, using our expressions for the line integral and the rate of change of flux,

$$h\,dE = -h\,dx\,\frac{dB}{dt}.$$

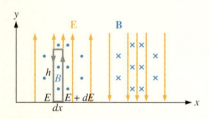

FIGURE 34-7 Cross section of Fig. 34-6 in the *x-y* plane. Also shown is a rectangular loop for evaluating the line integral in Faraday's law.

Dividing through by $h\,dx$, we have

$$\frac{dE}{dx} = -\frac{dB}{dt}.$$

(34-12a)

In deriving this equation, we considered changes in E with position at a fixed instant of time, as pictured in Fig. 34-7, so our derivative dE/dx means the rate of change of E with position while time is held fixed. Similarly, in evaluating the derivative of magnetic flux, we were concerned only with the time rate of change at the fixed position of our loop. Both our derivatives represent rates of change with respect to one variable while the other variable is held fixed, and are therefore partial derivatives (see Math Toolbox on p. 400 if you're not familiar with partial derivatives). Equation 34-12a should then be written more properly:

$$\frac{\partial E}{\partial x} = -\frac{\partial B}{\partial t}.$$

(34-12b)

Equation 34-12b—which is just Faraday's law applied to our electromagnetic wave—tells us that the rate at which the electric field changes with *position* is related to the rate at which the magnetic field changes with *time*.

Ampère's Law

Now imagine an observer looking down on Fig. 34-6 from above. This observer sees the magnetic field lines lying in the x-z plane, and electric field lines emerging perpendicular to the x-z plane as shown in Fig. 34-8. We can apply Ampère's law (Equation 34-9) to the infinitesimal rectangle shown, just as we applied Faraday's law to a similar rectangle in the x-y plane. Going counterclockwise around the rectangle, we get no contribution to the line integral of the magnetic field on the short sides, since they lie perpendicular to the field. Going down the left side, we get a contribution Bh to the line integral. Going up the right side, against the field, we get a negative contribution $-(B + dB)h$, where dB is the change in B from one side of the rectangle to the other. So the line integral in Ampère's law is

$$\oint \mathbf{B} \cdot d\boldsymbol{\ell} = Bh - (B + dB)h = -h\,dB.$$

The electric flux through the rectangle is simply $Eh\,dx$, so the rate of change of electric flux is

$$\frac{d\phi_E}{dt} = h\,dx\,\frac{dE}{dt}.$$

Ampère's law relates the line integral of the magnetic field to this time derivative of the electric flux, giving

$$-h\,dB = \varepsilon_0\mu_0 h\,dx\,\frac{dE}{dt}.$$

Dividing through by $h\,dx$ and noting again that we are really dealing with partial derivatives, we have

$$\frac{\partial B}{\partial x} = -\varepsilon_0\mu_0\frac{\partial E}{\partial t}.$$

(34-13)

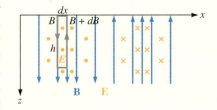

FIGURE 34-8 Cross section of Fig. 34-6 in the x-z plane, showing a rectangular loop for evaluating the line integral in Ampère's law.

Equations 34-12 and 34-13—derived from Faraday's and Ampère's laws—express fully the requirements that Maxwell's universal laws of electromagnetism pose on the field structure of Fig. 34-6. The two equations are remarkable in that each describes an induced field that arises from the changing of the other field. That other field, in turn, arises from the changing of the first field. Thus we have a self-perpetuating electromagnetic structure, whose fields exist and change without the need for charged matter. If Equations 34-10 and 34-11, which describe the fields in Fig. 34-6, can be made consistent with Equations 34-12 and 34-13, then we will have demonstrated that our electromagnetic wave does indeed satisfy Maxwell's equations and is thus a possible configuration of electric and magnetic fields.

Faraday, Ampère, and the Wave Fields

To see that Equation 34-12 is satisfied, we differentiate the electric field of Equation 34-10 with respect to x, and the magnetic field of Equation 34-11 with respect to t:

$$\frac{\partial E}{\partial x} = \frac{\partial}{\partial x}[E_p \sin(kx - \omega t)] = kE_p \cos(kx - \omega t)$$

and

$$\frac{\partial B}{\partial t} = \frac{\partial}{\partial t}[B_p \sin(kx - \omega t)] = -\omega B_p \cos(kx - \omega t).$$

Putting these expressions in for the derivatives in Equation 34-12 gives

$$kE_p \cos(kx - \omega t) = -[-\omega B_p \cos(kx - \omega t)].$$

The cosine term cancels from this equation, showing that the equation holds if

$$kE_p = \omega B_p. \tag{34-14}$$

To see that Equation 34-13 is also satisfied, we differentiate the magnetic field of Equation 34-11 with respect to x and the electric field of Equation 34-10 with respect to t:

$$\frac{\partial B}{\partial x} = kB_p \cos(kx - \omega t)$$

and

$$\frac{\partial E}{\partial t} = -\omega E_p \cos(kx - \omega t).$$

Using these expressions in Equation 34-13 then gives

$$kB_p \cos(kx - \omega t) = -\varepsilon_0 \mu_0 [-\omega E_p \cos(kx - \omega t)].$$

Again, the cosine term cancels, showing that this equation is satisfied if

$$kB_p = \varepsilon_0 \mu_0 \omega E_p. \tag{34-15}$$

Our analysis has shown that electromagnetic waves whose form is given by Fig. 34-6 and Equations 34-10 and 34-11 can exist, provided that the amplitudes E_p and B_p, and the frequency ω and wave number k, are related by Equations 34-14 and 34-15. Physically, the existence of these waves is possible because a change in either kind of field—electric or magnetic—induces the other kind of field, giving rise to a self-perpetuating electromagnetic field structure. Maxwell's theory thus leads to the prediction of an entirely new phenomenon—the electromagnetic wave. We will now explore some properties of these waves.

34-5 THE SPEED OF ELECTROMAGNETIC WAVES

In Chapter 16 we found that the speed of a sinusoidal wave is given by the ratio of the angular frequency and wave number:

$$\text{wave speed} = \frac{\omega}{k}.$$

To determine the speed of our electromagnetic wave, we solve Equation 34-14 for E_p:

$$E_p = \frac{\omega B_p}{k},$$

and use this expression in Equation 34-15:

$$kB_p = \varepsilon_0 \mu_0 \omega E_p = \frac{\varepsilon_0 \mu_0 \omega^2 B_p}{k}.$$

Solving for the wave speed ω/k then gives

$$\text{wave speed} = \frac{\omega}{k} = \frac{1}{\sqrt{\varepsilon_0 \mu_0}}. \tag{34-16}$$

This remarkably simple result shows that the speed of an electromagnetic wave in vacuum depends only on the electric and magnetic constants ε_0 and μ_0. All electromagnetic waves in vacuum, regardless of frequency or amplitude, share this speed. Although we derived this result for sinusoidal waves, the superposition principle ensures that it holds for any wave shape.

We can easily calculate the speed given in Equation 34-16:

$$\frac{1}{\sqrt{\varepsilon_0 \mu_0}} = \frac{1}{[(8.85 \times 10^{-12} \text{ F/m})(4\pi \times 10^{-7} \text{ H/m})]^{1/2}} = 3.00 \times 10^8 \text{ m/s}.$$

But this is precisely the speed of light!

As early as 1600, Galileo had tried to measure the speed of light by uncovering lanterns on different mountaintops. He was able to conclude only that "If not instantaneous, it is extraordinarily rapid." The first evidence for a finite speed of light came in the 1670s from observations by the Danish astronomer Ole Römer. He noted that eclipses of Jupiter's moons occurred at different times than predicted as Jupiter's distance from Earth changed. Römer interpreted the changes as being due to the different times for light traveling at a finite speed from Jupiter to Earth. Römer's data implied a speed of 2.3×10^8 m/s, about three quarters of the actual value. In 1728, James Bradley used changes in the apparent positions of stars resulting from Earth's orbital motion to calculate a value of 2.95×10^8 m/s for the speed of light—in error by less than 2%. As we saw in Chapter 1, the exact value $c = 2.99792458 \times 10^8$ m/s was adopted in

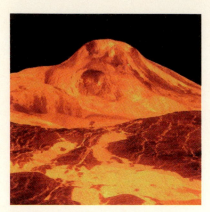

FIGURE 34-9 This image of Venus was made using data that include surface altitude determined by very accurate measurements of the travel time for radar signals reflected from Venus' surface. Radar, like all electromagnetic waves, travels at the speed of light.

1983 to define the meter. So we no longer measure the speed of light; instead, that speed provides an accurate way to measure distances (Fig. 34-9).

Do other properties of light besides its speed fit our picture of electromagnetic waves? In 1678—some 200 years before Maxwell—the Dutch physicist Christian Huygens had suggested that light is a wave. In 1801, Thomas Young experimented with a 2-slit system to show that light undergoes interference, giving conclusive evidence for the wave nature of light (see Fig. 16-25). But neither experiment nor theory could say what kind of wave light might be. Then, in the 1860s, came Maxwell. Using a theory developed from laboratory experiments involving electricity and magnetism, with no reference whatever to optics or light, Maxwell showed how the interplay of electric and magnetic fields could result in electromagnetic waves. The speed of those waves—calculated from the quantities ε_0 and μ_0 that were determined in laboratory experiments having nothing to do with light—was precisely the known speed of light. Maxwell was led inescapably to the conclusion that light is an electromagnetic wave.

Maxwell's identification of light as an electromagnetic phenomenon is a classic example of the unification of knowledge toward which science is ever striving. With one simple calculation, Maxwell brought the entire science of optics under the umbrella of electromagnetism. Maxwell's work stands as a crowning intellectual triumph, an achievement whose implications are still expanding our view of the universe.

34-6 PROPERTIES OF ELECTROMAGNETIC WAVES

Our demonstration that electromagnetic waves satisfy Maxwell's equations places definite constraints on the properties of those waves. The wave frequency ω and wave number k are not both arbitrary, but must be related through

$$\frac{\omega}{k} = c, \tag{34-17a}$$

where $c = 1/\sqrt{\varepsilon_0 \mu_0}$ is the speed of light. In Chapter 16 we related the angular frequency ω and wave number k to the more familiar frequency f and wavelength λ through the equations $\omega = 2\pi f$ and $k = 2\pi/\lambda$. Therefore we can also write Equation 34-17a in the form

$$f\lambda = c. \tag{34-17b}$$

Furthermore, Equation 34-14 shows that

$$E = \frac{\omega}{k}B = cB. \tag{34-18}$$

Thus, the field magnitudes in the wave are not independent but are in the ratio of the speed of light. Also, Fig. 34-6 and Equations 34-10 and 34-11 have the electric and magnetic fields perpendicular to each other and to the direction of wave propagation, with the electric and magnetic fields in phase. It is in fact the case that only waves with this form—**E** and **B** in phase and with **E**, **B**, and the propagation direction all perpendicular—can satisfy Maxwell's equations in vacuum (see Problem 11).

EXAMPLE 34-2	*Laser Light*

A laser beam with wavelength 633 nm is propagating in the $+z$ direction. Its electric field is parallel to the x axis and has amplitude 6.0 kV/m. Find the wave frequency, and the direction and amplitude of the magnetic field.

Solution

Equation 34-17b relates the wavelength and frequency to the speed of light. Solving for f gives

$$f = \frac{c}{\lambda} = \frac{3.00 \times 10^8 \text{ m/s}}{633 \times 10^{-9} \text{ m}} = 4.74 \times 10^{14} \text{ Hz} .$$

If we imagine reorienting the wave of Fig. 34-6 so it propagates along the z direction, then rotate it about the z direction so the electric field is parallel to the x axis, we find that the

magnetic field is parallel to the y axis. The magnetic field amplitude follows from Equation 34-18:

$$B_p = \frac{E_p}{c} = \frac{6.0 \times 10^3 \text{ V/m}}{3.00 \times 10^8 \text{ m/s}} = 2.0 \times 10^{-5} \text{ T} .$$

EXERCISE An electromagnetic wave is propagating in the $-y$ direction, with its magnetic field parallel to the x axis. The magnetic field amplitude is 8.0 μT. Write an expression for the wave's electric field vector at the point where the magnetic field points in the $+x$ direction and is at its peak value.

Answer: $\mathbf{E} = -2.4\hat{\mathbf{k}}$ kV/m

Some problems similar to Example 34-2: 24–27

Got It!

Light and radio waves are both electromagnetic waves; of the two, light has a much higher frequency. Does that mean light waves travel faster than radio waves? Regardless of your answer, are any other properties of these waves different?

34-7 THE ELECTROMAGNETIC SPECTRUM

Although an electromagnetic wave's frequency and wavelength must be related by Equation 34-17b, one or the other of these quantities is completely arbitrary. That means we can have electromagnetic waves of any frequency, or, equivalently, any wavelength. Direct measurement shows that visible light occupies a wavelength range from about 400 nm to 700 nm, corresponding to frequencies from 7.5×10^{14} Hz to 4.3×10^{14} Hz. The different wavelengths or frequencies correspond to different colors, with red at the long-wavelength, low-frequency end of the visible region and blue at the short-wavelength, high-frequency end (see the enlargement in Fig. 34-10).

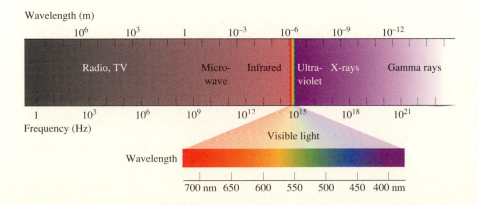

FIGURE 34-10 The electromagnetic spectrum ranges from radio waves to gamma rays, with visible light occupying only a narrow range of wavelengths and frequencies. Note the logarithmic scale, in which intervals between numbered tick marks correspond to factors-of-1000 changes in frequency and wavelength.

FIGURE 34-11 Subtle variations in body temperature are color coded in this infrared image of a human face. Regions at different temperatures emit different frequencies of infrared; these are coded into the image's visible colors according to the temperature scale at the top.

FIGURE 34-12 X rays are high-frequency, short wavelength electromagnetic waves that readily penetrate soft tissues but not bone. Here, an X-ray image reveals rheumatoid arthritis in a human knee joint.

Figure 34-10 shows the entire **electromagnetic spectrum**, including frequencies and wavelengths that differ by many orders of magnitude either way from those of visible light. The invisible electromagnetic waves beyond the narrow visible range were unknown in Maxwell's time. A brilliant confirmation of Maxwell's theory came in 1888, when the German physicist Heinrich Hertz succeeded in generating and detecting electromagnetic waves of much lower frequency than visible light. Hertz intended his work only to verify Maxwell's modification of Ampère's law, but the practical consequences have proven enormous. In 1896, the Italian scientist Guglielmo Marconi demonstrated that he could generate and detect the so-called Hertzian waves. In 1901, he transmitted electromagnetic waves across the Atlantic Ocean, creating a public sensation. From the pioneering work of Hertz and Marconi, spurred by the theoretical efforts of Maxwell, came the entire technology of radio, television, and microwaves that so dominates modern society. We now consider all electromagnetic waves in the frequency range from a few Hz to about 3×10^{11} Hz as radio waves, with ordinary AM radio at about 10^6 Hz, FM radio at 10^8 Hz, and microwaves used for radar, cooking, and satellite communications at 10^9 Hz and above.

Between radio waves and visible light lies the infrared frequency range. Electromagnetic waves in this region are emitted by warm objects, even when they are not hot enough to glow visibly. For this reason, infrared cameras are used to determine subtle body temperature differences in medical diagnosis, to examine buildings for heat loss, and to study the birth of stars in clouds of interstellar gas and dust (Fig. 34-11).

Beyond the visible region are the ultraviolet rays responsible for sunburn, then the highly penetrating X rays (Fig. 34-12), and finally the gamma rays whose primary terrestrial source is radioactive decay. All these phenomena, from radio to gamma rays, are fundamentally the same: They are all electromagnetic waves, differing only in frequency and wavelength. All travel with speed c, and all consist of electric and magnetic fields produced from each other through the induction processes described by Faraday's and Ampère's laws. Naming the different types of electromagnetic waves is just a matter of convenience; there are no gaps in the continuous range of allowed frequencies and wavelengths. Practical differences arise because waves of different wavelengths interact differently with matter; in particular, shorter wavelengths tend to be generated and absorbed most efficiently by smaller systems.

APPLICATION	*The New Astronomy*

Figure 34-10 shows that visible light occupies only a small part of the electromagnetic spectrum. For centuries our only information about the universe beyond Earth—except for an occasional meteorite—came from visible light. Processes like those occurring on the visible surface of our Sun and many other stars produce predominantly visible light. Optical astronomy, utilizing visible light, gave a good picture of the universe to the extent that it consists of objects not too different from the visible part of the Sun. The restriction to optical astronomy was in part imposed by Earth's atmosphere. Transparent to visible light, the atmosphere is largely opaque to other forms of electromagnetic radiation, although "windows" of relative transparency exist in parts of the radio and infrared bands. The

discovery by Bell Telephone Laboratories electrical engineer Karl Jansky in 1931 that radio waves from outer space can be detected on Earth led to the development of radio astronomy. For decades, radio astronomy has given a picture of the universe that complements the optical view, showing phenomena that are simply not detectable by optical means (Fig. 34-13).

The onset of the space age in the late 1950s finally opened the rest of the electromagnetic spectrum to astronomers. Before this time there were surprisingly few suggestions that anything interesting might be found beyond the visible range. But satellites carrying infrared, ultraviolet, X-ray, and gamma-ray detectors have literally revolutionized our view of the universe

(a)   (b)

FIGURE 34-13 (a) Some of the 27 dish antennas that comprise the Very Large Array (VLA) radio telescope in New Mexico. Each dish is 25 m in diameter, and the array can be configured to occupy an area larger than metropolitan Washington, DC. (b) The galaxy Centaurus A, a powerful radio emitter, imaged with the VLA. The two lobes are jets of material ejected from the galaxy's central core. The VLA was tuned to a wavelength of 20 cm for this observation.

ena that were once bizarre conjectures of theoreticians are now observed regularly. For example, astronomers are convinced that massive black holes—with the mass of a million Suns—lurk at the centers of galaxies, including our own. Closer to home, observations of the Sun with ultraviolet and X-ray instruments have brought new understandings of the star that sustains us. And by turning space-borne infrared detectors toward Earth, we have learned much about the structure and resources of our own planet (Fig. 34-15).

FIGURE 34-14 This false-color infrared image of the Andromeda galaxy was taken with the Infrared Astronomical Satellite. Yellow indicates regions of brightest infrared emission, corresponding to places where new stars are probably forming.

(Fig. 34-14). Exotic objects such as neutron stars and black holes are now objects of astronomical study. The opening of the entire electromagnetic spectrum has brought a new richness to astronomy, showing that our universe contains some of the most unusual objects that the laws of physics permit. Phenom-

FIGURE 34-15 This satellite image documents destruction of the Brazilian rainforest by human activities. False colors enhance the distinction between healthy forest (dark green) and cleared areas (light green, pink).

34-8 POLARIZATION

The fields of an electromagnetic wave in vacuum (and in common materials such as air and glass) are perpendicular to the propagation direction, but within the plane perpendicular to the wave propagation the orientation of one field is still arbitrary. **Polarization** is a wave property that specifies the electric field

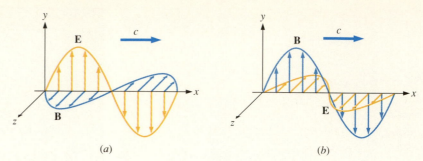

FIGURE 34-16 Field vectors for two electromagnetic waves. Both are propagating in the $+x$ direction, but have different polarizations. The polarization direction is that of the electric field, so the wave in (a) is vertically polarized (i.e., its electric field is in the y direction), while the wave in (b) is horizontally polarized (i.e., in the z direction). Unpolarized light would be a mix of such waves with their electric fields oriented at random in many different directions.

16.4

Polarizer

direction; since the two fields are perpendicular, polarization also determines the magnetic field direction (Fig. 34-16).

Electromagnetic waves used in radio, TV, and radar originate from antennas whose configuration gives the waves a definite polarization. The light waves produced by most lasers are also polarized. In contrast, visible light from hot sources such as the Sun or a light bulb is **unpolarized**, consisting of a mixture of electromagnetic waves with random field orientations.

Unpolarized light may be polarized either by reflection off surfaces or when it passes through substances whose molecular or crystal structure has a preferred direction called the **transmission axis**. Many crystals and synthetic materials such as the plastic Polaroid have this property. For example, sunlight reflecting off the hood of a car becomes partially polarized in the horizontal direction. Polaroid sunglasses, with their transmission axis vertical, block this reflected glare with only a modest reduction in overall light intensity.

A polarizing material passes unattenuated only that component of the wave's electric field that lies along its preferred direction. If θ is the angle between the field and the polarizer's preferred direction, then the field component in the preferred direction is $E\cos\theta$. As we will show shortly, the intensity S of an electromagnetic wave is proportional to the square of the field strength; as a result a wave of intensity S_0 incident on a polarizer emerges with intensity given by the so-called **Law of Malus**:

$$S = S_0\cos^2\theta. \qquad (34\text{-}19)$$

This equation shows that electromagnetic waves will be blocked completely if $\theta = 90°$, a situation that occurs when unpolarized light passes through one polarizer to give it a definite polarization, then through another oriented at 90° to the first (Fig. 34-17).

When unpolarized light passes through a polarizer its intensity is cut in half. You can see this from Equation 34-19 because the unpolarized light includes a mix of waves with random polarization angles θ. Averaging over all possible angles in Equation 34-19 amounts to taking the average of $\cos^2\theta$ over a full cycle. We've seen on several occasions that the average of the square of a sinusoidal function is $\frac{1}{2}$ (recall Fig. 16-16), so a polarizer does indeed cut the intensity of unpolarized light in half.

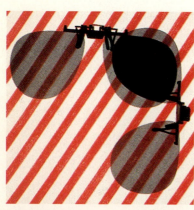

FIGURE 34-17 Two pairs of Polaroid sunglasses with their transmission axes at right angles. Where they overlap, no light can get through.

| **EXAMPLE 34-3** | *Multiple Polarizers* |

Unpolarized light with intensity S_0 is incident on a "stack" of three polarizers. The first has its polarization axis vertical, the second is at $25°$ to the vertical, and the third is at $70°$ to the vertical (Fig. 34-18). What is the intensity of light emerging from this stack?

Solution

We've just seen that the first polarizer cuts the unpolarized intensity in half, giving $\frac{1}{2}S_0$ for the intensity incident on the second polarizer. Equation 34-19 shows that the second and third polarizers each reduce the intensity by a factor $\cos^2\theta$, where θ is the angle between the incident polarization direction—established by one polarizer—and the next polarizer's axis. For the second polarizer this angle is $25°$; for the third it is $70° - 25° = 45°$. Thus light emerges from the stack with intensity

$$S = (\tfrac{1}{2})(\cos^2 25°)(\cos^2 45°)S_0 = 0.205S_0.$$

Interestingly, this is greater than the intensity we would get passing light through a vertical polarizer followed by a single polarizer at $70°$. And, as the exercise below shows, it is much

FIGURE 34-18 A stack of polarizers. Arrows on the sheets indicate directions of the polarization axes.

greater than what we would get by interchanging the second and third polarizers. Can you see why?

EXERCISE Rework Example 34-3 with the second and third polarizers interchanged.

Answer: $S = 0.029S_0$

Some problems similar to Example 34-3: 35, 36, 72

Polarization can tell us much about sources of electromagnetic waves or about materials through which the waves travel. Many astrophysical processes produce polarized waves; measuring the polarization then gives clues to the mechanisms operating in distant objects. Polarization of light as it passes through materials helps geologists to understand the composition and formation of rocks (Fig. 34-19) and helps engineers to locate stresses in structures (Fig. 34-20).

FIGURE 34-19 Photomicrograph of a thin section of rock placed between crossed polarizers. Individual mineral crystals, here a few mm in size, rotate the light's electric field, altering the transmitted light intensity.

FIGURE 34-20 Plastic model of a Gothic cathedral, photographed between polarizing sheets. The resulting patterns reveal stresses, helping architects and engineers understand the response of the building to wind and weight loading—topics related to our study of static equilibrium in Chapter 14.

| APPLICATION | *Liquid Crystal Displays and Electro-Optic Modulation* |

They're everywhere, from watches, calculators, and gas pumps to laptop computers—the **liquid-crystal displays** (LCDs) that convert electrical signals into letters, numbers, pictures, and graphs that we can read. Fundamental to their operation is the polarization of light.

In Fig. 34-17 we saw that two polarizers oriented at right angles block all light from getting through the pair. But if we could rotate the light's polarization direction through 90° while it's between the polarizers, then light from the first polarizer would reach the second aligned with the latter's preferred direction, and so would pass through. That's just what happens in an LCD—but it happens selectively, under control of an electrical signal, allowing the transmission of light to be turned on or off.

In Chapter 23 we introduced liquid crystals as systems of long dipole-like molecules that align with an applied electric field. Figure 34-21 shows how this property is used to make a display. The liquid crystal is contained between two transpar-

ent striated plates that cause the molecules to align with their striations. The two plates' striations are mutually perpendicular, so the molecules' alignment gradually changes through the device. The electric field of a light wave interacts with the molecules, with the result that the polarization direction of the light rotates through 90° as it passes through the liquid crystal. The whole system is between crossed polarizers and so, because of the 90° rotation of the polarization, light incident on the system gets through (Fig. 34-21a).

Now suppose a voltage is applied between the plates confining the liquid crystal. This results in an electric field that aligns the molecules as shown in Fig. 34-21b. In this new orientation the liquid crystal no longer rotates the polarization of the light, so the situation is just like that of Fig. 34-17, and the light is blocked. There's one such system for each of the segments that make up the numbers and letters of a liquid crystal display, and each segment can be turned on or off to make the desired symbol

FIGURE 34-21 Polarization plays a central role in the operation of a liquid-crystal display.

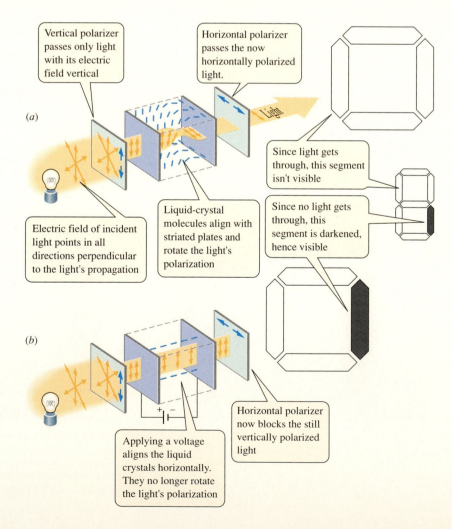

Vertical polarizer passes only light with its electric field vertical

Horizontal polarizer passes the now horizontally polarized light.

(a)

Light

Electric field of incident light points in all directions perpendicular to the light's propagation

Liquid-crystal molecules align with striated plates and rotate the light's polarization

Since light gets through, this segment isn't visible

Since no light gets through, this segment is darkened, hence visible

(b)

Applying a voltage aligns the liquid crystals horizontally. They no longer rotate the light's polarization

Horizontal polarizer now blocks the still vertically polarized light

In Fig. 34-21 we show light incident from the back of the display and emerging from the front for viewing. In many LCDs the light at the back is replaced by a mirror that reflects ambient light. Then there's no power needed to provide light, and the LCD operates at minimal electrical power. Low power consumption is one of the key advantages of liquid-crystal displays.

Liquid crystals are rather slow in their response, so they aren't good for rapid switching of light. But some solid crystals have much the same property as the liquid-crystal system of Fig. 34-21, in that they selectively rotate the polarization of light depending on the presence or absence of an electric field (in this case, it takes a field to produce the rotation). This is known as the **electro-optic effect**.

Placing an electro-optic crystal between crossed polarizers makes an **electro-optic modulator** (EOM) that can switch light on timescales as short as 10^{-10} s. Such devices, used as electronic shutters to interrupt laser beams and thus produce very short pulses of light, have been used to make some of the most accurate measurements of the speed of light. Applying a varying voltage to the EOM causes the light transmission to vary continuously. Voice or music, for example, can be transmitted on a laser beam by applying the amplified signal from a microphone to an EOM. EOMs are used in a wide variety of scientific and commercial applications. An especially important use is the conversion of electrical signals to light pulses on the highest-speed fiber optic communications systems.

34-9 PRODUCING ELECTROMAGNETIC WAVES

We have shown that electromagnetic waves can exist, and have explored some of their properties. But how do these waves originate?

All that's necessary is to produce a changing electric or magnetic field. Once a changing field of either type exists, Faraday's and Ampère's laws ensure the production of the other type—and so on, to give a propagating electromagnetic wave. Ultimately, changing fields of either type occur when we alter the motion of electric charge. Therefore:

▍ *Accelerated* electric charge is the source of electromagnetic waves.

In a radio transmitter, the accelerated charges are electrons moving back and forth in an antenna, driven by an alternating voltage from an *LC* circuit (Fig. 34-22). In an X-ray tube, high-energy electrons decelerate rapidly as they

FIGURE 34-22 (*a*) Simplified diagram of a radio transmitter. Oscillations of an *LC* circuit drive alternating current (green arrows) in the antenna, while a power source replenishes energy carried away in electromagnetic waves. (*b*)–(*d*) A changing electric field arises from the changing charge distribution in the antenna. The changing electric field induces a changing magnetic field, and the field structure propagates away from the antenna at the speed of light. Frames (*b*) and (*d*) are half a wave period apart in time, at times when charge separation in the antenna is a maximum.

slam into a target; their deceleration is the source of the electromagnetic waves, now in the X-ray region of the spectrum. In the magnetron tube of a microwave oven, electrons circle in a magnetic field; their centripetal acceleration is the source of the microwaves that cook your food. And the altered movement of electrons in atoms—although described accurately only by quantum mechanics—is the source of most visible light. If the motion of the accelerated charges is periodic, then the wave frequency is that of the motion; more generally, systems are most efficient at producing (and receiving) electromagnetic waves whose wavelength is comparable to the size of the system. That's why TV antennas are on the order of 1 m in size, while nuclei—some 10^{-15} m in diameter—produce gamma rays.

Calculation of electromagnetic waves emitted by accelerated charges presents challenging but important problems for physicists and communications engineers. Figure 34-23a shows a "snapshot" of the electric field produced by a single point charge undergoing simple harmonic motion, while Fig. 34-23b shows the field of an oscillating dipole—a configuration approximated by many systems from antennas to atoms and molecules. Both figures show that the waves are strongest in the direction at right angles to the acceleration of the charge distribution and that there is no radiation in the direction of the acceleration. This accounts for, among other phenomena, the directionality of radio and TV antennas, which transmit and receive most effectively perpendicular to the long direction of the antenna.

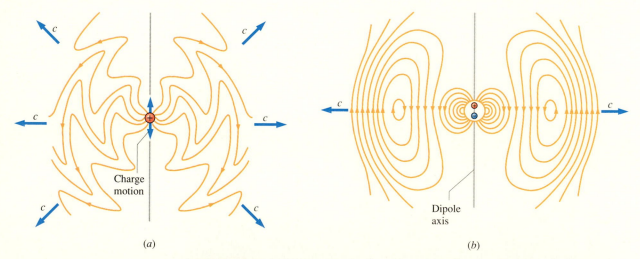

(a) (b)

FIGURE 34-23 "Snapshots" showing the electric fields of oscillating charge distributions. (a) A single point charge executing simple harmonic motion in the horizontal plane. Note that the field close to the charge approximates the radial field of a stationary point charge, but that farther out the "kinks" in the field—resulting from the accelerated motion—are more prominent and become essentially perpendicular to the radial direction. These transverse kinks move outward at the speed of light, and constitute the wave fields. (b) The electric field of an oscillating dipole. Note that the field forms closed loops, detached from the dipole. These are the outward propagating wave fields. The larger field loops shown were formed when the oscillating dipole had the opposite orientation. Not shown in either figure are the equally important magnetic fields.

The fields shown in Fig. 34-23 seem to bear little resemblance to the plane-wave fields of Fig. 34-7 that we used to demonstrate the possibility of electromagnetic waves. We could produce true plane waves only with an infinite sheet of accelerated charge—an obvious impossibility. But far from the source, the curved field lines evident in Fig. 34-23b, for example, would appear straight, and the wave would begin to approximate a plane wave. So our plane-wave analysis is a valid approximation at great distances—typically many wavelengths—from a localized wave source. Closer to the source more complicated expressions for the wave fields apply, but these, too, satisfy Maxwell's equations.

34-10 ENERGY IN ELECTROMAGNETIC WAVES

We showed in previous chapters that electric and magnetic fields contain energy. Here we have considered electromagnetic waves, in which a combination of electric and magnetic fields propagates through space. As the wave moves, it must transport the energy contained in those fields.

We define **intensity**, S, as the rate at which an electromagnetic wave transports energy across a unit area. This is the same definition we used for wave intensity in Chapter 16, and the units are also the same: power per unit area, or W/m^2. We can calculate the intensity of a plane electromagnetic wave by considering a rectangular box of thickness dx and cross-sectional area A with its face perpendicular to the wave propagation (Fig. 34-24). Within this box are wave fields **E** and **B** whose energy densities are given by Equations 26-3 and 32-11:

$$u_E = \tfrac{1}{2}\varepsilon_0 E^2$$

$$u_B = \frac{B^2}{2\mu_0}.$$

FIGURE 34-24 A box of length dx and cross-sectional area A at right angles to the propagation of an electromagnetic wave.

If dx is sufficiently small, the fields don't change much over the box, so the total energy in the box is just the sum of the electric and magnetic energy densities multiplied by the box volume $A\,dx$:

$$dU = (u_E + u_B)\,A\,dx = \frac{1}{2}\left(\varepsilon_0 E^2 + \frac{B^2}{\mu_0}\right)A\,dx.$$

This energy moves with speed c, so all the energy contained in the box length dx moves out of the box in a time $dt = dx/c$. The rate at which energy moves through the cross-sectional area A is then

$$\frac{dU}{dt} = \frac{1}{2}\left(\varepsilon_0 E^2 + \frac{B^2}{\mu_0}\right)\frac{A\,dx}{dx/c} = \frac{c}{2}\left(\varepsilon_0 E^2 + \frac{B^2}{\mu_0}\right)A.$$

So the intensity S, or rate of energy flow per unit area, is

$$S = \frac{c}{2}\left(\varepsilon_0 E^2 + \frac{B^2}{\mu_0}\right).$$

We can recast this equation in simpler form by noting that, for an electromagnetic wave, $E = cB$ and $B = E/c$. Using these expressions to replace one of

(a)

(b)

FIGURE 34-25 (a) The intensity of candle light is only a few W/m². (b) This laser produces a modest 3W of power, but because its beam diameter is less than 0.5 mm, its intensity is some 20 MW/m²—2000 times that of bright sunlight.

the E's in the term E^2 with B and similarly one of the B's in the term B^2 with E, we have

$$S = \frac{c}{2}\left(\varepsilon_0 c E B + \frac{EB}{\mu_0 c}\right) = \frac{1}{2\mu_0}(\varepsilon_0 \mu_0 c^2 + 1)\, EB.$$

But $c = 1/\sqrt{\varepsilon_0 \mu_0}$, so $\varepsilon_0 \mu_0 c^2 = 1$, giving

$$S = \frac{EB}{\mu_0}. \tag{34-20a}$$

Although we derived Equation 34-20a for an electromagnetic wave, it is in fact a special case of the more general result that nonparallel electric and magnetic fields are accompanied by a flow of electromagnetic energy. In general, the rate of energy flow per unit area is given by

$$\mathbf{S} = \frac{\mathbf{E} \times \mathbf{B}}{\mu_0}. \tag{34-20b}$$

Here a vector **S** is used to signify not only the magnitude of the energy flow, but also its direction. For an electromagnetic wave in vacuum, in which **E** and **B** must be at right angles, Equation 34-20b reduces to Equation 34-20a, with the direction of energy flow the same as the direction of wave travel. The vector intensity **S** is called the **Poynting vector** after the English physicist J. H. Poynting, who suggested it in 1884. Poynting's name is especially fortuitous, for the Poynting vector points in the direction of energy flow. Problem 76 explores an important application of the Poynting vector to fields that do not constitute an electromagnetic wave.

Equations 34-20 give the intensity at the instant when the fields have magnitudes E and B. In an electromagnetic wave the fields oscillate, and so does the intensity. We're usually not interested in this rapid oscillation. For example, an engineer designing a solar collector doesn't care that sunlight intensity oscillates at about 10^{14} Hz. What she really wants is the *average* intensity, $\bar{S}$. Because the instantaneous intensity of Equation 34-20a contains a product of sinusoidally varying terms, which are in phase, the average intensity is just half the peak intensity:

$$\bar{S} = \frac{\overline{EB}}{\mu_0} = \frac{E_p B_p}{2\mu_0} \quad \text{(average intensity)} \tag{34-21a}$$

(This follows because, as we've seen on several occasions, the average of $\sin^2 \omega t$ over one cycle is $\frac{1}{2}$.) Typical values for $\bar{S}$ in visible light range from a few W/m² in the faint light of a candle to many MW/m² in the most intense laser beams (Fig. 34-25).

We wrote Equation 34-21a in terms of both the electric and magnetic fields, but we can use the wave condition $E = cB$ to eliminate either field in terms of the other:

$$\bar{S} = \frac{E_p^{\,2}}{2\mu_0 c} \tag{34-21b}$$

and

$$\bar{S} = \frac{cB_p^{\,2}}{2\mu_0}. \tag{34-21c}$$

Got It!

Lasers 1 and 2 emit light of the same color, and the electric field in the beam from laser 1 is twice as strong as that in laser 2's beam. How do their (a) magnetic fields, (b) intensities, and (c) wavelengths compare?

EXAMPLE 34-4	*Solar Energy*

The average intensity of sunlight on a clear day at noon is about 1 kW/m^2 (Fig. 34-26). What are the electric and magnetic fields in sunlight? How many solar collectors would you need to replace a 4.8-kW electric water heater in noonday sun, if each collector has an area of 2.0 m^2 and converts 40% of the incident sunlight to heat?

Solution

Solving Equation 34-21b for the electric field gives

$$E_p = \sqrt{2\mu_0 c\overline{S}}$$

$$= [(2)(4\pi \times 10^{-7} \text{ H/m})(3.0 \times 10^8 \text{ m/s})(1 \times 10^3 \text{ W/m}^2)]^{1/2}$$

$$= 0.87 \text{ kV/m}.$$

The peak magnetic field is then given by $B_p = E_p/c$, so

$$B_p = \frac{E_p}{c} = \frac{870 \text{ V/m}}{3.0 \times 10^8 \text{ m/s}} = 3 \times 10^{-6} \text{ T}.$$

At 1 kW/m^2, we would then need 4.8 m^2 of collector area if the collectors were 100% efficient. At 40% efficiency, we therefore need 4.8 m^2/0.40 = 12 m^2, for a total of 6 collectors.

FIGURE 34-26 In bright sunlight, energy is incident on each square meter of these solar collectors at the rate of about 1000 watts.

EXERCISE A laser produces an average power of 7.0 W in a light beam 1.0 mm in diameter. Find (a) the average intensity and (b) the peak electric field of the laser light.

Answers: (a) 8.9 MW/m^2; (b) 82 kV/m

Some problems similar to Example 34-4: 38–41, 43

Waves from Localized Sources

As an electromagnetic wave propagates through empty space, its total energy does not change. With plane waves the intensity—power per unit area—does not change either. But when a wave originates in a localized source such as an atom, a radio transmitting antenna, a light bulb, or a star, its wavefronts are not planes but expanding spheres (recall Fig. 16-18). The wave's total energy remains the same, but as it expands that energy is spread over the area of an ever larger sphere—whose area increases as the square of the distance from the source. Therefore, as we found in Chapter 16, the power per unit area—the intensity—decreases as the inverse square of the distance:

$$S = \frac{P}{4\pi r^2}. \tag{34-22}$$

Here S and P can be either the peak or average intensity and power, respectively, and r is the distance from a localized source. This intensity decrease occurs not because electromagnetic waves "weaken" and lose energy but because that energy gets spread ever more thinly.

Because the intensity of an electromagnetic wave is proportional to the *square* of the field strengths (Equations 34-21), Equation 34-22 shows that the *fields* of a spherical wave decrease as $1/r$. Contrast that with the $1/r^2$ decrease in the electric field of a stationary point charge, and you can see why the electromagnetic wave fields associated with an accelerated charge dominate in all but the immediate vicinity of the charge (see Fig. 34-23a).

EXAMPLE 34-5	*A Garage-Door Opener*

A radio-activated garage-door opener responds to signals with average intensity as weak as 20 μW/m^2. If the transmitter unit produces a 240-mW signal, broadcast in all directions, what is the maximum distance at which the transmitter will activate the door opener? What is the minimum value for the peak electric field to which the unit responds?

Solution

Since the waves spread out in all the directions, Equation 34-22 applies. Solving for r gives

$$r = \sqrt{\frac{P}{4\pi S}} = \sqrt{\frac{240\times10^{-3}\ \text{W}}{(4\pi)(20\times10^{-6}\ \text{W/m}^2)}} = 31\ \text{m}.$$

Solving Equation 34-21b gives the electric field corresponding to the unit's 20-μW/m^2 sensitivity:

$$E_p = \sqrt{2\mu_0 c \overline{S}}$$

$$= \sqrt{(2)(4\pi\times10^{-7}\ \text{H/m})(3.00\times10^8\ \text{m/s})(20\times10^{-6}\ \text{W/m}^2)}$$

$$= 0.12\ \text{V/m}.$$

The sensitivity of radio receiving equipment is often expressed in terms of the minimum electric field strength.

EXERCISE A stereo receiver's AM tuner section has a rated sensitivity of 2.1 mV/m. What is the maximum distance at which this unit can receive broadcasts from a radio station's 5.0-kW transmitter, assuming the signal is broadcast in all directions?

Answer: 261 km

Some problems similar to Example 34-5: 49, 51, 52, 61, 62

34-11 WAVE MOMENTUM AND RADIATION PRESSURE

We know from mechanics that moving objects carry both energy and momentum. The same is true for electromagnetic waves. Maxwell showed that the wave energy U and momentum p are related by

$$p = \frac{U}{c}. \tag{34-23}$$

If an electromagnetic wave is incident on an object and the object absorbs the wave energy (as, for example, a black object exposed to sunlight), then the object also absorbs the momentum given by Equation 34-23. If the wave's average intensity is $\overline{S}$, then it carries energy per unit area at the average rate $\overline{S}$ J/s/m^2. According to Equation 34-23 it therefore carries momentum per unit area at the rate $\overline{S}/c$. Newton's law in its general form $\mathbf{F} = d\mathbf{p}/dt$ tells us that the rate of change of an object's momentum is equal to the net force on the object. Therefore, if an object absorbs electromagnetic wave momentum $\overline{S}/c$ per unit area per unit time, it experiences a force per unit area of this magnitude. Since force per unit area is pressure, we call this quantity the **radiation pressure**:

$$P_{\text{rad}} = \frac{\overline{S}}{c}. \tag{34-24}$$

FIGURE 34-27 The bright dot at the center of this picture is a 20-micron particle levitated by a laser beam reflected upward by the prism shown at the bottom.

FIGURE 34-28 Two-stage sailing spacecraft proposed for interstellar travel. At the target star, laser light beamed all the way from the solar system would reflect from the large sail to the smaller one, bringing the latter to a stop. The diameter of the large sail is 1000 km.

The radiation pressure is doubled if an object reflects electromagnetic waves, in the same way that bouncing a basketball off a backboard changes the ball's momentum by $2mv$ and, therefore, delivers momentum $2mv$ to the backboard.

The pressure exerted by ordinary light is very small, but Dartmouth College physicists E. F. Nichols and G. F. Hull demonstrated its existence in a sensitive experiment performed in 1903. With high-energy laser light or with objects of low mass and large area, radiation pressure can be appreciable. Lasers exert enough light pressure to levitate small particles (Fig. 34-27), and the pressure of light has been suggested as a means of driving interstellar "sailing ships" (Fig. 34-28). Finally, the idea that electromagnetic waves carry momentum played a crucial role in Einstein's development of his famous equation $E = mc^2$.

EXAMPLE 34-6	*Star Wars*

A proposed ballistic missile defense system calls for a laser that can focus 25 MW of light on an attacking warhead. The weapon works by heating the warhead to the point of destruction, but it also delivers momentum that alters the warhead's trajectory. If the beam dwells on a 200-kg warhead for 15 s, what velocity change does it impart to the warhead? Estimate the distance by which the warhead will be knocked off course over its remaining 30 minutes of flight.

Solution
The energy delivered in a 25-MW beam acting for 15 s is (25 MW)(15 s) = 375 MJ. According to Equation 34-23, the associated momentum is

$$p = \frac{U}{c} = \frac{375 \times 10^6 \text{ J}}{3.00 \times 10^8 \text{ m/s}} = 1.25 \text{ kg·m/s}.$$

The change in the warhead's velocity is given by $m\Delta v = \Delta p$, where Δp is the change in its momentum. Assuming the warhead absorbs all the beam's momentum, we then have

$$\Delta v = \frac{\Delta p}{m} = \frac{1.25 \text{ kg·m/s}}{200 \text{ kg}} = 0.00625 \text{ m/s}.$$

This is insignificant compared with a typical warhead speed of 7 km/s. Even though we don't know the direction of the velocity change, we can estimate crudely the error Δx in the impact point by multiplying this change by the flight time:

$$\Delta x = \Delta v \, t = (0.00625 \text{ m/s})(30 \text{ min})(60 \text{ s/min}) = 11 \text{ m}.$$

This is totally negligible, especially for a nuclear warhead. Even with this enormously powerful laser, radiation pressure has an insignificant effect. It's the energy delivered, not the momentum, that matters here.

EXERCISE A laser delivers 5.0 MW/m². If the beam is directed upward, what is the maximum mass for a 100-μm-diameter particle to be suspended in the beam?

Answer: 1.3×10^{-11} kg

Some problems similar to Example 34-6: 56–59

CHAPTER SYNOPSIS

Summary

1. Maxwell's modification of Ampère's law adds a **displacement current** term $\varepsilon_0 d\phi_E/dt$, showing that changing electric flux is a source of magnetic field. This modified law completes the set of **Maxwell's equations**—the four equations that govern the behavior of electromagnetic fields.

2. The interplay of electric and magnetic fields described by Faraday's and Ampère's laws gives rise to **electromagnetic waves**. When they propagate through vacuum, these waves
   a. Travel at the speed of light, $c = 1/\sqrt{\varepsilon_0\mu_0} = 3.00 \times 10^8$ m/s.
   b. Have their electric and magnetic fields at right angles to each other and to the direction of wave propagation.
   c. Have their fields in phase, with magnitudes related by $E = cB$.
   d. Can have any frequency or wavelength, provided the two are related by the equivalent expressions $f\lambda = c$ or $\omega/k = c$.

3. Radio waves, television, microwaves, infrared, visible light, ultraviolet, X rays, and gamma rays are all forms of electromagnetic radiation. They differ only in frequency and wavelength, and together comprise the **electromagnetic spectrum**.

4. **Polarization** describes the orientation of a wave's electric field in the plane perpendicular to the propagation direction. When a polarized wave passes through a polarizing material, its intensity is reduced by a factor $\cos^2\theta$, where θ is the angle between the wave polarization and the preferred axis of the material.

5. Electromagnetic waves are produced by accelerated electric charges, as in the alternating current of a radio antenna.

6. Electromagnetic waves carry energy. The rate at which energy is transported per unit area is the wave **intensity**. The **Poynting vector**,

$$\mathbf{S} = \frac{\mathbf{E} \times \mathbf{B}}{\mu_0},$$

describes this energy transport for any configuration of electromagnetic fields. The average intensity has half the peak value, or

$$\bar{S} = \frac{E_p B_p}{2\mu_0}.$$

The intensity of a plane wave remains constant, while the intensity from a localized source decreases as the inverse square of the distance from the source.

7. An electromagnetic wave with energy U also carries momentum $p = U/c$. As a result, it exerts a **radiation pressure** $P_{rad} = \bar{S}/c$ on an object that absorbs the wave and twice this pressure on a reflecting object.

Terms You Should Understand

(Pairs are closely related terms whose distinction is important; number in parentheses is chapter section where term first appears.)

displacement current (34-2)
Maxwell's equations (34-3)
electromagnetic wave (34-4)
electromagnetic spectrum (34-7)
polarization (34-8)
intensity, Poynting vector (34-10)
radiation pressure (34-11)

Symbols You Should Recognize

$\varepsilon_0 d\phi_E/dt$ (34-2)
$S, \mathbf{S}, \bar{S}$ (34-10)
c (34-5)
P_{rad} (34-11)

Problems You Should Be Able to Solve

evaluating induced magnetic fields in symmetric situations (34-2)
relating frequency and wavelength of electromagnetic waves (34-6)
relating electric and magnetic field strengths in electromagnetic waves (34-6)
calculating light intensity emerging from one or more polarizers (34-8)
relating wave intensity and fields (34-10)
evaluating wave intensity and fields as a function of distance from localized sources (34-10)
calculating radiation pressure and its effects (34-11)

Limitations to Keep in Mind

The description of electromagnetic waves developed in this chapter applies strictly only in vacuum.

QUESTIONS

1. Why is Maxwell's modification of Ampère's law essential to the existence of electromagnetic waves?
2. The presence of magnetic monopoles would require modification of Gauss's law for magnetism. Which of the other Maxwell equations would also need modification?
3. There is displacement current between the plates of a charging capacitor, yet no charge is moving between the plates. In what sense is the word "current" appropriate here?
4. Is there displacement current in an electromagnetic wave? Is there ordinary conduction current?
5. List some similarities and differences between electromagnetic waves and sound waves.
6. What aspect of the electromagnetic wave considered in Section 34-4 ensures that Gauss's laws for electricity and magnetism are satisfied?
7. Explain why parallel electric and magnetic fields in vacuum could not constitute an electromagnetic wave.
8. The speed of an electromagnetic wave is given by $c = \lambda f$. How does the speed depend on frequency? On wavelength?
9. When astronomers observe a supernova explosion in a distant galaxy, they see a sudden, simultaneous rise in visible light and other forms of electromagnetic radiation. How is this evidence that the speed of light is independent of frequency?
10. Turning a TV antenna so its rods point vertically may change the quality of your TV reception. Why? Think about polarization.
11. Unpolarized light is incident on two sheets of Polaroid with their polarization directions at right angles, and no light

gets through. A third sheet is inserted between the other two, and now some light gets through. How can this be?
12. Why is it not possible to define exactly where the visible region of the spectrum ends?
13. Why did the field of X-ray astronomy flourish only after the advent of space flight?
14. The Sun emits most of its electromagnetic wave energy in the visible region of the spectrum, with the peak in the yellow-green. Our eyes are sensitive to the same range, with peak sensitivity in the yellow-green. Is this a coincidence?
15. Suppose your eyes were sensitive to radio waves rather than light. What things would look bright?
16. An *LC* circuit is made entirely from superconducting materials, yet its oscillations eventually damp out. Why?
17. If you double the field strength in an electromagnetic wave, what happens to the intensity?
18. The intensity of light falls off as the inverse square of the distance from the source. Does this mean that electromagnetic wave energy is lost? Explain.
19. When your picture is taken with a flash camera, why doesn't the momentum of the light flash knock you over?
20. Some long-distance power transmission lines use DC rather than AC, despite the need to convert between DC and AC at either end. Why might this be? What energy loss mechanism occurs with AC but not DC?
21. Electromagnetic waves do not readily penetrate metals. Why might this be?

PROBLEMS

Section 34-2 Ambiguity in Ampère's Law

1. A uniform electric field is increasing at the rate of 1.5 V/m·μs. What is the displacement current through an area of 1.0 cm^2 at right angles to the field?
2. A parallel-plate capacitor has square plates 10 cm on a side and 0.50 cm apart. If the voltage across the plates is increasing at the rate of 220 V/ms, what is the displacement current in the capacitor?
3. A parallel-plate capacitor of plate area A and spacing d is charging at the rate dV/dt. Show that the displacement current in the capacitor is equal to the conduction current flowing in the wires feeding the capacitor.
4. A capacitor with circular plates is fed with long, straight wires along the axis of the plates. Show that the magnetic field *outside* the capacitor, in a plane that passes through the interior of the capacitor and is perpendicular to the

axis, is given by $B = \dfrac{\mu_0 \varepsilon_0 R^2}{2rd} \dfrac{dV}{dt}$. Here R is the plate radius, d the spacing, dV/dt the rate of change of the capacitor voltage, and r the distance from the axis.

5. A parallel-plate capacitor has circular plates with radius 50 cm and spacing 1.0 mm. A uniform electric field between the plates is changing at the rate 1.0 MV/m·s. What is the magnetic field between the plates (a) on the symmetry axis, (b) 15 cm from the axis, and (c) 150 cm from the axis?
6. An electric field points into the page and occupies a circular region of radius 1.0 m, as shown in Fig. 34-29. There are no electric charges in the region, but there is a magnetic field forming closed loops pointing clockwise, as shown. The magnetic field strength 50 cm from the center of the

FIGURE 34-29 Problem 6.

FIGURE 34-30 Problem 14.

region is 2.0 μT. (a) What is the rate of change of the electric field? (b) Is the electric field increasing or decreasing?

Section 34-4 Electromagnetic Waves

7. At a particular point the instantaneous electric field of an electromagnetic wave points in the $+y$ direction, while the magnetic field points in the $-z$ direction. In what direction is the wave propagating?
8. The fields of an electromagnetic wave are $\mathbf{E} = E_p \sin(kz + \omega t)\hat{\mathbf{j}}$ and $\mathbf{B} = B_p \sin(kz + \omega t)\hat{\mathbf{i}}$. Give a unit vector in the direction of propagation.
9. The electric field of a radio wave is given by $\mathbf{E} = E\sin(kz - \omega t)(\hat{\mathbf{i}} + \hat{\mathbf{j}})$. (a) What is the peak amplitude of the electric field? (b) Give a unit vector in the direction of the magnetic field at a place and time where $\sin(kz - \omega t)$ is positive.
10. Show by differentiation and substitution that Equations 34-12b and 34-13 can be satisfied by fields of the form $E(x, t) = E_p f(kx \pm \omega t)$ and $B(x, t) = B_p f(kx \pm \omega t)$, where f is any function of the argument $kx \pm \omega t$.
11. Show that it is impossible for an electromagnetic wave in vacuum to have a time-varying component of its electric field in the direction of its magnetic field. *Hint:* Assume **E** does have such a component, and show that you cannot satisfy both Gauss and Faraday.

Section 34-5 The Speed of Electromagnetic Waves

12. A light-minute is the distance light travels in one minute. Show that the Sun is about 8 light-minutes from Earth.
13. Your intercontinental telephone call is carried by electromagnetic waves routed via a satellite in geosynchronous orbit at an altitude of 36,000 km. Approximately how long does it take before your voice is heard at the other end?
14. An airplane's radar altimeter works by bouncing radio waves off the ground and measuring the round-trip travel time (Fig. 34-30). If that time is 50 μs, what is the altitude?

15. Roughly how long does it take light to go 1 foot?
16. If you speak via radio from Earth to an astronaut on the moon, how long is it before you can get a reply?
17. "Ghosts" on a TV screen occur when part of the signal goes directly from transmitter to receiver, while part takes a longer route, reflecting off mountains or buildings (Fig. 34-31). The electron beam in a 50-cm-wide TV tube "paints" the picture by scanning the beam from left to right across the screen in about 10^{-4} s. If a "ghost" image appears displaced about 1 cm from the main image, what is the difference in path lengths of the direct and indirect signals?

FIGURE 34-31 Ghost images of highways appear on this TV traffic report (Problem 17).

18. A computer can fetch information from its memory in 3.0 ns, a process that involves sending a signal from the central processing unit (CPU) to memory and awaiting the return of the information. If signals in the computer's wiring travel at $0.60c$, what is the maximum distance between the CPU and the memory? Your answer shows why high-speed computers are necessarily compact.
19. Problem 69 shows that the speed of electromagnetic waves in a transparent dielectric is given by $1/\sqrt{\kappa \varepsilon_0 \mu_0}$, where κ is the dielectric constant described in Chapter 26. An experimental measurement gives 1.97×10^8 m/s for the speed of light in a piece of glass. What is the dielectric constant of this glass at optical frequencies?

Section 34-6 Properties of Electromagnetic Waves

20. What are the wavelengths of (a) a 100-MHz FM radio wave, (b) a 3.0-GHz radar wave, (c) a 6.0×10^{14}-Hz light wave, and (d) a 1.0×10^{18}-Hz X ray?

21. A 60-Hz power line emits electromagnetic radiation. What is the wavelength?

22. Antennas for transmitting and receiving electromagnetic radiation usually have typical dimensions on the order of half a wavelength. Look at a TV antenna, and estimate the wavelength and frequency of a TV signal.

23. A CB radio antenna is a vertical rod 2.75 m high. If this length is one-fourth of the CB wavelength, what is the CB frequency?

24. A microwave oven operates at 2.4 GHz. What is the distance between wave crests in the oven?

25. What would be the electric field strength in an electromagnetic wave whose magnetic field equalled that of Earth, about 50 μT?

26. Dielectric breakdown in air occurs at an electric field strength of about 3×10^6 V/m. What would be the peak magnetic field in an electromagnetic wave with this value for its peak electric field?

27. A radio receiver can detect signals with electric fields as low as 320 μV/m. What is the corresponding magnetic field?

Section 34-8 Polarization

28. An electromagnetic wave is propagating in the z direction. What is its polarization direction, if its magnetic field is in the y direction?

29. Polarized light is incident on a sheet of polarizing material, and only 20% of the light gets through. What is the angle between the electric field and the polarization axis of the material?

30. Vertically polarized light passes through a polarizer whose polarization axis is oriented at 70° to the vertical. What fraction of the incident intensity emerges from the polarizer?

31. A polarizer blocks 75% of a polarized light beam. What is the angle between the beam's polarization and the polarizer's axis?

32. An electro-optic modulator is supposed to switch a laser beam between fully off and fully on, as its crystal rotates the beam polarization by 90° when a voltage is applied. But a power-supply failure results in only enough voltage for a 72° beam rotation. What fraction of the laser light is transmitted when it is supposed to be fully on?

33. Unpolarized light of intensity S_0 passes first through a polarizer with its polarization axis vertical, then through one with its axis at 35° to the vertical. What is the light intensity after the second polarizer?

34. Vertically polarized light passes through two polarizers, the first at 60° to the vertical and the second at 90° to the vertical. What fraction of the light gets through?

35. Unpolarized light with intensity S_0 passes through a stack of five polarizing sheets, each with its axis rotated 20° with respect to the previous one. What is the intensity of the light emerging from the stack?

36. Unpolarized light of intensity S_0 is incident on a "sandwich" of three polarizers. The outer two have their transmission axes perpendicular, while the middle one has its axis at 45° to the others. What is the light intensity emerging from this "sandwich?"

37. Polarized light with average intensity S_0 passes through a sheet of polarizing material which is rotating at 10 rev/s. At time $t = 0$ the polarization axis is aligned with the incident polarization. Write an expression for the transmitted intensity as a function of time.

Section 34-10 Energy in Electromagnetic Waves

38. A typical laboratory electric field is 1000 V/m. What is the average intensity of an electromagnetic wave with this value for its peak field?

39. What would be the average intensity of a laser beam so strong that its electric field produced dielectric breakdown of air (which requires $E_p = 3 \times 10^6$ V/m)?

40. Estimate the peak electric field inside a 625-W microwave oven under the simplifying approximation that the microwaves propagate as a plane wave through the oven's 750-cm^2 cross-sectional area.

41. A radio receiver can pick up signals with peak electric fields as low as 450 μV/m. What is the average intensity of such a signal?

42. Show that the electric and magnetic energy densities in an electromagnetic wave are equal.

43. A laser blackboard pointer delivers 0.10 mW average power in a beam 0.90 mm in diameter. Find (a) the average intensity, (b) the peak electric field, and (c) the peak magnetic field.

44. The laser of Example 34-6 produces a spot 80 cm in diameter at its target. What are the rms electric and magnetic fields at the target?

45. The United States' safety standard for continuous exposure to microwave radiation is 10 mW/cm^2. The glass door of a microwave oven measures 40 cm by 17 cm and is covered with a metal screen that blocks microwaves. What fraction of the oven's 625-W microwave power can leak through the door window without exceeding the safe exposure to someone right outside the door? Assume the power leaks uniformly through the window area.

46. A 1.0-kW radio transmitter broadcasts uniformly in all directions. What is the intensity of its signal at a distance of 5.0 km from the transmitter?

47. Use the fact that sunlight intensity at Earth's orbit is 1368 W/m^2 to calculate the Sun's total power output.

48. About two-thirds of the solar energy at Earth's orbit reaches the planet's surface. At what rate is solar energy incident on the entire Earth? See the previous problem,

and compare your result with the roughly 10^{13} W rate at which humanity consumes energy.

49. During its 1989 encounter with Neptune, the Voyager 2 spacecraft was 4.5×10^9 km from Earth (Fig. 34-32). Its images of Neptune were broadcast by a radio transmitter with a mere 21-W average power output. What would be (a) the average intensity and (b) the peak electric field received at Earth if the transmitter broadcast equally in all directions? (The received signal was actually somewhat stronger because Voyager used a directional antenna.)

FIGURE 34-32 Neptune, photographed by the Voyager II spacecraft when it was 4.5×10^9 km from Earth. How long did it take the radio signal carrying this image to reach Earth? (Problem 49)

50. A quasar 10 billion light-years from Earth appears the same brightness as a star 50,000 light-years away. How does the power output of the quasar compare with that of the star?

51. At 1.5 km from the transmitter, the peak electric field of a radio wave is 350 mV/m. (a) What is the transmitter's power output, assuming it broadcasts uniformly in all directions? (b) What is the peak electric field 10 km from the transmitter?

52. The peak electric field at a point 25 m from a point source of electromagnetic waves is 4.2 kV/m. What is the peak magnetic field 1.0 m from the source?

53. A typical fluorescent lamp is a little over 1 m long and a few cm in diameter. How do you expect the light intensity to vary with distance (a) near the lamp but not near either end and (b) far from the lamp?

Section 34-11 Wave Momentum and Radiation Pressure

54. A camera flash delivers 2.5 kW of light power for 1.0 ms (Fig. 34-33). Find (a) the total energy and (b) the total momentum carried by the flash.

55. What is the radiation pressure exerted on a light-absorbing surface by a laser beam whose intensity is 180 W/cm²?

56. A laser beam shines vertically upward. What laser power is necessary for this beam to support a flat piece of aluminum foil with mass 30 μg and diameter equal to that of the beam? Assume the foil reflects all the light.

FIGURE 34-33 How much energy and momentum are in light from the camera flash? (Problem 54)

57. The average intensity of noonday sunlight is about 1 kW/m². What is the radiation force on a solar collector measuring 60 cm by 2.5 m if it is oriented at right angles to the incident light and absorbs all the light?

58. Serious proposals have been made to "sail" spacecraft to the outer solar system using the pressure of sunlight. How much sail area must a 1000-kg spacecraft have if its acceleration at Earth's orbit is to be 1 m/s²? Assume the sails are made from reflecting material. You neglect the Sun's gravity. (Why?)

59. A 65-kg astronaut is floating in empty space. If the astronaut shines a 1.0-W flashlight in a fixed direction, how long will it take the astronaut to accelerate to a speed of 10 m/s?

60. A "photon rocket" emits a beam of light instead of the hot gas of an ordinary rocket. How powerful a light source would be needed for a photon rocket with thrust equal to that of a space shuttle (35 MN)? Compare your answer with humanity's total electric power-generating capability, about 10^{12} W.

Paired Problems

(Both problems in a pair involve the same principles and techniques. If you can get the first problem, you should be able to solve the second one.)

61. Find the peak electric and magnetic fields 1.5 m from a 60-W light bulb that radiates equally in all directions.

62. At 4.6 km from a radio transmitter, the peak electric field in the radio wave measures 380 mV/m. What is the transmitter's power, assuming it broadcasts equally in all directions?

63. Unpolarized light is incident on two polarizers with their axes at 45°. What fraction of the incident light gets through?

64. Find the angle between two polarizers if unpolarized light incident on the pair emerges with 10% of its incident intensity.

65. What is the radiation force on the door of a microwave oven if 625 W of microwave power hits the door at right angles and is reflected?

66. What is the power output of a laser whose beam exerts a 55-mN force on an absorbing object oriented at right angles to the beam? The object is larger than the beam's cross section.

67. A 60-W light bulb is 6.0 cm in diameter. What is the radiation pressure on an opaque object at the bulb's surface?

68. A white dwarf star is approximately the size of Earth but radiates about as much energy as the Sun. Estimate the radiation pressure on an absorbing object at the white dwarf's surface.

Supplementary Problems

69. Maxwell's equations in a dielectric resemble those in vacuum (Equations 34-6 through 34-9), but with ϕ_E in Ampère's law replaced by $\kappa\phi_E$, where κ is the dielectric constant introduced in Chapter 26. Show that the speed of electromagnetic waves in such a dielectric is $c/\sqrt{\kappa}$.

70. Use appropriate data from Appendix E to calculate the radiation pressure on a light-absorbing object at the Sun's surface.

71. A radar system produces pulses consisting of 100 full cycles of a sinusoidal 70-GHz electromagnetic wave. The average power while the transmitter is on is 45 MW, and the waves are confined to a beam 20 cm in diameter. Find (a) the peak electric field, (b) the wavelength, (c) the total energy in a pulse, and (d) the total momentum in a pulse. (e) If the transmitter produces 1000 pulses per second, what is its average power output?

72. In a stack of polarizing sheets, each sheet has its polarization axis rotated 14° with respect to the preceding sheet. If the stack passes 37% of the incident, unpolarized light, how many sheets does it contain?

73. The peak electric field measured at 8.0 cm from a light source is 150 W/m², while at 12 cm it measures 122 W/m². Describe the shape of the source.

74. Show that Equations 34-12b and 34-13 may be combined to yield a wave equation like Equation 16-13. *Hint:* Take the partial derivative of one equation with respect to x and of the other with respect to t, and use the fact that

$$\frac{\partial}{\partial x}\left(\frac{\partial f}{\partial t}\right) = \frac{\partial}{\partial t}\left(\frac{\partial f}{\partial x}\right)$$

for any well-behaved function $f(x, t)$.

75. Studies of the origin of the solar system suggest that sufficiently small particles might be blown out of the solar system by the force of sunlight. To see how small such particles must be, compare the force of sunlight with the force of gravity, and solve for the particle radius at which the two are equal. Assume the particles are spherical and have density 2 g/cm³. Why do you not need to worry about the distance from the Sun?

76. A cylindrical resistor of length ℓ, radius a, and resistance R carries a current I. Calculate the electric and magnetic fields at the surface of the resistor, assuming the electric field is uniform throughout, including at the surface. Calculate the Poynting vector, and show that it points into the resistor. Calculate the flux of the Poynting vector (that is, $\int \mathbf{S} \cdot d\mathbf{A}$) over the surface of the resistor to get the rate of electromagnetic energy flow into the resistor, and show that the result is just I^2R. Your result shows that the energy heating the resistor comes from the fields surrounding it. These fields are sustained by the source of electrical energy that drives the current.

PART 4 *Cumulative Problems*

These problems combine material from chapters throughout the entire part or, in addition, from chapters in earlier parts, or they present special challenges.

1. An air-insulated parallel-plate capacitor has plate area 100 cm^2 and spacing 0.50 cm. The capacitor is charged and then disconnected from the charging battery. A thin-walled, nonconducting box of the same dimensions as the capacitor is filled with water at 20.00°C. The box is released at the edge of the capacitor and moves without friction into the capacitor (Fig. 1). When it reaches equilibrium the water temperature is 21.50°. What was the original voltage on the capacitor?

FIGURE 1 Cumulative Problem 1.

2. A wire of length ℓ and resistance R is formed into a closed rectangular loop twice as long as it is wide. It is mounted on a nonconducting horizontal axle parallel to its longer dimension, as shown in Fig. 2. A uniform magnetic field **B** points into the page, as shown. A long string of negligible mass is wrapped many times around a drum of radius a attached to the axle, and a mass m is attached to the string. When the mass is released it falls and eventually reaches a speed that, averaged over one cycle of the loop's rotation, is constant from one rotation to the next. Find an expression for that average terminal speed.

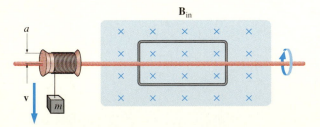

FIGURE 2 Cumulative Problem 2.

3. Five wires of equal length 25 cm and resistance 10 Ω are connected as shown in Fig. 3. Two solenoids, each 10 cm in diameter, extend a long way perpendicular to the page. The magnetic fields of both solenoids point out of the page; the field strength in the left-hand solenoid is increasing at 50 T/s while that in the right-hand solenoid is decreasing at 30 T/s. Find the current in the resistance wire shared by both triangles. Which way does the current flow?

FIGURE 3 Cumulative Problem 3.

4. A long solenoid of length ℓ and radius R has a total of N turns. The solenoid current is increasing linearly with time: $I(t) = bt$, where b is a constant. (a) Find an expression for the rate at which the magnetic energy in the solenoid is increasing. (b) Find an expression for the induced electric field at the inner edge of the solenoid coils. (c) Evaluate the Poynting vector at the inner edge of the coils, and show by integration that electromagnetic energy is flowing into the solenoid at a rate equal to the buildup of magnetic energy.

5. A coaxial cable consists of an inner conductor of radius a and an outer conductor of radius b; the space between the conductors is filled with insulation of dielectric constant κ (Fig. 4). The cable's axis is the z axis. The cable is used to carry electromagnetic energy from a radio transmitter to a broadcasting antenna. The electric field between the conductors points radially from the axis, and is given by $E = E_0(a/r) \cos(kz - \omega t)$. The magnetic field encircles the axis, and is given by $B = B_0(a/r) \cos(kz - \omega t)$. Here E_0, B_0, k, and ω are constants. Show, using appropriate closed surfaces and loops, that these fields satisfy Maxwell's equations. Your result shows that the cable acts as a "waveguide," confining an electromagnetic wave to the space between the conductors.

(a)   (b)

FIGURE 4 Cumulative Problem 5. (*a*) A coaxial cable. (*b*) Cross section, showing the electric and magnetic fields. The fields also vary with position z along the cable axis, according to the equations given.

Part Five   Optics

In "Prismatic Abstract," multiple light beams converge on a prism to create a dazzlingly colorful effect. Photographic artist Pete Saloutous exploited the principles of optics in creating this work.

Imagine a world without light. We see because light reflects off objects; our eyes form images because light refracts in our corneas and lenses; and the more subtle effect of interference lets us make some of the most precise measurements possible and is behind the operation of everyday technologies like compact discs. Although the behavior of light is ultimately grounded in Maxwell's equations of electromagnetism, we can learn much about light from the simpler perspective of optics. The next three chapters explore the behavior of light, images and optical instruments, and the phenomena of interference.

35 Reflection and Refraction

Reflection and refraction guide light in a wide range of natural and technological applications. Here the image of a bee's head appears in a large-scale version of the optical fibers that are essential to modern communications.

Maxwell's brilliant work shows that the phenomena of **optics**—that is, the behavior of light—are manifestations of the laws of electromagnetism. Except in the atomic realm, where it is necessary to use quantum mechanics, all optical phenomena can be understood in terms of electromagnetic wave fields as described by Maxwell's equations. But often we need not resort to the full electromagnetic or even wave description of light to understand optics.

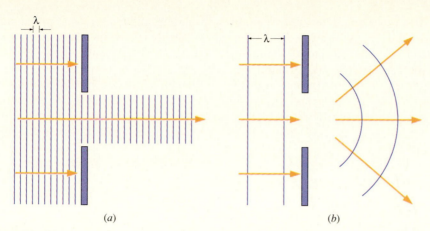

FIGURE 35-1 Light waves incident on an opaque barrier with a hole in it. (*a*) When the hole diameter is much greater than the wavelength λ, light emerges from the hole in essentially a straight line. In this case it is appropriate to treat the light as traveling in straight rays except where it actually hits something. (*b*) When the hole diameter is comparable to or smaller than the wavelength, then interference causes the wavefronts to bend, and the ray approximation is no longer adequate.

When the objects with which light—or, for that matter, any other wave—interacts are much larger than the wavelength, then to a good approximation the light travels in straight lines called rays (Fig. 35-1). Geometrical optics is the study of light under conditions when the ray approximation is valid. In this chapter we show how geometrical optics describes the behavior of light at interfaces between two different materials. In the next chapter we will show how geometrical optics explains optical systems including mirrors, lenses, the human eye, and many optical instruments.

35-1 REFLECTION AND TRANSMISSION

When a light ray propagating in one medium strikes the interface with a second medium, some light may be **reflected** back into the original medium and some **transmitted** into the new medium. Properties of the two materials, and the propagation direction of the incident light, determine the details of the reflection and transmission.

Some materials—notably metals—reflect nearly all the light incident on them. It's no coincidence that these materials are also good electrical conductors. The oscillating electric field of a light wave drives the metal's free electrons into oscillatory motion, and this accelerated motion, in turn, produces electromagnetic waves. The effect of all the oscillating electrons together is simply to reradiate the wave back into the original medium. Ultimately, that's why metals appear shiny.

In other materials, the atomic and molecular configurations are such that the material absorbs most of the incident light energy. Such materials therefore appear dark and opaque. Since an opaque material basically destroys light that's transmitted into it by converting the light to other forms of energy, these materials play little role in our exploration of optics.

FIGURE 35-2 A laser beam incident on a glass prism shows partial reflection and partial transmission at the interfaces between air and glass. Can you tell which is the incident beam?

Still other materials are essentially transparent, allowing light to propagate with little energy loss. These materials are generally insulators whose electrons, bound to individual atoms, cannot respond freely to the fields of an incident wave. The electrons do respond in a more limited way, however, effectively producing oscillating molecular dipoles. Although a microscopic description of this process is complicated, the net effect in most simple dielectric materials is a reduction in the propagation speed of electromagnetic waves from the speed in a vacuum c, as wave energy is absorbed and reradiated by the molecular dipoles.

In this chapter we explore the processes that occur at the interfaces between transparent and reflective materials and between different transparent materials. Reflection is the significant process in the former case, while in the latter both reflection and transmission generally occur (Fig. 35-2).

35-2 REFLECTION

Reflection returns some or all of the light incident on an interface to its original medium. Whether the reflection is essentially complete, as from a metal, or partial, as from a transparent material, it satisfies the same geometrical condition. Experiment—as well as analysis based on Maxwell's equations—shows that the incident ray, the reflected ray, and the normal to the interface between the two materials all lie in the same plane, and that the angle θ_1' that the reflected ray makes with the normal is equal to the angle θ_1 made by the incident ray (Fig. 35-3a). That is,

$$\theta_1' = \theta_1, \tag{35-1}$$

where the subscript 1 designates angles in the first medium. These angles are designated the **angle of reflection** and **angle of incidence**, respectively.

When a beam of parallel light rays reflects off a smooth surface, each ray in the beam reflects at the same angle, and the entire beam is thus reflected without distortion (Fig. 35-3b). This process is called **specular reflection**. But if the surface is rough, then individual rays, while still obeying Equation 35-1, reflect from differently oriented pieces of the surface. As a result the reflection is **diffuse**, with the original beam spreading in all directions (Fig. 35-3c). White wall paint is a good example of a diffuse reflector, while the aluminum or silver coating of a mirror is an excellent specular reflector. When we speak of reflection we generally mean specular reflection.

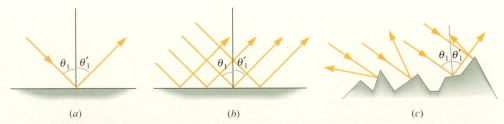

(a) (b) (c)

FIGURE 35-3 (a) The angles of reflection and incidence are equal. (b) In specular reflection, a smooth surface reflects a light beam undistorted. (c) A rough surface results in diffuse reflection. The law of reflection still holds for each individual ray.

| **EXAMPLE 35-1** | *The Corner Reflector* |

Two mirrors are joined at right angles, as shown in Fig. 35-4. Show that any light ray incident in the plane of the page will return antiparallel to its incident direction.

Solution

The law of reflection ensures that the two angles marked θ in Fig. 35-4 are equal, as are the two marked ϕ. But the mirrors make a 90° angle, and therefore $\phi = 90° - \theta$. The incident and first reflected ray make an angle 2θ, while the first reflected ray and the outgoing ray make an angle $2\phi = 180° - 2\theta$. Thus these two angles sum to 180°, which shows that the incident and outgoing rays are antiparallel. This result holds regardless of the actual value of the incidence angle θ.

Adding a third mirror at right angles to the other two makes a corner reflector that returns any beam in the direction from which it came—regardless of the orientation of the beam or the reflector (Fig. 35-5b). Corner reflectors, often made with prisms rather than mirrors, are widely used in optics. A corner

(a)   (b)

FIGURE 35-5 (a) Astronauts left this array of corner reflectors on the moon. Reflecting pulses of laser light off the array allows determination of the moon's position to within 15 cm. (b) Each reflector is functionally equivalent to three mutually perpendicular mirrors, although the reflectors are actually made from partial cubes of glass and work by the mechanism of total internal reflection that we'll explore in Section 35-4.

reflector left on the moon allows laser-based measurements of the moon's distance to within about 15 cm (Fig. 35-5a).

EXERCISE Two mirrors make an angle of 135°. A light ray is incident on the first mirror at an angle of 60° to the mirror's normal. Find (a) the angle of incidence on the second mirror and (b) the angle between the original incident beam and the final outgoing beam.

Answers: (a) 75°; (b) 90°

Some problems similar to Example 35-1: 2, 4, 5

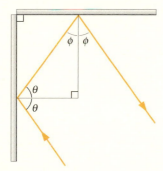

FIGURE 35-4 Two-dimensional corner reflector, made from two mirrors at right angles. Any ray incident in the horizontal plane returns parallel to its original direction.

35-3 REFRACTION

Figures 35-2 and 35-6 show that light passing between two transparent media is partially reflected and partially transmitted at their interface. In addition, the transmitted beam changes direction at the interface—a phenomenon known as **refraction**.

Physically, refraction occurs because the wave speeds in the two media are different. That difference results from delays introduced by the interaction of the electromagnetic wave fields with atomic electrons. But the wave frequency

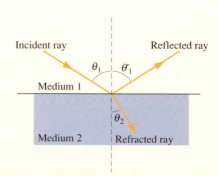

FIGURE 35-6 Reflection and refraction at an interface. This case, where the refracted ray is bent *toward* the normal, occurs when medium 2 has a higher refractive index and therefore a lower wave speed. Angles θ_1, θ_1', and θ_2 are, respectively, the angles of incidence, reflection, and refraction.

FIGURE 35-7 (a) Waves refract at an interface because wave speeds—and therefore the wave-lengths—are different in the two media. Two shaded triangles have a common hypotenuse, whose length is $vT/\sin\theta$, where v and θ are taken on either side of the interface. (b) Water waves refract as they pass over a triangular patch of shallower water. Note the decreased wavelength and different orientations of the waves within the shallow region. Reflection at the interfaces has resulted in waves propagating in several directions. (Wave speed in shallow water depends on depth, so a change in depth corresponds to a change in medium for light waves.)

15.1, 15.2

ActivPhysics

Reflection and Refraction

f—and thus the wave period $T = 1/f$—must be the same on both sides of the interface; otherwise, on one side of the interface more wave crests would pass in a given time than on the other side, implying the creation or destruction of waves at the interface. In Fig. 35-7a waves in medium 1 travel at speed v_1, and the wavelength—the distance between wave crests—is therefore $v_1 T$. We'll assume that in medium 2 the speed has some lower value v_2, so the wavelength $v_2 T$ is correspondingly shorter. The shaded triangles in Fig. 35-7a are right triangles with a common hypotenuse, and the ratio of the opposite side to the length of this hypotenuse defines the sines of the angles θ_1 and θ_2. Equating expressions for the hypotenuse length (opposite side divided by sine) in terms of these two angles gives

$$\frac{v_1 T}{\sin\theta_1} = \frac{v_2 T}{\sin\theta_2},$$

or

$$\frac{\sin\theta_2}{\sin\theta_1} = \frac{v_2}{v_1}.$$

We characterize the effect of a transparent medium on light through its **index of refraction**, n, defined as the ratio of the speed c of light in vacuum to the speed v of light in the medium:

$$n = \frac{c}{v}. \qquad \text{(index of refraction)} \qquad (35\text{-}2)$$

(Again there's an electromagnetic connection: the index of refraction is just the square root of the dielectric constant that we introduced in Chapter 26. But don't try to calculate refractive indices from the dielectric constants of

Table 26-1; the dielectric constant is frequency dependent, and its value at optical frequencies is significantly different from the DC values of Table 26-1.) Table 35-1 lists some indices of refraction. Note that for gases the index of refraction is essentially 1, the same as that of vacuum.

Using the definition 35-2 in our ratio of the sines of the angles of refraction and incidence, and then cross multiplying, gives **Snell's law**:

$$n_1 \sin\theta_1 = n_2 \sin\theta_2. \qquad \text{(Snell's law)} \qquad (35\text{-}3)$$

This law was developed geometrically in 1621 by Willebrord van Roijen Snell of the Netherlands and described analytically in the 1630s by René Descartes of France, where it is known to this day as Descartes' law. It allows us to predict what will happen to light at an interface, provided we know the refractive indices of the two media. As a special case, note that when $n_1 = n_2$ there's no change in angle; in general, the more different the refractive indices of the two media, the greater the change in angle.

Snell's law applies whether light goes from a medium of lower to higher refractive index or the reverse, as you can see by reversing the path of the light in Fig. 35-7. In the former case the refracted ray bends *toward* the normal, while in the latter case it bends *away* from the normal. Example 35-2 treats both cases.

TABLE 35-1 *Indices of Refraction**

SUBSTANCE	INDEX OF REFRACTION, n
Gases	
Air	1.000293
Carbon dioxide	1.00045
Liquids	
Water	1.333
Ethyl alcohol	1.361
Glycerine	1.473
Benzene	1.501
Diiodomethane	1.738
Solids	
Ice (H_2O)	1.309
Polystyrene	1.49
Glass	1.5–1.9
Sodium chloride (NaCl)	1.544
Diamond (C)	2.419
Rutile (TiO_2)	2.62

*At 1 atm pressure and temperatures ranging from 0°C to 20°C, measured at a wavelength of 589 nm (the yellow line of sodium).

EXAMPLE 35-2	*In and Out: Parallel Rays*

In Fig. 35-8, a light ray propagating in air strikes a transparent slab of thickness d and refractive index n with incidence angle θ_1. Show that the ray emerges from the slab propagating parallel to its original direction.

Solution
Applying Snell's law to the upper interface gives

$$\sin\theta_2 = \frac{\sin\theta_1}{n},$$

where we've taken $n_1 = 1$ for air and $n_2 = n$ for the slab. At the lower interface, with the light going from the slab to air, we have $n_1 = n$ and $n_2 = 1$, so Snell's law gives

$$\sin\theta_4 = n\sin\theta_3.$$

But the slab faces are parallel so, as the figure suggests, $\theta_3 = \theta_2$. Combining our two versions of Snell's law then gives

$$\sin\theta_4 = n\left(\frac{\sin\theta_1}{n}\right) = \sin\theta_1,$$

showing that the incident and outgoing rays are indeed parallel. They are, however, displaced by the distance x shown in the figure. You can find that displacement in Problem 54.

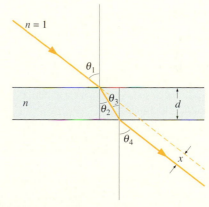

FIGURE 35-8 A light beam incident on a transparent slab emerges with its original direction, but is displaced from its original path.

EXERCISE A piece of glass with $n = 1.52$ is immersed in diiodomethane. A light ray strikes the glass with incidence angle 40°. Find the angle the refracted beam in the glass makes with the normal to the interface.

Answer: 47.3°

Some problems similar to Example 35-2: 11, 13, 54

EXAMPLE 35-3	CD Music

The laser beam that "reads" information from a compact disc is 0.737 mm wide at the point where it strikes the underside of the disc and forms a converging cone with half-angle 27°, as shown in Fig. 35-9. It then travels through a 1.2-mm-thick layer of transparent plastic with refractive index 1.55 before reaching the very thin, reflective information layer near the disc's top surface. What is the beam diameter (d in Fig. 35-9) at the information layer?

Solution

Figure 35-9 shows that $d = D - 2x$, where $D = 0.737$ mm is the beam diameter as it hits the disc. From the figure we also see that $x = t \tan \theta_2$, where $t = 1.2$ mm is the thickness of the plastic. Finally, Snell's law gives $\theta_2 = \sin^{-1}\left(\dfrac{\sin \theta_1}{n}\right)$, where we set the refractive index of air to 1. Putting these relations together gives

$$d = D - 2t \tan\left[\sin^{-1}\left(\frac{\sin \theta_1}{n}\right)\right]$$

$$= 737 \ \mu\text{m} - (2)(1200 \ \mu\text{m}) \tan\left[\sin^{-1}\left(\frac{\sin 27°}{1.55}\right)\right]$$

$$= 1.8 \ \mu\text{m},$$

which is a little larger than the "pits" cut into the CD to store its information. This narrowing of the beam plays a crucial role in keeping CDs noise free. The tiniest dust speck would blot out information at the μm-scale information layer, but at the point where the laser beam actually enters the disc—the closest dust can get to the information layer—it would take mm-size dust to cause problems.

In Chapter 37 we'll see more on how optical phenomena are at the heart of CD technology.

EXERCISE Figure 35-10 shows a polystyrene cylinder whose height is equal to its diameter. What is the maximum incidence angle θ_1 at which a light ray striking the top center of the cylinder will emerge through the bottom without first striking the side? The medium surrounding the cylinder is air.

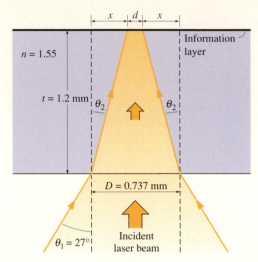

FIGURE 35-9 Section through a compact disc, showing convergence of the laser beam to a narrow spot at the information layer. All CDs share a common refractive index of 1.55 (Example 35-3).

FIGURE 35-10 What is the maximum θ_1 for which the beam will emerge through the bottom of the polystyrene cylinder?

Answer: 42°

Some problems similar to Example 35-3: 14, 15, 17, 18

We derived Snell's law using Fig. 35-7a, which shows the wavelength change that occurs as the wave speed changes between two media. There we argued that the frequencies in the two media must be the same. But wavelength, frequency, and wave speed are related by $\lambda f = v$; therefore, since wave speeds depend inversely on refractive index (Equation 35-2), so do the wavelengths. Thus the wavelengths in the two media are related by

$$\frac{\lambda_2}{\lambda_1} = \frac{n_1}{n_2}. \tag{35-4}$$

| EXAMPLE 35-4 | *Light and Diamond* |

Light with wavelength 589 nm in vacuum enters a diamond. Find the light speed and wavelength in the diamond.

Solution

Equation 35-2 gives

$$v = \frac{c}{n} = \frac{3.00 \times 10^8 \text{ m/s}}{2.419} = 1.24 \times 10^8 \text{ m/s} .$$

The refractive index of vacuum is 1, so Equation 35-4 gives

$$\lambda_{\text{diamond}} = \frac{\lambda_{\text{vacuum}}}{n} = \frac{589 \text{ nm}}{2.419} = 243 \text{ nm} .$$

The wavelength and speed are reduced by the same factor, as they must be to keep the wave frequency unchanged.

EXERCISE Microwaves propagate through glass at about 1.4×10^8 m/s. What is the refractive index of glass at microwave frequencies?

Answer: 2.14

Some problems similar to Example 35-4: 8, 9, 19

Got It!

Figure 35-11 shows the path of a light ray through three different media. Rank the media according to their indices of refraction.

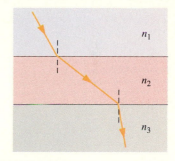

FIGURE 35-11 How do the refractive indices compare?

Multiple Refractions

Engineered optical systems often use several layers of refractive material to minimize reflective losses and certain types of distortion. To describe the path of a light ray in such a system, we need only apply Snell's law at each of the interfaces using the appropriate pair of refractive indices (Fig. 35-12a). When the layers are parallel the angular deflection of a light ray is the same as if it had gone through a single interface from the first medium to the last (see Problem 48). In many natural situations, including the human eye and Earth's atmosphere, the index of refraction is a continuously varying function of position (Fig. 35-12b). We can approximate such a case as a sequence of thin layers, each with a different refractive index. Going toward the limit of infinitely many infinitesimal layers gets us arbitrarily close to the exact solution, in which the

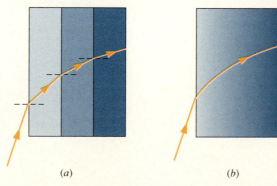

(a)   (b)

FIGURE 35-12 (a) Light propagating through a series of slabs with increasing refractive indices. At each interface the light bends more toward the normal. (b) Light follows a curved path in a medium with a continuously increasing refractive index.

FIGURE 35-13 (*a*) Mirages on a hot road. (*b*) The mirage occurs because hot air near the highway surface has a lower refractive index, resulting in a curved path for the light. The apparent position of the vehicle is not its actual position. Can you see why it appears upside down?

light follows a curved path. The mirage shown in Fig. 35-13 results from the fact that the refractive index of air is temperature dependent, causing the path of light rays to bend continuously.

35-4 TOTAL INTERNAL REFLECTION

Figure 35-14 shows light propagating inside a glass block and striking the interface with the surrounding air. Since air's refractive index is lower than that of glass, rays are bent *away* from the normal as they leave the glass. As the incidence angle increases, so does the angle of refraction—and Snell's law shows that the latter is always greater than the incidence angle. So at some incidence angle (see ray 3 in Fig. 35-14) the angle of refraction reaches 90°. What then?

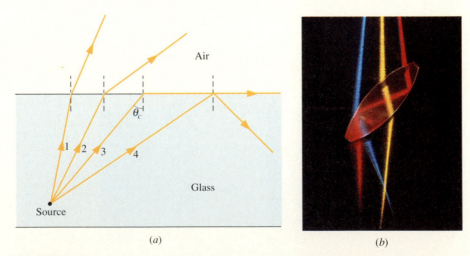

FIGURE 35-14 (*a*) Light propagating inside a glass block is refracted away from the normal at the glass-air interface. Ray 3, incident at the critical angle, just skims along the interface. At higher incidence angles, as with ray 4, the light undergoes total internal reflection at the interface. Not shown are weak reflected rays at angles less than the critical angle. (*b*) Three light beams strike a lens and are refracted at the air-glass interface. The red beam strikes the subsequent glass-air interface at an incidence angle greater than the critical angle and undergoes total internal reflection.

If the incidence angle is increased further, we find that the light is *totally* reflected at the interface. (There is always *some* reflection back into the glass, but now it's *all* reflected.) This phenomenon is called **total internal reflection**, and the incidence angle at which it first occurs is the **critical angle**, θ_c. We can find θ_c by setting $\theta_2 = 90°$ (i.e., $\sin\theta_2 = 1$) in Snell's law (Equation 35-3). The critical angle is then θ_1, and we have

$$\sin\theta_c = \frac{n_2}{n_1}. \qquad (35\text{-}5)$$

Since the sine of an angle cannot exceed 1, we must have $n_2 \leq n_1$ in order for this equation to have a solution. Thus, total internal reflection occurs only when light propagating in one medium strikes an interface with a medium of lower refractive index, and it occurs whenever the incidence angle exceeds the critical angle. You can simulate total internal reflection with ActivPhysics Activity 15.1, by clicking on the "Larger *n* region" button and varying the incidence angle until the refracted ray disappears.

Total internal reflection makes uncoated glass an excellent reflector when it's oriented appropriately (Fig. 35-15). Corner reflectors (Example 35-1) use total internal reflection in solid cubes of glass, rather than individual mirrors at right angles. For an observer inside a medium of higher refractive index, the existence of the critical angle affects the view of the outside world, as the example below shows. Finally, total internal reflection is the basis of the optical fibers now widely used in communications.

(a)   (b)

FIGURE 35-15 (*a*) A glass prism redirects light through total internal reflection. (*b*) Binoculars use a pair of prisms to "fold" the light path, allowing a more compact design.

EXAMPLE 35-5	*Whale Watch*

Planeloads of whale watchers fly above the ocean, as shown in Fig. 35-16. A whale looks upward, watching the planes. Within what range of viewing angles can the whale see the planes?

Solution

If the whale emitted light, that light would emerge from the ocean surface only for incidence angles less than the critical angle—that is, those angles within the cone shown in Fig. 35-16. Since the path of light is reversible, the whale can see objects above the surface only when it looks within this cone. For the water-air interface Equation 35-5 gives

$$\theta_c = \sin^{-1}\left(\frac{1}{1.333}\right) = 48.6°,$$

where we found water's refractive index in Table 35-1. The geometry of Fig. 35-16 then shows that this is also the half-angle of the cone, so the whale must look within 48.6° of the vertical to see out.

Can the whale see a plane only if it's *actually* within this angle? No. As the strongly refracted ray from the right-hand plane suggests, the entire outside world appears to the whale compressed into a cone of half-angle θ_c. If the whale looks beyond this cone, it will see instead reflections of objects below the surface.

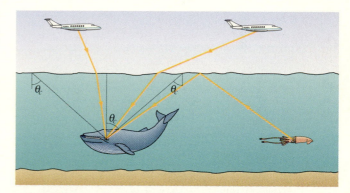

FIGURE 35-16 The whale sees the entire world above the surface in a cone of half-angle θ_c. Looking beyond this cone, it sees reflections of objects below the surface.

EXERCISE A diamond is submerged in water. What is the critical angle at the diamond-water interface?

Answer: 33.4°

Some problems similar to Example 35-5: 22, 23, 26, 28, 33

APPLICATION	*Optical Fibers and Lightwave Communication*

Refraction and total internal reflection are the basis for **optical fibers**, which revolutionized communications in the 1980s. Some three *billion* km of optical fiber were installed in the United States alone during that decade, and today they carry telephone conversations, television signals, computer data, and other information. Undersea fibers link the entire planet in a fiber-optic network.

A typical fiber consists of a glass core only 8 μm in diameter, surrounded by a so-called cladding consisting of glass with a lower refractive index than the core. The propagation of light in the fiber is a process whereby the core-cladding interface guides the light along the fiber. In some fibers this takes place by abrupt internal reflection at the interface (Fig. 35-17), while in others gradual refraction in the cladding guides the light. The glass used in optical fibers is so pure that a 1-km thick slab would appear as transparent as an ordinary window pane. Today's fibers use infrared light at wavelengths of 0.85 μm, 1.31 μm, and 1.55 μm. A recent development are optical fibers that actually regenerate signals as they travel long distances, extracting energy from an additional laser beam; these fibers are now used on some intercontinental and other ultra-long fiber communications systems.

FIGURE 35-18 The thin fiber-optic cable and the much bulkier cable of copper wires carry information at the same rate.

Optical fibers offer several advantages over copper wire or open-air transmission of electromagnetic waves. The main advantage is their very high **bandwidth**—the amount of information they carry per unit time. This large bandwidth results because the information carried on fiber is encoded on infrared radiation with a frequency on the order of 10^{14} Hz—much higher than the microwave frequencies used in conventional communication systems. A single fiber, for example, can carry tens of thousands of telephone conversations (Fig. 35-18).

Fibers are lighter and more rugged than copper cables and are less easy to "tap" illicitly. Because they are insulators, optical fibers are also less susceptible to electrical "noise" and are therefore used to carry information in electrically noisy environments like power plants or high-energy physics laboratories.

Optical fibers play a key role in the global Internet that links nearly all the world's computers. Computers on one floor or within a building may be connected in a small, relatively low-bandwidth network with copper wire. But connections between buildings are nearly always made with optical fibers, linking individual local networks into larger institution-wide networks. Entire institutions—corporations, universities, government laboratories, and the like—are then linked by very high speed fiber-optic "superhighways" that carry the Internet's burgeoning information "traffic."

Optical fiber technology requires more than just fiber: appropriate opto-electronic devices are needed at each end to convert information from light to electrical forms and vice versa. Such devices include miniature lasers made from semiconductors, and high-speed light-sensitive diodes. As the cost of these devices drops, high-speed optical fiber is becoming available to individual homes and offices, allowing nearly instant access to telecommunications, video resources, libraries, and databases.

(a)

(b)

FIGURE 35-17 (*a*) A beam of light undergoes a series of total internal reflections that guide it along an optical fiber. (*b*) A bundle of actual fibers.

35-5 DISPERSION

Refraction ultimately occurs because of the interaction of electromagnetic wave fields with atomic electrons. Although the details of that interaction require a quantum description, it should not be surprising that the index of refraction depends on frequency. After all, as we found in Chapter 15, the response of a system (here an atom) generally depends on the frequency of the driving force (here the oscillating fields of a wave).

Because the index of refraction depends on frequency, different frequencies—different colors for visible light—will refract through different angles. A light beam containing many colors will therefore spread into its constituent colors, a process called **dispersion** (Fig. 35-19). The classic example of dispersion is Newton's demonstration that white light is really a mixture of all the colors in the visible spectrum. Newton not only broke white light into its constituent colors, as in Fig. 35-19, but he also recombined the colors to produce the original white light.

In most materials in most frequency ranges, the index of refraction increases with increasing frequency and therefore decreases with increasing wavelength (Fig. 35-20). That means colors toward the violet end of the spectrum are refracted through the greatest angles, as is evident in Fig. 35-19.

Dispersion is the basis for the widely used technique of **spectral analysis** or **spectroscopy**, in which substances are identified and characterized by analyzing their **spectra**—meaning the range of wavelengths of electromagnetic radiation they emit, transmit, or absorb. Hot, dense objects, for example, emit a continuous spectrum of radiation, while diffuse gases radiate at only a few specific wavelengths (Fig. 35-21). The existence of those discrete spectra provided some of the strongest evidence for the nature of the atom, and today spectral analysis allows astronomers to identify and measure the abundances of elements in

FIGURE 35-19 Dispersion of white light by a glass prism. The refractive index of the glass is greater at higher frequencies—shorter wavelengths—and therefore results in greater refraction at the blue end of the spectrum. The incident beam comes in from the lower left edge, and a white reflected beam leaves the prism going straight downward. This beam has undergone an additional refraction that has again combined its colors.

FIGURE 35-20 Index of refraction as a function of wavelength for high-dispersion crown glass.

FIGURE 35-21 (Top) The solar spectrum is an essentially continuous band of wavelengths, produced by the hot, dense gases of the Sun's visible surface. Dark lines are discrete wavelengths absorbed by overlying gases in the solar atmosphere. (Bottom) Spectrum of a diffuse gas—in this case hydrogen—consists of light at discrete wavelengths. The pattern of lines allows identification of the emitting material.

distant astrophysical objects. Geologists use spectral analysis to identify minerals, and chemists use infrared spectra to study molecules. Spectral analysis is a powerful tool in nearly every branch of science. Although early spectroscopes used prisms, most modern instruments use instead diffraction gratings, whose operation we describe in Chapter 37.

Dispersion can be a nuisance in optical systems. Glass lenses, for example, focus different colors at different points, resulting in a distortion known as chromatic aberration. This effect can be minimized by making composite lenses of materials with different refractive indices.

EXAMPLE 35-6	*How Much Dispersion?*

White light strikes the prism in Fig. 35-22 normal to one surface. The prism is made of the glass whose refractive index is plotted in Fig. 35-20. Find the angle between outgoing red (700 nm) and violet (400 nm) light.

Solution

There is no refraction at the air-glass interface since the incident ray is normal to the surface. At the second interface medium 2 is air with $n_2 = 1$, and medium 1 is the glass whose indices at the two wavelengths we can read from Fig. 35-20: $n_{400} = 1.538$, $n_{700} = 1.516$. The geometry of Fig. 35-22 shows that the incidence angle at the second interface is the angle $\alpha = 40°$ at the top of the prism. The angles of refraction are then given by solving Snell's law for the two values of θ_2:

$$\theta_{400} = \sin^{-1}(n_{400}\sin\alpha) = \sin^{-1}[(1.538)(\sin 40°)] = 81.34°,$$

and

$$\theta_{700} = \sin^{-1}(n_{700}\sin\alpha) = \sin^{-1}[(1.516)(\sin 40°)] = 77.02°.$$

The angle between the two outgoing beams is therefore $\theta_{400} - \theta_{700} = 4.32°$, with the violet beam (400 nm) experienc-

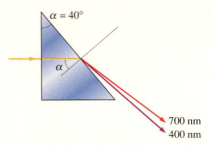

FIGURE 35-22 Example 35-6.

ing the greatest deflection. This 4.32° spread is called the *angular dispersion* of the beam.

EXERCISE If the angle α in Fig. 35-22 is increased, what is the maximum angular dispersion that can be achieved by refracting both the 400 nm and 700 nm beams? At what α does this occur?

Answer: $\Delta\theta = 9.7°$ at $\alpha = \sin^{-1}(1/n_{400}) = 40.556°$

Some problems similar to Example 35-6: 34–38

APPLICATION	*The Rainbow*

Nature provides a beautiful application of dispersion and internal reflection in the rainbow (Fig. 35-23), which occurs when sunlight strikes rain or other water droplets in the air. An observer standing between the Sun and the rain then sees the circular arc of colored bands. Figure 35-24 shows that the center of that arc lies on the line joining the Sun to the observer's eye. That means each observer sees a different rainbow! Furthermore, the rainbow's arc always subtends an angle of approximately 42°. How does the rainbow form, and why does it have this geometry?

Theories of the rainbow date back many centuries. By 1635 Descartes had produced a nearly complete explanation of the rainbow's shape and apparent location, but because he did not

FIGURE 35-23 Reflection, refraction, and dispersion all act to produce a rainbow.

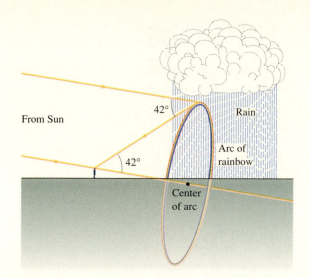

FIGURE 35-24 The rainbow is a circular arc located at 42° from the line that includes the Sun, the observer, and the center of the arc. The Sun is so far away that its rays are essentially parallel. Here part of the rainbow is blocked by the ground; however, observers in aircraft can sometimes see the entire circle.

FIGURE 35-25 A light ray passing through a spherical raindrop undergoes refraction at each interface and internal reflection at the back of the drop. (Not shown are the reflected rays at interfaces *A* and *C* and the transmitted ray at interface *B*.) The ray undergoes an overall deflection of $180° - \phi$.

FIGURE 35-26 Parallel rays striking a water drop. The minimum angle through which a ray is deflected is about 138°, corresponding to a maximum of about 42° in the angle ϕ of Fig. 35-25. Furthermore, rays tend to "bunch" at this angle, as suggested by the broad outgoing beam defined by three nearly parallel rays. This "bunching" causes an observer to see a bright arc at 42° from the antisolar direction in Fig. 35-24. The thicker ray undergoes the minimum deviation and has, correspondingly, the maximum ϕ.

know about dispersion he could not account for the colors. A few years later Newton, in his *Optics,* produced a full explanation.

Figure 35-25 shows a light ray passing through a spherical raindrop. The incidence angle θ in Fig. 35-25 is arbitrary, and

parallel light rays striking the curved surface of the drop will experience a range of values for θ. There will, therefore, be a range of angles ϕ between the incident and outgoing rays. As Fig. 35-26 shows, however, there is a maximum angle ϕ_{max} of about 42° and more light is returned at angles close to ϕ_{max} than at other angles. That's why a bright band—the rainbow—appears in the sky at an angle of about 42° to the direction of the Sun's rays. Problems 58 and 59 show how to find ϕ_{max}.

The "bunching" of light rays near ϕ_{max} explains why a bright band should appear, but why should it be colored? Because the refractive index varies with wavelength, so does ϕ_{max}. Each color will therefore appear brightest at a slightly different angle. For water, the refractive index in the visible region ranges from $n_{red} = 1.330$ to $n_{violet} = 1.342$. Using these values with the results of Problems 58 and 59 yields $\phi_{red} = 42.53°$ and $\phi_{violet} = 40.78°$. Thus the rainbow is seen as a band, itself subtending an angle of about 1.75°, with red at the top.

One occasionally sees a fainter and larger arc above the primary rainbow. This secondary rainbow results from two internal reflections, and as a result the order of its colors is reversed. Problem 60 explores the secondary rainbow.

35-6 REFLECTION AND POLARIZATION

Geometrical optics describes the paths of light rays at interfaces between different media, but doesn't tell how much of the light is reflected and how much transmitted at an interface. Application of Ampère's and Faraday's laws to electromagnetic wave fields at an interface does give that information. We won't undertake the calculation here, but will instead describe briefly the results.

FIGURE 35-27 Reflection in a plate-glass window is especially prominent at oblique incidence angles.

When light is incident on an interface, some fraction of it is reflected. The ratio of the reflected to the incident intensity is called the **reflection coefficient**. In general, the least reflection occurs with normal incidence—i.e., with light incident perpendicular to the interface. For glass with refractive index $n = 1.5$, for example, only about 4 percent of normally incident light is reflected—meaning the reflection coefficient in this case is 0.04. You usually don't notice this faint reflection, except when you're inside a lighted room at night; then there's no light coming through the windows from outside to overwhelm the reflection.

The reflection coefficient generally increases at oblique angles, approaching 1 as the incidence angle approaches 90°. You've undoubtedly experienced this effect when standing in front of a plate-glass window; as you look along the window at an oblique angle, the reflection of objects on your side of the window becomes more prominent (Fig. 35-27).

The Polarizing Angle

Reflection ultimately involves an interaction of the electric field of an incident light wave with electrons near the surface of the reflecting medium. It's not surprising, therefore, that the reflection coefficient depends on the orientation of a light wave's electric field relative to the interface. In Chapter 34 we defined the polarization direction of an electromagnetic wave as the direction of the wave's electric field; thus, the reflection coefficient depends on polarization.

For all but normal incidence, the incident, reflected, and refracted rays are in different directions and together define a plane (see Fig. 35-28, where that plane is the plane of the page). The electric field of the incident light may lie in that plane, or be perpendicular to it, or be at some angle in between. When the field lies in the plane defined by the three rays, there's a special angle at which incident light does not undergo any reflection, and instead is transmitted 100 percent across the interface. This special angle is called the **Brewster angle** or the **polarizing angle**, and designated θ_p. The name "polarizing angle" is apt because if light incident at this angle has a mix of polarizations, only the component with electric field perpendicular to the plane of Fig. 35-28 will be reflected; thus the reflected beam becomes completely polarized.

Why is there no reflection of the preferred polarization at this special angle? In Chapter 34 we saw that electromagnetic waves result from accelerated charge; Fig. 34-23 showed, nevertheless, that no electromagnetic radiation is emitted in the direction of a charged particle's acceleration. Knowing that, we can use Fig. 35-28 to understand the reason for the polarizing angle. For glass with refractive index $n = 1.5$, the polarizing angle is 56.3°. At this incidence angle, Snell's law gives an angle of refraction $\theta_2 = \sin^{-1}(\sin\theta_p/n) = 33.7°$. Now $56.3° + 33.7° = 90°$ and that means, as Fig. 35-28 shows, that any reflected ray would be perpendicular to the refracted ray. But the reflected ray really arises from the acceleration of the electrons in the glass—acceleration

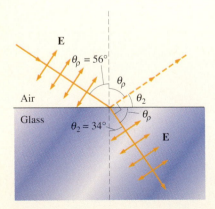

FIGURE 35-28 An electromagnetic wave polarized in the plane of the figure is incident on an interface. When the angles of incidence and refraction sum to 90°, the reflected ray would be in the direction of the electric field in the lower medium. Since there is no radiation in the direction of charged-particle acceleration, there is no reflected ray in this special case. Numbers shown are for an air-glass interface.

which, according to Newton's law of motion, is in the direction of the driving force. But here the driving force comes from the wave's electric field, and with polarization in the plane of Fig. 35-28, that field is itself parallel to the direction the reflected ray would take. But we've seen that there is no radiation in the direction of the acceleration, and hence for this special geometrical condition there is no reflected ray.

Figure 35-28 shows that the condition of no reflection will be met when the angles of incidence and refraction sum to 90°: $\theta_p + \theta_2 = 90°$, where we've set the incidence angle equal to the polarizing angle. This, in turn, implies that $\sin\theta_2 = \cos\theta_p$. But with the incidence angle $\theta_1 = \theta_p$, Snell's law gives $\sin\theta_2 = (n_1/n_2)\sin\theta_p$. Applying our condition $\sin\theta_2 = \cos\theta_p$, we then have $\cos\theta_p = (n_1/n_2)\sin\theta_p$. Multiplying both sides by $n_2/n_1\cos\theta_p$ then gives

$$\tan\theta_p = \frac{n_2}{n_1}. \tag{35-6}$$

(a)

For an air-glass interface with $n_1 = 1$ and $n_2 = 1.5$, this equation indeed gives $\theta_p = 56.3°$.

The polarizing angle is important in the construction of lasers, which require nearly loss-free transmission of light reflected from external mirrors back into the medium that generates the laser light. The common red-light helium-neon laser used in physics demonstrations and supermarket scanners, for example, includes a glass tube containing a helium-neon mixture; the ends of the tube are formed at the polarizing angle so that light with the correct polarization enters the tube without reflecting. Because zero reflection occurs only for light polarized in the plane of the incident and refracted rays, the laser itself produces light with this polarization. (We'll see more on how lasers work in Chapter 39 or, for extended-version readers, in Chapter 41).

A phenomenon similar to the polarizing angle occurs for reflection from metals and other surfaces, again because of the interaction of the electric field of the light wave with electrons. Somewhat similarly, scattering of sunlight by air molecules partially polarizes the light. Reduction of glare using Polaroid sunglasses or a camera's polarizing filter works by blocking the component of reflected or scattered light that has the enhanced polarization (Fig. 35-29).

(b)

FIGURE 35-29 Identical views of a storefront, photographed (a) without and (b) with a polarizing filter. Light reflecting obliquely from the glass is partially polarized. The polarizing filter eliminates this component, allowing the inside of the store to show much more clearly.

CHAPTER SYNOPSIS

Summary

1. **Geometrical optics** treats the behavior of light under the approximation that it travels in straight lines—**rays**—except at interfaces between different materials. This approximation is valid whenever waves interact with systems that are much larger than their wavelength.

2. **Reflection** occurs to some extent at nearly all interfaces. When a light ray reflects, the **angle of incidence** and **angle of reflection**—both measured from the normal to the interface—are equal. **Specular reflection** occurs from smooth surfaces, while rough surfaces produce **diffuse reflection**.

3. **Refraction** is the bending of light at an interface between transparent materials. The directions of the incident and refracted rays are related by **Snell's law**:

$$n_1\sin\theta_1 = n_2\sin\theta_2,$$

where θ_1 and θ_2 are the angles of incidence and refraction, respectively. The **indices of refraction** are n_1 and n_2, defined as the ratios of the speed of light in vacuum to the speed in a given medium.

4. **Total internal reflection** occurs when light propagating in a medium with refractive index n_1 is incident on an

interface at an angle greater than the **critical angle** θ_c given by

$$\sin \theta_c = \frac{n_2}{n_1}.$$

Total internal reflection is possible only when $n_2 < n_1$.

5. The index of refraction generally depends on frequency. This results in **dispersion** of the different colors as they refract through different angles. Dispersion is the basis of **spectroscopy**.

6. Some light is usually reflected at an interface, even with a transparent material. Generally, reflection increases as the angle of incidence becomes more oblique. But for the special case of light whose polarization is in the plane defined by incident and refracted rays, there's no reflection when the angle of incidence is equal to the **polarizing angle**, also called the **Brewster angle**. As a result, unpolarized light incident at this angle becomes completely polarized on reflection.

Terms You Should Understand

(Pairs are closely related terms whose distinction is important; number in parentheses is chapter section where term first appears.)

geometrical optics (introduction)
ray (introduction)
angles of incidence, reflection, refraction (35-2, 35-3)
specular reflection, diffuse reflection (35-2)
refraction, index of refraction (35-3)
Snell's law (35-3)

total internal reflection (35-4)
critical angle (35-4)
optical fiber (35-4)
dispersion (35-5)
spectroscopy (35-5)
reflection coefficient (35-6)
polarizing angle (35-6)

Symbols You Should Recognize

θ_1, θ_1', θ_2 (35-2, 35-3)
n (35-3)
θ_c (35-4)
θ_p (35-6)

Problems You Should Be Able to Solve

determining the directions of reflected and refracted rays at interfaces (35-2 and 35-3)
evaluating conditions for total internal reflection (35-4)
analyzing dispersion (35-5)
determining reflected and transmitted intensities at normal incidence (35-6)
evaluating the polarizing angle (35-6)

Limitations to Keep in Mind

Geometrical optics is an approximation valid only when light interacts with systems much larger than its wavelength.
Characterizing the optical properties of a material by a single number, the index of refraction, is a simplification. In many materials optical properties depend on direction, and include absorption as well as transmission.

QUESTIONS

1. Are light rays real? Discuss.
2. It's usually inappropriate to consider low-frequency sound waves as traveling in rays. Why? Why is it more appropriate for high-frequency sound and for light?
3. Describe why a spoon appears bent when it's in a glass of water.
4. Why do a diamond and an identically shaped piece of glass sparkle differently?
5. Specular reflection occurs with "smooth" surfaces. But on the microscopic scale all surfaces are rough. What do you suppose should be the criterion for "smoothness" in dealing with reflection?
6. White light goes from air through a glass slab with parallel surfaces. Will its colors be dispersed when it emerges from the glass?
7. Would light behave as in Fig. 35-15a if the refractive index of the prism were 1.25? Explain.

8. You send white light through two identical glass prisms, oriented as shown in Fig. 35-30. Describe the beam that emerges from the right-hand prism.

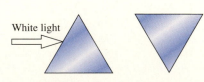
White light

FIGURE 35-30 Question 8.

9. Why can optical fibers carry much more information than copper wires?
10. What would happen if you scratched the outside of the reflecting surface of a prism used for total internal reflection?

11. Lightning produces a sudden burst of static in a nearby radio receiver. The static comprises a broad band of radio frequencies. But in a very distant receiver the "noise" from the flash arrives over an extended time. What does this say about the frequency dependence of the refractive index of Earth's atmosphere for radio waves?

12. What does a fish see as it looks around in directions above the horizontal? Explain.

13. In glass, which end of the visible spectrum has the lowest critical angle for total internal reflection?

14. Looking out the window of a lighted room at night, you see clear reflections of the room's interior. In the daytime those reflections would be much less obvious, yet the reflection coefficient has not changed. Explain.

15. Why can't you walk to the end of the rainbow?

16. What is wrong with the painting in Fig. 35-31? *Hint:* The rainbow subtends a half-angle of 42°.

FIGURE 35-32 Question 18.

FIGURE 35-31 What's wrong with this painting (*Niagara,* by Harry Fenn)? (Question 16)

17. Suppose the refractive index of water were not frequency dependent. Would anything like the rainbow occur?

18. Figure 35-32 shows a ball inside a transparent sphere inside an aquarium tank. The transparent sphere and the tank can each be filled with water. Explain which combination is occurring in each frame.

19. Why are polarizing sunglasses better than glasses that simply cut down on the total amount of light?

20. Does the transmitted intensity always exceed the reflected intensity at an air-glass interface?

21. Under what conditions will the polarizing angle be less than 45°?

PROBLEMS

ActivPhysics can help with these problems:
Activities 15.1, 15.2

Section 35-2 Reflection

1. Through what angle should you rotate a mirror in order that a reflected ray rotate through 30°?

2. The mirrors in Fig. 35-33 make a 60° angle. A light ray enters parallel to the symmetry axis, as shown. (a) How many reflections does it make? (b) Where and in what direction does it exit the mirror system?

3. To what angular accuracy must two ostensibly perpendicular mirrors be aligned in order that an incident ray returns on a path within 1° of its incident direction?

4. If a light ray enters the mirror system of Fig. 35-33 propagating in the plane of the page and parallel to one mirror, through what angle will it be turned?

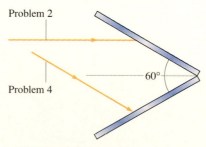

FIGURE 35-33 Problems 2, 4, 5.

5. Suppose the angle in Fig. 35-33 is changed to 75°. A ray enters the mirror system parallel to the axis. (a) How many reflections does it make? (b) Through what angle is it turned when it exits the system?

6. Two plane mirrors make an angle ϕ. For what value of ϕ will a light ray reflecting once off each mirror be turned through the same angle ϕ?

7. Two plane mirrors make an angle ϕ. A light ray enters the system and is reflected once off each mirror. Show that the ray is turned through an angle $360° - 2\phi$.

Section 35-3 Refraction

8. In which substance in Table 35-1 does the speed of light have the value 2.292×10^8 m/s?

9. Information in a compact disc is stored in "pits" whose depth is essentially one-fourth of the wavelength of the laser light used to "read" the information. That wavelength is 780 nm in air, but the wavelength on which the pit depth is based is measured in the $n = 1.55$ plastic that makes up most of the disc. Find the pit depth.

10. Light is incident on an air-glass interface, and the refracted light in the glass makes a 40° angle with the normal to the interface. The glass has refractive index 1.52. Find the incidence angle.

11. A light ray propagates in a transparent material at 15° to the normal to the surface. When it emerges into the surrounding air, it makes a 24° angle with the normal. What is the refractive index of the material?

12. Light propagating in the glass ($n = 1.52$) wall of an aquarium tank strikes the interior edge of the wall with incidence angle 12.4°. What is the angle of refraction in the water?

13. A block of glass with $n = 1.52$ is submerged in one of the liquids listed in Table 35-1. For a ray striking the glass with incidence angle 31.5°, the angle of refraction is 27.9°. What is the liquid?

14. A meter stick lies on the bottom of the rectangular trough in Fig. 35-34, with its zero mark at the left edge of the trough. You look into the long dimension of the trough at a 45° angle, with your line of sight just grazing the top edge of the tank, as shown. What mark on the meter stick do you see if the trough is (a) empty, (b) half full of water, and (c) full of water?

FIGURE 35-34 Problem 14.

15. You look at the center of one face of a solid cube of glass, on a line of sight making a 55° angle with the normal to the cube face. What is the minimum refractive index of the glass for which you will see through the opposite face of the cube?

16. The cylindrical tank in a public aquarium is 10 m deep, 11 m in diameter, and is full to the brim with water. If a flashlight shines on the tank from above, what is the minimum angle its beam can make with the horizontal if it is to illuminate part of the tank bottom?

17. You're standing 2.3 m horizontally from the edge of a 4.5-m-deep lake, with your eyes 1.7 m above the water surface. A diver holding a flashlight at the lake bottom shines the light so you can see it. If the light in the water makes a 42° angle with the vertical, at what horizontal distance is the diver from the edge of the lake?

18. You've dropped your car keys at night off the end of a dock into water 1.6 m deep. A flashlight held directly above the dock edge and 0.50 m above the water illuminates the keys when it's pointed at 40° to the vertical, as shown in Fig. 35-35. What is the horizontal distance x from the edge of the dock to the keys?

FIGURE 35-35 What is the horizontal distance x from the edge of the dock to the lost keys? (Problem 18)

19. A light ray is propagating in a crystal where its wavelength is 540 nm. It strikes the interior surface of the crystal with an incidence angle of 34° and emerges into the surrounding air at 76° to the surface normal. Find (a) the light's frequency and (b) its wavelength in air.

20. The prism in Fig. 35-36 has $n = 1.52$, $\alpha = 60°$, and is surrounded by air. A light beam is incident at $\theta_1 = 37°$. Find the angle δ through which the beam is deflected.

FIGURE 35-36 Problems 20, 51, 55.

Section 35-4 Total Internal Reflection

21. Find the critical angle for total internal reflection in (a) ice, (b) polystyrene, and (c) rutile. Assume the surrounding medium is air.
22. A drop of water is trapped in a block of ice. What is the critical angle for total internal reflection at the water-ice interface?
23. What is the critical angle for light propagating in glass with $n = 1.52$ when the glass is immersed in (a) water, (b) benzene, and (c) diiodomethane?
24. Total internal reflection occurs at an interface between a plastic and air at incidence angles greater than 37°. What is the refractive index of the plastic?
25. Light propagating in a medium with refractive index n_1 encounters a parallel-sided slab with index n_2. On the other side is a third medium with index $n_3 < n_1$. Show that the condition for avoiding internal reflection at *both* interfaces is that the incidence angle at the n_1-n_2 interface be less than the critical angle for an n_1-n_3 interface. In other words, the index of the intermediate material doesn't matter.
26. An aquarium measures 30 cm front to back, as shown in Fig. 35-37. It is made of glass with thickness much less than the size of the aquarium and is full of water with $n = 1.333$. You put your eye right up to the center of the aquarium's front wall and can still see the entire back wall. What is the maximum value of the aquarium's width w? *Hint:* You can ignore the glass; see the preceding problem.

FIGURE 35-37 Problem 26.

27. What is the minimum refractive index for which total internal reflection will occur as shown in Fig. 35-15a? Assume the surrounding medium is air and that the prism is an isosceles right triangle.
28. Where and in what direction would the main beam emerge if the prism in Fig. 35-15a were made of ice?
29. What is the speed of light in a material for which the critical angle at an interface with air is 61°?
30. The prism of Fig. 35-15a has $n = 1.52$. When it is immersed in a liquid, a beam incident as shown in the figure ceases to undergo total reflection. What is the minimum value for the liquid's refractive index?
31. A compound lens is made from crown glass ($n = 1.52$) bonded to flint glass ($n = 1.89$). What is the critical angle for light incident on the flint-crown interface?

32. Find a simple expression for the speed of light in a material in terms of the critical angle at an interface between the material and vacuum.
33. A scuba diver sets off a camera flash a distance h below the surface of water with refractive index n. Show that light emerges from the water surface through a circle of diameter $2h/\sqrt{n^2 - 1}$.

Section 35-5 Dispersion

34. Blue and red laser beams strike an air-glass interface with incidence angle 50°. If the glass has refractive indices of 1.680 and 1.621 for the blue and red light, respectively, what will be the angle between the two beams in the glass?
35. Suppose the red and blue beams of the preceding problem are now propagating in the same direction *inside* the glass. For what range of incidence angles on the glass-air interface will one beam be totally reflected and the other not?
36. White light propagating in air is incident at 45° on the equilateral prism of Fig. 35-38. Find the angular dispersion γ of the outgoing beam, if the prism has refractive indices $n_{red} = 1.582$, $n_{violet} = 1.633$.

FIGURE 35-38 Problem 36 (angles of dispersed rays are not accurate).

37. Two of the prominent spectral lines—discrete wavelengths of light—emitted by glowing hydrogen are hydrogen-α at 656.3 nm and hydrogen-β at 486.1 nm. Light from glowing hydrogen passes through a prism like that of Fig. 35-22, then falls on a screen 1.0 m from the prism. How far apart will these two spectral lines be? Use Fig. 35-20 for the refractive index.
38. Light from glowing sodium contains the two discrete wavelengths 589.0 nm and 589.6 nm. This light is passed through a prism like that of Fig. 35-22 and then allowed to fall on a screen 2.0 m distant. For wavelengths near 600 nm, the refractive index of the prism is $n = 1.546 - 4.47\times10^{-5}\lambda$, with λ the wavelength in nm. What must be the prism's apex angle α in order that the two sodium wavelengths be separated on the screen by 1.5 mm?

Section 35-6 Reflection and Polarization

39. Find the polarizing angle for diamond when light is incident from air.
40. What is the refractive index of a material for which the polarizing angle in air is 62°?
41. What is the polarizing angle for light incident from below on the surface of a pond?
42. For the interface between air (refractive index 1) and a material with refractive index n, show that the critical angle and the polarizing angle are related by $\sin \theta_c = \cot \theta_p$.

Paired Problems

(Both problems in a pair involve the same principles and techniques. If you can get the first problem, you should be able to solve the second one.)

43. Light propagating in air strikes a transparent crystal at incidence angle 35°. If the angle of refraction is 22°, what is the speed of light in the crystal?
44. A laser beam with wavelength 633 nm is propagating in air when it strikes a transparent material at incidence angle 50°. If the angle of refraction is 27°, what is the wavelength in the material?
45. A cylindrical tank 2.4 m deep is full to the brim with water. Sunlight first hits part of the tank bottom when the rising Sun makes a 22° angle with the horizon. Find the tank's diameter.
46. For what diameter tank in the preceding problem will sunlight strike some part of the tank bottom whenever the Sun is above the horizon?
47. Light is incident from air on the flat wall of a polystyrene water tank. If the incidence angle is 40°, what angle does the light make with the tank normal in the water?
48. A parallel-sided slab with refractive index n_2 separates two media with indices n_1 and n_3. Show that a light ray incident on the n_1-n_2 interface then enters the third medium with the same angle of refraction it would have had if the slab had not been present. (Assume total internal reflection does not occur at either interface.)
49. Light strikes a right-angled glass prism ($n = 1.52$) in a direction parallel to the prism's base, as shown in Fig. 35-39. The point of incidence is high enough that the refracted ray hits the opposite sloping side. (a) Through which side of the prism does the beam emerge? (b) Through what angle has it been deflected?

FIGURE 35-39 Problems 49, 50.

50. Repeat the preceding problem if the prism has refractive index 1.15.

51. Repeat Problem 20 for the case $n = 1.75$, $\alpha = 40°$, and $\theta_1 = 25°$.
52. The surfaces of a glass sheet are not quite parallel, but rather make an angle of 10°. The glass has refractive index $n = 1.52$. A light beam strikes one side of the sheet at incidence angle 35°, coming in on the thicker side of the normal. Through what angle is the beam direction changed when it emerges on the opposite side of the glass sheet?

Supplementary Problems

53. A cubical block is made from two equal-size slabs of materials with different refractive indices, as shown in Fig. 35-40. Find the index of the right-hand slab if a light ray is incident on the center of the left-hand slab and then describes the path shown.

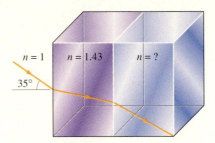

FIGURE 35-40 Problem 53.

54. Find an expression for the displacement x in Fig. 35-8, in terms of θ_1, d, and n.
55. Light is incident with incidence angle θ_1 on a prism with apex angle α and refractive index n, as shown in Fig. 35-36. Show that the angle δ through which the outgoing beam deviates from the incident beam is given by

$$\delta = \theta_1 - \alpha + \sin^{-1}\left\{n \sin\left[\alpha - \sin^{-1}\left(\frac{\sin\theta_1}{n}\right)\right]\right\}.$$

Assume the surrounding medium has $n = 1$.

56. Taking $n = 1.5$ and $\alpha = 60°$, plot the deviation δ of the preceding problem over the range $45° < \theta_1 < 50°$, and use your plot to find the incidence angle for minimum deviation. Trace the incident beam for this value of θ_1. Your result should be the symmetric path shown in Fig. 35-41; in fact, the minimum deviation always occurs with the incidence angle that gives this path, for any n and α.

FIGURE 35-41 Minimum deviation through an equilateral prism occurs when the path is symmetric (Problem 56).

57. Show that a three-dimensional corner reflector (three mirrors in three mutually perpendicular planes, or a solid cube in which total internal reflection occurs) turns an incident light ray through 180°, so it returns in the direction from which it came. *Hint:* Let $\mathbf{q} = q_x\hat{\mathbf{i}} + q_y\hat{\mathbf{j}} + q_z\hat{\mathbf{k}}$ be a vector in the direction of propagation. How does this vector get changed on reflection by a mirror in a plane defined by two of the coordinate axes?

58. Show that the angle ϕ that appears in Fig. 35-25 is given by $\phi = 4\sin^{-1}\left(\dfrac{\sin\theta}{n}\right) - 2\theta$, where θ is the angle of incidence.

59. (a) Differentiate the result of the preceding problem to show that the maximum value of ϕ occurs when the incidence angle θ is given by $\cos^2\theta = \frac{1}{3}(n^2 - 1)$. (b) Use this result and that of the preceding problem to find ϕ_{max} in water with $n = 1.333$.

60. Figure 35-42 shows the approximate path of a light ray that undergoes internal reflection twice in a spherical water drop. Find the maximum angle ϕ for this case, taking $n = 1.333$. This is the angle at which the secondary rainbow appears.

61. *Fermat's principle* states that the path of a light ray between two points is such that the time to traverse that

FIGURE 35-42 Problem 60.

path is an extremum (either a minimum or a maximum) when compared with the times for nearby paths. Consider two points A and B on the same side of a reflecting surface, and show that a light ray traveling from A to B via a point on the reflecting surface will take the least time if its path obeys the law of reflection. Thus, the law of reflection (Equation 35-1) follows from Fermat's principle.

62. Use Fermat's principle (see preceding problem) to show that a light ray going from point A in one medium to point B in a second medium will take the least time if its path obeys Snell's law. Thus, Snell's law follows from Fermat's principle.

36 Image Formation and Optical Instruments

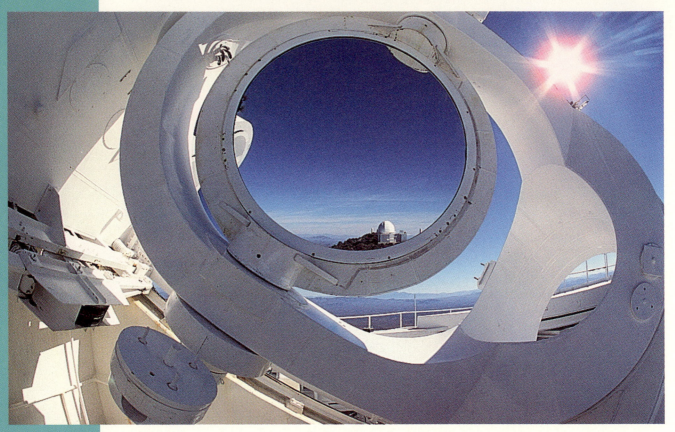

This mirror directs sunlight into the McMath–Pierce Solar Telescope in Arizona, which uses optical technology to form a detailed image of the Sun's surface.

Reflection and refraction alter the direction of light propagation, following laws we developed in the preceding chapter. Microscopes, telescopes, cameras, contact lenses, photocopiers, and your own eyes are among the many natural and technological systems that use reflection or refraction to form **images**—patterns of light that provide visual representations of reality. As you read this page, for example, refraction in your eye forms an image on the retina at the back of the eye. Your optic nerve conveys this image to your brain, which processes it to give you the conscience sense of seeing. In this chapter we study image formation using the approximation of geometrical optics—an approximation that remains valid as long as we consider only length scales much greater than the wavelength of light.

When we view an object with an optical system, we perceive light that seems to come straight from the object but whose path has actually been altered. As a result we see an image that may be different in size, orientation, or apparent position from the actual object. In some cases light actually comes from the image to our eyes; the image is then called a **real image**. In other cases light only apparently comes from the image location; the image is then called a **virtual image**.

36-1 PLANE MIRRORS

When you look at yourself in a flat mirror, you see an image that appears to be behind the mirror by the same distance that you are in front of it. The image is upright and the same size as you are, but appears reversed. Why?

Figure 36-1 shows how the image in a plane mirror comes about. In Fig. 36-1a we concentrate on a small object, in this case an arrowhead. (We'll frequently use arrows to represent objects in image-forming situations because they're both simple and sufficiently asymmetric that we can see whether images are inverted or upright.) We've drawn three light rays that leave the object, reflect off the mirror, and enter the observer's eye. The rays reflect at the mirror with equal angles of incidence and reflection. As Fig. 36-1a shows, light looks to the observer like it's coming from a point behind the mirror. That point is the location of the arrowhead's image. In this case the image is virtual because no light actually comes from behind the mirror.

Since two nonparallel lines define a point, we need only two rays to locate the arrowhead in Fig. 36-1a. We've repeated this image-location process in Fig. 36-1b, using as one of the rays the ray that reflects normally. The same procedure also locates the bottom of the arrow, and we could obviously fill in additional points to locate the entire arrow; the resulting image is shown in Fig. 36-1b.

Note that the triangles OPQ and $O'PQ$ in Fig. 36-1b share a common side and that the angles OPQ and $O'PQ$ are both right angles. And because the angles of incidence and reflection are equal, so are the angles OQP and $O'QP$. Therefore triangles OPQ and $O'PQ$ are congruent, showing that the distance PO' that the arrowhead's image lies behind the mirror is equal to the distance OP from the actual arrowhead to the mirror. A similar analysis applies to the rays from the bottom of the arrow, so we can conclude that the image appears as far behind the mirror as the object is in front. Furthermore, since extensions of the rays from the top and bottom of the arrow normal to the mirror pass through the top and bottom of the image, respectively, the image must be the same height as the object itself.

15.3
Plane Mirrors

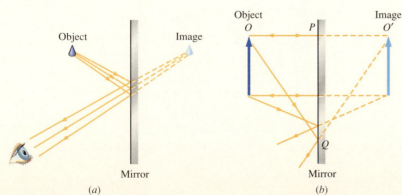

(a)

(b)

FIGURE 36-1 Image formation in a plane mirror. (*a*) Light reflected off the mirror seems to come from an image behind the mirror. Since there is really no light coming from behind the mirror, this is a virtual image. (*b*) Two rays from each point on an object serve to locate that point's image. Triangles OPQ and $O'PQ$ are congruent, showing that object and image are equidistant from the mirror.

EXAMPLE 36-1	*What Size Mirror to Buy?*

You want the smallest mirror that will show your full image. How tall must it be?

Solution

Figure 36-2 shows the situation. Because the angles of incidence and reflection are equal, light from your foot reflects from the mirror at a point that's vertically halfway between your eye and the floor. Similarly, light from the top of your head reflects midway between that point and your eye. The total distance from top to bottom of the mirror then needs to be half your eye-foot distance plus half the distance from your eye to the top of your head—for a total of half your height. Note that this result does not depend on how far from the mirror you stand. It does, however, require that you fix the mirror to the wall at just the right height.

EXERCISE You buy a mirror that's half your height, but you affix it to the wall 2 cm below its optimum location. How much of your image will be cut off?

FIGURE 36-2 A mirror half your height shows your entire image.

Answer: Top 4 cm

Some problems similar to Example 36-1: 1, 2

Image Reversal

Stand in front of a plane mirror and you appear reversed left to right; for example, your image's left hand is the image of your *right* hand (Fig. 36-3). It's not that the image of your left hand appears opposite your right hand, any more than the image of your head appears opposite your feet. A more accurate description is that the mirror reverses front to back. Figure 36-4 shows that the effect is to alter only one of the three coordinate axes, and that alters handedness, rotation, and all other phenomena connected with the right-hand rules we've been using.

Although image formation in a single plane mirror is straightforward, multiple reflections with more than one mirror can produce many—in some cases infinitely many—images; see Problem 2.

FIGURE 36-4 A plane mirror reverses front to back, reversing the direction of the *z* axis and making the image coordinate system a left-handed one. The rotation of the spinning top is reversed in the mirror, but its angular velocity vector still points upward because a left-hand rule applies in the "mirror world."

FIGURE 36-3 The image of a right hand is the image's left hand—but it's still the image of the *right* hand.

36-2 CURVED MIRRORS

In contrast to plane mirrors, curved mirrors form images that may be upright or inverted, virtual or real, and whose sizes need not be those of the original objects.

Parabolic Mirrors

A parabola—the curve generated by a quadratic function like $y = x^2$—has the property that a line drawn parallel to the parabola's axis makes the same angle to the normal as does a second line drawn to a special point called the **focus** or **focal point** (Fig. 36-5a). That means a concave mirror of parabolic shape will reflect rays parallel to the parabola's axis so they converge at the focus (Fig. 36-5b). This effect can be used to concentrate light to very high intensities (Fig. 36-6a). Conversely, a point source of light at the focus will emerge from the mirror in a beam of parallel rays (Fig. 36-6b).

(a)   (b)

FIGURE 36-5 A parabolic mirror reflects rays parallel to its axis to a common focus. (a) Ray diagram. (b) Photo shows rays reflecting from an actual mirror.

(a)

(b)

FIGURE 36-6 (a) This solar furnace uses a huge parabolic reflector to concentrate sunlight. It's used to study properties of materials at high temperatures. (b) A flashlight has a light bulb near the focus of a parabolic mirror and thus produces a beam of nearly parallel light rays.

Spherical Mirrors

Near the apex of the parabolic mirror in Fig. 36-6, you can't tell whether the mirror's shape is parabolic or spherical; parabola and sphere are essentially the same over a limited range. Because a sphere is a more regular shape and thus easier to form, most mirrors used for focusing are, in fact, spherical. The slight distortion caused by the deviation from a perfect parabola—called **spherical aberration**—is minimized by making the actual mirror only a tiny fraction of the entire sphere. In that case the focal length is much larger than the size of the mirror, and light rays from distant objects strike the mirror only if they're close to and nearly parallel to the mirror axis. Such rays are called **paraxial rays**, and it's only for them that the approximation of a parabola by a sphere results in accurate focusing. We assume paraxial rays in what follows. For clarity, however, our diagrams show mirrors much more curved than they should be for distortion-free focusing, and consequently not all the rays we draw will seem paraxial.

APPLICATION	*Hubble Trouble*

The Hubble Space Telescope was launched in 1990 as the flagship of a new generation of space-based astronomical observatories. Although Hubble is smaller than many ground-based telescopes, its vantage point above the atmosphere was to make it optically superior in resolving astronomical objects. Furthermore, Hubble can observe in infrared, visible, and ultraviolet wavelengths.

In the months following Hubble's launch, engineers and scientists checking the telescope were frustrated by their inability to achieve a clear focus. They came reluctantly to the conclusion that Hubble's 2.4-m-diameter primary mirror was flawed. Subsequent investigation showed that an error of 1.3 mm in the placement of instruments used during manufacture of the mirror had resulted in its being ground to the wrong curvature. The mirror itself is off by only 2.3 μm at its edge, but this is an enormous flaw in an optical system designed to be accurate to a small fraction of the wavelength of light. Hubble specifications called for 70% of the light energy from a point source to fall within an angular diameter of 0.1 second of arc; because of its spherical aberration, at best 16% of the light fell in the prescribed zone.

Although the Hubble mirror could not be replaced or repaired, in 1993 astronauts managed to install corrective mirrors that restored Hubble's optical system to better than its original design specifications. The result is a superb astronomical instrument with image quality greatly superior to that of ground-based telescopes (Fig. 36-7).

FIGURE 36-7 (Top) An astronaut installing new optical equipment on the Hubble Space Telescope. (Bottom) Hubble images of M-100, a galaxy 50 million light-years from Earth, taken before and after the repair.

We can see how spherical mirrors form images by tracing two rays from each of several points on the object, just as we did for plane mirrors. Some special rays make this process simple; their properties all follow from the law of reflection and the special properties of a spherical mirror in the paraxial approximation:

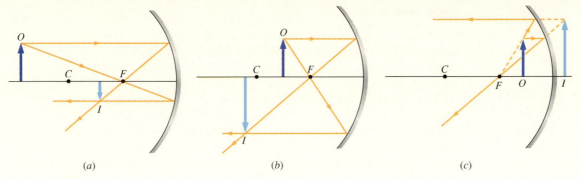

(a)   (b)   (c)

FIGURE 36-8 Image formation in a concave spherical mirror, using the fact that light passing through the focal point F reflects parallel to the mirror axis, and vice versa. O denotes the object and I its image. (a) With the object beyond the mirror's center of curvature C, the mirror forms a real image that is inverted and reduced in size. (b) With the object between the center of curvature and focus, the image is real, inverted, and magnified. (c) When the object is closer than the focus, the image is enlarged, upright, and virtual. Note, in this case, to get the ray including the object and the focus we had to extend the actual light path back toward the focus.

1. Any ray parallel to the mirror axis reflects through the focal point.
2. In the converse of (1), any ray that passes through the focal point reflects parallel to the axis.
3. Any ray that strikes the center of the mirror reflects symmetrically about the mirror axis.
4. Any ray through the center of curvature strikes the mirror normal to the mirror surface, and thus returns on itself.

Any two of these rays suffice to locate an image.

Figure 36-8 shows ray tracings, using our special rays (1) and (2) that go through the focal point, to find the image location in three cases. In all three cases symmetry ensures that the bottom of the image arrow will be on the axis, so we haven't bothered to trace it. In Fig. 36-8a we see that an object beyond the mirror's center of curvature (C) forms a smaller, inverted image. Light actually emerges from this image, so it's a *real* image. If you looked from the left in Fig. 36-8a you would actually see the image in space in front of the mirror (Fig. 36-9).

As the object moves closer to the mirror the real image grows; with the object between the center of curvature and the focus, the image is larger than the object, and farther from the mirror (Fig. 36-8b). As the object moves toward the focus the image grows larger and moves rapidly away from the mirror. With the object right at the focus the rays emerge in a parallel beam and there is no image. Finally, rays from an object closer to the mirror than the focus diverge after reflection. To an observer they appear to come from a point behind the mirror. Thus there is a virtual image, in this case upright and enlarged (Fig. 36-8c).

ActivPhysics Activity 15.4 provides a simulation of Fig. 36-8 in which you can move the object to any point.

FIGURE 36-9 A bear meets its real image, formed by the concave mirror at rear. The bear and its image are both in *front* of the mirror.

15.4
Spherical Mirrors

Got It!

Where would you place an object so its real image is the same size as the object?

FIGURE 36-10 Image formation using a convex mirror. The image is always virtual, upright, and reduced in size.

FIGURE 36-11 Reflecting spheres are convex mirrors, forming reduced, virtual, upright images.

Convex Mirrors

A convex mirror reflects on the outside of its spherical curvature, causing light to diverge rather than focus. Therefore, there is no possibility of forming a real image with a convex mirror. But Fig. 36-10 shows that the mirror can still form a virtual image. Although the focus has less obvious physical significance in this case, its location still controls the geometry of reflected rays. As Fig. 36-10 shows, we can still draw a ray parallel to the axis and another ray that would go through the focus if the mirror weren't in its way. By tracing these rays through the mirror we can see the directions in which they reflect. The reflected rays appear to diverge from a common point behind the mirror, showing that there is a virtual image, upright and reduced in size, behind the mirror. By considering different object positions, you can convince yourself that for a convex mirror the image always has these characteristics. Convex mirrors are widely used where an image of a broad region needs to be captured in a small space (Fig. 36-11).

The Mirror Equation

Drawing ray diagrams helps give an intuitive feel for image formation. More precise image locations and sizes follow from the **mirror equation**, which we now derive. This time we'll find the image using our special rays parallel to the axis and striking the mirror's center, as shown in Fig. 36-12a. The ray that strikes the center of the mirror reflects symmetrically about the axis; therefore, the two shaded triangles are similar. Then the **magnification**—the ratio image height h' to the object height h—is the same as the ratio of the image and object distances from the mirror. We'll consider the image height negative if the image is inverted; then from Fig. 36-12 we have

$$M = \frac{h'}{h} = -\frac{\ell'}{\ell}. \tag{36-1}$$

Here object and image are both in front of the mirror, so we take object and image distances ℓ and ℓ' as positive quantities; the negative sign in Equation 36-1 then shows that in this case the image is inverted. Also, for the object

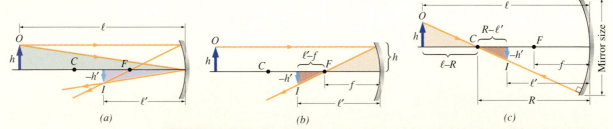

(a)   (b)   (c)

FIGURE 36-12 Finding the image I using the ray from the object O parallel to the axis, along with the ray that reflects from the center of the mirror. (a) Since the latter ray reflects symmetrically about the axis, the shaded triangles are similar, and allow us to relate the object and image heights h and h' to their distances ℓ and ℓ' from the mirror. (b) Another pair of similar triangles lets us relate the focal length f to the heights and distances. Length of the image arrow is marked $-h'$ because h' is considered negative for an inverted image. (c) The ray through the center of curvature strikes the mirror at 90°, so it reflects back on itself. Once again the shaded triangles are similar; now they let us relate the curvature radius to the heights and distances and thus, through the mirror equation, to the focal length. The curvature radius R proves to be twice the focal length f. The mirror size indicated should actually be considerably smaller than the curvature radius for the descriptions and equations of this section to be accurate.

location of Fig. 36-12, it's clear that $|M| < 1$, meaning the image is reduced rather than enlarged.

Figure 36-12*b* is the same as Fig. 36-12*a* except that now we show only the ray that reflects through the focus. We've also labeled the **focal length** *f*—i.e., the distance from the mirror to the focal point—and have shaded a different pair of similar triangles. From these triangles you can see that

$$\frac{-h'}{h} = \frac{\ell' - f}{f}.$$

Here we put the minus sign on h' because we've defined h' as a negative quantity for the inverted image, while comparing similar triangles requires a ratio of positive quantities. But Equation 36-1 shows that the ratio h'/h is the magnification $M = -\ell'/\ell$. So we have

$$-\frac{\ell'}{\ell} = \frac{\ell' - f}{f}$$

or, dividing both sides by ℓ' and doing a little algebra,

$$\frac{1}{\ell} + \frac{1}{\ell'} = \frac{1}{f}. \quad \text{(mirror equation)} \qquad (36\text{-}2)$$

Equations 36-1 and 36-2 together give us all the information we would get from ray tracing.

Although we derived the mirror equation using a real image, the equation applies to virtual images with the convention that a *negative* image distance ℓ' means the image is *behind* the mirror—as are virtual images (recall Fig. 36-8*c*). And we can handle *convex* mirrors as well by taking the focal length to be a *negative* quantity. Table 36-1 summarizes these sign conventions.

We can extract one more useful piece of information by considering the ray that passes through the center of curvature, *C*. This point is the center of the sphere of which the mirror is a part; therefore, its distance from the mirror is the mirror's curvature radius, *R*. That's why a ray through *C* hits normal to the mirror; it's following a radius, which intersects the sphere at right angles. So a ray through *C* reflects back on itself. We show this ray in Fig. 36-12*c*, along with yet another pair of similar triangles. From them, you can see that

$$\frac{R - \ell'}{\ell - R} = \frac{-h'}{h} = \frac{\ell'}{\ell},$$

where again we write $-h'$ because the inverted image has negative height according to our sign conventions, and we need a ratio of positive distances. The

TABLE 36-1 *Sign Conventions for Mirrors*

FOCAL LENGTH f^{1},	OBJECT DISTANCE, ℓ	IMAGE DISTANCE[2], ℓ'	TYPE OF IMAGE
+	$+, \ell > f$	+	Real, inverted, enlarged for $\ell < 2f$, reduced for $\ell > 2f$
+	$+, \ell < f$	−	Virtual, upright, enlarged
−	+	−	Virtual, upright, reduced

[1] + for concave mirror, − for convex.
[2] − means behind mirror.

last step in this equality follows from Equation 36-1. Solving the mirror equation for ℓ' and using the result in the equation above (see Problem 16) then yields

$$f = \frac{R}{2}. \tag{36-3}$$

Thus the focal length of a spherical mirror is half its curvature radius.

We emphasize that the results derived in this section are approximations based on the assumption that the spherical mirror is a good approximation to a parabola. Equivalent ways of describing this approximation are (1) that the curvature radius is large compared with the mirror's size as indicated in Fig. 36-12c and (2) that all rays reflecting from the mirror are close to and nearly parallel to the mirror axis (the paraxial approximation). Again, our drawings can't accurately reflect these conditions while at the same time showing clearly the ray paths.

15.7
Spherical Mirror Problems

EXAMPLE 36-2	*Sizing Up Hubble: A Concave Mirror*

During assembly of the Hubble Space Telescope, a technician stood 3.85 m in front of the telescope's concave mirror and viewed his image (Fig. 36-13). Given the telescope's focal length of 5.52 m, find (a) the location and (b) the magnification of the technician's image. (c) Repeat for the case when the technician stands 15.0 m from the mirror.

Solution

Since the mirror is *concave,* the focal length *f* is *positive.*
(a) We can solve the mirror equation (Equation 36-2) to get

$$\frac{1}{\ell'} = \frac{1}{f} - \frac{1}{\ell} = \frac{\ell - f}{f\ell},$$

or

$$\ell' = \frac{f\ell}{\ell - f} = \frac{(5.52 \text{ m})(3.85 \text{ m})}{3.85 \text{ m} - 5.52 \text{ m}} = -12.7 \text{ m}.$$

Why the *negative* answer? Because the technician is closer to the mirror than the focal point, so this is a *virtual* image located *behind* the mirror.
(b) Equation 36-1 gives the magnification:

$$M = -\frac{\ell'}{\ell} = -\frac{(-12.7 \text{ m})}{3.85 \text{ m}} = 3.30.$$

This *positive* answer indicates an *upright* image, enlarged about three times as shown in Fig. 36-13.
(c) If the technician stands 15.0 m from the mirror, similar calculations give

$$\ell' = \frac{f\ell}{\ell - f} = \frac{(5.52 \text{ m})(15.0 \text{ m})}{15.0 \text{ m} - 5.52 \text{ m}} = 8.73 \text{ m}$$

and

$$M = -\frac{\ell'}{\ell} = -\frac{8.73 \text{ m}}{15.0 \text{ m}} = -0.582.$$

FIGURE 36-13 A technician standing in front of the Hubble Space Telescope mirror. Where and how big is the technician's image?

Now ℓ' is *positive,* so the image is *real* and in *front* of the mirror; the *negative* magnification shows that it's *inverted,* and it's reduced in size.

EXERCISE You're scrutinizing your nose using a handheld concave mirror with curvature radius 2.2 m. How far from your face should you hold the mirror to see your nose doubled in size?

Answer: 55 cm

Some problems similar to Example 36-2: 3–9, 11–14

EXAMPLE 36-3 | *Jurassic Park: A Convex Mirror*

In the film *Jurassic Park,* horrified passengers in a car watch in the convex side-view mirror as a tyrannosaurus rex pursues them. Printed on the mirror is the warning "OBJECTS IN MIRROR ARE CLOSER THAN THEY SEEM." If the mirror's curvature radius is 12 m, and if the T-rex is actually 9.0 m from the mirror, by what factor does its image appear reduced in size?

Solution

This is a *convex* mirror, so its focal length is *negative*. Since the focal length is half the curvature radius, we have $f = -6.0$ m. Here we're asked for the magnification, given by Equation 36-1. For that we need the image distance ℓ'. We found an expression for ℓ' in the preceding example, and that expression holds here because the mirror equation from which it was derived works for both concave and convex mirrors. So Equation 36-1 becomes

$$M = -\frac{\ell'}{\ell} = -\frac{f\ell/(\ell - f)}{\ell} = -\frac{f}{\ell - f}$$

$$= -\frac{-6.0 \text{ m}}{9.0 \text{ m} - (-6.0 \text{ m})} = 0.40 .$$

The T-rex in the mirror appears only 40 percent of its actual size, and thus it "seems" further away. Note that the mirror's warning isn't really a comparison of the image and object *distances,* but of the apparent versus actual *size* of objects. Image distances in a convex mirror are always *smaller* than object distances, so the images are always *reduced* and we therefore perceive objects viewed in a convex mirror as being farther way.

EXERCISE An object 30 cm from a convex mirror appears reduced to one-fourth of its actual size when viewed in a convex mirror. What is the mirror's curvature radius?

Answer: 20 cm

Some problems similar to Example 36-3: 10, 15, 62

APPLICATION | *Nonimaging Optics and the Quest for Solar Energy*

Sometimes we want to concentrate light energy without necessarily forming an image. In solar energy systems producing heat, for example, concentrated sunlight leads to higher temperature and thus to higher efficiency; with photovoltaic systems inexpensive reflectors can be used to concentrate solar energy onto more expensive photovoltaic cells, reducing the cell area needed.

Nonimaging optics is a new field, and the design of concentrators is an ongoing challenge for optical engineers. Figure 36-14 shows a popular concentrator design, called a compound parabolic concentrator. Unlike a parabolic mirror designed for image formation, this concentrator is made from parabolic segments that do not form a single complete parabola. As Fig. 36-14a shows, all light entering the concentrator

(a)

(b)

FIGURE 36-14 (a) A compound parabolic concentrator. Rays (blue) entering at the angle θ to the concentrator axis are reflected to the edge of the exit aperture. Rays incident at smaller angles (red) pass through the exit aperture after at most one reflection; thus all light entering over the angular range 2θ is concentrated at the exit. (b) A section of parabolic concentrators used to concentrate solar energy on a black tube carrying a heat-transfer fluid. A piece of pipe lies at the exit of one concentrator; since light paths are reversible, the fact that the entire concentrator appears black confirms that the path of an incident light ray ends up on the black tube.

from a broad range of angles ends up at the narrow exit end. The light doesn't converge to form an image, but its intensity has nevertheless been increased because its energy is spread over a smaller area. Thus, the concentrator acts as a "funnel" for light. Figure 36-14b shows a section of a compound parabolic reflector used to concentrate light onto a tube carrying heat-transfer fluid in a solar thermal power plant. Because the concentrator accepts light from a broad angular range, it works without the need for expensive tracking equipment that would keep it facing the Sun.

Nonimaging concentrators have other applications as well. The first compound parabolic concentrator, built in 1965, was designed to collect light from a high-energy physics experiment. Nonimaging optics in the infrared were used on the Cosmic Background Explorer satellite that has provided crucial evidence in support of the Big Bang theory of the origin of the universe. And under development are solar-powered lasers that use nonimaging concentrators as the "pump" that supplies energy to run the laser.

36-3 LENSES

A **lens** is a transparent piece of material that uses refraction of light to form images. As with mirrors, lenses can be either concave or convex. But light goes through a lens, while it reflects off a mirror, so the roles of the two types are reversed. A convex lens focuses parallel rays to a **focal point**, and is therefore called a **converging lens** (Fig. 36-15). As we'll soon see, a convex lens can form real and virtual images, depending on the object location. A concave lens, in contrast, is a **diverging lens**; it refracts parallel rays so they appear to be diverging from a common focus. Like a *convex* mirror, a *concave* lens can form only virtual images (Fig. 36-16).

(a)

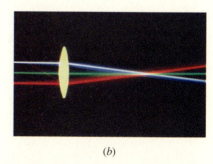

(b)

FIGURE 36-15 A convex lens brings parallel light rays to a focus *F*. The focal length *f* is the distance from the lens to the focal point. You can see in (a) that light is refracted at each surface of the lens, but as the lens is thin we can treat the two refractions as a single change of direction at the center plane of the lens. Because the lens surfaces are parallel at the center of the lens, any ray through the center passes undeflected.

(a)

(b)

FIGURE 36-16 Parallel light rays passing through a concave lens diverge so it looks as though they are coming from a common focus. But light doesn't actually come from the focus, so this type of lens can produce only virtual images.

In Section 36-4 we'll examine the detailed behavior of light as it refracts at the interfaces between a lens and the surrounding medium. First, we explore the simpler situation of a **thin lens**—one whose thickness is small compared with the curvature radii of its two surfaces. Although light refracts once as it enters a lens and once as it exits, in the thin-lens approximation the two lens surfaces are so close that it suffices to consider that the light bends just once, as it crosses the center plane of the lens. Unlike a mirror, light can go either way through a lens; that means the lens has two focal points, one on either side. For a thin lens, the focal length proves to be the same in either direction, so it doesn't matter which way we orient the lens. We'll justify the thin-lens approximation mathematically in Section 36-4.

Lens Images by Ray Tracing

As with mirrors, two rays serve to fix the image formed by a lens. For lenses, two special rays make ray tracing easy:

15.8
Thin Lens Ray Diagrams

1. Any ray parallel to the lens axis refracts through the focal point.
2. Any ray through the center of the lens passes undeflected.

Figure 36-17 shows ray tracings for different object placements in relation to a converging lens. In Fig. 36-17*a* we see that an object farther out than two focal lengths produces a smaller, inverted real image on the other side of the lens.

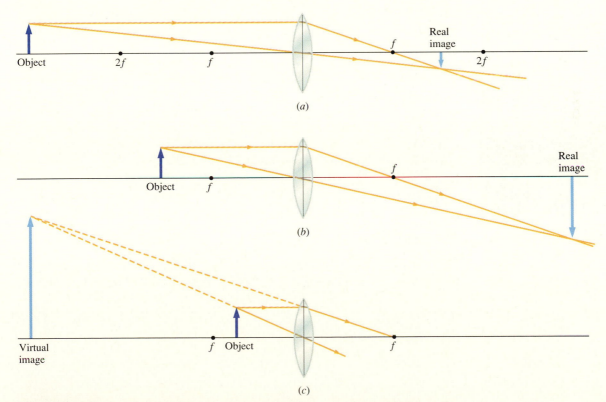

FIGURE 36-17 Image formation with a converging lens. Marks on axis are distances from the lens center, in units of the focal length *f*. In (*a*) and (*b*), an object distance greater than the focal length results in an inverted real image. The image is enlarged if the object distance is between one and two focal lengths, as shown in (*b*). In (*c*) a virtual image forms when the object is within the focal length. You can simulate these situations with ActivPhysics Activity 15.8.

FIGURE 36-18 Simplified diagram of a movie projector. The film lies just beyond the focal length, and thus the lens produces an enlarged real image on the distant screen. The audience normally views the image from the projector side of the screen. Note that the film is upside down.

Since light really emanates from this image, you could see it without actually looking through the lens. As the object moves toward the lens, the image moves away and grows in size. When the object is between one and two focal lengths from the lens, the image has moved beyond $2f$, and is enlarged. The image on a movie screen is formed in this way (Fig. 36-18). Moving the object closer than the focal point produces an enlarged virtual image that can be seen only by an observer looking *through* the lens (Fig. 36-17c).

Figure 36-19 shows ray tracings for a diverging lens. Like a convex mirror, this lens produces only virtual images that are upright and reduced in size. Like the virtual image of Fig. 36-17c these virtual images are visible only through the lens. You should convince yourself that the basic geometry of Fig. 36-19 does not change even if the object moves within the focal length.

FIGURE 36-19 A diverging lens always forms a reduced, upright, virtual image, visible only through the lens.

Got It!

Before going on to the mathematics of lenses, try answering the following simple questions about the image shown in Fig. 36-20. (Remember that answers to GOT IT! boxes are in the back of the book.)

a. What would happen if the bottom half of the lens were covered?
b. What would happen if the lens were removed?
c. What would happen if the screen's distance from the lens were doubled?
d. Would there be an image in the absence of the screen?

FIGURE 36-20 A simple optical system, consisting of an object (the candle), a converging lens, and a white screen on which the image appears. This system is a realization of Fig. 36-17b.

A Lens Equation

We can locate lens images quantitatively by deriving an equation like we did for mirrors. Note in Fig. 36-21 that the two unshaded right triangles formed by

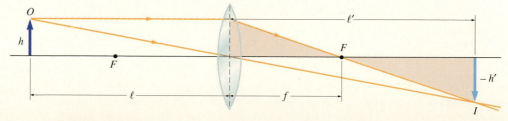

FIGURE 36-21 Ray diagram for deriving the lens equation. The unshaded pair of right triangles formed by the object and image arrows with their distances ℓ and ℓ' are similar, as are the pair of shaded triangles.

TABLE 36-2 *Sign Conventions for Lenses*

FOCAL LENGTH f^1,	OBJECT DISTANCE, ℓ	IMAGE DISTANCE[2], ℓ'	TYPE OF IMAGE
+	$+, \ell > f$	+	Real, inverted, enlarged for $\ell < 2f$, reduced for $\ell > 2f$
+	$+, \ell < f$	–	Virtual, upright, enlarged
–	+	–	Virtual, upright, reduced

[1] + for convex lens, − for concave.
[2] − means on same side of lens as object.

the object and image arrows with their distances ℓ and ℓ' are similar. Therefore, the magnification is

$$M = \frac{h'}{h} = -\frac{\ell'}{\ell}, \tag{36-4}$$

where again we take a negative height to signify an inverted image. Thus, the magnitude of the magnification is the ratio of object to image distance, as it was with mirrors. The two shaded triangles in Fig. 36-21 are also similar, and therefore

$$\frac{-h'}{\ell' - f} = \frac{h}{f}.$$

Combining this result with Equation 36-4 and doing some algebra then gives

$$\frac{1}{\ell} + \frac{1}{\ell'} = \frac{1}{f}, \qquad \text{lens equation} \tag{36-5}$$

which is identical to the mirror equation 36-2. Note that putting the object at infinity ($\ell = \infty$) gives $\ell' = f$ and that putting the object at the focus ($\ell = f$) gives $\ell' = \infty$. Thus, Equation 36-5 reflects our ray-tracing diagrams that show parallel rays (such as those from an object at a great distance) converging at the focal point, and vice versa.

Although we derived Equation 36-5 for the case of a real image, the equation holds for virtual images if we consider the image distance negative; in that case the image is on the same side of the lens as is the object. And it holds for diverging lenses if we consider the focal length negative. Table 36-2 summarizes these sign conventions for lenses, while Fig. 36-22 describes graphically the sizes and types of images formed at different object distances.

Got It!

You look through a lens at this page and see the words magnified. Is the image you observe real or virtual? Is the lens concave or convex?

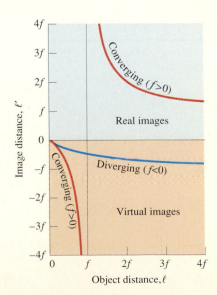

FIGURE 36-22 Image versus object distances for lenses. Real images correspond to positive image distances, virtual images to negative image distances. Red curves are for converging lenses, for which object distances below the focal length result in virtual images ($\ell' < 0$). The image moves toward infinity as the object approaches the focal point from either direction, and concurrently the image grows arbitrarily large. For diverging lenses (blue curve), images are always virtual and reduced in size.

EXAMPLE 36-4	*Slide Show*

You're projecting slides onto a wall 2.6 m from a slide projector whose single lens has focal length 12.0 cm. (a) How far should the slides be from the lens? (b) How big will be the image of a 35-mm slide?

Solution

We want the image focused on the screen, so the image distance ℓ' is 2.6 m. Solving Equation 36-5 for the object distance ℓ then gives

$$\ell = \left(\frac{1}{f} - \frac{1}{\ell'}\right)^{-1} = \left(\frac{1}{12 \text{ cm}} - \frac{1}{260 \text{ cm}}\right)^{-1} = 12.58 \text{ cm}.$$

This is just beyond the focal length, as Figs. 36-17 and 36-22 suggest it should be to get an enlarged, distant, real image.

Equation 36-4 then gives the height h' of the image formed of a 35-mm slide:

$$h' = -h\frac{\ell'}{\ell} = -(3.5 \text{ cm})\left(\frac{260 \text{ cm}}{12.58 \text{ cm}}\right) = -72 \text{ cm}.$$

The minus sign indicates inversion, showing that the slide must go in the projector upside down to get an upright image.

EXERCISE If the slide in Example 36-4 moves 1 mm farther from the lens, what will happen to the position of the image?

Answer: moves 36 cm toward projector

Some problems similar to Example 36-4: 18–30

EXAMPLE 36-5	*Fine Print*

You're using a magnifying glass (a converging lens) with 30 cm focal length to read a telephone book (Fig. 36-23). How far from the page should you hold the lens in order to see the print enlarged 3 times?

Solution

Here the image is virtual, so ℓ' is negative and $3\times$ magnification then corresponds to $\ell' = -3\ell$. So Equation 36-5 becomes

$$\frac{1}{\ell} - \frac{1}{3\ell} = \frac{2}{3\ell} = \frac{1}{f} = \frac{1}{30 \text{ cm}},$$

whence $\ell = (2)(30 \text{ cm})/3 = 20$ cm. Figure 36-23 confirms that the image appears further from the lens than the actual page.

EXERCISE A magnifying glass enlarges print by 50% when it's held 9.0 cm from a page. What is its focal length?

FIGURE 36-23 A magnifying glass is a converging lens used in its virtual-image mode.

Answer: 27 cm

Some problems similar to Example 36-5: 18–30

36-4 REFRACTION IN LENSES: THE DETAILS

We now examine in detail the refraction effects that cause lenses to form images. Every lens has two refracting surfaces, at least one of which is curved. We therefore look first at image formation due to refraction at a *single* curved interface. To understand the behavior of a lens we will then consider two such interfaces, using the image formed by the first as the object for the second.

Refraction at a Curved Surface

Figure 36-24 shows part of a spherically curved piece of transparent material with curvature radius R and refractive index n_2, surrounded by a medium with

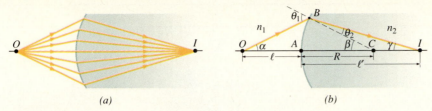

FIGURE 36-24 Refraction at an interface with a curved surface. (*a*) Rays from an object at *O* are refracted at the interface and converge to form an image at *I*. (*b*) Geometry for relating the object and image distances ℓ and ℓ'. All the labeled angles are considered small, even though the drawing doesn't show them as such.

index n_1 (typically air). A point-like object *O* lies on an axis through the center of curvature of the refracting surface. As we will now show, light rays diverging from this object at small angles to the optic axis are refracted to a common image point (Fig. 36-24*a*). The restriction to small angles is the same paraxial approximation we applied for spherical mirrors; again, our drawings don't always show the angles as being small.

In Fig. 36-24*b* we show a single ray in detail. Under the paraxial approximation the labeled angles in this figure are all small, and we can use the approximation $\sin x \simeq \tan x \simeq x$, where $x \ll 1$ with x an angle measured in radians. Then Snell's law—$n_1 \sin \theta_1 = n_2 \sin \theta_2$—becomes simply

$$n_1 \theta_1 = n_2 \theta_2 \,.$$

From triangles *BCI* and *OBC*, we see that $\theta_2 = \beta - \gamma$ and $\theta_1 = \alpha + \beta$. Using these results in our small-angle version of Snell's law gives

$$n_1(\alpha + \beta) = n_2(\beta - \gamma) \,.$$

Furthermore, in the small-angle approximation the arc *BA* is so close to a straight line that we can write $\alpha \simeq \tan \alpha \simeq BA/\ell$, with $\ell = OA$ the object's distance from the refracting surface. Similarly, $\beta \simeq BA/R$, and $\gamma \simeq BA/\ell'$. Thus, our expression of Snell's law becomes

$$n_1 \left(\frac{BA}{\ell} + \frac{BA}{R} \right) = n_2 \left(\frac{BA}{R} - \frac{BA}{\ell'} \right)$$

or, on canceling the arc length *BA* and rearranging,

$$\frac{n_1}{\ell} + \frac{n_2}{\ell'} = \frac{n_2 - n_1}{R} \,. \tag{36-6}$$

Notice that the angle α does not appear in this equation. Therefore, this relation between the object and image distances holds for *all* rays as long as they satisfy the small-angle approximation. That means all such rays come to a common focus at *I*, as in Fig. 36-24*a*, so an image must appear at this point.

Although we derived Equation 36-6 for a case of real image formation, as usual it holds for virtual images as well if we consider the image distance negative. It also works for surfaces that are concave toward the object—the opposite of Fig. 36-24—if we take *R* as a negative number. And it even works for flat surfaces, with $R = \infty$, as the exercise following Example 36-6 shows.

EXAMPLE 36-6	A Cylindrical Aquarium

Solution

An aquarium is made from a thin-walled tube of transparent plastic 70 cm in diameter (Fig. 36-25a). For a cat looking directly into the aquarium, what is the apparent distance to a fish 15 cm from the aquarium wall?

We can neglect the aquarium wall; since it's thin and has essentially parallel faces, light suffers no net deflection in passing through the wall. The surface is concave toward the object, so $R = -35$ cm. With $\ell = 15$ cm, $n_1 = 1.333$ for water, and $n_2 = 1$ for air, we can solve Equation 36-6 for the image distance ℓ':

$$\ell' = \frac{n_2}{\left(\dfrac{n_2 - n_1}{R} - \dfrac{n_1}{\ell}\right)} = \frac{1}{\left(\dfrac{1 - 1.333}{-35 \text{ cm}} - \dfrac{1.333}{15 \text{ cm}}\right)}$$

$$= -12.6 \text{ cm}.$$

The negative answer indicates that we have a virtual image, as shown in Fig. 36-25b.

A special case of Equation 36-6 is a flat surface, for which $R = \infty$. The right-hand side is then zero, but the equation still relates object and image distances. Looking down into a swimming pool from above, for example, you see virtual images of objects in the pool. The objects appear closer than they actually are, as Fig. 36-26 and the following exercise illustrate.

EXERCISE The bottom of a swimming pool looks to be 1.5 m below the surface. What is the pool's actual depth?

Answer: 2.0 m

Some problems similar to Example 36-6: 30, 31, 32, 34–36

(a)   (b)

FIGURE 36-25 (a) A cylindrical aquarium. (b) Top view, showing formation of a virtual image of a fish that is actually 15 cm from the aquarium wall.

FIGURE 36-26 Refraction at a flat surface produces an image I that appears closer than the object O. Objects at the bottom of a swimming pool, for example, appear closer than they really are.

Lenses, Thick and Thin

Figure 36-27 shows a lens of arbitrary thickness t made from material with refractive index n. For simplicity we will consider the lens surrounded by air or

FIGURE 36-27 A thick lens. The left-hand surface forms the virtual image I_1, which is also labelled O_2 because it forms the object for the right-hand surface. Together, the two surfaces then produce the final real image I_2. Absolute value signs are used on distances whose algebraic signs in Equation 36-6 are negative—i.e., the distance ℓ'_1 to the virtual image and the curvature radius R_2 of the right-hand surface, which is concave toward the incident light.

vacuum, with refractive index equal to 1. An object O_1 lies a distance ℓ_1 to the left of the left-hand surface. This surface focuses light from O_1 to form image I_1. Light from this image impinges on the right-hand surface, forming a second image I_2. We want to relate the original object O_1 and final image I_2.

We developed Equation 36-6, our description of refraction at a single curved surface, using Fig. 36-24. In that figure we considered formation of a real image, but we quickly noted that Equation 36-6 applies to virtual and real images alike. In considering a two-surface lens it proves more convenient to place the object where its image in the first surface is a virtual one, as in Fig. 36-27. Again, our final result will apply whatever the object placement.

At the left-hand surface in Fig. 36-27, the quantities in Equation 36-6 are $\ell = \ell_1$, $\ell' = \ell_1'$, $n_1 = 1$, $n_2 = n$, and $R = R_1$, so Equation 36-6 becomes

$$\frac{1}{\ell_1} + \frac{n}{\ell_1'} = \frac{n - 1}{R_1}. \qquad \text{(left-hand surface)}$$

The "object" O_2 for the right-hand surface is the image I_1 since light incident on that surface looks like it's coming from I_1. The distance ℓ in Equation 36-6 for this "object" is $\ell = \ell_2 = t - \ell_1'$, with t the lens thickness. Why minus? Because I_1 is a *virtual* image, so, according to our sign conventions, ℓ_1' is a *negative* quantity. The physical distance between the image and the lens is then the positive quantity $-\ell_1'$, and that quantity adds to the lens thickness to give the distance from this intermediate image to the right-hand surface. Also at the right-hand surface, $\ell' = \ell_2'$, $n_1 = n$, $n_2 = 1$, and $R = R_2$. Thus, at the right-hand surface Equation 36-6 reads

$$\frac{n}{t - \ell_1'} + \frac{1}{\ell_2'} = \frac{1 - n}{R_2}. \qquad \text{(right-hand surface)}$$

Now we'll let the lens become arbitrarily thin, taking the limit $t \to 0$. Then the first term in our right-surface equation becomes $-n/\ell_1'$. But the term n/ℓ_1' also occurs in the left-surface equation. So we can add the two equations to eliminate the intermediate-image distance ℓ_1'. Since we'll be left with only one object distance, ℓ_1, and one image distance, ℓ_2', we can drop the subscripts 1 and 2 on these quantities. The result is

$$\frac{1}{\ell} + \frac{1}{\ell'} = (n - 1)\left(\frac{1}{R_1} - \frac{1}{R_2}\right). \qquad (36\text{-}7)$$

The left-hand side of this equation is identical to the left-hand side of Equation 36-5. So we can equate the right-hand sides of the two equations to get an expression for the focal length of our lens:

$$\frac{1}{f} = (n - 1)\left(\frac{1}{R_1} - \frac{1}{R_2}\right). \qquad \text{(lensmaker's formula)} \qquad (36\text{-}8)$$

We emphasize that the radii R_1 and R_2 in Equation 36-8 may be positive or negative, depending on the lens curvature. In Fig. 36-27, for example, R_1 is positive because the left-hand surface is convex toward the incident light but R_2 is negative because the right-hand surface is concave toward the light. One of the radii can also be infinite if the corresponding surface is flat.

Lenses are made in a variety of shapes, as shown in Fig. 36-28. Those that are thicker at the center are converging lenses, for which Equation 36-8 gives a positive focal length. Those that are thinner at the center are diverging lenses, with negative focal length. These characterizations of lens behavior reverse if

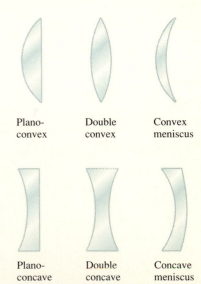

| Plano-convex | Double convex | Convex meniscus |

| Plano-concave | Double concave | Concave meniscus |

FIGURE 36-28 Common lens types.

the medium surrounding the lens has a higher refractive index than the lens itself (see Problem 73, which generalizes Equation 36-8 for an arbitrary refractive index in the surrounding medium).

EXAMPLE 36-7	A Plano-Convex Lens

The plano-convex lens in Fig. 36-28 is made from material with refractive index n. Given that the curved surface has curvature radius R, find an expression for the focal length of the lens. Show that the focal length is the same whether light is incident on the lens from right or left.

Solution

With light incident from the left in Fig. 36-28, $R_1 = R$. A flat surface has infinite curvature, so $R_2 = \infty$. Then Equation 36-18 gives

$$f = \left[(n-1) \left(\frac{1}{R} - \frac{1}{\infty} \right) \right]^{-1} = \frac{R}{n-1},$$

since $1/\infty = 0$. If light is incident from the right, then $R_1 = \infty$ and, since the curved surface is now concave toward the light,

$R_2 = -R$. Then

$$f = \left[(n-1) \left(\frac{1}{\infty} - \frac{1}{-R} \right) \right]^{-1} = \frac{R}{n-1},$$

showing that the two focal points are indeed equidistant from the lens.

EXERCISE A double-convex lens (see Fig. 36-28) is made of glass with $n = 1.75$. If the curvature radii of the two sides are equal, what should that radius be in order to give the lens a 25-cm focal length?

Answer: 37.5 cm

Some problems similar to Example 36-7: 38–47

Lens Aberrations

Lenses exhibit several types of optical defects. We've already described **spherical aberration** in connection with mirrors; this defect occurs because a spherical mirror only approximates the ideally focusing shape of a parabola. Lenses with spherical surfaces also exhibit spherical aberration (Fig. 36-29a). In our analysis of image formation by a curved refracting surface we had to make the small-angle approximation; otherwise, we would not have reached the conclusion that all rays from an object come to a common focus. Spherical aberration can be minimized by ensuring that incident rays are as close and as parallel as possible to the lens axis. That's easy for distant objects but harder with objects close to the lens. Using only the central portion of the lens eliminates those rays at larger angles (Fig. 36-29b), leading to a sharper focus. That's why a camera focuses over a wider range when it is "stopped down," the outer part of its lens

(a) (b)

FIGURE 36-29 (a) Spherical aberration occurs because rays striking the lens farther from the axis and at greater angles focus closer to the lens than do near-axial rays. The result is a "smearing" of the image. (b) This defect can be minimized by using only the central portion of the lens, at the expense of a dimmer image.

FIGURE 36-30 Chromatic aberration occurs because the refractive index varies with wavelength, causing different colors to focus at different points.

covered by an adjustable diaphragm. The trade-off, of course, is that there is less light available.

Another optical defect is **chromatic aberration** (Fig. 36-30), which occurs because the refractive index varies with wavelength. High-quality optical systems minimize this defect by using composite lenses made from different materials whose differing refractive indices allow several colors to focus at the same point. Chromatic aberration is unique to lenses since mirrors reflect light of all colors in exactly the same way. Figure 36-31 shows both spherical and chromatic aberration in a simple magnifying lens.

FIGURE 36-31 This magnifying lens shows both spherical aberration (distortion of the lines) and chromatic aberration (colors).

APPLICATION	*Fresnel Lenses*

Large-diameter lenses are impractically thick. Lighthouses, in particular, cannot use single lenses to concentrate their intense beams because too much light would be absorbed in the glass. To solve this problem the French scientist Augustin Fresnel, in 1822, devised a way of making large lenses thin. A Fresnel lens consists of concentric rings shaped to approximate segments of an ordinary lens surface, with steplike jumps to keep the overall structure thin (Fig. 36-32*a*). Fresnel lenses are still used in lighthouses (Fig. 36-32*b*), while Fresnel lenses made from plastic sheets appear in home, office, and car windows to provide magnification or wide-angle views (Fig. 36-32*c*). They are also used in some overhead projectors.

(*a*)   (*b*)   (*c*)

FIGURE 36-32 (*a*) A Fresnel lens approximates a normal curved lens by a series of short segments. (*b*) A giant Fresnel lens from a nineteenth-century lighthouse. (*c*) Fresnel lens made from sheet plastic mounted on the rear window of a van provides a wide-angle view of what's behind the van.

36-5 OPTICAL INSTRUMENTS

Numerous optical instruments make use of mirrors and/or lenses. We've already considered some simple applications using a single lens or mirror, including film projectors, magnifying glasses, and side-view mirrors. More complicated systems use several optical elements (lenses or mirrors), but the principles we have developed still apply. In particular, we trace light through a sequence of optical elements by using the image formed by one element as the object for the next.

The Eye

15.3

The Eye

Our eyes are our primary optical instruments. Each eye is a complex optical system with several refracting surfaces and mechanisms to vary both the focal length and the amount of light admitted. Figure 36-33a shows that the eye is essentially a fluid-filled ball about 2.3 cm in diameter. Light enters through the hard, transparent cornea and passes through a fluid called the aqueous humor before entering the lens. On exiting the lens, light traverses the vitreous humor that fills the main body of the eyeball and finally strikes the retina. The retina is covered with light-sensitive cells of two types (Fig. 36-33b). Cells called cones are sensitive to different colors but require moderate amounts of light, while rod cells function at low intensity but lack color discrimination. Both types of cells produce electrochemical signals that carry information to the brain.

A properly functioning eye produces a well-focused real image on the retina. Contrary to general belief, most of the refraction is provided not by the lens but by the cornea. The lens, however, is the adjustable element. Muscles pulling on the lens alter its focal length, allowing fine adjustment of the eye's focus—a process known as accommodation. Other muscles automatically adjust the iris, enlarging or contracting the pupil opening to compensate for different light levels.

When the muscles are relaxed, the lens is relatively flat, with focal length about 1.7 cm. But for nearsighted (myopic) people, the image still forms in front

(a)

(b)

FIGURE 36-33 (a) The human eye. (b) Scanning-electron-microscope image showing rods and a cone (blue structure) in a human retina.

FIGURE 36-34 (a) A myopic eye focuses light from distant objects in front of the retina. (b) A diverging lens corrects the problem, creating a virtual image closer to the eye. The eye's image of this virtual image then falls on the retina.

of the retina, causing distant objects to appear blurred (Fig. 36-34a). Diverging corrective lenses produce closer intermediate images that the myopic eye can then focus (Fig. 36-34). In farsighted (hyperopic) eyes, the image of nearby objects would focus behind the retina, and converging corrective lenses are used (Fig. 36-35). Actually, everyone is farsighted, in the sense that there is a minimum distance, called the **near point**, below which the eye cannot focus sharply. In the typical human the near point is at about 25 cm, and this value is taken as a standard. By the time one is 50, however, the typical near point is at about 40 cm, and it may be several meters.

ActivPhysics Activity 15.13 simulates both near- and farsighted eyes.

Prescriptions for corrective lenses usually specify the corrective power, P, in **diopters**, where the diopter measure of a lens is the inverse of its focal length in meters. Thus, a 1-diopter lens has $f = 1$ m, while a 2-diopter lens has $f = 0.5$ m and is more powerful in that it refracts light more sharply.

Since even the closest objects we view are many centimeters from the eye, it doesn't matter whether a corrective lens is a few cm away from the eye, as with glasses, or right on the cornea, as with a contact lens. Today's soft contact lenses are more than 50 percent water and provide a natural-seeming addition to the cornea that corrects simple vision defects. Contact lenses can be quite thin because, as Equation 36-8 shows, it's the curvature radii rather than the lens thickness that establishes the focal length. On the other hand, glasses remain somewhat more versatile, with variable-focus lenses available that can compensate for defects in both near and far vision. A more radical approach is laser surgery to reshape the cornea; when successful, it can eliminate the need for corrective lenses.

The human eye is a complex biological system, in many ways a direct extension of the brain. Although we understand its optical behavior, we understand much less clearly how visual perception actually occurs. Recently biologists and computer scientists have teamed to produce an electronic "silicon retina" whose behavior mimics many subtle aspects of the natural eye and may therefore shed light on the process of perception.

FIGURE 36-35 (a) A farsighted eye cannot focus light from the standard 25-cm near point at N. (b) A converging corrective lens makes light from an object at the near point appear to come from the point N′ which is at the minimum distance where this eye can focus.

EXAMPLE 36-8	*Lost Your Glasses!*

You're on vacation and have lost your reading glasses; without them, your eyes can't focus closer than 70 cm. Fortunately you can buy nonprescription reading glasses in a pharmacy, where they come in increments of 0.25 diopter. Which glasses should you buy?

Solution

Correcting your vision problem means bringing the range for clear focus to the standard 25-cm near point. According to Fig. 36-35b, that means you want a lens that will produce a virtual image at 70 cm of an object at 25 cm. The required power follows directly from Equation 36-5:

$$P = \frac{1}{f} = \frac{1}{\ell} + \frac{1}{\ell'} = \frac{1}{0.25 \text{ m}} + \frac{1}{-0.70 \text{ m}} = 2.57 \text{ diopters},$$

where the image distance is negative because the image is virtual. You'll do fine with 2.5-diopter glasses.

EXERCISE A nearsighted person cannot see clearly beyond 80 cm. Prescribe a lens power that will image the most distant objects at 80 cm, giving clear vision at all distances.

Answer: −1.25 diopters, the minus sign signifying a diverging lens

Some problems similar to Example 36-8: 48–50, 69

Got It!

You and your roommate have gotten your boxes of disposable contact lenses mixed up. One box is marked "−1.75 diopter," the other "+2.25 diopter." You're farsighted and your roommate is nearsighted. Which contact lenses are yours?

Cameras

A camera is much like the eye, except that in place of the eye's retina it focuses its image on film or a light-sensitive electronic device (Fig. 36-36). Most still cameras use the long-established technology of film based on light-sensitive chemicals, but light-sensitive silicon chips known as charge-coupled devices (CCDs) have already made film obsolete in astronomical imaging and, with the proliferation of digital cameras, are rapidly challenging conventional photography.

All but the simplest cameras have movable lenses that adjust, either manually or automatically, for optimum focus with different distances to the subject. Most

FIGURE 36-36 Optical system of a simple camera.

(a)	*(b)*

FIGURE 36-37 Photos taken from the same point with (*a*) short and (*b*) long focal length settings of an adjustable zoom lens.

cameras also have an adjustable diaphragm that covers part of the lens to regulate the amount of incoming light; "stopping down" the lens by closing the diaphragm admits less light but permits a broader range of distances to be in reasonably clear focus. Many cameras have zoom lenses whose focal lengths can be altered to provide different views of the same subject (Fig. 36-37).

Magnifiers and Microscopes

We wouldn't need optical instruments to examine small objects if we could bring our eyes arbitrarily close to the objects. But we've seen that the average human eye can't focus much closer than about 25 cm. We therefore use lenses to put enlarged images of small objects at distances where our eyes can focus.

What matters is not so much the actual size of the image, but how much of our field of view it occupies. Consequently we define the **angular magnification** m as the ratio of the angle subtended by an object seen through a lens to the angle subtended as seen by the naked eye when the object is at the standard 25-cm near point. Figure 36-38*a* shows that the former angle, α, is given

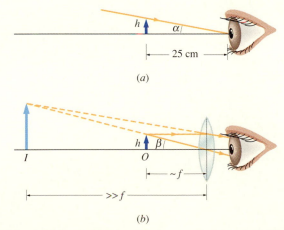

(a)

(b)

FIGURE 36-38 (*a*) An object of height h subtends a small angle $\alpha \simeq h/25$ cm at the standard 25-cm near point. (*b*) Putting the object near the focus of a converging lens gives an image that subtends an angle $\beta \simeq h/f$. The angular magnification is the ratio $m = \beta/\alpha$.

by $\alpha = h/25$ cm, where h is the object height and α is in radians. The maximum magnification would occur with the image itself at the near point (see Problem 53), but it's more comfortable to view a very distant image. We therefore place the object at just inside the focal length, forming an enlarged virtual image at a great distance. Figure 36-38*b* shows this geometry, from which we see that the angle β subtended by the image is essentially h/f. Then the magnification is

$$m = \frac{\beta}{\alpha} = \frac{h/f}{h/25 \text{ cm}} = \frac{25 \text{ cm}}{f} \qquad \text{(simple magnifier)}. \qquad (36\text{-}9)$$

This angular magnification is achieved only with the eye very close to the lens, which is not the way we normally hold an ordinary magnifying glass. But it is the way we place our eyes on instruments such as microscopes and telescopes that use simple magnifying lenses for their eyepieces.

Single lenses can produce angular magnifications of about 4 before aberrations compromise the image quality. Higher power magnification therefore requires more than one lens. A **compound microscope** is a two-lens system in which a lens of short focal length called the **objective** forms a magnified real image. This image is then viewed with a second lens, the **eyepiece**, positioned as a simple magnifier (Fig. 36-39). The object being viewed is positioned just beyond the focal length of the objective lens, and its image falls just inside the focal length of the eyepiece. If both focal lengths are small compared with the distance between the lenses, then the object distance for the objective lens is approximately the objective focal length f_o, and the resulting image distance is approximately the lens spacing L. The real image formed by the objective lens is larger than the object by the ratio of the image to object distance, or $-L/f_o$. The eyepiece enlarges the image further, by a factor of its angular magnification 25 cm$/f_e$. Then the overall magnification of the microscope is

$$M = M_o m_e = -\frac{L}{f_o}\left(\frac{25 \text{ cm}}{f_e}\right), \qquad \text{(compound microscope)} \qquad (36\text{-}10)$$

where, as usual, the minus sign signifies an inverted image.

Optical microscopes work well as long as the approximation of geometrical optics holds—that is, when the object being viewed is much larger than the wavelength of light. Viewing smaller objects requires waves of shorter wavelength than visible light. In the widely used electron microscope, those "waves" are electrons, whose wavelike nature we will examine in Chapter 39.

15.11
Two-Lens Systems

15.14
Microscope

FIGURE 36-39 Image formation in a compound microscope. Figure is not to scale; distance L should be much greater than either focal length, and the image I_1 should be very near the eyepiece's focus, resulting in greater magnification.

You can "build" and experiment with 2-lens optical systems, such as the microscope, using ActivPhysics Activity 15.11.

Telescopes

A telescope collects light from distant objects, either forming an image or supplying light to instruments for analysis. Telescopes are classified as **refracting** or **reflecting**, depending on whether the main light-gathering element is a lens or mirror. Small hand-held telescopes and binoculars are refracting instruments, as are telephoto camera lenses and some older astronomical telescopes. Modern astronomical telescopes are invariably reflectors.

The simplest refracting telescope consists of a single lens that images distant objects at essentially its focal point. Film or a CCD placed at that point captures the image, or an eyepiece is used to view the image. The world's largest refracting telescope, at Yerkes Observatory in Wisconsin, has a 1-m-diameter lens with 12-m focal length. Photographers can think of it as a 12,000-mm telephoto lens. Figure 36-40 shows the imaging process in an astronomical refracting telescope. The focal points of the objective and eyepiece lenses are nearly coincident, so the real image of a distant object that forms at the objective's focus is then seen through the eyepiece as a greatly enlarged virtual image. The angular magnification is the ratio of the angle β subtended by the final image to the angle α subtended by the actual object; Fig. 36-40 shows that this ratio is

15.12
The Telescope

$$m = \frac{\beta}{\alpha} = \frac{f_o}{f_e}. \qquad \text{(refracting telescope)} \qquad (36\text{-}11)$$

Since a real image is inverted and a virtual image is upright, a two-lens refracting telescope gives an inverted image. This is fine for astronomical work, but telescopes designed for terrestrial use have an extra lens, a diverging eyepiece (see Problem 72) or a set of reflecting prisms (as in binoculars; see Fig. 35-15) to produce an upright image.

Reflecting telescopes offer many advantages over refractors. Mirrors have reflective coatings on their front surfaces, eliminating chromatic aberration because light need not pass through glass. Having only one optically active

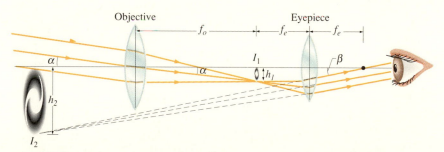

FIGURE 36-40 Image formation in a refracting telescope. A distant object, in this case a galaxy, is imaged first at the focus of the objective lens (image I_1). An eyepiece with its focus at nearly the same point then gives an enlarged virtual image (I_2). The angles α and β are given by $\alpha = h_1/f_o$ and $\beta = h_1/f_e$ in the small-angle approximation, leading to Equation 36-11 for the angular magnification.

FIGURE 36-41 One of the two identical Keck telescopes that occupy adjacent observatory domes atop Hawaii's Mauna Kea. Each telescope has a hexagonal mirror 10 m across, consisting of 36 individually adjustable segments; here the entire mirror is visible through the slit in the dome. The two Keck telescopes are being joined optically with four smaller telescopes to form an instrument with the resolving power of an 85-m telescope. (More on how this works in the next chapter.)

surface allows much larger reflectors to be built since the mirrors can be supported across their entire back surfaces—unlike lenses, which can be supported only at the edges. Where the largest refracting telescope ever built has a 1-m-diameter lens, the newest reflectors boast diameters in the 10-m range (Fig. 36-41). These designs incorporate segmented and/or flexible mirrors whose shape can be adjusted under computer control for optimum focusing; with so-called adaptive optics, such systems may adjust rapidly enough to compensate for the atmospheric turbulence that has traditionally limited the resolution of ground-based telescopes.

The simplest reflecting telescope is a curved mirror with a CCD or film at its focus. Superb image quality results, in principle limited only by wave effects we will discuss in the next chapter. More often the telescope is used as a "light bucket," collecting light from stars or other distant sources too small to image even with today's large optical telescopes. Then a secondary mirror sends light to a focus at a point that's convenient for telescope-mounted instrumentation. Optical fibers may also be used to bring light collected by the primary mirror to fixed instruments. Alternatively, an eyepiece can be mounted to examine the image visually. Figure 36-42 shows several common designs for reflecting telescopes.

Magnification is not a particularly important quantity in astronomical telescopes, which are used more for spectral and other analysis than for direct imaging. More important is the light-gathering power of the instrument, which is determined simply by the area of its objective lens or primary mirror. The 10-m Keck telescope, for instance, has 100 times the light-gathering power of the 1-m Yerkes refractor and more than 17 times the power of the 2.4-m Hubble Space Telescope.

(a)

(b)

(c)

FIGURE 36-42 Common arrangements for reflecting telescopes. (a) Placing a CCD or film at the prime focus gives the best image quality. (b) In the Cassegrain design light reflects from a secondary mirror and passes through a hole in the primary. The secondary may be either plane or convex. This design is widely used in large telescopes, with the added option of placing detectors directly at the prime focus as in (a). (c) The Newtonian design, used primarily in small telescopes, has an angled secondary mirror to direct light to the eyepiece.

CHAPTER SYNOPSIS

Summary

1. Reflection and refraction can result in **images**, which are **real** or **virtual** depending on whether light actually comes from the image location.
2. Plane mirrors form only virtual images, equal in size to the object being imaged. Concave curved mirrors have a **focal point** at which parallel rays converge after reflection. Concave mirrors form either real or virtual images, depending on the object location. For convex mirrors, parallel rays reflect as though diverging from the focal point; such mirrors can form only virtual images. The **mirror equation** quantifies the relation between object and image distances ℓ and ℓ':

$$\frac{1}{\ell} + \frac{1}{\ell'} = \frac{1}{f},$$

where the **focal length** f is the distance from the mirror's apex to its focus. The **magnification** is the ratio $M = -\ell'/\ell$, where a negative ℓ' corresponds to a virtual image and therefore to a positive M and so to an upright image.

3. **Lenses**, made from transparent materials, refract light to a focus. In the **thin-lens** approximation, the curvature radii of the lens surfaces are much greater than the thickness of the lens. Such a lens has two foci, located equal distances f on either side of the lens. The object and image distances and magnification then obey equations identical to those for mirrors. The focal length of a thin lens is given by the **lensmaker's equation**:

$$\frac{1}{f} = (n-1)\left(\frac{1}{R_1} - \frac{1}{R_2}\right),$$

where R_1 and R_2 are the curvature radii of the lens surfaces, taken as positive when the surface is convex toward the incident light. A positive focal length f corresponds to a **converging lens** and a negative f to a **diverging lens**. Lens **aberrations** are focusing defects resulting from imperfect lens shapes and the variation of refractive index with wavelength.

4. Optical instruments use lenses and/or mirrors to form useful images. The power of instrument lenses is often given in **diopters**—the inverse of the focal length in meters.

Terms You Should Understand

(Pairs are closely related terms whose distinction is important; number in parentheses is chapter section where term first appears.)

real image, virtual image (introduction)
focal point, focal length (36-2)
magnification (36-2)
thin lens (36-3)
converging lens, diverging lens (36-3)
spherical aberration, chromatic aberration (36-2, 36-3)
near point (36-5)
diopter (36-5)
angular magnification (36-5)
compound microscope (36-5)
reflecting telescope, refracting telescope (36-5)

Symbols You Should Recognize

ℓ, ℓ' (36-2, 36-3)
f (36-2, 36-3)
M, m (36-2–36-4)

Problems You Should be Able to Solve

finding mirror images and their sizes using ray tracing and algebraic techniques (36-2)
finding lens images and their sizes using ray tracing and algebraic techniques (36-3)
analyzing image formation at a single refracting surface (36-4)
designing lenses and analyzing their focal properties (36-4)
designing and analyzing simple optical instruments (36-5)

Limitations to Keep in Mind

Our description of imaging is based on the approximation of geometrical optics, valid only when objects and optical components are much larger than the wavelength of light.
Formulas for image formation in spherical mirrors and lenses are approximations, valid under the conditions that the curved surfaces used form only a small portion of a full sphere, and that all light rays are nearly parallel to the mirror or lens axis.

QUESTIONS

1. How can you see a virtual image, when it really "isn't there"?
2. You're trying to photograph yourself in a mirror, using an autofocus camera that sets its focus by bouncing ultrasound waves off the subject at which the camera is pointed. Why might the photo come out blurred?
3. You lay a magnifying glass (which is just a converging lens) on a printed page. Looking toward the glass, you move it away from the page. Explain the changes in what you see, especially as you move the lens beyond its focal length.
4. Under what circumstances will the image in a concave mirror be the same size as the object?

5. Describe the shapes of the mirrors making the images in Fig. 36-43.

FIGURE 36-43 Question 5.

6. If you're handed a converging lens, what can you do to estimate quickly its focal length?
7. What is the meaning of a negative object distance? Of a negative focal length?
8. A diverging lens always makes a reduced image. Could you use such a lens to start a fire by focusing sunlight? Explain.
9. Is there any limit to the temperature you can achieve by focusing sunlight? Think about the second law of thermodynamics.
10. Can a concave mirror make a reduced real image? A reduced virtual image? An enlarged real image? An enlarged virtual image? Give conditions for each image that is possible.
11. If you placed a screen at the location of a virtual image, would you see the image on the screen? Why or why not?
12. Where must the object be placed to form a reduced real image with a concave mirror or converging lens?
13. Where should you place a flashlight bulb in relation to the focus of its reflector?
14. Is the image on a movie screen real or virtual? How do you know?
15. Does a fish in a spherical bowl appear larger or smaller than it actually is?
16. A block of ice contains a hollow, air-filled space in the shape of a double-convex lens. Describe the optical behavior of this space.
17. The refractive index of the human cornea is about 1.38. If you can see clearly in air, why can't you see clearly underwater? Why do goggles help?
18. The compound microscope and the refracting telescope are both two-lens optical systems. How do their designs reflect their different uses?
19. Cheap binoculars sometimes show blurred images with different colors evident across the blurred region. Why?
20. Do you want a long or short focal length for the objective lens of a telescope? Of a microscope?
21. Give several reasons reflecting telescopes are superior to refractors.
22. Given that star images cannot be resolved even with telescopes, why are large telescopes superior to ordinary cameras for studying stars?

PROBLEMS

ActivPhysics can help with these problems: All activities in Section 15

Sections 36-1 and 36-2 Plane and Curved Mirrors

1. A shoe store uses small floor-level mirrors to let customers view prospective purchases. At what angle should such a mirror be inclined so that a person standing 50 cm from the mirror with eyes 140 cm off the floor can see her feet?
2. Two plane mirrors occupy the first four meters of the positive x and y axes, as shown in Fig. 36-44. Find the locations of all images of an object at $x = 2$ m, $y = 1$ m.
3. (a) What is the focal length of a concave mirror if an object placed 50 cm in front of the mirror has a real image 75 cm from the mirror? (b) Where and what type will the image be if the object is moved to a point 20 cm from the mirror?

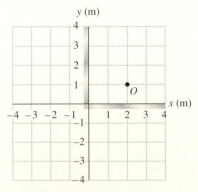

FIGURE 36-44 Problem 2.

4. A candle is 36 cm from a concave mirror with focal length 15 cm and on the mirror axis. (a) Where is its image? (b) How do the image and object sizes compare? (c) Is the image real or virtual?

5. An object is five focal lengths from a concave mirror. (a) How do the object and image heights compare? (b) Is the image upright or inverted?

6. The McMath-Pierce solar telescope at Kitt Peak National Observatory (Fig. 36-45) uses a single concave mirror to produce a real solar image 80 cm in diameter. (a) What is the focal length of the mirror? (b) How far is the image from the mirror? *Hint:* Consult Appendix E.

FIGURE 36-45 The McMath-Pierce solar telescope (Problem 6).

7. A virtual image is located 40 cm behind a concave mirror with focal length 18 cm. (a) Where is the object? (b) By how much is the image magnified?

8. (a) Where on the axis of a concave mirror would you place an object in order to produce a full-size image? (b) Will the image be real or virtual?

9. A 12-mm-high object is 10 cm from a concave mirror with focal length 17 cm. (a) Where, (b) how high, and (c) what type is its image?

10. Repeat the preceding problem for a convex mirror, assuming all numbers stay the same.

11. An object's image in a 27-cm-focal-length concave mirror is upright and magnified by a factor of 3. Where is the object?

12. What is the curvature radius of a concave mirror that produces a 9.5-cm-high virtual image of a 5.7-cm-high object, when the object is located 22 cm from the mirror?

13. When viewed from Earth, the moon subtends an angle of 0.5° in the sky. How large an image of the moon will be formed by the 3.6-m-diameter mirror of the Canada-France-Hawaii telescope, which has a focal length of 8.5 m?

14. At what two distances could you place an object from a 45-cm-focal-length concave mirror in order to get an image 1.5 times the object's size?

15. You look into a reflecting sphere 80 cm in diameter and you see an image of your face at one-third its normal size (Fig. 36-46). How far are you from the sphere's surface?

FIGURE 36-46 Problem 15.

16. Carry out the algebraic steps described just before Equation 36-3 to complete the proof that the focal length of a spherical mirror is half its curvature radius.

Section 36-3 Lenses

17. A light bulb is 56 cm from a convex lens, and its image appears on a screen located 31 cm on the other side of the lens. (a) What is the focal length of the lens? (b) By how much is the image enlarged or reduced?

18. By what factor is the image magnified when an object is placed 1.5 focal lengths from a converging lens? Is the image upright or inverted?

19. A lens with 50-cm focal length produces a real image the same size as the object. How far from the lens are image and object?

20. By holding a magnifying glass 25 cm from my desk lamp, I find I can focus an image of the lamp's bulb on a wall 1.6 m from the lamp. What is the focal length of my magnifying glass?

21. A simple camera uses a single converging lens to focus an image on its film. If the focal length of the lens is 45 mm, what should be the lens-to-film distance for the camera to focus on an object 80 cm from the lens?

22. A real image is 4 times as far from a lens as is the object. What is the object distance, measured in focal lengths?

23. How far from a page should you hold a lens with 32-cm focal length in order to see the print magnified 1.6 times?

24. A converging lens has a focal length of 4.0 cm. A 1.0-cm-high arrow is located 7.0 cm from the lens with its lowest point 5.0 mm above the lens axis. Make a full-scale drawing of the situation, and use ray tracing to locate the image. Confirm using the lens equation.
25. The largest refracting telescope in the world, at Yerkes Observatory, has a 1-m-diameter lens with focal length 12 m (Fig. 36-47). If an airplane flew 1 km above the telescope, where would its image occur in relation to the images of the very distant stars?

FIGURE 36-47 The Yerkes Observatory boasts the world's largest refracting telescope (Problem 25).

26. A small magnifying glass can focus sunlight sufficiently to ignite paper. Calculate the actual size of the solar image produced by a magnifying glass consisting of an optically perfect lens with 25-cm focal length. *Hint:* You'll need to consult Appendix E.
27. A lens has focal length $f = 35$ cm. Find the type and height of the image produced when a 2.2-cm-high object is placed at distances (a) $f + 10$ cm and (b) $f - 10$ cm.
28. How far apart are object and image produced by a converging lens with 35-cm focal length when the object is (a) 40 cm and (b) 30 cm from the lens?
29. A candle and a screen are 70 cm apart. Find two points between candle and screen where you could put a convex lens with 17 cm focal length to give a sharp image of the candle on the screen.
30. An object is placed two focal lengths from a diverging lens. (a) What type of image forms, (b) what is the magnification, and (c) where is the image?

Section 36-4 Refraction in Lenses: The Details

31. You're standing in a wading pool and your feet appear to be 30 cm below the surface. How deep is the pool?
32. A tiny insect is trapped 1.0 mm from the center of a spherical dew drop 4.0 mm in diameter. As you look straight into

the drop at the insect, what is its apparent distance from the edge of the drop?
33. Use Equation 36-6 to show that an object at the center of a glass sphere will appear to be its actual distance—one radius—from the edge. Draw a ray diagram showing why this result makes sense.
34. You're underwater, looking through a spherical air bubble (Fig. 36-48). What is its actual diameter if it appears, along your line of sight, to be 1.5 cm in diameter?

FIGURE 36-48 Problem 34.

35. Rework Example 36-6 for a fish 15 cm from the *far* wall of the tank.
36. Consider the inverse of Example 36-6: You're inside a 70-cm-diameter hollow tube containing air, and the tip of your nose is 15 cm from the wall of the tube. The tube is immersed in water, and a fish looks in. To the fish, what is the apparent distance from your nose to the tube wall?
37. Two specks of dirt are trapped in a crystal ball, one at the center and the other halfway to the surface. If you peer into the ball on a line joining the two specks, the outer one appears to be only one-third of the way to the other. What is the refractive index of the ball?
38. My magnifying glass is a double convex lens with equal curvature radii of 32 cm. If the lens glass has $n = 1.52$, what is its focal length?
39. A contact lens is in the shape of a convex meniscus (see Fig. 36-28); the inner surface is curved to fit the eye, with a curvature radius of 7.80 mm. The lens is made from plastic with refractive index $n = 1.56$. If it's to have a focal length of 44.4 cm, what should be the curvature radius of its outer surface?
40. For what refractive index would the focal length of a plano-convex lens be equal to the curvature radius of its one curved surface?
41. An object is 28 cm from a double convex lens with $n = 1.5$ and curvature radii 35 cm and 55 cm. Where and what type is the image?
42. A double convex lens has equal curvature radii of 35 cm. An object placed 30 cm from the lens forms a real image at 128 cm. What is the refractive index of the lens?
43. A plano-convex lens has curvature radius 20 cm and is made from glass with $n = 1.5$. Use the generalized lens-maker's formula given in Problem 73 to find the focal length when the lens is (a) in air, (b) submerged in water ($n = 1.333$), and (c) embedded in glass with $n = 1.7$. Comment on the sign of your answer to (c).

44. A slide projector has a double convex lens with equal curvature radii, and is made from glass with $n = 1.58$. (a) What should be the curvature radius to give a focal length of 104 mm? (b) With this focal length, how far from the focal point should the slide be placed to focus its image on a screen 3.1 m distant from the lens?

45. Two plano-convex lenses are geometrically identical, but one is made from crown glass ($n = 1.52$), the other from flint glass. An object at 45 cm from the lens focuses to a real image at 85 cm with the crown-glass lens and at 53 cm with the flint-glass lens. Find (a) the curvature radius (common to both lenses) and (b) the refractive index of the flint glass.

46. A double convex lens with equal 38-cm curvature radii is made from glass with refractive indices $n_{red} = 1.51$, $n_{blue} = 1.54$ at the edges of the visible spectrum. If a point source of white light is placed on the lens axis at 95 cm from the lens, over what range will its visible image be smeared?

47. An object placed 15 cm from a plano-convex lens made of crown glass focuses to a virtual image twice the size of the object. If the lens is replaced with an identically shaped one made from diamond, what type of image will appear and what will be its magnification? See Table 35-1.

Section 36-5 Optical Instruments

48. You find that you have to hold a book 55 cm from your eyes for the print to be in sharp focus (Fig. 36-49). What power lens is needed to correct your farsightedness?

55 cm

FIGURE 36-49 Problem 48.

49. Grandma's new reading glasses have 3.8-diopter lenses to provide full correction of her farsightedness. Her old glasses were 2.5 diopters. (a) Where is the near point for her unaided eyes? (b) Where will be the near point if she wears her old glasses?

50. A particular eye has a focal length of 2.0 cm instead of the 2.2 cm that would be required for a sharply focused image on the retina. (a) Is this eye nearsighted or farsighted? (b) What power of corrective lens is needed?

51. A camera's zoom lens covers the focal length range from 38 mm to 110 mm. You point the camera at a distant object and photograph it first at 38 mm and then with the camera zoomed out to 110 mm. Compare the sizes of its images on the two photos.

52. A camera can normally focus as close as 60 cm, but it has provisions for mounting additional lenses at the outer end of the main lens to provide closeup capability. What type and power of auxiliary lens will allow the camera to focus as close as 20 cm? The distance between the two lenses is negligible.

53. The maximum magnification of a simple magnifier occurs with the image at the 25-cm near point. Show that the angular magnification is then given by $m = 1 + \dfrac{25 \text{ cm}}{f}$, where f is the focal length.

54. A compound microscope has objective and eyepiece focal lengths of 6.1 mm and 1.7 cm, respectively. If the lenses are 8.3 cm apart, what is the magnification of the instrument?

55. A 300-power compound microscope has a 4.5-mm-focal-length objective lens. If the distance from objective to eyepiece is 10 cm, what should be the focal length of the eyepiece?

56. To the unaided eye, the planet Jupiter has an angular diameter of 50 arc seconds. What will be its angular size when viewed through a 1-m-focal-length refracting telescope with an eyepiece whose focal length is 40 mm?

57. A Cassegrain telescope like that shown in Fig. 36-42b has 1.0-m focal length, and the convex secondary mirror is located 0.85 m from the primary. What should be the focal length of the secondary in order to put the final image 0.12 m behind the front surface of the primary mirror?

58. The Hubble Space Telescope is essentially a Cassegrain reflector like that shown in Fig. 36-42b. The focal lengths of the concave primary and convex secondary mirrors are 5520.00 mm and −679.00 mm, respectively. The secondary is located at 4906.071 mm from the apex of the primary. Draw a ray diagram showing the path of initially parallel incident rays through the telescope, and use appropriate equations to determine where in Fig. 36-42b such rays are finally focused.

Paired Problems

(Both problems in a pair involve the same principles and techniques. If you can get the first problem, you should be able to solve the second one.)

59. (a) How far from a 1.2-m-focal length concave mirror should you place an object in order to get an inverted image 1.5 times the size of the object? (b) Where will the image be?

60. (a) How far from a 48-cm-focal length concave mirror should you place an object in order to get an upright image 1.5 times the size of the object? (b) Where will the image be?

61. Find the focal length of a concave mirror if an object 15 cm from the mirror has a virtual image 2.5 times the object's actual size.

62. An object is held 6.0 cm from the surface of a reflecting ball, and its image appears three-quarters full size. What is the ball's diameter?

63. How far from a 1.6-m focal length concave mirror should you place an object to get an upright image magnified by a factor of 2.5?

64. How far from a 25-cm-focal-length lens should you place an object to get an upright image magnified by a factor of 1.8?

65. An object and its lens-produced real image are 2.4 m apart. If the lens has 55-cm focal length, what are the possible values for the object distance and magnification?

66. An object and its converging-lens-produced virtual image are 2.4 m apart. If the lens has 55-cm focal length, what are the possible values for the object distance and magnification?

67. An object is 68 cm from a plano-convex lens whose curved side has curvature radius 26 cm. The refractive index of the lens is 1.62. Where and of what type is the image?

68. Both surfaces of a double concave lens have curvature radii of 18 cm, and the refractive index is 1.5. If a virtual image appears 14 cm from the lens, where is the object?

Supplementary Problems

69. My contact lens prescription calls for +2.25-diopter lenses with an inner curvature radius of 8.6 mm to fit my cornea. (a) If the lenses are made from plastic with $n = 1.56$, what should be the outer curvature radius? (b) Wearing these lenses, I hold a newspaper 30 cm from my eyes. Where is its image as viewed through the lenses?

70. Show that the powers of closely spaced lenses add; that is, placing a 1-diopter lens in front of a 2-diopter lens gives the equivalent of a single 3-diopter lens.

71. Show that identical objects placed equal distances on either side of the focal point of a concave mirror or converging lens produce images of equal size. Are the images of the same type?

72. Galileo's first telescope used the arrangement shown in Fig. 36-50, with a double convex eyepiece placed slightly before the focus of the objective lens. Use ray tracing to show that this system gives an upright image, which makes the design useful for terrestrial observing.

73. Generalize the derivation of the lensmaker's formula (Equation 36-8) to show a lens of refractive index n_{lens} in an external medium with index n_{ext} has focal length given

FIGURE 36-50 Problem 72.

by

$$\frac{1}{f} = \left(\frac{n_{lens}}{n_{ext}} - 1\right)\left(\frac{1}{R_1} - \frac{1}{R_2}\right).$$

74. An object is located 40 cm from a lens made from glass with $n = 1.5$. In air, the resulting image is real and is located 32 cm from the lens. When the entire system is immersed in a liquid, the image becomes virtual and is reduced to half the object's size. Use the result given in the preceding problem to find the refractive index of the liquid.

75. A Newtonian telescope like that of Fig. 36-42c has a primary mirror with 20-cm diameter and 1.2-m focal length. (a) Where should the flat diagonal mirror be placed to put the focus at the edge of the telescope tube? (b) What shape should the flat mirror have to minimize blockage of light to the primary?

76. A parabola is described by the equation $y = x^2/4f$, where f is the distance from vertex to focus. (a) Find the equation for the circle (which would be a sphere in three dimensions) that closely approximates the parabola $y = x^2$. *Hint:* How are the focal length and curvature radius of a spherical mirror related? (b) Solve the circle's equation for y, and use the binomial theorem to show that the equation reduces to that of the parabola for $x \ll 1$.

77. Just before Equation 36-7 are two equations describing refraction at the two surfaces of a lens with thickness t. Combine these equations to show that the object distance ℓ and image distance ℓ' for such a lens are related by

$$\frac{1}{\ell} + \frac{1}{\ell'} - \frac{[(n-1)\ell - R_1]^2 t}{\ell R_1[t(\ell + R_1) + n\ell(R_1 - t)]}$$
$$= (n - 1)\left(\frac{1}{R_1} - \frac{1}{R_2}\right).$$

78. Use the result of the preceding problem to find an expression for the focal length of a transparent sphere with radius R and refractive index n, as measured from the exit surface.

Interference and Diffraction 37

Interference of light plays an important role in extracting information from compact discs. Here adjacent tracks of the microscopic "pits" that code the CD's information produce interference that results in a rainbow of colors. This chapter explores many applications of wave interference and diffraction, including the operation of CDs and the newer DVDs.

The preceding chapters considered reflection and refraction, and their application in image formation, from the approximation of geometrical optics—an approximation valid when we can ignore the wave nature of light. We now turn to **physical optics**, which treats optical phenomena for which the wave nature of light plays an essential role. Two related phenomena, interference and diffraction, are central to much of physical optics.

37-1 COHERENCE AND INTERFERENCE

We introduced wave **interference** in Chapter 16, showing how wave displacements add in a manner that may be constructive or destructive depending on the relative phases of the waves. Interference of electromagnetic waves—including light—occurs in the same way; since the electric and magnetic fields separately obey the superposition principle, the net electric and magnetic fields at any point are the vector sums of the fields of individual waves. The superposition of those fields may result in an increase (constructive interference) or decrease (destructive interference) in the field strength.

Coherence

Although interference occurs any time two waves interact, we get steady interference patterns only when the interacting waves are **coherent**, meaning that they maintain a constant phase relation. Ordinary sources such as light bulbs or the Sun produce light that is coherent only over very short distances. That's because light from these sources consists of short wavetrains emitted with random phases (Fig. 37-1a). Lasers, in contrast, produce relatively long wavetrains that maintain coherence over many wavelengths (Fig. 37-1b). The typical length of a wavetrain is the **coherence length** for a given light source.

Regardless of the source, it takes two or more waves to interfere. It's virtually impossible for two different light sources to maintain coherence, so such interference usually results only when light from a single source splits, travels different paths, and then recombines. Even light from an ordinary light bulb will produce interference patterns provided the light in the two recombining light beams originated at nearly the same place and time. With the more coherent light of a laser, interference occurs over a wider range of path differences.

Coherence requires that the interfering light beams have exactly the same frequency and therefore wavelength or color. With a white-light source such as sunlight, interference occurs separately for each different color. Again, lasers make things easier: their light is very nearly **monochromatic**, consisting of a narrow band of wavelengths. Very slight deviations from exactly equal

(a)   (b)

FIGURE 37-1 (a) Light bulbs emit many short wavetrains whose phases are related in random ways. Coherent interference can occur only with light originating from a spatially limited region, so that a single wavetrain interferes with itself. The coherence length is defined as the average length of a wavetrain, here about 3 wavelengths. (b) Lasers produce light with much greater coherence length, making it easier to produce coherent interference with laser light.

(a)   (b)

FIGURE 37-2 Two waves that started out in phase but have traveled different paths before coming together. (a) Here the lower wave travels half a wavelength farther, so the two come together 180° out of phase, and their superposition results in destructive interference. (b) When the lower wave travels a full wavelength farther, the two come together in phase and the interference is constructive.

frequency and wavelength may still permit observable interference, and yield the optical equivalent of the beats we introduced in Section 16-5.

Destructive and Constructive Interference

Consider light waves that originate together at a single source, travel two different paths, and then come together again. Suppose one wave's path is exactly half a wavelength longer. Then when the waves recombine they will be out of phase by half a wavelength (Fig. 37-2a), and thus their superposition will have lower amplitude (zero, if the two interfering waves have exactly the same amplitude). If, on the other hand, the path lengths don't differ, or differ by a full wavelength, then the two waves recombine in phase (Fig. 37-2b) and their superposition has greater amplitude. These two cases correspond, respectively, to **destructive interference** and **constructive interference**. It doesn't really matter whether path lengths for the waves in Fig. 37-2a differ by half a wavelength, or $1\frac{1}{2}$ wavelengths, or $2\frac{1}{2}$ wavelengths—as long as the difference is an odd multiple of a half wavelength, the waves recombine out of phase and destructive interference results. Thus:

> **Destructive interference results when light paths differ by an odd integer multiple of a half wavelength.**

Similarly, it doesn't matter whether the path lengths in Fig. 37-2b are the same, or differ by 1, 2, 3, or any other integer number of wavelengths. Thus

> **Constructive interference results when light paths differ by an integer multiple of the wavelength.**

We'll use these two statements to build an understanding of interference in both simple and complicated situations.

There's one caveat to our statements: the path difference can't be greater than the coherence length that we defined in connection with Fig. 37-1. Otherwise the interfering waves will be from different wavetrains, and won't be coherent. Once again, laser light has the advantage here, because of its greater coherence length.

Of course, light paths don't have to differ by half or full multiples of the wavelength. In intermediate cases interfering waves superpose to make a composite wave whose amplitude may be enhanced or diminished, depending on the relative phase as determined by the path-length difference.

APPLICATION	CD Music, Continued

Example 35-3, "CD Music," showed how refraction in a compact disc (CD) helps focus the laser beam that "reads" information from the disc. Interference, too, plays a crucial role in "reading" a CD.

Information on a CD—whether an audio signal or computer data—is stored digitally in a sequence of "pits"—depressions stamped into a reflective metallic "information layer" (Fig. 37-3). The pits vary in length, but their depth is the same—very nearly one-quarter wavelength of the laser light used to "read" the CD. From the transparent underside of the disc, where the laser beam enters, each pit appears as an elevated bump. Since the bumps stick down one-quarter

FIGURE 37-4 Structure of a CD, showing that pits appear as bumps to the laser beam shining from below the disc. Path lengths for the laser beam differ by a half wavelength depending on whether the light reflects off a bump, resulting in destructive interference when the beam strikes a bump and the surrounding disc surface.

FIGURE 37-3 Information on a standard CD is stored in "pits" stamped into a metallic "information layer." The pits form a continuous spiral track that winds around the disc. Adjacent turns of the spiral are 1.6 μm apart, and the minimum pit length is 0.83 μm. Depth of the pits is approximately a quarter wavelength of the laser light used to "read" the CD.

wavelength, light reflecting off a bump follows a round-trip path that's shorter by half a wavelength compared with the path of light reflecting off the undisturbed information layer (Fig. 37-4). The laser beam that "reads" the disc is somewhat wider than the pit, so the reflected beam includes light both from the undisturbed disc as well as from the bump. The two interfere destructively, just as we showed in Fig. 37-2, making a much less intense reflected beam when a bump is present. As the disc spins, the result is a pattern of fluctuating light intensity conveying the information associated with the pattern of pits on the CD. A photodetector converts that pattern to electrical signals for processing by the electronic circuitry of the CD player or computer.

The new DVD discs introduced in the late 1990s work in exactly the same way, except—as we'll see later in this chapter—they can store much more information.

37-2 DOUBLE-SLIT INTERFERENCE

In Chapter 16 we looked briefly at interference patterns produced by a pair of coherent sources. With light, such a pair can be made by passing light from a single source through two narrow slits. In 1801 Thomas Young used this approach in a historic experiment that confirmed the wave nature of light. Young first admitted sunlight to his laboratory through a hole small enough to ensure coherence of the incoming light. The light then passed through a pair of narrow, closely spaced slits, after which it illuminated a screen. Each slit acts as a source of cylindrical wavefronts that interfere in the region between slits and screen (Fig. 37-5a). Constructive and destructive interference result in alternating regions of light and dark on the screen, as shown in Fig. 37-5b for laser light passed through double slits. The light and dark regions in Fig. 37-5b are called **interference fringes**.

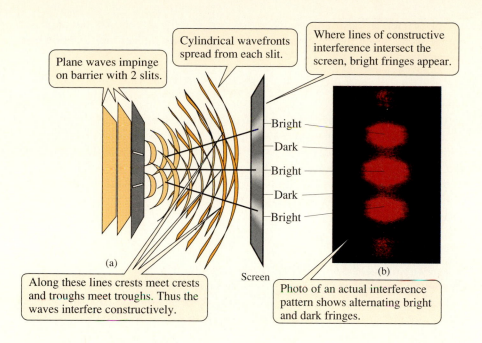

Plane waves impinge on barrier with 2 slits.

Cylindrical wavefronts spread from each slit.

Where lines of constructive interference intersect the screen, bright fringes appear.

Bright
Dark
Bright
Dark
Bright

(a)

Screen

(b)

Along these lines crests meet crests and troughs meet troughs. Thus the waves interfere constructively.

Photo of an actual interference pattern shows alternating bright and dark fringes.

FIGURE 37-5 Double-slit interference results when light from a single source passes through a pair of closely spaced slits. Photo in (b) was made by replacing the screen with film in an actual double-slit apparatus using laser light.

We can understand the occurrence and location of fringes in double-slit interference by applying our criteria for constructive and destructive interference. The bright fringes represent constructive interference, and therefore they occur where the difference in the path length for light traveling from the two different slits is a multiple of the wavelength. When the distance L from slits to screen is much larger than the slit spacing d, Fig. 37-6 then shows that the path difference from the two slits to a point on the screen is $d\sin\theta$, where θ is the angular position of a point on the screen, measured from an axis perpendicular to slits and screen. So our criterion for constructive interference—that this difference be an integer number of wavelengths—becomes

$$d\sin\theta = m\lambda. \qquad \text{(bright fringes)} \qquad (37\text{-}1a)$$

The integer m is the **order** of the fringe, with the central bright fringe being the zeroth order fringe and with higher order fringes on either side.

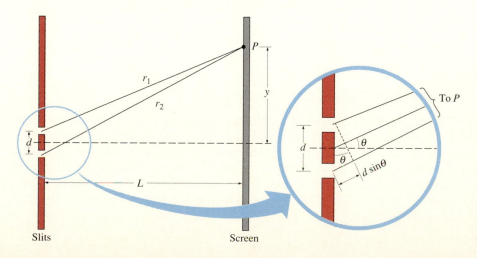

P

r_1

r_2

y

d

L

Slits

Screen

To P

d

θ

θ

$d\sin\theta$

FIGURE 37-6 Geometry for finding locations of the interference fringes. A bright fringe occurs when the path-length difference $r_2 - r_1$ is an integer multiple of the wavelength. In the blowup you can see that for $L \gg d$ the paths to P are nearly parallel, and the difference in their lengths is $d\sin\theta$. Can you find similar triangles to prove that the angles labeled θ are indeed the same?

FIGURE 37-7 Interference patterns made with the same double-slit system, using two different wavelengths of light. The different fringe spacings reflect the wavelength dependence in Equations 37-2.

16.5

Two-Source Interference

Waves interfere destructively when they arrive at the screen 180° out of phase, which occurs when their path lengths differ by an odd integer multiple of a half wavelength:

$$d \sin \theta = (m + \tfrac{1}{2})\lambda, \quad \text{(dark fringes)} \qquad (37\text{-}1b)$$

where m is any integer.

In a typical double-slit experiment, L may be on the order of 1 m, d a fraction of 1 mm, and λ the sub-μm wavelength of visible light. Then we have the additional condition that $\lambda \ll d$. This makes the fringes very closely spaced on the screen, so the angle θ in Fig. 37-6 is small even for large orders m. Then $\sin \theta \simeq \tan \theta = y/L$, and Equations 37-1 and the geometry of Fig. 37-6 show that a fringe's position y on the screen, measured from the central maximum, is given by

$$y_{\text{bright}} = m\frac{\lambda L}{d} \quad \text{and} \quad y_{\text{dark}} = (m + \tfrac{1}{2})\frac{\lambda L}{d}. \qquad (37\text{-}2a, b)$$

These equations show that the fringe spacing depends on wavelength, as confirmed experimentally in Fig. 37-7. Measurement of fringe spacing enabled Young to determine the wavelength of light. You can simulate two-slit interference—including the effect of varying wavelength—with ActivPhysics Activity 16.5.

Got It!

If you increase the slit separation in a 2-slit system, do the interference fringes become closer together or farther apart?

EXAMPLE 37-1	*Laser Wavelength*

A pair of narrow slits are 0.075 mm apart and are located 1.5 m from a screen. Laser light shining through the slits produces an interference pattern whose third-order bright fringe is 3.8 cm from the screen center. Find the wavelength of the light.

Solution

Here we have $m = 3$, $L = 1.5$ m, and $d = 0.075$ mm, so we can solve Equation 37-3a for the wavelength λ:

$$\lambda = \frac{y_{\text{bright}}d}{mL} = \frac{(0.038 \text{ m})(0.075 \times 10^{-3} \text{ m})}{(3)(1.5 \text{ m})} = 633 \text{ nm}.$$

This is in fact the wavelength of the red light from low-power helium-neon lasers commonly used in physics demonstrations.

EXERCISE What slit spacing will produce bright fringes 1.8 cm apart on a screen 85 cm from the slits, if the slits are illuminated with 589-nm light?

Answer: 27.8 μm

Some problems similar to Example 37-1: 2–5, 10

Intensity in the Interference Pattern

Geometric arguments allowed us to find the positions of the maxima and minima of a two-slit interference pattern. We can find the actual intensity variation by algebraically superposing the interfering waves. You might think this could be done by adding the wave intensities. But no! It's the electric and magnetic fields that obey the superposition principle, not the wave intensities. Intensity is proportional to the *square* of either field; if we added intensities we could never get the cancellation that occurs in destructive interference.

Consider again a point P on the screen of a double-slit apparatus (Fig. 37-8). Since light from the two slits reaches P over paths of different lengths, we expect that the waves will have different intensities when they reach P. In the approximation $d \ll L$, however, the path-length difference $d \sin \theta$ is so small that we can neglect this effect. So we consider that the two waves arriving at P have electric fields of equal amplitude, given by expressions like Equation 34-10. Since we're considering a fixed point P, it's convenient to take the origin at P to eliminate the kx term in Equation 34-10. Then we can express the electric fields of the two waves at P in the form

$$E_1 = E_p \sin \omega t \quad \text{and} \quad E_2 = E_p \sin(\omega t + \phi),$$

where E_p is the common amplitude, ω the common frequency, and ϕ the phase difference that occurs because of the different path lengths. We don't bother with vector notation because both waves, since they originate in the same source, are polarized in the same direction and thus their electric fields will simply add algebraically. Then net electric field at P is

$$E = E_1 + E_2 = E_p[\sin \omega t + \sin(\omega t + \phi)].$$

In Appendix A we find the trigonometric identity $\sin \alpha + \sin \beta = 2 \sin[(\alpha + \beta)/2]\cos[(\alpha - \beta)/2]$, which, with $\alpha = \omega t$ and $\beta = \omega t + \phi$, transforms our expression for E into

$$E = 2E_p \sin\left(\omega t + \frac{\phi}{2} \right)\cos\left(\frac{\phi}{2} \right).$$

We also used $\cos(-x) = \cos x$ to eliminate a minus sign in the argument of the cosine. Thus, the electric field at P oscillates with the wave frequency ω, and its overall amplitude is scaled by the factor $\cos(\phi/2)$.

What is the phase difference ϕ? We've already seen that the path-length difference is $d \sin \theta$, with d the slit spacing and θ the angle subtended at the slits between P and the slit centerline. If this difference is one-half wavelength λ, then we have a phase difference of $180°$ or π radians. More generally, the phase difference in radians is whatever fraction of the full cycle (2π radians) the path difference $d \sin \theta$ is of the wavelength:

$$\phi = 2\pi \frac{d \sin \theta}{\lambda}.$$

Using this result in our expression for the electric field at P gives

$$E = 2E_p \sin\left(\omega t + \frac{\phi}{2} \right)\cos\left(\frac{\pi d \sin \theta}{\lambda} \right).$$

This equation describes a field whose peak amplitude is not E_p but $2E_p \cos(\pi d \sin \theta/\lambda)$. The average intensity then follows from Equation 34-20b:

$$\bar{S} = \frac{[2E_p \cos(\pi d \sin \theta/\lambda)]^2}{2\mu_0 c} = 4\bar{S}_0 \cos^2\left(\frac{\pi d \sin \theta}{\lambda} \right), \tag{37-3}$$

where $\bar{S}_0 = E_p^2/2\mu_0 c$ is the average intensity of either wave alone. Equation 37-3 shows that the intensity varies between zero and $4\bar{S}_0$ as the angular position changes. We can also write Equation 37-3 in terms of position y on the screen. Under the approximation $d \gg \lambda$ even high-order fringes will occur at

FIGURE 37-8 Waves leaving the slits with the same phase arrive at P displaced by the path-length difference $d \sin \theta$. This corresponds to phase difference $\phi = d \sin \theta(2\pi/\lambda)$.

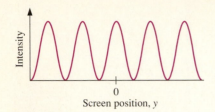

FIGURE 37-9 Intensity as a function of position y on the screen of a double-slit system. This intensity variation is reflected in Fig. 37-7, although saturation effects of the bright laser light on the film prevents the photo from showing accurately the sinusoidal dependence.

small angles θ, so we can write $\sin\theta \simeq \tan\theta \simeq y/L$, giving

$$\bar{S} = 4\bar{S}_0 \cos^2\left(\frac{\pi d}{\lambda L}y\right). \tag{37-4}$$

Now, $\cos^2\alpha$ has its maximum value, 1, when its argument is an integer multiple of π. Thus, the maxima of Equation 37-4 occur when $dy/\lambda L$ is an integer m, or when $y = m\lambda L/d$. This is just the condition of Equation 37-2a, showing that our intensity calculation is fully consistent with the simpler geometrical analysis. But the intensity calculation tells us more: it gives not only the positions of bright and dark fringes, but also the intensity variation in between, as shown in Fig. 37-9.

37-3 MULTIPLE-SLIT INTERFERENCE AND DIFFRACTION GRATINGS

Systems with multiple slits play a crucial role in optical instrumentation and in the analysis of materials. As we will soon see, gratings manufactured with several thousand slits per centimeter make possible high-resolution spectroscopic analysis. At a much smaller scale the regularly spaced rows of atoms in a crystal act much like a multiple-slit system for X rays, and the resulting X-ray patterns reveal the crystal structure.

Figure 37-10 shows waves from three evenly spaced slits interfering at a screen. Maximum intensity requires that all three waves either be in phase, or differ in phase by an integer number of wavelengths. Our criterion for the maximum in a two-slit pattern, $d\sin\theta = m\lambda$, ensures that waves from two adjacent slits will add constructively. Since the slits are evenly spaced with distance d between each pair, waves coming through a third slit will be in phase with the other two if this criterion is met. So the criterion for a maximum in an N-slit system is still Equation 37-1a:

FIGURE 37-10 Waves from three evenly spaced slits interfere constructively when they arrive at the screen in phase.

$$d\sin\theta = m\lambda. \quad \text{(maxima in multi-slit interference)}$$

With more than two waves, however, the criterion for destructive interference is more complicated. Somehow all the waves need to sum to zero. Figure 37-11 shows that this happens for three waves when each is out of phase with the others by one-third of a cycle. Thus, the path-length difference $d\sin\theta$ must be either $(m + \frac{1}{3})\lambda$ or $(m + \frac{2}{3})\lambda$, where m is an integer. The case $(m + \frac{3}{3})\lambda$ is excluded because then the path lengths differ by a full wavelength, giving constructive interference and thus a maximum in the interference pattern. More generally we can write

$$d\sin\theta = \frac{m}{N}\lambda, \tag{37-5}$$

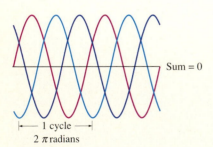

FIGURE 37-11 Waves from three slits must be out of phase by one-third of a cycle in order to interfere destructively. With this phase relationship the three waves sum to zero at every point.

for destructive interference in an N-slit system, where m is an integer *but not an integer multiple of N*. The reason for the exclusion is that when m is an integer multiple of N then m/N is an integer; then the path-length difference is an integer number of wavelengths, resulting in constructive rather than destructive interference. Mathematically, when m/N is an integer, Equation 37-5 becomes equivalent to Equation 37-1a that gives the condition for constructive interference.

FIGURE 37-12 Interference patterns for multiple slit systems with the same slit spacing. The bright fringes stay in the same place, but become narrower and brighter as the number of slits increases. Intensity plots do not have the same vertical scale; peak intensity in fact scales as the *square* of the number of slits.

Figure 37-12 shows interference patterns and intensity plots from some multiple-slit systems. Note that the bright, or *primary,* maxima are separated by several minima and fainter, or *secondary,* maxima. Why this complex pattern? Our analysis of the three-slit system shows two minima between every pair of primary maxima; for example, we considered the minima at $d \sin \theta$ equal to $(m + \frac{1}{3})\lambda$ or $(m + \frac{2}{3})\lambda$, which lie between the maxima at $d \sin \theta$ equal to $m\lambda$ and $(m + 1)\lambda$. More generally, Equation 37-5 shows that there are $N - 1$ minima between each pair of primary maxima given by Equation 37-2a. The secondary maxima that lie between these minima result from interference that is neither fully destructive nor fully constructive. The figure shows that the primary maxima become much brighter and narrower as the number of slits increases, while the secondary maxima become relatively less bright. With a large number N of slits, then, we should expect a pattern of bright but narrow primary maxima, with broad, essentially dark regions in between.

Diffraction Gratings

A set of many very closely spaced slits is called a **diffraction grating** and proves very useful in the spectroscopic analysis of light. Diffraction gratings commonly measure several cm across and have several thousand slits—usually called lines—per cm. Gratings are made by photoreducing images of parallel lines or by ruling with a diamond stylus on aluminum-plated glass (Fig. 37-13). Gratings like the slit systems we have been discussing are **transmission gratings** since light passes through the slits. **Reflection gratings** produce similar interference effects by reflecting incident light.

We've seen that the maxima of the multi-slit interference pattern are given by the same criterion, $d \sin \theta = m\lambda$, that applies to a two-slit system. For $m = 0$

FIGURE 37-13 A diffraction grating disperses white light into its constituent colors. The enlarged spectrum at rear results from reflecting the diffracted beam off a concave mirror.

2nd-order spectrum · 1st-order spectrum · Screen · Grating · Entrance slit · Light source

FIGURE 37-14 Essential elements of a grating spectrometer. The entrance slit regulates the amount of light entering the instrument. The grating disperses the light into its component wavelengths, which reach the screen to produce spectra of the different orders. An electronic detector would normally be used in place of the screen.

this equation implies that all wavelengths peak together at the central maximum, but for larger values of m the angular position of the maximum depends on wavelength. Thus, a diffraction grating can be used in place of a prism to disperse light into its component wavelengths, and the integer m is therefore called the **order** of the dispersion. Figure 37-14 shows a grating spectrometer working on this principle. Because the maxima in N-slit interference are very sharp for large N (recall Fig. 37-12), a grating with many slits diffracts individual wavelengths to very precise locations. Incidentally, the rainbow of colors produced by the compact disc in this chapter's opening photo results because adjacent "tracks" of CD pits (see Fig. 37-3) act as a diffraction grating.

EXAMPLE 37-2	A Grating Spectrometer

Light from glowing hydrogen contains discrete wavelengths ("spectral lines") called hydrogen-α and hydrogen-β, at 656.3 nm and 486.1 nm, respectively. Find the angular separation between these two wavelengths in a spectrometer using a grating with 6000 slits per cm. Consider both the first-order ($m = 1$) and second order ($m = 2$) dispersion.

Solution

With 6000 slits/cm, the slit spacing is $d = (1/6000)$ cm = 1.667 μm. Applying the criterion $d \sin \theta = m\lambda$ for the first-order spectrum, we then have

$$\theta_{1\alpha} = \sin^{-1}\left(\frac{\lambda}{d}\right) = \sin^{-1}\left(\frac{0.6563\ \mu m}{1.667\ \mu m}\right) = 23.2°$$

and

$$\theta_{1\beta} = \sin^{-1}\left(\frac{\lambda}{d}\right) = \sin^{-1}\left(\frac{0.4861\ \mu m}{1.667\ \mu m}\right) = 17.0°.$$

Thus, the angular separation is 6.2°. Repeating the same calculation with $m = 2$ gives $\theta_{2\alpha} = 51.9°$ and $\theta_{2\beta} = 35.7°$, for an angular spread of 16.2°. This wider spacing is characteristic of higher-order dispersion.

EXERCISE The bright yellow light emitted by glowing sodium vapor actually consists of two spectral lines at 589.0 nm and 589.6 nm. Find the angular separation of these lines in a second-order spectrum taken with a 4800-slit/cm grating.

Answer: 0.04°

Some problems similar to Example 37-2: 17–19

16.3

Multiple Slits Diffraction

In Example 37-2 the two lines are near the ends of the visible spectrum, and the values calculated for the angular positions show that there is no overlap between the first- and second-order visible spectra. But higher-order spectra do overlap, a fact that astronomers use in high-resolution spectroscopy when they wish to observe two different spectral lines simultaneously. Problem 21 explores this situation, while Fig. 37-15 shows the relative positions of the different orders.

Resolving Power

The detailed shapes and wavelengths of spectral lines contain a wealth of information about the systems in which light originates. Studying these details requires a high dispersion in order to separate nearby spectral lines or to analyze the intensity versus wavelength profile of a single line. Suppose we pass light

FIGURE 37-15 Positions of the different orders in a grating spectrum. The vertical separation between orders has been introduced for clarity; in actuality they would overlap. The central line of each spectrum is at 550 nm. Note the increased dispersion of the higher orders.

containing two spectral lines of nearly equal wavelengths λ and λ' through a grating. Figure 37-16 shows that we will just be able to distinguish two spectral lines if the peak of one line corresponds to the first minimum of the other; any closer and the lines will blur together. Suppose wavelength λ has its mth-order maximum at angular position θ. The criterion for this maximum is, as usual, $d \sin \theta_{\text{max}} = m\lambda$. We can equally well write this as $d \sin \theta_{\text{max}} = \dfrac{mN}{N} \lambda$, with N the number of slits in the grating. Equation 37-5 then shows that we get an adjacent minimum if we add 1 to the numerator mN. Thus, there is an adjacent minimum whose position satisfies

$$d \sin \theta_{\text{min}} = \frac{mN + 1}{N} \lambda .$$

Our criterion that the two wavelengths λ and λ' be just distinguishable is that the maximum for λ' fall at the location of this minimum for λ. But the maximum for λ' satisfies $d \sin \theta'_{\text{max}} = m\lambda' = \dfrac{mN}{N} \lambda'$, so for $\theta'_{\text{max}} = \theta_{\text{min}}$ we must have

$$(mN + 1)\lambda = mN\lambda' .$$

It's convenient to express this result in terms of the wavelength difference $\Delta\lambda = \lambda' - \lambda$. Solving our equation relating λ and λ' for λ' and using the result to write $\Delta\lambda$ in terms of λ alone then gives

$$\frac{\lambda}{\Delta\lambda} = mN . \qquad \text{(resolving power)} \qquad (37\text{-}6)$$

The quantity $\lambda / \Delta\lambda$ is the **resolving power** of the grating, a measure of its ability to distinguish closely spaced wavelengths. The higher the resolving power, the smaller the wavelength difference $\Delta\lambda$ that can be distinguished in the spectrum. Equation 37-6 shows that the resolving power increases with the number of lines, N, on the grating, and also with the order, m, of the spectrum we observe. Both these points make sense. We saw in Fig. 37-12 that the intensity of the maxima increases and their width decreases with increasing N, making different wavelengths easier to distinguish, while Fig. 37-15 shows that higher-order spectra are more dispersed and thus, again, make it easier to distinguish different wavelengths.

FIGURE 37-16 A plot of intensity versus wavelength for spectral lines of slightly different wavelengths, as observed with a spectrometer. The lines are just distinguishable if the maximum of one falls on the first minimum of the other; any closer and they blur together.

EXAMPLE 37-3	"Seeing" a Double Star

A certain double star system consists of a massive star essentially at rest, with a smaller companion in circular orbit. The stars are far too close to each other, and the system far too distant from Earth, for the pair to appear as anything but a single point in even the largest telescopes. Yet astronomers can "see" the companion star through the Doppler shift in the wavelengths of its spectral lines that occurs because of the star's motion. In the system under observation the hydrogen-α line from the companion star, at 656.272 nm when the source is at rest, is shifted to 656.215 nm (this corresponds to an orbital speed of about 26 km/s when the companion is moving toward Earth. If the telescope's spectrometer grating has 2000 lines/cm and measures 2.5 cm across, what order spectrum will resolve the hydrogen-α lines from the stationary star and its orbiting companion?

Solution

Here $N = (2000$ lines/cm$)(2.5$ cm$) = 5000$ lines. Solving Equation 37-6 for m then gives

$$m = \frac{\lambda}{N \, \Delta\lambda} = \frac{656.272 \text{ nm}}{(5{,}000)(656.272 \text{ nm} - 656.215 \text{ nm})} = 2.3 \,.$$

Since the order number must be an integer, the observation must be made in third order.

EXERCISE Putting atoms in a magnetic field causes what are normally single spectral lines to split, as the atoms produce light of two slightly different wavelengths. In a particular case, a spectral line of mercury at 435.8 nm splits into two lines separated in wavelength by 0.027 nm. To the nearest thousand, how many lines must a grating have to resolve these wavelengths in first order?

Answer: 16,000

Some problems similar to Example 37-3: 26–29

FIGURE 37-17 (a) X rays reflecting off the planes of atoms in a crystal. (b) X rays striking a lower plane travel an extra distance $2d \sin \theta$. The outgoing beam is enhanced by constructive interference when this distance is an integer multiple of the X-ray wavelength.

X-Ray Diffraction

The wavelengths of X rays, around 0.1 nm, are far too short to be dispersed with diffraction gratings produced mechanically or photographically. The regular spacing of atoms in a crystal, however, provides a "grating" of the appropriate scale. In the classical electromagnetic description, the reflection of electromagnetic waves occurs as electrons in each atom are set into oscillation by the electric field of the incident X-ray beam. The oscillating electrons then reradiate the energy they have absorbed, and the combined beam from all the atoms obeys the law of reflection. When the atomic spacing is regular, and comparable to the wavelength, however, interference enhances the reflected radiation at certain angles. Figure 37-17 shows an X-ray beam interacting with the layers of atoms in a crystal. In Fig. 37-17b we see that waves reflecting at one layer travel a total distance $2d \sin \theta$ farther than those reflecting at the layer above, where θ is the angle between the incident beams and the atomic planes. The outgoing beams will interfere constructively when this difference is an integer multiple of the wavelength:

$$2d \sin \theta = m\lambda \,. \qquad \text{(Bragg condition)} \qquad (37\text{-}7)$$

First derived by W. L. Bragg, this **Bragg condition** allows one to use a crystal with known spacing as a diffraction grating at X-ray wavelengths. More important is the converse: much of what we know about crystal structure comes from probing crystals with X rays and using the resulting patterns to deduce the positions of their atoms (Fig. 37-18).

FIGURE 37-18 This X-ray diffraction image of DNA, made by the British scientist Rosalind Franklin in 1952, was crucial in establishing the structure of the DNA molecule.

| **APPLICATION** | *Acousto-Optic Modulation* |

Anything containing regularly spaced structures—like the lines in a diffraction grating or the atoms in a crystal—can act as a diffraction grating for waves of suitable wavelength. A very useful technological development, the **acousto-optic modulator** (AOM), uses diffraction from the periodic variations in the structure of a quartz crystal caused by the presence of a sound wave in the crystal.

Figure 37-19 shows a diagram of an AOM. Attached to the crystal is a transducer—a loudspeaker-like device that converts an alternating voltage into sound waves that propagate through the quartz. The regular spacing of the acoustic wavefronts constitutes a diffraction grating whose spacing may be varied by changing the driving frequency. Light incident on the quartz is diffracted at preferred angles given by Equation 37-7, just as are X rays by regularly spaced atomic planes. Changing

the sound frequency allows the diffracted beam to be scanned in direction; turning the sound on and off switches the diffracted beam on and off. These effects are widely used in lightwave communication and other optical technologies; most laser printers, for example, use AOMs to control the laser beam that "paints" a picture of the printed page on a light-sensitive surface.

AOMs are also used to encode audio and video signals onto light, allowing transmission of these signals by light beams. Setting up the AOM's "diffraction grating" requires propagating sound waves into the crystal. Since this takes time, the AOM is not as fast a lightwave modulator as is the electro-optic modulator discussed in Chapter 34. The AOM is, however, much less expensive and less demanding of the associated electronic circuitry.

(a) (b)

FIGURE 37-19 (a) Schematic diagram of an acousto-optic modulator. The transducer drives sound waves into the crystal, and the acoustic wavefronts act like a moving diffraction grating to incident light. (b) An acousto-optic modulator. A laser beam enters the transparent crystal while electrical signals supplied by the cable in the foreground generate sound waves that act as a diffraction grating to modify the beam's propagation direction and frequency. Such modulators are used in laser printers and many other applications.

37-4 THIN FILMS AND INTERFEROMETERS

We treated the double-slit system first because it's one of the most straightforward examples of interference. Then we turned to the more complicated case of multiple-slit interference. But there are many ways to produce interfering light waves besides passing light through slits. All have in common the separation of light into two or more beams that travel on different paths, then recombine. Here we examine a few such examples.

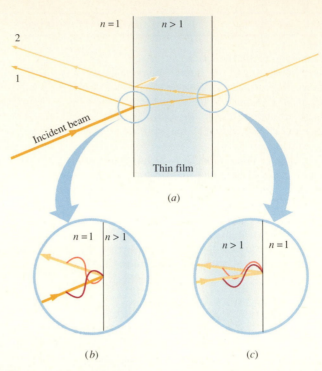

FIGURE 37-20 (*a*) Reflection and refraction at a thin layer of transparent material. (*b*) The reflected and incident beams are 180° out of phase when they reflect at an interface that goes from a lower to a higher refractive index. (*c*) There is no phase difference on reflection when the incident medium has a higher refractive index.

Interference in Thin Films

We saw in Section 35-6 that light is partially reflected and partially transmitted at an interface between two transparent media. Figure 37-20 shows what happens when light propagating in air strikes a thin film of transparent material. Partial reflection at both surfaces results in two beams propagating back into the air. Since these beams derive from the same source they're coherent, and therefore they interfere. (Additional reflections actually result in multiple beams, but they're successively fainter and we neglect them.)

In Section 17-5 we saw how waves on a string reflect when the end of the string is fixed, free, or tied to another string. Light incident from air on a medium with higher refractive index is like a wave on a string that's fixed or tied to a heavier string; the reflected wave undergoes a 180° phase change, as we showed in Figs. 17-5*a* and 17-6. Going the other way, from the medium with the higher refractive index to air, there's no phase change on reflection; here the situation is more like Fig. 17-5*b* or a string tied to a lighter string. So the incident light from air onto the film reflects with a 180° phase change, while light incident from inside the film on its rear surface reflects with no phase change. Figures 37-20*b* and *c* show these phase relations.

If the film in Fig. 37-20 were negligibly thin compared with the wavelength of the light, then the reflected beams 1 and 2 would be 180° out of phase because there's a 180° phase change at the first surface but none at the second. But if the film has thickness d, then there will be an additional phase difference due to the additional distance that beam 2 travels. We'll consider the case when light is incident on the film at nearly normal incidence, so that extra distance is essentially $2d$. In Section 37-1 we established that a path difference equal to an integer number of wavelengths results in constructive interference (recall Fig. 37-2), while an odd integer number of half wavelengths gives destructive interference. Here, though, we have an extra 180° phase change due to the reflections at the different surfaces of the film. Therefore a phase change of 180° due to the path length difference will combine with the reflection-caused 180° phase change to put the waves back in phase—giving constructive interference. Thus our usual criteria for constructive and destructive interference are reversed; for example, the condition for constructive interference is that the path difference be an integer multiple of a half wavelength, or

$$2d = (m + \tfrac{1}{2})\lambda_n,$$

where λ_n is the wavelength in the film, whose refractive index is n.

In Chapter 35 we saw that the wavelength in a material with refractive index n is reduced by a factor $1/n$ from its value λ in air or vacuum; thus $\lambda_n = \lambda/n$, and our criterion for constructive interference can be written

$$2nd = (m + \tfrac{1}{2})\lambda. \qquad \text{(constructive interference, thin film)} \qquad (37\text{-}8a)$$

Conversely, destructive interference occurs when

$$2nd = m\lambda. \qquad \text{(destructive interference, thin film)} \qquad (37\text{-}8b)$$

The destructive interference is not complete, however, since the two beams do not have the same intensity.

Suppose a film is illuminated with light of a single color—that is, a single wavelength λ. If the thickness of a film varies with position, then constructive interference described by Equation 37-8a will occur at different places. Thus there will be bright bands separated by darker areas in which destructive interference occurs. If, on the other hand, the film is illuminated with white light, then the interference conditions will be satisfied at different places for the different wavelengths of the colors that comprise white light. The bands of color you see in a soap bubble or oil slick result from such thin-film interference (Fig. 37-21). If part of the film is too thin to meet the constructive interference condition of Equation 37-8a for *any* visible wavelength, then that part of the film will appear dark. You can see this effect at the top of the soap film in Fig. 37-21.

FIGURE 37-21 Interference patterns formed in a soap film illuminated with white light. The film's thickness increases toward the bottom, resulting in constructive interference for a given wavelength occurring at several different vertical positions. The film is too thin at the top for constructive interference at any visible wavelength, so it appears dark.

EXAMPLE 37-4	A Soap Film

A rectangular wire loop 20 cm high is dipped into a soap solution and then held vertically, producing a soap film whose thickness varies linearly from essentially zero at the top to 1.0 μm at the bottom (Fig. 37-22). (a) If the film is illuminated with 650-nm light, how many bright bands (as shown in Fig. 37-21) will appear? (b) What region of the film would be dark if it were illuminated with white light? The refractive index of the film is that of water, $n = 1.33$.

Solution

The film thickness at the position of a bright band is given by Equation 37-8a. To find the number of bright bands, we can solve this equation for the value of m at which the thickness d would equal the maximum film thickness:

$$m = \frac{2nd}{\lambda} - \tfrac{1}{2} = \frac{(2)(1.33)(1.0 \ \mu m)}{0.65 \ \mu m} - \tfrac{1}{2} = 3.6 \ .$$

Since m must be an integer, the largest m corresponding to a thickness within the 1.0-μm maximum is 3. Since m ranges from zero to 3, there are a total of 4 bright bands.

At a given wavelength the minimum thickness for constructive interference occurs at $m = 0$. Taking the minimum wavelength for visible light at 400 nm = 0.40 μm, Equation 37-8a with $m = 0$ then gives

$$d = \frac{\lambda}{4n} = \frac{0.40 \ \mu m}{(4)(1.33)} = 0.0752 \ \mu m \ .$$

Since the film goes linearly from zero thickness to a maximum of 1.0 μm over its 20-cm height, this thickness occurs 0.0752

20 cm

1.0 μm

FIGURE 37-22 Cross section through a vertical soap film. The angle between the surfaces is greatly exaggerated.

of the way from the top edge, or at $(0.0752)(20 \text{ cm}) = 1.5$ cm. Above this level the film is too thin for constructive interference in visible light, so the top 1.5 cm of film appears darker.

EXERCISE At what distances from the top of the film will bright bands appear if the film in Example 37-4 is illuminated with 450-nm light?

Answer: 1.69 cm, 5.08 cm, 8.46 cm, 11.8 cm, 15.2 cm, 18.6 cm

Some problems similar to Example 37-4: 31–33, 35

We derived Equations 37-8 on the assumption that our thin film was surrounded by air. Frequently, though, a thin film is sandwiched between two different media—as when an oil film floats on water. If the intermediate medium has a higher refractive index than the medium at its rear (i.e., at the right in Fig. 37-20), then our results still apply. But if the rear medium has a greater refractive index than the thin film, then there will be an additional 180° phase change at the rear interface, and therefore, our interference conditions will reverse (see the application below, as well as Problems 67 and 68).

APPLICATION	Antireflection Coatings

Partial reflection limits the amount of light that can be transmitted from one transparent material into another, reducing the light-gathering power of cameras, binoculars, and other lens-based instruments and of nonimaging devices such as windows and solar collectors.

Coating lenses, photovoltaic cells, and other critical light-gathering components with thin layers of appropriate materials can reduce reflection through the use of destructive interfer-

ence. Normally such **antireflection coatings** have refractive indices between those of air and glass; consequently, there is a 180° phase change at *each* interface, and thus, Equation 37-8a rather than 37-8b gives the condition for *destructive* interference. Thus, the minimum thickness for an antireflection coating, given by Equation 37-8a with $m = 0$, is $d = \lambda/4n$. Since this result depends on wavelength, antireflection coatings are not equally effective for all colors. With composite lenses and multiple layers, however, reduced reflection is possible over the visible spectrum.

A perfect antireflection coating might seem impossible, even in principle, since the reflected rays 1 and 2 in Fig. 37-20 do not have the same intensity. But the situation is actually more complicated. There are additional rays resulting from multiple reflections inside the transparent material, and these also interfere. A full solution of the problem involving the application of Maxwell's equations at both interfaces shows that the reflection becomes exactly zero when the antireflection layer has not only the right thickness but also refractive index $n_2 = \sqrt{n_1 n_3}$, where n_1 and n_3 are the indices of the two media separated by the antireflection coating. A widely used antireflection coating, magnesium fluoride (MgF_2), has $n = 1.38$, a value that approximates the zero-reflection condition for high-index flint glasses in air.

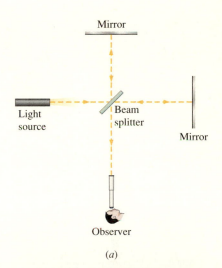

(a)   (b)

FIGURE 37-23 (a) Portion of a lens sitting on a flat glass plate. Newton's rings arise from the difference in path lengths between rays like 1 and 2. (b) Ring pattern produced in a test of a telescope's optical system.

Interference in thin layers is the basis of some very sensitive optical measuring techniques. The shape of a lens, for example, can be checked for accuracy to within a fraction of the wavelength of light using **Newton's rings**—interference patterns formed by interference of light reflected between the lens and a perfectly flat glass plate (Fig. 37-23; see also Problems 73–74).

The Michelson Interferometer

A number of optical instruments use interference for precise measurement of small distances. Among the simplest and most important of these is the **Michelson interferometer**, invented by the American physicist Albert Michelson and used in the 1880s by Michelson and his colleague Edward W. Morley in a famous experiment that paved the way for the theory of relativity. We discuss the Michelson-Morley experiment in the next chapter; here we describe the interferometer, which is still used for precision measurements.

Figure 37-24 shows the basic design of the Michelson interferometer. The key idea is that light from a monochromatic source is split into two beams by a half-silvered mirror called a **beam splitter**. The beam splitter is set at a 45° angle, so the reflected and transmitted beams travel perpendicular paths. Each then reflects off a flat mirror and returns to the beam splitter. The beam splitter again transmits and reflects half the light incident on it, with the result that some light from the originally separated beams is recombined. The recombined beams interfere, and the interference pattern is observed with a viewing lens at the bottom of Fig. 37-24.

(b)

FIGURE 37-24 (a) Schematic diagram of a Michelson interferometer. The beam splitter splits incident light into reflected and transmitted beams of equal intensity. (b) The observer looking through the viewer sees interference fringes arising from differing optical path lengths.

Suppose the path lengths in the two perpendicular arms of the interferometer were exactly the same or differed by a multiple of the wavelength. When the beams recombined they would then undergo fully constructive interference. If the path-length difference were a multiple of a half wavelength, on the other hand, the beams would interfere destructively. In reality, however, the mirrors are never exactly perpendicular, and therefore light reflecting from different parts of the mirrors experiences slightly different path lengths and thus recombines with different phase lags. To the observer, the result is a series of alternating light and dark fringes corresponding to constructive and destructive interference.

Now suppose one mirror is moved slightly. The path-length differences change and therefore, the interference pattern shifts. Moving the mirror a mere quarter wavelength adds an extra half wavelength to the round-trip path. That results in an additional 180° phase shift, moving dark fringes to where light ones were, and vice versa. An observer looking through the viewer as the mirror moves sees the fringes shift to their new positions. Shifts of a fraction of the distance between fringes are readily detected, allowing measurement of mirror displacements to within a small fraction of the wavelength.

A similar fringe shift occurs if a transparent material is placed in one path, retarding the beam because of its refractive index. This approach allows accurate measurements of the refractive indices of gases, which are so close to one that conventional methods would not be useful (see Problem 46).

Measurements with the Michelson interferometer depend not on the interference pattern itself but on *changes* in that pattern. Therefore, it doesn't really matter whether the path lengths are exactly equal. Whatever those lengths, there will be an interference pattern, and changes in that pattern will reflect changes in the relative optical paths.

EXAMPLE 37-5	*Sandstorm!*

A sandstorm has pitted the aluminum mirrors of a desert solar energy installation, and engineers want to know the depths of these pits so they can estimate the mirrors' useful lifetimes. They construct a Michelson interferometer with a sample from one of the pitted mirrors in place of one of the flat mirrors. With a 633-nm laser as the light source, the interference pattern shown in Fig. 37-25 results. What is the approximate depth of the pit?

Solution

The "bumps" in the interference pattern correspond to light that has traveled the extra distance into the bottom of the pit and back. A shift of one full cycle in the interference pattern would correspond to a full wavelength change in the path. As Fig. 37-25 shows, the pit causes a fringe shift corresponding to about 0.2 wavelengths. Since light makes a round trip through the pit, its actual depth is therefore 0.1λ, or 63 nm. Try measuring that with a meter stick!

EXERCISE A Michelson interferometer uses sodium light with wavelength 589.6 nm. As one mirror is moved, a photocell

FIGURE 37-25 Fringe pattern resulting from a pitted mirror. The size of the bump is a measure of the pit depth.

connected to a counter "watches" a fixed point through the viewer. If the counter records 4878 fringes passing that point, how far did the mirror move?

Answer: 1.438 mm

Some problems similar to Example 37-5: 44–46

37-5 HUYGENS' PRINCIPLE AND DIFFRACTION

The interference we've been studying in this chapter isn't the only optical phenomenon where the wave nature of light is important. There's also **diffraction**—the bending of light or other waves as they pass by objects. Interference and diffraction are closely related, and the double- and multiple-slit interference we've studied actually involves diffraction as well; hence the term "diffraction grating."

Diffraction, like other optical phenomena, is ultimately governed by Maxwell's equations of electromagnetism. But we can understand diffraction of light and other waves more readily using **Huygens' principle**, articulated in 1678 by the Dutch scientist Christian Huygens, who was first to suggest that light might be a wave. Huygens' principle states that

> All points on a wavefront act as point sources of spherically propagating "wavelets" that travel at the speed of light appropriate to the medium. At a short time Δt later, the new wavefront is the unique surface tangent to all the forward-propagating wavelets.

Figure 37-26 shows how Huygens' principle accounts for the propagation of plane and spherical (or cylindrical) waves. It is also possible to derive the laws of reflection and refraction from Huygens' principle.

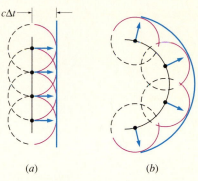

FIGURE 37-26 Application of Huygens' principle to (a) plane and (b) spherical waves. In each case the wavefront acts like a set of point sources that emit circular waves. A short time Δt later the wavelets have expanded to radius $c\,\Delta t$, and the new wavefront is the surface tangent to the wavelets. To follow the propagation further one could draw new wavelets originating on the new wavefront.

Diffraction

Figure 37-27 shows plane waves incident on an opaque barrier containing a hole. Since the waves are blocked by the barrier, Huygens' wavelets produced near each barrier edge cause the wavefronts to bend at the barrier (see blowups in Fig. 37-27). When the width of the hole is much greater than the wavelength, as in Fig. 37-27a, this diffraction is of little consequence, and the waves effectively propagate straight through the hole in a beam defined by the hole size. But when

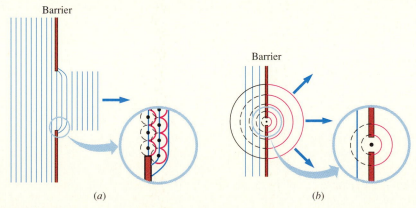

FIGURE 37-27 Plane waves incident on an opaque barrier with a hole in it are diffracted at the edges of the hole, as shown by Huygens' wavelet constructions. (a) When the hole size is much larger than the wavelength this diffraction is negligible, and the waves essentially propagate in a straight beam defined by the hole. (b) When the hole size and wavelength are comparable, the emerging wavefronts spread in a broad beam, approaching circular wavefronts as the hole size becomes negligibly small.

the hole size and wavelength are comparable, then wavefronts emerging from the hole spread in a broad pattern (Fig. 37-27*b*). Here diffraction is the dominant effect on the propagation. Thus diffraction, although it always occurs when waves pass an object, is significant only on length scales comparable to or smaller than the wavelength. That's why we could ignore diffraction and assume that light always travels through a single medium in straight lines, when we considered optical systems with dimensions much larger than the wavelength of light.

Diffraction ultimately limits our ability to image small objects and to focus light with arbitrary precision. Next, we'll see why this is so by examining quantitatively the behavior of light as it passes through a single slit. The result will help us understand optical challenges ranging from telescopic imaging of distant astrophysical objects, to "star wars" antimissile systems, to the development of the new digital video discs (DVDs) that hold a full-length movie on a disc identical in size to a standard audio CD.

37-6 SINGLE-SLIT DIFFRACTION

In treating double-slit and multiple-slit interference in Section 37-2, we assumed that plane waves passing through a slit emerged with circular wavefronts. According to Fig. 37-27, that assumption is vaild only if the width of the slits is very small compared with the wavelength, so the slit can be treated as a single, localized source of new waves. When the slit width is not small, Huygens' principle implies that we have to consider each point in the slit as a separate source—and then we can expect interference from waves originating at different points in the same slit. Thus a single wide slit is really like a multiple-slit system with infinitely many slits!

Figure 37-28 shows light incident on a slit of width a. Each point in the slit acts as a source of spherical wavelets propagating in all directions to the right of the slit. In Fig. 37-28 we focus on a particular direction described by the angle θ and will look at interference of light from the five points shown. Figure 37-28*b* concentrates on the points from which rays 1, 2, and 3 originate and shows that the path lengths for rays 1 and 3 differ by $\frac{1}{2}a\sin\theta$. These two beams will interfere destructively if this distance is half the wavelength; that is, if $\frac{1}{2}a\sin\theta = \frac{1}{2}\lambda$ or $a\sin\theta = \lambda$. But if rays 1 and 3 interfere destructively, so do rays 3 and 5, which have the same geometry, and so do rays 2 and 4, for the same reason. In fact, a ray leaving *any* point in the lower half of the slit will interfere destructively with the point located a distance $a/2$ above it. Therefore, an observer viewing the slit system at the angle θ satisfying $a\sin\theta = \lambda$ will see no light.

Similarly, the sources for rays 1 and 2 are $a/4$ apart and will therefore interfere destructively if $\frac{1}{4}a\sin\theta = \frac{1}{2}\lambda$, or $a\sin\theta = 2\lambda$. But then so will rays 2 and 3, and rays 3 and 4; in fact, any ray from a point in the lower three quarters of the slit will interfere destructively with a ray from the point $a/4$ above it, and therefore, an observer looking at an angle satisfying $a\sin\theta = 2\lambda$ will see no light.

We could equally well have divided the slit into six sections with seven evenly spaced points; we would then have found destructive interference if $\frac{1}{6}a\sin\theta = \frac{1}{2}\lambda$, or $a\sin\theta = 3\lambda$. We could obviously continue this process for any number of points in the slit, and therefore, we conclude that destructive

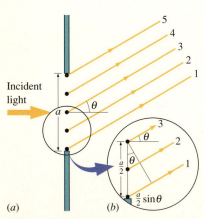

FIGURE 37-28 Each point in a slit acts as a source of Huygens' wavelets, which interfere in the region to the right of the slit. (*a*) Here we consider 5 points within the slit, and the relative phases of their light as viewed at an angle θ. (*b*) The path-length difference for two beams is their separation times $\sin\theta$.

interference occurs for all angles satisfying

$$a\sin\theta = m\lambda, \quad \text{(destructive interference, single-slit diffraction)} \quad (37\text{-}9)$$

16.1
Diffraction

with m any nonzero integer and a the slit width. Note that the case $m = 0$ is excluded; it produces not destructive interference but a central maximum in which all waves are in phase.

> ### Tip
>
> *Interference and Diffraction* Equation 37-9 for the *minima* of a single-slit diffraction pattern looks just like Equation 37-1a for the *maxima* of a multi-slit interference pattern, except that the slit width a replaces the slit spacing d. What's going on? Why does the same equation give the minima in one case and the maxima in another? Because we're dealing with two distinct but related phenomena. In the multi-slit case leading to Equation 37-1a, we considered each slit to be so narrow that it could be considered a single source, thus neglecting the interference of waves originating within the same slit. In the single-slit case leading to Equation 37-9, the diffraction pattern occurs precisely because of the interference of waves from different points within the same slit.

EXAMPLE 37-6	*Diffraction: A Narrow Slit*

For what slit width, in terms of wavelength, will the first minimum lie at an angular position of 45°?

Solution
With $m = 1$, Equation 37-9 gives

$$a = \frac{\lambda}{\sin\theta} = \frac{\lambda}{\sin 45°} = \sqrt{2}\lambda.$$

Here, with the slit width nearly equal to the wavelength, diffraction dominates the wave propagation. The incident light is therefore spread over a wide angular range. The exercise below shows that the beam spreading decreases as the slit width grows.

EXERCISE Light is incident on a slit whose width is 20 times its wavelength. Find the angular spread of the beam, taken as the angle between the first minima on either side of the central maximum.

Answer: 5.7°

Some problems similar to Example 37-6: 49–51

Intensity in Single-Slit Diffraction

We can find the intensity of the single-slit diffraction pattern by adding the electric fields of the individual beams, as we did for two-slit interference. Since there are infinitely many points along the slit, each constituting a point source of light, we might set up and evaluate a complicated integral. Instead we use the graphical method of phasors, introduced in Chapter 33 to deal with alternating currents and voltages of different phases. We have a similar situation here, as we want to combine waves with different phases that arise from different points on the slit. The phasor method is entirely equivalent to integrating over the electric field contributions from the individual sources in the slit.

We use phasors to represent the electric fields of the individual waves originating from points in the slit when they arrive at the screen. The length of a phasor gives the field amplitude E, and its direction gives its phase. Consider N such points, spaced evenly across a slit of width a, like the five points in Fig. 37-28.

These divide the slit into $N - 1$ sections, each of width $a/(N - 1)$. At some angle θ, the phase difference, in radians, of light coming from two adjacent points will be the ratio of their path-length difference to the wavelength, times the 2π radians that comprise a complete cycle. As usual, the path-length difference is the point spacing times $\sin\theta$ (see Fig. 37-28b). Here the point spacing is $a/(N - 1)$, so we have

$$\Delta\phi = \left(\frac{a}{N - 1}\sin\theta\right)\left(\frac{2\pi}{\lambda}\right) \tag{37-10}$$

for the phase difference between waves from adjacent points. All the waves have essentially the same amplitude, so we now need to add N equal-length phasors each of which differs in phase by $\Delta\phi$ from its nearest neighbors.

Consider first the case $\theta = 0$. Then Equation 37-10 gives $\Delta\phi = 0$ so, as Fig. 37-29a shows, the phasors are all in the same direction and sum to give a large total amplitude. This is the central maximum of the diffraction pattern. If, on the other hand, $\Delta\phi$ is such that the N phasors sum to a closed path, then the total amplitude—the net displacement from the beginning to the end of the phasor diagram—is zero (Fig. 37-29b). Here we have a point of zero intensity in the interference pattern. More generally, the phasors sum to give an amplitude E_θ that is neither zero nor as great as a simple sum of the individual amplitudes, as shown in Fig. 37-29c.

Now each point in the slit differs in phase by $\Delta\phi$ from its immediate neighbors. Thus, the second point differs by $\Delta\phi$ from the first point, the third point by $2\Delta\phi$ from the first point, and so forth until the Nth point differs in phase by $(N - 1)\Delta\phi$ from the first point. But this last quantity $(N - 1)\Delta\phi$ is just the total phase difference ϕ from one end of the slit to the other. Using Equation 37-10, we can write this phase difference as

$$\phi = (N - 1)\left(\frac{a}{N - 1}\sin\theta\right)\left(\frac{2\pi}{\lambda}\right) = \frac{2\pi}{\lambda}a\sin\theta. \tag{37-11}$$

To consider all points in the slit, we take the limit as $N \rightarrow \infty$; then the end-to-end chain of phasors in Fig. 37-29c becomes a circular arc of radius R, as shown in

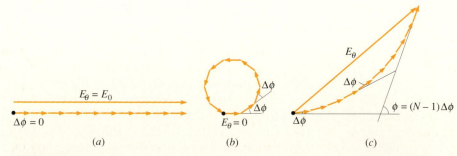

FIGURE 37-29 Phasor addition to find the amplitude in single-slit diffraction. (a) When $\theta = 0$ in Fig. 37-28 all waves have the same phase, and their amplitudes add to produce the central maximum. (b) When the phasor sum is a closed loop, the net phasor displacement is zero and so is the wave amplitude. (c) In general, the phasors add to produce an amplitude that is neither zero nor as great as in the central maximum. The angle between the first and last phasor is $N\Delta\phi$. Here $N = 10$ in all three frames, and in general we designate the resultant amplitude by E_θ. When the phasors all have the same phase, E_0 is the resultant amplitude.

Fig. 37-30. If we stretched this arc into a line, its length would be the electric field amplitude E_0 of Fig. 37-29a; therefore, E_0 is also the length of the arc. From the geometry of Fig. 37-30 we see that

$$\sin(\phi/2) = \frac{E_\theta/2}{R}.$$

We can also write the angle $\phi/2$ using the definition of an angle in radians as the ratio of arc length to radius. Here $\phi/2$ subtends an arc of length $E_0/2$, so we have

$$\phi/2 = \frac{E_0/2}{R}.$$

Dividing these two equations gives the ratio of the amplitudes:

$$\frac{E_\theta}{E_0} = \frac{\sin(\phi/2)}{\phi/2}.$$

But the wave intensity $\overline{S}$ is proportional to the square of the amplitude, so

$$\overline{S}_\theta = \overline{S}_0\left[\frac{\sin(\phi/2)}{\phi/2}\right]^2, \tag{37-12}$$

where ϕ is given by Equation 37-11.

Figure 37-31 shows plots of Equation 37-12 for several values of the slit width a in relation to the wavelength λ. Note that for wide slits—slit width a large compared with the wavelength λ—the central peak is very narrow and the secondary peaks are much smaller. Here diffraction is negligible, and the beam essentially propagates through the slit in the ray approximation of geometrical optics. But as the slit narrows the diffracted beam spreads, until, with $a = \lambda$, it covers an angular width of some $120°$.

The intensity given by Equation 37-12 will be zero when the numerator on the right-hand side is zero—that is, when the argument of the sine function is an integer multiple of π. That occurs when $\frac{\phi}{2} = \frac{\pi a}{\lambda}\sin\theta = m\pi$, or when $a\sin\theta = m\lambda$. Thus, we recover our earlier result of Equation 37-9 for the angular positions where destructive interference gives zero intensity.

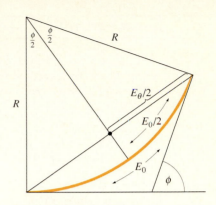

FIGURE 37-30 In the limit $N \to \infty$ the chain of phasors becomes a circular arc of length E_0.

(a)   (b)   (c)

FIGURE 37-31 Intensity in single-slit diffraction, as a function of the angle θ from the centerline, for three values of slit width a, expressed in units of wavelength λ. When the slit is much wider than the wavelength, most of the light is concentrated in a narrow central peak. But as the slit width approaches the wavelength, the diffracted beam becomes very wide.

EXAMPLE 37-7	*Other Peaks*

Secondary maxima lie approximately midway between the minima of the diffraction pattern. Use this fact to find the intensity at the first of these secondary maxima, in terms of the central-peak intensity $\bar{S}_0$.

Solution

The first and second minima are at positions given by $a \sin\theta_1 = \lambda$ and $a \sin\theta_2 = 2\lambda$, respectively. With the first of the secondary maxima midway between them, its angular position is given by $a \sin\theta = \frac{3}{2}\lambda$. (Here we assume θ is small, so $\sin\theta \simeq \theta$, and midway in angular position θ is therefore also midway in $\sin\theta$.) Then $\frac{a}{\lambda}\sin\theta = \frac{3}{2}$, so Equation 37-11 gives $\frac{\phi}{2} = \frac{\pi a}{\lambda}\sin\theta = \frac{3\pi}{2}$. Using this result in Equation 37-12, we

have

$$\bar{S} = \bar{S}_0 \left[\frac{\sin(\phi/2)}{\phi/2} \right] = \bar{S}_0 \left[\frac{\sin(3\pi/2)}{3\pi/2} \right] = \frac{4\bar{S}_0}{9\pi^2} = 0.045\bar{S}_0 \,.$$

Thus, the intensity at the first secondary maximum is only about 4.5% of the central-peak intensity. Note that this result is independent of the slit width, provided the secondary peak exists (which it may not; see Fig. 37-31c).

EXERCISE Show that the intensity in single-slit diffraction has half its maximum value when $\sin\theta = 1.3916\dfrac{\lambda}{\pi a}$.

Some problems similar to Example 37-7: 49–53

In analyzing single-slit diffraction we considered only parallel rays in the diffraction region. To make the diffraction pattern actually appear on a screen, we have to focus those rays with a lens between slit and screen (Fig. 37-32). Diffraction associated with parallel rays is called **Fraunhofer diffraction**. In general, Fraunhofer diffraction occurs when the distance from the diffracting system is large compared with the wavelength. Fraunhofer diffraction is an approximation to the more general case of **Fresnel diffraction**, whose analysis is more complicated because it also accounts for nonparallel rays that exist near a diffracting system.

Multiple Slits and Other Diffracting Systems

In treating multiple-slit systems in Section 37-2, we assumed the slits were so narrow compared with the wavelength that the central diffraction peak spread into the entire space beyond the slit system. When the slit width is not negligible, then the waves from each slit produce a single-slit diffraction pattern. The result is a pattern that combines single-slit diffraction with multiple-slit interference (Fig. 37-33).

FIGURE 37-32 Single-slit diffraction pattern produced by focusing light from a single slit onto a screen. Note the bright central peak and secondary maxima separated by regions of destructive interference.

FIGURE 37-33 When the slit width is not negligible, the double-slit pattern shows the regular variations of double-slit interference within a single-slit diffraction pattern.

(a)   (b)

FIGURE 37-34 (a) Light waves diffract as they pass by the straight edge of an opaque barrier. Photo shows the resulting interference pattern on a screen some distance beyond the barrier, whose shadow is the black region at left. (b) Diffraction of monochromatic light by a pair of crosshairs in a circular aperture. Diffraction effects from both the crosshairs (straight fringes) and the aperture itself (circular fringes) are evident.

Diffraction occurs any time light passes a sharp, opaque edge like the edges of the slits we've been considering. Close examination of the shadow produced by a sharp edge shows parallel fringes resulting from interference of the diffracting wavefronts (Fig. 37-34a). More complex diffraction patterns result from objects of different shape (Fig. 37-34b). The presence of such diffraction patterns limits our ability to form sharp optical images, as we show in the next section.

37-7 THE DIFFRACTION LIMIT

Diffraction imposes a fundamental limit on the ability of optical systems to distinguish closely spaced objects. Consider two point sources of light illuminating a slit. The sources are so far from the slit that waves reaching the slit are essentially plane waves, but the different source positions mean the waves reach the slit at different angles. We assume the sources are incoherent, so they don't

(a)

(b)

(a)

(b)

(c)

FIGURE 37-35 Two distant sources at different angular positions produce diffraction patterns whose central peaks have the same angular separation θ as the sources. (a) For sufficiently large θ the central peaks are clearly distinct, but (b) at small angular separations they overlap.

produce a regular interference pattern. Then light diffracting at the slit produces two single-slit diffraction patterns, one for each source. Because the sources are at different angular positions, the central maxima of these patterns don't coincide, as shown in Fig. 37-35.

If the angular separation between the sources is great enough, then the central maxima of the two diffraction patterns will be entirely distinct. In that case the eye or any other optical system can clearly distinguish the two sources (Fig. 37-35a). But as the sources get closer the central maxima begin to overlap (Fig. 37-35b). The two sources remain distinguishable as long as the total intensity pattern shows two peaks. Since the sources are incoherent the total intensity is just the sum of the individual intensities. Figure 37-36 shows how that sum loses its two-peak structure as the diffraction patterns merge. In general, two peaks are barely distinguishable if the central maximum of one coincides with the first minimum of the other. This condition is called the **Rayleigh criterion**, and when it is met we say that the two sources are just barely **resolved**.

What does all this have to do with optical instruments and images? Simply this: All optical systems are analogous to the single slit we've been considering. Every system has an aperture of finite size through which light enters the system. That aperture may be an actual slit or hole, like the diaphragm that "stops down" a camera lens, or it may be the full size of a lens or mirror. So all optical systems ultimately suffer loss of resolution if two sources—or two parts of the same extended object—have too small an angular separation. Thus, diffraction fundamentally limits our ability to probe the structure of objects that are either very small or very distant.

Figure 37-35 shows that the angular separation between the diffraction peaks is equal to the angular separation between the sources themselves. Then the Rayleigh criterion is just met if the angular separation between the two sources is equal to the angular separation between a central peak and the first minimum. We found earlier that the first minimum in single-slit diffraction occurs at angular position given by

$$\sin\theta = \frac{\lambda}{a},$$

with a the slit width and with θ measured from the central peak. In most optical systems the wavelength is much less than the size of any apertures, so we can use the small-angle approximation $\sin\theta \simeq \theta$ with θ in radians. Then the Rayleigh criterion—the condition that two sources be just resolvable—for single-slit diffraction becomes

$$\theta_{min} = \frac{\lambda}{a}. \qquad \text{(Rayleigh criterion, slit)} \qquad (37\text{-}13a)$$

FIGURE 37-36 Since the two sources are incoherent, the total intensity is just the sum of the intensities of the two diffraction patterns. Here we see the intensities in two patterns and their sum (light blue). (a) Fully resolved; (b) barely resolved (Rayleigh criterion); (c) unresolved.

FIGURE 37-38 Diffraction patterns produced by a pair of point sources imaged through a circular hole. The angular separation of the sources decreases going from left to right, and when the sources are too close they cannot be resolved.

(a)

(b)

FIGURE 37-37 (a) Diffraction pattern produced by a circular aperture. (b) Three-dimensional plot of intensity versus position.

16.6
Resolving Power

Most optical systems have circular apertures rather than slits. The diffraction pattern from such an aperture is a series of concentric rings (Fig. 37-37). A more involved mathematical analysis shows that the angular position of the first ring relative to the central peak, and therefore, the minimum resolvable source separation for a circular aperture, is

$$\theta_{min} = \frac{1.22\lambda}{D}, \quad \text{(Rayleigh criterion, circular aperture)} \quad (37\text{-}13b)$$

with D the aperture diameter. In most optical systems the angle θ is very small, and Equations 37-13a and b hold for this case. For larger angles, θ_{min} in both equations should be replaced with $\sin\theta_{min}$. Figure 37-38 shows the loss of resolution as two nearby sources are imaged through a circular aperture. You can simulate this effect using ActivPhysics Activity 16.6.

Equations 37-13 show that increasing the aperture size allows smaller angular differences to be resolved. In optical instrument design, that means larger mirrors, lenses, and other components. An alternative is to decrease the wavelength used, which may or may not be an option depending on the source. In high-quality optical systems diffraction is often the limiting factor preventing perfectly sharp image formation; such systems are said to be **diffraction limited**. For example, the diffraction limit sets a minimum size for objects resolvable with optical microscopes; that's why electron microscopes—with shorter effective wavelength—are used to image the smallest biological structures. Large ground-based telescopes are an exception to the diffraction limit; their image quality is limited by atmospheric turbulence, although this can be reduced with adaptive optics that we mentioned in Chapter 36. From its vantage point above the atmosphere, the Hubble Space Telescope is the first large diffraction-limited astronomical telescope.

Got It!

You're a biologist trying to resolve details of structures within a cell, but they look fuzzy even at the highest power of your microscope. Why won't it help to substitute a new eyepiece with shorter focal length, as suggested by Equation 36-10?

EXAMPLE 37-8 | Asteroid Alert!

An asteroid appears on a collision course with Earth, at a distance of 20×10^6 km. What is the minimum size asteroid that the 2.4-m-diameter diffraction-limited Hubble Space Telescope could resolve at this distance, using 550-nm reflected sunlight?

Solution

Resolving the asteroid means being able to distinguish its opposite edges in the telescope image and therefore to assess its size. Suppose the asteroid's long dimension (it might not be spherical) is ℓ. Then at a distance $L \gg \ell$ it subtends an angle given very nearly by $\theta = \ell/L$. Using this result in Equation 37-13b with the mirror diameter and wavelength given, we have

$$\frac{\ell}{L} = \frac{1.22\lambda}{D},$$

or

$$\ell = \frac{1.22\lambda L}{D} = \frac{(1.22)(550\times10^{-9}\ \text{m})(20\times10^{9}\ \text{m})}{2.4\ \text{m}} = 5.6\ \text{km}.$$

This is a potentially dangerous object, comparable in size to the asteroid that scientists believe caused the extinction of the dinosaurs, and somewhat larger than the comet fragments that slammed into Jupiter in 1994, causing Earth-sized disturbances on the giant planet.

EXERCISE Two ants stand 1 cm apart. Assuming diffraction limited vision, at what distance could a human eye, with an iris aperture of 4 mm, tell that two are indeed separate creatures? Assume a wavelength of 550 nm, near the middle of the visible spectrum. (Other limitations of the human eye prevent its realizing this diffraction limit.)

Answer: 60 m

Some problems similar to Example 37-8: 55–59

APPLICATION | Movies on Disc: From CD to DVD

Standard audio CDs have a maximum playing time of 74 minutes—a value that's set by the optical diffraction limit. In the Application "CD Music, Continued" earlier in this chapter, we showed how a CD encodes information in "pits" 1.6 μm apart and as short as 0.83 μm in length. Standard CDs are read with 780-nm infrared laser light and the pit size and separation are chosen to ensure that diffraction effects at this wavelength don't cause the CD player's optical system to confuse adjacent pits. Ultimately, therefore, the 74-minute playing time is limited by the laser wavelength.

When CDs were first developed, inexpensive semiconductor lasers were available only in the infrared. But today's semiconductor lasers produce light well into the visible—with significantly shorter wavelengths than the infrared light used in CD players and CD-ROM drives. The new DVDs introduced in the late 1990s (read either "digital video disc" or "digital versatile disc") exploit these shorter-wavelength lasers.

Using laser wavelengths of 635 and 650 nm, DVDs can function with significantly lower pit sizes and spacings because of the lower diffraction limit at these shorter wavelengths (Fig. 37-39). Coupled with a two-layer structure (as opposed to the standard CD's single information layer) and more sophisticated data-compression schemes, the smaller pits and pit spacing give DVDs more than 12 times the information-storage capacity of standard CDs. That translates into more than two hours of high quality video and audio, as opposed to the CD's 74 minutes of audio alone. In computer use, it means a DVD can hold some 8.5 gigabytes of information as opposed to a CD's 680 megabytes (by comparison, a 3.5″ floppy disc holds only 1.44 MB). And DVDs can be made in two-sided versions, providing another doubling in capacity. Despite their high capacity, DVDs are physically the same size as CDs, and DVD players are compatible with standard CDs as well.

FIGURE 37-39 A comparison of a standard CD and a DVD. The smaller pit size and closer pit spacing of the DVD result in a much higher information storage capacity, and are made possible by the lower diffraction limit associated with the DVD player's shorter wavelength laser light. Both images are microphotographs of the same size region, about 8 μm on a side. The pits lie on a single spiral track that covers most of the disc.

(a)   (b)

| APPLICATION | *Interferometry* |

At a given wavelength, Equations 37-13 show that resolving small angular separations requires a large aperture. To see detail in distant astronomical objects, in particular, requires impractically large apertures. A technique called **interferometry** provides a way around this problem. Used most commonly at radio wavelengths, interferometry employs two or more individual apertures (i.e., radio telescopes), usually separated by a considerable distance. Signals from each are combined with their phases intact, so that they interfere. A two-telescope interferometer works like a two-slit system in reverse: the system is most sensitive to radiation from angular positions that produce constructive interference (Fig. 37-40). The resolution of an interferometer is approximately that of a single slit-like aperture whose width is the spacing between the telescopes.

Today's interferometers include installations with many radio telescopes at a single site (Fig. 37-41), as well as coordinated instruments spread across the globe. The first interferometric arrays of optical telescopes began operation in the 1990s, and astronomers look forward to radio interferometers using the Earth-moon distance as their "aperture."

FIGURE 37-40 A radio interferometer consists of two or more radio telescopes at different locations. The interferometer can detect very slight changes in phase of the incoming wavefronts and can therefore resolve very small differences in angular position of objects in the sky.

FIGURE 37-41 Radio images of the active galaxy M87 show a jet of material emerging from the galactic core. (*top*) Image made with the Very Large Array (VLA) in New Mexico. The VLA consists of 27 radio telescopes whose separation gives an effective aperture of 27 km. (*bottom two images*) Much higher resolution is obtained with the Very Long Baseline Array (VLBA), whose radio telescopes extend from Hawaii to Puerto Rico. The resulting effective aperture of 8,000 km has a much lower diffraction limit, hence the higher resolution.

CHAPTER SYNOPSIS

Summary

1. **Interference** and **diffraction** are wave phenomena that constitute the subject of physical optics. Interference effects require two or more beams of **coherent** light, which maintain a fixed phase relation. Such coherent beams can be produced by allowing light from a single source to travel two different paths. If the path lengths differ by an integer multiple of the wavelength, **constructive interference** results when the beams recombine. If the path lengths differ by an integer multiple of a half wavelength, the result is **destructive interference**.

2. When coherent light of wavelength λ passes through a pair of narrow slits with spacing d, the resulting interference pattern shows bright fringes at maxima given by

$$d \sin \theta = m\lambda ,$$

where m is an integer. The average intensity in the interference pattern is given by

$$\bar{S} = 4\bar{S}_0 \cos^2\left(\frac{\pi d \sin \theta}{\lambda}\right) ,$$

where $\bar{S}_0$ is the average intensity incident on the slits.

3. A multiple-slit system has primary interference maxima in the same position as a double-slit system, with multiple minima and secondary maxima in between. For large numbers of slits the primary maxima are narrow and bright and the region between them relatively dark. A system with many slits constitutes a **diffraction grating**, which disperses light into its component wavelengths. A grating produces spectra of many **orders**, corresponding to the integer m in the equation that locates the interference maxima. The **resolving power** of an N-slit grating is given by

$$\frac{\lambda}{\Delta\lambda} = mN ,$$

where $\Delta\lambda$ is the minimum separation between wavelengths that can be distinguished in the mth-order spectrum. Other periodic systems, like the layers of atoms in a crystal, can serve as diffraction gratings for appropriate wavelengths.

4. Interference occurs in thin transparent films when light reflected from the front and back surfaces of the film recombines. For a film whose refractive index is larger than that of its surroundings, the interference is constructive when the film thickness d, refractive index n, and the light wavelength λ are related by $2nd = (m + \frac{1}{2})\lambda$, where m is an integer.

5. The **Michelson interferometer** utilizes the interference of light traveling on two perpendicular paths to make precise distance measurements.

6. **Huygens' principle** treats each point on a wavefront as a source of expanding spherical wavelets. Superposition of the wavelets describes the propagation of the wavefront and shows that waves undergo **diffraction** when passing by an opaque edge.

7. When the width of a single slit is not negligible compared with the wavelength, interference of light passing through different parts of the slit produces a **diffraction pattern** with a central maximum surrounded by lesser maxima separated by points of zero intensity.

8. The **diffraction limit** in optical systems arises because diffraction effects merge the images of two objects when their angular separation is sufficiently small. The **Rayleigh criterion** gives the minimum angular separation that can

be resolved by a given aperture. For a circular aperture of diameter D, the Rayleigh criterion is

$$\theta_{min} = \frac{1.22\lambda}{D} ,$$

where θ_{min} is in radians. Large apertures are therefore necessary to resolve objects with small angular separation.

Terms You Should Understand

(Pairs are closely related terms whose distinction is important; number in parentheses is chapter section where term first appears.)

physical optics, geometrical optics (introduction and preceding chapters)
interference (37-1)
coherence (37-1)
monochromatic light (37-1)
interference fringe (37-2)
order (37-2, 37-3)
diffraction grating (37-3)
resolving power (37-3)
Bragg condition (37-3)
Michelson interferometer (37-4)
diffraction (37-5)
Huygens' principle (37-5)
diffraction limit (37-7)
Rayleigh criterion (37-7)

Symbols You Should Recognize

m (37-1, 37-2, 37-3) θ_{min} (37-7)

Problems You Should Be Able to Solve

analyzing double-slit interference patterns (37-2)
finding locations of maxima and minima in multiple-slit interference (37-3)
analyzing dispersion of light with diffraction gratings (37-3)
determining conditions for constructive and destructive interference in thin films (37-4)
analyzing experiments using Michelson interferometers (37-4)
finding minima and beam widths in single-slit diffraction (37-6)
finding the diffraction limit in optical systems (37-7)

Limitations to Keep in Mind

This chapter's analysis of double-slit interference assumes slit widths much smaller than the wavelength; otherwise the pattern combines diffraction and interference.
Our analysis of diffraction applies only in the Fraunhofer limit, in which the incident light is essentially parallel and the pattern is formed at large distances from the diffracting aperture.

QUESTIONS

1. A prism bends blue light more than red. Is the same true of a diffraction grating?
2. Why does an oil slick show colored bands?
3. Would a Michelson interferometer work even if its two arms were not exactly the same length?
4. Why does a soap bubble turn colorless just before it dries up and breaks?
5. Why don't you see interference effects between the front and back of your eyeglasses?
6. Hold two fingers very close together while looking through them at a source of light. Explain what you see.
7. Figure 37-42 shows the shadow cast by a ball bearing on the end of a needle, under monochromatic laser light.

FIGURE 37-42 Question 7.

Explain the existence of the fringes shown and of the white spot at the center of the ball bearing's shadow.
8. You can hear around corners but you can't see around corners. Why the difference?
9. In deriving the intensity in double-slit interference, we could not simply add the intensities from the two slits. Why not?
10. In sketching the intensity pattern for two incoherent sources imaged through a single slit, we could simply add the intensities from the two sources. Why?
11. The primary maxima in multiple-slit interference are in the same angular positions as those in double-slit interference. Why, then, do diffraction gratings have thousands of slits instead of just two?
12. In what way is a widely separated pair of small radio telescopes superior to a single large one? In what way is it inferior?
13. Describe the change in the diffraction pattern of a single slit as the slit is narrowed.
14. What pattern would result from passing red and blue light simultaneously through a double-slit system?
15. When the moon passes in front of a star, the intensity of the starlight fluctuates instead of dropping abruptly. Explain.
16. In analyzing crystal structure using X-ray diffraction, it is necessary to take data with the crystal in several different orientations. Why?
17. A double-slit system has one slit much narrower than the wavelength of the incident light, the other much wider. Describe the resulting intensity pattern.
18. Sketch roughly the diffraction pattern you would expect for light passing through a square hole a few wavelengths wide.

PROBLEMS

ActivPhysics can help with these problems:
Activities in Section 16, Physical Optics

Section 37-2 Double-Slit Interference

1. A double-slit system is used to measure the wavelength of light. The system has slit spacing $d = 15$ μm and slit-to-screen distance $L = 2.2$ m. If the $m = 1$ maximum in the interference pattern occurs 7.1 cm from screen center, what is the wavelength?
2. A double-slit experiment with $d = 0.025$ mm and $L = 75$ cm uses 550-nm light. What is the spacing between adjacent bright fringes?
3. A double-slit experiment has slit spacing 0.12 mm. (a) What should be the slit-to-screen distance L if the bright fringes are to be 5.0 mm apart when the slits are

illuminated with 633-nm laser light? (b) What will be the fringe spacing with 480-nm light?
4. With two slits separated by 0.37 mm, the interference pattern has bright fringes with angular spacing 0.065°. What is the wavelength of the light illuminating the slits?
5. The green line of gaseous mercury at 546 nm falls on a double-slit apparatus. If the fifth dark fringe is at 0.113° from the centerline, what is the slit separation?
6. What is the angular position θ of the second-order bright fringe in a double-slit system with 1.5-μm slit spacing if the light has wavelength (a) 400 nm or (b) 700 nm?
7. Light shines on a pair of slits whose spacing is three times the wavelength. Find the locations of the first- and second-order bright fringes on a screen 50 cm from the slits. *Hint:* Do Equations 37-2 apply?

8. A double-slit experiment has slit spacing 0.035 mm, slit-to-screen distance 1.5 m, and wavelength 500 nm. What is the phase difference between two waves arriving at a point 0.56 cm from the center line?

9. For a double-slit experiment with slit spacing 0.25 mm and wavelength 600 nm, at what angular position is the path difference equal to one-fourth of the wavelength?

10. A screen 1.0 m wide is located 2.0 m from a pair of slits illuminated by 633-nm laser light, with its center on the centerline of the slits. Find the highest-order bright fringe that will appear on the screen if the slit spacing is (a) 0.10 mm; (b) 10 μm.

11. Laser light at 633 nm falls on a double-slit apparatus with slit separation 6.5 μm. Find the separation between (a) the first and second and (b) the third and fourth bright fringes, as seen on a screen 1.7 m from the slits.

12. A tube of glowing gas emits light at 550 nm and 400 nm. In a double-slit apparatus, what is the lowest-order 550-nm bright fringe that will fall on a 400-nm dark fringe, and what are the corresponding orders?

Section 37-3 Multiple-Slit Interference and Diffraction Gratings

13. In a 5-slit system, how many minima lie between the zeroth-order and first-order maxima?

14. In a 3-slit system the first minimum occurs at an angular position of 5°. Where is the first maximum?

15. A 5-slit system with 7.5-μm slit spacing is illuminated with 633-nm light. Find the angular positions of (a) the first 2 maxima and (b) the 3rd and 6th minima.

16. On the screen of a multiple-slit system, the interference pattern shows bright maxima separated by 0.86° and 7 minima between each bright maximum. (a) How many slits are there? (b) What is the slit separation if the incident light has wavelength 656.3 nm?

17. Green light at 520 nm is diffracted by a grating with 3000 lines per cm. Through what angle is the light diffracted in (a) first and (b) fifth order?

18. Find the angular separation between the red hydrogen-α spectral line at 656 nm and the yellow sodium line at 589 nm if the two are observed in 3rd order with a 3500 line/cm grating spectrometer.

19. Light is incident normally on a grating with 10,000 lines per cm. What is the maximum order in which (a) 450-nm and (b) 650-nm light will be visible?

20. Visible light has wavelengths between 400 nm and 700 nm. What is the lowest pair of consecutive orders for which there will be some overlap between the visible spectra as dispersed by a grating?

21. A solar astronomer is studying the Sun's 589-nm sodium spectral line with a 2500 line/cm grating spectrometer whose fourth order dispersion puts the wavelength range

from 575 nm to 625 nm on a detector. The astronomer is interested in observing simultaneously the so-called calcium-K line, at 393 nm. What order dispersion will put this line also on the detector?

22. (a) What portions of the 4th and 5th order visible spectra overlap in a 3000 lines/cm grating spectrometer? (b) How would your answer change for a 1000 lines/cm grating? (c) For a 10,000 lines/cm grating?

23. Estimate the number of lines per cm in the grating used to produce Fig. 37-15.

24. A grating spectrometer's detector covers an angular range of 10° and can be swung at any angle within ±75° of the normal to the grating. If the grating has 1200 lines/cm, in what orders can the hydrogen-α spectral line at 656 nm and the calcium-K line at 393 nm both fall on the detector? (b) What will be the angular position of the K line under these conditions?

25. When viewed in 6th order, the 486.1-nm hydrogen-β spectral line is flanked by another line that appears at the position of 484.3 nm in the 6th order spectrum. Actually the line is from a different order of the spectrum. What are the possible visible wavelengths of this line?

26. (a) Find the resolving power of a grating needed to separate the sodium-D spectral lines, which are at 589.0 nm and 589.6 nm. (b) How many lines must the grating have to achieve this resolution in first order? (This is very low resolution by present-day spectroscopic standards.)

27. Echelle spectroscopy uses relatively coarse gratings in high order. Compare the resolving power of an 80 line/mm echelle grating used in 12th order with a 600 line/mm grating used in 1st order, assuming the two have the same width.

28. The International Ultraviolet Explorer satellite carries a spectrometer with a 2.0-cm-wide grating ruled at 102 lines/mm. What is the minimum wavelength difference it can resolve in 12th order when observing in the ultraviolet at 155 nm?

29. You wish to resolve the calcium-H line at 396.85 nm from the hydrogen-ε line at 397.05 nm in a 1st-order spectrum. To the nearest hundred, how many lines should your grating have?

30. X-ray diffraction in calcium chloride (KCl) results in a first-order maximum when the X rays graze the crystal plane at 8.5°. If the X-ray wavelength is 97 pm, what is the spacing between crystal planes?

Section 37-4 Thin Films and Interferometers

31. Find the minimum thickness of a soap film ($n = 1.33$) in which 550-nm light will undergo constructive interference.

32. Light of unknown wavelength shines on a precisely machined wedge of glass with refractive index 1.52. The

closest point to the apex of the wedge where reflection is enhanced occurs where the wedge is 98 nm thick. Find the wavelength.

33. Monochromatic light shines on a glass wedge with refractive index 1.65, and enhanced reflection occurs where the wedge is 450 nm thick. Find all possible values for the wavelength in the visible range.

34. White light shines on a 100-nm-thick sliver of fluorite ($n = 1.43$). What wavelength is most strongly reflected?

35. As a soap bubble ($n = 1.33$) evaporates and thins, the reflected colors gradually disappear. (a) What is its thickness just as the last vestige of color vanishes? (b) What is the last color seen?

36. An oil film ($n = 1.25$) floats on water, and a soap film ($n = 1.33$) is suspended in air. Find the minimum thickness for each that will result in constructive interference with 500-nm green light.

37. Light reflected from a thin film of acetone ($n = 1.36$) on a glass plate ($n = 1.5$) shows maximum reflection at 500 nm and minimum at 400 nm. Find the minimum possible film thickness.

38. What minimum thickness of a coating with refractive index 1.35 should you use on a glass lens to minimize reflection at 500 nm, the approximate center of the visible spectrum?

39. An oil film with refractive index 1.25 floats on water. The film thickness varies from 0.80 μm to 2.1 μm. If 630-nm light is incident normally on the film, at how many locations will it undergo enhanced reflection?

40. Microwave ovens operate at a frequency of 2.4 GHz. What is the minimum thickness for a plastic tray with refractive index 1.45 that will cause enhanced reflection of microwaves incident normal to the plate?

41. Two perfectly flat glass plates are separated at one end by a piece of paper 0.065 mm thick. A source of 550-nm light illuminates the plates from above, as shown in Fig. 37-43. How many bright bands appear to an observer looking down on the plates?

Incident light

Paper

FIGURE 37-43 An air wedge (Problems 41, 42, 43, 72).

42. An air wedge like that shown in Fig. 37-43 shows N bright bands when illuminated from above. Find an expression for the number of bands that will appear if the air is replaced by a liquid of refractive index n different from that of the glass.

43. You apply a slight pressure with your finger to the upper of a pair of glass plates forming an air wedge as in Fig. 37-43. The wedge is illuminated from above with 500-nm light, and you place your finger where, initially, there is a dark band. If you push gently so the band becomes light, then dark, then light again, by how much have you deflected the plate?

44. A Michelson interferometer uses light from glowing hydrogen at 486.1 nm. As you move one mirror, 530 bright fringes pass a fixed point in the viewer. How far did the mirror move?

45. What is the wavelength of light used in a Michelson interferometer if 550 bright fringes go by a fixed point when the mirror moves 0.150 mm?

46. One arm of a Michelson interferometer is 42.5 cm long and is enclosed in a box that can be evacuated. The box initially contains air, which is gradually pumped out to create a vacuum. In the process, 388 bright fringes pass a fixed point in the viewer. If the interferometer uses light with wavelength 641.6 nm, what is the refractive index of the air?

47. The evacuated box of the previous problem is filled with chlorine gas, whose refractive index is 1.000772. How many bright fringes pass a fixed point as the tube fills?

48. Your personal stereo is in a dead spot caused by direct reception from an FM radio station at 89.5 MHz interfering with the signal reflecting off a wall behind you. How much farther from the wall should you move in order that the interference be fully constructive?

Sections 37-6 and 37-7 Single-Slit Diffraction and the Diffraction Limit

49. For what ratio of slit width to wavelength will the first minima of a single-slit diffraction pattern occur at $\pm 90°$?

50. Light with wavelength 633 nm is incident on a 2.5 μm wide slit. Find the angular width of the central peak in the diffraction pattern, taken as the angular separation between the first minima.

51. A beam of parallel rays from a 29-MHz citizen's band radio transmitter passes between two electrically conducting (hence opaque to radio waves) buildings located 45 m apart. What is the angular width of the beam when it emerges from between the buildings?

52. Use trial-and-error with a calculator, or a more sophisticated root-finding method, to verify the number 1.3916 in the exercise following Example 37-7.

53. Find the intensity as a fraction of the central peak intensity for the second secondary maximum in single-slit

diffraction, assuming the peak lies midway between the second and third minima.

54. A proposed "star wars" antimissile laser is to focus 2.8-μm wavelength infrared light to a 50-cm-diameter spot on a missile 2500 km distant. Find the minimum diameter for a concave mirror that can achieve this spot size, given the diffraction limit. Your answer suggests one of the many technical difficulties faced by antimissile defense systems.

55. The movie *Patriot Games* has a scene in which CIA agents use spy satellites to identify individuals in a terrorist camp (Fig. 37-44). Suppose that a minimum resolution for distinguishing human features is about 5 cm. If the spy satellite's optical system is diffraction limited, what diameter mirror or lens is needed to achieve this resolution from an altitude of 100 km? Assume a wavelength of 550 nm.

FIGURE 37-44 CIA agents using satellite imaging to identify terrorists in the film *Patriot Games*. How big a mirror or lens must the satellite's optical system have? (Problem 55)

56. Suppose one of the 10-m-diameter Keck telescopes in Hawaii could be trained on San Francisco, 3400 km away. Would it be possible to read (a) newspaper headlines or (b) a billboard sign at this distance? Justify your answers by giving the minimum separation resolvable with Keck at 3400 km, assuming 550-nm light. (c) Repeat for the case of the Keck optical interferometer, formed from the two 10-m Keck telescopes and several smaller ones, with an effective aperture of 50 m.

57. A camera has an $f/1.4$ lens, meaning that the ratio of focal length to lens diameter is 1.4. Find the smallest spot diameter (defined as the diameter of the first diffraction minimum) to which this lens can focus parallel light with 580-nm wavelength.

58. The distance from the center of a circular diffraction pattern to the first minimum on a screen 0.85 m distant from the diffracting aperture is 15,000 wavelengths. What is the aperture diameter?

59. While driving at night, your eyes' irises have dilated to 3.1-mm diameter. If your vision were diffraction limited, what would be the greatest distance at which you could see as distinct the two headlights of an oncoming car, which are spaced 1.5 m apart? Take $\lambda = 550$ nm.

60. Two stars are 4.0 light-years apart, in a galaxy 20×10^6 light-years from Earth. What minimum separation of two radio telescopes, acting together as an interferometer, is needed to resolve them? The telescopes operate at 2.1 GHz and are pointing straight upward; the stars are directly overhead.

61. Under the best conditions, atmospheric turbulence limits the resolution of ground-based telescopes to about 1 arc second (1/3600 of a degree) as shown in Fig. 37-45. For what aperture sizes is this limitation more severe than that of diffraction at 550 nm? Your answer shows why large ground-based telescopes do not produce better images than small ones, although they do gather more light.

FIGURE 37-45 Jupiter, photographed with a ground-based telescope (left) and from space (right). Atmospheric turbulence limits the ground-based image quality while diffraction limits the space-based image (Problem 61).

62. Two objects are separated by approximately one wavelength of the light with which an observer is attempting to resolve them. Show that the Rayleigh criterion then requires that the distance to the objects be less than the diameter of the observing aperture, and thus the Fraunhofer diffraction approximation is violated. It is in fact impossible to resolve two objects as close as one wavelength.

Paired Problems

(Both problems in a pair involve the same principles and techniques. If you can get the first problem, you should be able to solve the second one.)

63. Find the total number of lines in a 2.5-cm-wide diffraction grating whose third-order spectrum has the 656-nm hydrogen-α spectral line at an angular position of 37°.

64. Light is diffracted by a 5000 line/cm grating, and a detector sensitive only to visible light finds a maximum intensity at 28° from the central maximum. (a) What is the wavelength of the light? (b) In what order is it seen at the 28° position?

65. A 400 line/mm diffraction grating is 3.5 cm wide. Two spectral lines whose wavelengths average to 560 nm are just barely resolved in the 4th-order spectrum of this grating. What is the difference between their wavelengths?

66. What order is necessary to resolve wavelengths of 647.98 nm and 648.07 nm using a 4500-line grating?

67. A thin film of toluene ($n = 1.49$) floats on water. What is the minimum film thickness if the most strongly reflected light has wavelength 460 nm?

68. Oil with refractive index 1.38 forms a 210-nm-thick film on a piece of glass. What color of visible light is most strongly reflected?

69. What diameter optical telescope would be needed to resolve a Sun-sized star 10 light-years from Earth? Take $\lambda = 550$ nm. Your answer shows why stars appear as point sources in optical astronomy.

70. Could the 305-m radio telescope at Arecibo in Puerto Rico resolve the star in the preceding problem, if it were observing at a wavelength of 1.0 cm?

Supplementary Problems

71. White light shines on a 250-nm-thick layer of diamond ($n = 2.42$). What wavelength *of visible light* is most strongly reflected?

72. An air wedge like that of Fig. 37-43 displays 10,003 bright bands when illuminated from above. If the region between the plates is then evacuated, the number of bands drops to 10,000. Calculate the refractive index of the air.

73. In Fig. 37-23 the mth Newton's ring appears a distance r from the center of the lens. Show that the curvature radius of the lens is given approximately by $R = r^2/(m + \frac{1}{2})\lambda$, where the approximation holds when the thickness of the air space is much less than the curvature radius.

74. Given the result of the preceding problem, how many bright Newton's rings would be seen with a 2.5-cm-diameter glass lens with curvature radius 7.5 cm, when illuminated with 500-nm light?

75. How many rings would be seen if the system of the preceding problem were immersed in water ($n = 1.33$)?

76. A thin-walled glass tube of length L containing a gas of unknown refractive index is placed in one arm of a Michelson interferometer using light of wavelength λ. The tube is then evacuated, and during the process m bright fringes pass a fixed point as seen in the viewer. Find an expression for the refractive index of the gas.

77. The signal from a 103.9-MHz FM radio station reflects off a building 400 m away, effectively producing two sources of the same signal. You're driving at 60 km/h along a road parallel to a line between the station's antenna and the building and located a perpendicular distance of 6.5 km from them. How often does the signal appear to fade when you're driving roughly opposite the transmitter and building?

78. A satellite dish antenna 45 cm in diameter receives TV broadcasts at a frequency of about 12 GHz. (a) What is the angular size of its beam, defined as the full width of the central diffraction peak? (b) How many communications satellites could fit in geosynchronous orbit above Earth's equator if antennas like this one were not to pick up signals from more than one satellite? Your answer shows why geosynchronous orbit is such valuable "real estate."

79. The component of a star's velocity in the radial direction relative to Earth is to be measured using the doppler shift in the hydrogen-β spectral line, which appears at 486.1 nm when the source is stationary relative to the observer. What is the minimum speed that can be detected by observing in 1st order with a 10,000 line/cm grating 5.0 cm across? *Hint:* See Equation 17-9.

80. Light is incident on a diffraction grating at an angle α to the normal. Show that the condition for maximum light intensity becomes $d(\sin\theta \pm \sin\alpha) = m\lambda$.

81. In a double-slit experiment, a thin glass plate with refractive index 1.56 is placed over one of the slits. The fifth bright fringe now appears where the second dark fringe previously appeared. How thick is the plate if the incident light has wavelength 480 nm?

82. An arrangement known as Lloyd's mirror (Fig. 37-46) allows interference between a direct and a reflected beam from the same source. Find an expression for the separation of bright fringes on the screen, given the distances d and D in Fig. 37-46, and the wavelength λ of the light. *Hint:* Think of two sources, one the virtual image of the other.

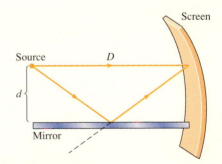

FIGURE 37-46 Lloyd's mirror (Problem 82).

PART 5 *Cumulative Problems*

These problems combine material from chapters throughout the entire part or, in addition, from chapters in earlier parts, or they present special challenges.

1. A *grism* is a grating ruled onto a prism, as shown in Fig. 1. The grism is designed to transmit undeviated one wavelength of the spectrum in a given order, as refraction in the prism compensates for the deviation at the grating. Find an equation relating the separation d of the grooves that constitute the grating, the wedge angle α of the prism, the refractive index n, the undeviated wavelength λ_0, and order m_0.

FIGURE 1 A grism (Cumulative Problem 1).

2. A double-slit system consists of two slits each of width a, with separation d between the slit centers ($d > a$). Light of intensity S_0 and wavelength λ is incident on the system, perpendicular to the plane containing the slits. Find an expression for the outgoing intensity as a function of angular position θ, taking into account both the slit width and the separation. Plot your result for the case $d = 4a$, and compare with Fig. 37-33.

3. A closed cylindrical tube whose glass walls have negligible thickness measures 5.0 cm long by 5.0 mm in diameter. It is filled with water, initially at 15°C, and placed with its long dimension in one arm of a Michelson interferometer. The water is not perfectly transparent, and it absorbs 3.2% of the light energy incident on it. The laser power incident on the water is 50 mW, and the wavelength is 633 nm. The refractive index of water in the vicinity of 15°C is given approximately by $n = 1.335 - 8.4 \times 10^{-5}T$, where T is the temperature in °C. As the water absorbs light energy, how long does it take the interference pattern to shift by one whole fringe?

4. A radio antenna broadcasts a 6.5-MHz signal in all directions. The antenna is on top of a tower 300 m above sea level, and the tower is located 2.0 km from the shore. An airplane is flying toward the tower, at twice the altitude of the tower top, as shown in Fig. 2. Radio waves from the tower reflect off the ocean surface with a 180° phase change, but there is no significant reflection off the land. As the plane heads toward the tower, at what horizontal distance or distances from the tower will its radio equipment detect a minimum in the radio signal amplitude due to destructive interference of the direct and reflected radio waves?

5. In one type of optical fiber, called a *graded-index fiber,* the refractive index varies in a way that results in light rays

FIGURE 2 Cumulative Problem 4 (figure is not to scale).

being guided along the fiber on curved trajectories, rather than undergoing abrupt reflections. Figure 3 shows a simple model that demonstrates this effect; it also describes the basic optical effect in mirages. A slab of transparent material has refractive index $n(y)$ that varies with position y perpendicular to the slab face. A light ray enters the slab at $x = 0$, $y = 0$, making an angle θ_0 with the normal just inside the slab. The refractive index at this point is $n(y = 0) = n_0$. (a) By writing $\sin\theta$ in Snell's law in terms of the components dx and dy of the ray path, show that that path (written in the form of x as a function of y) is given by

$$ x = \int_0^y \frac{n_0 \sin\theta_0}{\sqrt{[n(y)]^2 - n_0^2 \sin^2\theta_0}}\,dy . $$

(b) Suppose $n(y) = n_0(1 - ay)$, where $n_0 = 1.5$ and $a = 1.0$ mm^{-1}. If $\theta_0 = 60°$, find an explicit expression for x as a function of y, and plot your result to give the actual ray path. Explain the shape of your curve in terms of what happens when the ray reaches a point where $n(y) = n_0 \sin\theta_0$. What happens beyond this point?

FIGURE 3 Cumulative Problem 5.

Part Six

Modern Physics

A "quantum corral" of 48 iron atoms on a copper surface encloses circular ripples associated with the wave-like nature of electrons on the surface. The image was taken with a scanning-tunneling microscope, which utilizes the principles of quantum mechanics. The ideas of modern physics allow scientists to not only understand but also, increasingly, to image and even manipulate matter at the atomic scale.

How big is the universe? What are the fundamental constituents of matter? What holds atoms together, and why do they emit only certain colors of light? Is nature fundamentally predictable, or does uncertainty rule in the microscopic world? How do we know that the universe began in a "big bang"? How will it end? All these are questions for relativity and quantum physics—collectively called "modern physics" because they were developed after the year 1900. In Part 6 we give a brief account of Einstein's relativity, followed by a glimpse at quantum physics and its applications. We end with an overview of the latest developments in fundamental physics, from the nature of elementary particles to the origin and ultimate fate of the entire universe.

Readers of the extended version of this text will find an expanded Part 6 with much more on quantum physics and its applications.

38 Relativity

Albert Einstein's relativity revolutionized our understanding of space and time. Here Einstein explains his theory at a 1931 lecture in California.

By the last quarter of the nineteenth century, the basic laws of electromagnetism had been formulated. Maxwell's theory had demonstrated the electromagnetic nature of light, whose practical consequences we explored in the three preceding chapters. In the same era the work of Samuel Morse (telegraph), Alexander Graham Bell (telephone), Hertz and Marconi (radio), Thomas A. Edison (electric light, phonograph), and many others laid the foundation of electromagnetic technology. Yet at the same time the insights of Maxwell led to baffling questions and contradictions that shook the roots of physical understanding and even of common sense.

From the resolution of these contradictions arose relativity, a theory that radically altered the philosophical basis for our understanding of the physical world, and whose influence spilled over into all areas of twentieth-century thought. The theory of relativity stands as a monument of human intellect and imagination. It transcends the everyday world of common sense, and shows us a universe whose richness is almost beyond imagination. We'll approach relativity from a historical perspective, building on our understandings of electromagnetism through the work of Maxwell. From this, you'll gain a sense of the baffling nature of the questions that electromagnetism raised in the minds of nineteenth-century physicists, and how Einstein's resolution of those questions was at once profoundly bold and sweepingly simple.

38-1 SPEED *c* RELATIVE TO WHAT?

Maxwell's equations show that electromagnetic waves can exist, and that all such waves travel with speed c. Speed c relative to what? When we described the mechanics of a taut string in Chapter 16, we showed that waves could propagate along the string with a certain speed. A certain speed relative to what? Clearly, to the string. Similarly, sound waves propagate through the air with a certain speed relative to the air. If you move through the air, the speed of sound *relative to you* won't be the same as its speed relative to the air (Fig. 38-1).

The Ether Concept

Each type of wave has a characteristic speed *relative to the medium in which it propagates*. What about light? What is the medium through which light propagates? Light reaches us from the most distant galaxies, traveling through seemingly empty space. Yet all our other experience with waves suggests that there should be a medium—and that the speed c in Maxwell's equations should be the speed relative to that medium. Nineteenth-century scientists, supposing that light waves were like mechanical waves, believed that light propagated through a tenuous substance called the **ether** that permeated the entire universe. Electric and magnetic fields were visualized as stresses and strains in the ether. This

FIGURE 38-1 A plane moves at speed V through air, in which the sound speed is v_{sound}. waves approaching the plane from the front move at $v_{sound} + V$ relative to the plane, ose approaching from behind move at $v_{sound} - V$ relative to the plane.

mechanical view—that electromagnetic phenomena, including light, were disturbances of some substance—was deeply ingrained in the nineteenth-century scientists because of their previous experience that Newton's laws explained all known physical phenomena.

The ether had to have some unusual properties. It must offer no resistance to the motion of material bodies, or the planets would soon lose their energy and spiral into the Sun. At the same time the ether must be very stiff, for the speed of light is large. (If you make a spring stiffer, waves travel more quickly along it.) Indeed, the constants ε_0 and μ_0 must describe the mechanical properties of the ether that account for the high speed of light. These and other mechanical requirements make the ether a rather improbable substance, but without the ether we'd be left floundering for an answer to the question "speed c relative to what?"

The existence of electromagnetic waves traveling at speed c follows from Maxwell's equations. But this result could be true only in a frame of reference fixed with respect to the ether, for if we move relative to the ether then light should move at a speed other than c relative to us. Therefore Maxwell's equations—that is, our description of electromagnetism—were presumably correct only in the ether's frame of reference.

This situation put electromagnetism in a rather different position from mechanics. In mechanics, the concept of absolute motion is meaningless. You can eat your dinner, or throw a ball, or do any mechanical experiment as well on an airplane moving steadily at 1000 km/h as you can when the airplane is standing still on the ground. You need not take the uniform motion of the plane into account. This is the principle of **Galilean relativity**, which states that the laws of mechanics are valid in all inertial reference frames—i.e., frames of reference in uniform motion (see Section 3-5). But the laws of electromagnetism could be valid only in the ether's frame of reference, for it seemed that only in this frame could the prediction of electromagnetic waves moving at speed c be correct.

So the nineteenth-century physicists faced an unsatisfactory situation, in which the laws for one branch of physics (mechanics) seemed to work in all inertial frames, while those of another branch (electromagnetism) could not. But physicists had great faith in mechanical models and in the ether concept, for without ether the question "speed c relative to what?" seemed impossible to answer. Thus the late nineteenth century saw a flurry of activity as physicists sought to detect the ether. Ultimately they failed, paving the way for the new world view of relativity that, in Einstein's own words, "arose from necessity, from serious and deep contradictions in the old theory from which there seemed no escape."

38-2 MATTER, MOTION, AND THE ETHER

It was natural for nineteenth-century physicists to ask about Earth's motion relative to the ether. If Earth is moving through the ether, then the speed of light should be different in different directions, just as is the case for sound in Fig. 38-1. On the other hand, Earth might be at rest relative to the ether. Because other planets, stars, and galaxies move with respect to Earth, it's h

to imagine that ether is everywhere fixed with respect to Earth alone, for this violates the Copernican view that Earth does not occupy a privileged spot in the universe. But maybe Earth drags with it the ether in its immediate vicinity. If this "ether drag" occurs, then the speed of light must be independent of direction, but if ether drag does not occur then the speed of light measured on Earth must depend on direction. Through observation and experiment, nineteenth-century physicists sought to resolve the question of Earth's motion through the ether.

(a)

(b)

(c)

"Dragged" air

(d)

Aberration of Starlight

Imagine standing in a rainstorm with rain falling vertically. To keep dry, you hold your umbrella with its shaft straight up, as shown in Fig. 38-2a. But if you run, as in Fig. 38-2b, you'll keep driest if you tilt your umbrella forward. Why? Because then the direction of rainfall *relative to you* is not straight down but at an angle, as shown in Fig. 38-2c. This argument presupposes that you do not drag with you a large volume of air. If such an "air drag" occurred, raindrops entering the region around you would be accelerated quickly in the horizontal direction by the air moving with you, so they would now fall vertically relative to you, as in Fig. 38-2d. No matter which way you ran, as long as you dragged air with you, you would point your umbrella vertically upward to stay dry.

This umbrella example is exactly analogous to the observation of light from stars, with the rain being starlight and the umbrella a telescope. If Earth does not drag ether with it, the direction from which starlight comes will depend on the motion of Earth relative to the ether. But if "ether drag" occurs in analogy with Fig. 38-2d, then light from a particular star will always come from the same direction.

In fact we do observe a tiny change in the direction of starlight. As Earth swings around in its orbit, we must first point a telescope one way to see a particular star. Then, six months later, Earth's orbital motion is in exactly the opposite direction, and we must point the telescope in a slightly different direction. This phenomenon is called **aberration of starlight** and shows that *Earth does not drag the ether.*

The Michelson-Morley Experiment

If we reject the pre-Copernican notion that Earth alone is at rest relative to the ether, then aberration of starlight forces us to conclude that Earth moves through the ether. Furthermore, the relative velocity of the motion must change throughout the year, as Earth orbits the Sun.

In a series of experiments done in 1881–1887, the American scientists Albert A. Michelson and Edward W. Morley attempted to determine Earth's velocity

FIGURE 38-2 (a) Standing still in vertically falling rain, you hold your umbrella overhead to keep dry. (b) Running, you tilt the umbrella to compensate for the rain's motion relative to you. (c) The situation of (b) seen from the runner's frame of reference. (d) If you dragged a volume of air with you, you would hold the umbrella overhead whatever your state of motion.

FIGURE 38-3 Simplified diagram of a Michelson interferometer. An "ether wind" blowing in the direction shown should result in a longer time for the light beam on the horizontal arm.

FIGURE 38-4 Vector diagram showing resultant velocity **u** of light moving at right angles to an ether wind with speed v. The speed of light relative to the ether is c.

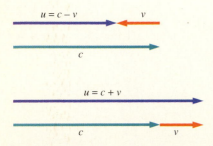

FIGURE 38-5 Vector diagrams showing the resultant velocities **u** for light moving with and against an ether wind whose speed is v.

relative to the ether. They used Michelson's interferometer (Fig. 38-3), whose operation we described in the preceding chapter. Recall that the interferometer produces a pattern of interference fringes that shifts if the round-trip travel time for light on one of its two perpendicular arms changes. The interference pattern reflects, among other things, possible differences in travel times that arise from differences in the speed of light in different directions—differences that should result from Earth's motion through the ether. Rotating the apparatus through 90° would interchange the directions of the arms and should therefore shift the interference pattern.

Now suppose that Earth moves at speed v relative to the ether. Then from the viewpoint of an observer on Earth, there is an "ether wind" blowing past Earth. Suppose that the Michelson-Morley apparatus is oriented with one light path parallel to the wind, the other perpendicular. Consider a light beam moving the distance L at right angles to the ether wind. The beam must be aimed slightly upwind, in order that it will actually move perpendicular to the wind. The light moves in this direction at speed c relative to the ether, but the ether wind sweeps it back so its path in the Michelson-Morley apparatus is at right angles to the wind. From Fig. 38-4, we see that its speed relative to the apparatus is

$$u = \sqrt{c^2 - v^2},$$

so the round-trip travel time is

$$t_{\text{perpendicular}} = \frac{2L}{u} = \frac{2L}{\sqrt{c^2 - v^2}}. \tag{38-1}$$

Light sent a distance L "upstream"—against the ether wind—travels at speed c relative to the ether but at speed $c - v$ relative to Earth. It therefore takes a time

$$t_{\text{upstream}} = \frac{L}{c - v}.$$

Returning, the light moves at $c + v$ relative to Earth, taking

$$t_{\text{downstream}} = \frac{L}{c + v}$$

(Fig. 38-5). So the round-trip time parallel to the wind is

$$t_{\text{parallel}} = \frac{L}{c - v} + \frac{L}{c + v} = \frac{2cL}{c^2 - v^2}. \tag{38-2}$$

The two round-trip travel times differ, with the trip parallel to the ether wind always taking longer (see Problems 1, 2, and 4). Light on the parallel trip is slowed when it moves against the ether wind, then speeds up when it moves with the wind. But slowing always dominates, because the light spends more time moving against the wind than with it.

The Michelson-Morley experiment of 1887 was sensitive enough to detect differences in the speed of light at least an order of magnitude smaller than Earth's orbital speed. The experiment was repeated with the apparatus oriented in different directions, and at different times throughout the year, and the same simple but striking result always emerged: there was never any difference in the travel times for the two light beams. In terms of the ether concept, the Michelson-Morley experiment showed that *Earth does not move relative to the ether.*

A Contradiction in Physics

Aberration of starlight shows that Earth does not drag ether with it. Earth must therefore move relative to the ether. But the Michelson-Morley experiment shows that it does not. This contradiction is a deep one, rooted in the fundamental laws of electromagnetism and in the analogy between mechanical waves and electromagnetic waves. The contradiction arises directly in trying to answer the simple question "with respect to what does light move at speed c?"

Physicists at the end of the nineteenth century made many ingenious attempts to resolve the dilemma of light and the ether, but their explanations were either inconsistent with experiment or lacked sound conceptual bases.

38-3 SPECIAL RELATIVITY

In 1905, at the age of 26, Albert Einstein (Fig. 38-6) proposed a theory that resolved the dilemma and at the same time altered the very foundation of physical thought. Einstein declared simply that the ether is a fiction. But then with respect to what does light move at speed c? With respect, Einstein declared, to anyone who cares to observe it. This statement is at once simple, radical, and conservative. Simple, because its meaning is clear and obvious. Anyone who measures the speed of light will get the value $c = 3.0 \times 10^8$ m/s. Radical, because it alters our commonsense notions of space and time. Conservative, because it asserts for electromagnetism what had long been true in mechanics: that the laws of physics do not depend on the motion of the observer. Einstein summarized his new ideas in the **special theory of relativity**, which is expressed in this simple sentence:

> The laws of physics are the same in all inertial reference frames.

Recall that inertial frames are those which are not accelerated—i.e., those in which the laws of *mechanics* were already valid. Einstein's statement encompasses *all* laws of physics, including mechanics and electromagnetism. The prediction that electromagnetic waves move at speed c must, then, be a universal prediction that holds in *all* inertial frames of reference. The *special* theory of relativity is special because it is valid only for the special case of inertial frames. Later we'll discuss the general theory of relativity, which removes this restriction.

Einstein's relativity readily explains the result of the Michelson-Morley experiment, for no matter what the speed of Earth relative to anything, an observer on Earth should measure the same speed for light in all directions. But at the same time, we will see that relativity flagrantly violates our common-sense notions of space and time.

38-4 SPACE AND TIME IN RELATIVITY

Consider a car driving past a pedestrian standing by the roadside (Fig. 38-7). Driver and pedestrian each measure the speed of the light from a blinking traffic signal. Relativity says they will get the same answer, $c = 3.0 \times 10^8$ m/s, even though the car is moving toward the source of light. How can this be? Consider how each observer might make the measurement. Let each be equipped with a

FIGURE 38-6 Albert Einstein was just 26 when he formulated the special theory of relativity.

FIGURE 38-7 As the car passes the pedestrian, a light pulse goes by. Both driver and pedestrian measure the same speed c for the light, even though they are in relative motion.

meter stick and an accurate, high-speed electronic stopwatch. Suppose that a light pulse passes the front ends of both meter sticks just as they coincide. Each observer measures the time it takes the light pulse to cross the meter stick, then divides the distance (one meter) by the measured time to get the speed of light. Since the stick on the car is moving toward the light source, common sense suggests that the light will pass the far end of this stick first, and therefore that the time on the car's stopwatch will be shorter. But this violates relativity, for if both observers use the same path length for their speed-of-light measurement, and if they get different times, then they will have measured different speeds for light. In fact, both stopwatches will read the same time, even though common sense tells us that the light passes the end of the "moving" meter stick "earlier."

How can this be? It follows logically from the statement of special relativity, which in turn is consistent with physical experiments. But how can it be? Something must be "wrong" with someone's meter stick or stopwatch. Maybe the motion of the car somehow affects the stopwatch on the car. But no: this suggestion violates the spirit of relativity, which says that steady, uniform motion is undetectable—that it makes no more sense to say that the car is moving and the pedestrian is at rest than to say the opposite. That's the whole point of relativity—any inertial reference frame is as good as any other for doing physics. So there can be nothing wrong with the clocks and meter sticks.

The only things left to go "wrong" are time and space. Time and space—the seemingly passive, universal backgrounds in which all physical events take place—must themselves depend on the observer. Two observers in different frames of reference, moving uniformly relative to each other, are measuring different quantities when they use clocks to record the passage of time and meter sticks to determine distances in space. In relativity it is the laws of physics—not measures of time and space—that must be the same for everyone. Time and space are altered in ways that allow the laws of physics, including Maxwell's theory and its prediction that light waves travel at speed c, to be the same for all observers in uniform motion.

In exploring space and time in relativity, we'll often speak of events. An **event** is an occurrence specified by giving its position—three spatial coordinates— and its time. Your birth, for example, is an event: it occurred at a certain place and a certain time, and both are needed to specify it fully.

Time Dilation

17.1
Relativity of Time

To understand how time gets altered, we'll consider the simple device shown in Fig. 38-8 (also simulated in ActivPhysics Activity 17.1), which we'll call a "light box." It consists of a box of length L with a light source at one end and a mirror at the other. A light flash leaves the source, reflects off the mirror, and returns to the source. We want to know the time interval between two distinct events: the light flash leaving the source and the flash returning to the source.

For concreteness, we'll imagine that the light box is in a spaceship moving past Earth at a uniform velocity. But don't think there's something special about space or spaceships going on here: the whole point of relativity is that all inertial frames are equivalent places for doing physics, so our spaceship is just one more inertial reference frame. We'll call that frame S'. There's also an accurate clock, C, in the spaceship, near the light source of the light box. Clock C reads zero just as the light flash is emitted.

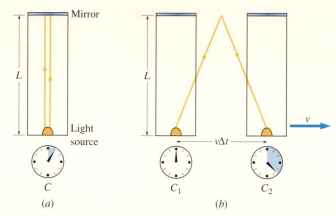

FIGURE 38-8 A "light box" provides a comparison of time intervals as measured in two different frames of reference. A light flash leaves the source, reflects off the mirror, and returns to the source. What's the time interval between emission and return of the flash? (a) In the reference frame in which the box is at rest—here the spaceship frame S'—the light travels a distance $2L$ at speed c. (b) In Earth's frame S, in which the box moves past with speed v, the light travels a greater distance. But its speed c is the same in *all* inertial frames, so the time interval is longer in this frame. Part (b) shows the box both when the light is emitted and again when it returns to the source.

In Fig. 38-8a we consider the light-box experiment viewed in the spaceship's frame of reference. Since the light box is at rest in this frame, the light travels a round-trip distance $2L$ from source to mirror and back, giving a round-trip travel time of

$$\Delta t' = \frac{2L}{c}.$$

Clearly, this is the time read on the spaceship's clock C.

Now consider the situation as viewed in Earth's frame of reference, which we'll call S. In this frame, spaceship and light clock are moving to the right with speed v, as shown in Fig. 38-8b. Suppose there are two clocks in Earth's frame, positioned so the light box passes clock C_1 just as the flash is emitted, and passes C_2 just as the flash returns to the source. The clocks are synchronized, and C_1 is set so it reads zero just as the light flash is emitted. What will C_2 read just as the light box passes it and the flash returns to the source? Whatever C_2 reads, that will be the time Δt between the emission and return of the light flash, as measured in the Earth frame S.

We can calculate the time interval in the Earth frame S, just as we did in the ship frame S', by figuring the total distance the light travels and dividing by its speed. Figure 38-8b—the situation viewed in the Earth frame S—shows that the box moves to the right a distance $v\Delta t$ in the time Δt between emission and return of the light flash. Meanwhile the light takes a diagonal path up to the mirror of the moving box, then back down to the source. The total length of this path is twice the diagonal from source to mirror, or, using the Pythagorean theorem, $2\sqrt{L^2 + (v\Delta t/2)^2}$. The time required for light to go this distance is just the distance divided by the speed of light, or

$$\Delta t = \frac{2\sqrt{L^2 + (v\Delta t/2)^2}}{c}. \tag{38-3}$$

Notice that we explicitly used relativity in writing Equation 38-3. We did not vectorially add the horizontal speed of the box to the vertical speed of light to get a new speed of light in frame S, for relativity says that the speed of light is the same in all frames of reference. Had we altered the speed, we would have had an increased path length in S, but an increased speed of light as well, and would have found that the time intervals in both frames were the same. But no! Relativity requires that we use the same speed c in both frames, even though the path lengths differ. That's why we get different answers for the time.

The unknown time Δt appears on both sides of Equation 38-3. Multiplying through by c and squaring gives

$$c^2(\Delta t)^2 = 4L^2 + v^2(\Delta t)^2.$$

We then solve for $(\Delta t)^2$ to get

$$(\Delta t)^2 = \frac{4L^2}{c^2 - v^2} = \frac{4L^2}{c^2}\left(\frac{1}{1 - v^2/c^2}\right).$$

Taking the square root of both sides, and noting that $2L/c$ is just the time $\Delta t'$ measured in the frame S' at rest with respect to the box, we have

$$\Delta t = \frac{\Delta t'}{\sqrt{1 - v^2/c^2}} \quad \text{or} \quad \Delta t' = \Delta t \sqrt{1 - v^2/c^2}. \tag{38-4}$$

Equation 38-4 describes the phenomenon of **time dilation**, in which the time interval between two events is always shortest in a frame of reference in which the two events occur at the same place. In our example, the two events are the emission and return of the light flash, and they occur at the same place in the spaceship frame S', namely at the bottom of the box, which is at rest relative to the spaceship. They don't occur at the same place in the Earth frame S because the box moves relative to this frame, so the bottom of the box isn't in the same place when the light flash returns as it was when the flash was emitted. Thus $\Delta t'$—the time interval measured in the spaceship frame S', where the events occur at the same place—is shorter than Δt, as you can see from Equation 38-4 and as illustrated in Fig. 38-9.

The shorter time measured by a single clock present at two events is called a **proper time**. Here the word "proper" doesn't mean "correct" or "right"— that would violate relativity, since a time measurement in any inertial frame has equal claim to being valid. Rather, "proper" is used in the sense of "proprietary," in that proper time is the time that *belongs* to this one particular clock. In the Earth frame S there is no single clock present at both the emission and return of the light flash, so the time measurement in this frame doesn't "belong" to any one clock.

Time dilation is sometimes characterized as saying that "moving clocks run slow," but this statement violates the spirit of relativity because it suggests that some frames are "really" moving and others not. The whole point of relativity is that all inertial frames are equally good for describing physical reality, so none can claim to be "at rest" while others are "moving." What the statement "moving clocks run slow" is trying to convey is what we've just seen: that the time interval between two events is shortest in a reference frame where the two

FIGURE 38-9 Clock C moves between clocks C_1 and C_2, which are at rest relative to each other and synchronized in their rest frame. Time between the event of C passing C_1 and the event of C passing C_2 is shorter in C's frame of reference.

occur at the same place. There's no significance whatever to our putting the light clock in a "moving" spaceship and comparing its time with Earth time. We could equally well have put the light box and its clock on Earth, and two separate clocks in the ship frame. Then Earth with its light box would be moving past the ship at speed v —so the Earth clock would measure the time $\Delta t'$ in Equation 38-4, and the two clocks in the ship frame would measure the longer time Δt. (That may sound like a contradiction, but it can't be because *there's nothing special about any inertial frame, including Earth's*. We'll return to this point in a few pages.)

We used a special "light box" to illustrate time dilation. But time dilation isn't something that happens only when we use light to determine time intervals. It's something that happens to time itself. Take away the light box in Fig. 38-8, and the clocks will show the same discrepancy. There's no use looking for a physical mechanism that slows things down. All manifestations of time—the oscillations of the quartz crystal in a digital watch, the swing of a pendulum clock, the period of vibration of atoms in an atomic clock, biological rhythms, and human lifetimes—all are affected in the same way.

EXAMPLE 38-1	*Star Trek*

A spaceship leaves Earth on a one-way trip that Earthbound observers judge will take 25 years. If the ship travels at $0.95c$ relative to Earth, how long is the journey as measured by a clock on board the ship?

Solution

The spaceship is like the light box we used in deriving Equation 38-4 for time dilation, in that in its frame the beginning and end of the journey occur at the same place—namely, on the ship. So the time interval on the ship clock is the $\Delta t'$ of Equation 38-4. For the observers on Earth, the beginning and end of the journey occur at different places, so their 25-year time interval is Δt. Applying Equation 38-4 then gives

$$\Delta t' = \Delta t \sqrt{1 - v^2/c^2} = (25\text{ y})(\sqrt{1 - (0.95c)^2/c^2})$$
$$= 7.8\text{ y}.$$

Thus at this high relative speed, the ship's time is considerably shorter. We'll soon explore what happens if the ship turns around and returns to Earth.

EXERCISE An extraterrestrial spacecraft whizzes through the solar system at $0.80c$. How long does it take to go the 8.3 light-minute distance from Earth to the Sun (a) according to an observer on Earth and (b) according to an alien aboard the ship?

Answers: (a) 10.4 min; (b) 6.23 min

Some problems similar to Example 38-1: 7, 8, 16

Why don't we notice time dilation as we travel about in our everyday lives? Because the factor v^2/c^2 in Equation 38-4 is so small for any velocities we have relative to Earth. Even in a jet airplane, we're moving at 1000 km/h or only about $10^{-6}\,c$. Then time in the airplane is different from that on Earth by only about 1 part in $(10^6)^2$, which amounts to a few milliseconds per century. This illustrates an important point: any results predicted by relativity should agree with our common sense, Newtonian ideas when relative velocities are small compared with the speed of light. Only at substantial fractions of c do relativistic effects become obvious. Since our intuitions and common sense are built on experience at low relative velocities, it's not surprising that effects at high velocities seem counter to common sense.

APPLICATION	*Mountains and Muons: Confirming Time Dilation*

Time dilation is dramatically obvious in experiments with sub-atomic particles moving, relative to us, at speeds near that of light. In a classic experiment, the "clocks" are the lifetimes of unstable particles called muons, which are created by the inter-action of cosmic rays with Earth's upper atmosphere. The experiment consists in counting the number of muons incident each hour on the top of a mountain—Mt. Washington in New Hampshire, about 2000 m above sea level. The measurement is then repeated at sea level (Fig. 38-10).

Using a detector that records only those muons moving at about $0.994c$ at the mountaintop altitude, the experiment shows an average of about 560 muons with this speed are incident on the mountaintop each hour. If the mountain weren't there, the muons would travel from the mountaintop altitude to sea level in a time given by

$$\Delta t = \frac{2000 \text{ m}}{(0.994)(3.0 \times 10^8 \text{ m/s})} = 6.7 \ \mu\text{s} .$$

The muon's decay rate is such that one should expect only about 25 of the original 560 muons to remain after this $6.7\text{-}\mu\text{s}$ interval, so that's approximately the number we might expect to detect each hour at sea level. More muons would survive a shorter time interval; in particular, we should expect 414 muons surviving $0.73 \ \mu\text{s}$ after passing the mountaintop altitude.

Why the value $0.73 \ \mu\text{s}$? Because the muons are moving at $0.994c$, and the $6.7\text{-}\mu\text{s}$ interval from mountaintop to sea level is measured in Earth's frame of reference—not in the muons' frame. In the muons' frame, time dilation should reduce that interval to

$$\Delta t' = \Delta t \sqrt{1 - v^2/c^2} = (6.7 \ \mu\text{s})(\sqrt{1 - 0.994^2})$$
$$= 0.73 \ \mu\text{s} .$$

FIGURE 38-10 The flux of high-speed muons is measured on a mountaintop and again at sea level. The muons decay at such a rate that only a small fraction of those incident at the mountaintop altitude should survive to reach sea level. But the number measured at sea level is much larger, showing the effect of time dilation on muon lifetimes.

Since the muons' decay is determined by *their* measure of time, we should therefore expect 414 muons surviving at sea level.

So what happens? When the experiment is done, muon counts of just over 400 per hour are observed at sea level. This is no subtle effect! The difference between 25 and 414 is dramatic. At $0.994c$, the nonrelativistic description is hopelessly inadequate. At this speed, the factor $\sqrt{1 - v^2/c^2}$ is one-ninth, and time dilation is obvious.

The Twin Paradox

The phenomenon of time dilation has startling consequences, for it allows us to travel into the future! The famous "twin paradox" illustrates this possibility. One of two twins boards a fast spaceship for a journey to a distant star (Fig. 38-11). The other stays behind on Earth. Imagine that there are clocks at Earth and star, like the two clocks in frame S of our light-box experiment (Fig. 38-12a). There is a clock on the spaceship, like the one clock in our light-box frame S'. When the ship arrives at the distant star, less time will have elapsed on the ship clock than on the Earth and star clocks (Fig. 38-12b). Now the ship turns around and comes home. Again, the situation is identical to our light-box example, so again

FIGURE 38-11 At departure, the twins are the same age.

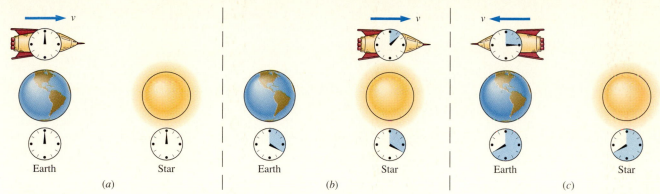

FIGURE 38-12 The traveling twin journeys to a distant star, then returns. Earth and star are at rest with respect to each other, and their clocks are synchronized. The figure is drawn from the Earth-star frame of reference. (*a*) Ship and Earth clocks agree as the ship leaves Earth. (*b*) When the ship reaches the star, less time has elapsed on its clock. (*c*) When the ship returns to Earth, its clock and Earth's no longer agree.

less time elapses on the ship clock, and the traveling twin arrives home younger than the earthbound twin (Fig. 38-13)! Depending on how far and how fast the traveling twin goes, the difference in ages could be arbitrarily large. The traveling twin could even return to Earth millions of years in the future, even though only hours had elapsed on the ship. But this is a one-way trip to the future! If the traveler doesn't like what he or she finds in the future, there's no going back!

FIGURE 38-13 It's not only clock readings, but all manifestations of time that differ. The traveling twin really has aged less!

EXAMPLE 38-2	*The Twin Paradox*

Earth and a star are 10 light-years (ly) apart, measured in a frame at rest with respect to Earth and star. Twin A boards a space ship and travels at $0.80c$ to the star, then immediately turns around and returns to Earth at $0.80c$. Twin B remains behind. Determine the round-trip travel time in the Earth-star frame of reference and in the ship frame. By how much will the twins' ages differ when they get back together?

Solution

At $0.80c$, the time to go 10 ly in the Earth-star frame is just

$$\Delta t = \frac{10 \text{ ly}}{0.80 \text{ ly/y}} = 12.5 \text{ y.}$$

The round-trip time is then 25 y. Equation 38-4 for $\Delta t'$ then gives the one-way travel time in the ship frame:

$$\Delta t' = \Delta t \sqrt{1 - v^2/c^2} = (12.5 \text{ y})(\sqrt{1 - 0.80^2}) = 7.5 \text{ y.}$$

Then the round-trip time in the ship frame is 15 y, so the twins' ages differ by 10 y when twin A returns.

Tip

Years, Light-Years, and the Speed of Light A light-year (ly) is the distance light travels in one year. By definition, therefore, the speed of light is 1 ly/y. It's often easiest in relativity to work in units where the speed of light is 1, whether those units be light-years and years, light-seconds and seconds, or whatever.

EXERCISE A spacecraft makes a round trip to a point a distance ℓ from Earth, as measured in Earth's frame of reference. The ship travels at 60% of the speed of light. If the round trip takes 1 hour by the ship's clock, (a) what is the distance ℓ and (b) how much time elapses on Earth between the ship's departure and its return?

Answers: (a) 0.375 light-hours (4×10^{11} m); (b) 1.25 h

Some problems similar to Example 38-2: 7, 9, 14, 15

This example of the twins is certainly unusual—but why is it a paradox? Consider the situation from the spaceship's reference frame. Doesn't the ship see Earth recede into the distance, turn around, and come back? And then shouldn't the earthbound twin be younger? This is the paradox. It's resolved by recognizing what's *special* about special relativity—namely that it applies only in inertial reference frames. The traveling twin must accelerate in order to return to Earth, and is therefore briefly in a noninertial reference frame. Although relativity precludes us saying that one twin is moving and the other is not, we can say that one twin's motion *changes* and the other's doesn't. This is obvious to the traveling twin, who experiences forces when the ship accelerates in turning around. But, of course, the earthbound twin doesn't feel anything unusual when the ship turns around. It isn't that all the unusual effects are crowded into the brief time during which the ship turns around; rather, it's that the ship on its round-trip journey occupies two distinct inertial reference frames, separated by that brief acceleration, while Earth remains in just one inertial frame. There's an asymmetry between the two twins, so there's no paradox.

Could it really happen? It could, and it has. Atomic clocks are now so accurate that experiments have been done to detect the minuscule time difference between a clock flown around the Earth and one left behind.

What if the traveling twin did not turn around? We could still argue that the ship clock runs slower than clocks on Earth and the star. But then isn't the situation symmetric? Couldn't the traveling twin argue that a clock on Earth should run slower than clocks on the ship? Yes—and we would find this to be true if we set up a series of clocks in the ship frame and measured time intervals on the Earth clock as Earth passed first one, then another of the ship clocks. But there's no contradiction. Unless the twins get together again, there is no way they can directly compare their clocks or their ages at the same place. Instead, they must compare one clock with a sequence of clocks that are all synchronized. And as we will soon see, clocks that are synchronized in one frame of reference are not synchronized in another frame! Only if the twins get back together can they compare just two clocks without having to worry about synchronization of distant clocks. And they can get back together only if at least one of them accelerates.

Got It!

Twins board their spaceships and head away from Earth in different directions, each traveling at the same high speed for the same distance. They turn around and head back, again at the same speed, and arrive at Earth together. How do their ages compare? Reconsider for the case where one twin travels twice as far, but goes twice as fast so that they still arrive home together.

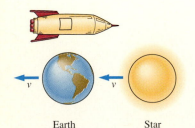

FIGURE 38-14 Earth-star trip viewed from the spaceship's frame of reference. Note that Earth-star distance is contracted, and the ship lengthened, relative to Fig. 38-12, which is drawn from the Earth-star frame.

Length Contraction

In Example 38-2 the spaceship moved 10 ly in 12.5 y at speed $0.80c$. These quantities are related by the simple expression $\Delta x = v\,\Delta t$, where Δx is the distance between Earth and star *measured in the Earth-star frame of reference* S. Now an observer in the ship frame of reference S' sees Earth and star moving past at speed v. First Earth passes the ship, then the star passes (Fig. 38-14). We found that the time interval between these two events, measured in the ship frame, is $\Delta t' = \Delta t \sqrt{1 - v^2/c^2} = 7.5$ y. Since Earth and star are moving past at $v = 0.80\,c$, the distance between Earth and star as measured in the ship frame must be

$$\Delta x' = v\,\Delta t' = v\,\Delta t \sqrt{1 - v^2/c^2} = \Delta x \sqrt{1 - v^2/c^2}, \qquad (38\text{-}5)$$

ActivPhysics
17.2
Relativity of Length

or 6.0 ly in our example. This equation shows that the distance between two points is always greatest in a frame (the Earth-star frame S, in this example) fixed with respect to those points. In any other frame the distance is smaller. This phenomenon of **length contraction** is also called **Lorentz contraction**, or Lorentz-Fitzgerald contraction, after the Dutch physicist H. A. Lorentz and the Irish physicist George F. Fitzgerald. In the 1890s, each independently proposed it as an ad hoc way of explaining the Michelson-Morley experiment. Only through Einstein's theory did the contraction acquire a solid conceptual basis.

Although we developed the Lorentz contraction using the distance between separate objects—Earth and a distant star—the effect occurs for any observer moving with respect to two fixed points. In particular, a rigid object like a meter stick or spaceship is shorter when measured by an observer with respect to whom it's moving. An object's length measured in a frame in which it's at rest is called its **proper length**; again, as with proper time, the word "proper" is used in the sense of being proprietary to that object. Proper length isn't the "right" or "real" length because a measurement made in any inertial frame has equal claim to be valid—and as length contraction shows, measurements made in different inertial frames yield different values for length.

As with time dilation, don't look for some physical mechanism that squashes moving objects. Rather, it is space itself that's different for different observers. In order to accept the simple fact that absolute motion is meaningless, we must alter our common-sense notions of time and space. Lorentz contraction and time dilation are manifestations of that alteration.

EXAMPLE 38-3	*A Stanford Electron*

At the Stanford Linear Accelerator Center (SLAC) (Fig. 38-15), subatomic particles are accelerated to high energies over a straight path whose proper length is 3.2 km. During a particular experiment, electrons are accelerated to 0.9999995 of the speed of light. In the SLAC frame, how long would it take electrons with this speed to travel the full length of the device? How long would the trip take in the rest frame of the electrons? How long would the accelerator be in the rest frame of the electrons?

Solution

The electron speed is so close to that of light that the travel time is, to a very good approximation,

$$\Delta t = \frac{\Delta x}{c} = \frac{3.2 \times 10^3 \text{ m}}{3.0 \times 10^8 \text{ m/s}} = 1.1 \times 10^{-5} \text{ s}.$$

FIGURE 38-15 The Stanford Linear Accelerator is 3.2 km (2 miles) long. But to electrons moving through it at 0.999 999 5c, the accelerator is only 3.2 m long.

In the rest frame of the electrons, the time to traverse the accelerator is given by Equation 38-4, or

$$\Delta t' = \Delta t \sqrt{1 - v^2/c^2} = (1.1 \times 10^{-5}\ \text{s})\sqrt{1 - 0.9999995^2}$$

$$= (1.1 \times 10^{-5}\ \text{s})(1.0 \times 10^{-3}) = 1.1 \times 10^{-8}\ \text{s} .$$

Our time calculation shows that the relativistic factor $\sqrt{1 - v^2/c^2}$ is 10^{-3}, so in the electron frame of reference the accelerator length is

$$\Delta x' = \Delta x \sqrt{1 - v^2/c^2} = 3.2 \times 10^{-3}\ \text{km} = 3.2\ \text{m} .$$

EXAMPLE 38-4	*A Stanford Student*

A physics student from New York flies to San Francisco to do an experiment at SLAC. She travels a distance of 4800 km on a plane going at 1000 km/h. How long does the trip take in a frame at rest with respect to Earth? How long does it take according to the student's watch? How far is it from New York to San Francisco in the airplane's frame of reference?

Solution

At 1000 km/h, the 4800-km trip takes

$$\Delta t = \frac{\Delta x}{v} = \frac{4800\ \text{km}}{1000\ \text{km/h}} = 4.8\ \text{hours} .$$

In the frame of the moving airplane, time and distance are altered by the relativistic factor $\sqrt{1 - v^2/c^2}$. The speed of the plane is 1000 km/h, or 278 m/s, so

$$\sqrt{1 - v^2/c^2} = \left[1 - \frac{(278\ \text{m/s})^2}{(3.0 \times 10^8\ \text{m/s})^2} \right]^{1/2}$$

$$\simeq 1 - (\tfrac{1}{2})(8.6 \times 10^{-13}) = 0.99999999999957 .$$

Here we used the binomial theorem, $(1 + x)^n \simeq 1 + nx$ for $|nx| \ll 1$ because most calculators don't carry enough significant figures to distinguish the result from 1 (see Appendix A). We really need not carry the calculation further. The time on the student's watch is the same as the time on the ground to about five parts in 10^{13}, and the distance in the plane's frame is the same as the ground distance to within this same factor. The student need take no account of time dilation and Lorentz contraction, except in a physics exam or high-energy experiment!

EXERCISE Our Milky Way galaxy is about 100,000 light-years in diameter. What is its diameter as measured by an intergalactic spacecraft traveling at 0.96c?

Answer: 28,000 ly

Some problems similar to Examples 38-3 and 38-4: 6, 11, 12

Examples 38-3 and 38-4 show that relativistic effects are significant only at high relative velocities—so high that the quantity v^2/c^2 is comparable with 1. In our daily lives we have no experience with such velocities. It is for this reason that relativity so offends our common-sense notions of space and time. Those notions are built on a groundwork of limited experience that does not include high velocities. If we did move regularly with respect to our surroundings at speeds near that of light, the relativity of space and time would be as obvious as our common-sense notions now seem.

For physicists working with high-energy elementary particles, relativistic effects *are* obvious. Unstable particles moving through the laboratory at high speeds live longer than they would at rest in the lab. And high-energy particle accelerators would not work if their design did not take relativity into account.

(a)

(b)

FIGURE 38-16 (a) In frame *S*, both sticks have the same speed *v* and both are contracted by the same amount. (b) Therefore their ends coincide at the same time.

Events and Simultaneity

Consider two identical sticks *A* and *B*, each of proper length *L*. Suppose these sticks are moving toward each other. For a frame of reference *S* in which both sticks are moving at the same speed *v*, the situation is shown in Fig. 38-16. Both sticks are Lorentz contracted, but since both are moving at the same speed *v* relative to the frame *S*, both are contracted by the same amount and therefore have the same length. What happens as the sticks pass each other? First, the

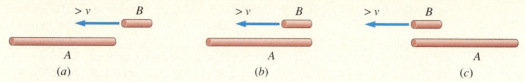

FIGURE 38-17 The passing sticks viewed in a frame S' at rest with respect to stick A. Relative speed is greater than the speed of either stick with respect to frame S. (a) First the left end of stick B passes the right end of A. (b) A while later, the right end of B passes the right end of A. This is event E_1. (c) Still later, the left end of B passes the left end of A. This is event E_2. Events E_1 and E_2 are not simultaneous in frame S'.

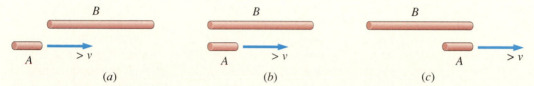

FIGURE 38-18 The passing sticks viewed in a frame of reference S'' at rest with respect to stick B. (a) First the right end of stick A passes the left end of B. (b) A while later, the left end of A passes the left end of B. This is event E_2. (c) Still later, the right end of A passes the right end of B. This is event E_1. The events E_1 and E_2 are not simultaneous in S'', and their time order is opposite what it was in S'.

right end of stick A and the left end of stick B pass (Fig. 38-16a). A little while later, the right end of A passes the right end of B. At the same time, *because the sticks have the same length in S*, the left end of A passes the left end of B (Fig. 38-16b). The passing of the two right ends of the sticks is an event that we designate E_1. Similarly, the passing of the two left ends we designate E_2. We have shown that, in the frame S, the two events E_1 and E_2 are **simultaneous**—they occur at the same time.

Now look at the situation from a frame of reference S' in which stick A is at rest. In this frame, stick B moves toward stick A. Since we are in S', we are at rest relative to stick A, and it has its proper length L. But stick B is contracted more than in frame S because of its higher relative velocity. The situation is shown in Fig. 38-17. As the figure indicates, the event E_1 occurs before E_2; the two events are not simultaneous in the frame S'. What happens in a frame S'' at rest with stick B? As Fig. 38-18 shows, the events E_1 and E_2 are again not simultaneous, and this time event E_2 occurs first.

Isn't this all just an illusion arising from the apparent length differences due to motion of the sticks? Isn't the picture in frame S (Fig. 38-16) "really" the right one? No! Relativity assures us that all inertial frames—including the frames S, S', and S'' of Figs. 38-16 through 38-18—are equally valid for describing physical reality. The length differences and the changes in time ordering of events E_1 and E_2 are not "apparent" and they are not "illusions." They arise from valid descriptions in different frames of reference, and each has equal claim to "reality." If you insist that one of the frames—say S—somehow has more validity, then you're reasserting the nineteenth-century notion that there is one favored reference frame in which alone the laws of physics are valid.

But how can observers disagree on the time order of events? Doesn't that violate causality? After all, if one event is a cause of another, we certainly expect the cause to precede the effect. It would be disturbing if some observer, with valid claim on "reality," found that cause and effect occurred in the reverse order. But there's no violation of causality. As we will soon show, the only events that can have their time order reversed are those that are so far apart in space, and so close in time, that not even light can travel fast enough to be at both events. There is no way that such events can influence each other, and therefore they can't be causally related. In a very real sense it doesn't matter which event occurs first, and indeed different observers will disagree on their relative time order. For example, an event on Earth now and another occurring five minutes from now on the Sun can't be causally related, since it takes light from the Sun eight minutes to reach Earth. For observers moving at high enough speeds relative to Earth and Sun, the solar event occurs first.

Only when events are close enough in space and separated enough in time so that light can travel from one to the other can the two be causally related. In that case, all observers will agree about their time order, although they may disagree about the actual time interval between the events. For example, an event on Earth now and another occurring fifteen minutes from now on the Sun could be causally related, and therefore all observers will agree that the terrestrial event occurs first. We'll explore these notions more quantitatively in the next section.

The Lorentz Transformations

Our demonstration that the time order of events may be relative deals implicitly with the coordinates—position and time—of specific events, and suggests that those coordinates may differ for different observers. Similarly, time dilation and length contraction arise as specific instances of the way positions and times in one frame of reference are related to their values in another frame. We now seek more general expressions—called **Lorentz transformations**—relating the space and time coordinates of an event in two frames of reference in relative motion. Consider coordinate axes in a frame of reference S and in another frame S' moving in the x direction with speed v relative to S. Suppose that the origins of the two coordinate systems coincide at time $t = t' = 0$. If an event has coordinates x, y, z, and t in S, what are its coordinates x', y', z', and t' in S'? Were it not for relativity, we would expect the coordinates y, z, and the time t to remain unchanged from one coordinate system to the other. For an event occurring at time $t = 0$, when the two origins coincide, we would expect the x coordinates to be the same too. But as S' moves in the positive x direction relative to S, a given x value in S would correspond to a value $x' = x - vt$ in S' (Fig. 38-19).

FIGURE 38-19 Nonrelativistic picture of two coordinate axes in relative motion. The x axis of frame S' moves to the right with speed v relative to S. At time $t = 0$ the two axes coincided. At a later time the x coordinates are related by $x' = x - vt$. Here $v = 1$ m/s, and the figure shows the two axes at $t = 2$ s.

How does relativity alter this coordinate transformation? Because of length contraction, we expect that the simple expression $x' = x - vt$ will need modification to be consistent with relativity. Any new expression we develop, however, must reduce to the nonrelativistic expression $x' = x - vt$ in the limit when $v \ll c$. A simple form that has this property is

$$x' = \gamma(x - vt), \qquad (38\text{-}6)$$

where γ depends on the relative speed v. We could also transform the other way. The only difference is the direction of relative motion; frame S' is moving in the positive x direction relative to S, while S is moving in the negative x direction relative to S'. Therefore, the transformation from x' and t' to x should look like Equation 38-6 except with v replaced by $-v$:

$$x = \gamma(x' + vt'). \qquad (38\text{-}7)$$

To see if we can make Equations 38-6 and 38-7 consistent with relativity, we impose the requirement that the speed of light be the same in both frames of reference. Suppose that a light flash goes off at the origin $x = 0$ at time $t = 0$. Since the origins of our frames coincide at this time, and since clocks at the origin in S and S' both read zero when the origins coincide, the light flash also occurs at $x' = 0$ and $t' = 0$ in frame S'. Call this event—the emission of the light flash—event E_1. Some time later, an observer at some position x in S observes the light flash. Call this event E_2. When does E_2 occur? Since light travels at speed c, we must have $x = ct$. Now in frame S' event E_2 has some coordinates x' and t'. But relativity requires that the speed of light in S' also be c, so we must have $x' = ct'$. Substituting $x = ct$ and $x' = ct'$ into Equations 38-6 and 38-7 gives

$$ct' = \gamma(ct - vt) = \gamma t(c - v)$$

and

$$ct = \gamma(ct' + vt') = \gamma t'(c + v).$$

Multiplying together the left-hand sides of these equations, and then the right-hand sides, and equating the results gives

$$c^2 t't = \gamma^2 tt'(c - v)(c + v),$$

so

$$c^2 = \gamma^2(c^2 - v^2),$$

or

$$\gamma = \frac{1}{\sqrt{1 - v^2/c^2}}. \qquad (38\text{-}8)$$

That we could find a value of γ depending on the relative velocity but not on the coordinates shows that our guess for the form of the transformation equations was correct. Equations 38-6 and 38-7, with γ given by Equation 38-8, are the relativistically correct transformations for the coordinates x and x'. Taking $v \ll c$ in Equation 38-8 shows that $\gamma \to 1$ in this limit, so our transformation equations correctly reduce to the nonrelativistic result at low relative velocities.

What about the y and z coordinates? They're measured at right angles to the relative motion, so there's no length contraction involved, and thus we have

FIGURE 38-20 As measured in frame S, lengths in S' are contracted and the clocks in S' are not synchronized.

simply $y' = y$ and $z' = z$. But the fourth coordinate—time—*does* transform. You can find the time transformation by solving Equations 38-6 and 38-7 simultaneously for t' and t (see Problem 25). The result is

$$t' = \gamma\left(t - \frac{vx}{c^2}\right) \tag{38-9}$$

and

$$t = \gamma\left(t' + \frac{vx'}{c^2}\right). \tag{38-10}$$

Because we know about time dilation, we should not be too surprised to find that measures of time differ between the two frames of reference. But why should the time in one frame depend not only on the time in the other frame but also on location in space? Because, as we have found, events that are simultaneous in one frame of reference are not simultaneous in a frame of reference moving relative to the first. In this case, our simultaneous events are the pointing of all clock hands to the same time. For the clocks in S', these events are simultaneous in S'. But they are not simultaneous in S. In fact, as Equation 38-9 shows, clocks that are farther to the right in S' read successively earlier times (Fig. 38-20). The term vx/c^2 in Equation 38-9 and its analog in Equation 38-10 account for this nonsynchronism of clocks in one frame as measured from the other frame.

Our earlier qualitative discussion of simultaneity can be made quantitative using the Lorentz transformations, as Example 38-5 and Problems 22 to 24 illustrate. Similarly, applying the Lorentz transformations to the coordinates describing the emission and return of the light flash in our light-box example results in a derivation of time dilation (see Problem 26). Table 38-1 summarizes the Lorentz transformations between coordinates in frames S and S', where S' is moving at speed v in the positive x direction relative to S.

TABLE 38-1 *The Lorentz Transformations*

S TO S′	**S′ TO S**
$y' = y$	$y = y'$
$z' = z$	$z = z'$
$x' = \gamma(x - vt)$	$x = \gamma(x' + vt')$
$t' = \gamma(t - vx/c^2)$	$t = \gamma(t' + vx'/c^2)$

$$\text{where } \gamma = \frac{1}{\sqrt{1 - v^2/c^2}}$$

EXAMPLE 38-5	*Galactic Fireworks*

Solution

Our Milky Way galaxy and the Andromeda galaxy are approximately at rest with respect to each other and are 2.0×10^6 light-years apart. At time $t = 0$ in the reference frame of these two galaxies, supernova explosions occur in both galaxies. Are these explosions simultaneous to the pilot of a spacecraft traveling at $0.80c$ from the Milky Way toward Andromeda? If not, what is the time interval between them, as measured in the spacecraft frame? Find also the spatial interval between the two explosions in the spacecraft frame.

The supernova explosions constitute two distinct events, and we're interested in their coordinates in the spaceship's frame of reference. Let the origin of the galaxy frame S be at the supernova in the Milky Way, and let the x axes of the galaxy frame S and spacecraft frame S' lie on the line connecting the two supernovae. Let the two frames coincide at time $t = t' = 0$. Then the space and time coordinates in S of the two supernova explosions are $x_1 = 0$, $t_1 = 0$, $x_2 = 2.0$ Mly, $t_2 = 0$. Similarly, the coordinates of the Milky Way explosion in the spacecraft frame S' are $x_1' = 0$, $t_1' = 0$. We seek the coordinates x_2' and t_2' of the Andromeda explosion in S'.

Referring to Table 38-1, we first calculate the relativistic factor γ:

$$\gamma = \frac{1}{\sqrt{1 - v^2/c^2}} = \frac{1}{\sqrt{1 - 0.80^2}} = 1.67 .$$

Then using the Lorentz transformations, we have

$$t_2' = \gamma(t_2 - vx_2/c^2)$$

$$= (1.67)\left[0 \text{ y} - \frac{(0.80 \text{ ly/y})(2.0 \text{ Mly})}{(1 \text{ ly/y})^2} \right] = -2.7 \text{ My} .$$

and $x_2' = \gamma(x_2 - vt_2) = (1.67)[2.0 \text{ Mly} - (0.80 \text{ ly/y})(0 \text{ ly})]$

$$= 3.3 \text{ Mly}.$$

Do these results make sense? In the spacecraft frame, the Andromeda supernova occurs nearly three million years before the Milky Way supernova! (The minus sign tells us that t_2' is earlier than the time $t_1' = 0$ of the Milky Way supernova.) Here is an example of events that are simultaneous in one frame but not in another. To make matters worse, consider an observer moving at $0.80c$ from Andromeda toward the Milky Way. For an observer in this frame, we reverse the sign of v in the transformation equations, obtaining a time of $+2.7$ million years. For this observer, the Milky Way supernova occurs first! How can this be? There's no contradiction, and no violation of cause and effect. In the spaceship's frame S' the two supernova events occur 3.3 Mly apart in space but only 2.7 My apart in time. Light from the "earlier" event cannot travel to the "later" event, so there can't be any causal influence between the two. It really doesn't matter which occurs first, and indeed which does depends on the observer.

Had we considered supernova events occurring not simultaneously but a long time apart— longer than 2 million years—in the galaxy frame of reference, we would find that all observers would agree on the time order of the two events, although not necessarily on the actual value of the time interval (see Problems 22 to 24).

EXERCISE Coordinate system S' is moving along the x axis of system S at $0.90c$. The two origins coincide when clocks in both systems read noon. An event occurs at $x = 5.0$ light-hours, $t = 3$ P.M. in frame S. Find the position and time of this event in S'.

Answers: 5.28 lh, 8:34 A.M.

Some problems similar to Example 38-5: 21–23, 53, 54

Relativistic Velocity Addition

If you're in an airplane moving at 1000 km/h relative to the ground and you walk toward the front of the plane at 5 km/h, common sense suggests that you move at 1005 km/h relative to the ground. But relativity implies that measures of time and distance vary among frames of reference in relative motion. For this reason the velocity of an object with respect to one frame does not simply add to the relative velocity between frames to give the object's velocity with respect to another frame. In the airplane your speed with respect to the ground is actually a little less than 1005 km/h as you stroll down the aisle of the plane, though the difference is insignificant at such a low speed.

The correct expression for **relativistic velocity addition** follows from the Lorentz transformations. Consider a frame of reference S and another frame S' moving in the positive x direction with speed v relative to S. Let their origins coincide at time $t = t' = 0$, so the Lorentz transformations of Table 38-1 apply.

Suppose an object moves with velocity u' along the x' axis in S'. In our airplane example, S' would be the airplane frame of reference, u' the velocity at which you walk through the plane, and v the velocity of the plane relative to the ground, or S frame. We seek the velocity u of the object relative to the frame S (that is, your velocity relative to the ground as you walk in the plane).

In either frame, velocity is the ratio of change in position to change in time, or

$$u = \frac{\Delta x}{\Delta t}.$$

Designating the beginning of the interval Δt by the subscript 1 and the end by 2, we can use Equations 38-7 and 38-10 to write

$$\Delta x = x_2 - x_1 = \gamma[(x_2' - x_1') + v(t_2' - t_1')] = \gamma(\Delta x' + v \Delta t')$$

and

$$\Delta t = t_2 - t_1 = \gamma[(t_2' - t_1') + v(x_2' - x_1')/c^2] = \gamma(\Delta t' + v \Delta x'/c^2).$$

Forming the ratio of these quantities, we have

$$\frac{\Delta x}{\Delta t} = \frac{\Delta x' + v \Delta t'}{\Delta t' + v \Delta x'/c^2} = \frac{(\Delta x'/\Delta t') + v}{1 + v(\Delta x'/\Delta t')/c^2}.$$

But $\Delta x'/\Delta t'$ is the velocity u' of the object in frame S', and $\Delta x/\Delta t$ is the velocity u, so

$$u = \frac{u' + v}{1 + u'v/c^2}. \tag{38-11}$$

The numerator of this expression is just what we would expect from common sense. But this simple sum of two velocities is altered by the second term in the denominator, which is significant only when both the object's velocity u' and the relative velocity v between frames are comparable with c. Solving Equation 38-11 for u' in terms of u, v, and c gives the inverse transformation:

$$u' = \frac{u - v}{1 - uv/c^2}. \tag{38-12}$$

EXAMPLE 38-6	*Collision Course!*

Two spacecraft approach Earth from opposite directions, each moving at $0.80c$ relative to Earth, as shown in Fig. 38-21. How fast do the spacecraft move relative to each other?

Solution

Call the Earth frame of reference S', and let S be the frame of spacecraft A. Then S' is moving at speed $v = 0.80c$ relative to S, while spacecraft B is moving at $u' = 0.80c$ relative to S'. Then the velocity of B relative to A is given by Equation 38-11:

$$u = \frac{u' + v}{1 + u'v/c^2} = \frac{0.80c + 0.80c}{1 + (0.80c)(0.80c)/c^2} = \frac{1.6c}{1.64} = 0.98c.$$

The relative speed remains less than the speed of light. This result is quite general: Equations 38-11 and 38-12 imply that as long as an object moves at a speed $v < c$ relative to some frame of reference, its speed relative to any other frame of reference will also be less than c (see Problem 20).

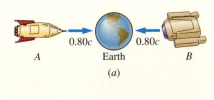

0.80c   0.80c

A   Earth   B

(a)

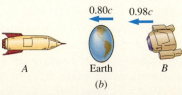

0.80c   0.98c

A   Earth   B

(b)

FIGURE 38-21 (a) Two spaceships approaching Earth at $0.80c$. (b) The situation in the frame of reference of the left-hand spaceship. Note changes in lengths of the ships and Earth.

EXAMPLE 38-7	*Catch That Light*

A light wave moves past Earth at the speed of light c. You try to chase the light wave by hopping on a fast spacecraft, moving at $0.95c$ relative to Earth. What is the speed of the light relative to the spacecraft?

Solution

Call the Earth frame S, so $u = c$, and the spacecraft frame S', so $v = 0.95c$. Then u', the speed of light relative to the spacecraft, is given by Equation 38-12:

$$u' = \frac{u - v}{1 - uv/c^2} = \frac{c - 0.95c}{1 - 0.95c^2/c^2} = \frac{0.05c}{0.05} = c.$$

We really didn't need to calculate this result, since a fundamental premise of relativity is that the speed of light is the same for all observers. The equations of relativistic velocity addition reflect this basic fact. No matter what the relative velocity v between two frames, light moving at c in one frame moves at c in any other frame. You cannot even begin to catch up with light!

EXERCISE A spacecraft whizzes by Planet X at $0.75c$, in excess of the Galactic Federation speed limit. A space cop takes off from Planet X at $0.90c$ relative to the planet. Find the relative speed of cop and speeder.

Answer: $0.46c$

Some problems similar to Examples 38-6 and 38-7: 17–19

Got It!

You're driving down the highway, and your speedometer reads exactly 30 km/h. A car passes you, going in the same direction at exactly 20 km/h relative to you. Does its speedometer—which measures the car's speed relative to the road—read more or less than 50 km/h?

38-5 ENERGY AND MOMENTUM IN RELATIVITY

Conservation of momentum and conservation of energy are cornerstones of Newtonian mechanics, where they hold in any inertial—i.e., uniformly moving—frame of reference. But both the momentum and energy of a particle are functions of its velocity, and we've just seen that relativity alters the Newtonian picture of how velocities transform from one reference frame to another. How, then, can momentum and energy be conserved in all frames of reference?

Momentum

Conservation of momentum is a cornerstone of mechanics, and, for consistency with relativity, we should hope that momentum conservation would hold in all inertial reference frames. In Newtonian mechanics the momentum of a particle with mass m and velocity **u** is m**u**. (Here we use **u** for particle velocities, reserving **v** for the relative velocity between two reference frames.) But if a system's momentum—the sum of its individual particles' momenta m**u**—is conserved in one frame of reference, then relativistic velocity addition suggests that it won't be conserved in another. The problem here lies not with momentum conservation but with the Newtonian expression for momentum. The expression m**u** is actually an approximation valid only for speeds u much less than c. The measure of momentum valid at any speed is

$$\mathbf{p} = \frac{m\mathbf{u}}{\sqrt{1 - u^2/c^2}} = \gamma m\mathbf{u}, \tag{38-13}$$

FIGURE 38-22 The ratio of the relativistic momentum to the Newtonian expression mu. Curve is a plot based on Equation 38-13, while crosses and circles represent measurements from three different experiments done with electrons. The relativistic momentum goes asymptotically to infinity as $u \rightarrow c$.

where γ is the familiar relativistic factor. The momentum given in expression 38-13 is conserved in all reference frames, and at low velocities it reduces to the Newtonian expression $\mathbf{p} = m\mathbf{u}$.

As $u \rightarrow c$ the factor γ grows arbitrarily large, and so does the relativistic momentum (Fig. 38-22). Since force is the rate of change of momentum, that means a very large force is required to produce even the slightest change in the velocity of a rapidly moving particle. This effect shows one answer to a common question about relativity: Why is it impossible to accelerate an object to the speed of light? The answer is that the object's momentum would approach infinity, and no matter how close to c it was moving it would still require infinite force to give it the last bit of speed needed to reach c.

EXAMPLE 38-8	Pushing Toward c

Compare the momentum changes required to accelerate an electron (a) from rest to $0.01c$ and (b) from $0.98c$ to $0.99c$.

Solution
The momentum at each speed is given by Equation 38-13:

$$p(0) = \frac{mu}{\sqrt{1 - u^2/c^2}} = 0,$$

$$p(0.01c) = \frac{m(0.01c)}{\sqrt{1 - 0.01^2}} = 0.010 \, mc,$$

$$p(0.98c) = \frac{m(0.98c)}{\sqrt{1 - 0.98^2}} = 4.92 \, mc,$$

and

$$p(0.99c) = \frac{m(0.99c)}{\sqrt{1 - 0.99^2}} = 7.02 \, mc.$$

In both cases the speed has increased by the same $0.01c$, but the momentum changes are very different: Going from rest to $0.01c$ gives $\Delta p = 0.010mc$, while from $0.98c$ to $0.99c$ the momentum change is $\Delta p = (7.02mc - 4.92mc) = 2.1mc$—a factor of 210 greater than the slow-speed case. To accomplish the second velocity change in the same time would therefore require a force 210 times greater. Note, incidentally, that we would have had sufficient accuracy using the Newtonian expression $p = mu$ in the low-speed case.

EXERCISE An electron is moving at $0.35c$. If its speed is doubled, by what factor does its momentum increase?

Answer: 2.62

Some problems similar to Example 38-8: 28, 30–32

Energy and Mass

The most widely known result of relativity is the famous equation $E = mc^2$. Here we develop a general expression for relativistic energy and show just what $E = mc^2$ means. In the process we'll gain new insights into energy, momentum, and mass.

In Chapter 7 we derived the work-energy theorem, and thus developed the concept of kinetic energy, by calculating the work needed to accelerate a mass m from rest to some final speed. We did that by integrating force over distance, using Newton's law $F = dp/dt$ to write force in terms of momentum. We'll do the same thing here, now using our relativistic expression for momentum. So we have

$$\frac{dp}{dt} = \frac{d}{dt}\left[\frac{mu}{\sqrt{1 - u^2/c^2}}\right] = \frac{m(du/dt)}{(1 - u^2/c^2)^{3/2}}.$$

Then the kinetic energy gained as a particle accelerates from rest in our reference frame to a final speed u is

$$K = \int \frac{dp}{dt} dx = \int \frac{dp}{dt} u \, dt = \int \frac{m(du/dt)}{(1 - u^2/c^2)^{3/2}} u \, dt = \int_0^u \frac{mu}{(1 - u^2/c^2)^{3/2}} du \, ,$$

where we used $u = dx/dt$ to replace dx with $u \, dt$. The integral is readily evaluated using the fact that $u \, du = \frac{1}{2} d(u^2)$, giving

$$K = \frac{mc^2}{\sqrt{1 - u^2/c^2}} - mc^2 = \gamma mc^2 - mc^2 \qquad (38\text{-}14)$$

for a particle's kinetic energy in a reference frame where the particle has speed u. Once again, γ here is the relativistic factor $1/\sqrt{1 - u^2/c^2}$.

For speeds low compared with c, Equation 38-14 should reduce to the familiar Newtonian expression $K = \frac{1}{2}mu^2$. And it does, as you can show in Problem 43. But Equation 38-14 suggests, more generally, that kinetic energy is the difference of two energies—the velocity-dependent quantity γmc^2 and the term mc^2 that depends only on mass but not velocity. Pursuing this interpretation, we identify γmc^2 as the particle's **total energy** and mc^2 as its **rest energy**. Rearranging Equation 38-14 lets us write the total energy as the sum of the kinetic energy and the rest energy:

$$E = K + mc^2$$

or, more simply

$$E = \gamma mc^2 = \frac{mc^2}{\sqrt{1 - u^2/c^2}} \quad \text{(total energy)} . \qquad (38\text{-}15)$$

What does all this mean? Put $u = 0$ in Equation 38-15 and $E = mc^2$—showing that the total energy of a stationary particle is not zero but is directly proportional to its mass. That means a particle has energy simply by virtue of having mass, or, as Einstein first recognized, that *mass and energy are equivalent*. The proportionality between mass and energy is a whopping big c^2—about 9×10^{16} J/kg in SI units.

Although we developed Equation 38-15 by considering kinetic energy, the mass-energy equivalence expressed in $E = mc^2$ is universal. Energy, like mass, exhibits inertia. A hot object is slightly harder to accelerate than a cold one because of the inertia of its thermal energy. A stretched spring is more massive than an unstretched one, because of its extra potential energy. When a system loses energy, it loses mass as well.

To the general public, $E = mc^2$ is synonymous with nuclear energy. This equation does describe mass changes that occur in nuclear reactions, but it applies equally well to chemical reactions and all other occurrences in which energy enters or leaves a system. If you weigh a nuclear power plant just after it's been refueled, then weigh it again a month later, you will find it weighs slightly less. If you weigh a coal-burning power plant and all the coal and oxygen that go into it for a month, and then weigh all the carbon dioxide and other combustion products that come out, you will find a discrepancy between the mass of what goes in and what comes out. If both plants produce the same amount of energy, the mass discrepancy will be the same for both. The only difference lies in the amount of mass released as energy in each individual reaction. The

Energetic neutrons

^{235}U → +200 MeV

(a)

C + O_2 → CO_2 +4.1 eV

(b)

FIGURE 38-23 Energy release in (a) nuclear and (b) chemical reactions. A mass change occurs in both cases, but is more than 10^7 times greater in the nuclear reaction.

FIGURE 38-24 Pair creation events observed in a bubble chamber at Brookhaven National Laboratory. Each pair of curved paths represent the trajectories of a particle-antiparticle pair, bending in opposite directions in a magnetic field. The energy required to create the particles—mc^2 per particle—ultimately came from a single high-energy proton that collided with a stationary proton.

fissioning of a single uranium nucleus involves about 50 million times as much energy, and therefore mass, as the reaction of a single carbon atom with oxygen to make carbon dioxide (Fig. 38-23). That's why a coal-burning power plant consumes many hundred-car trainloads of coal each week, while a nuclear plant is refueled every 18 months with a few truckloads of uranium. Incidentally, neither process converts very much of the fuel mass to energy; if we could convert *all* the mass in a given object to energy, ordinary matter would be an almost limitless source of energy. Such conversion is in fact possible, but only in the annihilation of matter and antimatter. The opposite conversion also occurs, with a particle-antiparticle pair appearing where before there had been only energy (Fig. 38-24).

EXAMPLE 38-9	*Annihilation*

A positron is an antimatter particle with the same mass as the electron but the opposite electric charge. When an electron and a positron meet, they annihilate and produce a pair of gamma rays (bundles of electromagnetic wave energy) of equal energy. Find the energy of each gamma ray.

Solution

Here two electron masses annihilate to give two gamma rays. So the energy of each gamma ray is the energy equivalent of one electron mass, or

$$E = mc^2 = (9.11 \times 10^{-31} \text{ kg})(3.00 \times 10^8 \text{ m/s})^2$$
$$= 8.20 \times 10^{-14} \text{ J},$$

or 511 keV. The detection of 511-keV gamma rays from laboratory or astrophysical sources is a sure indication that electron-positron annihilation is occurring.

EXERCISE The Sun radiates energy at the rate of 3.85×10^{26} W. Find the rate at which it loses mass.

Answer: 4.28×10^9 kg/s

Some problems similar to Example 38-9: 38, 39

Given the fame of $E = mc^2$, it's easy to overlook the fact that the rest energy mc^2 is generally only part of a particle's total energy. Only when a particle is at rest does mc^2 equal its total energy. For a particle moving at velocity u that's small compared with c, the total energy is only slightly greater than mc^2; the extra is very nearly the Newtonian kinetic energy, $\frac{1}{2}mu^2$. (Here "at rest" and "moving" are, of course, relative to some inertial reference frame.) But when a particle moves with nearly the speed of light, the relativistic factor $\gamma = 1/\sqrt{1 - u^2/c^2}$ becomes much greater than 1, and the total energy γmc^2 is many times the rest energy. Such a particle is termed **relativistic**.

EXAMPLE 38-10	*A Relativistic Electron*

An electron has a total energy of 2.50 MeV. Find (a) its kinetic energy and (b) its speed.

Solution

In Example 38-9 we found that the rest energy of an electron is 511 keV, or 0.511 MeV. The electron's kinetic energy is the difference between its total energy and its rest energy, or

$$K = E - mc^2 = 2.5 \text{ MeV} - 0.511 \text{ MeV} = 1.99 \text{ MeV} .$$

Equation 38-15 shows that the total energy is just γ times the rest energy. Thus we have

$$\gamma = \frac{1}{\sqrt{1 - u^2/c^2}} = \frac{E}{mc^2} = \frac{2.5 \text{ MeV}}{0.511 \text{ MeV}} = 4.89 .$$

Solving for u then gives

$$u = c\sqrt{1 - \frac{1}{4.89^2}} = 0.979c = 2.94 \times 10^8 \text{ m/s} .$$

EXERCISE Find (a) the rest energy of a proton, and (b) its kinetic energy when moving at $0.98c$.

Answers: (a) 938 Mev; (b) 3.78 GeV

Some problems similar to Example 38-10: 40–42

Got It!

The rest energy of a proton is 938 MeV. Without doing any calculations, give a quick estimate of the speed of a proton with total energy 1 TeV ($=10^{12}$ eV).

The Energy-Momentum Relation

In Newtonian physics the equations $p = mu$ and $K = \frac{1}{2}mu^2$ yield the relation $p^2 = 2K/m$. Similarly, Problem 44 shows that in relativity we can combine the equations $p = \gamma mu$ and $E = \gamma mc^2$ to get

$$E^2 = p^2 c^2 + (mc^2)^2 , \qquad (38\text{-}16)$$

where now the energy-momentum relation involves E rather than K because in relativity the energy includes both kinetic and rest energies. For a particle at rest, $p = 0$ and Equation 38-16 shows that the total energy is just the rest energy. For highly relativistic particles the rest energy is negligible and the total energy becomes very nearly $E = pc$. Some "particles"—like the photons that, in quantum physics, are "bundles" of electromagnetic energy, have no mass. These particles exist only in motion at the speed of light, and for them Equation 38-16 gives the exact relation $E = pc$.

38-6 WHAT IS NOT RELATIVE

In relativity, space and time are not absolute but depend on the reference frame of the observer. But the speed of light c remains the same for all observers. Are there other such **relativistic invariants**? Yes—and these invariants, not the shifting measures of space and time, are at the basis of relativity's objective description of physical reality.

One invariant is mass. Rearranging Equation 38-16 gives

$$m = \frac{\sqrt{E^2 - p^2 c^2}}{c^2}. \tag{38-17}$$

If you look at a particle in a frame in which it's at rest, its total energy E is its rest energy mc^2, and Equation 38-17 obviously gives m for its mass. If you go to a frame in which the particle is moving, it now has kinetic energy and momentum as well. But Equation 38-16 shows that E and p change together in a way that keeps the mass m the same. Thus a particle's mass is invariant—it's a quantity that observers in different frames of reference will agree on.

We've been talking about the mass, energy, and momentum of a single particle. But the same concepts apply to systems of particles. In the case of a multiparticle system, Equation 38-17 can be taken as a definition of the system's mass in terms of its energy and momentum. The system's energy and momenta are the sums of the energies and momenta of its constituent particles—but, remarkably, Equation 38-17 then shows that the system's mass isn't just the sum of the masses of its constituent particles. That's because any relative motion between the system particles, or any potential energy associated with internal forces, also contributes to the system's mass.

In Newtonian physics the distance between two points is the same no matter who measures it. In relativity, distance depends on the observer. So does time. But there is a quantity, analogous to distance but incorporating time as well, that remains invariant. Called the **spacetime interval**, it is a kind of "four-dimensional distance," not between two points in space but between two events in space and time. The spacetime interval Δs is given by an expression that looks like a modified Pythagorean theorem:

$$(\Delta s)^2 = c^2(\Delta t)^2 - [(\Delta x)^2 + (\Delta y)^2 + (\Delta z)^2], \tag{38-18}$$

where the Δ quantities describe the differences between the space and time coordinates of two events. The invariance of Equation 38-18 follows directly from the Lorentz transformations (see Problem 45).

The invariance of the spacetime interval suggests that something absolute underlies the shifting sands of relativistic space and time. That absolute is **spacetime**—a four-dimensional framework linking space and time. The points in spacetime are events, specified by four coordinates. The time interval or space interval between two events depends on the particular frame of reference of the observer, but the spacetime interval—a four-dimensional "distance" that takes all four coordinates into account—is the same for all observers.

In more advanced treatments of relativity, it is convenient to consider four-dimensional vectors called **four-vectors**. The displacement between two events in spacetime—specified by the four quantities Δx, Δy, Δz, and Δt—is a four-vector, with "length" given by Equation 38-18. Similarly, the three quantities $p_x c$, $p_y c$, $p_z c$, and E form a four-vector called the energy-momentum four vector or, more imaginatively, the momenergy vector. Either of Equations 38-16 or 38-17 shows that the "length" of this four-vector is just the rest energy mc^2.

It's possible to break an ordinary 2- or 3-dimensional vector into components in many different ways. Although the values of the individual components depend on your choice of coordinate system, the length or magnitude of the vector does not (Fig. 38-25). Similarly, the individual space and time components of a four-vector depend on your choice of reference frame—that is, on your velocity. But the "length" of a four-vector—for example, the spacetime interval between two events or the rest energy (equivalently, the mass) of a system—does not. Be careful, though, with the analogy in Fig. 38-25. The "length" of a vector in ordinary 2- or 3-dimensional space is given by the Pythagorean theorem. But the "length" of a four-vector involves a modified Pythagorean theorem, in which the Pythagorean sum of squares of the three spatial components is *subtracted* from the square of the time component. That's obvious in Equation 38-18, and you can also see it with Equations 38-16 or 38-17 if you isolate the rest energy term $(mc^2)^2$ on one side and recognize p^2 as the sum $p_x^2 + p_y^2 + p_z^2$. This unusual way of forming a vector's "length" reflects the fact that the underlying geometry of spacetime is not the Euclidean geometry you learned in high school.

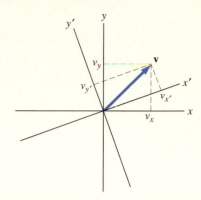

FIGURE 38-25 Although the x and y components of an ordinary vector depend on the choice of coordinate system, the length of the vector does not.

38-7 ELECTROMAGNETISM AND RELATIVITY

Historically, relativity arose from deep questions regarding the propagation of electromagnetic waves. We've seen that relativity alters concepts such as space, time, energy, and momentum, that are fundamental to Newtonian physics. For that reason, Newtonian physics becomes an approximation valid at small speeds. But relativity is built on the premise that Maxwell's equations of electromagnetism are correct in all reference frames—including the prediction of electromagnetic waves propagating at speed c. Even the title of Einstein's famous 1905 paper shows how intimately related are electromagnetism and relativity. The title isn't "The Theory of Relativity"—it's "On the Electrodynamics of Moving Bodies." Maxwell's equations are relativistically correct, and require no modification.

Although electric and magnetic fields in any frame of reference obey the same Maxwell equations, this doesn't mean the fields themselves are invariants. If, for example, you sit in the rest frame of a point charge, you see a spherically symmetric point-charge field. If you move relative to the charge, you see a magnetic field as well, associated with the moving charge. So electric and magnetic fields aren't absolutes; what one observer sees as a purely electric field another may see as a mix of electric and magnetic fields, and vice versa. You can think of the electric and magnetic fields as components of a more fundamental electromagnetic field; how that electromagnetic field breaks out into electric and magnetic fields depends on your frame of reference (see Problem 65). Although the fields are different in different reference frames, there is one important electromagnetic quantity that's invariant, namely electric charge.

We'll illustrate the deep relationship that relativity imposes on electricity and magnetism by considering the force on a charged particle near a current-carrying wire. For simplicity, we assume the wire contains equal line charge densities of positive and negative charge, and that both lines of charge are moving

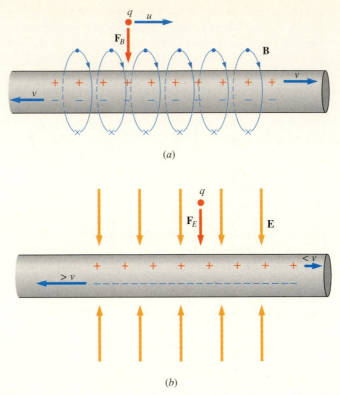

(a)

(b)

FIGURE 38-26 (a) A current-carrying wire with equal densities of positive and negative charge moving in opposite directions. A magnetic field surrounds the wire, so a charge q moving parallel to the wire experiences a magnetic force toward the wire. (b) The same situation, viewed in the rest frame of the charged particle. The line of negative charges is Lorentz contracted more, and therefore the positively charged particle "sees" a net negative charge on the wire. Thus the particle experiences an electric force toward the wire. There's still a magnetic field, but we don't show it since the stationary particle doesn't experience a magnetic force.

in opposite directions with the same speed v relative to the wire (Fig. 38-26a). The resulting current produces a magnetic field that encircles the wire, and the charged particle moving with velocity **u** as shown in Fig. 38-26a experiences a *magnetic* force $\mathbf{F} = q\mathbf{u} \times \mathbf{B}$ toward the wire. Because the positive and negative line charge densities are equal the wire is neutral, so there's no *electric* force on the charged particle.

Now look at the situation in the reference frame of the charged particle. The positive charges in the wire have a lower speed relative to the particle than do the negative charges, so in the particle's frame the distances between negative charges are contracted *more* than the distances between positive charges, as shown in Fig. 38-26b. But charge is invariant, so that means the charge per unit length is *greater* for the negative charges. So in the frame of the charged particle, *the wire carries a net negative charge!* That results in an *electric* field pointing toward the wire, and therefore the charged particle experiences an *electric* force toward the wire (see Fig. 38-26b). Of course there's still a magnetic field as well, but since the particle is at rest in its own frame, the magnetic force $q\mathbf{u} \times \mathbf{B}$ is zero.

We've given two quite different descriptions of the force on the charged particle. In the wire's frame, the particle experienced a purely magnetic force that we could determine knowing the magnetic force $q\mathbf{u}\times\mathbf{B}$ and how magnetic fields arise from currents. In contrast, describing the force in the particle's reference frame required no knowledge of magnetism whatever. We only needed to know the electric force $q\mathbf{E}$, and how electric fields arise from electric charges. This illustrates a profound point: that electricity and magnetism aren't independent phenomena that happen to be related. Rather, they're two aspects of a single phenomenon—electromagnetism. Given the relativity principle, it's not even logically possible to have electricity without magnetism, or vice versa. Relativity provides the complete unification of electromagnetism that we've hinted at throughout our study of these phenomena.

38-8 GENERAL RELATIVITY

The *special* theory of relativity is special because it's restricted to uniform motion. Following the development of special relativity, Einstein attempted to formulate a theory that would express the laws of physics in the same form in all frames of reference, including those in accelerated motion. But Einstein recognized that it is impossible to distinguish the effects of uniform acceleration from those of a uniform gravitational field (Fig. 38-27). Consequently, Einstein's general theory became a theory of gravity. Building on the notion of four-dimensional spacetime, Einstein introduced geometrical curvature of spacetime to account for gravity, and the result, published in 1916, is his **general theory of relativity**. The theory's predictions differ substantially from those of Newton's theory of gravity only in regions of very strong gravitational fields—far stronger than those found anywhere in our solar system. For that reason general relativity has become a cornerstone of modern astrophysics, describing the physics of such bizarre objects as neutron stars, black holes, and so-called gravitational lenses (Fig. 38-28). General relativity also addresses cosmological questions of the origin and ultimate fate of the universe. But we're not without terrestrial uses for general relativity; the satellite-based Global Positioning System (GPS), which can locate objects to within a few meters anywhere on Earth, would be off by several kilometers if the effects of general relativity weren't included in GPS calculations.

FIGURE 38-28 In general relativity, gravitation is synonymous with the curved geometry of spacetime. As a result, even light follows curved paths. This Hubble Space Telescope image shows the "Einstein Cross," whose four bright spots are images of the same distant quasar. The four images result from different light paths through the curved spacetime associated with a massive galaxy between the quasar and Earth. This process is called gravitational lensing.

9.8 m/s²   9.8 m/s²

Earth

(a)   (b)

FIGURE 38-27 Why general relativity is a theory of gravity. (*a*) In a spaceship at rest on Earth, objects accelerate downward at $g = 9.8$ m/s². (*b*) Far from any gravitating body, a spaceship accelerates at 9.8 m/s². Relative to the accelerating reference frame of the spaceship, objects accelerate toward the back of the ship at 9.8 m/s². Situations (*a*) and (*b*) are impossible to distinguish, so a theory dealing with nonuniform motion must also consider gravity.

CHAPTER SYNOPSIS

Summary

1. The **ether** was a hypothetical medium whose properties were supposed to explain the propagation of electromagnetic waves. In particular, such waves were supposed to have speed c relative to the ether.

2. The **Michelson-Morley experiment** and the observation of the aberration of starlight led to a contradiction in physics: Earth's motion through the ether could not be detected, yet Earth did not drag ether with it.

3. Einstein's **special theory of relativity** (1905) resolved the contradiction by asserting that uniform motion is undetectable by any experiment, mechanical or electromagnetic. Einstein did away with the ether, declaring simply that **the laws of physics are the same in all inertial frames of reference**. Mechanics and electromagnetism alike are included in Einstein's theory, so Maxwell's prediction of electromagnetic waves moving at speed c is correct in all frames of reference.

4. The simple statement that the laws of physics are the same in all frames of reference requires profound changes in our common-sense notions of space and time. These changes are described by the **Lorentz transformations**, which relate space and time measurements made in different frames of reference:

$$y' = y$$
$$z' = z$$
$$x' = \gamma(x - vt)$$
$$t' = \gamma(t - vx/c^2),$$

where

$$\gamma = \frac{1}{\sqrt{1 - v^2/c^2}}.$$

Particular manifestations of these transformations include **time dilation**, **Lorentz contraction**, and the relativistic velocity addition formulas.

5. Relativistic transformations result in new expressions for momentum:

$$\mathbf{p} = \frac{m\mathbf{u}}{\sqrt{1 - u^2/c^2}} = \gamma m\mathbf{u}$$

and total energy:

$$E = \frac{mc^2}{\sqrt{1 - u^2/c^2}} = \gamma mc^2,$$

for a particle of mass m moving with velocity $\mathbf{u}$. Kinetic energy is the difference between the total energy E and the **rest energy** mc^2. The existence of rest energy shows that matter and energy may be converted into each other.

6. Relativity links space and time into a four-dimensional framework called **spacetime**. Although individual space and time measurements depend on one's frame of reference, the **spacetime interval** between two events does not.

7. Maxwell's equations of electromagnetism are fully consistent with relativity. But the fields themselves are different in different frames of reference; what appears as a magnetic field in one frame may be partially electric in another, and vice versa. Relativity imposes a logical relationship between electricity and magnetism, in that neither phenomenon is possible without the other.

8. The **general theory of relativity** is Einstein's generalization of relativity to include accelerated reference frames. Because the effects of acceleration mimic those of gravity, general relativity is a theory of gravity.

Terms You Should Understand

(Pairs are closely related terms whose distinction is important; number in parentheses is chapter section where term first appears.)

ether (38-1)
Michelson-Morley experiment (38-2)
special relativity, general relativity (38-2, 38-8)
time dilation, length contraction (38-4)
proper time, proper length (38-4)
Lorentz transformations (38-4)
rest energy (38-5)
spacetime (38-6)
spacetime interval (38-6)

Symbols You Should Recognize

γ (38-4)

Problems You Should Be Able to Solve

finding times and distances for different observers in relative motion (38-4)
transforming space and time coordinates among coordinate systems in relative motion (38-4)
transforming velocities (38-4)
applying mass-energy equivalence (38-5)
evaluating relativistic energy and momentum (38-5)

Limitations to Keep in Mind

The special theory of relativity applies only to inertial frames of reference—i.e., those that are not accelerated.

QUESTIONS

1. Why was the Michelson-Morley experiment a more sensitive test of motion through the ether than independent measurements of the speed of light in two perpendicular directions?

2. Why was it necessary to repeat the Michelson-Morley experiment at different times throughout the year?

3. Why do we reject the idea that the ether frame of reference is the Earth frame?

4. What is special about the special theory of relativity?

5. Does relativity require that the speed of sound be the same for all observers? Why or why not?

6. How would the world be different if the speed of light were 160 km/h (100 miles per hour)? Would our "common-sense" notions change?

7. A friend argues that the speed of light cannot be the same for all observers, for if one of them is moving toward a light source then that one will clearly measure a higher speed. How would you refute this argument?

8. Time dilation is sometimes described by saying that "moving clocks run slow." In what sense is this true? In what sense does the statement violate the spirit of relativity?

9. If you're in a spaceship moving at $0.95c$ relative to Earth, do you perceive time to be passing more slowly than it would on Earth? Think! Is your answer consistent with the principle of relativity?

10. In our light-box example for time dilation, we found that a time interval between two events measured in frame S' was shorter than in frame S. But you could equally well say that frame S is moving relative to frame S', so clocks in S should "run slow" compared with those in S'. An observer in each frame should judge the clocks in the other frame to "run slow." Is this a contradiction?

11. To try to circumvent the difficulty of accelerating an object to the speed of light, you build a series of conveyor belts, all running in the same direction, and each moving 10 m/s relative to the one next to it (Fig. 38-29). You step from the ground onto the first conveyor belt, then to the next, and so forth. By the time you reach the 3×10^7th conveyor belt, you should be moving at c relative to the ground. Why doesn't this scheme work?

FIGURE 38-29 A series of conveyor belts, each moving at 10 m/s relative to its neighbors (Question 11).

12. If you took your pulse while traveling in a high-speed spacecraft, would it be faster, slower, or the same as on Earth?

13. The Andromeda galaxy is 2 million light-years from our Milky Way. Although nothing can go faster than light, it would still be possible to travel to Andromeda in much less than 2 million years. How is this possible?

14. Is matter converted to energy in a nuclear reactor? In a burning candle? In your body?

15. An unstretched rubber band is weighed on an extraordinarily sensitive scale. It is then stretched and weighed again. Is there a difference in the weight? Why or why not?

16. The rest energy of an electron is 511 keV. What is the approximate speed of an electron whose total energy is 1 GeV ($= 10^9$ eV)? You need not do any calculations!

17. An atom in an excited state emits a burst of light. What happens to the mass of the atom?

18. In some of the hottest parts of the universe, the thermal energy of particles may be many millions of electron volts. At such temperatures, the number of particles within a closed volume may vary. How is this possible? *Hint:* The mass of the electron is equivalent to about 0.5 MeV.

19. The quantity $\mathbf{E} \cdot \mathbf{B}$ is invariant. What does this say about how different observers will measure the angle between $\mathbf{E}$ and $\mathbf{B}$ in a light wave?

PROBLEMS

Section 38-2 Matter, Motion, and the Ether

1. Consider an airplane flying at 800 km/h airspeed between two points 1800 km apart. What is the round-trip travel time for the plane (a) if there is no wind? (b) if there is a wind blowing at 130 km/h perpendicular to a line joining the two points? (c) if there is a wind blowing at 130 km/h along a line joining the two points? Ignore relativistic effects. (Why are you justified in doing so?)

2. What would be the difference in light travel times on the two legs of the Michelson-Morley experiment if the ether existed and if Earth moved relative to it at (a) its orbital speed relative to the Sun (Appendix E)? (b) $10^{-2}c$? (c) $0.5c$? (d) $0.99c$? Assume each light path is exactly 11 m in length, and that the paths are oriented parallel and perpendicular to the ether wind.

3. Figure 38-30 shows a plot of James Bradley's data on the aberration of light from the star γ Draconis, taken in 1727–1728. (a) From the data, determine the magnitude of Earth's orbital velocity. (b) The data very nearly fit a perfect sine curve. What does this say about the shape of Earth's orbit?

FIGURE 38-30 Bradley's data on the aberration of starlight (Problem 3). Aberration is measured in seconds of arc ($1'' = 1/3600°$).

4. Show that the time of Equation 38-2 is larger than that of Equation 38-1 as long as $0 < v < c$.

5. Suppose the speed of light differed by 100 m/s in two perpendicular directions. How long should the arms be in a Michelson interferometer if this difference is to cause the interference pattern to shift one-half cycle (i.e., a light fringe shifts to where a dark one was) relative to the pattern if there were no speed difference? Assume 550-nm light.

Section 38-4 Space and Time in Relativity

6. Two stars are 50 ly apart, measured in their common rest frame. How far apart are they to a spaceship moving between them at $0.75c$?

7. How long would it take a spacecraft traveling at 65% of the speed of light to make the 5.8×10^9 km journey from Earth to Pluto according to clocks (a) on Earth and (b) on the spacecraft?

8. A spaceship goes by at half the speed of light, and you determine that it is 35 m long. What is its length as measured in its rest frame?

9. Earth and Sun are 8.3 light-minutes apart, as measured in their rest frame. (a) What is the speed of a spacecraft that makes the trip in 5.0 min according to its on-board clocks? (b) What is the trip time as measured by clocks in the Earth-Sun frame?

10. The Andromeda galaxy is two million light-years from Earth, measured in the common rest frame of Earth and Andromeda. Suppose you took a fast spaceship to Andromeda, so it got you there in 50 years as measured on the ship. If you sent a radio message home as soon as you reached Andromeda, how long after you left Earth would it arrive, according to timekeepers on Earth?

11. How fast would you have to move relative to a meter stick for its length to measure 99 cm in your frame of reference?

12. Electrons in a TV tube move through the tube at 30% of the speed of light. By what factor is the tube foreshortened in the electrons' frame of reference, relative to its length at rest?

13. You wish to travel to a star N light-years from Earth. How fast must you go if the one-way journey is to occupy N years of your life?

14. The nearest star beyond our solar system is about 4 light-years away. If a spaceship can get to the star in 5 years, as measured on Earth, (a) how long would the ship's pilot judge the journey to take? (b) How far from Earth would the pilot find the star to be?

15. Twins A and B live on Earth. On their 20th birthday, Twin B climbs into a spaceship and makes a round-trip journey at $0.95c$ to a star 30 light-years distant, as measured in the Earth-star frame of reference. What are their ages when the twins are reunited?

16. Radioactive oxygen-15 decays at such a rate that half the atoms in a given sample decay every two minutes. If a tube containing 1000 O-15 atoms is moved at $0.80c$ relative to Earth for 6.67 minutes according to clocks at rest with respect to Earth, how many atoms will be left at the end of that time?

17. Two distant galaxies are receding from Earth at $0.75c$, in opposite directions. How fast does an observer in one galaxy measure the other to be moving?

18. Two spaceships are having a race. The "slower" one moves past Earth at $0.70c$, and the "faster" one moves at $0.40c$ relative to the slower one. How fast does the faster ship move relative to Earth?

19. Muons traveling vertically downward at $0.994c$ relative to Earth are observed from a rocket traveling upward at $0.25c$. What speed does the rocket's crew measure for the muons?

20. Use relativistic velocity addition to show that if an object moves at speed $v < c$ relative to some uniformly moving frame of reference, then its speed relative to any other uniformly moving frame must also be less than c.

21. Earth and Sun are 8.33 light-minutes apart. Event A occurs on Earth at time $t = 0$, and event B on the Sun at time $t = 2.45$ min, as measured in the Earth-Sun frame. Find the time order and time difference between A and B for observers (a) moving on a line from Earth to Sun at $0.750c$,

(b) moving on a line from Sun to Earth at 0.750c, and (c) moving on a line from Earth to Sun at 0.294c.

22. Two civilizations are evolving on opposite sides of a galaxy, whose diameter is 10^5 light-years (Fig. 38-31). At time $t = 0$ in the galaxy frame of reference, civilization A launches its first interstellar spacecraft. Civilization B launches its first spacecraft 50,000 years later. A being from a more advanced civilization C is traveling through the galaxy at 0.99c, on a line from A to B. Which civilization does C judge to have first achieved interstellar travel, and how much in advance of the other?

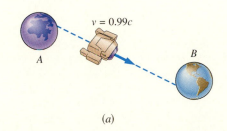

$v = 0.99c$

A

B

(a)

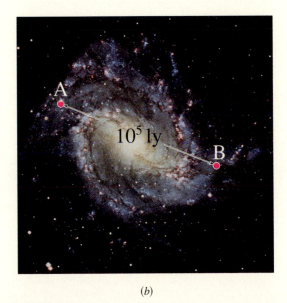

10^5 ly

(b)

FIGURE 38-31 Civilizations A and B are on opposite sides of a galaxy, a distance of 10^5 light-years in the galaxy frame of reference (Problems 22, 23, 24).

23. Repeat the preceding problem, now assuming that civilization B lags A by 1 million years in the galaxy frame of reference.

24. Could there be observers who would judge the events in the two preceding problems to be simultaneous? If so, how must each be moving relative to the galaxy?

25. Derive the Lorentz transformations for time, Equations 38-9 and 38-10, from the transformations for space.

26. In the light box of Fig. 38-8, let event A be the emission of the light flash and event B its return to the source. Assign suitable space and time coordinates to these events in the frame in which the box moves with speed v. Apply the Lorentz transformations to show that the time $\Delta t'$ between the two events in the box frame is given by Equation 38-4.

27. Two spaceships are each 25 m long, as measured in their rest frames (Fig. 38-32). Ship A is approaching Earth at 0.65c. Ship B is approaching Earth from the opposite direction at 0.50c. Find the length of ship B as measured (a) in Earth's frame of reference and (b) in ship A's frame of reference.

A

B

$v = 0.65c$

$v = 0.50c$

FIGURE 38-32 Problem 27 (the drawing is in Earth's frame of reference).

Section 38-5 Energy and Momentum in Relativity

28. By what factor does the momentum of an object change if you double its speed when its original speed is (a) 25 m/s and (b) 1.0×10^8 m/s?

29. At what speed will the momentum of a proton (mass 1 u) equal that of an alpha particle (mass 4 u) moving at 0.5c?

30. At what speed will the Newtonian expression for momentum be in error by 1%?

31. A particle is moving at 0.90c. If its speed increases by 10%, by what factor does its momentum increase?

32. Compare the momentum changes needed to boost a spacecraft (a) from 0.1c to 0.2c and (b) from 0.8c to 0.9c.

33. Find (a) the total energy and (b) the kinetic energy of an electron moving at 0.97c.

34. A TV tube accelerates electrons from rest through a potential difference of 30 kV. Find their final speed.

35. At what speed will the relativistic and Newtonian expressions for kinetic energy differ by 10%?

36. Find (a) the speed and (b) the momentum of a proton whose kinetic energy is 500 MeV.

37. Among the most energetic cosmic rays ever detected are protons with energies around 10^{20} eV. Find the momentum of such a proton, and compare with that of a 25-mg insect crawling at 2 mm/s (Fig. 38-33).

(a)   (b)

FIGURE 38-33 (a) A cosmic ray proton slams into the upper atmosphere, creating a shower of elementary particles. How does the proton's energy compare with that of (b) a crawling bug? (Problem 37)

38. A large city consumes electrical energy at the rate of 10^9 W. If you converted all the rest mass in a 1-g raisin to electrical energy, for how long could it power the city?

39. In a nuclear fusion reaction, two deuterium nuclei (^{2}H) combine to give a helium nucleus (^{3}He) plus a neutron. The energy released in the process is 3.3 MeV. By how much do the combined masses of the helium nucleus and neutron differ from the combined masses of the original deuterium nuclei?

40. Find the kinetic energy of an electron moving at (a) $0.0010c$, (b) $0.60c$, and (c) $0.99c$. Use suitable approximations where possible.

41. Find the speed of an electron with kinetic energy (a) 100 eV, (b) 100 keV, (c) 1 MeV, (d) 1 GeV. Use suitable approximations where possible.

42. How much energy would it take to accelerate a proton (a) from rest to $0.10c$, (b) from $0.50c$ to $0.55c$, (c) from $0.95c$ to $0.99c$, and (d) from $0.99c$ to c.

43. Use the binomial approximation (Appendix A) to show that Equation 38-14 reduces to the Newtonian expression for kinetic energy in the limit $u \ll c$.

44. Show that Equation 38-16 follows from the expressions for relativistic momentum and total energy.

Section 38-6 What Is Not Relative

45. Show from the Lorentz transformations that the spacetime interval of Equation 38-18 has the same value in all frames of reference.

46. A spaceship travels at $0.80c$ from Earth to a star 10 light-years distant, as measured in the Earth-star frame (Fig. 38-34). Let event A be the ship's departure from Earth and event B its arrival at the star. (a) Find the distance and time between the two events in the Earth-star

FIGURE 38-34 Problem 46.

frame. (b) Find the distance and time between the two events in the ship frame. *Hint:* The distance in the ship frame is the distance an observer has to move *with respect to that frame* to be at both events—not the same as the Lorentz-contracted distance between Earth and star. (c) Compute the square of the spacetime interval in both frames to show explicitly that it is invariant.

47. Use Equation 38-18 to calculate the square of the spacetime interval between the events (a) of Problem 22 and (b) of Problem 23. Comment on the signs of your answers in relation to the possibility of a causal relation between the events.

48. A light beam is emitted at event A and arrives at event B. Show that the spacetime interval between the two events is zero.

Paired Problems

(Both problems in a pair involve the same principles and techniques. If you can get the first problem, you should be able to solve the second one.)

49. An extraterrestrial spacecraft passes Earth and 4.5 s later, according to its clocks, it passes the moon. Find its speed.

50. How fast would you have to go to reach a star 200 light-years distant in a 75-year human lifetime?

51. An electron moves down a 1.2-km-long particle accelerator at $0.999992c$. In the electron's frame, (a) how much time does the trip take and (b) how long is the accelerator?

52. An advanced civilization has developed a spaceship that goes, with respect to the galaxy, only 50 km/s slower than light. (a) How long, according to the ship's crew, does it take to cross the 100,000-ly diameter of the galaxy? (b) What is the galactic diameter as measured in the ship's frame of reference?

53. Event A occurs at $x = 0$ and $t = 0$ in a frame of reference S. Event B occurs at $x = 3.8$ light-years, $t = 1.6$ years in S. Find (a) the distance and (b) the time between A and B in a frame S' moving at $0.80c$ along the x axis of S.

54. Two nuclear reactions occur in a particle accelerator, at points 45 cm apart and separated by 6.8 ns in time, as measured in the accelerator's frame of reference. Find (a) the distance and (b) the time between the events as measured in the reference frame of a proton moving at $0.92c$ on a line between the two.

55. When a particle's speed doubles, its momentum increases by a factor of 3. What was the original speed?
56. Find the speed of an electron whose momentum is 50% greater than the value given by the Newtonian expression $p = mu$.

Supplementary Problems

57. How fast would you have to travel to reach the Crab Nebula, 6500 light-years from Earth, in 20 years? Give your answer to 7 significant figures.
58. At what speed are a particle's kinetic and rest energies equal?
59. A cosmic ray proton with energy 20 TeV is heading toward Earth. What is Earth's diameter measured in the proton's frame of reference?
60. Plot the ratio of relativistic to Newtonian kinetic energy as a function of speed, and from your graph find the speed where the relativistic kinetic energy is 50% greater than the Newtonian value.
61. When the speed of an object increases by 5%, its momentum goes up by a factor of 5. What was the original speed?
62. Use the Lorentz transformations to show that if two events are separated in space and time so that a light signal leaving one event could not reach the other, then there is an observer for whom the two events are simultaneous. Show that the converse is true as well: that if a light signal could get from one event to the other, then no observer will find them simultaneous.

63. A source emitting light with frequency f moves toward you at speed u. By considering both time dilation and the effect of wavefronts "piling up" as shown in Fig. 17-20, show that you measure a Doppler shifted frequency given by

$$f' = f\sqrt{\frac{c + u}{c - u}}.$$

Use the binomial approximation to show that this result can be written in the form of Equation 17-10 for $u \ll c$.

64. Use Equation 9-7 to estimate the size to which you would have to squeeze each of the following before escape speed at its surface approximated the speed of light: (a) Earth; (b) the Sun; (c) the Milky Way galaxy, containing about 10^{11} solar masses. Your answers show why general relativity is not needed for most astronomical calculations.
65. Consider a line of positive charge with line charge density λ, as measured in a frame S at rest with respect to the charges. (a) Show that the electric field a distance r from this charged line has magnitude $E = \lambda/2\pi \varepsilon_0 r$, and that there is no magnetic field (no relativity needed here). Now consider the situation in a frame S' moving at speed v parallel to the line of charge. (b) Show that the line charge density as measured in S' is given by $\lambda' = \gamma\lambda$, with $\gamma = 1/\sqrt{1 - v^2/c^2}$. (c) Use the result of (b) to find the electric field in S'. Since the line of charge is moving with respect to S', there is a current in S'. (d) Find an expression for this current and (e) for the magnetic field it produces. Show that the quantities (f) $\mathbf{E} \cdot \mathbf{B}$ and (g) $E^2 - B^2$ are the same in both frames of reference. (In fact, these quantities are always invariant.)

39 Light and Matter: Waves or Particles?

This pattern, produced by diffraction of an electron beam from a titanium-nickel alloy, shows that electrons exhibit a wave-like character. Quantum theory has blurred the distinction between particles and waves, showing that particles have a wave-like aspect and that light has a particle-like aspect.

Newtonian mechanics and Maxwell's electromagnetism constitute the core of **classical physics**, providing a rich and deep understanding of physical reality. Application of classical physics has led to many of the technological developments essential to modern civilization. Although these theories were firmly established by the middle of the nineteenth century, they remain central to the work of many contemporary scientists and engineers.

Nevertheless, at the end of the nineteenth century a few seemingly minor phenomena defied classical explanation. Most physicists felt that it was only a matter of time before these, too, came under the classical umbrella. But that was not to be. We've already seen how questions about the speed of light led to a radical restructuring of our fundamental notions of space and time. Other questions, especially those concerning the behavior of matter at the atomic scale, were to bring an even more radical transformation of physical thought.

The essence of the new physics lies in the ideas of **quantum physics**, developed first to explain the workings of the atom and subsequently pushed into the nuclear realm and then to the subnuclear world of elementary particles. In this chapter we introduce some phenomena that led to the ideas of quantum physics and explore early attempts to explain them. The next chapter gives a fuller account of the complete theory called quantum mechanics, and subsequent chapters explore its application to atoms, molecules, and nuclei, and to technologies based on quantum systems.

39-1 TOWARD THE QUANTUM THEORY

Are matter and energy continuously divisible? The essential difference between classical and quantum physics is that the former answers this question "in general, yes," while the latter says definitively "no." Instead, most physical quantities in quantum physics are **quantized**, coming only in certain discrete values.

The idea that physical quantities might come in discrete "chunks" is not new to modern quantum theory. Some 2400 years ago the Greek philosopher Democritus proposed that all matter consists of indivisible atoms of different types. By the start of the twentieth century a more scientifically grounded atomic theory was widely accepted. J. J. Thomson's discovery of the electron in 1897 showed that atoms might be divisible after all, but at the same time it revealed a finer division of matter into discrete "chunks." Millikan's 1909 oil-drop experiment (Section 23-2) showed that electric charge is similarly quantized, seemingly available only in integer multiples of the elementary charge e. Discovery of the proton and later the neutron further solidified the notion that matter is comprised of fundamental building blocks with definite and unchanging values for their various physical properties.

Quantization of matter into particles with discrete properties is not incompatible with classical physics as long as those particles behave according to classical laws—in particular, that they move continuously through space and can have *any* amount of energy. Add electromagnetism to the picture and the classical viewpoint requires that the fields be continuous, exerting forces on charged particles and sometimes changing, in a gradual and continuous way, the particles' energies.

The startling fact of quantum physics is that this classical behavior does not occur at the atomic scale; instead, energy itself is in many instances quantized. Reconciling the implications of that fact with our common-sense notions of matter and motion has proved impossible; instead, the quantum world speaks a different language, one in which deeply ingrained ideas about causality and the solid reality of matter seem no longer to apply. Here we look at three distinct phenomena that force us to accept the idea that energy can be quantized.

FIGURE 39-1 As the temperature of an electrically heated coil increases, it emits increasing amounts of radiation, with the maximum emission occurring at ever shorter wavelengths. The color, correspondingly, changes from a dull red to a bright yellow-white.

39-2 BLACKBODY RADIATION

Heat an object hot enough and it begins to glow, emitting electromagnetic radiation in the form of light. As we saw in Section 19-6, the total power radiated is proportional to the fourth power of the temperature. There's also a change in wavelength with increasing temperature: The first visible glow is a dull red, changing with higher temperatures to orange and then yellow colors corresponding to ever shorter wavelengths (Fig. 39-1).

An object that is a perfect absorber of electromagnetic radiation is called a **blackbody**, since it absorbs all light incident on it and thus appears black. When a blackbody is heated, it emits electromagnetic radiation at a range of wavelengths; this is called **blackbody radiation**. Many objects—such as the Sun, the glowing coil shown in Fig. 39-1, or a wood stove—behave approximately as blackbodies. An excellent approximation to a blackbody can be achieved using a hollow piece of *any* material with a small hole in it. As Fig. 39-2 shows, any radiation entering the hole undergoes multiple reflections from the interior walls and is eventually absorbed. The hole, therefore, is a nearly perfect absorber, so when the material is heated the radiation emerging from it will be blackbody radiation.

Experimental study of blackbody radiation shows several characteristic features:

1. The radiation covers a continuous range of wavelengths, with the total power radiated at all wavelengths combined being given by the Stefan-Boltzmann law that we introduced in Chapter 19:

$$P_{\text{blackbody}} = \sigma A T^4, \qquad (39\text{-}1)$$

where A is the area of the radiating surface, T is its temperature, and $\sigma = 5.67 \times 10^{-8}$ W/m²·K⁴ is the Stefan-Boltzmann constant.

FIGURE 39-2 A piece of material containing a cavity and a small hole. Any radiation incident on the hole undergoes multiple reflections inside the cavity, and its energy is eventually absorbed. Very little radiation escapes the hole, which therefore appears black. The hole is a nearly perfect blackbody.

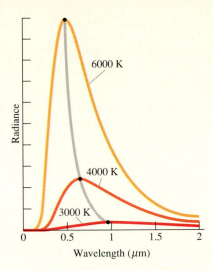

FIGURE 39-3 Radiance of a black-body as a function of wavelength. Gray curve shows the shift in peak radiance toward shorter wavelengths. Total radiation, given by the area under each curve, increases rapidly with increasing temperature.

2. The wavelength at which the most radiation occurs is inversely propor-tional to the temperature; specifically

$$\lambda_{max} T = 2.898 \times 10^{-3} \text{ m·K}. \tag{39-2}$$

This relation is called **Wien's displacement law**.

3. The distribution of wavelengths depends only on the temperature, not on the material of which the blackbody is made.

It's convenient to characterize a blackbody in terms of its **radiance**, or power emitted per unit area of blackbody surface per unit wavelength interval. Radi-ance is a function of wavelength and also of blackbody temperature. Figure 39-3 shows the radiance of a blackbody at several temperatures.

Microscopically, blackbody radiation is associated with the thermal motions of atoms and molecules, so it's not surprising that the radiation increases with temperature. In the late 1800s physicists tried to apply the laws of electromag-netism and statistical mechanics to explain the experimental observations of blackbody radiation. They met with some success in describing such aspects as the T^4 dependence of the total energy radiated, and the shifting of the radiation distribution toward shorter wavelengths with increasing temperature, but they could not reproduce the actual observed distribution at all wavelengths.

In 1900 the German physicist Max Planck formulated an equation that fit the observed radiance-versus-wavelength curves of blackbody radiation at all wave-lengths. Planck's equation describes the radiation arising from the thermal vibration of molecules, and in terms of radiance it reads

$$R(\lambda, T) = \frac{2\pi hc^2}{\lambda^5 (e^{hc/\lambda kT} - 1)}. \tag{39-3}$$

Two familiar quantities in Equation 39-3 are Boltzmann's constant $k = 1.38 \times 10^{-23}$ J/K, introduced in Chapter 20, and the speed of light c. A new quantity is the constant h, whose value Planck chose to make the equation fit experimen-tal data.

Planck first presented his law as a purely empirical equation describing black-body experiments. Later he began searching for a physical interpretation. Through arguments involving entropy and statistical mechanics, Planck showed that his equation had a remarkable implication:

> The energy of a vibrating molecule is quantized, meaning it can have only certain discrete values. Specifically, if *f* is the vibration frequency, then the energy must be an integer multiple of the quantity *hf*:

$$E = nhf, \qquad n = 0, 1, 2, 3, \ldots \tag{39-4}$$

where h is the constant Planck introduced in Equation 39-3. Today we know h as one of the fundamental constants of nature and call it **Planck's constant**. Its value is approximately 6.63×10^{-34} J·s, and it is precisely because h is so small that quantum phenomena usually become obvious only when we delve into the atomic and molecular realm. Planck's quantization of the energy of vibrating molecules implies further that a molecule can absorb or emit energy only in discrete "bundles" of size hf, and that in doing so it jumps abruptly from one of its allowed energy levels to another (Fig. 39-4). (Later developments showed that Planck was correct about the size of the energy jumps, but that the factor n in Equation 39-4 should actually be $n + \frac{1}{2}$.)

Planck himself was a very conservative scientist and was reluctant to accept or elaborate on his theory's evident disagreement with classical physics; nevertheless, his revolutionary work won Planck the 1918 Nobel Prize. Other

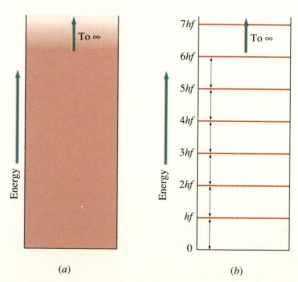

FIGURE 39-4 (*a*) In classical physics, a vibrating molecule can have any energy, and its energy can change through absorption or emission of any amount of energy. (*b*) Allowed energies in Planck's theory are integer multiples of *hf*, and can change only through absorption or emission of discrete energy "bundles" of size *hf*. Two-headed arrows indicate these allowed transitions. Energy-level diagrams like this are used frequently in quantum physics, and usually, as here, the horizontal axis has no physical significance.

FIGURE 39-5 Radiance versus wavelength for blackbody radiation from an object at 6000 K. Classical theory agrees with experiment and with Planck's theory at the longer wavelengths, but predicts an excess of radiation at short wavelengths—the "ultraviolet catastrophe."

physicists subsequently emphasized the contrast between Planck's work and the classical treatment of blackbody radiation. That earlier treatment, based on the assumption that energy is shared equally among all possible vibrational modes, had led to the **Rayleigh-Jeans law** for the radiance of a blackbody:

$$R(\lambda, T) = \frac{2\pi ckT}{\lambda^4}. \qquad (39\text{-}5)$$

Not only did the Rayleigh-Jeans prediction contradict experimental measurements, but it also led to the absurd conclusion that every glowing object should emit electromagnetic energy at an infinite rate, with that energy concentrated at the shortest wavelengths (Fig. 39-5). Since the shortest wavelength radiation known at the time was ultraviolet, this phenomenon was called the **ultraviolet catastrophe**. In Planck's equation the exponential term in the denominator grows rapidly with decreasing wavelength, diminishing the radiance and averting the ultraviolet catastrophe. Problems 13, 86, and 87 show that Planck's law reduces to the Rayleigh-Jeans law for longer wavelengths, and that it also leads to Wien's displacement law and the Stefan-Boltzmann law.

EXAMPLE 39-1	*Light-Bulb Efficiency, Blackbody Radiation, and the Universe*

A standard incandescent light bulb's filament temperature is about 3000 K. (a) Find the peak wavelength emitted by the bulb. (b) How does the radiance at 550 nm—the approximate center of the visible spectrum—compare with that at the peak wavelength?

Solution

Equation 39-2 relates the peak wavelength and temperature, so for the light bulb we have

$$\lambda_{max} = \frac{2.898 \times 10^{-3}\ \mathrm{m \cdot K}}{T} = \frac{2.898 \times 10^{-3}\ \mathrm{m \cdot K}}{3000\ \mathrm{K}} = 966\ \mathrm{nm}\ .$$

That this wavelength lies in the infrared shows why incandescent lamps are not particularly efficient—they emit more heat (infrared radiation) than light. To compare the radiance at 550 nm with that at the 966-nm peak, we take the ratio of the radiances given by Planck's law, Equation 39-3:

$$\frac{R(\lambda_2, T)}{R(\lambda_1, T)} = \frac{2\pi hc^2/[\lambda_2^5(e^{hc/\lambda_2 kT} - 1)]}{2\pi hc^2/[\lambda_1^5(e^{hc/\lambda_1 kT} - 1)]} = \frac{\lambda_1^5(e^{hc/\lambda_1 kT} - 1)}{\lambda_2^5(e^{hc/\lambda_2 kT} - 1)}$$

$$= 0.38\ ,$$

where we used $\lambda_2 = 550$ nm, $\lambda_1 = 966$ nm, $T = 3000$ K, and the values for the constants c, h, and k. Thus the light bulb's brightness in the center of the visible spectrum is only about 40% of its peak brightness. That is, incidentally, why color film intended for daylight applications gives incorrect color rendition when used indoors under ordinary light bulbs.

EXERCISE The universe is bathed in "cosmic microwave background radiation," which originated about 10^5 years after the "big bang" event that began the universe (more on this in Chapter 45). Intensity of the cosmic background radiation peaks at a wavelength of about 1 mm. To what temperature does this blackbody radiation correspond?

Answer: 3 K

Some problems similar to Example 39-1: 2, 4, 7, 10, 11

Got It!

Two identical blackbodies are heated until blackbody A's temperature is twice that of blackbody B. Compare their total radiated power and the wavelengths at which they emit the most energy.

39-3 PHOTONS

Planck's hypothesis quantized the energies of vibrating molecules and showed that they could exchange energy with electromagnetic radiation only in discrete "bundles" of size hf. But what about the radiation itself? Is its energy also quantized?

The Photoelectric Effect

Experiments begun by Hertz in 1887 and continued by other physicists in the early 1900s showed that some metals emit electrons when illuminated with light. This **photoelectric effect** is studied by putting a metal surface in an evacuated glass container that also contains a second conducting electrode (Fig. 39-6). Making the second electrode positive with respect to the emitting surface attracts the electrons, and the resulting current provides a measure of the number of electrons ejected. If the second electrode is sufficiently negative with respect to the emitting surface, on the other hand, then the electron energy is not enough to overcome the potential difference and the current ceases. The value of this so-called stopping potential thus provides a measure of the ejected electrons' maximum kinetic energy.

Classical physics suggests that the photoelectric effect should occur because an electron in a metal experiences a force in the oscillating electric field of a light

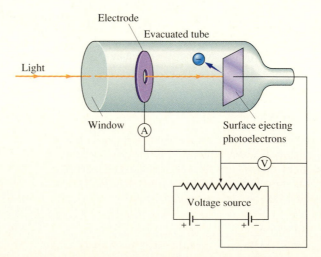

FIGURE 39-6 Apparatus for studying the photoelectric effect. Light passes through a hole in a metal electrode and strikes a metal surface, where it ejects electrons. Moving the contact on the resistor varies the voltage between the electrode and the metal surface, allowing measurement of the number and energy of the ejected electrons. Note that the circuit allows the electrode to be made either positive or negative with respect to the electron-emitting surface.

FIGURE 39-7 Current versus voltage for the photoelectric experiment of Fig. 39-6, shown for two different values of light intensity at the same light frequency. Magnitude of the stopping potential V_s measures the maximum energy of the ejected electrons. In contrast to the classical prediction, that energy is independent of light intensity.

wave. As the electron continuously absorbs energy from the wave, the amplitude of its oscillatory motion should grow until eventually it has enough energy to escape from the metal. Because the energy in a wave is spread throughout the entire wave, it might take a while for a single tiny electron to absorb enough energy to escape. Increasing the light intensity should increase the electric field in the light wave, so should result in the electron's being ejected sooner and with more energy. Changing the wave frequency should have little effect on the behavior of the free electrons in the metal.

The photoelectric effect does occur, but not in the way classical physics suggests. Figure 39-7 shows results from a photoelectric experiment, in the form of current versus voltage as read by the meters in Fig. 39-6. These results, along with observations made by varying the frequency of the incident light, show three major disagreements with the classical prediction:

1. Current begins immediately, showing that electrons are ejected immediately, even in dim light.
2. The maximum electron energy, as measured by the stopping potential V_s, is independent of the light intensity.
3. Below a certain cutoff frequency *no* electrons are emitted, no matter how strong the light. Above the cutoff frequency electrons are emitted with a maximum energy that increases in proportion to the light-wave frequency.

17.3
Photoelectric Effect

In 1905, the same year in which he formulated the special theory of relativity, Albert Einstein offered an explanation for the photoelectric effect. Einstein suggested that the energy of an electromagnetic wave is not spread throughout the wave, but is instead concentrated in "bundles" called **quanta** or **photons**. Einstein applied to these photons the same energy-quantization condition that Planck had already proposed for molecular vibrations: that photons in light with frequency f have energy hf, where again h is Planck's constant:

$$E = hf. \quad \text{(energy of a photon)} \quad (39\text{-}6)$$

The more intense the light, the more photons—but the energy of each photon is unrelated to the light intensity.

Einstein's idea readily explains the nonclassical aspects of the photoelectric effect. He held that for each material there is a minimum energy—called the work function, ϕ, required to eject an electron. (Table 39-1 lists work functions for selected elements.) Since the energy in a photon of light with frequency f is hf, the photons in low-frequency light have less energy than the work function and are therefore unable to eject electrons—no matter how many photons there are. At the cutoff frequency the photon energy equals the work function, and the photons have just enough energy to eject electrons. As the frequency increases still further the electrons emerge with maximum kinetic energy K equal to the difference between the photon energy and the work function:

$$K_{\text{max}} = hf - \phi. \quad (39\text{-}7)$$

TABLE 39-1 *Work Functions ϕ for Selected Elements*

NAME (SYMBOL)	ϕ, eV
Silver (Ag)	4.26
Aluminum (Al)	4.28
Cesium (Cs)	2.14
Copper (Cu)	4.65
Potassium (K)	2.30
Sodium (Na)	2.75
Nickel (Ni)	5.15
Silicon (Si)	4.85

FIGURE 39-8 Results of a photoelectric experiment. White line shows stopping potential as a function of light frequency and wavelength. The stopping potential in volts is a direct measure of the electron energy in eV. Slope of the straight line is numerically equal to Planck's constant, which confirms Equation 39-7. Can you tell from the graph which material in Table 39-1 is the emitting surface?

Thus, the electrons' maximum kinetic energy depends only on the photon energy—that is, on the light frequency but not on its intensity (Fig. 39-8). Finally, the immediate ejection of electrons follows because an individual photon delivers its entire bundle of energy to an electron all at once. Einstein received the 1921 Nobel Prize for his explanation of the photoelectric effect; his special and general theories of relativity were at that time sufficiently controversial that they bore no mention in the Nobel citation. In 1914 Millikan, who had earlier demonstrated the quantization of electric charge, carried out meticulous photoelectric experiments that confirmed Einstein's hypothesis and helped earn Millikan the 1923 Nobel Prize.

Got It!

If you replot Fig. 39-8 for a substance with a different work function, will the slope of the line change? Will the point at which it intersects the horizontal axis change?

| APPLICATION | *The Photomultiplier* |

The photoelectric effect is put to good use in the photomultiplier, an extremely sensitive device for detecting light. Figure 39-9 shows a typical photomultiplier, consisting of an evacuated tube containing a number of electrodes held at increasing electric potentials. Light is incident at one end on the lowest voltage electrode, called the photocathode. Photoelectrons ejected by the photoelectric effect are attracted to the higher potential of the next electrode, called a dynode. Each electron is accelerated with enough energy that it releases several additional electrons when it hits the dynode. These accelerate to the next dynode, again releasing several electrons. This process continues through a chain of typically 10 dynodes, resulting in "electron multiplication" factors as high as 10^9. For each electron ejected at the photocathode, that means there may be as many as a billion electrons striking the final electrode. These produce a measurable pulse of electric current.

Typical photomultipliers have so-called quantum efficiencies around 20%, meaning that only 20% of the incident photons actually dislodge photoelectrons. But in most applications that inefficiency is more than compensated for by the enormous gain due to electron multiplication.

Photomultipliers are widely used in research, industry, environmental monitoring, laser-based distance measurements, radiation detectors, and medical instrumentation. In some applications they have been superseded by solid-state devices based on semiconductor technology, but for very sensitive and high-speed measurements, the photomultiplier remains at present the detector of choice.

FIGURE 39-9 Part of a photomultiplier tube, showing the multiplying effect as electrons hit each dynode. Many electrons strike the last dynode (not shown), producing a current pulse in external circuitry.

EXAMPLE 39-2	*Designing a Photomultiplier*

A photomultiplier is to respond to visible light with wavelengths shorter than 575 nm. Choose a suitable photocathode material from Table 39-1, and find the maximum kinetic energy of an electron that can be ejected by visible light.

Solution

Equation 39-7 shows that a photosensitive material will just barely eject electrons when the photon energy hf is equal to the work function ϕ. With $\lambda = c/f$, this gives a maximum work function

$$\phi = \frac{hc}{\lambda} = \frac{(6.63 \times 10^{-34} \text{ J·s})(3.00 \times 10^8 \text{ m/s})}{(575 \times 10^{-9} \text{ m})}$$

$$= 3.46 \times 10^{-19} \text{ J}.$$

The only material in Table 39-1 with a work function below this value is cesium, with $\phi = 3.43 \times 10^{-19}$ J.

The minimum wavelength—and therefore maximum photon energy—in visible light is about 400 nm. Using this value in Equation 39-7 with $f = c/\lambda$ gives

$$K = \frac{hc}{\lambda} - \phi$$

$$= \frac{(6.63 \times 10^{-34} \text{ J·s})(3.00 \times 10^8 \text{ m/s})}{400 \times 10^{-9} \text{ m}} - 3.43 \times 10^{-19} \text{ J}$$

$$= 1.54 \times 10^{-19} \text{ J}$$

for the maximum photoelectron energy.

Although cesium has the lowest elemental work function, photocathodes made with appropriate compounds operate throughout the visible and even into the infrared regions of the spectrum.

EXERCISE Which materials in Table 39-1 exhibit the photoelectric effect for at least some wavelengths of visible light? Comment on where they occur in the periodic table (see inside back cover).

Answer: cesium, potassium, sodium—all alkali metals from column 1 of the periodic table

Some problems similar to Example 37-2: 20–22, 25

Waves or Particles?

The work of Planck and Einstein applies quantization of energy to two quite different systems, vibrating molecules in Planck's case and electromagnetic waves in Einstein's. In positing the existence of photons, Einstein gave the first inklings of the **wave-particle duality**—the seemingly dual nature of light, which acts in some situations like a wave and in others, as in the photoelectric effect, more like a localized particle. We now turn to another phenomenon that demonstrates light's particle-like aspect. Later we will see how the wave-particle duality encompasses not only light but matter as well.

The Compton Effect

17.4
Compton Scattering

In 1923 the American physicist Arthur Holly Compton, at Washington University in St. Louis, did an experiment that dramatically confirmed the particle-like aspect of electromagnetic radiation. Although Compton's work came much later in the history of quantum theory than did Einstein's 1905 interpretation of the photoelectric effect, we include it here because it so strongly corroborates Einstein's photon hypothesis.

Compton was studying the interaction of electromagnetic radiation—specifically, X rays—with electrons. Classically, an electron subject to an electromagnetic wave should undergo oscillatory motion, driven by the wave's oscillating electric field. Since accelerated charge is the source of electromagnetic waves, the electron should itself produce electromagnetic waves *of the same frequency as the incident waves* (Fig. 39-10a). From the considerations of Section 34-9, the electron should radiate in essentially all directions, with maximum radiation perpendicular to its oscillatory motion.

Compton and his co-workers subjected a graphite target to an X-ray beam, and measured the intensity of scattered X rays as a function of wavelength for different scattering angles. Remarkably, they found the greatest concentration of scattered X rays at a wavelength *longer* than that of the incident radiation (Fig. 39-11). They interpreted their results as implying that particle-like photons had collided with free electrons in the graphite, losing energy to the electrons and therefore, since $E = hf$, emerging with lower frequency and correspondingly longer wavelength (Fig. 39-10b). (Another peak, which appears at the incident wavelength, is due to scattering from the atoms' tightly bound inner electrons. This scattering occurs without significant energy loss.)

We can understand the Compton effect by treating the interaction as a relativistic elastic collision between the incident photon and a stationary electron. In Chapter 38 we found that momentum and energy for light are related by $p = E/c$. Since $E = hf$ and $f = c/\lambda$, a photon of wavelength λ therefore has energy $E = hc/\lambda$ and momentum $p = h/\lambda$. Using Equations 38-13 and 38-15

(a) (b)

FIGURE 39-10 Scattering of electromagnetic radiation by an electron. (a) In the classical picture, the electron oscillates at the wave frequency and so reradiates at this frequency. Vertical lines are the incident wavefronts; circles are reradiated wavefronts. (b) In the quantum picture, the electron undergoes a particle-like collision with a quantum of electromagnetic radiation. As a result the photon energy decreases and its wavelength increases.

for relativistic momentum and energy, we can write conservation of energy and momentum for the collision of Fig. 39-10*b* as

$$\frac{hc}{\lambda_0} + mc^2 = \frac{hc}{\lambda} + \gamma mc^2 \qquad \text{(energy)}$$

$$\frac{h}{\lambda_0} = \frac{h}{\lambda}\cos\theta + \gamma mu\cos\phi \qquad (x \text{ momentum})$$

$$0 = \frac{h}{\lambda}\sin\theta - \gamma mu\sin\phi \qquad (y \text{ momentum}).$$

Here λ_0 is the wavelength of the incident photon, m the electron mass, u the speed of the electron after collision, and θ and ϕ the angles shown in Fig. 39-10*b*. Energy and momentum on the left-hand sides of these equations are values before the collision; right-hand sides are the corresponding quantities after the collision. Where two terms appear, the first is for the photon and the second for the electron. Even though the electron is initially at rest, it does have rest energy mc^2 that appears on the left-hand side of the energy equation. Solving these equations (see Problem 83) gives the **Compton shift**—i.e., the change in wavelength $\Delta\lambda = \lambda - \lambda_0$:

$$\Delta\lambda = \frac{h}{mc}(1 - \cos\theta). \qquad (39\text{-}8)$$

This equation is in excellent agreement with data from scattering experiments (Fig. 39-11).

The term h/mc in Equation 39-8 is called the **Compton wavelength** of the electron and gives the wavelength shift for a photon scattering at $\theta = 90°$. Its value is readily calculated:

$$\lambda_c = \frac{h}{mc} = 0.00243 \text{ nm}.$$

Equation 39-8 shows that the largest wavelength shift will be $2\lambda_C$, occurring at $\theta = 180°$. For the shift to be noticeable it should be a significant fraction of the incident wavelength, which therefore cannot be too many times the Compton wavelength. For X rays, λ is in the range from approximately 0.01 nm to 10 nm, and therefore, detection of the Compton shift in X rays is already a difficult experiment. It would be totally impossible with visible light. Scattering from more massive particles than electrons only makes matters worse, since a particle's Compton wavelength is inversely proportional to its mass.

Today, Compton scattering with gamma rays is a widely used technique for studying the structure of matter. For example, abnormalities in human bone can be detected through Compton scattering of gamma rays emitted by a small radioactive source embedded in bone. And the inverse Compton effect—the scattering of a rapidly moving electron off a photon—is a common process in high-energy astrophysical systems and is used in the laboratory to produce beams of gamma radiation.

The wavelength shift in Compton scattering simply admits no classical explanation. Coming after a decade of experimental and theoretical work that pointed increasingly to quantization as the essence of the atomic world, Compton's experimental results were for many physicists the convincing evidence for the reality of quanta.

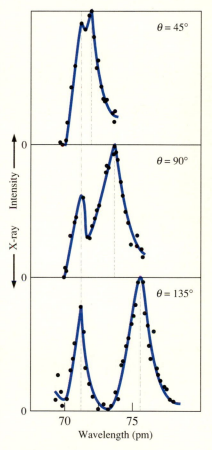

FIGURE 39-11 Compton's results for scattering of X rays with average wavelength about 71 pm, for scattering angles of 45°, 90°, and 135° (recall that the prefix *p* means 10^{-12}). The right-hand peak is from photons scattering off free electrons, and shows clearly the wavelength shift that is the hallmark of the Compton effect. The left-hand peak, at the incident wavelength, is from photons scattering off tightly bound atomic electrons, which cannot absorb significant energy from the photons.

EXAMPLE 39-3	*Compton Scattering*

A 0.0650-nm X-ray photon scatters off an electron at 75° to its initial direction. Find (a) the wavelength of the scattered photon and (b) the energy gained by the electron.

Solution

The wavelength shift is given by Equation 39-8:

$$\Delta\lambda = \lambda_c\,(1 - \cos\theta) = (0.00243\text{ nm})(1 - \cos 75°)$$

$$= 0.00180\text{ nm},$$

where we used $h/m_e c = \lambda_c$. The wavelength of the scattered photon is, therefore,

$$\lambda = \lambda_0 + \Delta\lambda = 0.0650\text{ nm} + 0.0018\text{ nm} = 0.0668\text{ nm}.$$

The photon's energy loss is the electron's gain; with $E = hf = hc/\lambda$ for the photon, we have

$$\Delta E = \frac{hc}{\lambda_0} - \frac{hc}{\lambda} = hc\left(\frac{1}{\lambda_0} - \frac{1}{\lambda}\right)$$

$$= (6.63\times10^{-34}\text{ J·s})(3.00\times10^8\text{ m/s})$$

$$\times \left(\frac{1}{6.5\times10^{-11}\text{ m}} - \frac{1}{6.68\times10^{-11}\text{ m}}\right)$$

$$= 8.25\times1\times10^{-17}\text{ J} = 515\text{ eV}.$$

EXERCISE A gamma ray undergoes a 45° Compton scattering. If its wavelength increases by 25%, find (a) its initial wavelength and (b) its initial energy in electron-volts.

Answers: (a) 2.85 pm; (b) 436 keV

Some problems similar to Example 39-3: 27, 30, 31

Got It!

Will the Compton wavelength shift be greater or less for photons of a given wavelength scattering off protons rather than electrons?

39-4 ATOMIC SPECTRA AND THE BOHR ATOM

In Chapter 34 we found that accelerated charges are the source of electromagnetic radiation. By 1900 it was known that atoms contain negative electrons as well as regions of positive charge; by 1911 experiments by Ernest Rutherford and his colleague Hans Geiger and student Ernest Marsden had localized the positive charge in the tiny but massive nucleus. According to classical physics, electrons should orbit the nucleus under the influence of the electric force, radiating electromagnetic wave energy as they accelerate in their orbits. In fact, a classical calculation shows atomic electrons will quickly radiate away all their energy and spiral into the nucleus. Thus, the very existence of atoms seems at odds with classical physics.

The Hydrogen Spectrum

A more subtle problem involving radiation from atoms dates to 1804, when William Wollaston noticed apparent lines of demarcation between some of the colors dispersed by a prism. Ten years later the German optician Josef von Fraunhofer dispersed the solar spectrum sufficiently that he could see hundreds of narrow, dark lines against the otherwise continuous spectrum of colors (Fig. 39-12a). Studies of light emitted by diffuse gases excited by electric discharges show similar **spectral lines**, these bright against an otherwise dark background (Fig. 39-12b). We know that bright-line spectra—called **emission spectra**—are produced when atoms emit light of discrete frequencies. **Absorption spectra**, in contrast, arise when atoms in a diffuse gas absorb certain discrete frequencies of light from a continuous source. What Fraunhofer saw in sunlight were absorption lines arising from the diffuse gas overlying the bright

(a)

(b)

FIGURE 39-12 (*a*) The solar spectrum is a continuous band of colors overlain by dark lines where specific colors have been absorbed in the solar atmosphere. Prominent line in the red is hydrogen-alpha, and the pair in the yellow are the 589-nm sodium doublet. (*b*) Spectral lines emitted by a diffuse sample of hydrogen gas. This is the Balmer series of lines, with Hα the bright line in the red.

visible surface of the Sun. We emphasize the word *diffuse:* Discrete spectra generally arise only when a gas is sufficiently diffuse that light from one atom stands a strong chance of escaping the gas before it interacts with other atoms. In dense gases, or in solids and liquids, multiple interactions result in the continuous spectrum characteristic of blackbody radiation.

Every element produces its own unique set of spectral lines, so analysis of spectra, even from the remote reaches of the cosmos, allows us to identify and characterize the material emitting the light. Spectral analysis led to the discovery of helium in the Sun's atmosphere before that element had been identified on Earth; hence the name, from the Greek word *helios* for Sun. Measuring the Doppler shift of spectral lines lets us "see" stars orbiting black holes in our own and distant galaxies, and also gives direct evidence for the expansion of the universe and thus its origin in a "big bang" explosion. Back on Earth, the technique of atomic absorption spectroscopy uses spectral lines to determine the elemental composition of substances, helping identify pollutants or trace the flow of elements in biological samples.

But why should atoms emit and absorb light of discrete frequencies? In 1884, several decades before a satisfactory answer was given, a Swiss schoolteacher named Johann Balmer realized that the wavelengths of the first four lines in the visible spectrum of hydrogen (Fig. 39-12*b*) were related by the equation

$$\frac{1}{\lambda} = R_H\left(\frac{1}{2^2} - \frac{1}{n^2}\right),$$

where $n = 3, 4, 5, 6 \ldots$ and R_H is a constant called the **Rydberg constant** for hydrogen, with value approximately $1.097 \times 10^7 \text{ m}^{-1}$. Other series of lines in the hydrogen spectrum were soon found, and Balmer's equation was generalized to read

$$\frac{1}{\lambda} = R_H\left(\frac{1}{n_2^2} - \frac{1}{n_1^2}\right), \tag{39-9}$$

where $n_1 = n_2 + 1, n_2 + 2, \ldots$. The Balmer series of lines has $n_2 = 2$; the Lyman series, in the ultraviolet, has $n_2 = 1$, and the infrared Paschen series has $n_2 = 3$. There are in fact infinitely many such series, corresponding to $n_2 = 1, 2, 3, \ldots$.

But why should atoms emit light in discrete spectral lines? And why should the hydrogen lines form patterns with the simple regularity described by Equation 39-9?

FIGURE 39-13 Danish physicist Niels Bohr (left) was a powerful influence on many of the twentieth century's greatest physicists. In his later years he was also active in humanitarian causes, especially the effort to control nuclear weapons. Here Bohr is discussing quantum mechanics with Max Planck.

18.1
Bohr Model

The Bohr Atom

In 1913 the great Danish physicist Niels Bohr (Fig. 39-13) proposed an atomic theory that accounted for the spectral lines of hydrogen. In the **Bohr atom** the electron moves in a circular orbit about the nucleus, held in orbit by the electric force. Classically, any orbital radius and correspondingly any energy and any angular momentum should be possible. But Bohr introduced quantization to the atom, stating that only certain orbits were possible—namely, those whose angular momentum is an integer multiple of Planck's constant divided by 2π. Angular momentum quantization implies energy quantization, which, as we soon show, leads to Equation 39-9 for the hydrogen spectral lines.

Bohr's theory asserted that an electron in any of the allowed orbits does not radiate energy, in direct contradiction to the predictions of classical electromagnetism. But an electron can jump from one orbit to another, emitting or absorbing a photon whose energy is equal to the energy difference between the two orbital levels. We can therefore find the expected photon energies—and thus the corresponding wavelengths—if we know the allowed energy levels.

To find the quantized atomic energy levels in Bohr's model, consider a hydrogen atom consisting of a fixed proton and an electron in circular orbit about the proton, held in its orbit by the electric force. Treating the proton as fixed is a good approximation because its mass is nearly 2000 times the electron's. We consider only electron speeds much less than that of the light, which is an excellent approximation in hydrogen and hydrogen-like atoms.

In Example 13-2 we found that the angular momentum of a particle with mass m and speed v, moving in a circular path of radius r, is mvr. Thus, Bohr's quantization condition reads

$$mvr = n\hbar, \qquad (39\text{-}10)$$

where $n = 1, 2, 3, \ldots$ and where we define $\hbar \equiv h/2\pi$. (This quantity, called "h bar," is useful because the combination $h/2\pi$ occurs frequently in quantum mechanics.) We need to relate the electron's angular momentum to its energy so we can find out what Equation 39-10 implies about energy quantization.

We studied circular orbits for the inverse-square force of gravity in Chapter 9, where we found that the kinetic and potential energy in a circular orbit are related by $K = -\frac{1}{2}U$, where the zero of potential energy is taken at infinity. The total energy $K + U$ is therefore $\frac{1}{2}U$. These results hold for any $1/r^2$ force, including the electric force. In the electric case the potential energy U is the point-charge potential of the proton, ke/r, multiplied by the electron charge, $-e$. Then the total energy is

$$E = \tfrac{1}{2}U = -\frac{ke^2}{2r}.$$

The minus sign has a simple interpretation: it means the electron is *bound* to the proton, in that it would take energy to separate them. Solving this equation for r then gives

$$r = -\frac{ke^2}{2E}. \qquad (39\text{-}11)$$

Since the kinetic energy is $K = -\frac{1}{2}U = -E$, we also have $\frac{1}{2}mv^2 = -E$ or

$$v = \sqrt{-2E/m}.$$

(Remember that E is negative, so the quantity under the square root is positive.) Using our expressions for r and v in the quantization condition of Equation 39-10 gives

$$m\sqrt{-2E/m}\left(-\frac{ke^2}{2E}\right) = n\hbar .$$

Solving for the energy E, we find

$$E = -\frac{k^2 e^4 m}{2\hbar^2 n^2} .$$

It is convenient to define the **Bohr radius**, a_0, given by

$$a_0 = \frac{\hbar^2}{mke^2} = 0.0529 \text{ nm} ;$$

with this definition our expression for the energy becomes

$$E = -\frac{ke^2}{2a_0}\left(\frac{1}{n^2}\right). \tag{39-12a}$$

Equation 39-12a gives us the allowed energy levels under Bohr's quantization condition. Evaluating this expression for the case $n = 1$ gives $E_1 = -2.18 \times 10^{-19}$ J $= -13.6$ eV; it's then convenient to write Equation 39-12a numerically in the form

$$E = -\frac{13.6 \text{ eV}}{n^2} , \tag{39-12b}$$

where in both forms $n = 1, 2, 3, \ldots$. The lowest energy state, $n = 1$, is called the **ground state**, while the others are **excited states**.

Now we have the allowed energy levels. What about spectra? Recall that Bohr's theory allows the electron to jump between energy levels, emitting or absorbing a photon whose energy hf is equal to the energy difference between the levels. So imagine an electron going from a higher level n_1 to a lower level n_2. The energy difference, according to Equation 39-12a, is

$$\Delta E = -\frac{ke^2}{2a_0}\left(\frac{1}{n_1^2} - \frac{1}{n_2^2}\right) = \frac{ke^2}{2a_0}\left(\frac{1}{n_2^2} - \frac{1}{n_1^2}\right),$$

and this is equal to the energy of the emitted photon. But the photon energy is $\Delta E = hf = hc/\lambda$, and therefore $1/\lambda = \Delta E/hc$ or, using our expression for ΔE,

$$\frac{1}{\lambda} = \frac{ke^2}{2a_0 hc}\left(\frac{1}{n_2^2} - \frac{1}{n_1^2}\right).$$

This looks just like Equation 39-9 for the hydrogen spectral lines, except that $ke^2/2a_0 hc$ replaces the Rydberg constant R_H. Evaluating this quantity gives

$$\frac{ke^2}{2a_0 hc} = \frac{(9.0 \times 10^9 \text{ N·m}^2/\text{C}^2)(1.6 \times 10^{-19} \text{ C})^2}{(2)(5.29 \times 10^{-11} \text{ m})(6.63 \times 10^{-34} \text{ J·s})(3.00 \times 10^8 \text{ m/s})}$$

$$= 1.09 \times 10^7 \text{ m}^{-1} ,$$

which is very close to the experimentally observed Rydberg constant. Using more accurate values for the fundamental constants would show that the two are the same to within about 1 part in 2000, with nearly all of the discrepancy resulting from our approximation that the proton is stationary.

FIGURE 39-14 Energy-level diagram for the Bohr model of the hydrogen atom, showing transitions responsible for the first three series of spectral lines. Each series arises in jumps to a common final state; $n = 2$ gives the visible and ultraviolet lines of the Balmer series.

18.2

Atomic Physics: Spectroscopy

Bohr's theory of quantized angular momentum thus accounts brilliantly for the observed spectrum of hydrogen. We can understand the origin of the various spectral line series using Fig. 39-14, which is an **energy level diagram** for the Bohr model of hydrogen. The diagram shows the allowed energy levels as horizontal lines, and the various possible transitions among levels as vertical arrows. Transitions with a common final state are grouped together in sets, and each such set represents a different series of spectral lines.

Knowing the energy levels of Equation 39-12, we can also find the radii of the allowed electron orbits, as given by Equation 39-11:

$$r = -\frac{ke^2}{2E} = \left(\frac{ke^2}{2}\right)\left(\frac{2a_0 n^2}{ke^2}\right) = n^2 a_0. \qquad (39\text{-}13)$$

Thus, the lowest energy orbit has a radius of one Bohr radius, with higher energy orbits growing rapidly with increasing n. A hydrogen atom in its ground state— $n = 1$—has a diameter of two Bohr radii, or about 0.1 nm. As we will see in the next chapter, the Bohr model's precise electron orbits are not compatible with the fully developed theory of quantum mechanics; nevertheless, Equation 39-13 gives correctly the approximate size of atoms.

EXAMPLE 38-4	*Big Atoms*

The tendency of physical systems toward states of lowest energy holds for atoms, and therefore, hydrogen atoms are normally found in their ground state, with diameter approximately 0.1 nm. But in the diffuse hydrogen gas of interstellar space, atoms can be found in highly excited states with sizes approaching a fraction of a millimeter. Transitions between adjacent energy levels of these so-called Rydberg atoms result in emission of photons at radio frequencies, and therefore, radio astronomers can study the enormous atoms. One of the longest wavelengths observed is the so-called 272α, corresponding to a transition from $n = 273$ to $n = 272$. (a) What is the diameter of a hydrogen atom in the $n = 273$ state? (b) At what wavelength and frequency should a radio telescope be set to observe this transition?

Solution

(a) Equation 39-13 gives the atomic diameter:

$$d = 2r = 2n^2 a_0 = (2)(273^2)(0.0529 \times 10^{-9} \text{ m}) = 7.9 \ \mu\text{m}.$$

This is just about the size of a red blood cell!

(b) To find the transition wavelength we use Equation 39-9 with $n_1 = 273$ and $n_2 = 272$;

$$\lambda = \left[R \left(\frac{1}{n_2^2} - \frac{1}{n_1^2} \right) \right]^{-1}$$

$$= \left[(1.097 \times 10^7 \text{ m}^{-1}) \left(\frac{1}{272^2} - \frac{1}{273^2} \right) \right]^{-1} = 92 \text{ cm}.$$

Using $f = c/\lambda$ gives the corresponding frequency of 325 MHz, which happens to lie in a gap between the VHF and UHF bands used for television broadcasting.

EXERCISE Find the wavelength of the spectral line Hα, corresponding to the transition from $n = 3$ to $n = 2$.

Answer: 656.3 nm

Some problems similar to Example 39-4: 38, 45, 74

Equation 39-12 shows, and Fig. 39-14 suggests, that there are infinitely many electron energy levels between the ground state at -13.6 eV and the zero of energy. It's possible to give an atomic electron enough energy to bring it above the $E = 0$ level, but then it is no longer bound to its proton and is therefore not part of the atom. The process of removing an electron is called **ionization**, and Equation 39-12b and Fig. 39-14 show that it takes 13.6 eV to ionize a hydrogen atom in its ground state. This quantity is called the **ionization energy**.

Limitations of the Bohr Model

Bohr's theory proved astoundingly successful in explaining the hydrogen spectrum. It also explains the spectra of hydrogen-like atoms—those with all but one of their electrons removed—with the appropriate change in the value of the nuclear charge. And it has some success in predicting the spectra of atoms such as lithium and sodium that have a single valence electron beyond a group of more tightly bound electrons. But it cannot account for the spectrum of more complicated atoms, even the simple two-electron helium. And even with hydrogen, there are subtle spectral details that the Bohr model does not address. Furthermore, like Planck's original quantum hypothesis, Bohr's quantization of atomic energy levels lacked a convincing theoretical basis. We will see in the next chapter how the much more comprehensive theory of quantum mechanics overcomes these limitations.

39-5 MATTER WAVES

In classical physics, light is purely a wave phenomenon. Einstein's introduction of energy "bundles"—photons—gave light a particle-like quality as well. In 1923, ten years after Bohr published his atomic theory, a French prince named Louis de Broglie (pronounced "de Broy") set forth a remarkable hypothesis in his doctoral thesis. If light has both wave-like and particle-like properties, he reasoned, why shouldn't particles of matter also exhibit both properties?

We saw in Chapter 34 that light with energy E also carries momentum $p = E/c$. Combined with Equation 39-6, that means a photon of light with frequency f has momentum $p = hf/c$. Since $f\lambda = c$, the photon's momentum and wavelength are therefore related by

$$\lambda = \frac{h}{p}. \quad \text{(de Broglie wavelength)} \qquad (39\text{-}14)$$

De Broglie proposed that this same relation should hold for particles of matter; at nonrelativistic speed, for example, an electron should have associated with it a "wavelength" given by h/mv.

De Broglie used his matter-wave hypothesis to explain why atomic electron orbits should be quantized. He proposed that Bohr's allowed orbits were those in which standing waves could exist (Fig. 39-15), in much the same way that a violin string can support only certain frequencies of standing waves. Suppose that n full wavelengths of a de Broglie electron wave fit around the circumference of the electron's circular orbit. Then we must have

FIGURE 39-15 The allowed electron orbits in the Bohr atom are those that can fit an integral number of de Broglie wavelengths around the circular Bohr orbit. Blue curve represents the de Broglie standing wave for the $n = 3$ level; red curve does not satisfy the Bohr condition. See ActivPhysics Activity 18.3 for a simulation of this figure.

$$n\lambda = n\frac{h}{p} = n\frac{h}{mv} = 2\pi r,$$

with r the orbit radius. Multiplying both sides by $mv/2\pi$ then gives $mvr = nh/2\pi = n\hbar$, which is Bohr's quantization condition. Thus, de Broglie's hypothesis provides a natural explanation for the quantization of atomic energy levels.

If matter has wave-like properties, why don't we notice them all the time? As in optics, wave effects become important only in systems whose size is comparable to or smaller than a wavelength. For macroscopic objects, and even for electrons moving at substantial speed in, say, a TV tube, the electron wavelength is so small that wave properties are unnoticeable. But when an electron is confined to small spaces—like the lower energy orbits of the Bohr atom—its wave properties are dominant.

18.3

Standing Electron Waves

EXAMPLE 39-5	*Baseballs and Electrons*

Find the de Broglie wavelength of (a) a 150-g baseball pitched at 45 m/s and (b) an electron moving at 1.0×10^6 m/s. Compare with the sizes of home plate and an atom, respectively.

Solution

Equation 39-14 gives the wavelengths:

$$\lambda_{\text{baseball}} = \frac{h}{p} = \frac{h}{mv} = \frac{6.63\times10^{-34}\ \text{J}\cdot\text{s}}{(0.15\ \text{kg})(45\ \text{m/s})} = 9.8\times10^{-35}\ \text{m}.$$

This is about 34 orders of magnitude smaller than home plate, showing that the baseball's wave properties are totally negligible in the systems with which it interacts. For the electron, on the other hand,

$$\lambda_{\text{electron}} = \frac{h}{mv} = \frac{6.63\times10^{-34}\ \text{J}\cdot\text{s}}{(9.11\times10^{-31}\ \text{kg})1(1.0\times10^6\ \text{m/s})} = 0.73\ \text{nm},$$

several times the size of an atom. Thus we should expect wave effects to be prominent when electrons with this speed interact with atoms.

EXERCISE At what speed would a proton's de Broglie wavelength equal the approximately 0.1-nm diameter of the hydrogen atom?

Answer: 4 km/s

Some problems similar to Example 39-5: 47, 48, 50, 52

| APPLICATION | *The Electron Microscope* |

In Chapter 37 we saw that the resolution of an optical system—the minimum separation between two distinguishable points—is on the order of the wavelength of the illuminating light. That limits the resolution of conventional optical microscopes to some hundreds of nanometers. But the wavelengths of electrons can be made much smaller than that of visible light, and therefore **electron microscopes** can achieve resolutions on the order of 1 nm or less. The associated magnifications approach 10^6.

Figure 39-16 shows a simplified diagram of an electron microscope used for high magnification of small objects. An electric field accelerates electrons to energies of 50–100 keV, with corresponding wavelengths on the order of 0.005 nm. The electron beam then enters the microscope's "optical" system, consisting of magnetic fields that focus the electron beam in the way glass lenses focus light in a conventional microscope. Aberration in these lenses, rather than diffraction effects, ultimately limits the resolution of the electron microscope at electron energies above about 100 keV. An object placed in the beam scatters electrons, resulting in darker regions on a screen, film, or other detector (Fig. 39-17a).

The electron microscope must be evacuated to prevent scattering of the electron beam by air molecules. This complicates the preparation of biological samples, which must be desiccated and mounted on substrates that themselves do not significantly scatter the beam. Nevertheless, the electron microscope has become an indispensable tool not only in biology, but also in chemistry and metallurgy.

A related instrument, the scanning electron microscope, affords dramatic three-dimensional imaging at magnifications in the range from about 10 to 10^5 (Fig. 39-17b).

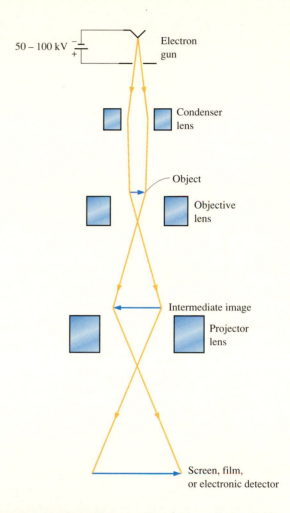

FIGURE 39-16 An electron microscope. The electron gun accelerates electrons to energies of 50–100 keV. The optical system then uses magnetic fields as lenses to focus the beam and form an image.

(a)

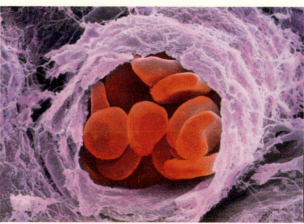

(b)

FIGURE 39-17 (a) Conventional electron micrograph of a tuberculosis bacterium. (b) Scanning electron micrograph showing blood cells in an arteriole.

Incident electron beam

Scattered beam

$\phi = 50°$

$\theta = 65°$

$d = 91$ pm

ActivPhysics

17.5
Electron Interference

FIGURE 39-18 In reflecting an electron beam off a nickel crystal, Davisson and Germer found maximum reflection at $\theta = 50°$—the angle one would expect for diffraction of waves with the de Broglie wavelength of the incident electrons.

Electron Diffraction

In 1927 the American physicists Clinton J. Davisson and Lester H. Germer gave a convincing experimental verification of de Broglie's matter-wave hypothesis. Davisson and Germer were not actually searching for de Broglie's waves, but were studying the interaction of an electron beam with a nickel crystal. They found that the intensity of the scattered beam peaked when the angle between the incident and scattered beams was 50° (Fig. 39-18). Davisson and Germer noted the similarity between this phenomenon and the peaks observed in X-ray diffraction from the planes of atoms in a crystal, and concluded that they were seeing the results of interference among the electron waves scattered from the nickel atoms. As the example below shows, their measurement also confirmed quantitatively de Broglie's formula for the electron wavelength.

EXAMPLE 39-6	*Electron Diffraction*

The electron energy in Davisson and Germer's experiment was 54 eV, and the spacing between atomic planes in their nickel crystal was 0.091 nm. Show that their detection of a scattered peak at $\phi = 50°$ in Fig. 39-18 is consistent with de Broglie's expression for the electron wavelength (Equation 39-14).

Solution

We need the electron momentum to find the wavelength. Writing $K = \frac{1}{2}mv^2$ and then solving for the momentum gives $p = mv = \sqrt{2mK}$. Then the de Broglie wavelength is

$$\lambda = \frac{h}{p} = \frac{h}{\sqrt{2mK}}$$

$$= \frac{6.63 \times 10^{-34} \text{ J·s}}{\sqrt{(2)(54 \text{ eV})(1.6 \times 10^{-19} \text{ J/eV})(9.11 \times 10^{-31} \text{ kg})}}$$

$$= 1.67 \times 10^{-10} \text{ m} = 0.167 \text{ nm}.$$

If we treat the scattering of the electron waves like the X-ray scattering of Fig. 37-17, we can apply the Bragg condition (Equation 37-7): $2d \sin\theta = m\lambda$. Comparison of Figs. 37-17 and 39-18 shows that the angle θ in the Bragg condition is 65° for Davisson and Germer's experiment. Using the calculated de Broglie wavelength and the 0.091-nm atomic spacing, we expect a first-order diffraction peak at

$$\theta = \sin^{-1}\left(\frac{\lambda}{2d}\right) = \sin^{-1}\left(\frac{0.167 \text{ nm}}{(2)(0.091 \text{ nm})}\right) = 67°,$$

in close agreement with the experimental result that $\theta = 65°$.

EXERCISE Find the angle ϕ in Fig. 39-18 if 100-eV electrons are incident on the same crystal.

Answer: 95°

Some problems similar to Example 39-6: 53, 54

Shortly after Davisson and Germer's work, the Scottish physicist George Thomson passed an electron beam through a small aperture and observed electron diffraction directly (Fig. 39-19). Thomson was the son of J. J. Thomson, who had discovered the electron in 1897. In a familial manifestation of the wave-particle duality, father Thomson identified the electron as a particle while his son found it to be a wave. Since Thomson's work, similar diffraction experiments have shown the wave nature of neutrons and other particles.

FIGURE 39-19 Diffraction pattern produced by passing an electron beam through a circular aperture shows that electrons have a wave-like character, and looks very much like the diffraction pattern created when light passes through a circular aperture (see Fig. 37-37).

FIGURE 39-20 Heisenberg's "quantum microscope." (*a*) A short-wavelength, high-energy photon interacts with an electron. Diffraction is minimal, and the electron's position is accurately determined. But the high-energy photon disturbs the electron's trajectory, leaving its momentum very uncertain. (*b*) Using a long-wavelength, low-energy photon minimizes the disturbance and allows accurate determination of the electron's momentum. But now diffraction is significant, and the measurement does not give an accurate position.

(*a*)

(*b*)

39-6 THE UNCERTAINTY PRINCIPLE

In classical physics it is, in principle, possible to know the exact position and velocity of a particle and, therefore, to predict with certainty its future behavior. But not so in quantum mechanics! In 1927 the German physicist Werner Heisenberg presented his famous **uncertainty principle**, which states that certain pairs of quantities cannot be measured simultaneously with arbitrary precision. Position and momentum constitute one such pair; if we measure a particle's position to within an uncertainty Δx, then we cannot simultaneously determine its momentum to an accuracy better than Δp, where

$$\Delta x \, \Delta p \geq \hbar. \qquad \text{(uncertainty principle)*} \qquad (39\text{-}15)$$

Why this limitation? The fundamental reason is quantization. To measure some property of a system requires interacting with that system—for example, shining light on it. That interaction invariably involves energy, and the interaction energy disturbs the system slightly. As a result of this disturbance, values inferred from the measurement are no longer quite the right ones. In classical physics the energy can be made as small as possible, resulting in an arbitrarily small disturbance. But in quantum theory the minimum energy is a single quantum, like a photon of light, and thus, the disturbance cannot be made arbitrarily small.

So why not use lower frequency light, whose photon energy hf is smaller? Because lower frequency means longer wavelength and, as we found in Chapter 37, precise imaging requires light whose wavelength is much shorter than the system being imaged. Heisenberg summarized this dilemma with the "thought experiment" shown in Fig. 39-20. He imagined a "quantum microscope" observing a single photon interacting with an electron. Using a short-wavelength photon allows precise localization of the electron (Fig. 39-20*a*). But short wavelength means high frequency and thus, through Equation 39-6, high photon energy. The high-energy photon imparts considerable momentum to the electron, and thus, any knowledge we had of the electron's momentum is seriously degraded by the very act that determines its position. We might instead choose a long-wavelength, low-energy photon, as in Fig. 39-20*b*. Then there is little disturbance, and we can get an accurate assessment of the electron's momentum. But now the photon wavelength is so long that diffraction effects severely limit our ability to tell just where the electron was. So we can measure the electron's position accurately, but only at the expense of very imprecise knowledge of its momentum. Or we can measure its momentum accurately, but then we'll know very little about its position. With a photon of intermediate wavelength we could measure both quantities, but neither with perfect accuracy. Equation 39-15 establishes the minimum uncertainties possible.

*Activ*Physics

17.6
Uncertainty Principle

---

* The right-hand side of this inequality is approximate; the important point is that the product cannot exceed a quantity on the order of $\hbar$. The absolute minimum value for the uncertainty is $\frac{1}{2}\hbar$ and occurs only when the waveform of the particle's de Broglie wave has a certain shape.

EXAMPLE 39-7	*Microelectronics*

A beam of aluminum atoms is used to "dope" a semiconductor chip to give it the proper electrical characteristics. If the electron velocity is required to be $(4.50000 \pm 0.00001) \times 10^4$ m/s, how accurately can the electrons be localized?

Solution

With a velocity spread of $\pm 0.00001 \times 10^4$ m/s, the maximum tolerable uncertainty in the atoms' velocity is $\Delta v = 0.00002 \times 10^4$ m/s, or 0.20 m/s. Then corresponding momentum uncertainty is

$$\Delta p = m\Delta v = (26.98 \text{ u})(1.66 \times 10^{-27} \text{ kg/u})(0.20 \text{ m/s})$$

$$= 8.96 \times 10^{-27} \text{ kg·m/s},$$

where we found the atomic weight of aluminum in Appendix D. Equation 39-15 then gives the uncertainty in position:

$$\Delta x \geq \frac{\hbar}{\Delta p} = \left(\frac{1.055 \times 10^{-34} \text{ J·s}}{8.96 \times 10^{-27} \text{ kg·m/s}} \right) = 12 \text{ nm}.$$

This is about 100 atomic diameters and shows that the uncertainty principle will ultimately limit our ability to manufacture very small microelectronic structures by the means now used.

EXERCISE You measure the velocity of a 3.0-g ping-pong ball to an accuracy of ± 0.01 mm/s. Find the minimum uncertainty in its position, and compare with the size of the ball.

Answer: 1.8×10^{-27} m, totally negligible compared with the ball's size

Some problems similar to Example 39-7: 55, 57, 58

The uncertainty principle is intimately connected with de Broglie's wave hypothesis. Suppose we pass an electron beam through a slit, as shown in Fig. 39-21. Then we know the vertical component of the electrons' position to within the slit width. But the electrons have a wave-like aspect, and they diffract in passing through the slit. If the slit width is much greater than the wavelength, then, as we found in Chapter 37, the diffracted beam does not spread much (Fig. 39-21a). We can therefore be quite sure of the electrons' vertical momentum; in this case it is very close to zero. But the slit is wide, so there is considerable uncertainty in the vertical position. We can find the position more accurately by making the slit smaller, but then the diffracted beam widens and the uncertainty in the vertical momentum increases. So the wave nature of matter imposes a trade-off: the more accurately we know the position of a particle, the less accurately can we know its momentum, and vice versa.

It sounds like the uncertainty principle only diminishes our knowledge. But actually, the principle provides useful estimates of the size and energy scales in atomic and subatomic systems, as the example below illustrates.

Wide slit   Narrow slit

(a)   (b)

FIGURE 39-21 (a) An electron beam passing through a wide slit shows little diffraction, and therefore the vertical component of an electron's momentum is known to be very nearly zero. With a wide slit, though, an electron's position is not well known. (b) A narrow slit localizes the electron, increasing the accuracy with which its vertical position is known. But diffraction increases the uncertainty in its vertical momentum.

EXAMPLE 39-8	*Atomic and Nuclear Energies*

Use the uncertainty principle to estimate the minimum energy possible for (a) an electron confined to a one-dimensional region of atomic dimensions (about 0.1 nm) and (b) a proton confined to a region of nuclear dimensions (about 1 fm).

Solution

The key to making these estimates is to understand that momentum is a vector quantity. We could know its exact magnitude and still be uncertain about it if we didn't know its direction. In this example we have particles confined in one-dimensional regions. The uncertainty in position is at most the width Δx of the region. The particles can't be at rest because then we would know their exact momentum—zero—in violation of the uncertainty principle. The minimum momentum p they could have, in fact, is such that the uncertainty associated with not knowing which way the particles are going satisfies the uncertainty principle. Not knowing the direction amounts to being uncertain by the difference between the momentum in one direction and the momentum in the other, or $\Delta p = p - (-p) = 2p$. Then the uncertainty principle (Equation 39-15) gives $2p\Delta x \geq \hbar$, or

$$p \geq \frac{\hbar}{2\Delta x}.$$

Using $p = mv$ and $K = \frac{1}{2}mv^2$, we can write the kinetic energy as $K = p^2/2m$, and therefore, we must have

$$K \geq \frac{1}{2m}\left(\frac{\hbar}{2\Delta x}\right)^2.$$

Using the electron mass for m and $\Delta x = 0.1$ nm then gives a minimum energy of about 1 eV, roughly the energy scale for atomic energy levels. Using the proton mass and the nuclear size gives a minimum energy of 5 MeV. The ratio of these energies, a factor of about 10^7, indicates the dramatic difference between chemical energy sources—that ultimately involve rearrangement of atomic electrons—and nuclear sources that disrupt the nucleus itself.

EXERCISE An electron is somewhere in a molecule 15 nm long. What is its minimum kinetic energy?

Answer: 42 μeV

Some problems similar to Example 39-8: 60, 84

Energy-Time Uncertainty

A second pair of variables that defy simultaneous accurate measurement are the energy of a system and the time it takes to measure that energy. The uncertainty ΔE in the energy is related to the measurement time Δt through the inequality

$$\Delta E\, \Delta t \geq \hbar. \qquad (39\text{-}16)$$

One effect of energy-time uncertainty is to render atomic and nuclear energy levels inexact and, therefore, to broaden spectral lines. If an atom were forever in a fixed energy state, we could take infinitely long to measure its energy and therefore make ΔE arbitrarily small. But excited states of atoms have characteristic lifetimes (typically $\sim 10^{-8}$ s), which limits the measurement time and therefore sets a minimum uncertainty in the energy level. Problems 64 and 65 explore energy-time uncertainty.

Observers, Uncertainty, and Causality

The uncertainty principle moves the observer from a passive onlooker to an active participant in physical events. To observe is necessarily to disturb, and quantum theory is therefore very much concerned with the role of the observer and the process of measurement. The uncertainty principle is fundamentally a statement about what can and cannot be learned through measurement.

Position and momentum cannot be measured simultaneously with perfect accuracy. Surely, though, a particle has well-defined values of both, even though we cannot know them? The answer seems to be no. The standard interpretation of quantum mechanics suggests that it makes no sense to talk about what cannot be measured, and recent experiments have ruled out "hidden variables" that might be active at a lower level to guide the particle in a deterministic path. Its wave aspect makes a particle a "fuzzy" thing, and it really makes no sense to think of the particle as a tiny ball with definite momentum and position. For that reason it also makes no sense to think of the particle's future as being fully determined in the sense that Newton's laws determine the future path of, say, a baseball. We are left with uncertainty—or indeterminacy, as Heisenberg's word also translates—as a fundamental fact of our universe.

39-7 THE WAVE-PARTICLE DUALITY, COMPLEMENTARITY, AND THE CORRESPONDENCE PRINCIPLE

FIGURE 39-22 Niels Bohr, Werner Heisenberg, and Wolfgang Pauli in an intense discussion of quantum mechanics.

One of the most disturbing aspects of quantum theory is the wave-particle duality—the seemingly contradictory fact that matter and light have both wave-like and particle-like properties. If this bothers you, you're in good company: it bothered the early quantum physicists, too (Fig. 39-22). Werner Heisenberg, who first stated the uncertainty principle, described his frustration in trying to understand the quantum world:

> I remember discussions with Bohr which went through many hours till very late at night and ended almost in despair; and when at the end of the discussion I went alone for a walk in the neighboring park I repeated to myself again and again the question: Can nature possibly be as absurd as it seems to us in these atomic experiments?*

Bohr dealt with the wave-particle duality through his **principle of complementarity**. The wave and particle pictures, he said, are complementary aspects of the same reality. If we do an experiment to measure a wave-like property—for example, the diffraction of electrons—then we find wave properties but not particle properties. If we do an experiment to measure a particle-like property—for example, localizing an electron—then we will not find wave properties. The two measurements require different experiments, and we can't perform both simultaneously on the same entity. So we will never catch wave and particle in an outright contradiction, and the answer to the question "Which is it, wave or particle?" has to be that it's both, and that which you find depends on what experiment you choose to perform.

Bohr articulated a second principle that helps reconcile the seeming contradiction between classical and quantum physics. His **correspondence principle** states that the predictions of classical and quantum physics should agree in situations where the size of individual quanta is negligible. Taking $h \rightarrow 0$ in Planck's law, for example, gives the classical Rayleigh-Jeans law (see Problem 13). Or, for large values of n, the energies of adjacent atomic states in the Bohr model become so close that the levels appear essentially as a continuum of allowed energies—as expected in classical physics. Or consider a 1000-W radio beam; the photon energy hf is so small that the beam contains an enormous number of photons, and we can consider the energy distributed essentially continuously over the beam. But in a 1000-W X-ray beam, the photon energy is much higher and the number of photons correspondingly fewer; it's therefore difficult to avoid the fundamental fact of energy quantization in this beam. Visible light lies somewhere in between; we can often treat its energy as being continuously distributed, except when it interacts with systems as small as individual atoms.

* Heisenberg, Werner. *Physics and Philosophy: The Revolution in Modern Science* (New York: Harper & Brothers, 1962).

CHAPTER SYNOPSIS

Summary

1. **Quantum physics** describes matter and energy on the small scale. The essential difference between quantum and classical physics is that some physical quantities in quantum physics come only in certain discrete values, or quanta.

2. The wavelength distribution of **blackbody radiation** could not be explained using classical physics. Only by introducing the notion that the energies of vibrating molecules are quantized in units of hf could Planck give a mathematical description of the blackbody spectrum. The constant h, called **Planck's constant**, sets the fundamental scale for quantum phenomena.

3. Einstein's explanation of the **photoelectric effect** introduced quantization of electromagnetic wave energy in the form of **photons** of energy hf, where f is the wave frequency. Einstein's proposal gave the first hint of the **wave-particle duality**.

4. Experiments with the **Compton effect** reinforced the view that light has a particle-like aspect, by demonstrating that a photon and electron undergo particle-like collisions from which the photon emerges with lower energy and longer wavelength.

5. Light from glowing gases shows discrete **spectral lines** whose wavelength, for hydrogen, follows the simple formula

$$\frac{1}{\lambda} = R_H \left(\frac{1}{n_2^2} - \frac{1}{n_1^2} \right),$$

where n_1 and n_2 are any integers with $n_1 > n_2$. Bohr was able to explain the hydrogen spectrum by postulating that electrons circle the nucleus in only certain allowed orbits whose angular momentum mvr is quantized according to the relation $mvr = nh/2\pi$, where n is an integer. Bohr's theory shows that the resulting electron energies are given by

$$E = -\frac{13.6 \text{ eV}}{n^2},$$

and that the radius of the electron orbit is

$$r = a_0 n^2,$$

where a_0 is the **Bohr radius** given by $a_0 = \hbar^2/mke^2 = 0.0529$ nm.

6. De Broglie proposed that particles of matter have a wave-like aspect, with their wavelength given by

$$\lambda = \frac{h}{p},$$

where p is the particle's momentum. Bohr's quantization condition for atomic electron orbits follows under the assumption that the allowed orbits are those that can accommodate an integer number of de Broglie electron wavelengths. Experiments by Davisson and Germer and by

G. Thomson demonstrated wave interference and diffraction effects in electron beams.

7. Heisenberg's **uncertainty principle** states that it is impossible to measure simultaneously and with perfect accuracy both the position and momentum of a particle; in particular,

$$\Delta p \, \Delta x \geq \hbar,$$

where Δp and Δx are the uncertainties in position and momentum, respectively, and where $\hbar = h/2\pi$. Similarly, it is impossible to measure energy with perfect accuracy in a finite time.

8. Bohr's **principle of complementarity** reconciles the seeming contradiction inherent in the wave-particle duality. The principle states that wave and particle are complementary aspects of reality, and that any given experiment detects one aspect or the other, but never both simultaneously. Bohr's **correspondence principle** states that quantum and classical physics must agree in situations where individual quanta are negligibly small.

Terms You Should Understand

(Pairs are closely related terms whose distinction is important; number in parentheses is chapter section where term first appears.)

classical physics, quantum physics (introduction)
blackbody radiation (39-2)
Planck's constant (39-2)
ultraviolet catastrophe (39-2)
photoelectric effect, Compton effect, Compton wavelength (39-3)
wave-particle duality (39-3, 39-7)
spectral line (39-4)
Bohr atom (39-4)
ground state, excited state (39-4)
de Broglie wavelength (39-5)
uncertainty principle (39-6)
principle of complementarity (39-7)
correspondence principle (39-7)

Symbols You Should Recognize

$h, \hbar$ (39-2, 39-4)
λ_c (39-3)
a_0 (39-4)
$\Delta p, \Delta x, \Delta E, \Delta t$ (39-6)

Problems You Should Be Able to Solve

evaluating radiance of blackbodies (39-2)
solving for electron energies and other quantities in the photoelectric effect (39-3)
analyzing Compton scattering (39-3)

evaluating energy levels and orbital radii in the Bohr atom (39-4)

finding spectral lines for hydrogen and hydrogen-like atoms (39-4)

evaluating de Broglie wavelengths (39-5)

applying the uncertainty principle (39-6)

Limitations to Keep in Mind

The theories outlined in this chapter represent first attempts at explaining quantum phenomena, and are neither fully accurate nor universally applicable.

QUESTIONS

1. Can you explain the workings of the telephone using classical physics? What about the electron microscope? Why or why not?

2. Why does classical physics predict that atoms should collapse?

3. Looking at the night sky, you see one star that appears reddish, another yellowish, and another bluish. Compare their temperatures.

4. Any object warmer than absolute zero radiates energy. So why can't you see warm objects in a dark room?

5. Imagine an atom that, unlike hydrogen, had only three energy levels. If these levels were evenly spaced, how many spectral lines would result? How would their wavelengths compare?

6. What colors of visible light have the highest energy photons?

7. Why is the immediate ejection of electrons in the photoelectric effect surprising from a classical viewpoint?

8. Suppose the Compton effect were significant at radio wavelengths. What problems might this present for radio and TV broadcasting?

9. How are the wave-particle duality and the uncertainty principle related?

10. Why is the Compton effect seen only with X rays and gamma rays?

11. For what photon scattering angle θ will the electron in a Compton scattering event gain the most energy?

12. How many spectral lines are in the entire Balmer series?

13. Why are the lines of the Lyman series in the ultraviolet while some Balmer lines are visible?

14. Why does the photoelectric effect suggest that light has particle-like properties?

15. Energy-time uncertainty limits the precision with which we can know the mass of unstable particles (those that decay after a finite time). Why?

16. Why is the Compton effect a particularly convincing demonstration that electromagnetic radiation has particle-like properties?

17. Equation 39-14 suggests that a particle that's not moving should have infinite wavelength. Speculate on what this might mean.

18. If you measure a particle's position with perfect accuracy, what do you know about its momentum?

19. How might our everyday experience be different if Planck's constant had the value 1 J·s?

20. Why is the energy given by Equation 39-12 negative?

PROBLEMS

ActivPhysics can help with these problems:
Activities 17.3–17.6

Section 39-2 Blackbody Radiation

Note: Most answers in this section are very sensitive to the exact values used for the constants h and c.

1. If you increase the temperature of a blackbody by a factor of 2, by what factor does its radiated power increase?

2. The surface temperature of the star Rigel is 10^4 K. (a) What is the power per square meter radiated by Rigel? (b) At what wavelength is the maximum power radiated? In what spectral region is this?

3. At what wavelength does Earth, approximated as a 286-K blackbody, radiate the most energy?

4. Spacecraft instruments measure the radiation from a passing asteroid and find that it peaks at 40 μm wavelength. Assuming the asteroid is a blackbody, what is its surface temperature?

5. According to Planck's theory, what is the minimum nonzero energy of a molecule with vibration frequency 3.4×10^{14} Hz?

6. An oscillating molecule drops from one allowed energy level to the next lower one. If its energy changes by 2.2 eV, what is its oscillation frequency?

7. The Sun approximates a blackbody at 5800 K. Find the wavelength at which the Sun emits the most energy.

8. What is the power per unit area emitted by a 3000-K incandescent lamp filament in the wavelength interval from 500 nm to 502 nm? *Hint:* The radiance given by Equation 39-3 is the power per unit area *per unit wavelength interval.*

9. Find the temperature range over which a blackbody will radiate the most energy at a wavelength in the visible range (400 nm–700 nm).

10. By what factors does the radiance of a blackbody at 600 nm increase as the temperature is increased from 1000 K to (a) 1500 K and (b) 2000 K?

11. Treating the Sun as a 5.8-kK blackbody, how does its ultraviolet radiance at 200 nm compare with its visible radiance at its peak wavelength of 500 nm?

12. For a 2.0-kK blackbody, by what percentage is the Rayleigh-Jeans law in error at wavelengths of (a) 1.0 mm, (b) 10 μm, and (c) 1.0 μm?

13. Use the series expansion for e^x (Appendix A) to show that Planck's law (Equation 39-3) reduces to the Rayleigh-Jeans law (Equation 39-5) when $\lambda \gg hc/kT$.

Section 39-3 Photons

14. Find the energy in electron-volts of (a) a 1.0-MHz radio photon, (b) a 5.0×10^{14}-Hz optical photon, and (c) a 3.0×10^{18}-Hz X-ray photon.

15. What is the wavelength of a 6.5-eV photon? In what spectral region is this?

16. A microwave oven uses electromagnetic radiation at 2.4 GHz. (a) What is the energy of each microwave photon? (b) At what rate does a 625-W oven produce photons?

17. A red laser at 650 nm and a blue laser at 450 nm emit photons at the same rate. How do their total power outputs compare?

18. (a) Find the Compton wavelength for the proton. (b) Find the energy, in electron-volts, of a gamma ray whose wavelength equals the proton's Compton wavelength.

19. Find the rate of photon production by (a) a radio antenna broadcasting 1.0 kW at 89.5 MHz, (b) a laser producing 1.0 mW of 633-nm light, and (c) an X-ray machine producing 0.10-nm X rays with a total power of 2.5 kW.

20. What is the work function of a material whose photoelectric cutoff wavelength is 390 nm?

21. Electrons in a photoelectric experiment emerge from an aluminum surface with maximum kinetic energy of 1.3 eV. What is the wavelength of the illuminating radiation?

22. What is the maximum work function for a surface to emit electrons when illuminated with 900-nm infrared light?

23. (a) Find the cutoff frequency for the photoelectric effect in copper. (b) Find the maximum energy of the ejected electrons if the copper is illuminated with light of frequency 1.8×10^{15} Hz.

24. The stopping potential in a photoelectric experiment is 1.8 V when the illuminating radiation has wavelength 365 nm. (a) What is the work function of the emitting surface? (b) What would be the stopping potential for 280-nm radiation?

25. Which materials in Table 39-1 exhibit the photoelectric effect *only* for wavelengths shorter than 275 nm?

26. What is the initial wavelength of a photon that loses half its energy when it Compton scatters from an electron and emerges at 90° to its initial direction of motion?

27. A photon with wavelength 15 pm Compton scatters off an electron, its direction of motion changing by 110°. What fraction of the photon's initial energy is lost to the electron?

28. When light shines on a potassium surface, the maximum speed with which electrons leave the surface is 4.2×10^5 m/s. Find the wavelength of the light.

29. The maximum electron energy in a photoelectric experiment is 2.8 eV. When the wavelength of the illuminating radiation is increased by 50%, the maximum electron energy drops to 1.1 eV. Find (a) the work function of the emitting surface and (b) the original wavelength.

30. A 150-pm X-ray photon undergoes Compton scattering with an electron, and emerges at 135° to its original direction of motion. Find (a) the wavelength of the scattered photon and (b) the kinetic energy of the electron.

31. An electron is initially at rest. What will be its kinetic energy after a 0.10-nm X-ray photon scatters from it at 90° to its original direction of motion?

32. Find the scattering angle for which a 7.5-MeV photon undergoing Compton scattering with an electron loses half its energy.

33. What is the minimum photon energy for which it is possible for the photon to lose half its energy undergoing Compton scattering with an electron?

Section 39-4 Atomic Spectra and the Bohr Atom

34. Calculate the wavelengths of the first three lines in the Lyman series for hydrogen.

35. Which spectral line of the hydrogen Paschen series ($n_2 = 3$) has wavelength 1282 nm?

36. The wavelengths of a spectral line series tend to a limit as $n_1 \rightarrow \infty$. Evaluate the series limit for (a) the Lyman series and (b) the Balmer series in hydrogen.

37. What is the maximum wavelength of light that can ionize hydrogen in its ground state? In what spectral region is this?

38. A Rydberg hydrogen atom makes a downward transition to the $n = 225$ state. If the photon emitted has energy 9.32 μeV, what was the original state?

39. It takes 4.9 eV of energy to excite electrons to a higher energy level in mercury atoms. What will be the wavelength of the photon emitted when an excited electron drops back to the lower level?

40. A hydrogen atom is in its ground state when its electron absorbs 48 eV in an interaction with a photon. What is the energy of the resulting free electron?

41. At what energy level does the Bohr hydrogen atom have diameter 5.18 nm?

42. How much energy does it take to ionize a hydrogen atom in its first excited state?

43. Ultraviolet light with wavelength 75 nm shines on a gas of hydrogen atoms in their ground states, ionizing some of the atoms. What is the energy of the electrons freed in this process?

44. A hydrogen atom in its ground state absorbs a 12.75-eV photon and ends up in an excited state. It then drops to the next lower state. What is the energy of the photon emitted in this downward transition?

45. Helium with one of its two electrons removed acts very much like hydrogen, and the Bohr model successfully describes it. Find (a) the radius of the ground-state electron orbit and (b) the photon energy emitted in a transition from

the $n = 2$ to the $n = 1$ state in this singly ionized helium. *Hint:* The nuclear charge is $2e$.

46. What is the radius of the first Bohr orbit in O^{7+} (i.e., oxygen with 7 of its 8 electrons removed)?

Section 39-5 Matter Waves

47. Find the de Broglie wavelength of (a) Earth, in its 30-km/s orbital motion, and (b) an electron moving at 10 km/s.

48. How slowly must an electron be moving for its de Broglie wavelength to equal 1 mm?

49. A proton and an electron have the same de Broglie wavelength. How do their speeds compare, assuming both are much less than that of light?

50. Find the de Broglie wavelength of electrons with kinetic energies of (a) 10 eV, (b) 1.0 keV, and (c) 10 keV.

51. Through what potential difference should you accelerate an electron from rest so its de Broglie wavelength will be the size of a hydrogen atom, about 0.1 nm?

52. What is the minimum electron speed that would make an electron microscope superior to an optical microscope using 450-nm light?

53. A Davisson-Germer type experiment using nickel gives peak intensity in the reflected beam when the incident and reflected beams have an angular separation of 100°. What is the electron energy?

54. A beam of 75-eV electrons impinges on an unknown crystal, producing a first-order diffraction peak at 65° from the incident beam. What is the spacing of the atomic planes in the crystal?

Section 39-6 The Uncertainty Principle

55. A proton is confined to a space 1 fm wide (about the size of the atomic nucleus). What is the minimum uncertainty in its velocity?

56. Is it possible to measure an electron's velocity to an accuracy of ± 1.0 m/s while simultaneously finding its position to an accuracy of ± 1.0 μm? What about a proton?

57. A proton has velocity $\mathbf{v} = (1500 \pm 0.25)\hat{\imath}$ m/s. What is the uncertainty in its position?

58. An electron is moving in the $+x$ direction with speed measured at 5.0×10^7 m/s, to an accuracy of $\pm 10\%$. What is the minimum uncertainty in its position?

59. Find the minimum energy for a neutron in a uranium nucleus whose diameter is 15 fm.

60. An electron is trapped in a "quantum well" 20 nm wide. What is the minimum speed it could have?

61. A proton is moving along the x axis with speed $v = (1500 \pm 0.25)$ m/s, but its direction ($+$ or $-x$) is unknown. Find the uncertainty in its position. *Hint:* What is the difference between the two extreme possibilities for the *velocity?*

62. An electron beam is accelerated at the back of a TV tube and then heads toward the center of the screen with horizontal velocity 2.2×10^7 m/s. As the electrons leave the acceleration region, their vertical position is known to within ± 45 nm. Find the minimum angular spread in the beam, as set by the uncertainty principle.

63. An electron is moving at 10^6 m/s, and you wish to measure its energy to an accuracy of $\pm 0.01\%$. What is the minimum time necessary for the measurement?

64. A cosmic-ray particle interacts with an energy-measuring device for a mere 12 fs. What is the minimum uncertainty in its energy?

65. The lifetimes of unstable particles set energy-time uncertainty limits on the accuracy with which the rest energies—and hence the masses—of those particles can be known. The particle known as the neutral pion has rest energy 135 MeV and lifetime 8×10^{-17} s. Find the uncertainty in its rest energy (a) in eV and (b) as a fraction of the rest energy.

66. An electron has momentum of magnitude p, but the direction is unknown. (a) Argue that the uncertainty in its momentum is therefore $2p$. (b) Compare the minimum uncertainty in its position with its de Broglie wavelength. Why might the two be related?

Paired Problems

(Both problems in a pair involve the same principles and techniques. If you can get the first problem, you should be able to solve the second one.)

67. (a) Find the wavelength at which a 2200-K blackbody has its peak radiance. (b) How does the radiance for 690-nm red light compare with the peak radiance?

68. The radiance of a blackbody peaks at 660 nm. (a) What is its temperature? (b) How does the radiance at 400 nm compare with that at 700 nm?

69. A photocathode ejects electrons with maximum energy 0.85 eV when illuminated with 430-nm blue light. Will it eject electrons when illuminated with 633-nm red light, and if so what will be the maximum electron energy?

70. A photocathode ejects electrons with maximum energy 2.8 eV when illuminated with 360-nm ultraviolet light. Will it eject electrons when illuminated with 840-nm infrared light, and if so what will be the maximum electron energy?

71. Find the initial wavelength of a photon that loses 20% of its energy when it Compton scatters from an electron through an angle of 90°.

72. A photon Compton scatters from an electron, emerging at 140° to its initial direction and with an energy of 85 keV. Find the photon's initial energy.

73. (a) Find the energy of the highest energy photon that can be emitted as the electron jumps between two adjacent

allowed energy levels in the Bohr hydrogen atom. (b) Which energy levels are involved?

74. Find (a) the wavelength and (b) the energy in eV of the photon emitted as a Rydberg hydrogen atom drops from the $n = 180$ level to the $n = 179$ level.

75. In which of the following will wave properties be more evident: an electron moving at 10 Mm/s or a proton moving at 1 km/s?

76. Give the ratio of de Broglie wavelength to diameter for (a) a 100-g ball, 10 cm in diameter, moving at 5 m/s; (b) a 4-u alpha particle, 3.4 fm in diameter, moving at 20 Mm/s; and (c) a 238-u uranium nucleus, 13 fm in diameter, moving at 1.4 km/s. (d) For which will wave properties be negligible, and for which will they be completely dominant?

77. An electron is known to be within about 0.1 nm of the nucleus of an atom. What is the uncertainty in its velocity?

78. Typically, an atom remains in an excited state for about 10^{-8} s before it drops to a lower energy state, emitting a photon in the process. What is the uncertainty in the energy of this transition?

Supplementary Problems

79. An ▬▬▬▬▬ accelerated from rest through a 4.5-kV potential difference. What is its de Broglie wavelength?

80. A photomultiplier consists of a photocathode, 8 dynodes, and an anode that collects the electrons from the final dynode. The photocathode's active surface is a circle 2.5 cm in diameter, and its quantum efficiency is 7.5% for light with wavelength 633 nm. Each dynode releases 3 electrons for each electron incident on it. Find the current at the anode if the cathode is illuminated with 633-nm light with intensity 6.5 mW/m².

81. A photon's wavelength is equal to the Compton wavelength of a particle with mass m. Show that the photon's energy is equal to the particle's rest energy.

82. Show that the frequency range occupied by the hydrogen spectral line series involving transitions ending at the nth level is given by $\Delta f = cR_H/(n + 1)^2$.

83. Solve the energy and momentum conservation equations given in the discussion of the Compton effect to verify Equation 39-8. Show also that the electron's recoil angle ϕ can be written $\phi = \tan^{-1}\dfrac{\cot(\theta/2)}{1 + \lambda_c/\lambda_0}$.

84. (a) Show that the energy required for an electron to escape from the surface of a nucleus of radius R and charge Ze is kZe^2/R, where k is the Coulomb constant. Evaluate for the oxygen nucleus, with $Z = 8$ and $R = 3$ fm. (b) Use the uncertainty principle to find the minimum energy for an electron confined to this nucleus. (You'll need to use the relativistic energy-momentum relation, which you can simplify to $E = pc$ because the energy is much higher than the rest energy.) (c) Compare your answers to (a) and (b) to show that your nucleus cannot contain an electron.

85. A photon undergoes a 90° Compton scattering off a stationary electron, and the electron emerges with *total* energy $\gamma m_e c^2$, where γ is the relativistic factor introduced in Chapter 38. Find an expression for the initial photon energy.

86. Show that Wien's displacement law follows from Planck's law (Equation 39-3). *Hint:* Differentiate Planck's law with respect to wavelength.

87. The total power per unit area emitted by a blackbody is given by integrating the power per area per unit wavelength (Equation 39-3) over all wavelengths. Carry out this integration to show that the radiated power per unit area is proportional to the fourth power of the temperature. Show that your result also implies that the Stefan-Boltzmann constant is given by $\sigma = 2\pi^5 k^4/15c^2h^3$. *Hint:* Use the quantity $u = hc/\lambda kT$ as your integration variable, and consult a table of definite integrals.

40 Quantum Mechanics

29.8 A
x 3.0

337 Å

338 Å

electron is

Scanning tunneling microscope image of a portion of a DNA molecule shows a region only 34 nm on a side—so small that 6 million of these regions would fit across the width of this page. This new type of microscope exploits the phenomenon of quantum tunneling, made possible by the wave-like aspect of matter.

The ideas developed in the preceding chapter comprise the core of what is called the **old quantum theory**. The old quantum theory introduced the basic concepts of quantum physics, and it was remarkably successful in explaining a number of quantum phenomena—for example, blackbody radiation, the photoelectric effect, and the hydrogen spectrum. On the other hand, it could not treat even the simplest of multielectron atoms, and it left some subtle spectral features unexplained. Furthermore, the old quantum theory was a hodgepodge of separate but loosely related propositions, each developed to explain a particular phenomenon; it lacked coherence and clear guiding principles.

Is there a more coherent theory that predicts the behavior of systems at the atomic and subatomic scales, and that offers a satisfying description of how such systems really work? The answer is at once an emphatic yes and a disappointing no. Yes, because the theory of **quantum mechanics** developed in the 1920s predicts with remarkable precision many of the observed properties of atomic systems, including their energies, the wavelengths of spectral lines they emit, and even the lifetimes of excited atoms. No, because quantum mechanics cannot give us a satisfying visual *picture* of the atomic and subatomic world. The uncertainty principle and wave-particle duality are essential aspects of quantum mechanics. Any picture we formulate of electrons or photons whizzing around like miniature baseballs with precise positions and momenta is simply inappropriate at the atomic scale. But quantum mechanics does provide a self-consistent mathematical description that lets us explore and predict the behavior of atoms, the organization of chemical elements, the physics of semiconductors and superconductors, the extraordinary behavior of matter at low temperature, the formation of white dwarf stars, the operation of lasers, and a host of other phenomena that are crucial in our technological world and for which classical physics is at best inaccurate and at worst totally inappropriate. In this chapter we explore the mathematical structure and physical interpretation of quantum mechanics. In Chapters 41 and 42, we will apply quantum mechanics first to the atom and then to more complex systems like semiconductor materials that involve the quantum-mechanical interaction of many atoms.

40-1 PARTICLES, WAVES, AND PROBABILITY

Photons and Light Waves

In Maxwell's electromagnetic theory (Chapter 34) we had a seemingly complete description of light as an electromagnetic wave. Now we find, through the photoelectric and Compton effects, that light sometimes manifests itself as particles. What is the connection between the wave and particle descriptions?

In a photoelectric experiment, the *energy* of the ejected electrons depends on the wavelength of the light, not on its intensity. But the *number* of ejected electrons does depend on the intensity. Since an electron is ejected when it absorbs a photon, we conclude that the number of photons in the incident light is proportional to the light intensity. Now, the intensity of an electromagnetic wave depends on the square of the electric or magnetic field (recall Equations 34-21b, c). The fields, in turn, obey Maxwell's equations, so one aspect of a photoelectric experiment—namely, the number of electrons ejected—is related to Maxwell's description of light as an electromagnetic wave.

Can we quantify further the relation between wave intensity and number of photons? Yes, but only in a statistical sense. The ejection of individual electrons in a photoelectric experiment is quite random. The uncertainty principle prevents us from following a photon trajectory and therefore from predicting when and where an electron will be ejected. All we can say is that electrons are more likely to be ejected where the wave intensity is greater. Specifically, the probability that an electron will be ejected is directly proportional to the intensity of the incident electromagnetic waves—that is, to the square of the electric or magnetic field. More generally, the probability of finding a photon in a beam of electromagnetic waves is directly proportional to the wave intensity (Fig. 40-1).

Light wave

Photon beam

FIGURE 40-1 The probability of finding a photon is proportional to the intensity of the electromagnetic wave. Here we show an electromagnetic wave whose intensity varies with position, and the corresponding distribution of photons. Actually, the picture itself is only suggestive, for by localizing the individual photons we would have lost any knowledge of their wave properties. It is better to think of the wave intensity as determining the probability of finding a photon *when you look for it*. In the standard interpretation of quantum mechanics, it makes no sense to ask where the photons are, or even whether they are there at all, when you're not trying to detect them.

FIGURE 40-2 Computer simulation showing the development of a two-slit interference pattern from random photon events. The probability of a photon landing at a particular point is proportional to the electromagnetic wave intensity. The statistical pattern becomes evident only with a large number of photons. (*a*) 50 photons; (*b*) 250 photons; (*c*) 1000 photons; (*d*) 10,000 photons.

In this quantum-mechanical description, the fields still evolve according to Maxwell's equations. For example, the fields of an electromagnetic wave undergoing double-slit interference develop regions of maximum and minimum wave intensity—the bright and dark bands of the interference pattern. But the wave fields determine only the *probability* that individual photons will be detected in the interference pattern. That's why a very short exposure or low-intensity beam results not in a weak version of the interference pattern but in a seemingly arbitrary pattern. Only with large numbers of photons does the statistical pattern emerge (Fig. 40-2).

In quantum mechanics, then, the relation between the wave and particle aspects of light is this: As long as we don't try to detect the light, it propagates as a wave governed by Maxwell's equations. We can use those equations, and the theory of physical optics (Chapter 37) that is based on them, to predict the behavior of the waves. But when we detect the light, we do so through interactions involving individual photons. Those interactions are random events whose probability depends on the wave intensity—that is, on the square of the wave fields.

Electrons and Matter Waves

In the preceding chapter we introduced de Broglie's remarkable hypothesis that matter, as well as light, exhibits both wave and particle properties. We don't experience the wave properties of everyday matter because the de Broglie wavelength of objects such as baseballs, cars, and people is so much smaller than the systems with which they interact (recall Example 39-5). But at the atomic scale the wave properties of matter are important, and that's precisely why the quantum description becomes necessary.

The wave-particle duality puts matter and light on essentially the same footing, and the statistical interpretation of that duality is the same for each. Figure 40-3 shows a beam of electrons and its associated de Broglie matter wave. Just as the probability of finding a photon is proportional to the wave intensity—that is, the *square* of the electromagnetic field amplitude—so the probability of finding a

Matter wave

Electron beam

FIGURE 40-3 A beam of electrons and its associated matter wave. Where the wave amplitude is larger there is a higher probability of finding an electron. The situation is exactly analogous to that of Fig. 40-1, and the same statistical interpretation and reservations apply; in particular, it makes no sense to talk about individual electrons unless we set out to detect them.

FIGURE 40-4 This interference pattern was produced by passing a beam of electrons through a double-slit system with slit spacing comparable to the electrons' de Broglie wavelength. It looks just like the corresponding double-slit pattern for light (e.g., Fig. 40-2 or 37-5).

particle is proportional to the square of the matter wave amplitude. And as with light, the particle nature of matter manifests itself only when we try to detect a particle; leave the particle alone, and its behavior is governed by its wave nature. That's why electrons exhibit the same interference phenomena as does light (Fig. 40-4).

We know the equations that determine the behavior of light waves: they're Maxwell's equations of electromagnetism. But what equation describes matter waves? Answering that question culminated the development of quantum mechanics during the first three decades of this century. Two quite separate approaches were taken to the quantum description of matter. In 1925, at the age of 24, the German physicist Werner Heisenberg collaborated with Max Born and Pascual Jordan to develop **matrix mechanics**—a quantum theory formulated in the abstract mathematical language of matrices. At nearly the same time, the Austrian physicist Erwin Schrödinger (Fig. 40-5) formulated his theory of **wave mechanics**. Schrödinger soon showed that his theory and Heisenberg's were, in fact, equivalent. Heisenberg and Schrödinger received the 1932 and 1933 Nobel Prizes in Physics (the latter shared with Paul Dirac), respectively, for their contributions to the development of quantum mechanics. Because Schrödinger's work has a more obvious physical interpretation in terms of waves, we will continue our discussion of quantum mechanics entirely in terms of Schrödinger's formulation.

17.7
Wave Packets

40-2 THE SCHRÖDINGER EQUATION

Schrödinger's challenge was to find an equation describing matter waves. The resulting **Schrödinger equation** gives the full space and time dependence of those waves. Here we'll consider only the spatial variation of matter waves, and for now we'll restrict ourselves to one dimension. We follow Schrödinger in using the symbol ψ for the **wave function** of the matter waves.

We can understand the Schrödinger equation by considering a sinusoidal wave of the form $\psi(x) = A \sin kx$, where, as usual in describing waves, $k = 2\pi/\lambda$ with λ the wavelength. Differentiating this expression twice gives

FIGURE 40-5 Erwin Schrödinger.

$$\frac{d^2\psi(x)}{dx^2} = -Ak^2 \sin kx = -k^2\psi(x).$$

But $k = 2\pi/\lambda$ and, for matter waves, de Broglie's hypothesis gives $\lambda = h/p$, with h Planck's constant and p the particle's momentum. Thus we can write k in terms of momentum as $k = 2\pi p/h = p/\hbar$. (We defined $\hbar = h/2\pi$ in the preceding chapter.) Now in classical physics a particle of mass m has kinetic energy K and momentum p related by $K = p^2/2m$. Furthermore, kinetic energy is the difference between the total energy E and potential energy U; thus $E - U = p^2/2m$. Putting this all together, we can write the quantity k^2 in our differentiated wave expression as

$$k^2 = \frac{p^2}{\hbar^2} = \frac{2m(E - U)}{\hbar^2}.$$

Making this substitution and doing a little algebra then gives

$$-\frac{\hbar^2}{2m}\frac{d^2\psi(x)}{dx^2} + U(x)\psi(x) = E\psi(x). \qquad (40\text{-}1)$$

Equation 40-1 is the **time-independent Schrödinger equation**, giving the spatial variation of matter waves in one dimension. A solution of the full time-dependent equation consists of a solution to Equation 40-1 multiplied by a sinusoidal oscillation with frequency $f = E/h$, where E is the particle energy. Incidentally, we wrote $U(x)$ in Equation 40-1 because potential energy is often a function of position. We developed the time-independent Schrödinger equation by merging de Broglie's matter-wave hypothesis $\lambda = h/p$ with the Newtonian relation $K = p^2/2m$; for that reason, we expect the equation to hold only for nonrelativistic particles.

How do we know that the Schrödinger equation is right? The answer is clear: the equation provides a description of physical reality in remarkable agreement with experiments. And not only a few arcane experiments; as we'll see, Schrödinger's equation goes a long way toward explaining the structure of atoms, their chemical properties, and indeed the entire science of chemistry. Furthermore, the Schrödinger description agrees with Newtonian mechanics for macroscopic systems so large that the "graininess" or quantization at the atomic level is not obvious.

The Meaning of ψ

What is the meaning of the wave function ψ? That's a much deeper question than "Is the Schrödinger equation right?" and it's a question that physicists and philosophers continue to debate. In the standard interpretation, first articulated by Max Born in 1926, ψ itself is not an observable quantity. It manifests itself only in the statistical distributions of particle detections. More specifically, the probability per unit volume—also called the **probability density**—that we will find a particle is given by ψ^2. For a particle confined to one dimension, the probability density becomes the probability per unit length, and we interpret ψ to mean that the probability of finding the particle in a small interval dx at position x is

$$\text{Probability} = \psi^2(x)\,dx. \qquad (40\text{-}2)$$

Here we write $\psi^2(x)$ to indicate that the probability may vary from place to place. We can interpret Equation 40-2 in two ways. At face value, it is the probability that a single experiment, with a detector at position x set up to find particles in an interval of width dx, will detect the particle. Or, if we do many such experiments, the equation gives the fraction of the experiments in which we will find a particle in our detector.

But *what is* ψ? How can it be unobservable yet govern the behavior of matter? There can't be a direct causal link between the wave function and individual particles, since ψ determines only the *probability* that a particle will behave in a certain way. Think about this! In quantum mechanics the outcome of an experiment is not fully determined. The Schrödinger equation describes only the probability of a given outcome. How can this be? The quantum world is so different, according to the standard interpretation, that our macroscopic language, concepts, and pictorial models are simply inadequate. In particular, macroscopic causality is replaced by a microscopic indeterminancy in which quantum events are truly random; only the overall statistical pattern of those events is governed by physical laws.

The philosophical implications of quantum mechanics have been a matter of debate since the theory was formulated (Fig. 40-6). A central theme in this debate is the possibility of "hidden variables," physical quantities that might be hidden from us by the uncertainty principle but that might nevertheless govern the microscopic world in a fully deterministic way. Experiments of the early 1980s have placed severe restrictions on such hidden-variable theories, but fascinating discussions on the interpretation of quantum mechanics continue. Here, however, we turn to the Schrödinger equation to see how it is used in analyzing quantum-mechanical systems.

FIGURE 40-6 Niels Bohr and Albert Einstein discussing the implications of quantum mechanics. Einstein never accepted the probabilistic interpretation of quantum mechanics, as indicated in his paraphrased remark "God does not play dice."

Normalization and Other Constraints on the Wave Function

In one dimension, the quantity $\psi^2\,dx$ represents the probability of finding the particle associated with the wave function $\psi(x)$ in the interval dx at some position x. But the particle *must* be *somewhere*. Therefore, if we sum the probabilities of finding the particle in all such intervals dx along the x axis, the result must be 1; there must be a 100% chance that we'll find the particle somewhere. Since the probability density may vary with position, that sum becomes an integral, and we have

$$\int_{-\infty}^{+\infty} \psi^2\,dx = 1 . \qquad \text{(normalization condition)} \qquad (40\text{-}3)$$

Equation 40-3 is called the **normalization condition**; once we have a solution $\psi(x)$ to the Schrödinger equation 40-1, this condition can be used to set the overall amplitude of the function ψ.

The Schrödinger equation contains the second derivative of ψ. In order that this term be well-defined, both ψ itself and its first derivative must be continuous. (An exception to the continuity condition on $d\psi/dx$—possible only in unrealistic example situations—occurs if the potential energy U becomes infinite.)

40-3 THE INFINITE SQUARE WELL

We now solve the Schrödinger equation for a particularly simple system—a particle trapped in one dimension between two perfectly rigid walls. Although unrealistic in some respects, this system nevertheless is a surprisingly good approximation to some real quantum systems, including some new electronic devices and simple nuclei such as deuterium. More importantly, its analysis illustrates the general procedure for applying the Schrödinger equation and shows how energy quantization emerges from Schrödinger's theory.

In classical physics, a particle trapped between rigid walls moves back and forth with constant speed (Fig. 40-7). In the absence of friction or other losses, the particle's energy remains constant at its initial value. And in classical physics, that value can be anything.

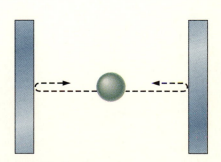

FIGURE 40-7 A particle trapped between rigid walls. Classically, it moves back and forth with constant total energy. That energy can have any value.

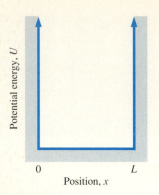

FIGURE 40-8 Infinite square well potential-energy curve describes a particle constrained to move in one dimension between rigid walls a distance L apart.

We can describe the particle's situation in terms of its potential-energy curve. (We will be talking a great deal about potential-energy curves for a while, and you might find it helpful to review Section 8-4 at this point.) Since the particle is subject to no forces while it's between the walls, its potential energy is constant in this region, and we can fix the arbitrary zero of potential energy by setting that constant to zero. If the walls are perfectly rigid then the particle should not be able to penetrate them at all, no matter what its energy. This means that the potential energy becomes abruptly infinite at the walls. Then the potential-energy curve for our particle looks like Fig. 40-8; you can see from the figure why this curve is called an **infinite square well**.

We now consider the quantum-mechanical description of the particle in a square well. The particle has a wave function whose time-independent part is given by the Schrödinger equation (Equation 40-1):

$$-\frac{\hbar^2}{2m}\frac{d^2\psi}{dx^2} + U(x)\psi = E\psi,$$

where the potential energy $U(x)$ is that of the square well in Fig. 40-8:

$$U = 0 \text{ for } 0 < x < L$$

$$U = \infty \text{ for } x < 0 \text{ or } x > L.$$

Physically, we set $U = \infty$ so there is no chance that the particle can penetrate the rigid walls. Quantum-mechanically, this means that the function ψ must be exactly zero in the region where $U = \infty$, giving zero probability of finding the particle beyond the walls. Whatever the wave function ψ looks like, then, it must go to zero at $x = 0$ and at $x = L$, and must remain zero beyond those points. All we need to calculate, then, is ψ inside the well, where $0 < x < L$.

Within the well, $U = 0$ and the Schrödinger equation becomes

$$-\frac{\hbar^2}{2m}\frac{d^2\psi}{dx^2} = E\psi. \tag{40-4}$$

To find solutions, recall de Broglie's hypothesis that the allowed orbits in the Bohr atom are those for which standing waves just "fit" around the orbit. We have a similar situation with the infinite square well, in which the allowed solutions should be standing waves with nodes at the ends of the well—exactly analogous to standing waves on a string with both ends clamped that we discussed in Chapter 17. So we want a sinusoidal wave for $\psi(x)$, with $\psi = 0$ at $x = 0$ and at $x = L$. The first condition is satisfied if we take a wave of the form $\psi = A \sin kx$, with A and k both constants. The second condition is satisfied if $k = n\pi/L$, where n is any integer—a condition equivalent to saying that an integer number of half wavelengths fit in the well. So we propose a solution of the form

$$\psi(x) = A \sin\left(\frac{n\pi x}{L}\right), \tag{40-5}$$

20.2

Particle in a Box

with the constant A still undetermined. Equation 40-5 represents standing waves with nodes at the ends of the square well, but does it satisfy the Schrödinger equation? We can find out by substituting Equation 40-5 into Equation 40-4. We need not only ψ but also its second derivative; twice differentiating Equation 40-5 gives

$$\frac{d^2\psi}{dx^2} = -A\frac{n^2\pi^2}{L^2}\sin\left(\frac{n\pi x}{L}\right).$$

Substituting ψ and $d^2\psi/dx^2$ into Equation 40-4 gives

$$\left(-\frac{\hbar^2}{2m}\right)\left[-A\frac{n^2\pi^2}{L^2}\sin\left(\frac{n\pi x}{L}\right)\right] = EA\sin\left(\frac{n\pi x}{L}\right),$$

which reduces to

$$E = \frac{n^2\pi^2\hbar^2}{2mL^2} = \frac{n^2h^2}{8mL^2}, \tag{40-6}$$

where we used $\hbar = h/2\pi$. What is Equation 40-6 telling us? It says that our proposed solution 40-5 can indeed satisfy the Schrödinger equation—provided the particle energy E has a value given by Equation 40-6, with n an integer.

Our standing-wave solutions and the associated condition 40-6 show how the quantization of energy arises naturally from the mathematics of the Schrödinger equation. Physically, of course, the reason for quantization remains as de Broglie had postulated: matter waves in a confined system must be standing waves with an integer number of half wavelengths. Although de Broglie's hypothesis and the Schrödinger equation lead to exactly the same conclusion in this simple case of an infinite square well, we will see that with more complicated potential-energy functions only the Schrödinger equation can give us the full story.

The integer n that appears in the wave function 40-5 and in the energy quantization condition of Equation 40-6 is called the **quantum number** for the particle in the square well. The physical state of a quantum-mechanical system is called a **quantum state**. In this simple case, one quantum number suffices to specify the quantum state, which then tells us everything quantum mechanics has to say about the situation. As far as the Schrödinger equation is concerned, it looks like all integer values of n are allowed. The choice of negative or positive n has no physical significance since ψ^2 has the same value with either sign of ψ; for this reason, negative n's are redundant. But $n = 0$ implies $\psi = 0$ everywhere, giving no chance of finding the particle anywhere. This case cannot satisfy the normalization condition of Equation 40-3, leaving only positive integer values of n.

With only nonzero n's allowed, Equation 40-6 shows that the particle's energy is always positive; zero energy is not allowed. The lowest possible energy is $h^2/8mL^2$, obtained with $n = 1$. This lowest energy is called the **ground-state energy**; the corresponding wave function is the **ground-state wave function**. A nonzero ground-state energy is a common feature of quantum systems, and one with no classical counterpart.

Why isn't it possible to have a zero-energy particle in the square well? The answer lies in the uncertainty principle. If the energy were exactly zero, then the particle's momentum would also be exactly zero—and we would know its momentum precisely. In other words, the uncertainty Δp in the momentum would be zero. However, we know that the particle is confined to the well; therefore, the uncertainty Δx in its position is L. So $\Delta p\Delta x$ would be zero, in violation of the uncertainty principle.

EXAMPLE 40-1 | *Hints of Atomic Structure*

What is the ground-state energy of an electron in an infinite square well whose width L is the approximate diameter of a hydrogen atom, about 10^{-10} m? Give the answer in both joules and electron volts.

Solution

With $n = 1$ for the ground state, and with $L = 1 \times 10^{-10}$ m, Equation 40-6 gives:

$$E = \frac{h^2}{8mL^2}$$

$$= \frac{(6.63 \times 10^{-34} \text{ J·s})^2}{(8)(9.11 \times 10^{-31} \text{ kg})(1 \times 10^{-10} \text{ m})^2} = 6 \times 10^{-18} \text{ J} .$$

Since $1 \text{ eV} = 1.6 \times 10^{-19}$ J, this is

$$E = \frac{6 \times 10^{-18} \text{ J}}{1.6 \times 10^{-19} \text{ J/eV}} = 40 \text{ eV} ,$$

where in this approximate calculation we've rounded to one significant figure. Although a square-well potential is very different from the electrical potential energy associated with electrical forces in an atom, this example nevertheless provides a rough estimate of the energies we should expect to find for electrons in atoms—typically in the tens-of-eV range.

EXERCISE Find the width of an infinite square well containing a proton whose ground-state energy is 10 MeV.

Answer: 4.5 fm, about the size of a nucleus

Some problems similar to Example 40-1: 8, 9, 11, 13

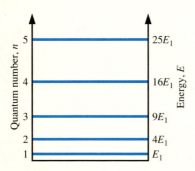

FIGURE 40-9 Energy-level diagram for a particle in an infinite square well. The levels are not evenly spaced because the energy (Equation 40-6) is proportional to n^2.

The ground-state energy is the minimum possible for a particle in a square well. But it can have any of the other energies given by Equation 40-6 with $n > 0$. Figure 40-9 shows an **energy-level diagram** for a particle in an infinite square well. In Chapter 39 we used a similar diagram (Fig. 39-14) to represent atomic energy levels in the Bohr atom.

Got It!

Electron A is confined to a square well 1 nm wide; electron B to a similar well only 1 fm wide. How do their ground-state energies compare?

Normalization, Probabilities, and the Correspondence Principle

In our solution for the infinite square well, we still don't know the value of the constant A in the wave function. This can be determined from the normalization condition 40-3:

$$\int_{-\infty}^{\infty} \psi^2 \, dx = 1 .$$

Inside the well, ψ is given by Equation 40-5. Outside the well, ψ is exactly zero. So we can write the normalization condition as an integral involving only Equation 40-5 and the interval $0 < x < L$:

$$\int_0^L A^2 \sin^2\left(\frac{n\pi x}{L}\right) dx = 1 .$$

If we divided $\int_0^L \sin^2(n\pi x/L) \, dx$ by the well width L, we would have the average of $\sin^2$ over an integer number of half cycles—or just $\frac{1}{2}$. So the integral of $\sin^2(n\pi x/L)$ from 0 to L is L times $\frac{1}{2}$, and therefore

$$A^2 \frac{L}{2} = 1 , \quad \text{or} \quad A = \sqrt{\frac{2}{L}} .$$

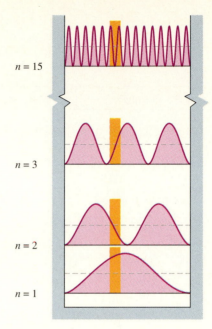

FIGURE 40-10 Wave functions for a particle in an infinite square well. Horizontal axis for each wave function is drawn at the corresponding energy level, although each axis represents $\psi = 0$ for its wave function. The break indicates that the $n = 15$ state has a much higher energy.

FIGURE 40-11 Classical (dashed) and quantum (solid) probability densities for a particle in an infinite square well. Orange bands represent the resolution of an instrument used to detect the particle; for large n the instrument averages over one or more cycles and therefore detects the particle with an average probability everywhere equal to the classical value.

The complete wave function is then given by

$$\psi_n = \sqrt{\frac{2}{L}} \sin\left(\frac{n\pi x}{L}\right), \qquad (40\text{-}7)$$

where the subscript n refers to the function associated with the nth quantum state. Figure 40-10 shows wave functions for the ground state and several excited states.

Where are we likely to find the particle? Classically, it would be moving back and forth at constant speed, and therefore would be equally likely to be anywhere in the well. Quantum-mechanically, the probability of finding it at some position x is proportional to the probability density ψ^2 at that point. Figure 40-11 shows the probability densities given by squaring the wave functions of Fig. 40-10. For $n = 1$ we're clearly more likely to find the particle near the middle of the well—in marked contrast to the classical prediction of equal probability everywhere. For other low-n states there are obvious regions of high and low probability, again in contrast to the classical prediction. But as the quantum number increases the maxima and minima of the probability density get closer together. Any instrument we use to find the electron has a finite resolution, and once the periodicity of the wave function drops below that resolution, then we measure an average probability, which is essentially constant over the interval (see Fig. 40-11).

We have here a manifestation of Bohr's correspondence principle: At large values of the quantum number n the interval between adjacent energy levels becomes small compared with the energy itself, and a measurement—in this case of the electron's position—gives results in agreement with classical physics. But classical physics is totally inadequate at low n, where the nonclassical zero-point energy and energy quantization are most evident.

EXAMPLE 40-2	Quantum Probability

A particle is in the ground state of an infinite square well that extends from $x = 0$ to $x = L$. What is the probability that it will be found in the leftmost quarter of the well?

Solution

In deriving the normalization condition 40-3, which expresses mathematically that the particle has probability 1 of being found *somewhere,* we integrated the probability density ψ^2 over all values of x. Similarly, to determine the probability of finding the particle in a particular region, we integrate ψ^2 over that region. Here the region is the left-hand fourth of the well, given by $0 < x < \frac{1}{4}L$. Then the probability is

$$P = \int_0^{L/4} \psi^2 \, dx \, .$$

The ground-state wave function, given by Equation 40-7 with $n = 1$, is $\psi_1 = \sqrt{2/L} \sin(\pi x/L)$, so we have

$$P = \frac{2}{L} \int_0^{L/4} \sin^2\left(\frac{\pi x}{L}\right) dx \, .$$

We can integrate using the table at the end of Appendix A; the result is

$$P = \frac{2}{L}\left(\frac{x}{2} - \frac{\sin(2\pi x/L)}{4\pi/L}\right)\Bigg|_0^{L/4} = \frac{2}{L}\left(\frac{L}{8} - \frac{L}{4\pi}\right) = \frac{1}{4} - \frac{1}{2\pi}$$
$$= 0.091 \, .$$

This is considerably lower than the probability $P = 0.25$ we would expect classically for finding the particle in any quarter of the well, and reflects the lower value of ψ^2 nearer the well ends. Problem 54 repeats the calculation of this example for arbitrary quantum numbers, and shows that classical and quantum probabilities agree at large n.

EXERCISE Determine the probability of finding a ground-state particle in the central quarter of an infinite square well (i.e., in the range $\frac{3}{8}L < x < \frac{5}{8}L$), and compare with the classical prediction.

Answer: 0.475, nearly twice the classical value

Some problems similar to Example 40-2: 23, 47, 48, 55

The infinite square well has given us insights into a number of important quantum phenomena shared by more realistic quantum systems such as atoms. These include the quantization of energy levels, the existence of a nonzero ground-state energy, probability distributions that differ from classical predictions, and the correspondence of quantum and classical physics at high quantum numbers. In the next chapter we will apply the Schrödinger equation to atoms, using the same techniques and finding many of the same phenomena we did here. First, however, we examine other simple quantum systems that give rise to new and surprising behavior quite at odds with classical physics.

40-4 THE HARMONIC OSCILLATOR

In Chapter 15 we studied the motion of a particle subject to a restoring force directly proportional to the particle's displacement from its equilibrium position. Using classical physics, we found that the particle undergoes simple harmonic motion, oscillating sinusoidally at an angular frequency ω determined by its mass and the nature of the restoring force. Although our paradigm system was a mass on a spring, we pointed out (see Fig. 15-26) that any other system whose potential energy is a quadratic function of displacement also constitutes a simple harmonic oscillator. That includes many systems at the atomic and molecular level. And our

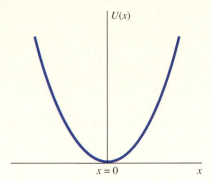

FIGURE 40-12 The potential-energy curve for a simple harmonic oscillator is a parabola.

most fundamental theories of physics—the so-called quantum field theories—treat the entire universe of particles and fields as essentially a huge ensemble of harmonic oscillators. Understanding the quantum-mechanical harmonic oscillator is therefore crucial in describing the behavior of matter at the smallest scales.

We now use the Schrödinger equation to investigate the behavior of a quantum-mechanical harmonic oscillator. If the system were a mass on a spring, it would have potential energy $U = \frac{1}{2}kx^2$ (Equation 8-4) and frequency $\omega = \sqrt{k/m}$, with k the spring constant and m the particle mass. Combining these equations, we can write the potential energy in terms of the frequency and mass:

$$U = \tfrac{1}{2}m\omega^2 x^2. \tag{40-8}$$

A quantum-mechanical harmonic oscillator is more likely to be an electron in some electric field configuration that gives rise to a restoring force, or an atom vibrating at the end of a molecular bond; for this reason Equation 40-8, which makes no explicit reference to springs and spring constants, is an appropriate form for the potential energy of a general harmonic oscillator.

Figure 40-12 shows a plot of the parabola described by Equation 40-8. This curve is analogous to Fig. 40-8 for the infinite square well. In contrast to the discontinuous potential-energy curve of the square well, we have in Equation 40-8 a single continuous function giving the harmonic oscillator's potential energy everywhere. Using this function for $U(x)$ in the Schrödinger equation (Equation 40-1) gives

$$-\frac{\hbar^2}{2m}\frac{d^2\psi}{dx^2} + \tfrac{1}{2}m\omega^2 x^2\psi = E\psi. \tag{40-9}$$

Solving this differential equation involves advanced mathematical techniques; here we give only the ground-state solution:

$$\psi_0 = \left(\frac{m\omega}{\pi\hbar}\right)^{1/4} e^{-m\omega x^2/2\hbar} \tag{40-10}$$

and let you show (Problem 28) that it satisfies Equation 40-9. In so doing you will also find that the ground-state energy is $\frac{1}{2}\hbar\omega$.

Many other functions also satisfy Equation 40-9, but most of them approach infinity at large values of x and, therefore, cannot be normalized (Fig. 40-13). The

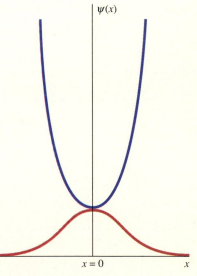

FIGURE 40-13 The functions $e^{+m\omega x^2/2\hbar}$ (blue) and $e^{-m\omega x^2/2\hbar}$ (red) both satisfy the Schrödinger equation 40-9 for the harmonic oscillator, but only the function with the negative exponent can be normalized. Aside from the normalization constant $\pi^{-1/4}$, this latter function is the ground-state wave function.

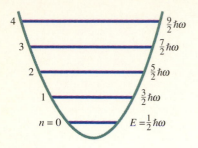

FIGURE 40-14 Energy-level diagram for a quantum-mechanical harmonic oscillator, superposed on the potential-energy curve. Energy levels are evenly spaced at intervals of $\hbar\omega$ ($= hf$), starting with the ground state at energy $\frac{1}{2}\hbar\omega$.

normalizable solutions—those that represent particle wave functions—occur only for certain discrete values of the energy E, given by

$$E_n = (n + \tfrac{1}{2})\hbar\omega, \qquad (40\text{-}11)$$

where $n = 0$ now corresponds to the ground state. Figure 40-14 shows an energy-level diagram for the harmonic oscillator; note the even spacing implied by Equation 40-11. Incidentally, the additive factor $\frac{1}{2}$ in Equation 40-11 shows that Planck was wrong in his hypothesis that the allowed harmonic oscillator energies should be multiples of $hf(= \hbar\omega)$. Planck's spectral distribution (Equation 39-3) is nevertheless correct, but he did not foresee the existence of nonzero ground-state energy.

The even spacing between energy levels of the harmonic oscillator as shown in Fig. 40-14 is in marked contrast to the situation in atoms (see Fig. 39-14) or in the infinite square well (see Fig. 40-9). A quantum harmonic oscillator emits or absorbs photons as it makes transitions among adjacent levels, and the even level spacing implies that all transitions between adjacent levels of a pure harmonic oscillator involve photons of the same energy.

EXAMPLE 40-3	*A High-Power Laser*

Among the most intense lasers are those powered by chemical reactions that produce harmonic-oscillator-like molecules in excited states. In a laser using hydrogen fluoride molecules, developed under the "Star Wars" antimissile program, much of the emission occurs as a result of transitions between the $n = 1$ harmonic oscillator level and the ground state. The infrared light emitted in these transitions has wavelength 2.7 μm. What is the ground-state energy, in electron-volts, of the HF molecules?

Solution

Equation 40-11 shows that the spacing between adjacent energy levels is $\hbar\omega$ and that the ground state energy is $E_0 = \frac{1}{2}\hbar\omega$. The energy difference $\hbar\omega$ between the $n = 1$ state and the ground state ($n = 0$) is the energy of a 2.7-μm photon. In Chapter 39 we saw that $E = hc/\lambda$ for a photon of wavelength λ. Thus, we

have $\hbar\omega = hc/\lambda$, or, for the ground-state energy,

$$E_0 = \tfrac{1}{2}\hbar\omega = \frac{hc}{2\lambda} = \frac{(6.63\times10^{-34}\text{ J·s})(3.0\times10^8\text{ m/s})}{(2)(2.7\times10^{-6}\text{ m})(1.6\times10^{-19}\text{ J/eV})}$$

$$= 0.23\text{ eV}.$$

This is the minimum possible energy for the hydrogen fluoride molecule. We will explore chemical lasers further in Chapter 42.

EXERCISE An electron is trapped in a harmonic oscillator potential. Find the classical angular frequency of oscillation if the spacing between adjacent energy levels is 2.8 eV.

Answer: 4.25×10^{15} s^{-1}

Some problems similar to Example 40-3: 24, 25, 27

With the infinite square well, we found that the probability density for the low-energy states bore no resemblance to what we would expect from classical physics. The same is true for the harmonic oscillator. If we consider a classical harmonic oscillator—for example, a mass moving back and forth on a spring—we find the mass moving slowly near the extremes of its displacement, but rapidly through its equilibrium position. We're most likely to find the particle where it spends the most time—namely, near the turning points—and least likely to find it where it's moving fastest—namely, near its equilibrium position. Figure 40-15a shows this classical probability density, along with the probability density $\psi_0^2(x)$ associated with the quantum-mechanical ground state. As with the infinite square well, the classical and quantum probability densities for the ground state are completely at odds. Quantum-mechanically, we're most likely to find the particle in its equilibrium position, where classical physics says the probability is lowest. As the energy level rises, however, the quantum description becomes more like the classical one, with the probability greatest near the classical turning points

20.3
Potential Wells

FIGURE 40-15 Probability densities for the $n = 0$, 1, 2, 3, and 10 states of the harmonic oscillator. Dashed curve is the classical prediction, in which the probability density is highest where the particle is moving most slowly—namely, near its turning points (marked with vertical lines). Quantum mechanically, the probability density in the ground state (a) is completely at odds with the classical prediction. But at higher energy levels (e) the probability density (averaged over the rapid fluctuations in ψ) approaches the classical prediction. Note the increase in amplitude of the motion with increasing energy, a feature common to both classical and quantum analysis. The classical probability drops abruptly to zero beyond the turning points, but in quantum mechanics there remains a nonzero chance of finding the particle beyond the classical turning points. The horizontal axis is position, in units of the classical turning point at the ground-state energy.

(Fig. 40-15, parts b–e). We found the same result for the infinite square well. Again, we see Bohr's correspondence principle at work: In the limit of large quantum numbers, quantum mechanics gives the same predictions as classical physics. Only when the energy is reasonably close to the ground-state energy do the new effects of quantum mechanics become dramatically obvious.

40-5 QUANTUM TUNNELING

One of the most remarkable features of Fig. 40-15 is the nonzero probability of finding a quantum-mechanical harmonic oscillator beyond its classical turning points—the points at which its kinetic energy has been converted entirely to potential energy. This unusual situation, which seems at first to violate energy conservation, has no counterpart in the classical description of matter.

Another example of penetration into a classically forbidden region is provided by a particle encountering a potential barrier (Fig. 40-16). Examples of such barriers include electrical potential differences associated with atomic nuclei, gaps between solid materials, and artificially constructed atomic barriers like that shown in the opening photo for Part 6 of this text. After exploring the behavior of particles at potential barriers, we'll look briefly at some specific examples.

Classically, a particle whose total energy lies below the barrier energy is confined to one side of the barrier. (It can't be inside the barrier because its kinetic energy $\frac{1}{2}mv^2$ would have to be negative—a clear impossibility.) If we solve the Schrödinger equation for this potential-energy curve, however, we find oscillatory solutions on either side of the barrier, joined according to the continuity conditions on ψ and $d\psi/dx$ by exponential functions within the barrier (see Problems 30, 31, and 56). Such a solution is shown superimposed on the barrier in Fig. 40-16. The probability density ψ^2 associated with this solution remains

FIGURE 40-16 A potential barrier of height U, showing the wave function for a particle incident from the left with energy E less than the barrier energy U. Classically, the particle should rebound and remain to the left of the barrier. But quantum-mechanically, there is a nonzero probability—reflected in the nonzero wave function on the right—that it will penetrate the barrier and emerge on the other side.

FIGURE 40-17 For a massive particle, the wave function drops rapidly in the barrier, giving a negligible probability of penetration. Compare with the case of a low-mass particle shown in Fig. 40-16.

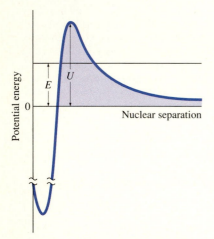

FIGURE 40-18 Potential-energy curve for a pair of deuterium nuclei, showing the barrier associated with the repulsive electrical force and the deep well arising from the strongly attractive nuclear force. Horizontal line is the energy E of a particle that, according to classical physics, cannot penetrate the Coulomb barrier. Quantum mechanically, though, it can tunnel through the barrier and fall into the potential well, releasing energy. Such tunneling is ultimately responsible for the fusion reactions that make the Sun shine.

20.4
Potential Barriers

nonzero through the barrier, and continues to give a nonzero probability of finding the particle on the far side—implying that a particle initially on one side of the barrier may later be found on the other side.

How likely is this phenomenon, called **quantum tunneling**? That depends on the relation of the particle energy to the barrier energy, and also on the width of the barrier. As you can show in Problem 30, the ψ function inside the barrier involves exponential functions of the form $e^{\pm\sqrt{2m(U_0-E)}x/\hbar}$, where U_0 is the barrier height. In general, these exponentials drop very rapidly across the barrier width unless the particle energy E is very close to the barrier energy or the particle mass m is very small. The probability that a particle will be found on the far side of the barrier is therefore very small when the mass m is large, so quantum tunneling is essentially a microscopic phenomenon (Fig. 40-17).

It looks as if tunneling violates the conservation-of-energy principle. But conservation of energy is saved by the uncertainty principle. Suppose we try to catch a particle in the act of tunneling. If we know the particle is within the barrier, the uncertainty in its position is no greater than the barrier width. Then the uncertainty principle sets a minimum value for the uncertainty in the particle's momentum, and this in turn implies a minimum energy. A quantitative analysis shows that minimum to be such that we can no longer be sure that the particle energy is below the barrier energy. But quantum tunneling really does occur, as we can confirm experimentally by measuring particle energies on both sides of the barrier. They can both be below the barrier height, and yet somehow particles have gone from one side to the other.

If we don't try to detect a particle within the barrier, its penetration is a purely wave phenomenon to which our particulate energy considerations don't apply. Again we see the wave-particle duality at work: If we don't observe the particle, its behavior is governed by the associated waves and may result in most unparticle-like phenomena such as tunneling. If we do try to catch it in the act of such behavior, it ceases to be wavelike and the surprising phenomena are not there.

Tunneling is important in a number of quantum-mechanical phenomena and technological devices. That the Sun shines—and therefore that we're alive—is a consequence of quantum tunneling of nuclei in the Sun's core. Classically, those nuclei don't have sufficient energy to get close enough to overcome their mutual electrical repulsion. But they can tunnel quantum mechanically through this "Coulomb barrier," then fuse together to release the enormous energy that powers the Sun (Fig. 40-18). An opposite nuclear process, alpha decay, occurs as alpha particles (composites of 2 protons and 2 neutrons) tunnel through a nuclear potential barrier that traps them inside large nuclei such as uranium. Measurement of the alpha particles' energy shows it to be below the barrier energy, confirming that tunneling occurs. Finally, we have artificially constructed potential barriers including "nanocrystallites" of nanometer-size grains of semiconductor crystal, "quantum dots" that trap single electrons, and "quantum corrals" as shown in Fig. 40-19 and on page 1007. Such devices offer great promise for the miniaturizing of electronic and optical components, including the possibility of computer memory cells in which a single electron carries the "yes" or "no" information of binary code. Quantum tunneling plays a central role in the movement of electrons through such devices. You can simulate quantum tunneling with ActivPhysics Activity 20.4.

FIGURE 40-19 A "quantum corral," produced by laying an oval ring of iron atoms (blue peaks) on a copper substrate. Electrons within the corral can escape by quantum tunneling through the iron barrier.

Conventional optical microscopes cannot resolve objects smaller than the wavelength of visible light—several hundred nanometers, or several thousand times the size of an atom. Electron microscopes achieve higher resolution using a beam of electrons in place of light. But quantum mechanics shows that even electrons have wavelike properties and an associated wavelength. For high resolution, the electron wavelength must be small; since $\lambda = h/p$, small electron wavelength implies a large momentum and correspondingly large energy. The high-energy electrons in an electron microscope beam penetrate deeply into the material under observation. The electron microscope is therefore good at resolving internal structure, but it reveals little about the surface of a material. Yet surface structure has become increasingly important to semiconductor engineers, as their microelectronic devices continue to shrink in size, and the ability to picture surface details is also necessary for biologists studying the workings of viruses, and for chemists concerned with reactions that take place on the surfaces of materials.

The **scanning-tunneling microscope** (STM), developed in the 1980s by Heinrich Rohrer and Gerd Binnig of IBM Zurich Research Laboratory in Switzerland, makes explicit use of quantum tunneling to resolve surface details down to the size of a single atom (Fig. 40-20). Their work brought Rohrer and Binnig the 1986 Nobel Prize in Physics, shared also with Ernst Ruska for his 1931 invention of the conventional electron microscope.

To see how the tunneling microscope works, consider the electrons in a material whose surface is to be studied. Classically, those electrons are bound inside the material by a steep potential-energy jump at the surface; this jump is associated with the electrostatic attraction between the electrons and the rest of the material. The material surface therefore marks the edge of a finite potential well in which classical physics suggests the electrons should be bound. But quantum-mechanically, electron wave functions extend beyond the surface, into the classically forbidden region (Fig. 40-21a). Since there are many electrons in the material, we can picture the material as surrounded by a "cloud" of electrons that have quantum-mechanically tunneled beyond the material surface (Fig. 40-21b). As Problem 30 shows, the wave function decreases exponentially with distance into the barrier—that is, with distance from the material surface. Statistically, this means that the density of the electron cloud also decreases exponentially.

In the scanning-tunneling microscope, a very sharp needle (Fig. 40-22; sometimes so sharp that a single atom sits at its tip!)

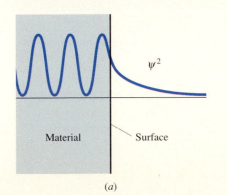

FIGURE 40-20 Heinrich Roher (left) and Gerd Binning with an early model of their scanning-tunneling microscope.

FIGURE 40-22 Conventional scanning electron microscope image of a tunneling microscope needle. Such a needle may be only one atom wide at its tip.

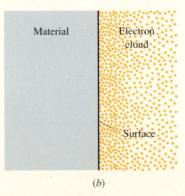

(a)   (b)

FIGURE 40-21 (a) The probability density ψ^2 extends beyond a material surface, giving a finite chance of finding electrons outside the material. (b) Statistically, an electron "cloud" surrounds the material. The cloud density decreases exponentially with distance from the surface.

is brought so close—typically less than 1 nm—to the surface under study that the electron clouds of needle and surface overlap. Application of a small potential difference between needle and surface then results in an electric current. In potential-energy terms, electrons carrying this current are tunneling through a potential barrier between surface and needle (Fig. 40-23); the current is therefore called a **tunneling current**. Because the wave function and therefore the electron cloud density decrease exponentially with distance from the surface, the tunneling current is very sensitive to the surface-needle separation. By scanning the needle parallel to the surface and recording the current, an image of the surface can be constructed. (In practice, the image construction method is more complicated. Feedback mechanisms are used to move the needle

vertically to keep the current and therefore the surface-needle separation constant. Then the needle itself traces out the contours of the surface.)

Successful operation of the scanning-tunneling microscope requires precise control of the needle position to within 0.01 nm, or less than one-tenth of a typical atomic diameter. Needle motion is controlled with so-called piezoelectric materials, whose size changes slightly in response to an applied electric field. External disturbances are isolated and damped through an elaborate combination of springs and eddy-current damping. The result is a device that has already made significant contributions in physics, biology, chemistry, and semiconductor engineering. Figure 40-24 shows some images produced with scanning-tunneling microscopes.

FIGURE 40-23 (*a*) The scanning-tunneling microscope detects currents flowing between the tip of a needle-like probe and the surface under study. Because the electron cloud density falls exponentially with distance from the surface, the current is a sensitive measure of that distance. By scanning the needle parallel to the surface, an image of the irregular surface can be constructed. (*b*) Potential energy curve for the region from material to needle tip. The electron current flows by tunneling through the potential barrier.

(*a*)

(*b*)

(*a*)

(*b*)

FIGURE 40-24 (*a*) This STM image shows a molecule of 8 cesium and 8 iodine atoms, on a copper surface. The molecule was assembled by moving the individual atoms with the STM tip. (*b*) Scientists at the University of California developed the world's smallest batteries. Two of the batteries appear in this STM image; each consists of a pile of copper (Cu) and gold (Ag) atoms only about 80 nm apart. The region shown measures 140 nm on a side, with axes labeled in angstrom units (1Å = 0.1 nm).

40-6 FINITE POTENTIAL WELLS

Both the infinite square well and the ideal harmonic oscillator involve potential wells of infinite depth. A particle is necessarily bound to such a well; it cannot escape to arbitrarily large distances. Its quantized energy states are therefore called **bound states**. But what if we have a well of finite depth, as shown in Fig. 40-25? We will state without mathematical analysis some properties of the resulting quantum states; aspects of the mathematics are covered in Problems 30 and 55–57. Provided the well is not too shallow, there are, in general, bound states with quantized energy levels whose wave functions resemble those of the infinite square well (Fig. 40-26). The main difference between these and the states of the infinite square well lies in the nonzero probability of quantum tunneling into the classically forbidden region outside the well.

The bound states, with their quantized energy levels, describe particles whose energies lie below the well height. Although those particles can penetrate somewhat into the classically forbidden region beyond the edges of the well, the exponential decrease in ψ ensures that they are very unlikely to be found far from the well. Particles whose energies exceed the well height, however, should be free to move anywhere. Quantum-mechanically, this freedom is reflected in a wave function that is everywhere oscillatory, without exponential decay (see Problem 30). Furthermore, particles in these **unbound states** can have any energy whatsoever as long as it exceeds the well height; unbound energy values are not quantized. Rather, there is a **continuum** of allowed energies above the well height, in contrast to the discrete, quantized levels below. Figure 40-27 is an energy-level diagram reflecting the presence of quantized, bound states as well as the continuum of unbound states. We will again find the presence of both bound and continuum states when we study the atom in the next chapter.

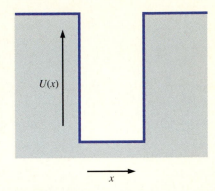

FIGURE 40-25 Potential-energy curve for a finite square well.

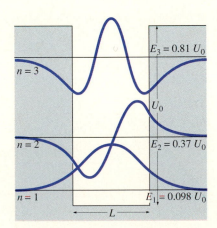

FIGURE 40-26 Bound-state wave functions for a finite square well, superimposed on the associated energy levels. Within the well the solutions are oscillatory, while outside they decay exponentially. For this particular combination of well depth and width (given by $\sqrt{mU_0 L^2/2\hbar^2} = 4$), there are only three bound states.

FIGURE 40-27 A complete energy-level diagram for a finite square well shows discrete bound states as well as a continuum of unbound states extending to arbitrarily high energy.

40-7 QUANTUM MECHANICS IN THREE DIMENSIONS

Our study of one-dimensional quantum systems has revealed important features of the quantum world, such as energy quantization. But atoms and most other quantum systems are inherently three dimensional. The wave function then depends on all three spatial variables, and the Schrödinger equation must be generalized to involve these variables. We will not use the full three-dimensional Schrödinger equation here but will simply point out some new features that arise in three-dimensional quantum systems.

We characterized the quantum states of a one-dimensional square well by a single integer quantum number n. That number arose mathematically from the condition that the wave function within the well include an integer number of half wavelengths. Solving the Schrödinger equation under these conditions gave a unique energy level associated with each value of n. In a similar way, the potential-energy function for a one-dimensional harmonic oscillator led to allowed energy levels described in terms of a single quantum number.

In three dimensions, constraints analogous to those imposed by the boundaries of a one-dimensional well or the potential-energy function of a one-dimensional harmonic oscillator lead to independent quantum numbers for each dimension (Fig. 40-28). For each set of three quantum numbers there is an associated energy level. For a particle of mass m confined to a cubical box of side L, for example, a generalization of the one-dimensional square well leads to the energy levels

$$E = \frac{h^2}{8mL^2}(n_x^2 + n_y^2 + n_z^2),\qquad (40\text{-}12)$$

where the n's are the quantum numbers associated with each spatial dimension. As in one dimension, the allowed values for the n's are positive integers. Thus, the ground state has $n_x = n_y = n_z = 1$. But what's the first excited state? It could be $n_x = 2$, $n_y = n_z = 1$. But it could equally well be $n_x = n_y = 1$, $n_z = 2$, or $n_x = n_z = 1, n_y = 2$, since all three of these combinations give the same energy.

When two or more quantum states have the same energy, those states are said to be **degenerate**. The first excited state of a particle confined to a cubical box, for example, is threefold degenerate, meaning there are three distinct states with the same energy. Degeneracy is often associated with symmetry of the quantum-mechanical system. In the cubical box, for example, it is the fact that the sides have

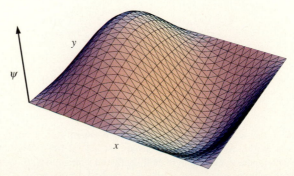

FIGURE 40-28 The wave function for a particle confined to a square region in two dimensions. The function (which satisfies the two-dimensional Schrödinger equation) is $\sin(n_x \pi x/L) \sin(n_y \pi y/L)$, with $n_x = 2$ and $n_y = 1$. Note that the two different quantum numbers correspond to a full wavelength in the x direction and a half wavelength in the y direction. An analogous expression would hold for a particle confined in three dimensions.

FIGURE 40-29 Energy-level diagrams showing the ground state and first excited state for a particle in a three-dimensional box. At left the box is cubical and the excited level is threefold degenerate. At right, the sides L_x, L_y, and L_z are slightly unequal, and chosen so that $L_x < L_y < L_z$ in such a way that the ground-state energy is unchanged. Introducing this asymmetry to the box removes the degeneracy, splitting the first excited state into three closely spaced but distinct levels.

equal lengths that allows different combinations of quantum numbers to give exactly the same energy. Making all three sides different would remove the degeneracy, splitting what was a single energy level into three levels (Fig. 40-29). The same thing happens in more realistic quantum systems. For example, imposing a magnetic field on an otherwise spherically symmetric atom breaks the symmetry and may split energy levels that were previously degenerate (Fig. 40-30). Detection of this splitting in optical spectra allows measurement of magnetic fields on the Sun and in other remote objects.

40-8 RELATIVISTIC QUANTUM MECHANICS

Like Newtonian physics, quantum mechanics based on the Schrödinger equation is a nonrelativistic theory—that is, it is not consistent with the requirement of special relativity that the laws of physics be the same in all inertial frames. It is therefore an approximation valid only at speeds much less than that of light. For many applications, including most in atomic, molecular, solid-state, and low-temperature physics, the nonrelativistic approximation $v \ll c$ is valid and the Schrödinger equation applies. But when particle speeds become significant compared with the speed of light, the Schrödinger equation becomes inadequate and must be replaced with a truly relativistic wave equation. And even for slowly moving particles, the requirement of relativistic invariance leads to some remarkable new phenomena.

The Dirac Equation and Antiparticles

In 1928, the English physicist Paul Dirac formulated a relativistic wave equation applicable to electrons. In the process, Dirac encountered several unexpected mathematical requirements with deep physical implications.

In Schrödinger's quantum mechanics, as well as in Newtonian physics, the nonrelativistic assumption $v \ll c$ is inherent in the expression $E = \frac{1}{2}mv^2$ for the energy of a free particle. The corresponding relation between energy and momentum, $E = p^2/2m$, is at the heart of the Schrödinger equation. Actually this expression is an approximation to the more general form we developed in Chapter 38:

$$E^2 = m^2c^4 + p^2c^2 .$$

Dirac's relativistic wave equation incorporated this relativistic energy-momentum relation in place of the Newtonian expression used in the Schrödinger equation. But Dirac faced a problem: the relativistic expression implies two possible values for the energy E, depending on which sign is used when one takes the square root. Dirac found that it was not possible to dismiss the negative root, and that therefore quantum states with negative energy must be possible. Through a rather involved argument, Dirac showed that the existence of these negative-energy states implied the possibility of a particle identical in mass to the electron but carrying a positive charge. The 1932 discovery of this particle, called the **positron**, was a vindication of Dirac's brilliant theorizing. Although Dirac's theory applies specifically to the electron, we now know that every particle has an

FIGURE 40-30 (a) Spectral line at 404.66 nm produced by mercury atoms undergoing transitions from the $n = 7$ level to the $n = 6$ level. The upper level is actually threefold degenerate. (b) Application of a 2.38-T magnetic field breaks the symmetry and removes the degeneracy, giving three lines where before there was one. Plots were taken with a grating spectrometer.

FIGURE 40-31 Pair creation. A high-energy photon produced an electron-positron pair whose trajectories spiral in opposite directions about a magnetic field. This image was made with the ALEPH detector at the Large Electron-Positron Collider at CERN.

associated **antiparticle**, identical in mass but opposite in electrical, magnetic, and related properties.

Since a particle-antiparticle pair has no net electric charge, the creation of such a pair does not violate charge conservation. Einstein's mass-energy equivalence therefore implies that such **pair creation** is possible given an energy of at least $2mc^2$, where m is the mass of each particle. In a typical pair-creation event, a photon of energy at least twice the rest energy of a given particle interacts with a massive nucleus (needed to conserve momentum) and disappears. In its place appear the particle and its antiparticle (Fig. 40-31). In 1997 scientists at the Stanford Linear Accelerator Center succeeded for the first time in pair creation from photons alone, by allowing very high-energy photons to interact with ordinary laser photons. The opposite process, annihilation, also occurs as a particle and its antiparticle disappear to form a pair of photons.

The concept of antiparticles has proved essential in high-energy physics and in studies of the very early universe. Essentially, particle-antiparticle pairs form when available energy—in a particle accelerator beam, or in the thermal energy just following the big bang—exceeds the threshold $2mc^2$ for formation of a pair of particles of mass m. In such a high-energy environment the equivalence of mass and energy is far more obvious than in the cool, low-energy world of our everyday experience.

Electron Spin

Another unexpected mathematical result of Dirac's work was his discovery that the relativistic wave function could not be a scalar; instead, it proved necessary to work with matrices. Dirac showed that this mathematical situation implied physically that the electron must possess an intrinsic angular momentum of magnitude $(\sqrt{3}/2)\hbar$—something that other quantum physicists had already inferred from experiments, but without any theoretical grounding. The existence of this angular momentum, called **spin**, has enormous significance in quantum mechanics, as we will examine extensively in the next chapter.

For now, we note simply that electron spin is grounded ultimately in the requirement of relativistic invariance, and is a relativistic effect that applies to *all* electrons, regardless of speed. In this sense, the intrinsic angular momentum of an electron is like its rest energy: both are consequences of relativistic invariance and both exist even for electrons at rest.

CHAPTER SYNOPSIS

Summary

1. Experimentally, matter and light are observed to manifest both particle-like and wave-like behavior. Quantum mechanics describes this situation through a statistical linkage between the **wave function**, ψ, and the probability of finding the corresponding particle. The probability of finding a particle is proportional to the square of the associated wave amplitude.
   a. For light, the appropriate wave equation derives from Maxwell's equations for the electric and magnetic fields. The corresponding particle is the photon, and the probability of finding a photon is proportional to the light intensity—that is, to the square of the electric or magnetic field strength.
   b. For matter moving nonrelativistically, the appropriate wave equation is the **Schrödinger equation**. The associated particles are most likely to be found where the square of the wave function is greatest.

2. The **time-independent Schrödinger equation**,

$$-\frac{\hbar^2}{2m}\frac{d^2\psi}{dx^2} + U(x)\psi = E\psi,$$

governs the spatial behavior of matter waves associated with particles moving in one dimension.
   a. The wave function ψ is itself an unobservable quantity. Its square, however, gives the **probability density**, or probability per unit volume (unit length in one dimension) of finding the particle. That is, in one dimension the

probability of finding the particle in some small interval dx at position x is given by

$$\text{Probability} = \psi^2(x)\,dx\,.$$

b. In the standard interpretation of quantum mechanics, the link between ψ and the finding of an actual particle is always statistical; the strict causality and determinism of classical physics are replaced by a fundamental indeterminism in the microscopic world. Only the probability that a particular outcome will occur is governed by exact laws.

c. A particle must be found *somewhere*. Mathematically, this is expressed through the **normalization condition**:

$$\int_{-\infty}^{\infty} \psi^2(x)\,dx = 1\,.$$

d. The wave function ψ is further constrained to be a continuous function, and its derivative $d\psi/dx$ must also be continuous except where the potential energy $U(x)$ goes discontinuously to infinity.

3. Solving the Schrödinger equation involves finding functions that satisfy the equation for a given potential energy $U(x)$. Of these functions, only those that are normalizable and that meet the continuity conditions are acceptable solutions. In general, potential-energy functions that would confine a classical particle to a finite region result in **energy quantization**—that is, in solutions to the Schrödinger equation for only certain discrete values of the total energy E.

4. A particle in an infinite square well provides a simple application of the Schrödinger equation. Here the discrete energy levels are given by

$$E_n = \frac{n^2 h^2}{8mL^2}\,,$$

where n, a positive integer, is the **quantum number**. The lowest possible energy, in this case $h^2/8mL^2$, is the **ground-state energy**. In general, zero energy is not permitted for a quantum-mechanical particle in a confined space—a manifestation of the uncertainty principle.

5. For the quantum-mechanical harmonic oscillator, a good approximation to many atomic and molecular systems, the energy levels are given by

$$E_n = (n + \tfrac{1}{2})\hbar\omega\,,$$

where the ground-state energy $\frac{1}{2}\hbar\omega$ corresponds to $n = 0$.

6. Solutions $\psi(x)$ to the Schrödinger equation can have nonzero values in regions where particle motion is not possible in classical physics. As a result, a particle may tunnel through a finite-width barrier even though classical physics suggests it has insufficient energy. Such **quantum tunneling** is important in many nuclear reactions including fusion and alpha decay, and is exploited in technologies including some microelectronic devices and the scanning-tunneling microscope.

7. For potential wells of finite depth, particle energies are quantized only for levels below the well top. Any energy above that level is possible, giving rise to a **continuum** of states.

8. In three dimensions, three quantum numbers are needed to specify the quantum state of a particle. Different combinations of quantum numbers, representing distinct quantum states, may give the same quantized energy. When this happens, the associated energy level is said to be **degenerate**.

9. For electrons moving at relativistic speeds, the Schrödinger equation must be replaced with Dirac's relativistic wave equation. In deriving that equation, Dirac predicted the existence of **antiparticles**, a prediction soon confirmed experimentally with the discovery of the positron in 1932. Dirac's work also shows that relativistic invariance requires that all electrons have an intrinsic angular momentum called **spin**.

Terms You Should Understand

(Pairs are closely related terms whose distinction is important; number in parentheses is chapter section where term first appears.)

old quantum theory, quantum mechanics (introduction)
Schrödinger equation (40-2)
wave function (40-2)
probability density (40-2)
normalization (40-2)
quantum number (40-3)
square-well potential (40-3, 40-6)
quantum tunneling (40-5)
bound state, unbound state (40-6)
continuum (40-6)
degeneracy (40-7)
positron, antimatter (40-8)
spin (40-8)

Symbols You Should Recognize

ψ (40-2)

Problems You Should Be Able to Solve

solving the Schrödinger equation for simple potentials (40-3)
evaluating energy levels for the infinite square well (40-3)
evaluating energy levels for the quantum harmonic oscillator (40-4)
finding energy levels and degeneracy in three-dimensional square wells (40-7)

Limitations to Keep in Mind

The Schrödinger equation is nonrelativistic, and therefore only applies when particle velocities are much less than the speed of light.

Quantum mechanics defies explanation in terms of Newtonian particles with well-defined trajectories. This is especially true at low quantum numbers; at higher quantum numbers, the correspondence principle shows how the quantum and classical views converge.

QUESTIONS

1. Explain qualitatively why a particle confined to a finite region cannot have zero energy.
2. Does quantum tunneling violate energy conservation? Explain.
3. In describing the quantum-mechanical behavior of matter, we settled on a wave equation (the Schrödinger equation) to express the underlying physical law. Does this mean that the wave properties of matter are in some way more important than the particle properties? Explain.
4. Bohr's correspondence principle states that quantum and classical mechanics must agree in a certain limit. What limit? Give an example.
5. Our analysis of the infinite square well shows that it is possible to know exactly the energy of a quantum-mechanical particle confined to a finite region. How is this not a violation of the uncertainty principle? After all, knowing the energy amounts to knowing the speed and therefore the magnitude of the momentum.
6. Why is the blue curve of Fig. 40-13 not a suitable wave function? After all, it satisfies the Schrödinger equation.
7. Consider two states of a particle in a cubical box that correspond to the same energy (for example, the states n_x, n_y, $n_z = 1, 1, 2$ and $n_x, n_y, n_z = 2, 1, 1$). Is there any physically measurable property that could be used to distinguish these states? Explain.
8. In terms of de Broglie's matter wave hypothesis, how does making the sides of a box different lengths remove some of the degeneracy associated with a particle confined to that box?
9. A particle is confined to a two-dimensional box whose sides are in the ratio $1:2$. Are any of its energy levels degenerate? If so, give an example. If not, why not?
10. Is there any ratio of sides of a two-dimensional box for which a particle confined to the box has no degenerate energy levels? If so, give a ratio. If not, why not?
11. What did Einstein mean by his remark, loosely paraphrased, that "God does not play dice"?
12. Some philosophers argue that the strict determinism of classical physics is inconsistent with human free will, but that the indeterminacy of quantum mechanics does leave room for free will. Others claim that physics has no bearing on the question of free will. What do you think?

PROBLEMS

ActivPhysics can help with these problems: Activities in Section 20, Quantum Mechanics

Section 40-2 The Schrödinger Equation

1. What are the units of the wave function $\psi(x)$ in a one-dimensional situation?
2. The wave function for a particle is given by $\psi = Ae^{-x^2/a^2}$, where A and a are constants. (a) Where is the particle most likely to be found? (b) Where is the probability per unit length half its maximum value?
3. The solution to the Schrödinger equation for a particular potential is $\psi = 0$ for $|x| > a$, and $\psi = A\sin(\pi x/a)$ for $-a \le x \le a$, where A and a are constants. In terms of a, what value of A is required to normalize ψ?
4. Find an expression for the normalization constant A for the wave function given by $\psi = 0$ for the case $|x| > b$ and by $\psi = A(b^2 - x^2)$ for $-b \le x \le b$.
5. Use a table of definite integrals or symbolic math software to help evaluate the normalization constant A in the wave function of Problem 2.
6. Suppose ψ_1 and ψ_2 are both solutions of the Schrödinger equation for the same energy E. Show that $a\psi_1 + b\psi_2$ is also a solution, where a and b are arbitrary constants.
7. Describe the potential, in terms of x and the particle energy E, that would result in the wave function of Problem 4.

Section 40-3 The Infinite Square Well

8. Determine the ground-state energy for an electron in an infinite square well of width 10.0 nm.
9. What is the width of an infinite square well in which a proton cannot have an energy less than 100 eV?
10. One reason we don't notice the effects of quantum mechanics in everyday life is that Planck's constant h is too small. Treating yourself as a particle in a room-sized one-dimensional infinite square well, how big would h have to be if your minimum possible energy corresponded to a speed of 1.0 m/s? Take the width of the room to be 2.6 m and your mass to be 60 kg.
11. A particle is confined to an infinite square well of width 1.0 nm. If the energy difference between the ground state and the first excited state is 1.13 eV, is the particle an electron or a proton?
12. An electron is trapped in an infinite square well 25 nm wide. Find the wavelengths of the photons emitted in the following transitions: (a) $n = 2$ to $n = 1$; (b) $n = 20$ to $n = 19$; (c) $n = 100$ to $n = 1$.
13. An electron drops from the $n = 7$ to the $n = 6$ level of an infinite square well 1.5 nm wide. Find (a) the energy and (b) the wavelength of the photon emitted.
14. Show explicitly that the difference between adjacent energy levels in an infinite square well becomes arbitrarily small

compared with the energy of the upper level, in the limit of large quantum number n.

15. An electron is in a narrow molecule 4.4 nm long, a situation that approximates a one-dimensional infinite square well. If the electron is in its ground state, what is the maximum wavelength of electromagnetic radiation that can cause a transition to an excited state?

16. A snail of mass 3 g crawls between two rocks 15 cm apart at a speed of 0.5 mm/s. Treating this system like a particle in an infinite square well, determine the approximate quantum number. Does the correspondence principle permit the use of the classical approximation in this case?

17. Repeat Example 40-1 for a proton trapped in a nuclear-size square well of width 1 fm. Comparison with the result of Example 40-1 gives a rough estimate of the energy difference between nuclear and chemical reactions.

18. The ground-state energy for an electron in square well A is equal to the energy of the first excited state for an electron in square well B. How do the widths of the two wells compare? Both are infinitely deep.

19. An electron drops from the $n = 5$ level to the $n = 2$ level of an infinite square well, emitting a 1.4-eV photon in the process. Find the width of the square well.

20. Electrons in an ensemble of many square-well systems are initially in the $n = 4$ state. The wells all have the same width, 0.10 nm. Find the wavelengths of all spectral lines emitted as the electrons cascade downward to the ground state through all possible transitions.

21. Sketch the probability density for the $n = 2$ state of an infinite square well extending from $x = 0$ to $x = L$, and determine where the particle is most likely to be found.

22. An infinite square well extends from $-L/2$ to $L/2$. (a) Find expressions for the normalized wave functions for a particle of mass m in this well. You will need separate expressions for even and odd quantum numbers. (b) Find the corresponding energy levels.

23. A particle is in the ground state of an infinite square well. What is the probability of finding the particle in the left-hand third of the well?

Sections 40-4, 40-5, and 40-6 The Harmonic Oscillator, Quantum Tunneling, and Finite Potential Wells

24. A particle in a harmonic oscillator potential has a ground-state energy of 0.14 eV. What would be the classical frequency f of the oscillator?

25. What is the ground-state energy for a particle in a harmonic oscillator potential whose classical angular frequency is $\omega = 1.0 \times 10^{17}$ s^{-1}?

26. A 10-g mass is attached to an ideal spring of spring constant $k = 150$ N/m, and set oscillating with amplitude 12 cm. To what quantum number does this oscillation correspond?

27. A hydrogen chloride molecule may be modeled as a hydrogen atom (mass 1.67×10^{-27} kg) on a spring; the other end

of the spring is attached to a rigid wall (the massive chlorine atom). If the minimum photon energy that will promote this molecule to its first excited state is 0.358 eV, find the "spring constant."

28. Show that the two functions $Ae^{\pm m\omega x^2/2\hbar}$ both satisfy the Schrödinger equation for a one-dimensional harmonic oscillator (Equation 40-9), and determine the corresponding energies.

29. For what quantum numbers is the spacing between adjacent energy levels in a harmonic oscillator less than 1% of the actual energy?

30. Figure 40-32 shows a finite square well. Consider a particle of mass m whose energy E lies between 0 and U_0. (a) Write two forms of the Schrödinger equation for this particle, corresponding to the interior ($U = 0$) and exterior ($U = U_0$) of the well. (b) Show that the equation for the well interior is satisfied by functions of the form $A\sin(\sqrt{2mE}\,x/\hbar)$ and $A\cos(\sqrt{2mE}\,x/\hbar)$, where A is an arbitrary constant. (c) Show that the equation for the well exterior is satisfied by functions of the form $Ae^{\pm\sqrt{2m(U_0-E)}x/\hbar}$. *Note:* You are not solving the whole problem because your solutions aren't necessarily consistent at the well boundaries. Problems 55–57 treat the full solution.

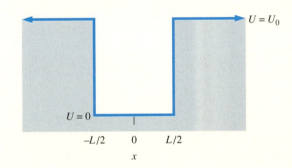

FIGURE 40-32 Problems 30, 31.

31. When the square well width L of Fig. 40-32, the well depth U_0, and the particle mass m are related by $\sqrt{mU_0 L^2/2\hbar^2} = 4$, a detailed analysis shows that the ground-state energy is given by $E_1 = 0.098U_0$ and the corresponding wave function is approximately

$$\psi_1 = 17.9 \frac{1}{\sqrt{L}} e^{7.60x/L} \qquad (x \leq -\tfrac{1}{2}L)$$

$$\psi_1 = 1.26 \frac{1}{\sqrt{L}} \cos(2.50x/L) \qquad (-\tfrac{1}{2}L \leq x \leq \tfrac{1}{2}L)$$

$$\psi_1 = 17.9 \frac{1}{\sqrt{L}} e^{-7.60x/L} \qquad (x \geq \tfrac{1}{2}L).$$

Show that both ψ_1 and its first derivative are continuous at the well edge. (That is, show that the second and third expressions above have the same value and the same derivative at $x = \frac{1}{2}L$.) Because the functions given are approximate, your results won't be exactly the same, but should agree to two significant figures. *Hint:* Be sure to set your calculator to radians before computing the trig functions! (b) Plot the function $\sqrt{L}\psi_1$ as a function of x/L, indicating the well edges on your plot.

32. For the state described in the preceding problem, find the probability that the particle will be found somewhere in the classically forbidden region outside the well. *Hint:* Symmetry allows you to calculate the probability for $x \geq \frac{1}{2}L$, then double the result. Also see Example 40-2, and consult the integral table in Appendix A if necessary.

33. The probability that a particle of mass m and energy $E < U$ will tunnel through the potential barrier of Fig. 40-16 is approximately

$$P = e^{-2\sqrt{2m(U - E)}L/h},$$

where L is the barrier width. Evaluate this probability (a) for a 2.8-eV electron incident on a 1.0-nm wide barrier 4.0 eV high and (b) for a 1200-kg car moving at 15 m/s striking a 1.0-m-thick stone wall requiring, classically, 150 kJ to breach it.

Section 40-7 Quantum Mechanics in Three Dimensions

34. If all sides of a cubical box are doubled, what happens to the ground-state energy of a particle in that box?

35. A very crude model for an atomic nucleus is a cubical box about 1 fm on a side. What would be the energy of a gamma ray emitted if a proton in such a nucleus made a transition from its first excited state to the ground state?

36. (a) Use Equation 40-12 to draw an energy-level diagram for the first six energy levels of a particle in a cubical box, in terms of $h^2/8mL^2$, and (b) give the degeneracy of each.

37. The generalization of the Schrödinger equation to three dimensions is

$$-\frac{\hbar^2}{2m}\left(\frac{\partial^2\psi}{\partial x^2} + \frac{\partial^2\psi}{\partial y^2} + \frac{\partial^2\psi}{\partial z^2}\right) + U(x, y, z)\psi = E\psi.$$

(a) For a particle confined to the cubical region $0 \leq x \leq L$, $0 \leq y \leq L$, $0 \leq z \leq L$, show by direct substitution that the equation is satisfied by wave functions of the form $\psi(x, y, z) = A\sin(n_x\pi x/L)\sin(n_y\pi y/L)\sin(n_z\pi z/L)$, where the n's are integers and A is a constant. (b) In the process of working part (a), verify that the energies E are given by Equation 40-12.

Section 40-8 Relativistic Quantum Mechanics

38. (a) Find the minimum energy of a photon needed to create an electron-positron pair. (b) Find the corresponding photon frequency and wavelength, and use Fig. 34-10 to determine in what part of the electromagnetic spectrum it lies.

39. Early in the history of the universe, the temperature was so high that thermal energy resulted in frequent pair creation.

Find the approximate temperature such that the thermal energy kT would be enough to create (a) an electron-positron pair and (b) a proton-antiproton pair.

40. A muon and its antiparticle annihilate, producing a pair of 106-MeV gamma-ray photons. Find the mass of the muon.

Paired Problems

(Both problems in a pair involve the same principles and techniques. If you can get the first problem, you should be able to solve the second one.)

41. An alpha particle (mass 4 u) is trapped in a uranium nucleus of diameter 15 fm. Treating the system as a one-dimensional square well, what would be the minimum energy for the alpha particle?

42. A laser emits 1.96-eV photons. If this emission is due to electron transitions from the $n = 2$ to $n = 1$ states of an infinite square well, what is the width of the well?

43. What is the probability of finding a particle in the central 80% of an infinite square well, assuming it's in the ground state?

44. A particle detector has a resolution 15% of the width of an infinite square well. What is the chance that the detector will find a particle in the ground state of the square well if the detector is centered on (a) the midpoint of the well or (b) a point one-fourth of the way across the well?

45. A harmonic oscillator emits a 1.1-eV photon as it undergoes a transition between adjacent states. What is its classical oscillation frequency?

46. An ensemble of harmonic oscillators undergoes transitions from the first excited state to the ground state, and the spectrum from the ensemble shows a line at 5.6 μm. Find the frequency of the oscillators.

47. An electron is confined to a cubical box. For what box width will a transition from the first excited state to the ground state result in emission of a 950-nm infrared photon?

48. A proton is confined to a cubical box 2.5 fm on a side. Find the minimum energy of a gamma ray that could be emitted in transitions among any of the first 6 energy levels.

Supplementary Problems

49. For what quantum state is the probability of finding a particle in the left-hand quarter of an infinite square well equal to 0.303?

50. A 9-W laser beam shines on an ensemble of 10^{24} electrons, each in the ground state of a one-dimensional infinite square well 0.72 nm wide. The photon energy is just enough to raise an electron to its first excited state. How many electrons can be so excited if the beam shines for 10 ms?

51. A large number of electrons are all confined to infinite square wells 1.2 nm wide. They are undergoing transitions among all possible states. How many (a) visible (400 nm to 700 nm) and (b) infrared lines will there be in the spectrum emitted by this ensemble of square-well systems?

52. Show that the turning points for a classical harmonic oscillator of mass m with energy given by Equation 40-11 are located at $x = \pm\sqrt{\dfrac{\hbar}{m\omega}(2n+1)}$.

53. Consider an infinite square well with a steplike potential in the bottom, as shown in Fig. 40-33. Without solving any equations, sketch what you think the wave function should look like for a particle whose energy is (a) less than the step height and (b) greater than the step height.

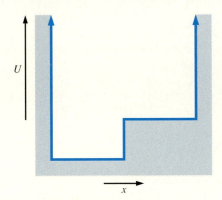

FIGURE 40-33 Problem 53.

54. A particle is in the nth quantum state of an infinite square well. (a) Show that the probability of finding it in the left-hand quarter of the well is

$$P = \frac{1}{4} - \frac{\sin(n\pi/2)}{2n\pi}.$$

(b) Show that for even n the probability is exactly $\frac{1}{4}$.
(c) Show that for odd n the probability approaches the classical value $\frac{1}{4}$ as $n \to \infty$.

55. The next three problems illustrate the matching procedure necessary in solving the Schrödinger equation for potential wells of finite depth. Consider a particle of mass m in the semi-infinite well shown in Fig. 40-34. It will be convenient to work with dimensionless forms of the particle energy E and well depth U_0, in units of $\hbar^2/2mL^2$: $\varepsilon \equiv 2mL^2E/\hbar^2$ and $\mu \equiv 2mL^2U_0/\hbar^2$. Assume that $E < U_0$ or, equivalently, $\varepsilon < \mu$. Show by substitution that the following wave functions satisfy the Schrödinger equation in the regions indicated:

$$\psi_1 = A\sin(\sqrt{\varepsilon}\,x/L) \qquad (0 \le x \le L)$$
and
$$\psi_2 = Be^{-\sqrt{\mu-\varepsilon}\,x/L} \qquad (x \ge L),$$

where A and B are arbitrary constants.

FIGURE 40-34 Problems 55–57.

56. The wave functions of the preceding problem satisfy the Schrödinger equation, but they don't necessarily meet the conditions that ψ and its derivative be continuous; in particular, both quantities could change abruptly at the well edge $x = L$. To achieve continuity of ψ, evaluate ψ_1 and ψ_2 at $x = L$ and set them equal. To achieve continuity of $d\psi/dx$, first differentiate ψ_1 and ψ_2, then evaluate the derivatives at $x = L$, and then set them equal. The two continuity conditions should result in these equations:

$$A\sin(\sqrt{\varepsilon}) = Be^{-\sqrt{\mu-\varepsilon}}$$

and

$$\sqrt{\varepsilon}\,A\cos(\sqrt{\varepsilon}) = -\sqrt{\mu-\varepsilon}\,Be^{-\sqrt{\mu-\varepsilon}}$$

Use the first of these equations to eliminate the quantity $Be^{-\sqrt{\mu-\varepsilon}}$ from the second equation, and show that the result is

$$\tan(\sqrt{\varepsilon}) = -\sqrt{\frac{\varepsilon}{\mu-\varepsilon}}.$$

This equation determines the values of ε that give the quantized energy levels.

57. The equation derived in the preceding problem cannot be solved algebraically since the unknown ε appears in a trig function and under the square root. It can be solved graphically, by plotting both sides on the same graph and determining where they intersect. It can also be solved by trial and error on a calculator, by using a calculator with a root-finding routine, or by computer. Use one of these methods to find all possible values of ε for (a) $\mu = 2$, (b) $\mu = 20$, and (c) $\mu = 50$. *Note:* The number of solutions varies with μ; there may be no solutions, meaning no bound states are possible, or there may be one or more bound states.

41 Atomic Physics

Quantum mechanics reveals the atom as a nucleus surrounded by a "cloud" of probability indicating the likelihood of finding the atomic electrons at particular locations. This computer-generated model of a beryllium atom shows that the electron "cloud" consists of two distinct shells.

In the preceding chapter we developed some key ideas of quantum mechanics, largely by exploring the Schrödinger equation in application to simple systems. Now we turn to the atom and show how the Schrödinger equation, along with two important results from relativistic quantum mechanics, leads us to a description of atomic spectra and to an explanation of the periodic table and the chemical properties of the elements. We will deal most thoroughly with the simplest atom, hydrogen. Even with this case, we must skip many mathematical details that are beyond the scope of this book. Nevertheless, we will make plausible a number of important atomic properties, many of which have analogs in the simpler systems of the preceding chapter.

41-1 THE HYDROGEN ATOM

Like the particle in a box that we considered in the preceding chapter, the electron in a hydrogen atom is also a particle confined by a force associated with a three-dimensional potential-energy function. For the electron, the potential energy is that of the electrostatic attraction between the electron and proton. In Chapter 25 we found that the electric potential associated with a point charge q is $V(r) = kq/r$, where r is the distance from the point charge and where the zero of potential is at infinity. In treating the hydrogen atom, we consider the more massive proton (1836 times the electron mass) to be fixed at the origin. The proton's charge is $+e$, so the electric potential associated with the proton becomes $V(r) = ke/r$. Recall that electric potential is the potential energy *per unit charge;* to get the potential energy of an electron in the field of the proton, we multiply the potential by the electron charge, $-e$, to get

$$U(r) = \frac{-ke^2}{r}. \tag{41-1}$$

This is the potential-energy function to be used in the Schrödinger equation for the electron in a hydrogen atom. Figure 41-1 shows a plot of Equation 41-1.

The Schrödinger Equation in Spherical Coordinates

Substituting Equation 41-1 for $U(x, y, z)$ in the three-dimensional Schrödinger equation (see Problem 40-37) gives the equation we must solve to find the structure of the hydrogen atom. But our potential energy is expressed in terms of the radial distance r, while the Schrödinger equation is in rectangular coordinates x, y, z. We could write r in terms of x, y, z, but the resulting equation would be unenlightening because rectangular coordinates do not reflect the natural

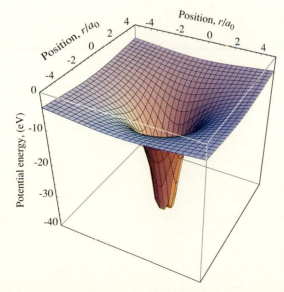

FIGURE 41-1 Potential energy of the electron in a hydrogen atom (in eV), as a function of position in a plane containing the proton. Horizontal axes are distances from the proton in units of the Bohr radius $a_0 = 0.0529$ nm.

FIGURE 41-2 Spherical coordinates locate a point P in terms of its distance r from the origin, the angle θ from the z axis, and the angle ϕ that measures position relative to the x axis.

spherical symmetry of the potential-energy function. Instead, it is easiest to work in **spherical coordinates**, where the position of a point is specified by giving its distance r from the origin along with two angles θ and ϕ that specify its orientation (Fig. 41-2). Converting to spherical coordinates is straightforward but tedious; the result is the equation

$$-\frac{\hbar^2}{2mr^2}\left[\frac{\partial}{\partial r}\left(r^2\frac{\partial\psi}{\partial r}\right) + \frac{1}{\sin\theta}\frac{\partial}{\partial\theta}\left(\sin\theta\frac{\partial\psi}{\partial\theta}\right) + \frac{1}{\sin^2\theta}\frac{\partial^2\psi}{\partial\phi^2}\right] - \frac{ke^2}{r}\psi = E\psi,$$

(41-2)

where we have used Equation 41-1 for the potential energy function, and where ψ is now a function of the spherical coordinates r, θ, ϕ of Fig. 41-2.

Although Equation 41-2 looks forbidding, it can be solved readily using more advanced techniques than you have probably encountered. For total energy E less than zero, corresponding to bound states in the potential well of Fig. 41-1, it turns out that most solutions become infinite at large values of r and are therefore not normalizable. As with simpler one-dimensional systems, only certain values of the energy E result in acceptable bound-state solutions. For total energy greater than zero, the electron is unbound and any energy proves possible, as with the finite square well.

The Hydrogen Ground State

In general, solutions to Equation 41-2 depend on all three variables r, θ, ϕ. But some solutions, including the ground state, are spherically symmetric—that is, they depend only on r. Here we show that the ground state has the form of a simple exponential, and in the process derive the ground-state energy. Consider the function

$$\psi = Ae^{-r/a_0}, \tag{41-3}$$

where A and a_0 are as yet undetermined constants, the latter with the units of length. For this spherically symmetric function, nothing depends on the angular variables θ and ϕ, so derivatives with respect to those variables are strictly zero. We're then dealing with a function of only one variable, so we can write total instead of partial derivatives. Equation 41-2 then becomes

$$-\frac{\hbar^2}{2mr^2}\frac{d}{dr}\left(r^2\frac{d\psi}{dr}\right) - \frac{ke^2}{r}\psi = E\psi, \tag{41-4}$$

where ψ is now a function of r only. Substituting the proposed solution 41-3 into this equation gives good practice in differentiating (see Problem 19); the result, after some algebra, is

$$-\frac{\hbar^2}{2ma_0^2} + \frac{\hbar^2}{mra_0} - \frac{ke^2}{r} = E_1, \tag{41-5}$$

where we write E_1 to designate the ground-state energy. Equation 41-5 must be satisfied for all values of r. But this can be the case only if the two r-dependent terms cancel; that is, if

$$\frac{\hbar^2}{mra_0} = \frac{ke^2}{r}$$

or

$$a_0 = \frac{\hbar^2}{mke^2} = 5.29\times10^{-11}\text{ m} = 0.0529\text{ nm}.$$

Substituting this result into Equation 41-5 then gives the ground-state energy:

$$E_1 = -\frac{\hbar^2}{2ma_0^2} = -2.18 \times 10^{-18} \text{ J} = -13.6 \text{ eV},$$

where the minus sign shows that the atom is a bound system. Equivalently, it would take 13.6 eV or more to bring the electron to an unbound state with $E > 0$—that is, to ionize the atom. Our results show that Equation 41-3 is indeed a solution to the Schrödinger equation for the hydrogen atom, corresponding to energy $E_1 = -13.6$ eV.

In deriving expressions for a_0 and E_1, we've shown how Schrödinger's theory gives two of the fundamental parameters of atomic physics. We've actually seen these parameters before: a_0 is the **Bohr radius**, defined in Chapter 39 as the orbital radius of the smallest allowed electron orbit in Bohr's atomic theory, while our E_1 is the same as the electron energy in the lowest Bohr orbit. But while the ground-state energies of the Bohr and Schrödinger theories agree, the interpretation of these two theories differs dramatically, and, as we will see later, only the Schrödinger theory describes correctly the finer details of atomic structure. In particular, the Bohr theory still clings to the classical notion of electrons in circular orbits about the nucleus. But the Schrödinger theory is truly quantum-mechanical, with the electron represented by the wave function ψ and associated probability distribution. The Bohr radius a_0 no longer represents an actual orbital radius but instead determines the atomic size only in a statistical sense. Since the wave function falls off exponentially as e^{-r/a_0}, the probability of finding the electron becomes negligible at distances from the nucleus much greater than a_0.

The Radial Probability Distribution

Where are we most likely to find the electron in a hydrogen atom that is in its ground state? Figure 41-3 shows the probability density ψ^2 associated with Equation 41-3. At first glance it looks as if the electron is most likely to be at $r = 0$, right inside the nucleus! With the one-dimensional systems of the preceding chapter, that interpretation would have been correct. But here we have to be more careful. What we really want to know is this: at what radial distance from the nucleus are we most likely to find the electron? The probability density ψ^2 is the probability *per unit volume* of finding the electron at some point. What we want is the probability *per unit radial distance*. In a one-dimensional system, probability per unit length and per unit volume are proportional (see Equation 40-2 and preceding discussion). But in three dimensions they need not be. Figure 41-4 shows why.

Consider a thin spherical shell of radius r and thickness dr centered on the nucleus. The area of the shell is $4\pi r^2$, so its volume is $dV = 4\pi r^2 dr$. Since the ground-state wave function depends only on radius, so does the probability density, and the probability of finding the electron in the shell is given by

$$\text{Probability} = \psi^2 dV = 4\pi r^2 \psi^2 dr.$$

We define the **radial probability density**, $P(r)$, as the probability per unit radius of finding the electron at radial distance r. Then the probability itself is $P(r)dr$, and we see from the equation above that

$$P(r) = 4\pi r^2 \psi^2, \tag{41-6}$$

(a)

(b)

FIGURE 41-3 Probability density ψ^2 for the hydrogen ground state of Equation 41-3. (a) is a plot of ψ^2 as a function of distance from the proton, while (b) shows the "electron cloud," whose intensity is proportional to the probability density.

FIGURE 41-4 The volume of a spherical shell of radius r and thickness dr is $dV = 4\pi r^2 dr$. The probability per unit radial distance is thus $4\pi r^2$ times the probability per unit volume.

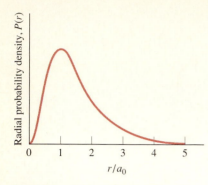

FIGURE 41-5 Radial probability density for the hydrogen ground state. The electron is most likely to be found one Bohr radius from the nucleus.

or $4\pi r^2$ times the probability per unit volume. Physically, the two probability densities differ because spherical shells of fixed thickness dr increase in volume as their radii increase.

When we ask at what radius are we most likely to find the electron, we are asking where the radial probability density of Equation 41-6 is a maximum. Multiplying ψ^2 from Equation 41-3 by $4\pi r^2$ to get the radial probability density for the hydrogen ground state gives

$$P_1(r) = 4\pi r^2 A^2 e^{-2r/a_0}.$$

Figure 41-5 is a plot of this radial probability density. As the figure shows, and as you can prove in Problem 20, the maximum occurs at $r = a_0$, showing that the electron is most likely to be found at the distance suggested by Bohr for the ground-state orbit. But in contrast to Bohr's theory, the electron is *not* in orbit, and it need not be found at a distance of one Bohr radius. In quantum mechanics, the atom is a fuzzy thing. A hydrogen atom is roughly two Bohr radii in diameter, but that statement is necessarily vague because of the nonzero probability of finding the electron either closer to or farther from the nucleus.

EXAMPLE 41-1	*Normalization and the Probability Distribution*

Determine the normalization constant A of Equation 41-3 and, using the result, calculate the probability that the electron in the hydrogen ground state will be found beyond the Bohr radius.

Solution

The electron must be *somewhere,* so integrating the probability density ψ^2 over all space must give 1; mathematically this gives the normalization condition

$$\int_{\text{all space}} \psi^2 dV = 1.$$

With the spherically symmetric ground-state function, we found that $\psi^2 dV$ can be written $4\pi r^2 \psi^2 dr$. Using Equation 41-3 for ψ, the normalization condition then becomes

$$\int_{r=0}^{r=\infty} 4\pi A^2 r^2 e^{-2r/a_0} dr = 1.$$

We could evaluate the integral using integration by parts; however, the result is given in the integral table at the end of Appendix A. Replacing x by r and a by $-2/a_0$ in the table's expression for $\int x^2 e^{ax} dx$, we have

$$\int_{r=0}^{r=\infty} 4\pi A^2 r^2 e^{-2r/a_0} dr$$

$$= 4\pi A^2 \left\{ \frac{r^2 e^{-2r/a_0}}{(-2/a_0)} - \frac{2}{(-2/a_0)} \left[\frac{e^{-2r/a_0}}{(-2/a_0)^2} \left(-\frac{2}{a_0} r - 1 \right) \right] \right\} \Bigg|_{r=0}^{r=\infty}$$

$$= 1.$$

The expression in curly brackets vanishes at $r = \infty$, and at $r = 0$ the exponentials are just 1, so we have

$$4\pi A^2 [0 - (-\tfrac{1}{4} a_0^3)] = 1,$$

or

$$A = \frac{1}{\sqrt{\pi a_0^3}}.$$

The probability of finding the electron beyond the Bohr radius is given by integrating the probability density over the region of interest—from $r = a_0$ to $r = \infty$:

$$P(r > a_0) = \int_{a_0}^{\infty} 4\pi r^2 \psi^2 dr.$$

This is the same integral that arose in evaluating the normalization constant A, except that now we evaluate between a_0 and infinity. Making this change in the limits, we are led to

$$P(r > a_0) = 4\pi A^2 a_0^3 (\tfrac{1}{2} e^{-2} + \tfrac{3}{4} e^{-2}) = 5\pi A^2 a_0^3 e^{-2}.$$

With $A^2 = 1/\pi a_0^3$, this becomes

$$P(r > a_0) = 5e^{-2} \simeq 0.677.$$

Thus in about two of three cases, the electron will be found beyond the Bohr radius. In Problem 4, you can do a similar calculation to show that there is about one chance in four of finding the electron beyond two Bohr radii.

EXERCISE Find the probability that the electron in the hydrogen ground state will be found within $\tfrac{1}{2} a_0$ of the nucleus.

Answer: 0.080

Some problems similar to Example 41-1: 4, 51, 52

Excited States of Hydrogen

So far we have examined only the ground state of the hydrogen atom. But Equation 41-2 admits many more normalizable solutions, corresponding to the excited states of hydrogen.

In general, each energy level is associated with one spherically symmetric wave function and a number of nonsymmetric ones. For historical reasons, the spherically symmetric states are called **s-states**. The distinct energy levels are labeled by the quantum number n, called the **principal quantum number**. The ground state, for example, is the $1s$ state. The energy of the nth level, derivable from the Schrödinger equation, turns out to agree exactly with the earlier Bohr theory:

$$E_n = -\frac{1}{n^2}\frac{\hbar^2}{2ma_0^2} = \frac{E_1}{n^2} = \frac{-13.6\text{ eV}}{n^2}. \qquad (41\text{-}7)$$

The spherically symmetric state with energy E_2—that is, the $2s$ state—has wave function given by

$$\psi_{2s} = \frac{1}{4\sqrt{2\pi a_0^3}}\left(2 - \tfrac{r}{a_0}\right)e^{-r/2a_0}. \qquad (41\text{-}8)$$

By substituting this function into Equation 41-4, you can verify that the energy E_2 is given by Equation 41-7 (see Problem 61). The radial probability densities for the first three spherically symmetric states are plotted in Fig. 41-6; notice that the excited states correspond to larger, more "smeared-out" atoms.

Although we are discussing hydrogen specifically, our results are easily generalized to any single-electron atom—that is, to an atom of atomic number Z ionized $Z - 1$ times. For such an atom the potential energy function becomes $-kZe^2/r$, and our calculations go through as before except that the factor e^2 is replaced by Ze^2. Then the energy levels become

$$E_n = -\frac{Z^2}{n^2}\frac{\hbar^2}{2ma_0^2} = \frac{Z^2E_1}{n^2} = -\frac{(13.6\text{ eV})Z^2}{n^2}, \qquad (41\text{-}9)$$

reflecting the tighter binding of the more highly charged nucleus (see Problem 63).

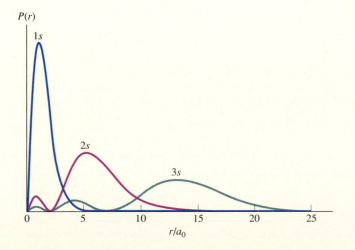

FIGURE 41-6 Radial probability densities for the spherically symmetric $1s$, $2s$, and $3s$ states of hydrogen. For higher energy states the probability density peaks at larger radii, showing that the excited atom is larger in a fuzzy, statistical sense.

Got It!

Approximately where are you most likely to find the electron in the 2s state of hydrogen?

Orbital Quantum Numbers and Angular Momentum

So far we have discussed only spherically symmetric states and the principal quantum number n that determines the energy. In these states, it turns out that the angular momentum associated with the electron's motion about the nucleus—also called the **orbital angular momentum**—is exactly zero. (This makes sense because the electron probability distribution is spherically symmetric, so there's no preferred axis about which angular momentum might be defined. But it contrasts with the Bohr theory prediction—which is incorrect—that the ground-state orbital angular momentum should be $\hbar$.) But there are other solutions to the Schrödinger equation for the hydrogen atom, solutions that involve all three variables and are therefore not spherically symmetric. These solutions correspond to nonzero values of the orbital angular momentum.

For a given principal quantum number n, there are in fact n distinct solutions with different angular momenta that share the nth energy level given by Equation 41-7. (The fact that these different states have essentially the same energy is related to the $1/r$ nature of the Coulomb potential; so-called fine structure effects that we will discuss shortly can actually lead to small differences in energy among states with the same n.) The different angular momentum states are labeled with the quantum number ℓ, called the **orbital quantum number**. The range for ℓ is from 0 to $n - 1$, so the ground state ($n = 1$) corresponds to the single value $\ell = 0$. Higher energy levels, however, are degenerate. The orbital quantum number ℓ determines the magnitude, L, of the angular momentum associated with the electron's motion:

$$L = \sqrt{\ell(\ell + 1)}\hbar.$$

(41-10)

EXAMPLE 41-2	*Orbital Angular Momentum*

What are the possible values for the magnitude of the orbital angular momentum of an electron in the $n = 3$ energy level of hydrogen?

Solution

At the nth energy level, there are n different angular momentum values associated with the ℓ values $\ell = 0, 1, \ldots n - 1$. Here $n = 3$ so $\ell = 0, 1,$ or 2. The corresponding magnitudes of the orbital angular momentum are, from Equation 41-10,

$$\ell = 0: L = \sqrt{0(0 + 1)}\hbar = 0,$$

$$\ell = 1: L = \sqrt{1(1 + 1)}\hbar = \sqrt{2}\hbar,$$

$$\ell = 2: L = \sqrt{2(2 + 1)}\hbar = \sqrt{6}\hbar.$$

EXERCISE Find the maximum value for the orbital angular momentum of a hydrogen atom in the $n = 5$ state.

Answer: $2\sqrt{5}\hbar$

Some problems similar to Example 41-2: 5, 6, 8, 9

For historical reasons states with the ℓ values 0, 1, 2, 3, 4, 5, . . . are given the letter labels $s, p, d, f, g, h, \ldots$. These are combined with the principal quantum-number n to specify both the energy and angular momentum of a state. Thus, the ground state is 1s, while the $n = 2$ state with $\ell = 1$ is the 2p state. (The lowercase letters s, p, d, . . . are used in discussing individual electrons, while the corresponding capital letters denote orbital angular momentum states of an entire atom. For one-electron hydrogen, the two are the same.)

In the quantization of orbital angular momentum, we have another nonclassical result of quantum mechanics, for in classical physics an electron of a given energy should be able to have any angular momentum, up to a maximum value that corresponds to a circular orbit (Fig. 41-7). At large n, however, the number of allowed angular momentum values (namely, n) is so large that there appears to be a nearly continuous range of values. Again, we see Bohr's correspondence principle at work: for large quantum numbers, the quantum description approaches the classical limit. But for small n, the quantum-mechanical discreteness of both energy and angular momentum is clearly evident, and stands in marked contrast to the classical prediction.

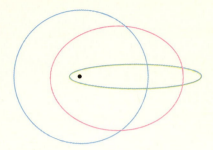

FIGURE 41-7 Classical electron orbits with the same energy but different angular momenta. Any value of L from zero to the maximum L of the circular orbit is allowed in classical physics.

EXAMPLE 41-3	*Quantizing the Moon*

Estimate the orbital angular momentum of the moon, and find the corresponding ℓ value. Then find the difference in angular momenta between this ℓ value and the next lower value, and compare with the angular momentum itself.

Solution

Although the Earth-moon system is not an electrostatically bound hydrogen atom, we can still use Equation 41-10 to estimate the quantized angular momentum values. From Chapter 12, we have $L = mvr$ for an object of mass m moving at speed v in a circle of radius r. For the moon, these three quantities are given in Appendix E, and we have

$$L = mvr = (0.0735 \times 10^{24} \text{ kg})(1.0 \times 10^3 \text{ m/s})(0.385 \times 10^9 \text{ m})$$

$$= 2.8 \times 10^{34} \text{ J·s}.$$

(Check that the units can be written as J·s.) Since $\hbar$ is so small, ℓ in Equation 41-10 will clearly be very large, so we can approx-

imate $\ell + 1 \simeq \ell$ to avoid solving a quadratic equation. With this approximation, Equation 41-10 becomes $L \simeq \ell\hbar$, or

$$\ell = \frac{L}{\hbar} = \frac{2.8 \times 10^{34} \text{ J·s}}{1.06 \times 10^{-34} \text{ J·s}} = 3 \times 10^{68}.$$

This huge ℓ value indicates that the system is essentially at the classical limit. With $L \simeq \ell\hbar$, the difference between adjacent angular momentum values is just $\hbar$. This is such a tiny fraction ($\simeq 10^{-68}$) of the angular momentum that quantization is totally unobservable.

EXERCISE A hydrogen atom in an interstellar gas cloud is in the $n = 293$ state. What is the difference between the maximum possible value for its angular momentum and the next lower value, as a fraction of the maximum value?

Answer: 0.34%

Some problems similar to Example 41-3: 11, 14

Space Quantization

Angular momentum is a vector, and in quantum mechanics it turns out that the angular momentum vector is quantized not only in magnitude but also in direction—a phenomenon called **space quantization**. Space quantization of orbital angular momentum gives rise to a third quantum number, designated m_ℓ. Space quantization becomes evident when an atom is in a magnetic field that establishes a preferred axis along which the angular momentum component can be measured; for this reason m_ℓ is called the **orbital magnetic quantum number**.

Space quantization requires, in particular, that the component L_z of orbital angular momentum along any chosen axis can have only values given by

$$L_z = m_\ell \hbar, \tag{41-11}$$

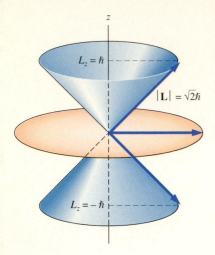

z
$L_z = \hbar$
$|\mathbf{L}| = \sqrt{2}\hbar$
$L_z = -\hbar$

FIGURE 41-8 The three possible orientations for the orbital angular momentum vector in the $\ell = 1$ state, where $L = \sqrt{2}\hbar$. More generally, there are $2\ell + 1$ possible orientations. Only the *z* component is fixed for a given state; *x* and *y* components are uncertain.

where m_ℓ takes on integer values from $-\ell$ to ℓ, with ℓ the orbital quantum number. Thus an $\ell = 1$ state can have one of the three possible m_ℓ values -1, 0, or $+1$, corresponding to angular momentum components $-\hbar$, 0, or $+\hbar$ along some axis. Since the magnitude of the angular momentum in an $\ell = 1$ state is $\sqrt{2}\hbar$ (see Example 41-2), none of these values corresponds to full alignment with the given axis. Instead, we can think geometrically of the angular momentum vectors as being constrained to lie at angles $\cos^{-1}(L_z/L)$ to the axis; for $\ell = 1$ these angles are $\pm45°$ and $90°$ (see Fig. 41-8). Although knowing the angle is useful for diagramming the angular momentum vector, we emphasize that the quantum numbers ℓ and m_ℓ tell everything there is to know about quantized orbital angular momentum. Quantum physicists, therefore, are usually not concerned with the orientation of angular momentum vectors.

41-2 ELECTRON SPIN

Detailed observation of the hydrogen spectrum shows that spectral lines exhibit a fine splitting; where a lower resolution spectrum shows one spectral line, at higher resolution there appears a closely spaced pair of lines. This splitting could not be explained using the three quantum numbers n, ℓ, and m_ℓ. In 1925 the Austrian physicist Wolfgang Pauli suggested that a fourth quantum number, capable of taking only two values, might be needed. Soon Samuel Goudsmit and George Uhlenbeck (Fig. 41-9), graduate students at the University at Leiden, realized that the spectral splitting could be explained if this fourth quantum number were associated with an intrinsic angular momentum, or **spin**, carried by the electron. Their proposal was without firm theoretical grounding; later, as we indicated in Chapter 40, Paul Dirac showed that electron spin follows from the requirement of relativistic invariance. Spin is an inherently quantum-mechanical property with no classical analog. Although spin can be visualized crudely by imagining the electron to be a small sphere spinning about an axis, this classical picture is really inappropriate.

Spin angular momentum is quantized in a manner similar to that of orbital angular momentum. But unlike the orbital quantum number ℓ that takes a range of integer values, the electron spin quantum number s has only the single value $s = \frac{1}{2}$. The electron is therefore called a **spin-$\frac{1}{2}$ particle**. The magnitude of the spin angular momentum is related to the spin quantum number in the same way that the magnitude of orbital angular momentum is related to the orbital quantum number ℓ; that is, by an equation like Equation 41-10:

$$S = \sqrt{s(s + 1)}\hbar . \qquad (41\text{-}12)$$

Since s takes only the value $\frac{1}{2}$, Equation 41-12 shows that the magnitude of the electron spin angular momentum is $S = \frac{\sqrt{3}}{2}\hbar$.

Spin angular momentum also exhibits space quantization. That is, the component of spin along a chosen axis can take only values given by an equation

FIGURE 41-9 George Uhlenbeck (left) and Samuel Goudsmit (right) who, as graduate students, first proposed electron spin.

analogous to Equation 41-11:

$$S_z = m_s \hbar, \tag{41-13}$$

where the quantum number m_s has only the two possible values $-\frac{1}{2}$ and $+\frac{1}{2}$ (m_s is thus analogous to the orbital magnetic quantum number m_ℓ, whose values range from $-\ell$ to $+\ell$ in increments of 1). Figure 41-10 shows space quantization of electron spin.

Magnetic Moment of the Electron

Together, the electron's spin and electric charge mean the electron behaves like a miniature current loop, with an intrinsic magnetic dipole moment. The dipole moment vector **M** associated with the spin angular momentum vector **S** is given by

$$\mathbf{M} = -\frac{e}{m}\mathbf{S}, \tag{41-14}$$

with e/m the charge-to-mass ratio of the electron. (The minus sign reflects the electron's negative charge, and we consider here *only* the magnetic moment arising from electron spin. An additional magnetic moment results from orbital motion in states with nonzero ℓ.) Since the component of **S** on any axis can take only the values $\pm\frac{1}{2}\hbar$, the components of the magnetic moment can only be

$$M_z = \pm\frac{e\hbar}{2m}. \tag{41-15}$$

The quantity $e\hbar/2m$ is a fundamental unit for measuring magnetic moments called the **Bohr magneton**; its value is 9.27×10^{-24} A·m².

The ratio of magnetic moment to spin angular momentum is twice what we would expect classically for a charged particle in circular motion. Like spin itself, the factor of 2 is a relativistic effect first explained by Dirac. Actually, the factor is not quite 2 but approximately 2.00232, a result first established by Lamb and Retherford in 1947, and justified theoretically by Schwinger using the theory of quantum electrodynamics.

The Stern-Gerlach Experiment

In 1922, Otto Stern and Walther Gerlach at the University of Hamburg performed an experiment that demonstrated the quantization of atomic angular momentum vectors. The **Stern-Gerlach experiment** used a nonuniform magnetic field to separate a beam of silver atoms according to the component of their angular momentum along the field direction. (Recall from our study of magnetism that a magnetic dipole in a nonuniform magnetic field experiences a net force. Atoms with different angular momentum components and therefore differently oriented magnetic moments are thus deflected by different amounts.) The experiment was repeated in 1927 by Phipps and Taylor in a way that gives an unambiguous verification of quantized electron spin. Phipps and Taylor used hydrogen atoms in the ground state; as we have seen, this state has zero orbital angular momentum, so the only angular momentum effects are due to electron spin. Classically, a beam of hydrogen should be spread into a continuous band corresponding to

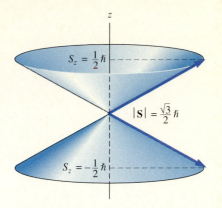

FIGURE 41-10 Space quantization of electron spin. The component of spin angular momentum on a given axis can take only the value $\pm\frac{1}{2}\hbar$.

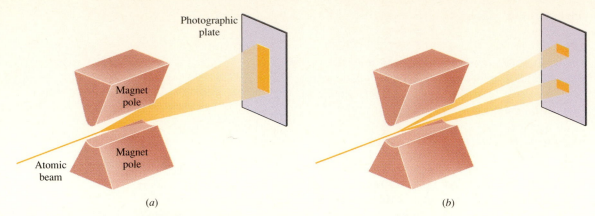

FIGURE 41-11 In the Stern-Gerlach experiment, as repeated by Phipps and Taylor, a beam of ground-state hydrogen atoms passes through a nonuniform magnetic field, established here by concave and convex magnet pole pieces. Atoms deflect according to the orientations of their magnetic moments, which in turn depend on electron spin orientation. (*a*) Classically, any spin is possible, and the beam should spread into a band. (*b*) In reality, the beam splits in two, corresponding to the two possible spin angular momentum components $+\frac{1}{2}\hbar$ and $-\frac{1}{2}\hbar$.

angular momentum components from $\frac{-\sqrt{3}}{2}\hbar$ to $\frac{+\sqrt{3}}{2}\hbar$. But when the experiment is done, the beam always splits into two beams, corresponding to the two angular momentum components $\pm\frac{1}{2}\hbar$. Figure 41-11 shows the experiment.

Spins of Other Particles

Spin angular momentum as proposed by Goudsmit and Uhlenbeck and confirmed theoretically by Dirac applies specifically to the electron. We now know that particles in general carry quantized intrinsic angular momentum or spin. Some, like electrons, protons, and neutrons, have a spin quantum number $s = \frac{1}{2}$. According to Equation 41-12, these spin-$\frac{1}{2}$ particles have spin angular momentum of $\frac{\sqrt{3}}{2}\hbar$ and, according to Equation 41-13, the spin component on a given axis takes only the values $\pm\frac{1}{2}\hbar$. Other particles—including photons and some composite nuclei—are spin-1 particles, for which $s = 1$. There are profound and fascinating differences in behavior between particles with half-integer and integer spins, as we describe shortly. Whatever the spin quantum number s, particle spins obey the quantization relations of Equations 41-12 and 41-13, with values of m_s ranging from $-s$ to s in increments of 1. (Photons are an exception to this rule; the fact that they are massless turns out to eliminate the state $m_s = 0$ that would otherwise be possible for a spin-1 particle.)

EXAMPLE 41-4	*Spin of the Deuteron*

The spins of its constituent proton and neutron combine to make the deuteron a spin-1 particle. Determine the magnitude of its spin angular momentum and the possible values of its spin component on a given axis.

Solution

Equation 41-12 gives the magnitude of the spin angular momentum; with $s = 1$ for the spin-1 deuteron, we have

$$S = \sqrt{s(s + 1)}\hbar = \sqrt{2}\hbar.$$

The possible values of m_s range from $-s$ to s in increments of 1; thus, $m_s = -1, 0,$ or 1. The corresponding angular momentum components, given by Equation 41-13, are $-\hbar$, 0, and $\hbar$.

EXERCISE Find a minimum angle between the spin vector and the z axis for a spin-$\frac{1}{2}$ particle.

Answer: 55°

Some problems similar to Example 41-4: 16, 24

Got It!

The nucleus of oxygen-17 has spin $\frac{5}{2}$. How many possible orientations are there for O-17's spin angular momentum vector?

Total Angular Momentum and Spin-Orbit Coupling

The total angular momentum of the atom, designated **J**, is the vector sum of the orbital and spin angular momenta; that is,

$$\mathbf{J} = \mathbf{L} + \mathbf{S}. \tag{41-16}$$

A detailed quantum-mechanical analysis shows that the magnitude of **J** is quantized according to a rule similar to those for orbital and spin angular momenta:

$$J = \sqrt{j(j + 1)}\hbar. \tag{41-17}$$

For an atom with a single electron, the quantum number j takes the values

$$j = \ell \pm \tfrac{1}{2} \qquad \text{for } \ell \neq 0 \tag{41-18a}$$

$$j = \tfrac{1}{2} \qquad \text{for } \ell = 0. \tag{41-18b}$$

The state of an atom with total angular momentum of magnitude J is often given by listing the principal quantum number, the capital letter designating the orbital angular momentum ($S, P, D, F, G, \ldots$), and, as a subscript, the j value. Thus a hydrogen atom with $n = 3$, $\ell = 2$, and $j = \frac{3}{2}$ would be designated $3D_{3/2}$.

Total angular momentum also exhibits space quantization, with the possible values for the component of **J** on some axis given by

$$J_z = m_j \hbar, \tag{41-19}$$

where the quantum number m_j takes values from $-j$ to j, with a difference of 1 between successive m_j values (i.e., $m_j = -j, -j + 1, -j + 2, \ldots, j$).

Although derivation of these so-called **angular momentum coupling rules** is not at all straightforward, we can understand them in terms of simple vector diagrams. Recall that the magnitude of the electron's spin angular momentum is $\frac{\sqrt{3}}{2}\hbar$ and its component on a given axis takes only the two values $\pm\frac{1}{2}\hbar$. When $\ell = 0$, there is no orbital angular momentum, and Equations 41-17 and 41-18b then show that the magnitude of the total angular momentum is just that of the spin. When $\ell \neq 0$, spin and orbital angular momentum add in accordance with Equation 41-18a to produce a total angular momentum vector **J** whose magnitude is given by Equation 41-17 (see Fig. 41-12).

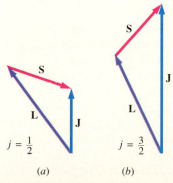

$j = \tfrac{1}{2}$

$j = \tfrac{3}{2}$

(a)   (b)

FIGURE 41-12 Two possibilities for coupling of spin and orbital angular momentum. Case shown is for $\ell = 1$, so Equation 41-18a gives $j = \frac{1}{2}$ or $j = \frac{3}{2}$.

| **EXAMPLE 41-5** | *Total Angular Momentum* |

Find the possible magnitudes of the total angular momenta for a hydrogen atom in an $\ell = 2$ state. For each possible J value, how many possible values are there for the component of **J** on a given axis?

Solution

For $\ell = 2$, Equation 41-18a gives $j = \frac{3}{2}$ or $j = \frac{5}{2}$. Then, from Equation 41-17, we have

$$J = \sqrt{j(j+1)}\hbar = \sqrt{\frac{3}{2}(\frac{3}{2}+1)}\hbar = \frac{\sqrt{15}}{2}\hbar$$

or

$$J = \sqrt{j(j+1)}\hbar = \sqrt{\frac{5}{2}(\frac{5}{2}+1)}\hbar = \frac{\sqrt{35}}{2}\hbar .$$

Figure 41-13 shows vector diagrams for the angular momentum coupling in these two cases.

For the $j = \frac{3}{2}$ state, there are four possible values of m_j, ranging from $-j$ to j; they are $-\frac{3}{2}, -\frac{1}{2}, \frac{1}{2},$ and $\frac{3}{2}$, corresponding to four values of the total angular momentum component given by Equation 41-19. For the $j = \frac{5}{2}$ state, there are six possible m_j's and correspondingly six values for the angular momentum component.

EXERCISE Find the maximum possible angular momentum for a hydrogen atom in an $\ell = 3$ state.

Answer: $\dfrac{3\sqrt{7}}{2}\hbar$

Some problems similar to Example 41-5: 25–27

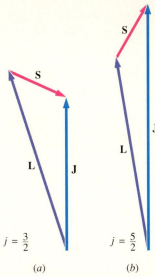

FIGURE 41-13 Vector diagrams for spin-orbit coupling with $\ell = 2$ (Example 41-5).

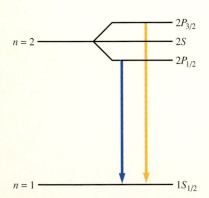

FIGURE 41-14 Energy-level diagram showing spin-orbit splitting in hydrogen. Left side shows $n = 1$ and $n = 2$ levels if there were not splitting; at right the $2P$ state splits because of the spin-orbit interaction, while the zero-angular-momentum S states are unaffected. Allowed transitions to the ground state are indicated by arrows; they result in spectral lines of slightly different wavelengths. In hydrogen the $2P$ levels actually differ by only about 10^{-5} of the separation between the $n = 1$ and $n = 2$ levels.

The two possible values of j for a given ℓ correspond to two distinct quantum states, and these states actually have slightly different energies. This energy difference is associated with the orientation of the electron's magnetic moment in a magnetic field that, in the electron's reference frame, is created by the motion of the positively charged nucleus—a field whose direction is that of the electron's orbital angular momentum **L**. As Equation 41-14 shows, the electron's negative charge means that its spin **S** and magnetic moment **M** have opposite directions. Because a magnetic dipole has the highest energy when it's oriented opposite the magnetic field, this means that the more-nearly parallel alignment of **S** and **L**—corresponding to $j = \ell + \frac{1}{2}$—has the higher energy.

In hydrogen, the magnitude of the energy difference between the $j = \frac{1}{2}$ and $j = \frac{3}{2}$ states of the first excited level is only 5×10^{-5} eV, far smaller than the separation between this level and the ground state. Because the $n = 2$, $\ell = 1$ state is actually two states of slightly different energy, hydrogen atoms undergoing transitions from these states to the ground state emit two spectral lines slightly separated in wavelength. The term **fine structure** describes the splitting associated with spin-orbit and related effects. In the present example, the split spectral line is called a **doublet**. (Transitions from the $n = 2$, $\ell = 0$ state to the ground state are forbidden; otherwise the line would be a triplet.) Figure 41-14 is an energy-level diagram showing the effect of spin-orbit splitting in hydrogen, while Fig. 41-15 shows the doublet spectral line resulting from the analogous transition in sodium. (The effect is difficult to observe in hydrogen.)

The spin-orbit effect results from a magnetic field internal to the atom itself. But splitting of energy levels also occurs in an external magnetic field, and is

FIGURE 41-15 A high-resolution spectrometer provides a close look at the doublet spectral lines resulting from the $3P_{1/2} \rightarrow 3S_{1/2}$ and $3P_{3/2} \rightarrow 3S_{1/2}$ transitions in sodium. The sodium doublet also shows prominently near the center of the solar spectrum in Fig. 39-12a.

FIGURE 41-16 This image of the galaxy NGC 784 was made with the Very Large Array radio telescope tuned to the 21-cm line of hydrogen. The image shows the distribution of neutral hydrogen in the galaxy.

called the **Zeeman effect**. We showed an example of Zeeman splitting in Fig. 40-30.

Since it has zero orbital angular momentum, the ground state does not exhibit spin-orbit splitting. But interaction of the electron spin with the magnetic dipole moment of the nucleus does result in an even finer splitting known as **hyperfine structure**. The transition between the two hyperfine levels of the hydrogen ground state—corresponding physically to a change in the orientation of the electron spin vector—involves a photon of 21-cm wavelength. Radio astronomers use the 21-cm hydrogen radiation to map regions of interstellar hydrogen in the cosmos (Fig. 41-16).

41-3 THE PAULI EXCLUSION PRINCIPLE

In trying to understand why atomic electrons distributed themselves as they did, Pauli was led in 1924 to his famous **exclusion principle** which, loosely, states the following:

Two electrons cannot be in the same quantum state.

Since the quantum state of an electron includes its spin orientation specified by the spin quantum number m_s, the exclusion principle means that at most two electrons can occupy a state whose other quantum numbers n, ℓ, and m_ℓ are the same.

The Pauli exclusion principle has profound implications for the behavior of multielectron systems, for it requires that most electrons remain in high-energy states (Fig. 41-17). If the exclusion principle did not hold, atomic electrons would collapse to the ground state and there would be no such thing as chemistry or life! And, as the following application indicates, the exclusion principle even manifests itself in objects at the astrophysical scale.

FIGURE 41-17 (a) If the exclusion principle did not hold, the lowest energy configuration of a square well containing five electrons would have them all in the ground state. (b) The exclusion principle permits only two electrons, their spin components opposite, to occupy a single energy level.

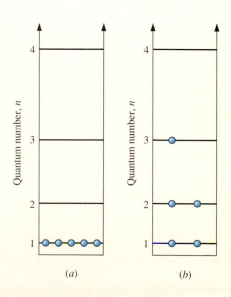

| APPLICATION | *White Dwarfs and Neutron Stars* |

The endpoint of a star's life comes when nuclear fusion ceases in the stellar interior. Then there is no longer sufficient pressure to maintain the star against its own gravity, and collapse ensues. For stars with more than about five times the Sun's mass, no known force can halt the collapse, and the star beomes a black hole—a singular configuration governed by the general theory of relativity. But for less massive stars, including our Sun, collapse is ultimately halted by a quantum-mechanical pressure arising from the exclusion principle.

In the eventual collapse of the Sun, some five billion years hence, all the electrons will drop to the lowest possible energy states. But the exclusion principle asserts that only two electrons, with their spin components opposite, can occupy each energy level. So most of the electrons end up in high-energy states (imagine putting 10^{57} electrons in the square well of Fig. 41-17). Consequently they have substantial momentum and therefore exert considerable pressure. It is this **degenerate electron pressure**—quite independent of temperature, unlike the pressure of an ordinary gas—that sustains the star against further collapse. In its final configuration, the resulting **white dwarf** star is only about the size of the Earth (Fig. 41-18a).

In stars more massive than 1.4 times the Sun's mass, degenerate electron pressure is insufficient to counter the gravitational force, and the star collapses further until electrons and protons join to form neutrons. The neutrons, too, develop a degenerate pressure that halts the collapse. At this point the former star is only a few kilometers in diameter, and is called a **neutron star** (Fig. 41-18b). For the most massive stars, even degenerate neutron pressure cannot withstand gravity, and the star collapses to become a black hole (Fig. 41-18c).

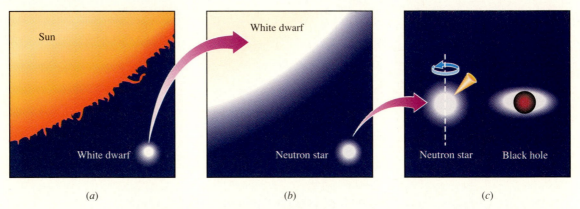

(a)   (b)   (c)

FIGURE 41-18 Comparative sizes of the Sun, a white dwarf, a neutron star, and a black hole. The white dwarf and neutron star are supported against gravitational collapse by pressure that arises as a consequence of the exclusion principle. Neutron star is shown in rapid rotation, emitting a searchlight-like beam of radiation; black hole is surrounded by a disk of accreting gas.

The Pauli exclusion principle quickly became a fundamental rule in the developing quantum mechanics of the late 1920s. But physicists remained dissatisfied by the need to invoke this seemingly ad hoc rule with no theoretical basis. Late in the 1930s, following detailed analysis of relativistic quantum theories, Pauli finally showed that the exclusion principle, like the existence of spin, is ultimately grounded in the requirement of relativistic invariance. More specifically, Pauli found that particles whose spin quantum number *s* is a half-integer (collectively called **fermions**) must necessarily obey the exclusion principle. On the other hand, particles with integer spin (called **bosons**) *do not* obey the exclusion principle. Photons, for example, are spin-1 particles and therefore an arbitrarily large number of them can occupy exactly the same quantum state. The laser, with its intense, coherent beam of light, is possible because the many photons comprising the beam are essentially all in the same state. In 1995 physicists at the University of Colorado first succeeded in producing an assemblage of bosonic matter all in the same quantum state. This so-called Bose-Einstein condensation had been a

FIGURE 41-19 Velocity distribution of atoms in a Bose-Einstein condensation shows a large peak at the near-zero velocity of atoms all in their common ground state. Such a condensation, first produced in 1995, is possible only with bosons—particles with integer spin that don't obey the exclusion principle. The three peaks mark the evolution from a normal gas (at left) to the condensate (right).

goal of physicists since 1924, when the Indian physicist Satyendra Nath Bose first suggested the possibility. The Bose-Einstein condensation represents a truly new state of matter, in which thousands of atoms join quantum mechanically to behave as a single entity (Fig. 41-19). Since 1995 physicists around the world have created these condensations, and are just beginning to explore their possible applications.

41-4 MULTIELECTRON ATOMS AND THE PERIODIC TABLE

The Chemical Elements

Our modern understanding of the chemical elements developed in the late eighteenth century. At that time, the French chemist Antoine Lavoisier found that burning resulted from a combination of substances with the newly discovered gas oxygen. Lavoisier made the first reasonable list of elements, defined as substances that cannot be decomposed by chemical means (Fig. 41-20).

Chemists proceeded to discover formulas for various compounds, such as H_2O for water. From the formulas, they could determine the relative atomic masses of various elements. In 1864, the British chemist John Newlands discovered that similar chemical properties occurred periodically when elements were arranged in order of atomic mass. The periodic arrangement was clarified further by the German chemist Lothar Meyer and the Russian chemist Dmitri Mendeleev. In 1869 Mendeleev set up a table with the approximately 60 elements then known.

FIGURE 41-20 The chemical elements exhibit an amazing variety of forms, colors, and other physical and chemical properties. All these are ultimately traceable to atomic structure, which in turn follows from the principles of quantum physics. On glass dishes, clockwise from top left, are sodium, chromium, sulfur, iodine, and iron. Also shown are zinc and copper strips and a magnesium ribbon. Invisible gaseous elements—primarily oxygen and nitrogen—comprise the surrounding air.

FIGURE 41-21 The periodic table. A larger version, with atomic weights, is printed inside the back cover; names of the elements are listed in Appendix D. Blue represents nonmetals, pink represents metals, and yellow represents semimetals.

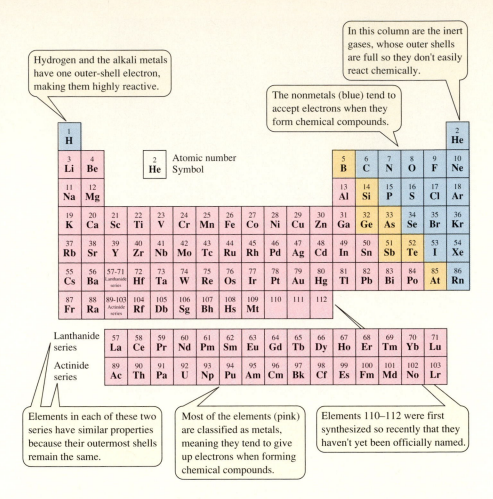

Hydrogen and the alkali metals have one outer-shell electron, making them highly reactive.

In this column are the inert gases, whose outer shells are full so they don't easily react chemically.

The nonmetals (blue) tend to accept electrons when they form chemical compounds.

Elements in each of these two series have similar properties because their outermost shells remain the same.

Most of the elements (pink) are classified as metals, meaning they tend to give up electrons when forming chemical compounds.

Elements 110–112 were first synthesized so recently that they haven't yet been officially named.

He left blanks where necessary to maintain the periodic occurrence of similar chemical properties. Elements filling the blanks were soon discovered, validating Mendeleev's periodic table and strongly suggesting an underlying order in the composition of atoms. Then, early in the twentieth century, studies of X-ray spectra led to a table organized by atomic number Z (the number of protons in the nucleus), rather than by atomic weight. When this was done, a number of elements missing from earlier periodic tables were identified. The modern periodic table is shown in Fig. 41-21 and is printed with atomic weights inside the back cover.

Explaining the Periodic Table

The orderly arrangement of elements in the periodic table has greatly enhanced our understanding of chemistry and our ability to formulate new and useful compounds. But why does nature exhibit this regularity? The answer lies in the Schrödinger equation and the exclusion principle.

Solution of the Schrödinger equation for multielectron atoms is complicated by the interactions among the electrons; simple, analytic solutions like those for hydrogen are generally not available. But qualitatively, we still find energy levels characterized by the principal quantum number n. Each such level is called a **shell**; for historical reasons, the shells $n = 1, 2, 3, \ldots$ are also labeled with the letters $K, L, M, \ldots$. As with hydrogen, an electron at the nth energy level can

TABLE 41-1 *Atomic Shell Structure Notation*

QUANTUM NUMBER	SHELL-STRUCTURE NOTATION	ALLOWED VALUES	LETTER LABELS	NUMBER OF STATES
n	Shell	1, 2, 3, . . .	$K, L, M, \ldots$	Infinite
ℓ	Subshell	0, 1, 2, . . . , $n-1$	$s, p, d, f, \ldots$	n
m_ℓ	Orbital	$-\ell, -\ell+1, \ldots, \ell-1, \ell$	–	$2\ell + 1$
m_s	–	$-\frac{1}{2}, +\frac{1}{2}$		2

take on any of the n values $\ell = 0, 1, 2, \ldots, n - 1$ for the orbital quantum number ℓ that determines its orbital angular momentum. The different angular momentum states within a shell are termed **subshells**; subshells with the values $\ell = 0, 1, 2, 3, \ldots$ are labeled with the letters $s, p, d, f, \ldots$. Finally, for each subshell there are $2\ell + 1$ possible values of the magnetic orbital quantum number m_ℓ, ranging from $-\ell$ to ℓ. A state characterized by all three quantum numbers n, ℓ, and m_ℓ is called an **orbital**. Table 41-1 summarizes shell-structure notation in relation to quantum numbers. For completeness, the table lists also the spin quantum number m_s.

The structure of a multielectron atom is determined by the quantum states of its constituent electrons—that is, by their distribution among the shells, subshells, and orbitals. That distribution, in turn, is determined by the exclusion principle. According to the exclusion principle, no two electrons can be in exactly the same quantum state—that is, they cannot have the same values for all four quantum numbers n, ℓ, m_ℓ, and m_s. Since an atomic orbital is characterized by the three quantum numbers n, ℓ, and m_ℓ, the exclusion principle implies that at most two electrons can occupy a single orbital.

We're now ready to understand the ground-state electronic structure of multi-electron atoms. The simplest is helium (He), with two electrons. The K shell ($n = 1$) is the lowest possible energy level. As Table 41-1 shows, only the zero-angular-momentum s subshell is permitted within the K shell, and within that subshell only the single orbital corresponding to $m_\ell = 0$ is allowed. But that orbital can accommodate two electrons. So in the ground state of helium, both electrons occupy the s subshell of the K shell. Atomic physicists describe this situation with the shorthand notation $1s^2$, where 1 stands for the principal quantum number n (i.e., for the K shell), s for the subshell, and the superscript 2 for the number of electrons in that subshell. The corresponding notation for hydrogen would be $1s^1$.

After helium comes lithium (Li), with three electrons. From our analysis of helium, we know that the K shell is full with two electrons. So the third electron goes into the L shell, or $n = 2$ energy level. Of the subshells in the L shell, the s subshell turns out to have slightly lower energy than the others, so the third electron occupies the s subshell. Then the electronic configuration of lithium is $1s^2 2s^1$; i.e., a helium-like core with a single outer electron in the s subshell of the $n = 2$ level.

Beryllium (Be), with four electrons, fills the $1s$ and $2s$ subshells; its designation is $1s^2 2s^2$. The fifth electron of boron then goes into the $2p$ subshell, giving the structure $1s^2 2s^2 2p^1$. Table 41-1 shows that a p subshell ($\ell = 1$) allows three m_ℓ values—i.e., three orbitals, capable of holding a total of six electrons. As we advance in atomic number, electrons continue to fill the p subshell. Finally, at

TABLE 41-2 *Electronic Configurations and Ionization Energies of Elements 1–18*

ATOMIC NUMBER, Z	ELEMENT	ELECTRONIC CONFIGURATION	IONIZATION ENERGY, eV
1	H	$1s^1$	13.60
2	He	$1s^2$	24.60
3	Li	$1s^2 2s^1$	5.390
4	Be	$1s^2 2s^2$	9.320
5	B	$1s^2 2s^2 2p^1$	8.296
6	C	$1s^2 2s^2 2p^2$	11.26
7	N	$1s^2 2s^2 2p^3$	14.55
8	O	$1s^2 2s^2 2p^4$	13.61
9	F	$1s^2 2s^2 2p^5$	17.42
10	Ne	$1s^2 2s^2 2p^6$	21.56
11	Na	$1s^2 2s^2 2p^6 3s^1$	5.138
12	Mg	$1s^2 2s^2 2p^6 3s^2$	7.644
13	Al	$1s^2 2s^2 2p^6 3s^2 3p^1$	5.984
14	Si	$1s^2 2s^2 2p^6 3s^2 3p^2$	8.149
15	P	$1s^2 2s^2 2p^6 3s^2 3p^3$	10.48
16	S	$1s^2 2s^2 2p^6 3s^2 3p^4$	10.36
17	Cl	$1s^2 2s^2 2p^6 3s^2 3p^5$	13.01
18	Ar	$1s^2 2s^2 2p^6 3s^2 3p^6$	15.76

Gaps mark ends of periodic-table rows.

neon ($Z = 10$), the $2p$ subshell is full. Only with the next element, sodium ($Z = 11$), does the $n = 3$ shell begin to fill. Table 41-2 lists electronic configurations for the elements hydrogen ($Z = 1$) through argon ($Z = 18$), along with their ionization energies (that is, the energy required to remove the outermost electron).

The chemical behavior of an atom is determined primarily by its outermost electrons, both because these electrons interact most directly with nearby atoms and because they are most weakly bound to their nuclei. A look at Table 41-2 shows that the configurations of outer electrons for the elements lithium through neon (Ne) are the same as the corresponding configurations for sodium (Na) through argon (Ar). The chemical properties of the corresponding atoms are therefore similar. Both lithium and sodium, for example, have a single electron in their outermost shell. As the relatively low ionization energy suggests, this electron is loosely bound, so it readily interacts with other atoms, giving these elements their extreme reactivity. Neon and argon, in contrast, both have completely filled outermost shells. All the outer-shell electrons have essentially the same energy; the corresponding ionization energy is high; and there is little tendency for these electrons to interact with those of other atoms. As a result, argon and neon are **inert**—that is, they don't readily form chemical compounds, and at normal temperatures their lack of interaction makes them gases. Other element pairs also share similar properties. Fluorine and chlorine, for instance, each need only one more electron to achieve the energetically favorable situation of an inert gas with its filled outer shell. Consequently these elements readily accept electrons. Materials such as common salt, NaCl, owe their high melting points to the strong bond that results when reactive sodium gives up its outer electron to electron-accepting chlorine and the resulting positive and negative ions bind strongly by the electrostatic force. We will consider such molecular bonding in the next chapter.

As we move beyond argon ($Z = 18$), things become slightly more complicated. In these multielectron atoms, shielding effects of the inner electrons result in the $4s$ state having lower energy than the $3d$ states. Potassium (symbol K; $Z = 19$) thus has the electronic configuration $1s^2 2s^2 2p^6 3s^2 3p^6 4s^1$ rather than $1s^2 2s^2 2p^6 3s^2 3p^6 3d^1$ as would occur if the $n = 4$ state had a higher energy than any $n = 3$ state. After potassium comes calcium, with two electrons in its single $4s$ orbital. But the $4p$ orbitals do have higher energy than the $3d$ orbitals, so elements beyond calcium begin filling the $3d$ orbitals. The next ten elements, scandium through zinc, have chemical properties that vary only slightly because both their outermost electrons remain $4s$ electrons; collectively, they are called **transition elements**. (Chromium ($Z = 24$) and copper ($Z = 29$) are minor exceptions; in these an extra electron goes into the $3d$ orbitals, leaving only one $4s$ electron). Finally, elements 31 (gallium) through 36 (krypton) repeat the pattern of aluminum through argon shown in Table 41-2, as their $4p$ orbitals fill with electrons. Krypton, with its outer p subshell full, is again an inert gas.

EXAMPLE 41-6	*Electronic Configuration of Iron*

Given that the $4s$ states fill before the $3d$ states in the transition element iron (Fe), but with no other anomalies, determine the electronic configuration of iron.

Solution

Consulting the periodic table (Fig. 41-21 or inside back cover), we find that iron has atomic number $Z = 26$—that is, it has eight more electrons than the last entry, argon, in Table 41-2. So its inner structure is like that of argon, and we need determine only the configuration of the remaining eight electrons. In argon, the $3p$ orbitals are full; the next $n = 3$ orbitals are the d orbitals, but we are told the $4s$ orbital fills first. So two electrons fill the

single $4s$ orbital, leaving six in the $3d$ orbitals. (Since the d subshell has five orbitals, accommodating up to ten electrons, the iron d subshell can handle all the remaining electrons.) So the electronic structure of iron is $1s^2 2s^2 2p^6 3s^2 3p^6 3d^6 4s^2$. As a check, note that the superscripts add to give a total of 26 electrons.

EXERCISE What is the electronic configuration of the transition element titanium ($Z = 22$)?

Answer: $1s^2 2s^2 2p^6 3s^2 3p^6 3d^2 4s^2$

Some problems similar to Example 41-6: 33–35

We can now understand the organization of the periodic table, shown in Fig. 41-21. The vertical columns contain elements with the same outer-electron configurations, resulting in similar chemical properties. Each row starts with an element whose outermost shell contains a single s electron; these are hydrogen and the highly reactive alkali metals. Each row ends with an inert gas, its outermost p subshell full. Traversing the first row involves filling the $1s$ orbital only; since this orbital holds at most two electrons, there are only two elements in the first row. The second row has eight elements, associated with the filling of the $2s$ and $2p$ orbitals as shown in Table 41-2. The third row is like the second, but with the $3s$ and $3p$ orbitals filling. Because the $4s$ orbitals have lower energy than the $3d$ orbitals, the third row ends with an inert gas whose $3p$ orbitals are full, and the fourth row begins as the $4s$ orbital begins to fill. But then come the elements $Z = 21$ through $Z = 30$, in which the $3d$ orbitals are filling; these make for ten additional elements in the fourth row. The fifth row is a repeat of the fourth, as first the $5s$ orbitals fill, then the higher energy $4d$ orbitals, then the remaining $5p$ orbitals. The sixth row begins with element 55, cesium, as the $6s$ orbitals fill before either the $5d$ or $4f$ orbitals. Two elements later, at lanthanum, the $4f$ orbitals begin to fill. Since their outer electrons remain $6s$ electrons, these elements have very similar properties. Together, they constitute the **lanthanide**

FIGURE 41-22 Ionization energy as a function of atomic number Z. Shaded bands mark the rows of the periodic table.

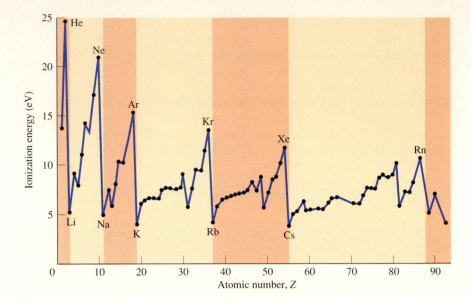

series. Because its elements are so similar chemically, and to avoid stretching the sixth row of the periodic table, the lanthanide series is usually printed separately below the main table. Row seven repeats row six, with the **actinide series** again printed below. The actinides include elements 89 (actinium) through 103 (lawrencium, named for the inventor of the cyclotron). Elements beyond 92 (uranium) are radioactive with half-lives short compared with the age of the Earth. They are not found naturally but are produced in particle accelerators, fission reactors, and nuclear weapons. Elements 104 through 112 follow the actinide series and partially complete the seventh row of the periodic table. This group, some of which were first synthesized only in the 1990s, are generally very short lived.

The periodic ordering of the chemical elements is reflected not only in the periodic table, but also in their ionization energies. Figure 41-22 is a plot of ionization energy versus atomic number. A peak at each inert gas is followed by an abrupt drop to the easily ionized alkali metal that starts the next row of the periodic table. Then the ionization energies generally increase across each row, showing that the atoms become more reluctant to give up their electrons and more willing to accept electrons as they combine chemically with other atoms.

Note the crucial role the exclusion principle plays in this discussion of the chemical elements. Without that principle, every atom in its ground state would have all its electrons in the $1s$ orbital. There would be no qualitative distinction among the elements, and the science of chemistry would not exist. Nor would there be any chemists or physicists; life itself would be impossible without the rich diversity of chemical compounds formed from the different elements.

41-5 TRANSITIONS AND ATOMIC SPECTRA

Emission and absorption of photons with specific energies provide the most direct experimental manifestation of quantized atomic energy levels, and give rise to the spectral lines that permit precise analysis of atomic systems from the laboratory to distant astrophysical objects. Even the simple hydrogen atom exhibits a myriad of quantum states. In multielectron atoms, the possibilities for electronic

18.2
Atomic Physics: Spectroscopy

excitation are even more numerous. The spectra of atoms reflect this rich array of available quantum states (Fig. 41-23).

Selection Rules

Of the conceivable transitions between energy levels, not all are equally likely. So-called **selection rules** determine which are **allowed transitions**—those most likely to occur. One selection rule limits allowed transitions to those for which the orbital quantum number ℓ changes by $\Delta\ell = \pm 1$; this and other selection rules are related to conservation of angular momentum. Quantum mechanics also provides a way of calculating transition probabilities or, equivalently, the mean lifetimes of excited states. For outer electrons, excited states that de-excite by allowed transitions have typical lifetimes on the order of 10^{-9} s.

Transitions that are not allowed by selection rules are called **forbidden transitions**; most are not strictly impossible but just extremely unlikely. States that can lose energy only by forbidden transitions are **metastable states**; their lifetimes are many orders of magnitude longer than the nanosecond timescale for allowed transitions (Fig. 41-24). Forbidden spectral lines are especially valuable probes of the solar corona and other low-density astrophysical gases in which collisions are rare, and atoms can, therefore, remain in metastable states.

Optical Spectra

Spectral lines in or near the visible region result from transitions involving electrons in the incompletely filled outer shells of atoms. The alkali metals, with a single s electron in their outer shells, therefore produce spectra qualitatively similar to that of hydrogen. However, the more complicated structure of a multielectron atom causes shifts in energy levels not observed in hydrogen. Figure 41-25 shows energy-level diagrams and allowed transitions for hydrogen

FIGURE 41-23 Optical spectra of oxygen, carbon and neon. The distinctly different line patterns are characteristic of each element, and reflect underlying differences in electronic structure.

FIGURE 41-24 Phosphorescent materials emit light after their atoms are excited to long-lived metastable states. Such materials are used in a variety of applications from TV picture tubes to day-glo paints.

FIGURE 41-25 Simplified energy-level diagrams for (a) hydrogen and (b) sodium, neglecting spin-orbit splitting. The letters S, P, D, ... correspond to the $\ell = 0, 1, 2, \ldots$ states. Only the outermost electron of sodium participates in optical transitions, so the lowest level shown is the $3s$ level of this electron in the sodium ground state. Note that the $4s$ and $4p$ levels are widely separated in sodium, with the $3d$ level between them; this explains why the $4s$ orbital fills before the $3d$. Allowed transitions are shown in purple.

(a)   (b)

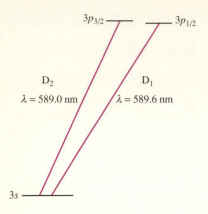

$3p_{3/2}$ — — $3p_{1/2}$

D_2 D_1

$\lambda = 589.0$ nm $\lambda = 589.6$ nm

$3s$ —

FIGURE 41-26 A magnified portion of the energy-level diagram for sodium, showing that the spin-orbit effect splits the $3p$ level into two closely spaced states. Transitions from these states result in the slightly different wavelengths of the sodium D doublet.

and sodium. The diagram for an atom with more than one electron in its outer shell is even more complicated.

Each of the transitions shown in Fig. 41-25 corresponds to a spectral line of specific wavelength. At higher resolution, many spectral lines are observed to be doublets or triplets of closely spaced lines. The p levels, for example, are split into two by the spin-orbit splitting; therefore, transitions between s and p levels result in doublet lines (Fig. 41-26).

Got It!

Which of the transitions shown in Fig. 41-25b results in a photon of the shortest wavelength?

EXAMPLE 41-7	*The Sodium Doublet*

Use the sodium-D wavelengths shown in Fig. 41-26 to calculate the energy difference between the $3p$ states of sodium.

Solution

Using $E = hf$ and $f\lambda = c$, we can write $E = hc/\lambda$ for the energy of each transition. Then the energy splitting is

$$\Delta E = \frac{hc}{\lambda_1} - \frac{hc}{\lambda_2} = hc\left(\frac{1}{\lambda_1} - \frac{1}{\lambda_2}\right)$$

$$= (6.63 \times 10^{-34} \text{ J·s})(3.00 \times 10^8 \text{ m/s})$$

$$\times \left(\frac{1}{589.0 \times 10^{-9} \text{ m}} - \frac{1}{589.6 \times 10^{-9} \text{ m}}\right)$$

$$= 3.44 \times 10^{-22} \text{ J} = 2.15 \times 10^{-3} \text{ eV}.$$

This fine-structure splitting is only about 0.1% of the sodium-D photon energies but is readily measured with spectroscopic techniques.

EXERCISE The transition $6s \rightarrow 4p$ in sodium results in doublet spectral lines at 1.6374 μm and 1.6388 μm. What is the splitting between the two energy levels of the $4p$ state?

Answer: 1.038×10^{-22} J $= 648$ μeV

Some problems similar to Example 41-7: 39–41

FIGURE 41-27 An X-ray tube. Current from the small battery heats the filament, which boils off electrons. Potential difference between filament and target—tens of kilovolts—accelerates electrons toward the target. X-ray production occurs when the electrons decelerate rapidly on hitting the target.

X-ray Spectra

With sufficient energy, it is possible to eject an electron from an inner atomic orbital. Then an electron in a higher-energy orbital drops into the unoccupied state, emitting a photon whose energy is equal to the difference between the two atomic energy levels. For inner-shell electrons, those energies are typically measured in keV, corresponding to wavelength in the range 0.01 to 1 nm—which lies in the X-ray region of the spectrum.

A simple way to produce X-ray photons is to bombard a metal target with a beam of high-energy electrons (Fig. 41-27). Most of the electrons decelerate rapidly on hitting the target, resulting in a broad band of X-ray radiation (recall from Chapter 34 that accelerated charges produce electromagnetic radiation). But some electrons undergo collisions with inner-shell electrons in the target atoms, knocking them out and initiating the chain of transitions that produce X-ray photons of specific energies—the **characteristic X rays** whose spectrum is characteristic of the target material. In analogy with the hydrogen spectrum, X-ray spectral lines are labeled $K\alpha$, $K\beta$, $K\gamma$, . . . for transitions from the L, M, N, . . . shells to the K shell. Transitions from the M, N, O, . . . shells to the L shell, respectively, are labelled $L\alpha$, $L\beta$, $L\gamma$, . . . , and so forth. Figure 41-28 shows X-ray spectra obtained by bombarding molybdenum with high-energy electrons.

We can estimate the energy of an X-ray transition by considering that the electron undergoing the transition experiences not the full electric field of the nucleus but a field reduced by the effect of any electrons closer to the nucleus. In the $L \rightarrow K$ transition that gives rise to the $K\alpha$ line, for instance, the electron sees an effective electric charge of $(Z - 1)e$ from the Z protons in the nucleus (charge Ze) and the remaining K-shell electron (charge $-e$). So the electron making the transition is in roughly the same circumstances as the electron in a single-electron atom of atomic number $Z - 1$. More generally, an electron in the nth energy level sees an effective nuclear charge Z_{eff} that is reduced approximately by the number of electrons in the shells below the upper level. For transitions across several levels, Z_{eff} is more difficult to calculate, since the shielding of the nucleus changes from shell to shell. Once Z_{eff} is known, the approximate transition energy can be obtained by subtracting the energy levels given by Equation 41-9:

$$\Delta E = E_{n_1} - E_{n_2} \simeq -\frac{(13.6 \text{ eV})Z_{eff}^2}{n_1^2} - \left[-\frac{(13.6 \text{ eV})Z_{eff}^2}{n_2^2} \right]$$

$$= (13.6 \text{ eV})Z_{eff}^2 \left(\frac{1}{n_2^2} - \frac{1}{n_1^2} \right).$$

(41-20)

FIGURE 41-28 X-ray spectra obtained by bombarding molybdenum with electrons. Curves are labeled with the electron energy in keV. The highest energies are sufficient to eject inner-shell electrons from molybdenum atoms. As other electrons drop into the vacant states, X-ray photons corresponding to the $K\alpha$ and $K\beta$ spectral lines are emitted. Also shown is the broad band of X radiation associated with the electrons' rapid deceleration as they hit the target.

EXAMPLE 41-8	*An X-ray Transition*

Estimate the energy and wavelength of the $L\alpha$ line in molybdenum ($Z = 42$).

Solution

The transition $L\alpha$ is from the M shell ($n = 3$) to the L shell ($n = 2$). Interior to the M shell are the K shell, holding two electrons, and the L shell, holding at most two s electrons and six p electrons. Since the M-shell electron is making a transition to the L shell, one of the L-shell electrons is missing. So there are nine electrons interior to the M shell, and $Z_{eff} = 42 - 9 = 33$. Applying Equation 41-20 then gives

$$\Delta E = (13.6 \text{ eV})Z_{eff}^2 \left(\frac{1}{n_2^2} - \frac{1}{n_1^2} \right)$$

$$= (13.6 \text{ eV})(33^2)\left(\frac{1}{2^2} - \frac{1}{3^2} \right) = 2057 \text{ eV} .$$

The corresponding wavelength is then

$$\lambda = \frac{hc}{\Delta E} = \frac{(6.63 \times 10^{-34} \text{ J·s})(3.00 \times 10^8 \text{ m/s})}{(2057 \text{ eV})(1.6 \times 10^{-19} \text{ J/eV})} = 0.60 \text{ nm} .$$

EXERCISE Estimate the energy of the $K\alpha$ transition in tungsten ($Z = 74$).

Answer: 54 keV

Some problems similar to Example 41-8: 43, 46, 57, 58

We emphasize that the use of an effective Z in Equation 41-20 is only a crude approximation; detailed analysis of inner-shell electron energies requires more realistic approximations to the Schrödinger problem for the multielectron atom. Nevertheless, Equation 41-20 and related forms do provide reasonable fits to experimental data, especially when Z_{eff} is adjusted to help the fit. For example, the difference between the actual atomic number Z and Z_{eff} for $L\alpha$ transitions turns out to be about 7.4 instead of the estimate of 9 we made in Example 41-8. The use of empirical equations relating the wavelengths of X-ray spectral lines to atomic number was historically important in establishing the periodic table. In 1914, Henry G. J. Moseley in Great Britain made a plot of the square root of X-ray frequencies ($f = c/\lambda$) versus atomic number Z. Equation 41-20 shows that such a plot should be a straight line, and that is what Moseley found experimentally

FIGURE 41-29 Moseley's plot showing square root of $K\alpha$-line frequencies as a function of atomic number Z.

FIGURE 41-30 X-ray applications. (a) X-ray spectrum produced by excitation of air pollutants trapped on a filter. The labeled spectral-line peaks show the presence of lead (Pb) and arsenic (As) in the samples. (b) X-ray image shows metal pins used to repair a human hip joint.

(Fig. 41-29). He was then able to determine the atomic numbers of many elements where before only the atomic weights were known. Recasting the periodic table in terms of atomic number rather than atomic weight made possible our modern understanding of the chemical elements.

Atomic X-ray transitions find wide practical use (Fig. 41-30). Astrophysicists studying high-energy processes in the cosmos use X-ray spectrometers on satellites to identify X-ray emission lines produced by highly excited atoms. Geologists bombard rocks with high-energy electrons, knocking out inner-shell electrons and causing X-ray emission. Analysis of the resulting X-ray spectra helps identify abundances of elements within the rock. Airborne pollutants are readily measured by passing air through a filter, exciting the atoms caught on the filter, and analyzing the resulting X-ray spectrum. Absorption of X-rays is also significant. The element barium exhibits particularly strong X-ray absorption; in medical applications, barium is used to coat the intestinal tract, making it opaque to X rays.

Spontaneous and Stimulated Transitions

What causes an electron to jump from one energy level to another? In the case of an upward transition, the electron must absorb the appropriate amount of energy. Generally, that energy is supplied by a photon whose energy is equal to the energy difference between the two levels; the process is then called **stimulated absorption** (Fig. 41-31a). (Upward transitions can result from other processes as well, as, for example, in an energetic collision between two atoms or the interaction of a free electron with atomic electrons.)

For most downward transitions, however, we can identify no specific cause. An electron spontaneously jumps from a higher to a lower energy level and a photon

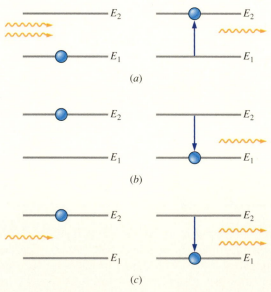

FIGURE 41-31 Interaction of photons with atomic electrons. Horizontal lines denote two atomic energy levels and the wave a photon with energy equal to the energy difference between the two levels. (a) In stimulated absorption, one photon is absorbed and the electron jumps to the higher level. (b) In spontaneous emission, a photon is emitted as the electron jumps spontaneously to the lower level. (c) In stimulated emission, the nearby passage of a photon stimulates a drop from the higher level. A second photon is emitted in the process. The two photons have the same energy, phase, and direction, and thus interfere constructively.

is emitted; the process is termed **spontaneous emission** (Fig. 41-31b). Although an individual spontaneous emission is a random event, quantum mechanics gives the probability per unit time for that event to occur; the inverse of that probability is the mean lifetime of the excited state.

In 1917 Einstein recognized a third possibility: excited atoms can be stimulated to drop into lower energy states by the nearby passage of a photon, again of energy appropriate to the transition. A photon is emitted in the process, and has the same energy and phase as the stimulating photon, and is moving in the same direction. This process, called **stimulated emission**, is the reverse of stimulated absorption (Fig. 41-31c).

Spontaneous emission, stimulated absorption, and stimulated emission play an important role in the transfer of radiation through gases. And stimulated emission is responsible for an important technological development: the laser.

18.4
The Laser

APPLICATION	*The Laser*

As Fig. 41-31c suggests, stimulated emission provides a way of increasing the number of photons of a given energy and phase. The **laser**, whose name derives from Light Amplification by Stimulated Emission of Radiation, exploits this effect to produce intense beams of visible light or other electromagnetic radiation. The key to laser action is to establish a large population of atoms in an excited state; since most atoms are normally in the ground state, this excited situation is termed a **population inversion** (Fig. 41-32). For most excited states, the atoms would normally decay very rapidly by spontaneous emission, and the population inversion could not be maintained. For this reason, the excited level is usually chosen to be a metastable one, whose lifetime is relatively long. Excitation is then a two-step process: first, atoms are excited to an energy state higher than the desired metastable state; then they quickly decay by spontaneous emission to the metastable state where they're "stuck" for a relatively long time by the lack of an allowed transition back to the ground state (Fig. 41-33). The excitation process is called **pumping** and the energy source for the excitation is the **pump**. Laser

pumps include flash lamps, sunlight, other lasers, electric discharges, chemical reactions, and, in the X-ray laser once proposed for ballistic missile defense, nuclear explosions.

If they were isolated, the excited atoms would remain in the metastable state for a long time, occasionally and randomly dropping to the ground state. The result would be a feeble, incoherent emission of photons. But when one atom drops to the ground state, it emits a photon that passes by many other excited atoms. The passing photon stimulates the others to emit as well. Now more photons are available, and more stimulated emission takes place. The process snowballs, resulting in a burst of photons of the same energy and phase. In a so-called **pulsed laser**, the pump is again applied to restore the population inversion, and another photon pulse is obtained. In a **cw laser** (for continuous wave), pumping and lasing occur continuously, resulting in continuous light output. You can simulate laser action with ActivPhysics Activity 18.4.

Working at Columbia University in the 1950s, Charles Townes and Arthur Schawlow realized that stimulated emission could be enhanced by enclosing the radiating medium in a cavity

FIGURE 41-32 A group of atoms, represented as pairs of energy levels. (a) Normally, most atoms are in the ground state, although random thermal energy may occasionally excite some to higher levels. (b) In a population inversion, most atoms are in excited states.

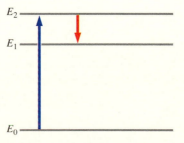

FIGURE 41-33 Pumping a laser usually involves a two-step process. First, atoms are excited from the ground state (E_0) to a higher energy level (E_2). Then they drop spontaneously to the metastable level E_1, where they remain until passing photons stimulate transitions back to the ground state.

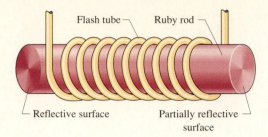

FIGURE 41-34 A ruby laser. The flash tube is the pump that excites chromium atoms in the ruby rod. Mirrors at either end form an optical cavity; one mirror is partially silvered to let the beam emerge.

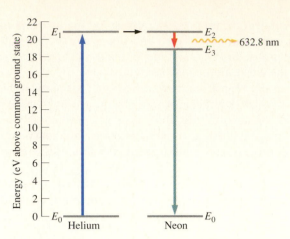

FIGURE 41-35 Energy-level diagram for the helium-neon laser. Electron-ion collisions excite helium to the metastable level E_1 at 20.61 eV. Collisions between helium and neon then transfer energy to the neon atoms, exciting them to the 20.66-eV level E_2. Laser emission occurs as the neon atoms drop by stimulated emission to the level E_3. Finally, they drop spontaneously to the ground state and are ready to begin the process again.

with reflective ends. Photons bouncing back and forth in the cavity would then cause a high level of stimulated emission. One end of the cavity could be made only partially reflective, so light could emerge to form the laser beam. The first working laser was built in 1960 by Theodore Maiman. In Maiman's laser, a flash tube surrounded a ruby rod; light from the flash excited chromium ions in the ruby to a metastable state; stimulated emission at a wavelength of 694 nm followed. Mirrors at either end of the rod provided the cavity, with one mirror partially silvered to allow the beam to emerge. Figure 41-34 is a schematic diagram of this ruby laser.

Since 1960, a myriad of laser types has been developed. Virtually anything can be used as the lasing medium, as long as a population inversion can be established. Media now in use include gases, solids, liquids, semiconductor materials, and ionized plasmas. Natural laser action has even been detected in interstellar gas clouds. Some lasers, particularly those using chemical dyes as the lasing medium, are tunable over a broad range of wavelengths.

A laser commonly used in elementary physics experiments is the **helium-neon laser**, containing a mixture of 80% helium and 20% neon. The operation of this device is slightly more complicated than the two-step excitation process of Fig. 41-33. Electric current passes through the gas mixture, exciting helium by electron-ion collisions. The metastable excited state, 20.61 eV above the ground state, is very close to a 20.66-eV excited state of neon. Collisions between helium and neon populate this 20.66-eV neon level. The neon atoms drop to a lower energy level by stimulated emission, resulting in red laser light at 632.8 nm. Finally, the neon atoms fall to the ground state by spontaneous emission, and the process continues. Figure 41-35 shows an energy-level diagram for the He-Ne laser. This laser, like most, is not very efficient. Note from the energy-level diagram how little of the neon excitation energy is emitted in the laser beam. Problem 47 explores the energetics of the He-Ne laser.

The special characteristics of a laser beam are as follows: (1) The light is extremely monochromatic. Because the beam results entirely from transitions between a pair of discrete energy levels, all photons have essentially the same energy. (2) The beam is coherent; in wave terms, waves emitted by all the atoms are in phase, and interfere constructively to form a single wave-

whose wavefronts are well-defined surfaces that advance in step. This coherence allows the beam to travel long distances with minimal spreading (Fig. 41-36), and also makes possible very precise focusing. The fundamental limitation of optical diffraction is often the only barrier to a perfectly focused laser spot. (3) The beam can be made very powerful since the stimulated-emission process extracts energy from a great many atoms simultaneously. Since photons are spin-1 particles that do not obey the exclusion principle, there is no limit to the number of photons with the same energy and phase that can participate in a laser beam. Although typical lasers used in introductory physics laboratories have power outputs in the milliwatt range,

FIGURE 41-36 Coherence of laser light allows laser beams to travel considerable distances with minimal spreading. Here a laser is aimed at the moon in a lunar ranging experiment.

(a)   (b)   (c)

large continuous-wave lasers with power outputs in excess of 1 megawatt are now available, and pulsed outputs of 100 MW and higher have been achieved (Fig. 41-37). Multiple laser systems, such as those used for laser fusion experiments, reach peak power outputs of 10^{14} W—about 100 times the output of all the world's electric generating plants.

In the several decades since the first laser, applications of lasers have become ubiquitous. Small lasers scan price codes at supermarket checkouts and extract audio signals and computer data from compact discs. A laser beam makes possible the versatility and quality of laser-printed computer output. Medical uses increase each year; eye surgery in particular has been revolutionized by the laser. Lasers help surveyors establish straight lines over long distances. On a grander scale, they provide extremely accurate measurement of the distance to the moon. Small solid-state lasers provide drivers for fiber-optic systems, encoding telephone signals and computer data as pulses of light. Lasers entertain us with dazzling light shows. In science and engineering, they are used to provide precise excitation of atomic and molecular states, allowing analysis of materials and elucida-

(d)

FIGURE 41-38 Laser applications include (a) cutting steel, (b) treating skin cancer, (c) reading bar codes at a supermarket checkout, and (d) measuring fluid flow velocity.

FIGURE 41-37 This high-power laser beam is used to separate uranium isotopes.

tion of fundamental physical principles. Experiments that were once difficult to perform—such as Young's double-slit interference experiment—become trivial with lasers. Through the process of **holography**, laser wavefronts are captured on photographic emulsions from which three-dimensional images are reconstructed. Applications of holography include measurement of very fine motions—on the order of a wavelength of light—in connection with measurements of stress in materials. At higher powers, laser beams serve as precision cutting tools, eating their way through metals. Ultrafast lasers provide picosecond pulses of light, probing phenomena that occur on extremely short timescales. And as laser applications proliferate, so do the wavelengths available from lasers. Currently, lasers are available with outputs in the radio, the infrared, the visible, the ultraviolet, and the X-ray regions of the spectrum. A gamma-ray laser, using population inversions of excited nuclear states, is under development. Figure 41-38 samples the diversity of laser applications.

CHAPTER SYNOPSIS

Summary

1. The hydrogen atom is described by the Schrödinger equation in spherical coordinates. Especially simple states are those that depend only on the radial coordinate r, among them the ground state. The probability of finding an electron at a given position is best described by the **radial probability density**, given by $P(r) = 4\pi r^2 \psi^2$. In general, solutions to the Schrödinger equation involve all three coordinates, and are described by three quantum numbers:

   a. The **principal quantum number** n determines the energy:

   $$E_n = -\frac{1}{n^2}\frac{\hbar^2}{2ma_0^2} = \frac{E_1}{n^2} = \frac{-13.6 \text{ eV}}{n^2}.$$

   b. The **orbital quantum number**, ℓ, determines the magnitude of the angular momentum:

   $$L = \sqrt{\ell(\ell+1)}\,\hbar,$$

   where allowed values of ℓ are integers ranging from 0 to $n-1$.

   c. The **orbital magnetic quantum number** m_ℓ determines the component of orbital angular momentum along a given axis:

   $$L_z = m_\ell\hbar,$$

   where m_ℓ takes integer values between $-\ell$ and $+\ell$. This quantization of the direction of the angular momentum vector **L** is called **space quantization**.

2. The electron carries an intrinsic angular momentum called **spin**. In general, the spin of a particle is described by a quantum number s that takes a single value; for electrons and other **spin-$\frac{1}{2}$ particles**, including protons and neutrons, $s = \frac{1}{2}$. The magnitude of the spin angular momentum is given by

   $$S = \sqrt{s(s+1)}\hbar;$$

   for electrons $S = (\sqrt{3}/2)\hbar$. Like orbital angular momentum, spin exhibits space quantization. The component S_z of spin on a given axis can take on only discrete values, spaced at intervals of $\hbar$:

   $$S_z = m_s\hbar,$$

   where for a spin-$\frac{1}{2}$ particle like the electron, m_s takes only the values $\pm\frac{1}{2}$. Spin is ultimately a consequence of relativistic invariance but was discovered experimentally before its theoretical explanation was known. The **Stern-Gerlach experiment** and related experiments provide evidence for electron spin and its space quantization.

3. The total angular momentum of a hydrogen atom is the vector sum of the orbital and spin angular momenta of its single electron. Magnetic interaction between magnetic dipole moments associated with electron spin and the orbital angular momentum—called **spin-orbit coupling**—results in **fine-structure splitting** of atomic energy levels. Such splitting occurs only for states with nonzero angular momentum; that is, for $\ell \neq 0$ states. **Hyperfine structure**, which arises from the interaction of the spin magnetic moment with the nuclear magnetic moment, occurs in $\ell = 0$ states as well.

4. Pauli's **exclusion principle** states that no two electrons can be in the same state. Since an electron has two possible spin orientations, the exclusion principle means that at most two atomic electrons can be in a state with the same quantum numbers n, ℓ, and m_ℓ. The exclusion principle applies only to particles of half-integer spin; integer-spin particles such as photons can have arbitrarily many particles in the same state.

5. In multielectron atoms, the exclusion principle governs the organization of electrons among the various quantum states, designated **shells**, **subshells**, and **orbitals**. Chemical properties of the elements are determined by the outermost atomic electrons. The regular arrangement of elements along each row in the **periodic table** reflects the filling of the outermost atomic shells; the inert gas at the end of each row has a completely full outer p subshell and does not readily participate in chemical reactions.

6. Transitions among atomic energy levels are governed by **selection rules**, many related to underlying conservation laws.

   a. The visible and near-visible spectra of atoms are the result of transitions involving the outermost electrons. At high resolution, spectral lines are often split by effects like the spin-orbit splitting.

   b. X-ray spectra result from transitions involving inner-shell electrons, typically when one such electron is ejected from the atom and another drops into its place.

   c. Atomic transitions include **stimulated absorption**, in which an atom absorbs a photon as an atomic electron jumps to a higher energy level; **spontaneous emission**, in which an electron jumps spontaneously to a lower level, resulting in emission of a photon; and **stimulated emission**, whereby a passing photon stimulates an excited atom to undergo a downward transition and emit a photon of the same energy and phase. Stimulated emission is the essential process in the **laser**, a source of intense and coherent electromagnetic radiation.

Terms You Should Understand

(Pairs are closely related terms whose distinction is important; number in parentheses is chapter section where term first appears.)

radial probability density (41-1)
orbital angular momentum (41-1)
principal quantum number, orbital quantum number, orbital magnetic quantum number (41-1)
space quantization (41-1)
Stern-Gerlach experiment (41-2)
fine structure, spin-orbit effect (41-2)
exclusion principle (41-3)
fermion, boson (41-3)
shell, subshell, orbital (41-4)
selection rules (41-5)
allowed transition, forbidden transition (41-5)
spontaneous emission, stimulated emission (41-5)
population inversion (41-5)

Symbols You Should Recognize

n, ℓ, m_ℓ (41-1)
s, m_s, j, m_j (41-1)
$s, p, d, f, \ldots$ (41-2)
$S, P, D, F, \ldots$ (41-2)
$K, L, M, \ldots$ (41-2)
symbols like $3D_{3/2}$ (41-2)
electronic configurations like $1s^2 2s^2 2p^6$ (41-4)

Problems You Should Be Able to Solve

evaluating energies and angular momenta associated with states of one-electron atoms (41-1)
finding probabilities for locating electrons in one-electron atoms (41-1)
evaluating spin angular momentum (41-2)
finding allowed orientations for spin and other angular momentum vectors (41-2)
finding allowed values for total angular momentum (41-2)
determining electronic structure of atoms (41-4)
relating energy-level diagrams and spectra (41-5)

Limitations to Keep in Mind

Closed-form solutions of the Schrödinger equation apply only to one-electron atoms.
Optical spectra of multielectron atoms are much more complex than those discussed in this chapter.
The analysis presented here for X-ray line spectra is only a crude approximation.

QUESTIONS

1. The electron in a hydrogen atom is very crudely like a particle confined to a three-dimensional box. In the atom, what plays the role of the confining box?
2. A friend who has not studied physics asks you the size of a hydrogen atom. How do you answer?
3. How many quantum numbers are required to specify fully the state of a hydrogen atom? Why? (Don't forget about spin.)
4. Both the Bohr and Schrödinger theories of the hydrogen atom predict the same ground-state energy. Do they agree about the angular momentum in the ground state? Explain.
5. Plotted as a function of radius r, the probability density ψ^2 for the hydrogen ground state shows a maximum right at the nucleus, $r = 0$ (see Fig. 41-3). Why is it not correct to infer that the electron is most likely to be found at $r = 0$?
6. Describe space quantization, and explain why it is most obviously at odds with classical physics for states with low orbital quantum number ℓ.
7. The potential-energy function for the electron in a hydrogen atom is spherically symmetric. Does this mean that all possible ψ functions for the electron depend only on radial distance from the proton? Explain.
8. Is it possible for a hydrogen atom to be in the $2d$ state? Explain.
9. Can the component of a quantized angular momentum measured on a given axis ever equal the magnitude of the angular momentum vector? Explain.
10. The electron is a spin-$\frac{1}{2}$ particle. Does this mean the intrinsic angular momentum of the electron is $\frac{1}{2}\hbar$? Explain.
11. How does the Stern-Gerlach experiment provide convincing evidence for space quantization?
12. Why is there no spin-orbit splitting in the ground state of hydrogen?
13. In what way is the exclusion principle responsible for the diversity of chemical elements?
14. Helium and lithium exhibit very different chemical behavior, yet they differ by only one unit of nuclear charge. Explain how the exclusion principle brings about the dramatic difference in chemical behavior.
15. Are forbidden transitions really forbidden? Explain and contrast with allowed transitions.
16. Why is an inner-shell vacancy generally required for an atom to emit an X-ray photon?
17. Optical spectra change qualitatively from one element to the next, while X-ray spectra exhibit only a gradual, continuous change from one element to the next. What might explain the difference?
18. Why is stimulated emission essential for laser action?
19. Why do most lasers use a metastable state of the lasing medium?

PROBLEMS

ActivPhysics can help with these problems:
Activities 18.2, 18.4

Section 41-1 The Hydrogen Atom

1. Using physical constants accurate to four significant figures (see the inside front cover), verify the numerical values of the Bohr radius a_0 and the hydrogen ground-state energy E_1.

2. A group of hydrogen atoms are in the same excited state. It is found that photons with at least 1.5 eV energy are required to ionize these atoms. What is the quantum number n for the initial excited state?

3. Find (a) the probability density and (b) the radial probability density at $r = \frac{1}{2}a_0$ for the hydrogen ground state. Comment on the relative sizes of these quantities.

4. Repeat the calculation in the second half of Example 41-1 to find the probability that an electron in the hydrogen ground state will be found beyond two Bohr radii.

5. What is the maximum possible magnitude for the orbital angular momentum of an electron in the $n = 7$ state of hydrogen?

6. Which of the following is not a possible value for the magnitude of the orbital angular momentum in hydrogen?
   (a) $\sqrt{12}\hbar$;   (b) $\sqrt{20}\hbar$;   (c) $\sqrt{30}\hbar$;
   (d) $\sqrt{40}\hbar$;   (e) $\sqrt{56}\hbar$.

7. The orbital angular momentum of the electron in a hydrogen atom has magnitude 2.585×10^{-34} J·s. What is the minimum possible value for its energy?

8. What is the orbital quantum number for an electron whose orbital angular momentum has magnitude $L = \sqrt{30}\hbar$?

9. Determine the principal and orbital quantum numbers for a hydrogen atom whose electron has energy -0.850 eV and orbital angular momentum of magnitude $\sqrt{12}\hbar$.

10. Find (a) the energy and (b) the magnitude of the orbital angular momentum for an electron in the $5d$ state of hydrogen.

11. A 1200-kg car rounds a turn of radius 150 m at a speed of 10 m/s. Assuming its angular momentum is quantized according to Equation 41-10, find the approximate value for ℓ.

12. A hydrogen atom is in the $6f$ state. What are (a) its energy and (b) the magnitude of its orbital angular momentum?

13. Give a symbolic description for the state of the electron in a hydrogen atom when the total energy is -1.51 eV and the orbital angular momentum is $\sqrt{6}\hbar$.

14. The maximum possible angular momentum for a hydrogen atom in a certain state is $30\sqrt{11}\hbar$. Find (a) the principal quantum number and (b) the energy.

15. Which of the following pairs of energy and magnitude of orbital angular momentum are possible for a hydrogen atom, and to what n and ℓ values do they correspond?
    (a) -0.544 eV, 3.655×10^{-34} J·s;
    (b) -1.51 eV, 3.655×10^{-34} J·s;
    (c) -1.51 eV, 5.842×10^{-34} J·s;
    (d) -3.4 eV, 1.492×10^{-34} J·s

16. A hydrogen atom is in a state with $\ell = 2$. What are the possible angles its orbital angular momentum vector can make with an arbitrary axis?

17. A hydrogen atom has energy $E = -0.850$ eV. What are the maximum possible values for (a) the magnitude of its orbital angular momentum and (b) the component of that angular momentum on a chosen axis?

18. An electron in hydrogen is in the $5f$ state. What possible values, in units of $\hbar$, could a measurement of the orbital angular momentum component on a given axis yield?

19. Substitute Equation 41-3 for ψ in Equation 41-4, and carry out the indicated differentiation to show that the result is Equation 41-5.

20. Differentiate the radial probability density for the hydrogen ground state, and set the result to zero to show that the electron is most likely to be found at one Bohr radius.

Section 41-2 Electron Spin

21. Verify the value of the Bohr magneton in Equation 41-15.

22. Theories of quantum gravity predict the existence of a spin-2 particle called the graviton. What would be the magnitude of the graviton's spin angular momentum?

23. Some very short-lived particles known as delta resonances have spin $\frac{3}{2}$. Find (a) the magnitude of their spin angular momentum and (b) the number of possible spin states.

24. What is the minimum angle between the spin vector and the z axis for (a) a spin-$\frac{3}{2}$ particle and (b) a spin-2 particle?

25. What are the possible j values for a hydrogen atom in the $3D$ state?

26. Repeat the preceding problem for the case where you know only that the principle quantum number is 3; that is, ℓ might have any of its possible values.

27. Draw vector diagrams similar to Fig. 41-12 for spin-orbit coupling of an electron in the $\ell = 3$ state.

28. A hydrogen atom is in the $4F_{5/2}$ state. What are (a) its energy in units of the ground-state energy, (b) the magnitude of its orbital angular momentum in units of $\hbar$, and (c) the magnitude of its total angular momentum in units of $\hbar$? (d) Which of (b) or (c) is greater? Why?

Section 41-3 The Pauli Exclusion Principle

29. Suppose you put five electrons into an infinite square well. (a) How do the electrons arrange themselves to achieve the lowest total energy? (b) Give an expression for this energy in terms of the electron mass m, the well width L, and Planck's constant.

30. A harmonic oscillator potential of natural frequency ω contains eight electrons and is in the state of lowest energy. (a) What is that energy? (b) What would be the energy if the electrons were replaced by spin-1 particles of the same mass?

31. A harmonic oscillator potential of natural frequency ω contains a number of electrons and is in the state of lowest energy. If that energy is $6.5\hbar\omega$, (a) how many electrons are in the potential well and (b) what is the energy of the highest energy electron?

Section 41-4 Multielectron Atoms and the Periodic Table

32. Use shell notation to characterize the outermost electron of rubidium.

33. Write the full electronic structure of scandium.

34. Write the full electronic structure of bromine.

35. Determine the electronic configuration of copper.

36. Using the table below, make a plot of atomic volume versus atomic number, for the elements from $Z = 30$ to $Z = 59$ listed in the table. Comment on the structure of your graph in relation to the periodic table, the electronic structures of atoms, and their chemical properties. (Volumes are in units of 10^{-30} m³.)

Z	V	Z	V	Z	V
30	7.99	40	26.1	50	11.2
31	12.5	41	20.2	51	8.78
32	6.54	42	18.8	52	6.88
33	4.99	43	17.5	53	5.28
34	3.71	44	16.2	54	4.19
35	2.85	45	12.8	55	95.9
36	2.57	46	12.0	56	51.6
37	70.3	47	11.2	57	49.0
38	37.2	48	10.5	58	46.5
39	28.3	49	17.2	59	44.0

Section 41-5 Transitions and Atomic Spectra

37. An electron in a highly excited state of hydrogen ($n_1 \gg 1$) drops into the state $n = n_2$. What is the lowest value of n_2 for which the emitted photon will be in the infrared ($\lambda > 700$ nm)?

38. A solid-state laser made from lead-tin selenide has a lasing transition at a wavelength of 30.0-μm. If its power output is 2.0 mW, how many lasing transitions occur each second?

39. The $4f \to 3p$ transition in sodium produces a spectral line at 567.0 nm. What is the energy difference between these two levels?

40. The $4p \to 3s$ transition in sodium produces a doublet spectral line at 330.2 and 330.3 nm. What is the energy splitting of the $4p$ level?

41. The $4s \to 3p$ transition in sodium produces a doublet spectral line at 1138.1 nm and 1140.4 nm. Combine this fact with the discussion in Example 41-7 to find an accurate value for the energy difference between the $3s$ and $4s$ states in sodium.

42. Show that the wavelength λ in nm of a photon whose energy E is given in eV is

$$\lambda = \frac{1240}{E}.$$

43. Estimate the wavelength of the $K\alpha$ X-ray line in calcium.

44. Estimate the minimum electron energy that could excite the $K\alpha$ line in molybdenum, and compare with Fig. 41-28. *Hint:* You need to eject a K-shell electron entirely from the atom.

45. What is the approximate minimum accelerating voltage for an X-ray tube with an iron target to produce the $L\alpha$ line? *Hint:* See Problem 44.

46. Use the X-ray spectrum shown in Fig. 41-30 to estimate Z_{eff} for the $L\beta$ transition in lead.

47. Use information from Fig. 41-35 (a) to find the energy of the level E_3 and (b) to determine the maximum fraction of the energy delivered to excite neon atoms in a He-Ne laser that actually ends up as laser light.

48. A selection rule for the infinite square well allows only those transitions in which n changes by an odd number. Suppose an infinite square well of width 0.200 nm contains an electron in the $n = 4$ state. (a) Draw an energy-level diagram showing all allowed transitions that could occur as this electron drops toward the ground state (include transitions from lower levels that could be reached from $n = 4$). (b) What are all the possible photon energies emitted in these transitions?

49. An ensemble of square-well systems of width 1.17 nm all contain electrons in highly excited states. They undergo all possible transitions in dropping toward the ground state, obeying the selection rule that Δn must be odd. (a) What wavelengths of visible light are emitted? (b) Is there any infrared emission? If so, how many spectral lines fall in the infrared?

50. For hydrogen, fine-structure splitting of the $2p$ state is only about 5×10^{-5} eV. What percentage is this difference of the

photon energy emitted in the $2p \rightarrow 1s$ transition? Your answer shows why it's hard to observe spin-orbit splitting in hydrogen.

Paired Problems

(Both problems in a pair involve the same principles and techniques. If you can get the first problem, you should be able to solve the second one.)

51. Find the probability that the electron in the hydrogen ground state will be found in the radial distance range $r = a_0 \pm 0.1a_0$.

52. A gas contains 1000 hydrogen atoms in their ground states. If you locate their electrons, in how many atoms, on average, will they be within 3 Bohr radii of the nucleus?

53. Find the spacing between adjacent orbital angular momentum values for hydrogen in the $n = 2$ state.

54. What is the most orbital angular momentum that could be added to an atomic electron initially in the $6d$ state, without changing its principal quantum number? What would be the new state?

55. (a) Write the alpha-numeric description of a hydrogen atom whose energy is one-ninth of the ground-state energy (recall that the latter is negative, so this is an excited state), whose orbital angular momentum has magnitude $\sqrt{6}\hbar$, and whose spin and orbital angular momentum are as nearly aligned as possible. (b) How many m_j values are there for this state?

56. A hydrogen atom is in an F state. (a) Find the possible values for the magnitude of its total angular momentum and, (b) for the state with the greatest angular momentum, the number of possible values for the component of $\mathbf{J}$ on a given axis.

57. Estimate the energy of the $L\alpha$ X-ray transition in arsenic.

58. Estimate the wavelength of the $L\alpha$ X-ray transition in lead.

Supplementary Problems

59. Verify that the normalization constant $1/4\sqrt{2\pi a_0^3}$ in Equation 41-8 is correct.

60. A hydrogen atom is in the $2s$ state. What is the probability that its electron will be found (a) beyond one Bohr radius and (b) beyond 10 Bohr radii?

61. Substitute the ψ_{2s} wave function (Equation 41-8) into Equation 41-4 to verify that the equation is satisfied and that the energy is given by Equation 41-7 with $n = 2$.

62. Show that the maximum number of electrons in the nth shell of an atom is $2n^2$.

63. (a) Verify Equation 41-9 by considering a single-electron atom with nuclear charge Ze instead of e. (b) Calculate the ionization energies for single-electron versions of helium, oxygen, iron, lead, and uranium.

64. Form the radial probability density $P_2(r)$ associated with the ψ_{2s} state of Equation 41-8, and find the most probable radial position for the electron.

Molecular and Solid-State Physics

42

Atoms bind together to form molecules, shown here in a computer simulation.

Quantum mechanics is remarkably successful in explaining the properties of atoms, and should also apply to more complex structures such as molecules and solids. In principle, we could treat a complicated molecule such as DNA by solving the Schrödinger equation for all its constituent electrons and nuclei. Recently, the availability of supercomputers has made calculations of molecular structure increasingly successful, although we are still far from computing the structure of DNA. Nevertheless, in most applications it remains appropriate to think of multiparticle systems in more qualitative terms, categorizing them according to the mechanisms that bind their constituent particles.

42-1 MOLECULAR BONDING

The binding of atoms into molecules involves both electrical forces and peculiarly quantum-mechanical effects associated with the exclusion principle. Although individual atoms are electrically neutral, the distribution of positive and negative charge within them nevertheless can result in attractive or repulsive forces. When atoms are squeezed closely together, interactions involving spins of their outermost electrons also result in attractive or repulsive interactions. For atoms with unfilled outer shells, it is energetically favorable for electrons to pair with opposite spins; this causes an attractive interaction. When the outer atomic shells are filled, two electrons originally in identical states in the separate atoms are forced into different states as the atoms are brought together. This occurs because the two atoms become effectively a single system to which the exclusion principle applies. One or more electrons are forced into higher-energy states, raising the overall energy and giving rise to a repulsive interaction. Finally, if atoms are pushed very closely together, the electrical repulsion of the nuclei becomes important. Ultimately, the balance of attractive and repulsive interactions determines the equilibrium configuration of a molecule. In energy terms, we can think of a stable molecule as a minimum-energy configuration of the electrons and nuclei making up two or more atoms (Fig. 42-1). Although such force and energy considerations ultimately govern all molecules, it is convenient to distinguish several molecular binding mechanisms, based on which of the attractive and repulsive interactions are most important.

Ionic Bonding

As we saw in the preceding chapter, elements near the left side of the periodic table have relatively few electrons in their outermost shells and correspondingly low ionization energies. In contrast, elements near the right side of the table exhibit nearly filled shells and consequently have strong affinities for electrons. When atoms from these different regions of the periodic table come together, relatively little energy is needed to transfer electrons between them. Sodium, for example, has an ionization energy of 5.1 eV, meaning that this much energy must be supplied to make an Na^+ ion. Chlorine, at the opposite end of the periodic table, has such a strong electron affinity that the energy of a Cl^- ion is actually 3.8 eV below that of a neutral Cl atom. Thus an expenditure of only 1.3 eV (5.1 eV $-$ 3.8 eV) is required to transfer the outermost electron from a sodium

FIGURE 42-1 Potential energy of a pair of hydrogen atoms as a function of the distance between their nuclei. The minimum of the curve marks the internuclear separation r_0 of an H_2 molecule at equilibrium.

FIGURE 42-2 Potential-energy curve for a Na^+ and Cl^- ion. The zero of potential energy corresponds to neutral Na and Cl atoms at infinite separation. Here the curve goes toward +1.3 eV at large separation, reflecting the energy needed to transfer an electron from Na to Cl to form the ion pair.

to a chlorine atom. The oppositely charged ions that result are strongly attracted, and they come together until they reach equilibrium at an internuclear separation of about 0.24 nm. The total energy of the pair is then 4.2 eV below that of neutral chlorine and sodium atoms at large separation (Fig. 42-2). Since it would therefore take 4.2 eV to separate the atoms, this quantity is called the **dissociation energy**.

Because the minimum-energy sodium-chlorine structure consists of ions bound by the electrostatic force, the binding mechanism is termed **ionic bonding**. Ionic bonding generally occurs in crystalline solids. Because the building blocks of an ionically bound substance are electrically charged, each can bind to several of the opposite charge, resulting in the regular pattern of a crystalline solid (Fig. 42-3). Because the electrostatic force between ions is relatively strong, ionic solids are tightly bound and therefore have high melting points (801°C for NaCl). And because all electrons are bound to individual nuclei, there are no free electrons and therefore ionic solids are excellent electrical insulators.

FIGURE 42-3 A sodium chloride crystal is a regular array of sodium and chlorine ions, bound by the electrostatic force. In the crystalline solid, the effect of multiple-ion interactions increases the internuclear spacing from 0.24 nm of the isolated ion pair (Fig. 42-2) to 0.28 nm.

Covalent Bonding

In an ionic bond, each electron is associated with only one ion. In a **covalent bond**, on the other hand, electrons are shared among atoms. Covalent bonds can occur between atoms whose outermost shells are not full, and whose outer electrons can therefore pair with opposite spins. The simplest example of a covalent bond is the hydrogen molecule, H_2. Since each hydrogen atom has only a single $1s$ electron, each could accommodate in its $1s$ shell a second electron with opposite spin. When two hydrogen atoms come together, quantum mechanics predicts a molecular ground state in which both electrons share a single orbital, with the greatest probability of finding the electrons between the nuclei (Fig. 42-4). Dissociation energies for covalent bonds are, like those of ionic bonds, on the order of a few eV.

With their outermost molecular orbitals full (i.e., containing paired electrons of opposite spin), covalently bonded molecules often have no room for another electron in their structures. For example, the addition of a third hydrogen atom to an H_2 molecule is not possible because the ground-state orbital already

FIGURE 42-4 Probability density for the ground state of molecular hydrogen (H_2). Binding of the molecule results from sharing of the two electrons, here evident in the enhanced probability density between the nuclei.

FIGURE 42-5 In the diamond crystal, each atom bonds covalently to its four nearest neighbors.

contains two electrons with opposite spins, and the exclusion principle requires that a third electron go into a higher energy state. The energy of that state is, in fact, higher than that of an H_2 molecule and a distant H atom; for this reason H_3 is not a stable molecule. Because their outermost molecular orbitals are full, covalent molecules interact only weakly, and as a result many common covalent materials—for example, H_2, CO, N_2, and H_2O—are either gases or liquids at ordinary temperatures. In other cases covalent bonds can form crystalline structures; an example is diamond, formed when each carbon atom bonds covalently to its four nearest neighbors (Fig. 42-5).

APPLICATION	*Buckminsterfullerine*

Molecular bonding has been with us for a long time, but nature continues to surprise us. As early as the 1960s, chemists suggested that carbon atoms might bond to produce large, hollow molecules. Such molecules were first identified in the 1980s, and in 1990 they were produced in abundance. They join diamond and graphite as a third form of pure carbon.

The paradigm of this new molecular family is C_{60}, named buckminsterfullerine after the American philosopher, designer, and engineer R. Buckminster Fuller, who is best known for the geodesic dome. With its 60 covalently bonded carbon atoms identically positioned at the corners of adjacent pentagons and hexagons, buckminsterfullerine resembles a soccer ball and is the roundest molecule possible (Fig. 42-6a). Other fullerines include asymmetric "buckybabies" such as C_{32} and C_{44}, the rugby-ball-shaped C_{70}, and giants such as C_{540} and C_{960}.

Buckminster Fuller argued that his geodesic domes are among the strongest of structures, and his namesake molecules are similarly stable and resilient. Yet despite their stability, fullerines show remarkable chemical reactivity. Other atoms and molecular complexes attach readily to the outside of the fullerine "cage," and individual atoms can be "trapped" inside. C_{60} compounds with potassium and rubidium even prove to be superconductors.

Closed forms such as C_{60} aren't the only possibility for these remarkable carbon compounds; in the early 1990s Japanese scientists developed "buckytubes"—tiny, tough, hollow cylinders made from carbon atoms (Fig. 42-6b). These prove useful in the scanning tunneling microscope that we discussed in

(a)   (b)

FIGURE 42-6 Two of the many fullerine forms. (a) Buckminsterfullerine, C_{60}, is a symmetric arrangement of 60 carbon atoms. (b) "Buckytubes" are tubelike structures made from carbon atoms.

Chapter 40, where they make strong yet flexible tips for the microscope's probe. By capping the ends of "buckytubes," researchers at the National Renewable Energy Laboratory in 1997 produced "nanotanks" capable of storing safely the hydrogen fuel that may someday power a new generation of pollution-free vehicles.

Fullerines are easy to produce by vaporizing carbon; they form, for example, in sooty candle flames and in interstellar space. Chemists and physicists are only just beginning to understand these remarkable manifestations of molecular bonding, which will almost certainly see widespread technological applications in the coming decades.

Hydrogen Bonding

If water consists of covalent molecules, why does it ever take on a solid form? The answer lies in **hydrogen bonds**, a type of weak bond that forms when the positively charged proton of a hydrogen nucleus attracts negative parts of other molecules. In ice, for example, hydrogen bonds link a proton in one H_2O

(a)   *(b)*

FIGURE 42-7 Hydrogen bonding. (*a*) Hydrogen bonds (shown as dashed lines) join water molecules in an ice crystal. (*b*) In DNA, hydrogen bonds hold two covalently bonded chains of atoms in the double-helix structure.

molecule to the oxygen atom in another (Fig. 42-7*a*). This bonding is most accurately described as an attraction between electric dipoles composed of OH structures in each molecule. These dipoles occur because the electrons in H_2O are statistically more likely to be found in the vicinity of the oxygen, giving the oxygen a slight negative charge and leaving the hydrogens slightly positive. This statistical effect results, on the average, in the transfer of much less than one negative elementary charge to the oxygen. For this reason, and because dipole-dipole forces fall much more rapidly with distance than point-charge forces, hydrogen bonds are much weaker than ionic or covalent bonds. A typical hydrogen-bond energy is 0.1 eV. Hydrogen bonds are often important in determining the overall physical configuration of complicated molecules. In DNA, for example, covalent bonds join individual atoms to form long chains; the chains are then linked into the famous double-helix structure by hydrogen bonds (Fig. 42-7*b*).

You might expect hydrogenlike bonds to occur any time part of a molecule is positively charged. But the bond is unique to hydrogen, for this reason: The core of a hydrogen atom consists only of the proton, about 10,000 times smaller than the atom. This tiny proton can easily nestle close to negatively charged structures, giving rise to a bond of nonnegligible strength. With any other atom, the core consists of a nucleus and the tightly bound inner electrons, giving a structure that is more atomic size than nuclear size. These larger cores cannot fit tightly between other atomic cores, and so cannot form hydrogenlike bonds of significant strength.

Van der Waals Bonding

In Section 23-6 we showed how the weak van der Waals force results from electrostatic interaction between induced dipole moments of otherwise nonpolar molecules. In gases of covalent molecules, the van der Waals force causes deviations from the perfect gas law that are most pronounced at high densities. But as the temperature drops, this weakly attractive force becomes effective in binding molecules into liquids and solids. Liquid and solid forms of oxygen (O_2) and nitrogen (N_2), for example, are held together by van der Waals bonding.

Metallic Bonding

In a metal, the outermost atomic electrons are not bound to individual nuclei, but move throughout the material. The metal forms a crystal lattice of positive ions, bound together by this "electron gas" spread through the metal. The free electrons give a metal its high electrical and thermal conductivities.

42-2 MOLECULAR ENERGY LEVELS

In a molecule, electrons and nuclei are bound into a single structure by electric forces. Like an atom or other quantum-mechanical bound system, the possible energy levels of a molecule are quantized. As in atoms, differences among molecular energy levels may be associated with different electronic configurations (Fig. 42-8). But molecules are more complex structures than atoms, and molecular energy can take additional forms.

In Chapter 21, we found that a complete description of the specific heats of gases required that we consider the rotational and vibrational motion of individual molecules. We even hinted at quantum mechanics, as we pointed out that each of these modes of molecular motion could absorb only certain discrete amounts of energy (see Fig. 21-23 and accompanying text). Here, in a quantum-mechanical treatment of molecular energetics, we consider rotational and vibrational energy states as well as electronic configuration.

Rotational Energy Levels

If a molecule is rotating, it has angular momentum whose magnitude, from Equation 13-4, is

$$J = I\omega,$$

where I is the rotational inertia and ω the angular speed, and where we use J rather than L for consistency with the molecular physics literature. In the preceding chapter, we found that atomic angular momentum is quantized, with magnitude given by

$$J = \sqrt{j(j + 1)}\hbar. \tag{42-1}$$

Not surprisingly, the angular momentum of molecular rotation, $J = I\omega$, also obeys the quantization condition of Equation 42-1, where the quantum number j takes on integer values 0, 1, 2, 3, ... But then the rotational energy, which from Equation 12-21 is

$$E_{rot} = \tfrac{1}{2}I\omega^2,$$

must also be quantized. Solving the equation $J = I\omega$ for ω allows us to write the energy as

$$E_{rot} = \tfrac{1}{2}I\left(\frac{J}{I}\right)^2 = \frac{J^2}{2I}.$$

Applying the quantization condition 42-1 for J, we then have the quantized rotational energy levels:

$$E_{rot} = \frac{\hbar^2}{2I}j(j + 1), \qquad j = 0, 1, 2, 3, \ldots. \tag{42-2}$$

(a)

(b)

FIGURE 42-8 Electron probability densities for (a) the ground state and (b) the first excited state of the hydrogen molecule (H_2). Note that the excited state has a greater internuclear separation, and that electrons in this state are less likely to be found in the region between the nuclei.

Molecules, like atoms, can jump between energy levels by absorbing or emitting photons. Conservation of angular momentum and the fact that the photon is a spin-1 particle limit the allowed transitions to those for which $\Delta j = \pm 1$, so the only allowed jumps are between adjacent rotational levels. Measuring the photon energy associated with rotational transitions allows us to infer the rotational inertia through Equation 42-2, and thus to learn about the structures of molecules.

EXAMPLE 42-1	*How Big Is a Molecule?*

The spectrum of a gas of molecules undergoing transitions among rotational states shows a series of sharp lines at wavelengths corresponding to the energy differences between adjacent states (see Fig. 42-12 in the next section). The energy difference between adjacent spectral lines is constant; for hydrogen chloride (HCl), this difference is 2.63 meV (2.63×10^{-3} eV). Use this experimental result to calculate the rotational inertia of the HCl molecule. Then, making the approximation that the massive chlorine atom stays nearly fixed while the hydrogen rotates rigidly about it (Fig. 42-9), find an expression for the rotational inertia in terms of the hydrogen mass and internuclear separation. By equating with your spectroscopically measured rotational inertia, determine the approximate internuclear separation in HCl.

FIGURE 42-9 Dumbbell model for a rotating HCl molecule. Since the chlorine atom is much more massive than the hydrogen, it is a good approximation to consider the chlorine fixed. Then the rotational inertia is due entirely to the hydrogen.

Solution

Equation 42-2 shows that the difference between the jth and $(j-1)$th energy levels is

$$\Delta E_{j \to (j-1)} = \frac{\hbar^2}{2I}\left[j(j+1) - (j-1)j\right] = \frac{\hbar^2 j}{I}.$$

This is the energy of a single spectral line corresponding to a transition between the jth and $(j-1)$th levels. An adjacent spectral line, corresponding to a transition between the $(j-1)$th and $(j-2)$th levels, would similarly have $\Delta E_{(j-1) \to (j-2)} = \hbar^2(j-1)/I$. So the difference $\Delta(\Delta E)$ between adjacent spectral lines is

$$\Delta(\Delta E) = \frac{\hbar^2 j}{I} - \frac{\hbar^2(j-1)}{I} = \frac{\hbar^2}{I},$$

a result that is independent of j. Solving for the rotational inertia I, and using our 2.63-meV value for $\Delta(\Delta E)$ gives

$$I = \frac{\hbar^2}{\Delta(\Delta E)} = \frac{(1.055 \times 10^{-34} \text{ J·s})^2}{(2.63 \times 10^{-3} \text{ eV})(1.6 \times 10^{-19} \text{ J/eV})}$$

$$= 2.65 \times 10^{-47} \text{ kg·m}^2.$$

For a point mass m moving in a circle of radius R, the rotational inertia given by Equation 12-15 is just $I = mR^2$, so

$$R = \sqrt{\frac{I}{M}} = \sqrt{\frac{2.65 \times 10^{-47} \text{ kg·m}^2}{1.67 \times 10^{-27} \text{ kg}}} = 1.26 \times 10^{-10} \text{ m}$$

$$= 0.126 \text{ nm},$$

slightly larger than an isolated hydrogen atom. (Here we used the proton mass as an excellent approximation to the hydrogen mass.) Physical chemists use the method of this example to determine molecular bond lengths.

In this example, we made the approximation that the more massive Cl atom remained essentially fixed, with the rotational inertia determined entirely by the hydrogen atom. A more accurate calculation—essential in a diatomic molecule whose constituent atoms had more comparable masses—would account for the motion of both atoms in determining the rotational inertia (see Problems 10 and 52). This calculation is also approximate in that the wavelength given is for a hypothetical molecule with no vibrational energy. The nonzero ground-state vibrational energy of a real molecule results in a slight stretching of the molecular bond and, therefore, in a shift in the rotational transitions.

EXERCISE The equilibrium bond length in hydrogen bromide (HBr) is 0.141 nm. Treating the bromine as essentially at rest, determine the energy of the first rotational excited state.

Answer: 2.09 meV

Some problems similar to Example 42-1: 6, 7, 9, 10, 51

FIGURE 42-10 Near its minimum, the molecular potential-energy curve approximates a parabola. For small-amplitude displacements from equilibrium, the molecule is thus a quantum-mechanical harmonic oscillator.

As Example 42-1 indicates, transitions among rotational energy states generally involve fairly low-energy photons, typically in the microwave region of the spectrum.

Vibrational Energy Levels

The equilibrium configuration of a molecule corresponds to the minimum of the molecular potential-energy curve. In the vicinity of that minimum the curve is generally well approximated by a parabola (Fig. 42-10; see also Fig. 15-26). In Chapter 15, we saw that parabolic potential-energy curves result in simple harmonic motion, and in Chapter 40 we examined the quantum harmonic oscillator by using a parabolic potential-energy curve in the Schrödinger equation. There, we found that quantized vibrational energy levels are given by

$$E_{vib} = (n + \tfrac{1}{2})\hbar\omega, \tag{42-3}$$

where the quantum number n takes on integer values 0, 1, 2, 3, . . . , and where ω is the natural frequency for classical harmonic oscillations of the molecule. The selection rule for harmonic oscillators limits allowed transitions to those with $\Delta n = \pm 1$, so $\hbar\omega$ is the energy of photons emitted or absorbed in allowed transitions among vibrational energy levels. (Actually, the small-amplitude approximation is often justified only for the lower quantum states, so Equation 42-3 and the selection rule $\Delta n = \pm 1$ may apply only to these states.) For typical diatomic molecules, ω is on the order of 10^{14} s^{-1}, in the infrared region of the spectrum. Consequently, study of molecular vibrations involves infrared spectroscopy.

As we found in Chapter 40, the minimum energy of a quantum harmonic oscillator is the ground-state energy $E_0 = \tfrac{1}{2}\hbar\omega$, from Equation 42-3 with $n = 0$. Thus a molecule can never have zero vibrational energy, although Equation 42-2 shows that it *can* have zero rotational energy.

EXAMPLE 42-2	*Molecular Energies*

An HCl molecule is in its vibrational ground state. Its classical vibration frequency is $f = 8.66 \times 10^{13}$ Hz. If its rotational and vibrational energies are nearly equal, what are its rotational quantum number and its angular momentum?

Solution

The rotational and vibrational energies are given by Equations 42-2 and 42-3, respectively, the latter with $n = 0$ to give the ground state. Equating these energies gives

$$\frac{\hbar^2}{2I}j(j+1) = \tfrac{1}{2}\hbar\omega = \tfrac{1}{2}hf,$$

or
$$j(j+1) = \frac{4\pi^2 If}{h},$$

where we used $\hbar = h/2\pi$. We could solve this quadratic equation for j, but from Example 42-1 we already know that transitions between rotational energy levels involve microwave photons that have much lower energy than the infrared photons associated with vibrational transitions. Therefore, we anticipate that a highly excited rotational state is necessary for the rotational energy to approximate the ground-state vibrational

energy. So j should be large, and we approximate $j + 1$ by j. Then we have approximately

$$j = \sqrt{\frac{4\pi^2 If}{h}} = \sqrt{\frac{(4\pi^2)(2.65 \times 10^{-47}\text{ kg·m}^2)(8.66 \times 10^{13}\text{ Hz})}{6.63 \times 10^{-34}\text{ J·s}}}$$
$$= 12,$$

to the nearest integer. Equation 42-1 then gives the molecular angular momentum:

$$J = \sqrt{j(j+1)}\hbar \simeq \sqrt{j^2}\hbar = j\hbar$$
$$= (12)(1.06 \times 10^{-34}\text{ J·s}) = 1.3 \times 10^{-33}\text{ J·s},$$

where again we use the approximation $j(j+1) \simeq j^2$.

EXERCISE What photon energy would be needed to increase by one both the vibrational and rotational quantum numbers of the molecule in Example 42-2?

Answer: 0.39 eV

Some problems similar to Example 42-2: 11, 14, 15, 41, 42

FIGURE 42-11 Energy-level diagram showing the ground state and first two vibrational excited states of a diatomic molecule. Each vibrational level (*n* value) is split into infinitely many levels corresponding to different rotational energies. Here four rotational states are shown for each vibrational level. Colored arrows indicate allowed photon-absorbing transitions from the vibrational ground state to the first excited state; these obey the selection rules $\Delta n = \pm 1$, $\Delta j = \pm 1$. The allowed transitions form two bands of higher and lower energy transitions, corresponding to $\Delta j = +1$ and $\Delta j = -1$. These bands form the two distinct sets of spectral lines shown in Fig. 42-12.

Molecular Spectra

A molecule in a state characterized by vibrational quantum number *n* and rotational quantum number *j* can undergo transitions obeying the selection rules $\Delta n = \pm 1$ and $\Delta j = \pm 1$. If molecules could not rotate, the molecular spectrum would consist of a single line at the classical vibration frequency, corresponding to transitions among adjacent vibrational states. But each vibrational level corresponds to an infinity of rotational states, with $j = 0, 1, 2, 3, \ldots$, and having different rotational energies. The resulting energy-level diagram, shown in Fig. 42-11, is therefore much more complicated. At typical temperatures, only the ground and first vibrational levels are significantly populated, but with energy distributed among many rotational levels. As a result, molecular spectra typically show a rich structure, with many lines corresponding to the different transitions of Fig. 42-11. Figure 42-12 is a spectrum of HCl, taken with a high-resolution infrared spectrometer that resolves the individual spectral lines. With a lower resolution instrument, the pattern shows up simply as a broad band, and we often speak of infrared absorption bands in describing the effect of molecules on infrared radiation. For example, absorption bands of atmospheric carbon dioxide limit the escape of infrared radiation from Earth. Carbon dioxide emissions from burning fossil fuels have increased atmospheric CO_2 levels to the point where this "trapping" of infrared radiation—the "greenhouse effect"—may result in a global warming trend, with serious climatic implications.

FIGURE 42-12 Absorption spectrum of HCl, taken with an infrared spectrophotometer. The lines shown result from transitions between the vibrational ground and first excited state. The left-hand group of lines are from transitions in which the rotational quantum number *j* increases by 1; for the right-hand group *j* decreases by 1. These two groups correspond, respectively, to the left-hand and right-hand sets of transitions in Fig. 42-11. Each line is split because of the presence of the two chlorine isotopes ^{35}Cl and ^{37}Cl, whose different masses lead to different rotational inertias and therefore different rotational energies. The higher peaks are from the more abundant isotope ^{35}Cl.

APPLICATION	*A Star Wars Laser*

High-power lasers have been proposed as one component of the ballistic missile defense called "star wars." The technological requirements for a missile defense laser are staggering: the laser must deliver at least 30 MW and be able to focus its beam on a target thousands of kilometers distant while the target itself moves at several km/s.

Thirty megawatts is an awesome laser power, equivalent to some 60 billion of the typical helium-neon lasers used in introductory physics demonstrations. How can this much power be generated, especially in a space-based system with no connection to terrestrial energy sources? A possible answer is the chemical laser, whose operation relies on molecular excited states.

In the preceding chapter we saw that laser action requires a population inversion, with a majority of atoms or molecules in excited states. Stimulated emission then results in an intense beam of coherent photons. In a chemical laser, a chemical reaction produces the lasing molecules; the population inversion occurs because the molecules form in vibrational excited states. Among the reactions considered for star wars chemical lasers are those involving hydrogen fluoride (HF) and deuterium fluoride (DF).

Figure 42-13 shows an HF or DF laser. Two sets of nozzles deliver supersonic streams of fluorine and hydrogen (or deuterium). They react to produce HF or DF; in the HF reaction the majority of molecules are in the $n = 2$ excited state. The excited molecules enter a region between curved mirrors, where stimulated emission produces the intense laser beam.

Emission from chemical lasers, like other molecular vibrational transitions, lies in the infrared region of the spectrum. The HF laser's emission is in a band centered at about 2.8 μm, while the DF band is centered at 3.8 μm. Atmospheric transmission of infrared radiation varies drastically between these two wavelengths; in consequence the HF laser would have to be space-based, while the DF laser could operate from the ground.

The technological challenges of a 30-MW missile-defense laser are daunting. The device must be readied for firing within minutes after months of dormancy. A space-based HF laser affords little opportunity for testing and maintenance. The exhaust of spent HF gas acts as a rocket thrust, disturbing the aim

FIGURE 42-13 Schematic diagram of a hydrogen fluoride chemical laser. Fluorine and hydrogen emerge from two sets of nozzles to form HF molecules in vibrational excited states. The molecules de-excite through stimulated emission between two mirrors, producing an intense beam of coherent photons. The diagonal mirror extracts the beam.

of the weapon and also contaminating optical surfaces. The fundamental limit of optical diffraction requires huge mirrors for aiming the beam. These technical reasons, coupled with changing political realities, have led to greatly decreased emphasis on space-based laser weapons. However, the U.S. Air Force is continuing development of an aircraft-mounted chemical laser to be used against limited missile attacks. And more benign industrial uses exist too; chemical lasers designed for materials processing, for example, can cut rapidly through inch-thick steel.

42-3 SOLIDS

We have already seen that molecular bonding mechanisms can join relatively few atoms to form a molecule, or many to form a solid. In the lowest energy state, the atoms of a solid are arranged in a regular, repeating pattern; the solid is then **crystalline**. Sometimes solids form without their atoms having the opportunity to arrange themselves into a crystal structure; such solids are termed **amorphous**. Glass is a common amorphous solid. Amorphous materials are difficult to analyze due to their inherent randomness, so we concentrate here on crystalline solids.

(a) (b) (c)

FIGURE 42-14 (a) In the cesium chloride crystal, each cesium ion is surrounded by eight chlorine ions, located at the corners of a cube. The unit cell is a cube with one-eighth of a chlorine ion at each corner and a cesium ion at the center. (b) A cesium chloride crystal is a periodic array of unit cells. (c) Sodium chloride is different; in this crystal each ion is surrounded by only six nearest neighbors of the opposite type.

Crystal Structure

The hallmark of a crystalline solid is the regular, repetitive arrangement of its atoms. If we examine a crystal with very fine resolution, we find that a basic arrangement of a few atoms repeats throughout the crystal (Fig. 42-14). This basic arrangement is the **unit cell**. Different crystalline materials have different unit cells; the structure of the unit cell is determined by the individual atoms and the bonding mechanism that acts between them. Sometimes the same underlying matter may assume different structures, depending on how the solid was formed; this is the case with diamond and graphite, both crystalline forms of pure carbon.

As with individual molecules, properties like atomic separation in a crystalline solid are determined by the interplay of attractive and repulsive interactions. The situation is complicated, however, because an individual atom experiences forces from many other atoms in the surrounding crystal. With ionic bonding, those forces are electrical attraction and repulsion as described by Coulomb's law; that makes ionic crystals most amenable to simple mathematical treatment.

For ionic crystals, we can consider the individual ions as point charges, and the attraction between them as resulting from the electrostatic force. Consider, for example, the NaCl structure shown in Fig. 42-14c. Each sodium ion is surrounded by six nearest chlorine ions, each some distance r away. The potential energy of a singly ionized positive sodium ion in the potential of each negative chlorine ion is $-ke^2/r$. So the contribution to the potential energy of the six nearest chlorine ions is $-6ke^2/r$, with the minus sign indicating an attractive interaction. But then there are 12 sodium ions a distance $\sqrt{2}r$ from the sodium in question; they give rise to a repulsive force and consequently a positive potential energy of value $+12ke^2/\sqrt{2}r$. At a distance of $\sqrt{3}r$ there are eight more chlorines, giving a potential energy $-8ke^2/\sqrt{3}r$. The result is that the electrostatic potential energy of the sodium ion can be written

$$U_1 = -\alpha \frac{ke^2}{r},$$

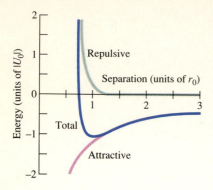

FIGURE 42-15 Potential-energy function for an ionic crystal, as given by Equation 42-4. Also shown are separate contributions of the attractive and repulsive terms. Equilibrium separation occurs where the total energy is a minimum.

where $\alpha = 6 - 12/\sqrt{2} + 8/\sqrt{3} - \ldots$, and is called the **Madelung constant**. Many terms in the series above are required to compute α accurately, showing that the effect of distant ions is significant in determining the energy of an ion in the crystal. For the NaCl structure, α is approximately 1.748.

As ions are brought closer together, they experience the repulsive effect of the exclusion principle, as we discussed in Section 40-1. This repulsion is described approximately by a potential energy of the form $U_2 = A/r^n$, where A and n are constants. So the total potential energy of an ion in the crystalline solid is

$$U = U_1 + U_2 = -\alpha\frac{ke^2}{r} + \frac{A}{r^n}.$$

At equilibrium the potential energy is a minimum (Fig. 42-15), corresponding to zero net force on the ion. Differentiating the potential energy with respect to r and setting dU/dr to zero to find the minimum, we have

$$0 = \frac{\alpha ke^2}{r_0^2} - \frac{nA}{r_0^{n+1}},$$

where r_0 designates the equilibrium separation. Solving for A gives $A = \alpha ke^2 r_0^{n-1}/n$, so the potential energy becomes

$$U = -\alpha\frac{ke^2}{r_0}\left[\frac{r_0}{r} - \frac{1}{n}\left(\frac{r_0}{r}\right)^n\right]. \tag{42-4}$$

The value of U at the equilibrium separation r_0 is designated U_0 and is called the **ionic cohesive energy**. The magnitude of U_0 represents the energy needed to remove an ion entirely from the crystal. The cohesive energy is sometimes given in kcal/mol, in which case its magnitude is the energy per mole needed to break an entire crystal into its constituent ions (see Problem 17).

EXAMPLE 42-3	*Potential-Energy Function for the NaCl Crystal*

For NaCl, the ionic cohesive energy is measured to be −7.84 eV. The equilibrium separation can be determined from density considerations (see Problem 16); its value is 0.282 nm. Use these values, and the Madelung constant $\alpha = 1.748$, to determine the exponent n in Equation 42-4 for NaCl.

Solution

Evaluating Equation 42-4 at $r = r_0$ gives

$$U_0 = -\alpha\frac{ke^2}{r_0}\left(1 - \frac{1}{n}\right).$$

Solving for n, we have

$$n = \left(1 + \frac{U_0 r_0}{\alpha ke^2}\right)^{-1}$$

$$= \left[1 + \frac{(-7.84\text{ eV})(1.6\times10^{-19}\text{ J/eV})(0.282\times10^{-9}\text{ m})}{(1.748)(9.0\times10^9\text{ N·m}^2/\text{C}^2)(1.6\times10^{-19}\text{ C})^2}\right]^{-1}$$

$$= 8.22.$$

The large value of this exponent shows how strongly resistant the NaCl crystal is to compression. In Problem 21 you can calculate the repulsive force involved in maintaining this "stiffness."

EXERCISE Find the potential energy of an ion in an NaCl crystal if the atomic separation throughout the crystal has been (a) compressed by 20% or (b) expanded by 20%.

Answers: (a) −4.36 eV; (b) −7.20 eV

Some problems similar to Example 42-3: 18–20, 43, 44

(a)   (b)   (c)

FIGURE 42-16 (a) Energy levels of the $1s$ and $2s$ states as a pair of atoms are brought close together. When the atoms are widely separated, both $1s$ levels and both $2s$ levels have the same energy. But the exclusion principle causes the levels to separate as the atoms approach. (b) With five atoms, each level splits into a group of five closely spaced levels. (c) In a crystalline solid, the large number of atoms results in essentially continuous energy bands, separated by gaps: r_0 marks the equilibrium separation, and E_g the band-gap energy.

Band Theory

Quantum-mechanical analysis of a solid containing 10^{23} atoms or so might seem a hopeless task. But the regularity of a crystalline solid makes that problem, while not an easy one, at least amenable to mathematical treatment. The physical regularity of the solid is reflected mathematically in the properties of the wave function; specifically, the wave function for a crystalline solid in equilibrium is itself periodic. This must be the case because equivalent points in two different unit cells have exactly the same physical properties.

Here, we will not attempt to solve the Schrödinger equation for a crystal, or even to write the solutions. But we can see qualitatively what some properties of those solutions must be. Consider two identical atoms, initially widely separated, as they are brought closer together. When the atoms are far apart, they are described by identical wave functions and associated energy-level diagrams; a given electron state, for example, has exactly the same energy in each atom. But as the atoms move closer, their wave functions begin to overlap to form a single wave function that characterizes the entire composite system. Because of the exclusion principle, two electrons that were in identical states in the two widely separated atoms can no longer be in the same state. This effect manifests itself as a separation of what were originally identical energy levels. Figure 42-16a shows schematically what happens to the energies of two states as a pair of atoms are brought close together. As more and more atoms come together, initially identical energy levels split into ever more finely spaced levels (Fig. 42-16b). In a crystalline solid, there are so many atoms that each energy level splits into an essentially continuous **band** of allowed energies (Fig. 42-16c). Often, the bands arising from distinct single-atom states are separated by **band gaps**, as shown in Fig. 42-16c. An electron can have any energy between the top and bottom of a band, but electron energies in the band gaps are forbidden. The situation is like that of a single atom, where electrons are allowed only certain discrete energies, except now the discrete levels have broadened into bands.

We're usually interested in the properties of a solid at or near its equilibrium state, designated r_0 in Fig. 42-16c. There the solid is characterized by an energy-level diagram in which the energy levels are those of Fig. 42-16c at the value $r = r_0$. Figure 42-17 shows such a diagram.

FIGURE 42-17 Energy-level diagram for the equilibrium separation of Fig. 42-16c, showing the band structure of the $1s$ and $2s$ states. The vertical axis still represents energy, but now the horizontal axis has no significance.

3p

3s

2p

2s

1s

FIGURE 42-18 Band structure of metallic sodium, with blue indicating filled states and tan unfilled states. The lowest bands correspond to the innermost electrons, whose wave functions do not overlap significantly and therefore do not result in much energy splitting. Splitting and shifting of the 3s and 3p levels, in contrast, is so great that the bands arising from these states overlap. Note that the 3s/3p band is only partially full.

Conductors, Insulators, and Semiconductors

Figures 42-16 and 42-17 are highly simplified; actually many more states contribute to the band structure of a solid. Furthermore, the splitting and shifting process depicted in Fig. 42-16 can result in overlapping bands. Figure 42-18 shows the band structure of sodium, in which the 3s and 3p bands overlap. Note that the high-energy band containing electrons—here the 3s/3p band—is not completely full, meaning that energy levels in the upper portion of the band are not occupied by electrons.

We can determine which of the allowed energy levels of a solid will be occupied in the same way we established the electronic structure of atoms: by placing a given total number of electrons in the lowest possible levels consistent with the exclusion principle. In some materials, like the sodium of Fig. 42-18, that filling process results in the highest-energy occupied band being only partially full. But in others, like the material shown in Fig. 42-19, the highest occupied band is completely full.

In the difference between Figs. 42-18 and 42-19 lies the difference between electrical conductors and insulators. We introduced electrical conductors in Chapter 23, and elaborated on conduction mechanisms in Chapter 27. Our discussion was necessarily sketchy and qualitative, because a full description of electrical conduction in solids requires quantum mechanics. We're now ready to understand, in quantum-mechanical terms, the distinctions among insulators, conductors, and semiconductors.

We defined a conductor as a material containing charges that are free to move in response to an electric field, thereby establishing an electric current. Classically, there's no problem with this: we apply an electric field, and if an electron is "free," it will be accelerated by the field and gain energy. But quantum-mechanically, an electron can gain energy only by jumping from one allowed energy level to another. When an electric field is applied to a solid, an electron's energy can increase only if there is a higher, unoccupied energy level for it to jump into.

Consider sodium, a conductor whose band structure is shown in Fig. 42-18. If there are N atoms in a sample of sodium, then each band arises from the splitting of a single atomic energy level into N distinct levels. Each band can therefore hold N times as many electrons as the corresponding atomic level. In sodium, all the 1s electrons (a total of $2N$) from the individual atoms are in the 1s band of the solid, all the 2s electrons (again, $2N$ of them) are in the 2s band, and all the 2p electrons (a total of $6N$) are in the 2p band. But the 3s shell of an individual sodium atom contains only one of its two possible electrons; in the solid, this means that the corresponding 3s band is only half full, as suggested in Fig. 42-18. Because energy levels within a band are very closely spaced, an electron near the top of the filled portion of the 3s band needs to gain only a very small amount of energy to move into one of the unoccupied states within the same band. Therefore, it is very easy for an electric field to promote electrons to higher energy states, and for this reason sodium is a good conductor.

FIGURE 42-19 Band structure for an insulator. Blue indicates occupied states, tan unoccupied states. The highest occupied band is completely full, so there are no nearby unoccupied states to which electrons can jump in response to an applied electric field.

Energy

In other materials, as shown in Fig. 42-19, the uppermost occupied band is completely full. An electron in this band cannot move to another state in the band because all such states are full. For an electron to gain energy at all, it must jump the band gap to the lowest unoccupied band. Since this jump requires a relatively large energy, an applied electric field of reasonable magnitude cannot increase electron energies, so no current flows and the material is an insulator.

Thus we see how quantum mechanics determines whether a material is a conductor or an insulator: if the highest-energy occupied band is only partially full it's a conductor, but if the highest-energy band is full, then there are no nearby states for electrons to move into, and the material is an insulator. Although being an insulator or conductor follows ultimately from a material's atomic structure, actually predicting electrical properties can be complicated by the occurrence of overlapping bands.

Quantum mechanics not only determines whether a material is a conductor or an insulator; it also answers some of the baffling questions raised in our semi-classical discussion of electrical conduction in Chapter 27. Next we'll look at what quantum mechanics has to say about conduction in metals.

Metallic Conductors

In Chapter 27 we found that classical physics could not account correctly for all aspects of metallic conduction. In particular, classical predictions of the temperature dependence of electrical conductivity proved quite inconsistent with actual measurements (see Fig. 27-7 and related text).

Since the conduction electrons in a metal are not bound to particular atoms, their behavior can be approximated by assuming that the metal contains a gas of free electrons. Quantum-mechanically, the conduction electrons are confined to the metal by a three-dimensional square well potential like the one we discussed in Chapter 40. By counting the quantum states available to electrons in such a potential, it can be shown that the number of states per unit energy interval increases with energy. (You can see this trend already in Fig. 40-29, showing the first few states in a three-dimensional box.) We will not carry out this counting procedure; the result, however, is given by

$$g(E) = \left(\frac{2^{7/2}\pi m^{3/2}}{h^3}\right) E^{1/2}, \tag{42-5}$$

where E is the energy, m the electron mass, and $g(E)$ the number of states per unit volume per unit energy interval at the energy E. The quantity $g(E)$ is called the **density of states**. The calculation leading to Equation 42-5 accounts for the two possible spin states at each energy level.

At absolute zero, electrons fill the lowest available energy states according to the exclusion principle, which permits only one electron in a given state. In terms of band theory, the result is the partially filled conduction band that we described in the preceding section. In terms of Equation 42-5, the number of electrons increases with energy up to the maximum energy needed to accommodate all the electrons. That maximum energy is called the **Fermi energy**. In a metal at absolute zero, the Fermi energy is the energy of the highest-filled state in the conduction band. All energy levels below the Fermi energy are fully occupied, and all those above the Fermi energy are empty, as shown in Fig. 42-20a.

FIGURE 42-20 Occupation of energy states by electrons in a metal. Curve is the density of states given by Equation 42-5, while shaded region indicates occupied energy levels. (*a*) At absolute zero, all states up to the Fermi energy E_F are completely full, and those above empty. (*b*) At higher temperatures, thermal excitation promotes a few electrons above the Fermi energy.

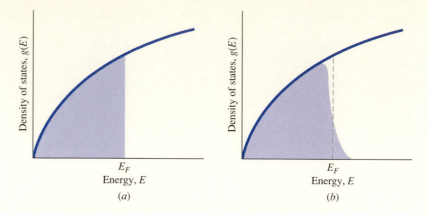

At temperatures above absolute zero, thermal excitation promotes some electrons to levels above the Fermi energy, leaving some levels just below that energy vacant (see Fig. 42-20*b*). But the Fermi energy in most metals is on the order of 1 to 10 eV, much higher than the thermal energy at typical temperatures. Therefore, the electron distribution is altered only slightly from its form at absolute zero. The result is that conduction in metals is provided almost entirely by electrons whose energies are near the Fermi energy, regardless of temperature. This situation causes the mean electron speed to be different from the classical thermal speed, and makes the temperature dependence of the electrical conductivity in a metal quite different from the classical prediction. Problem 26 compares the classical and quantum results for electron speed in a metal.

EXAMPLE 42-4	*Electron Density and the Fermi Energy*

Use Equation 42-5 to derive a relation between the Fermi energy E_F and the density n of conduction electrons in a metal. Then calculate the electron density in magnesium, whose Fermi energy is 7.13 eV.

Solution

Equation 42-5 gives $g(E)$, the number of states per unit volume per unit energy available to conduction electrons in a metal. Multiplying by some small energy interval dE then gives $g(E) \, dE$, the number of states per unit volume in the energy interval between E and $E + dE$. At absolute zero, all states up to the Fermi energy are occupied, and all those above the Fermi energy are vacant. Therefore, if we sum the number of states per unit volume in all the energy intervals from $E = 0$ to $E = E_F$, we will have the total number of occupied states per unit volume or, equivalently, the density of conduction electrons. Since $g(E)$ is a continuously varying function, that summation becomes an integration, and we have

$$n = \int_{E=0}^{E=E_F} g(E) \, dE = \int_0^{E_F} \left(\frac{2^{7/2} \pi m^{3/2}}{h^3} \right) E^{1/2} \, dE$$

$$= \left(\frac{2^{7/2} \pi m^{3/2}}{h^3} \right) \left(\frac{E^{3/2}}{3/2} \right) \Bigg|_0^{E_F} = \left(\frac{2^{9/2} \pi m^{3/2}}{3h^3} \right) E_F^{3/2}.$$

This is the desired relation between electron density and Fermi energy.

For magnesium, with $E_F = 7.13$ eV, we have

$$n = \left(\frac{2^{9/2} \pi m^{3/2}}{3h^3} \right) E_F^{3/2}$$

$$= \left(\frac{2^{9/2} \pi (9.11 \times 10^{-31} \text{ kg})^{3/2}}{(3)(6.63 \times 10^{-34} \text{ J} \cdot \text{s})^3} \right)$$
$$\times \, [(7.13 \text{ eV})(1.6 \times 10^{-19} \text{ J/eV})]^{3/2}$$

$$= 8.61 \times 10^{28} \text{ m}^{-3} .$$

This value is consistent with conduction electron densities we encountered in Chapter 27.

[If you try the above calculation on your calculator, you may get an error indication, showing that powers of 10 involved are outside the calculator's range. Try grouping the quantity $m^{3/2}/h^3$ as $(\sqrt{m}/h)^3$ to get some cancellation of the large negative exponents.]

EXERCISE Lithium has 4.6×10^{28} conduction electrons per cubic meter. Find the Fermi energy for lithium.

Answer: 4.71 eV

Some problems similar to Example 42-4: 22–24, 46

Semiconductors

In Chapter 27 we discussed semiconductors—the materials at the heart of our modern electronic world—and gave a simple classical explanation of their properties. Here we see how band theory gives a more correct quantum-mechanical explanation of semiconductors.

Our band diagram for an insulator (Fig. 42-19) is strictly correct only at absolute zero temperature. At this temperature, the highest occupied band—called the **valence band**—is completely full and the next band—the **conduction band**—is completely empty. At any other temperature, random thermal energy may give an occasional electron enough energy to leap the gap between the valence and conduction bands. Such electrons would then be available for acceleration by an electric field, and the material would have nonzero electrical conductivity. In good insulators, though, the band gaps are relatively large (several eV or more), so the number of electrons thermally excited to the conduction band is negligible. But in some materials, notably silicon and germanium, the band gap is relatively small (on the order of 1 eV; see Table 42-1). Even at room temperature, random thermal excitation in low-band-gap materials promotes a significant number of electrons to the conduction band. As a result these materials conduct electricity, although their conductivity is much lower than in metallic conductors. A material whose conductivity results from thermal excitation across a small band gap is a **semiconductor**. Figure 42-21 shows schematically the band-structure differences between conductors, insulators, and semiconductors. Note that a semiconductor really has the same band structure as an insulator and would, at absolute zero, be an insulator.

In Chapter 27 we described how the addition of small quantities of impurities—a process called doping—could radically alter the electrical conductivity of semiconductors. In terms of band theory, a dopant such as arsenic, with five valence electrons, adds so-called **donor levels** just a few tenths of an eV below the conduction band (Fig. 42-22a). Thermal energy readily promotes electrons from these levels to the conduction band, greatly increasing the conductivity of the material. In this case, as we defined it in Chapter 27, the doped material becomes an **N-type semiconductor**, N-type because its predominant charge carriers are *negative* electrons. If we add a dopant like aluminum with only three valence electrons, we create a **P-type semiconductor**, whose dominant charge carriers are *positive* **holes**, or absences of electrons in the crystal lattice. Figure 42-22b shows how the holes occur when electrons in the valence band are thermally excited into the nearby **acceptor levels**.

TABLE 42-1 *Band-Gap Energies for Selected Semiconductors (at 300 K)*

SEMI-CONDUCTOR	BAND-GAP ENERGY (eV)
Si	1.14
Ge	0.67
InAs	0.35
InP	1.35
GaP	2.26
GaAs	1.43
CdS	2.42
CdSe	1.74
ZnO	3.2
ZnS	3.6

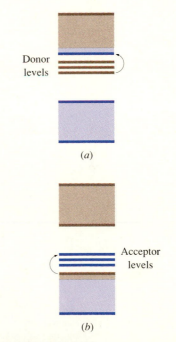

FIGURE 42-22 Band structure in doped semiconductors. In (a), the dopant introduces donor levels just below the conduction band. Arrow indicates that thermal energy can excite electrons from these levels to the conduction band, making an N-type semiconductor whose charge carriers are predominantly negative electrons. In (b), doping results in acceptor levels that lie just above the valence band, and are initially vacant. Thermal energy excites electrons to the acceptor levels, leaving positive holes that make the material a P-type semiconductor.

FIGURE 42-21 Band structures for a conductor, an insulator, and a semiconductor. Blue and tan show, respectively, occupied and unoccupied states. The semiconductor's small band gap allows thermal excitation of some electrons into the conduction band, giving the material a modest electrical conductivity.

(a)

(b)

Metallic contact
Insulating layer
P-type semiconductor
N-type semiconductor
Substrate (N-type)
Metallic contact

FIGURE 42-23 (a) Construction of a light-emitting diode. Photons are emitted at the *PN* junction, where electrons drop from the conduction band to the valence band. Much of the light never escapes the device, making it relatively inefficient. (b) Each segment of this display is a light-emitting diode.

The diodes, transistors, and integrated circuits we discussed in Chapter 27 are technological manifestations of our mastery of the physics of semiconductors. Theoretical understanding coupled with technological precision now enables us to fabricate materials with arbitrarily chosen band gaps. Light-emitting diodes (LEDs), commonly used as indicator lights and for digital displays, involve semiconductors whose band-gap energies correspond to visible-light photons. As electrons jump from the conduction to the valence band, the device emits visible light (Fig. 42-23). A related device is the semiconductor laser, in which a population inversion is established by pumping excess electrons into the conduction band. Stimulated emission then results in a coherent beam of light as electrons fall across the band gap. Semiconductor lasers are used in fiber-optic communications, and to extract information from compact discs.

Conversely, a material with a band gap corresponding to visible-light photons can absorb light energy, in the process promoting electrons to the conduction band and creating electron-hole pairs and driving an electric current. The result is the conversion of light energy into electricity. A semiconductor device designed for such conversion is called a **photovoltaic cell** (Fig. 42-24). Photovoltaic cells have long been used to generate electricity in remote applications such as automated weather stations, electric fences for livestock control, and spacecraft. But increasingly, they are also being used to power individual buildings and even to replace central power plants (Fig. 42-25). For years, solar cells have been too expensive to compete with more conventional means of large-scale electric power generation. But the cost of photovoltaics has dropped by roughly two orders of magnitude since the 1970s, and continues to fall by about

FIGURE 42-25 Photovoltaic cells are increasingly competitive with more traditional sources of electrical energy. These panels are part of a large-scale photovoltaic power plant.

Metallic contact
Antireflection coating
N-type semiconductor
P-type semiconductor
Metallic contact

FIGURE 42-24 Operation of a photovoltaic cell. Sunlight photons traverse the thin *N*-type layer to the *PN* junction, where they promote electrons to the conduction band and thus produce electron-hole pairs. Energy acquired from the solar photons drives current through the metallic contacts and on to an external load. Antireflection coating maximizes light transmission into the device.

10 percent each year. Already, the cost per watt of photovoltaic power is lower than that of the few nuclear plants that opened in the United States in the past decade. With vigorous ongoing research, and with production of photovoltaic cells increasing every year, photovoltaic solar energy may soon make significant contributions to our electrical energy supply.

EXAMPLE 42-5	*Engineering the Band Gap*

A digital videodisc (DVD) is read with 650-nm laser light. What should be the band gap in a diode laser operating at this wavelength?

Solution

We want 650-nm photons to be emitted as electrons drop across the band gap, so the gap energy should be that of a 650 nm photon. Equating photon energy to frequency and then wavelength, we have

$$E = hf = \frac{hc}{\lambda} = \frac{(6.63 \times 10^{-34} \text{ J·s})(3.00 \times 10^8 \text{ m/s})}{650 \times 10^{-9} \text{ m}}$$

$$= 3.06 \times 10^{-19} \text{ J} = 1.9 \text{ eV}.$$

Of the materials listed in Table 42-1, CdSe (cadmium selenide) comes closest to this band-gap energy. By carefully adjusting the composition of the semiconductor material, the desired gap energy can be achieved more precisely.

EXERCISE What is the maximum wavelength of light that can produce electron-hole pairs in gallium phosphide (GaP)?

Answer: 550 nm

Some problems similar to Example 42-5: 28–31

42-4 SUPERCONDUCTIVITY

In Chapter 27 we introduced **superconductivity**—the complete loss of electrical resistance in certain materials at low temperature. First discovered in mercury by H. Kamerlingh Onnes in 1911, superconductivity was for decades limited to a few individual elements and alloys at temperatures below about 20 K. Then, in 1986, a series of rapid breakthroughs led to materials that are superconducting at temperatures on the order of 100 K (Fig. 42-26). The first of these breakthroughs—surmounting the "30 K barrier"—brought the 1987 Nobel Prize in physics to J. Georg Bednorz and K. Alex Müller of IBM's Zurich Research Laboratory (Fig. 42-27). Physicists achieved superconducting temperatures of 125 K in the late 1980s, and by the late 1990s the highest superconducting temperatures had reached 164 K. Research on superconductivity is one of

FIGURE 42-26 Resistivity versus temperature for a thin film of $YBa_2Cu_3O_x$ superconductor. The superconducting transition occurs at 93 K, well above the 77 K boiling point of liquid nitrogen.

FIGURE 42-27 1987 Nobel laureates Bednorz and Müller discovered a new class of high-temperature superconductors.

TABLE 42-2 *Superconducting Transition Temperatures*

MATERIAL	T_c, K
Al	1.20
Sn	3.72
Pb	7.19
Nb	9.26
La_3In	10.4
NbN	16.0
V_3Si	17.1
Nb_3Sn	18.1
$Rb-C_{60}$	28
La-Sr-Cu-O	40
Y-Ba-Cu-O	93
Ti-Ba-Ca-Cu-O	125

the most exciting areas of contemporary physics. A host of new superconducting materials have been found—including even doped versions of the remarkable compound buckminsterfullerine described earlier in this chapter. The ultimate goal of a room-temperature superconductor, once thought beyond reach, may one day be achieved.

Table 42-2 shows **transition temperatures**, T_c, for selected superconductors. In general, materials that are relatively poor conventional conductors make the best superconductors; note that aluminum, the best conventional conductor in Table 42-2, has the lowest T_c. Pure elements tend to have lower transition temperatures than alloys and other compounds. Niobium has the highest elemental T_c, and niobium compounds are the most widely used of the more established low-temperature superconductors. The newer high-T_c superconductors are complex structures whose essential ingredient is a metal-oxide compound; they come in many forms with a range of T_c's (Fig. 42-28).

To date, cooling requirements of conventional superconductors have limited their use to specialized applications including large, high-field electromagnets and very sensitive magnetic-field measuring devices (Fig. 42-29). But the discovery of high-temperature superconductors, which can be cooled with inexpensive liquid nitrogen ($T = 77$ K, and a price per liter less than that of orange juice), may propel superconducting technology into everyday use. Potential uses for superconductivity include electromagnets for energy storage, for magnetic resonance imaging in medicine, for fusion and high-energy physics research, and for separation of materials for purification and recycling; electric power

FIGURE 42-28 Structure of yttrium-barium-copper-oxide (Y-Ba-Cu-O) superconductor, whose transition temperature is 93 K. Conduction occurs along planes defined by the green pyramids, which represent copper oxide.

(a)

(b)

FIGURE 42-29 Applications for conventional superconductors include (a) superconducting cables and (b) electromagnets, such as this one being constructed for use in a particle accelerator.

generation and transmission; electric motors and magnetically levitated mass transportation vehicles; superconducting computers; magnetometers for brainwave imaging; and even mechanical devices such as low-friction bearings. Figure 42-30 shows an application of high-temperature superconductors.

To make use of a superconductor's ability to carry electric current without loss, it is advantageous to form the material into flexible wires, tapes, or thin films. High-temperature superconductors are ceramics that are brittle and are difficult to form into elongated structures. Considerable effort by materials scientists and others has resulted in wires and ribbon-shaped conductors of high-temperature superconductor (Fig. 42-31). High-temperature superconducting films have also been deposited successfully on silicon, paving the way for semiconductor chips whose interconnections will be superconducting. Already high-T_c thin film electronic components for microwave systems are commercially available.

FIGURE 42-30 This SQUID (superconducting quantum interference device) uses high-temperature superconductors to measure very small magnetic fields.

Superconductivity and Magnetism

The hallmark of a superconductor is its zero electrical resistance; currents established in superconductors—called **supercurrents**—have persisted for years without diminishing. Another property is the so-called **Meissner effect**, wherein a superconductor excludes magnetic flux from its interior. Faraday's law suggests that a *changing* magnetic flux should not be possible inside a superconductor, for such a change would result in a finite electric field and therefore an infinite current in the zero-resistance superconductor. But the Meissner effect goes beyond this, disallowing even static magnetic fields from the interior of the superconductor. Figure 42-32*a* shows a piece of potentially superconducting material immersed in a uniform magnetic field. The material is not yet superconducting, and magnetic field lines pass through it. But as it cools below its transition temperature, the material becomes superconducting and the magnetic flux is excluded, as shown in Fig. 42-32*b*. Figure 42-32*c* shows that the overall field can be considered the sum of the original applied field and a dipole-like field generated by supercurrents induced within the superconductor.

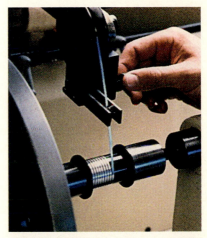

FIGURE 42-31 Coil being wound from a ribbon of high-temperature superconductor at the American Superconductor Corporation.

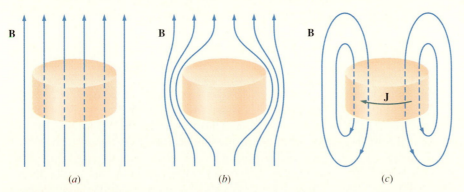

FIGURE 42-32 The Meissner effect. (*a*) Above the superconducting transition temperature, magnetic field lines penetrate the material. (*b*) In the superconducting state, magnetic flux is excluded from the material. (*c*) The overall field in (*b*) is the sum of the original uniform field and the field associated with persistent supercurrents **J** (arrow) in the superconductor.

FIGURE 42-33 A small cubical magnet floats above a wafer of high-temperature superconductor. The superconductor is cooled in a bath of liquid nitrogen at 77 K.

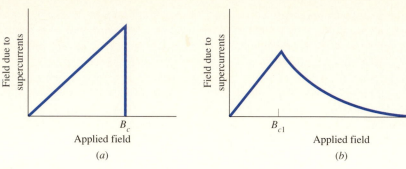

FIGURE 42-34 Responses of (a) type I and (b) type II superconductors to applied magnetic fields. Horizontal axis is the applied field, while vertical axis is the magnitude of the fields generated by the induced supercurrents. For low fields, the induced field is equal in magnitude to the applied field but opposite in direction; the two cancel to give no net field inside the superconductor. Superconductivity ceases abruptly in a type I superconductor at the critical field B_c, while a type II superconductor gradually allows magnetic flux to penetrate while remaining superconducting above the lower critical field B_{c1}. Only at the upper critical field B_{c2} does superconductivity cease altogether.

The Meissner effect is dramatically illustrated by placing a magnet on a piece of potentially superconducting material. When the material is cooled below its transition temperature, the magnet levitates and floats above the superconductor (Fig. 42-33). This levitation is a result of the repulsive interaction between the magnetic moments of the permanent magnet and the superconductor, the latter resulting from persistent currents that flow in the superconductor to cancel the flux from the permanent magnet. As we indicated in Section 31-5, the complete exclusion of magnetic flux in a superconductor makes it a diamagnetic material with magnetic susceptibility -1.

As the strength of an externally applied magnetic field increases, so do the supercurrents and resulting magnetic field of the superconductor. But beyond a certain field strength, called the **critical field**, the external magnetic field begins to alter the superconducting state, and the superconductor no longer excludes magnetic flux. This alteration can happen in two ways, as shown in Fig. 42-34; a given superconducting material is classified as **type I** or **type II superconductor** depending on which response it exhibits. In a type I superconductor, magnetic fields beyond the critical field abruptly destroy the superconducting state (Fig. 42-34a), and the material becomes a normal conductor. A type II superconductor, in contrast, has a lower and an upper critical field, as shown in Fig. 42-34b. At the lower critical field, the material begins to allow flux penetration but maintains its zero-resistance state until the much higher upper critical field is reached. In between the lower and upper critical fields in a type II superconductor, the material develops distinct regions of normal conductivity, centered on magnetic field lines and forming a regular array (Fig. 42-35). As the field increases, the fraction of the material in the superconducting state diminishes, until at the upper critical field superconductivity stops altogether. Type I superconductors tend to be pure elements like lead or mercury, while type II superconductors are generally alloys or more complicated mixtures. High-temperature superconductors are all of type II.

FIGURE 42-35 Structure of a type II superconductor between the upper and lower critical fields shows a regular array of concentrated magnetic flux regions (red). Each region is a zone of normal (nonsuperconducting) material. The flux regions are about 1 μm apart.

FIGURE 42-36 Critical current density J_c versus applied magnetic field in thin film, single crystal, and polycrystalline forms of $YBa_2Cu_3O_7$ superconductor at 77 K. Those data show that thin films, especially, can carry large currents before the associated magnetic field destroys their superconductivity.

The destruction of superconductivity by magnetic fields has enormous practical significance, for a major application of superconductivity is to carry large electric currents without energy loss, especially in electromagnets. But electric current is the source of magnetic field (Chapter 30), so there is a maximum current in a given superconductor at which the magnetic field of that current destroys the superconducting state. For type I superconductors, the critical field is so low as to make these materials useless in high-current applications such as electromagnets. But the upper critical fields of type II superconductors are generally much higher, so most practical applications of superconductors use type II materials. Critical fields of the high-temperature superconductors are as high as 100 T, and current densities greater than 10^6 A/cm^2 have been achieved in thin-film superconducting samples at 77 K (Fig. 42-36).

APPLICATION	*Maglev Transportation*

Maglev (short for magnetically levitated) vehicles using superconducting magnets offer the promise of high-speed transportation in densely populated urban corridors. Maglev systems are currently under development in several countries, especially Germany and Japan. Japanese maglev prototypes (Fig. 42-37a) have already carried passengers at speeds in excess of 400 km/h on test tracks up to 43 km in length. Several commercial lines are on the drawing boards, including a maglev run from Hiroshima's city center to the airport.

Engineers worldwide are pursuing a number of different maglev configurations. In the Japanese system, on-board superconducting magnets are the heart of the maglev system, and

(a)

(b)

FIGURE 42-37 (a) The Japanese MLU002 prototype uses superconducting magnets for levitation, propulsion, and guidance. (b) Principle of magnetic levitation in the Japanese system. The on-board superconducting magnet induces currents in the figure-8-shaped coils in the guideway walls. The direction of the induced currents is such that the upper part of the coil exerts an attractive force with an upward component, while the downward part exerts a repulsive force, also with an upward component. Coils on opposite sides of the guideway are electrically connected in such a way that the horizontal forces act to keep the vehicle centered in the guideway.

illustrate not only superconductivity but many other principles of physics. The vehicle "flies" in a guideway with wire coils embedded in its vertical sides. Moving with the vehicle, the on-board superconducting magnets induce currents in these guideway coils, making them momentarily into electromagnets. Figure 42-37b shows how forces between the on-board magnets and the guideway coils levitate the vehicle, which rides just centimeters above the bottom of the guideway.

Magnetic forces not only lift the maglev vehicle, but also propel it forward. A separate set of coils in the guideway is supplied with alternating current, alternately pulling and pushing the vehicle's on-board magnet. In effect, the system becomes a linear electric motor, much like a conventional motor that has been "unwound" to produce straight-line motion.

How does the maglev vehicle, traveling at hundreds of kilometers per hour, stay safely in its guideway? That, too, is accomplished with magnetic forces. Levitation coils on opposite sides of the guideway are electrically connected, forming a single current loop. When the vehicle is centered in the guideway, there's no net lateral force from the magnetic field associated with current in this loop. But if the vehicle strays from dead center, the induced current results in magnetic forces that drive the vehicle back toward the guideway center.

The maglev vehicles now under development generally use superconducting magnets to minimize power consumption, but as a result they require on-board refrigeration systems. Development of a room-temperature superconductor would make maglev a much more attractive transportation alternative.

Theories of Superconductivity

Superconductivity is a purely quantum-mechanical phenomenon; classical physics is totally inadequate to explain its existence. A successful theory of conventional low-T_c superconductors, called the **BCS theory** after its originators, was formulated in 1957 by John Bardeen, Leon Cooper, and John Robert Schrieffer; the trio shared the 1972 Nobel Prize in Physics.

In the BCS theory, superconductivity results from a quantum-mechanical pairing of electrons that leads to a lower-energy state in which electron pairs move through the crystal lattice with no energy loss to the ions. This absence of energy loss is what gives the superconductor its zero electrical resistance. (Contrast with the collisional origin of electrical resistance discussed in Section 27-2.) The electron pairing itself involves an interaction with the ion lattice; one electron of the pair slightly deforms the lattice, and the second electron is attracted by the slight positive charge of the deformed lattice site, resulting in a new minimum-energy state (Fig. 42-38). But the paired electrons are not physically close; typically, a million other electrons, each paired with another distant electron, may lie between two electrons constituting a pair (Fig. 42-38c). The result of this long-range pairing is a coherence in the motion of the conduction electrons that extends throughout the superconductor. Like dancers in a well-choreographed dance, the electrons all move together in a way that precludes energy loss to the ion lattice.

High-temperature superconductors are not fully understood, although they almost certainly involve quantum-mechanical pairing of charge carriers—probably holes rather than electrons. The mechanism of the pairing is less clear; one promising candidate involves magnetic interactions, although other mechanisms are under investigation. Superconductivity should provide a continuing challenge to theorists and experimentalists for some time to come.

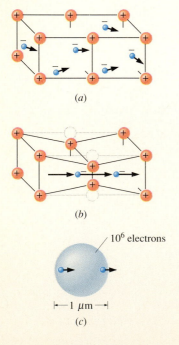

(a)

(b)

10^6 electrons

|←— 1 μm —→|

(c)

FIGURE 42-38 Electron pairing in the BCS theory. (a) A normal conductor, with uncorrelated electrons. (b) In a superconductor, one electron passes through the lattice, deforming it slightly. About 10^{-12} s later a second electron passes and experiences the potential of the distorted lattice. The motions of the two electrons are therefore correlated. (c) Part (b) is not to scale; paired electrons are typically 1 μm apart, and there are about a million other electrons in their vicinity. The long-range coherent motion of all these electron pairs gives rise to superconductivity.

CHAPTER SYNOPSIS

Summary

1. The bonding of atoms into molecules involves a balance between attractive and repulsive forces, producing stable structures when the potential energy is a minimum.
   a. **Ionic bonding** occurs between atoms near opposite ends of the periodic table, when one atom gives up an electron to the other. The resulting ions are strongly attracted.
   b. In **covalent bonding**, two or more atoms share their outermost electrons in common orbitals. Molecules such as O_2, N_2, and H_2O are covalently bonded.
   c. **Hydrogen bonds** are weak bonds involving the protons of hydrogen-containing molecules.
   d. **Van der Waals bonding** is very weak and results from electric forces between molecular dipoles.
   e. **Metallic bonding** results from sharing of a free electron gas among all the ions of the crystal lattice.
2. The energy-level structure of molecules is very rich. In addition to atomic-like excitation of electrons to different energy levels, internal motion also contributes to the quantization of molecular energy:
   a. The **rotational energy** of molecules is quantized, with the allowed levels given by

   $$E_{rot} = \frac{\hbar^2}{2I} j(j + 1),$$

   where I is the molecule's rotational inertia and the integer j the rotational quantum number. Spacings between rotational energy levels are measured in milli-electron-volts.
   b. The **vibrational energy** is also quantized, with quantized energy levels being approximately those of the quantum harmonic oscillator:

   $$E_{vib} = (n + \tfrac{1}{2})\hbar\omega.$$

   Ground-state vibrational energies are on the order of 0.1 eV.
   c. Selection rules limit rotational and vibrational transitions to those for which $\Delta j = \pm 1$ and $\Delta n = \pm 1$.
3. **Solids** may be held together by any of the molecular bonding mechanisms. The minimum-energy configuration for a solid consists of a regular array of atoms that compromise a **crystal**. The energy needed to disassemble a crystalline solid is its **cohesive energy**.
   a. In solids, individual atomic electron energy levels spread into **energy bands**, separated by **band gaps**.
   b. Electrical **conductors** have their highest bands partially filled with electrons, making it easy for electric fields to promote electrons to higher energy and thus allowing current to flow. The **Fermi energy** is the energy of the highest occupied level in a metallic conductor.
   c. **Insulators** have their highest band completely full, so excitation of electrons is impossible unless they gain enough energy to jump the substantial band gap.

   d. **Semiconductors** are like insulators, but with smaller band gaps. Random thermal energy promotes some electrons across this gap to the **conduction band**, giving these materials modest conductivity. Doping with impurities introduces additional levels that result in high concentrations of electrons (*N*-type semiconductor) or holes (*P*-type semiconductor).
4. The sudden loss of all electrical resistance below a **critical temperature** marks the onset of **superconductivity**. All superconductors exhibit the **Meissner effect**, wherein magnetic flux is excluded from the interior of the material. In a **type I superconductor** the exclusion is complete up to a **critical field** at which superconductivity ceases. In a **type II superconductor**, flux exclusion is complete up to a lower critical field, above which flux penetration gradually occurs until superconductivity ceases at an upper critical field. The **BCS theory** invokes the correlated motion of electron pairs to explain low-temperature superconductors.

Terms You Should Understand

(Pairs are closely related terms whose distinction is important; number in parentheses is chapter section where term first appears.)

ionic, covalent, hydrogen, van der Waals, and metallic bonding (42-1)
unit cell (42-3)
band, band gap (42-3)
conductor, insulator, semiconductor (42-3)
conduction band, valence band (42-3)
Fermi energy (42-3)
superconductor, type I and type II (42-4)
transition temperature (42-4)
Meissner effect (42-4)
critical field (42-4)
BCS theory (42-4)

Symbols You Should Recognize

α (42-3) r_0 (42-3) E_F (42-3) T_c (42-4)

Problems You Should Be Able to Solve

evaluating molecular rotational and vibrational energy levels and associated spectra (42-2)
determining molecular bond lengths from rotational spectra (42-2)
evaluating ionic energies and related quantities (42-3)
relating Fermi energy and electron density (42-3)
relating band-gap energies and light emission or absorption (42-3)

Limitations to Keep in Mind

Models developed in this chapter to describe the properties of molecules and solids are only approximations; in particular, most real molecules do not act like simple harmonic oscillators for vibrational excitations with $n > 1$.

QUESTIONS

1. Why is the exclusion principle crucial to the existence of stable molecules?
2. Why do ionically bonded materials have high melting points?
3. Ionic bonds clearly result from electrostatic attraction between ions. In what way do covalent bonds also involve electrostatic attraction?
4. Does it make sense to distinguish individual NaCl molecules in a salt crystal? What about individual H_2O molecules in an ice crystal? Explain.
5. Why is the hydrogen bond unique to hydrogen?
6. Is it useful to think of the highest-energy electrons "belonging" to individual atoms in an ionically bonded molecule? In a covalently bonded molecule?
7. What are the approximate relative magnitudes of the energies associated with electronic excitation of a molecule, with molecular vibration, and with molecular rotation?
8. You meet a scientist who uses a microwave technology to study molecules. What form of molecular energy is she most concerned with?
9. Radio astronomers have discovered many complex organic molecules in interstellar space. Why were these discoveries made with radio telescopes as opposed to optical telescopes?
10. In Fig. 21-23, why are the rotational states excited at lower temperatures than the vibrational states?
11. How can diamond be an insulator while graphite, also pure carbon, is a conductor of electricity? Suggest an explanation based on band theory.
12. Would you expect solid hydrogen to conduct electricity? Why or why not?
13. The Fermi energy in metals is much greater than the thermal energy at typical temperatures. Why does this make the mean speed of conduction electrons nearly independent of temperature?
14. Why does the size of the band gap determine whether a material is an insulator or a semiconductor?
15. How would you expect the conductivity of an undoped semiconductor to depend on temperature? Why?
16. Compare the relative contributions made by electrons and by holes to the electrical conductivity of undoped semiconductors, P-type semiconductors, and N-type semiconductors.
17. Name some technological innovations that might result from a room-temperature superconductor.
18. Suppose a room-temperature superconductor were discovered, but that it had a very low critical field. In what way would this limit its practical applicability?
19. Why can a magnet float motionless above a superconductor but not above an ordinary conductor? After all, bringing the magnet near either conductor induces eddy currents whose fields repel the magnet.
20. How do type I and type II superconductors differ?
21. What is the economic advantage of high temperature superconductors? After all, they still require cooling far below room temperature.
22. What role does the ion lattice play in the electron pairing described by the BCS theory?

PROBLEMS

Section 42-2 Molecular Energy Levels

1. Find the energies of the first four rotational states of the HCl molecule described in Example 42-1.
2. Find the wavelength of electromagnetic radiation needed to excite oxygen molecules (O_2) to their first rotational excited state. The rotational inertia of an oxygen molecule is 1.95×10^{-46} kg·m^2.
3. A molecule drops from the $j = 2$ to the $j = 1$ rotational level, emitting a 2.50-meV photon. If the molecule then drops to the rotational ground state, what energy photon will be emitted?
4. Calculate the wavelength of a photon emitted in the $j = 5$ to $j = 4$ transition of a molecule whose rotational inertia is 1.75×10^{-47} kg·m^2.
5. Photons of wavelength 1.68 cm excite transitions from the rotational ground state to the first rotational excited state in a gas. What is the rotational inertia of the gas molecules?
6. A molecule absorbs a photon and jumps to the next higher rotational state. If the photon energy is three times what would be needed for a transition from the rotational ground state to the first rotational excited state, between what two levels is the transition?
7. Find an expression for the energy of a photon required for a transition from the $(j - 1)$th level to the jth level in a molecule with rotational inertia I.
8. A molecule of rotational inertia I undergoes a transition from the jth rotational level to the $(j - 1)$th level. Show that the wavelength of the emitted photon is $\lambda = 4\pi^2 Ic/hj$.
9. The rotational spectrum of diatomic oxygen (O_2) shows spectral lines spaced 0.356 meV apart in energy. Find the atomic separation in this molecule. *Hint:* See Example 42-1, but remember that the oxygen atoms have equal mass.
10. Use the result given in Problem 52 to find the bond length in carbon monoxide (CO), using the fact that excitation of the first rotational state requires photons of wavelength 2.59 mm.
11. For the HCl molecule of Example 42-2, determine (a) the energy of the vibrational ground state and (b) the energy of a photon emitted in a transition between adjacent vibrational levels. Assume the rotational quantum number does not change.
12. The classical vibration frequency for diatomic hydrogen (H_2) is 1.32×10^{14} Hz. What is the spacing between its vibrational energy levels?

13. The energy between adjacent vibrational levels in diatomic nitrogen is 0.293 eV. What is the classical vibration frequency of this molecule?

14. Diatomic deuterium (D_2) has classical vibration frequency 9.35×10^{13} Hz and rotational inertia 9.17×10^{-48} kg·m². Find (a) the energy and (b) the wavelength of a photon emitted in a transition between the $n = 1, j = 1$ state and the $n = 0, j = 2$ state.

15. An oxygen molecule is in its vibrational and rotational ground states. It absorbs a photon of energy 0.19653 eV and jumps to the $n = 1, j = 1$ state. It then drops to the $n = 0, j = 2$ level, emitting a 0.19546-eV photon. Find (a) the classical vibration frequency and (b) the rotational inertia of the molecule.

Section 42-3 Solids

16. Use the 2.16 g/cm³ density of NaCl to calculate the ionic spacing r_0 in the NaCl crystal. *Hint:* Consult Appendix D for atomic weights.

17. Express the 7.84-eV ionic cohesive energy of NaCl in kilocalories per mole of ions.

18. Lithium fluoride, LiF, has the same crystal structure as NaCl and therefore essentially the same Madelung constant α. Its ionic cohesive energy is -10.5 eV, and the value of n in Equation 42-4 is 6.25. What is the equilibrium ionic separation in LiF?

19. Determine the constant n in Equation 42-4 for potassium chloride (KCl), for which $r_0 = 0.315$ nm and $U_0 = -7.21$ eV. The crystal structure is the same as for NaCl.

20. A salt crystal contains 10^{21} sodium-chlorine pairs. How much energy would it take to compress the crystal to 90% of its normal size?

21. (a) Differentiate Equation 42-4 to obtain an expression for the force on an ion in an ionic crystal. (b) Use your result to find the force on an ion in NaCl if the crystal could be compressed to half its equilibrium spacing (see Example 42-3 for relevant parameters). Compare with the electrostatic attraction between the ions at a separation of $\frac{1}{2} r_0$. Your result shows how very "stiff" this ionic crystal is.

22. The Fermi energy in aluminum is 11.6 eV. What is the density of conduction electrons in aluminum?

23. Determine the Fermi energy for calcium, which has 4.6×10^{28} conduction electrons per cubic meter.

24. Metal A has twice the Fermi energy of metal B. How do their conduction electron densities compare?

25. Suppose the charge carriers in a material were protons, with density 10^{28} m^{-3}—comparable to that of electrons in a metal. What would be the order of magnitude of the Fermi energy?

26. The Fermi energy for copper is 7.0 eV. Compare the electron speed associated with this energy with the classical thermal speed for an electron at room temperature (300 K). Your result shows how the quantum and classical descriptions of metallic conduction differ.

27. The *Fermi temperature* is defined by equating the thermal energy kT to the Fermi energy, where k is Boltzmann's constant. Calculate the Fermi temperature for silver ($E_F = 5.48$ eV), and compare with room temperature.

28. What is the wavelength of light emitted by a gallium phosphide (GaP) light-emitting diode? *Hint:* See Table 42-1.

29. What is the shortest wavelength of light that could be produced by electrons jumping the band gap in a material from Table 42-1? What is the material?

30. Which material in Table 42-1 would provide the longest wavelength of light in a light-emitting diode? What is the wavelength?

31. A common light-emitting diode is made from a combination of gallium, arsenic, and phosphorous (GaAsP) and emits red light at 650 nm. What is its band gap?

32. Photons with energy less than a semiconductor's band gap are not readily absorbed by the material, so measurement of light absorption versus wavelength gives the band gap. An absorption spectrum for silicon shows no absorption for wavelengths greater than 1090 nm. Use this information to calculate the band gap in silicon, and verify its entry in Table 42-1.

33. The Sun radiates most strongly at about 500 nm, the peak of its Planck curve. The semiconductor zinc sulfide has a band gap of 3.6 eV. (a) What is the maximum wavelength absorbed by ZnS? (b) Would ZnS make a good photovoltaic cell? Why or why not?

34. Light-emitting diodes are available commercially with output wavelengths in the range from about 550 nm to 1300 nm. What is the corresponding range of band-gap energies?

35. A blue-green semiconductor laser being developed for long-playing compact discs emits at 447 nm (see Application: From CD to DVD, in Chapter 37). What is the band gap in this laser?

Section 42-4 Superconductivity

36. Pure aluminum, which superconducts below 1.20 K, exhibits a critical field of 9.57 mT. What is the maximum current that can be carried in a 30-gauge (0.255 mm diameter) aluminum superconducting wire without the field from that current exceeding the critical field? *Hint:* Where is the field greatest? Consult Example 30-4.

37. The critical magnetic field in niobium-titanium superconductor is 15 T. What current is required in a 5000-turn solenoid 75 cm long to produce a field of this strength?

38. A danger in superconducting magnet systems is "quenching"—a sudden loss of superconductivity associated with rapid resistive heating and loss of coolant. If the solenoid of the preceding example has a diameter of 5.0 cm, (a) how much magnetic energy will be released if it quenches? (b) What volume of liquid helium could be vaporized? The heat of vaporization is 2.6 kJ per liter of liquid helium.

Paired Problems

(Both problems in a pair involve the same principles and techniques. If you can get the first problem, you should be able to solve the second one.)

39. The atomic spacing in diatomic hydrogen (H_2) is 74 pm. Find the energy of a photon emitted in a transition from the first rotational excited state to the ground state.

40. Find the wavelength of electromagnetic radiation needed to excite a transition from the rotational ground state to the first rotational excited state in N_2, whose atomic spacing is 109 pm.

41. What wavelength of infrared radiation is needed to excite a transition between the $n = 0, j = 3$ state and the $n = 1$, $j = 2$ state in KCl, for which the rotational inertia is 2.43×10^{-45} kg·m^2 and the classical vibration frequency is 8.40×10^{12} Hz?

42. Find the wavelengths emitted in all possible allowed transitions between the first three rotational states in the $n = 1$ level to any states in the $n = 0$ level in diatomic hydrogen. The rotational inertia of the hydrogen molecule is 4.60×10^{-48} kg·m^2, and its classical vibration frequency is 3.69×10^{14} Hz.

43. Lithium chloride, LiCl, has the same structure and therefore the same Madelung constant as NaCl. The equilibrium separation in LiCl is 0.257 nm, and $n = 7$ in Equation 42-4. Find the ionic cohesive energy of the LiCl crystal.

44. Find the energy per unit cell needed to decrease the ionic separation in NaCl by 15%.

Supplementary Problems

45. What would be the Fermi energy in a one-dimensional infinite square well 10 nm wide and holding 100 electrons? Assume two electrons (with opposite spins) per energy level.

46. Find the fraction of conduction electrons in a metal at absolute zero that have energies less than half the Fermi energy.

47. Figure 21-23 shows that diatomic hydrogen acts like a monotomic gas at temperatures below about 100 K. Use the fact that the rotational inertia of H_2 is 4.6×10^{-48} kg·m^2 to show that this low-temperature behavior makes sense. *Hint:* Compare the thermal energy kT with the minimum energy needed to excite molecular rotation.

48. Repeat the preceding problem, now considering vibrational energy levels. The classical vibration frequency for H_2 is about 4.4×10^{13} Hz, and Fig. 21-23 shows that vibrational effects on the specific heat become important at about 2000 K.

49. The transition from the ground state to the first rotational excited state in diatomic oxygen (O_2) requires about 356 μeV. At what temperature would the thermal energy kT be sufficient to set diatomic oxygen into rotation? Would you ever find diatomic oxygen exhibiting the specific heat of a monatomic gas at normal pressure?

50. The density of rubidium iodide (RbI) is 3.55 g/cm^3, and its ionic cohesive energy is -145 kcal/mol. Determine (a) the equilibrium separation and (b) the exponent n in Equation 42-4 for RbI.

51. The HCl bond-length calculation of Example 42-1 assumed the molecule was not vibrating. But because the ground-state vibrational energy is not zero, the bond is actually stretched slightly by its vibration. To estimate this stretching (a) use the classical vibration frequency of 8.66×10^{13} Hz to find the ground-state vibrational energy. (b) Use the result of part (a) and the fact that the effective "spring constant" for HCl is about 480 N/m to estimate the stretching of the molecular bond due to ground-state vibration. Assume that half the energy is potential energy of the stretched "spring." (c) Compare your result with the bond length found in Example 42-1.

52. Consider a diatomic molecule consisting of different atoms with masses m_1 and m_2, separated by a distance R. (a) Find the molecule's center of mass. (b) Use your result to show that the rotational inertia about the center of mass is

$$I = \left(\frac{m_1 m_2}{m_1 + m_2} \right) R^2 .$$

The quantity $m_1 m_2 / (m_1 + m_2)$ is called the reduced mass. (c) Use this result to compute a more accurate value for the HCl bond length of Example 42-1.

53. The Madelung constant (Section 42-3) is notoriously difficult to calculate because it is the sum of an alternating series of nearly equal terms. But it can be calculated for a hypothetical one-dimensional crystal consisting of a line of alternating positive and negative ions, evenly spaced (Fig. 42-39). Show that the potential energy of an ion in this "crystal" can be written

$$U = -\alpha \frac{ke^2}{r_0} ,$$

where the Madelung constant α has the value $2 \ln 2$. *Hint:* Study the series expansions listed in Appendix A.

FIGURE 42-39 Problem 53.

Nuclear Physics

Gamma rays are high-energy photons produced in quantum transitions within the atomic nucleus. This image, taken with gamma rays, reveals the presence of cancer in this person's vertebrae.

The last two chapters showed how modern physics describes the structure of atoms and how atoms join to form molecules and solids. Now we turn inward, to the atomic nucleus. This chapter deals with nuclear structure and the phenomenon of radioactivity; in the next chapter we consider nuclear energy and its applications. In the final chapter we go even deeper into the structure of matter, probing the innards of the neutrons and protons that make up atomic nuclei.

43-1 DISCOVERY OF THE NUCLEUS

When J. J. Thomson discovered the electron in 1897, it was not known how electrons and the equivalent positive charge were distributed in atoms. Thomson himself favored the **plum-pudding model**, with electrons embedded throughout a "pudding" of positive charge (Fig. 43-1a). In a 1909–1911 series of experiments suggested by Ernest Rutherford, Rutherford's colleague Hans Geiger and undergraduate student Ernest Marsden shot positively charged alpha particles (helium nuclei) from a radioactive source into thin gold foils. They expected the particles to pass straight through the gold or to be deflected at most very slightly. Most behaved as expected, but a few bounced back in nearly the direction they had come from (Fig. 43-2), implying collision with a much more massive object. It was, to paraphrase Rutherford, as though you shot a cannonball at a piece of tissue paper and the ball bounced back. Rutherford realized that a rebounding alpha particle had collided with a small but massive object exerting a large repulsive force. He therefore concluded that all the atom's positive charge and nearly all its mass must be concentrated in a small, massive core: the **nucleus**. A new atomic model emerged, with the tiny, positively charged nucleus at the center, surrounded by electrons in distant orbits (Fig. 43-1b). We now know, of

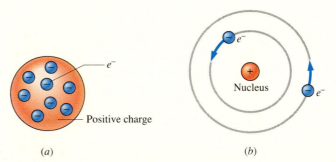

(a) (b)

FIGURE 43-1 (a) Thomson's "plum-pudding" model of the atom had electrons immersed in a "pudding" of positive charge. (b) In Rutherford's nuclear model, electrons orbit the tiny but massive nucleus. Figure is not to scale; the nucleus is actually about 10^{-5} times the size of the atom.

FIGURE 43-2 In the Rutherford-Geiger-Marsden experiment, alpha particles recoiling from a gold foil revealed the existence of the massive atomic nucleus. Most alphas went right through the foil, showing that the nucleus is tiny; those few that bounced back showed that the nucleus nevertheless contains nearly all the atomic mass.

course, that the distinct orbits of Rutherford's atomic model are not consistent with modern quantum mechanics. But the essence of his discovery remains: The atom consists of a tiny, massive nucleus surrounded by electrons that are, at least statistically, distant from the nucleus.

By 1920 Rutherford had proposed that nuclei other than hydrogen must contain electrically neutral as well as positively charged entities. The discovery of the neutron in 1932 confirmed this proposal. Today we know that protons and neutrons, collectively called **nucleons**, are the constituents of nuclei.

43-2 BUILDING NUCLEI: ELEMENTS, ISOTOPES, AND STABILITY

The Nuclear Force

What holds the nucleus together, given the mutual electrical repulsion of its protons? There must be another force—the **nuclear force**—acting attractively between the constituents of the nucleus. Through much of the twentieth century the nuclear force was thought to be a fundamental force of nature, but today we recognize it as a manifestation of a more fundamental force that acts between particles called quarks that themselves make up the nucleons. We will study quarks and their interactions in Chapter 45.

19.1
Particle Scattering

The nuclear force acts attractively between pairs of nucleons—neutrons and protons, protons and protons, neutrons and neutrons—without regard for their electric charge. The force is very strong at separations less than a few fm (10^{-15} m), but it falls off with distance much more rapidly than the inverse-square falloff of the electric force (the nuclear force drops roughly exponentially with distance). The attractive nuclear force between two protons therefore dominates at short distances, but the repulsive electric force becomes dominant at separations of more than a few fm. The structure and characteristics of nuclei are determined, to a first approximation, by the interplay between the weaker but longer range electric force and the stronger but shorter range nuclear force.

The Atomic Number

A nucleus consists of protons and neutrons. The **atomic number**, Z, is the number of protons in a nucleus. The number of protons determines the nucleus's electric charge (Ze) and, therefore, the number of electrons in a neutral atom. Since an atom's chemical properties depend on its electron configuration, the atomic number Z ultimately determines chemical behavior. That's why all atoms with the same Z belong to the same element.

Isotopes and Nuclear Symbols

Although nuclei of a given element have the same number of protons, they don't have to have the same number of neutrons since the neutrons don't affect electric charge and therefore have negligible influence on chemical behavior. Nuclei of the same element but with different numbers of neutrons are different **isotopes**

FIGURE 43-3 Isotopes of a given element have the same number of protons but different numbers of neutrons. Their chemical properties are the same but their nuclear properties may differ significantly.

(Fig. 43-3). We specify a given isotope by its **mass number**, A, or total number of nucleons (protons plus neutrons). Specification of the atomic number Z and mass number A then fully describes a nucleus. For this description we use a shorthand notation, giving the element symbol used in the periodic table, preceded by a subscript for the atomic number and a superscript for the mass number. Thus, $^{4}_{2}\text{He}$ is the common isotope of helium, and $^{235}_{92}\text{U}$ is the uranium isotope that undergoes fission in nuclear reactors. The atomic number and element symbol are redundant since helium *means* the element with two protons in its nucleus and uranium *means* 92 protons. Sometimes, therefore, we write expressions like helium-4, He-4, or ^{4}He to mean the same thing as $^{4}_{2}\text{He}$.

Most elements have several naturally occurring isotopes. Hydrogen, for example, is most abundant as the isotope $^{1}_{1}\text{H}$, whose nucleus consists of a single proton. But about one in every 6500 hydrogen atoms is deuterium ($^{2}_{1}\text{H}$), whose nucleus contains a neutron and a proton. Although the chemical properties of isotopes are similar, their nuclear properties may differ dramatically. Uranium provides an extreme case: the more common isotope $^{238}_{92}\text{U}$ cannot sustain a nuclear fission chain reaction, while the rare (0.72% of natural uranium) isotope $^{235}_{92}\text{U}$ can. Uranium enriched to a high proportion of U-235 is therefore a dangerous material, suitable for making nuclear weapons—as we will see in the next chapter.

Stable Nuclei

Of the infinitely many possible combinations of neutrons and protons, only about 400 are known to be **stable** under normal conditions, meaning they don't spontaneously decay into other nuclei. Figure 43-4 shows the stable nuclei, along with many unstable ones, on a chart of atomic number Z versus neutron number $N = A - Z$. This figure shows that the lighter stable nuclei have roughly the same number of neutrons and protons, but that heavier nuclei have more neutrons than protons. Figure 43-5 suggests the reason more neutrons are necessary in the heavier nuclei. As the nuclear size grows, the more distant protons become quite far apart compared with the range of the nuclear force.

FIGURE 43-4 A chart of the nuclei, plotting all known nuclei according to the number of protons (atomic number Z) versus number of neutrons ($A - Z$). The nuclei are color coded by half-life. The more massive nuclei all lie below the line $N = Z$, showing that they are richer in neutrons; this is necessary to counter the electrical repulsion of the protons.

They feel essentially only the repulsive electric force, and alone they would tend to tear the nucleus apart. Extra neutrons, subject only to the nuclear force, help bind the protons into the nucleus. But even this effect has a limit, and the result is that there are no stable nuclei for $Z > 83$.

Electrical repulsion explains why neutrons are necessary and why their numbers increase with increasing atomic number. But why aren't there stable nuclei with a great many more neutrons than protons? One reason lies in the exclusion principle: extra neutrons have to go into higher-energy states (recall Fig. 41-17), making individual nucleons more likely to escape the nucleus. Furthermore, the neutron itself is an unstable particle; an isolated neutron decays spontaneously into a proton, an electron, and an elusive particle called the neutrino. This decay also happens in nuclei with too many neutrons, but is suppressed in stable nuclei. We'll take a closer look at neutron decay and related processes in Section 43-5. ActivPhysics Activity 19.2 lets you create your own nuclei and test their stability.

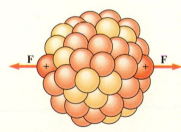

FIGURE 43-5 Two widely separated protons in a large nucleus experience significant electrical repulsion, but the attractive nuclear force between them is negligible because of its short range. More neutrons are therefore needed to keep the nucleus from flying apart.

43-3 PROPERTIES OF THE NUCLEUS

Nuclear Masses

A nucleus of mass number A contains Z protons and $A - Z$ neutrons. Although the proton and neutron have slightly different masses, those masses are the same to three significant figures. Thus we can approximate the mass of a nucleus as A proton masses. Mass spectrometers (see Problem 29-27) give more accurate measurements of nuclear masses. It's convenient to express nuclear masses in

TABLE 43-1 *Selected Masses*

	MASS, kg	MASS, u	MASS, MeV/c²
Electron	$9.109\ 39 \times 10^{-31}$	$0.000\ 548\ 579$	$0.510\ 999$
Proton	$1.672\ 62 \times 10^{-27}$	$1.007\ 276$	938.272
Neutron	$1.674\ 93 \times 10^{-27}$	$1.008\ 665$	939.566
${}_{1}^{1}\text{H}$ atom	$1.673\ 53 \times 10^{-27}$	$1.007\ 825$	938.783
α particle (${}_{2}^{4}\text{He}$ nucleus)	$6.644\ 66 \times 10^{-27}$	$4.001\ 506$	3727.38
${}_{6}^{12}\text{C}$ atom	$1.992\ 65 \times 10^{-26}$	12	$11\ 177.9$
Unified mass unit (u)	$1.660\ 54 \times 10^{-27}$	1	931.494

terms of the **unified mass unit**, u, defined as one-twelfth of the mass of a neutral carbon-12 atom. The unified mass unit is very nearly 1.66057×10^{-27} kg, slightly less than the mass of the proton or neutron. High-energy physicists, cognizant of the Einstein mass-energy equivalence, often express nuclear and particle masses in MeV/c^2—a value numerically equal to the particle's rest energy in MeV. Table 43-1 lists some important masses in these various units.

Since the unified mass unit u is defined as one-twelfth the mass of the ${}_{6}^{12}\text{C}$ atom and is less than the proton or neutron mass, that means the carbon-12 atom's mass is less than that of its constituent nucleons alone—and this is before we consider the slight additional mass of its six electrons. Why is this? The answer lies in the Einstein mass-energy equivalence. The nucleus is a bound system, meaning that its total energy is *less,* by some amount $\Delta E,$ than that of its constituents when they are widely separated. Its mass is therefore less as well, by the amount $\Delta E/c^2.$ The "missing" energy ΔE was released when the nucleus formed.

As we emphasized in Chapter 38, there's nothing uniquely nuclear about mass-energy equivalence, and therefore, other bound systems exhibit this mass decrease as well. A water molecule, for example, has less mass than the total of two separate hydrogen atoms and one oxygen atom. The nuclear difference lies in the great strength of the nuclear force and the correspondingly large energies involved in nuclear interactions. That makes the mass difference much more obvious in nuclear situations. We will consider nuclear mass differences and energies more in the next section.

Nuclear Size

Rutherford interpreted data from the alpha-particle experiments to show that the nuclear size must be on the order of 1 fm (see Problem 1). Since Rutherford's time physicists have used high-energy electrons and other probes to learn the details of nuclear structure. The results of these studies show that nuclei are approximately spherical and give the **nuclear radius** R defined as the radius at which the nuclear density has fallen to half its central value. A rough, empirical formula for this radius is

$$R = R_0 A^{1/3}, \tag{43-1}$$

where $R_0 = 1.2$ fm and A is the mass number.

Equation 43-1 shows that the nuclear radius is proportional to the cube root of the mass number A. That means the nuclear volume is proportional to A itself—just what we would expect if the nucleus consisted of a tight, spherical bundle of protons and neutrons, as shown in Fig. 43-5. To the extent that the nuclear mass is the sum of the masses of its constituent particles, that means all nuclei have essentially the same ratio of mass to volume—i.e., the same density. The following example shows that nuclear density is enormously greater than the densities of ordinary materials.

| **EXAMPLE 43-1** | *Gibraltar in a Teaspoon: Neutron Stars* |

(a) Calculate the density of nuclear matter, and determine the mass of 1 teaspoon (about 5 ml) of it. (b) The largest known agglomerations of nuclear matter are neutron stars, which sometimes form in the supernova explosions that mark the deaths of massive stars. Find the radius of a neutron star with mass equal to the Sun's.

Solution

Taking the mass of the nucleus to be approximately A proton masses, we can use Equation 43-1 to find the density of nuclear matter:

$$\rho = \frac{m}{V} = \frac{Am_p}{\frac{4}{3}\pi R^3} = \frac{Am_p}{\frac{4}{3}\pi R_0^3 A} = \frac{3m_p}{4\pi R_0^3} = \frac{(3)(1.67\times10^{-27}\text{ kg})}{(4\pi)(1.2\times10^{-15}\text{ m})^3}$$

$$= 2.3\times10^{17}\text{ kg/m}^3,$$

about 10^{14} times the density of ordinary matter. A teaspoon of nuclear matter (5 mL or 5×10^{-6} m³) then has a mass of about 10^{12} kg, about that of the Rock of Gibraltar!

Appendix E shows that the Sun's mass is about 2×10^{30} kg. The number of nucleons in this mass is then given by $A = 2\times10^{30}$ kg$/1.67\times10^{-27}$ kg/nucleon or 1.2×10^{57}. Equation 43-1 then gives the radius:

$$R = R_0 A^{1/3} = (1.2\times10^{-15}\text{ m})(1.2\times10^{57})^{1/3} = 13\text{ km},$$

about the size of a city. Another approach would have been to get the volume and then the radius from the density calculated in part (a). Our answer is approximate because Equation 43-1 does not scale exactly to gravitationally bound nuclei of stellar mass, but it is good within a factor of about two.

EXERCISE Compare the radius of a uranium-238 nucleus with that of an oxygen-16 nucleus.

Answer: $R_{\text{U-238}} = 2.46 R_{\text{O-16}}$

Some problems similar to Example 43-1: 8, 9, 11, 12

Nuclear Spin

In Chapter 41 we discussed the important role of electron spin in atomic structure, and we noted that protons and neutrons, like electrons, are spin-$\frac{1}{2}$ particles—meaning they have intrinsic angular momentum of magnitude $\sqrt{s(s+1)}\hbar$, where $s = \frac{1}{2}$. The spins of the individual nucleons, combined with any angular momentum associated with their motions within the nucleus, give the nucleus a quantized angular momentum I that obeys the same rules we've seen for other angular momenta in quantum mechanics:

$$I = \sqrt{i(i+1)}\hbar, \tag{43-2}$$

where i, the nuclear spin quantum number, is an even or odd multiple of a half integer. The component of the nuclear angular momentum on a given axis is also quantized, just like other angular momenta, according to the rule $I_z = m_i\hbar$, where m_i ranges from $-i$ to i in steps of 1.

The values of i may be integer or half integer, depending on whether the number of nucleons is even or odd, respectively. We noted in Chapter 41 that particles with half-integer spins obey the Pauli exclusion principle, while those with integer spins do not. That can lead to profound differences in the behavior of isotopes of the same element; in particular, the spin-$\frac{1}{2}$ nucleus $_2^3$He exhibits

FIGURE 43-6 This fountain of liquid helium-4 occurs because the spin-0 helium-4 nuclei can all occupy the same quantum state, resulting in a "superfluid" that flows without viscosity. The phenomenon does not occur with spin-1/2 helium-3 until a far lower temperature where He-3 atoms pair to form spin-1 particles.

very different behavior from that of spin-zero $^4_2$He, even though both are isotopes of helium (Fig. 43-6).

The angular momentum of the nucleus results in a nuclear magnetic moment, usually expressed in units of the **nuclear magneton**,

$$\mu_N = \frac{e\hbar}{2m_p} = 5.05 \times 10^{-27} \text{ J/T},$$

where m_p is the proton mass. This quantity is smaller than the Bohr magneton that we introduced in Chapter 41 by the ratio of the electron to the proton mass, showing that magnetic effects involving nuclei are much smaller. Since the proton is positive, its magnetic moment is in the same direction as the spin. There is no fully satisfactory theory to predict the exact value of this moment, but experiments show that its projection on a given axis takes the values $\pm 2.793\mu_N = \pm 1.4106 \times 10^{-26}$ J/T. Listings of fundamental constants often call this quantity "the magnetic moment of the proton," although the actual magnitude is greater by a factor of $\sqrt{3}$ because of space quantization (recall Fig. 41-10). Interestingly, the neutral neutrons as well as the charged protons contribute to the nuclear magnetic moment, giving a tantalizing hint that the neutron may itself contain charged particles whose total charge sums to zero. We'll see in Chapter 45 that this is indeed the case.

Interaction of the nuclear magnetic moment with magnetic fields—either externally applied or produced by atomic electrons—alters very slightly the energy levels of the atom (Fig. 43-7a). In hydrogen, for example, the spin-$\frac{1}{2}$ proton can have either of two orientations in the magnetic field of the electron, and the result is **hyperfine splitting** of the hydrogen ground state into two levels spaced a mere 5.9 μeV apart (Fig. 43-7b). Transitions between these levels

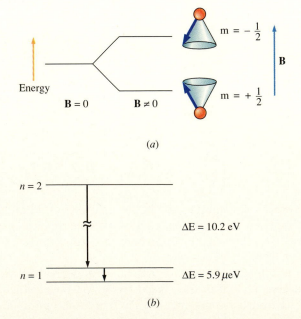

FIGURE 43-7 (a) A nonzero magnetic field **B** splits the energy level of the spin-$\frac{1}{2}$ proton into two levels. The upper level corresponds to magnetic quantum number $m = -\frac{1}{2}$, meaning that the spin is aligned as nearly as possible anti-parallel to the field. (b) The two possible orientations of the proton's magnetic moment in the magnetic field of the electron split the hydrogen ground state into two levels 5.9 μeV apart.

result in a spectral line at a radio wavelength of 21 cm. This transition is used in hydrogen masers, which serve as very accurate atomic clocks, and radio astronomers use it to detect interstellar clouds of neutral hydrogen.

EXAMPLE 43-2 | *Flipping Protons*

A 2.5-T magnetic field points in the z direction. A proton has its magnetic moment oriented so the z component is opposite the field direction. It then flips into the lower energy state, where the z component of its magnetic moment is aligned with the field. Find the energy and frequency of the photon it emits in flipping between the two states.

Solution

In Chapter 29 (Equation 29-12) we found that the potential energy of a magnetic dipole of moment $\boldsymbol{\mu}$ in a magnetic field $\mathbf{B}$ is $U = -\boldsymbol{\mu} \cdot \mathbf{B}$. The value $\mu_p = 1.41 \times 10^{-26}$ J/T for the proton magnetic moment is already the projection on an axis defined by the magnetic field direction, so we can write $U = \pm\mu_p B$ for the energies of the two states. Then the magnitude of the energy difference between the two states is

$$\Delta U = \mu_p B - (-\mu_p B) = 2\mu_p B$$
$$= (2)(1.41 \times 10^{-26} \text{ J/T})(2.5 \text{ T})$$
$$= 7.05 \times 10^{-26} \text{ J} = 0.44 \ \mu\text{eV} .$$

Using $E = hf$ shows that a photon with this energy has a frequency of 106 MHz, which happens to lie in the FM broadcasting band.

EXERCISE Use the hyperfine splitting energy of 5.9 μeV in the hydrogen ground state to estimate the magnetic field experienced by the proton in the hydrogen atom.

Answer: 33.5 T

Some problems similar to Example 43-2: 13–15

APPLICATION | *Nuclear Magnetic Resonance*

Putting nuclei in an external magnetic field results in two possible energy states, as shown in Fig. 43-7a, depending on the orientation of the nuclear magnetic moments relative to the field. Applying electromagnetic radiation with photon energy equal to the energy difference between the states will cause nuclei to "flip" from the lower to the higher energy state. But nuclei also experience magnetic fields from the atomic electrons that surround them, and those fields are extremely sensitive to the details of the surrounding electron distribution—i.e., to the molecular structures in which nuclei are embedded. Measuring the exact energy needed to flip nuclear spins in a magnetic field thus provides information about molecular structure.

Figure 43-8 shows an **NMR spectrometer**, widely used for structural analysis of chemical compounds. The sample under analysis is placed in a uniform magnetic field, often provided by superconducting coils. A smaller coil carries current at a frequency f—usually a few hundred MHz—corresponding to the photon energy hf that would flip the spin of an isolated proton in the spectrometer's magnetic field. This coil emits electromagnetic waves of frequency f. If nuclei in the sample absorb the associated photons, they flip into their higher energy states and subsequently drop back, emitting radiation of frequency f in the process. A receiver coil mounted perpendicular to the transmitter coil detects this radiation.

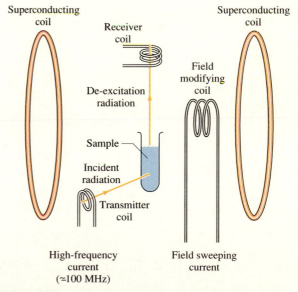

FIGURE 43-8 Schematic diagram of a nuclear magnetic resonance spectrometer. At certain values of magnetic field, nuclei absorb photons from the transmitter coil, flip to their higher energy states, and then drop back down, emitting photons detected by the receiver coil. The exact fields needed for such resonance provide information about molecular structure.

Nuclei in a sample are surrounded by electrons that produce additional magnetic fields, and therefore, they will not generally absorb the photons of frequency f in the steady magnetic field of the spectrometer. So the magnetic field is varied slightly, by sending a varying current through yet another coil. When this additional field exactly cancels the field due to surrounding electrons, nuclei can absorb photons. They flip to the upper spin state, drop back down, and produce a signal detected in the receiver coil. Scanning the field through a range of values allows detection of nuclei in different electron environments (Fig. 43-9). Comparison with data from known molecular substructures leads to determination of the structure of the sample molecules.

Nuclear magnetic resonance with protons (H nuclei) is the basis of the widely used medical technique called **MRI**, for **magnetic resonance imaging**. In MRI, a person is placed inside a large solenoid. A gradient in the solenoid's magnetic field means that the nuclear magnetic resonance frequency for protons varies with position, and thus, the resonance signal can be used to provide positions of protons giving the magnetic resonance signal. A computer then uses the position information to construct an image.

Most of the MRI signal from the human body comes from fat and water. Fat gives the strongest signal because it has a higher density of protons. Thus, MRI is good for imaging soft tissues that don't show in X rays (see Fig. 43-10).

FIGURE 43-10 MRI image of a healthy human ankle. MRI allows physicians to see both bone (shown here in beige) and soft tissue (bright colors).

FIGURE 43-9 NMR spectrum of ethyl bromide (CH_3CH_2Br). Each peak corresponds to a different magnetic field in the NMR spectrometer, reflecting the different electron configurations surrounding the various nuclei.

Magnetic field strength

43-4 BINDING ENERGY

19.2

Nuclear Binding Energy

We've seen that the mass of a nucleus is less than the sum of the masses of its constituent nucleons—a reflection of the fact that the nucleus is a bound system whose energy is lower than that of its widely separated constituents. Disassembling a nucleus thus requires that we supply energy. The more tightly bound the nucleus, the more energy is needed. Quantitatively, conservation of energy and the Einstein mass-energy equivalence allow us to write

$$m_N c^2 + E_b = Z m_p c^2 + (A - Z) m_n c^2 \qquad (43\text{-}3)$$

for a nucleus of mass m_N, atomic number Z, and mass number A, where m_p and m_n are the proton and neutron masses, respectively. We can regard E_b in Equation 43-3 as the energy it would take to disassemble the nucleus; consequently, E_b is called the **binding energy**. Alternatively, E_b is the energy released when the constituent particles come together to form the nucleus.

EXAMPLE 43-3	*Powering the Sun*

Use information from Table 43-1 to calculate the binding energy of the helium-4 nucleus.

Solution

Table 43-1 gives the masses of the proton, neutron, and $^4_2$He nucleus. Using these values in Equation 43-3 and solving for E_b gives

$$E_b = Zm_pc^2 + (A - Z)m_nc^2 - m_Nc^2$$

$$= (2)(938.272 \text{ MeV}) + (4 - 2)(939.566 \text{ MeV})$$

$$- 3727.38 \text{ MeV} = 28.3 \text{ MeV} .$$

Here we used Table 43-1's values for mass in MeV$/c^2$; multiplying by c^2 then gives an energy in MeV with the same numerical value.

The formation of helium through a sequence of nuclear reactions is the process that powers the Sun, and our 28.3-MeV result is very close to the actual 26.7 MeV released for each helium-4 nucleus so formed.

EXERCISE Find the binding energy of the deuterium nucleus ($^2_1$H), whose mass is 2.0136 u.

Answer: 2.19 MeV

Some problems similar to Example 43-3: 16, 19, 20–22

In practice one often knows the atomic mass rather than the nuclear mass, but Equation 43-3 still applies because the constituent electrons then appear on both sides of the equation and so cancel—provided, as is always the case, that the binding energy of the electrons to the atom is negligible compared with the nuclear binding energy.

Nuclear binding energies play a crucial role in determining the origin of the nuclei as well as their behavior. Figure 43-11 shows the **curve of binding energy**, which is a plot of the binding energy *per nucleon* (E_b/A), as a function of mass number. The larger the binding energy per nucleon, the more tightly

FIGURE 43-11 The curve of binding energy. Nuclei lighter than those at the peak of the curve can release energy by fusing to produce heavier nuclei. Nuclei heavier than those at the peak of the curve can release energy by fissioning into two smaller nuclei. The common iron isotope ^{56}Fe is near the top of the curve and is therefore among the most tightly bound nuclei. ActivPhysics Activity 19.2 will locate any stable nucleus on the curve of binding energy.

bound the nucleus. Note that the curve has a broad peak in the vicinity of $A = 60$, indicating that nuclei with mass numbers near this value are most tightly bound. That means it's energetically favorable for light nuclei to join together in the process called **nuclear fusion**. But the heaviest nuclei have lower binding energies per nucleon than do the middle-weight ones, so it's energetically favorable for them to split or **fission** into lighter nuclei. We discuss these nuclear energy processes at length in the next chapter.

EXAMPLE 43-4	*Tritium and Helium*

The mass of a helium-3 nucleus ($^3_2$He) is 3.01493 u, and that of a tritium nucleus ($^3_1$H) is 3.01550 u. Calculate the binding energy per nucleon for each.

Solution

Helium-3 consists of two protons and a neutron, so $Z = 2$ and $A = 3$; using masses from Table 43-1 in Equation 43-3 then gives

$$E_b = Zm_pc^2 + (A - Z)m_nc^2 - m_Nc^2$$
$$= (2)(938.272 \text{ MeV}) + (3 - 2)(939.566 \text{ MeV})$$
$$- 2808.39 \text{ MeV} = 7.721 \text{ MeV} ,$$

where we first converted the He-3 mass to MeV/c^2 by multiplying its value in u by the factor 931.494 MeV/c^2/u given in Table 43-1. For tritium, with $Z = 1$ and $A = 3$, we similarly have

$$E_b = Zm_pc^2 + (A - Z)m_nc^2 - m_Nc^2$$
$$= (1)(938.272 \text{ MeV}) + (3 - 1)(939.566 \text{ MeV})$$
$$- 2808.92 \text{ MeV} = 8.484 \text{ MeV} .$$

Since each nucleus contains three nucleons, the binding energies per nucleon for helium-3 and tritium are 2.57 MeV and 2.83 MeV, respectively.

EXERCISE The isotope iron-56 is near the peak of the binding-energy curve of Fig. 43-11, with 8.8 MeV/nucleon binding energy. Use the curve, along with data from Table 43-1, to find the nuclear mass of iron-56.

Answer: 55.92 u

Some problems similar to Example 43-4: 17, 54

Nucleosynthesis and the Origin of the Elements

Where did the chemical elements come from? The curve of binding energy holds many of the answers. Since it's energetically favorable for light nuclei to fuse together, they will do so if given enough energy to overcome the repulsive electric force. Cosmologists believe that this happened in the high-temperature conditions from about 1 minute to about 30 minutes after the big bang event that started the universe, in the process forming the helium that we now find composing about 25% of the observed matter in the universe. Also formed were minute amounts of deuterium, lithium, beryllium, and boron—but nothing with $Z > 5$ emerged under those primordial conditions. Billions of years later stars formed, and in the interiors of the hotter stars conditions proved suitable for a process in which three helium nuclei join to make carbon. From there nuclear fusion reactions lead to the formation of isotopes up to those near the $A = 60$ peak of the binding energy curve. In fact, the nuclei of essentially all the elements with $A < 60$ except hydrogen and helium—including most of the materials that make our bodies—were formed in the interiors of massive stars.

Once fusion builds nuclei at the peak of the binding energy curve, it then requires energy to make still more massive nuclei with their lower binding energies; therefore this process does not occur spontaneously. So where did the more massive nuclei—like silver, gold, or uranium—come from? And how did the lighter nuclei get out of the stars? Astrophysicists believe the heavier elements were formed in supernova explosions, runaway nuclear reactions in which massive stars explode, spewing their contents into interstellar space (Fig. 43-12).

FIGURE 43-12 An exploding supernova produces nuclei with $A > 60$, and spews them and the lighter nuclei formed in the star's interior into interstellar space. Photo shows Supernova 1987A (lower right) in the Large Magellanic Cloud, a satellite galaxy of our Milky Way. This was the brightest supernova seen from Earth in several centuries.

Energy released in the process helped to form the heavier elements. The expanding cloud from the supernova explosion thus distributes both lighter and heavier elements into the interstellar medium, where eons later they may find themselves incorporated into new stars, planets, and living beings.

43-5 RADIOACTIVITY

In 1896 Henre Becquerel of Paris noticed that a photographic plate that had been stored near uranium compounds appeared fogged when he developed it. He reasoned that some sort of rays must have emanated from the uranium to fog the film. Becquerel had discovered **radioactivity**, the phenomenon wherein some substances spontaneously emit either particles or photons of very high energy.

In 1898, also in Paris, the Polish student Marie Curie (Fig. 43-13) decided to do her doctoral thesis on Becquerel's rays. Working with her husband, Pierre Curie, she found that not only uranium but also thorium gave off rays, and she coined the term "radioactivity." She soon discovered the more intensely radioactive substances polonium, which she named for her native Poland, and radium. The Curies shared the 1903 Nobel Prize in Physics with Becquerel for their work on radioactivity, and Marie Curie later won the 1911 Nobel Prize in Chemistry for discovering polonium and radium. Today we know a great many radioactive isotopes.

FIGURE 43-13 Marie Curie and Pierre Curie in their Paris laboratory in 1906.

Decay Rates and Half Lives

Different radioactive isotopes emit radiation at different rates, their nuclei decaying in the process into other forms. The greater the number of radioactive nuclei present, the greater the number of decays per unit time, and therefore the more rapidly the number of nuclei will decrease. We can express this mathematically by writing

$$\frac{dN}{dt} = -\lambda N, \tag{43-4}$$

where N is the number of nuclei and λ, the **decay constant**, gives the probability that any one nucleus will decay in a 1-second time interval. We've seen a similar equation before, in the analysis of RC and RL circuits. We solve it the same way, by multiplying through by dt/N and integrating. Taking the integration from time 0, when there are N_0 nuclei, to an arbitrary time t, when there are N nuclei, gives

$$\int_{N_0}^{N} \frac{dN}{N} = -\lambda \int_0^t dt.$$

Evaluating the integrals, we have

$$\ln\left(\frac{N}{N_0}\right) = -\lambda t,$$

or, exponentiating each side and using $e^{\ln x} = x$,

$$N = N_0 e^{-\lambda t}. \tag{43-5a}$$

Thus, the number of nuclei drops exponentially with time (Fig. 43-14).

FIGURE 43-14 Decay of a radioactive sample occurs exponentially. N_0 is the number of nuclei initially in the sample. After one half-life, half the initial number remain; after two half-lives, only one-fourth remain, and so on. ActivPhysics Activity 19.4 simulates radioactive decay.

It's often convenient to express Equation 43-5a in terms of the **half-life**, $t_{1/2}$, defined as the time it takes half the nuclei in a given sample to decay. Since $2 = e^{\ln 2}$, we can write $e^{-\lambda t} = 2^{-\lambda t/\ln 2}$; then Equation 43-5a becomes

$$N = N_0 2^{-t/t_{1/2}}, \tag{43-5b}$$

where

$$t_{1/2} = \frac{\ln 2}{\lambda} = \frac{0.693}{\lambda} \tag{43-6}$$

19.4
Radioactivity

is the half-life. Table 43-2 lists some significant radioisotopes and their half-lives.

The rate at which nuclei decay determines how vigorously radioactive a sample is. That rate is given by $|dN/dt|$, with the absolute value signs because dN/dt is itself negative as the number of nuclei decreases. The rate $|dN/dt|$ is called the **activity** of the sample. Equation 43-4 shows that dN/dt is proportional to N, so the activity also decays exponentially according to Equations 43-5. The SI unit of activity is the **becquerel** (Bq), equal to one decay per second. An older unit, the **curie** (Ci), is defined as 3.7×10^{10} Bq and is approximately the activity of one gram of radium-226.

TABLE 43-2 *Selected Radioisotopes*

ISOTOPE	HALF-LIFE	DECAY MODE	COMMENTS
Carbon-14 ($^{14}_{6}$C)	5730 years	β^-	Used in radiocarbon dating.
Iodine-131 ($^{131}_{53}$I)	8.04 days	β^-	Fission product abundant in fallout from nuclear weapons and reactor accidents; damages thyroid gland.
Oxygen-15 ($^{15}_{8}$O)	2.03 minutes	β^+	Short-lived oxygen isotope produced with cyclotrons and used for medical diagnosis in positron emission tomography (PET).
Potassium-40 ($^{40}_{19}$K)	1.25×10^9 years	β^-	Comprises 0.012% of natural potassium; dominant radiation source within the normal human body. Used in radioisotope dating.
Plutonium-239 ($^{239}_{94}$Pu)	24,110 years	α	Fissile isotope used in nuclear weapons; produced by neutron capture in $^{238}_{92}$U.
Radium-226 ($^{226}_{88}$Ra)	1600 years	α	Highly radioactive isotope discovered by Marie and Pierre Curie and which results from decay of $^{238}_{92}$U.
Radon-222 ($^{222}_{86}$Rn)	3.82 days	α	Radioactive gas formed naturally in decay of $^{226}_{88}$Ra; seeps into buildings, where it may cause serious radiation exposure.
Strontium-90 ($^{90}_{38}$Sr)	29 years	β^-	Fission product that behaves chemically like calcium; readily absorbed into bones.
Tritium ($^{3}_{1}$H)	12.3 years	β^-	Hydrogen isotope used in biological studies and to enhance yields of nuclear weapons.
Uranium-235 ($^{235}_{92}$U)	7.04×10^8 years	α	Fissile isotope comprising 0.72% of natural uranium; used as reactor fuel and in primitive nuclear weapons.
Uranium-238 ($^{238}_{92}$U)	4.46×10^9 years	α	Predominant uranium isotope; cannot sustain a chain reaction.

EXAMPLE 43-5	*Indoor Radon*

The radioactive gas radon-222 is a serious pollutant in many homes. The gas results from the decay of radium-226, which itself arises in the decay series of the uranium-238 that occurs naturally in rocks, soil, and building materials. It seeps through cracks in basement walls and floors, where it reaches a steady concentration with infiltration balancing radioactive decay. The mean activity of Rn-222 in American homes is about 0.05 Bq per liter of air, although this quantity varies widely depending on the underlying bedrock and on the home's foundation construction. How many Rn-222 nuclei are there per liter of air in the average home?

Solution

We're given the activity $|dN/dt|$ in a 1-L sample, so we can solve Equation 43-4 for N. Table 43-2 gives 3.82 days for the half-life, so we can use Equation 43-6 to write $\lambda = 0.693/t_{1/2}$. Then we have

$$N = -\frac{1}{\lambda}\frac{dN}{dt} = \frac{t_{1/2}}{0.693}\left|\frac{dN}{dt}\right|$$

$$= \left[\frac{(3.82\ \text{d})(86{,}400\ \text{s/d})}{0.693}\right](0.05\ \text{Bq})$$

$$= 2.4\times10^4\ \text{Rn-222 atoms in each liter}.$$

Here we took away the minus sign and put the absolute value sign on dN/dt because the latter is a negative quantity.

Even the modest activity level calculated here presents a serious health threat: Statistically, the increased risk of fatal cancer from a lifetime exposure to this level of radon activity reduces the human lifespan by an average of 40 days, comparable to the statistical risk associated with drowning and much greater than the risk from living next door to a nuclear power plant.

EXAMPLE 43-6	*Fallout from Chernobyl*

The 1986 accident at the Chernobyl nuclear power plant in the then-Soviet Ukraine spread radioactive fallout over eastern Europe and Scandinavia (Fig. 43-15). A particularly dangerous isotope released was iodine-131, which is absorbed by the thyroid gland and can cause thyroid cancer. Following the accident, I-131 levels in milk in Romania rose to 2900 Bq per liter. How long did Romanians have to wait before I-131 levels dropped below their government's 185-Bq/L safety standard?

Solution

We could solve this problem formally using Equation 43-5b. But what the equation says is that after n half-lives, the activity drops to $1/2^n$ of its initial value. So the number of half-lives

needed to bring the I-131 activity from 2900 Bq/L to 185 Bq/L is given by

$$\frac{1}{2^n} = \frac{185}{2900}$$

or, inverting and taking logarithms of both sides,

$$\ln(2^n) = \ln\left(\frac{2900}{185}\right).$$

But $\ln 2^n = n\ln 2$, so

$$n = \frac{\ln(2900/185)}{\ln 2} = 3.97\ \text{half-lives}.$$

Table 43-2 gives 8.04 days for I-131's half-life, so Romanians had to wait 32 days before their milk was considered safe. Other countries, with less stringent safety standards and lower initial levels, had much shorter waits. We discuss the Chernobyl accident further in the next chapter.

> **Tip**
>
> *Powers of Two* After n half lives, the activity of a radioactive sample is down by a factor of $1/2^n$. In estimating activity levels it's useful to note that $2^{10} = 1024$, showing that activity drops by very nearly a factor of 1000 every 10 half-lives.

EXERCISE A patient undergoing a PET scan receives a dose of radioactive oxygen-15 ($t_{1/2} = 2$ minutes). How long will it take for the activity of the dose to decline to one one-millionth of its original level?

Answer: about 40 min (20 half-lives)

Some problems similar to Examples 43-5 and 43-6: 28, 31, 35, 36, 42

FIGURE 43-15 The Chernobyl accident spread radioactive fallout over much of Europe. Here a crane works to build a concrete tomb around the damaged reactor.

APPLICATION	*Radiocarbon Dating*

Archaeologists, art historians, geologists, and others use radioactive decay to date ancient materials. The radioactive isotope carbon-14 forms continually in the atmosphere through the interaction of cosmic rays with nitrogen. Carbon-14 has a 5730-year half-life. Since it acts chemically just like ordinary carbon, it gets incorporated into living organisms. While alive, an organism maintains a steady concentration of ^{14}C, established by the balance between radioactive decay and the uptake of new ^{14}C. At death, carbon uptake ceases but radioactive decay continues, so the concentration of ^{14}C decreases in relation to that of the stable isotope ^{12}C. By measuring ^{14}C activity in an ancient sample of organic material (charcoal, bone, etc.) and comparing the $^{14}C/^{12}C$ ratio with that found in living material, the age of the sample can be determined (Fig. 43-16).

Until recently, samples of 1 to 5 grams were required to count enough decays for accurate radiocarbon dating. A newer technique ionizes carbon in the sample, then uses an electrostatic accelerator to produce an ion beam. A mass spectrometer, using the fact that different mass ions move on trajectories of different radii in a magnetic field, then separates ^{14}C and ^{12}C for an accurate determination of their abundance ratio. This technique counts all the ^{14}C atoms, not just those that decay, and as a result it requires only about 1 mg of carbon. The new technique therefore provides dates for objects too valuable or too small for the earlier technique (Fig. 43-17).

The $^{14}C/^{12}C$ ratio varies slightly with time, as a result of fluctuations in cosmic ray intensity that are in turn related to solar activity. By counting growth rings in ancient trees, scientists have established a precise measure of age independent of carbon dating, thus allowing calibration of the radiocarbon method to account for fluctuations in the $^{14}C/^{12}C$ ratio. Radiocarbon dating is quite accurate to about 20,000 years and can be used back to about 50,000 years. For longer time spans, up to the billions of years characterizing the ages of rocks, ratios of longer-lived isotopes provide age information. Much knowledge of our own past, and our planet's and our solar system's, comes from radioisotope dating.

FIGURE 43-16 The principle of radiocarbon dating. (*a*) Carbon-14 formed in the atmosphere is incorporated into a living organism through the food chain. While it lives, the organism exhibits a mild, steady level of ^{14}C radioactivity (arrows). (*b*) At death, ^{14}C uptake ceases. (*c*) Much later, ^{14}C activity has decayed considerably. (*d*) Archaeologists excavate the long-dead remains. By measuring ^{14}C activity, they can infer the time since death. Note that the archaeologists, with their active ^{14}C intake, are more radioactive than their ancient ancestor.

FIGURE 43-17 This piece of the Shroud of Turin, smaller than a postage stamp, was used in accelerator-based radiocarbon dating to establish the shroud's age at 700–800 years, thus refuting the belief that the shroud was the burial cloth of Jesus.

EXAMPLE 43-7	*Archaeology*

Archaeologists unearth charcoal from an ancient campfire. They find that carbon-14 activity per unit mass of the sample is 7.4% that expected in living wood. Find the sample's age.

Solution

The sample contains 0.074 of its original ^{14}C radioactivity. This is the factor $1/2^n$, with n the number of half-lives elapsed. So $0.074 = 1/2^n$, or

$$2^n = \frac{1}{0.074} = 13.5 .$$

Taking logarithms of both sides and noting that $\ln(2^n) = n \ln 2$, we have

$$n \ln 2 = \ln 13.5 ,$$

or

$$n = \frac{\ln 13.5}{\ln 2} = 3.76 .$$

Since ^{14}C has a 5730-year half life, the sample's age is $(3.76)(5730 \text{ years}) = 21{,}500$ years.

EXERCISE Find the age of a bone sample whose ^{14}C activity is 19% that of living bone.

Answer: 13,700 years

Some problems similar to Example 43-7: 40, 45

Got It!

What's the approximate age of an ancient bone whose ^{14}C activity is one-fourth that of a contemporary bone?

Types of Radiation

Passing the emanations from radioactive materials through a magnetic field shows that there are three kinds of radiation, one positively charged, one negatively charged, and one neutral (Fig. 43-18). Earlier researchers gave these the names alpha, beta, and gamma radiation, respectively. Today we know that alpha radiation consists of helium-4 nuclei, that beta radiation consists of high-energy electrons (or sometimes positrons), and that gamma radiation is a stream of very-high energy photons. These radiations also differ in penetrating power: alpha particles are stopped by a sheet of paper and beta particles by several cm of matter, but some gamma rays can penetrate substantial thicknesses of concrete or lead. Radioactive nuclei—radioisotopes—are characterized not only by which type of radiation they emit, but also by the radiation energy. Uranium-238, for example, emits alpha particles with energies around 4 MeV, while carbon-14, widely used in radioactive dating, emits beta particles with maximum kinetic energy of 156 keV.

FIGURE 43-18 There are three types of radiation, which separate in a magnetic field into a positive (α particles), negative (β particles), and neutral (γ rays).

Alpha Decay

Alpha emitters are nuclei with too much positive charge, so the nuclear force cannot keep the repulsive electric force in check. They decay into a smaller nucleus with less charge and less mass by emitting a bundle of two protons and two neutrons—a helium-4 nucleus, also called an alpha particle. The process, therefore, drops the atomic and mass numbers by two and four, respectively, and symbolically we write

$$^{A}_{Z}X \longrightarrow \, ^{A-4}_{Z-2}Y + \, ^{4}_{2}\text{He} . \tag{43-7}$$

Here X is the original nucleus—the **parent**—and Y the resulting **daughter** nucleus. Since an alpha particle is a helium nucleus, we write $^{4}_{2}\text{He}$ for the alpha particle, although this is sometimes abbreviated α. Note that the sums of the atomic numbers on both sides of this equation are equal, as are the sums of the mass numbers; this must be true in any nuclear reaction that conserves charge and the number of nucleons. Energy is released in the alpha decay, and appears as the alpha particle's kinetic energy. That the alpha's energy is less than that needed to overcome the nuclear potential barrier is one of the most direct confirmations of quantum tunneling; alpha decay is possible only because the escaping particle tunnels through the potential barrier.

Beta Decay

Beta emitters typically have too many neutrons to be stable nuclei. But they emit electrons, not neutrons. So how does that help? And where did the electron come from, if the nucleus contains only protons and neutrons? The answer lies in the **weak force**, a relative of electromagnetism, that can bring about the decay of a neutron into a proton and an electron. That's just what happens in beta decay, and the result is one less neutron and one more proton. The ejected electron carries off one unit of negative charge, so the atomic number of the remaining nucleus must increase by one to conserve charge. The total number of nucleons remains the same, however, so the mass number does not change.

In the 1920s physicists studying beta decay found a disturbing thing: the ejected electrons did not account for all the energy released in the process. Where was the rest of it? In 1930, Pauli proposed the existence of a hitherto undetected particle, which he called the **neutrino**, for "little neutral one." The neutrino is an elusive particle indeed: it has little or possibly no mass, meaning that it travels near or at the speed of light, and it interacts only through the weak force. Neutrinos emitted in nuclear reactions in the Sun's core, for example, travel right through the solid Earth with little probability of interaction. Neutrinos have nevertheless been detected through rare nuclear reactions, and today they are used to probe conditions in the core of the Sun and in the earliest second of the universe, as we describe in Chapter 45 (Fig. 43-19).

FIGURE 43-19 Computer reconstruction of a neutrino detection at the Sudbury Neutrino Observatory in Ontario. The bright cone-like structure is radiation from particles moving faster than light does in the water that fills the detector.

The neutrino, like other particles, has an antiparticle. Ordinary beta decay, in fact, produces antineutrinos. A typical beta decay can be written

$$_Z^A X \longrightarrow {}_{Z+1}^A Y + e^- + \bar{\nu}, \qquad (43\text{-}8a)$$

where e^- is the electron and $\bar{\nu}$ is the antineutrino.

EXAMPLE 43-8	*Nuclear Decays*

Write equations for (a) the α decay of uranium-238 and (b) the β decay of carbon-14.

Solution

Alpha decay drops the atomic number by two and the mass number by four. A glance at the periodic table shows that uranium has atomic number 92, and that the element two numbers lower is thorium. Thus the daughter nucleus is $_{90}^{234}\text{Th}$, and the reaction reads

$$_{92}^{238}\text{U} \longrightarrow {}_{90}^{234}\text{Th} + {}_2^4\text{He} .$$

Beta decay increases the atomic number by one, changing carbon ($Z = 6$) to nitrogen ($Z = 7$). Since the mass number stays the same, this reaction reads

$$_6^{14}\text{C} \longrightarrow {}_7^{14}\text{N} + e^- + \bar{\nu} .$$

Here the daughter nucleus is the abundant, stable form of nitrogen.

EXERCISE What is the daughter nucleus formed when oxygen-15 decays by positron emission?

Answer: $_7^{15}\text{N}$

Some problems similar to Example 43-8: 26–28, 65

A second type of beta-decay reaction turns a proton into a neutron, emitting a positron (anti-electron, e^+) and a neutrino:

$$_Z^A X \longrightarrow _{Z-1}^A Y + e^+ + \nu. \tag{43-8b}$$

Since the positron carries positive charge, this reaction drops the atomic number by one rather than increasing it as in beta-decay reactions producing electrons. Reaction 43-8b occurs in some short-lived isotopes of carbon and oxygen, and gamma rays from the subsequent annihilation of the positrons are used to produce images in the medical diagnostic procedure known as positron emission tomography (PET).

A third beta-decay process is **electron capture**, in which a nucleus captures an inner-shell atomic electron, converting a proton into a neutron and lowering the atomic number by one. A neutrino is also ejected:

$$_Z^A X + e^- \longrightarrow _{Z-1}^A Y + \nu. \tag{43-8c}$$

Gamma Decay

Alpha and beta decays usually leave the daughter nucleus in an excited state. Nuclei can also be excited by collisions with high-energy particles or through photon absorption. As with atoms, nuclear excited states decay through photon emission. But—as a simple uncertainty-principle estimate (Example 39-8) shows—the energies associated with nuclear states must be on the order of MeV. The high-energy photons emitted in nuclear transitions are called gamma rays. A photon carries neither mass nor charge, so gamma emission, unlike alpha or beta, does not change the type of nucleus; we therefore write

$$_Z^A X^* \longrightarrow _Z^A X + \gamma, \tag{43-9}$$

where X^* designates an excited state.

Since each nucleus has a distinctive energy-level pattern, gamma-ray spectral lines can be used to identify nuclear species (Fig. 43-20). Even nonradioactive materials can be identified this way, by first exciting their nuclei through

FIGURE 43-20 Gamma-ray spectra of cobalt-60. The actual decay is a two-step process, with ^{60}Co beta decaying to excited ^{60}Ni*, which then emits gamma radiation as it drops to its stable ground state. Two different sets of data are shown, one taken with a sodium-iodide (NaI) detector, the other with a much higher resolution germanium-lithium (Ge(Li)) semiconductor detector.

exposure to a neutron beam from a nuclear reactor and measuring the gamma-ray spectrum that results from the subsequent de-excitation.

Decay Series and Artificial Radioactivity

A few radioisotopes—like ^{40}K and ^{238}U in Table 43-2—have half-lives comparable with the age of the Earth. It's not surprising that we find these occurring naturally. But most half-lives are much shorter. So why do we find short-lived isotopes at all? Some, like cosmic ray-produced carbon-14, result from naturally occurring nuclear reactions. Many others arise in the decay of long-lived isotopes, while some are produced artificially in nuclear reactors and particle accelerators.

Figure 43-21 shows the **decay series** for uranium-238, whose 4.46-billion-year half-life ensures that there's still plenty of it around. Even though they have

FIGURE 43-21 The decay of uranium-238 results in a series of shorter-lived nuclei that occur wherever uranium is present. Red arrows represent alpha decay, blue arrows beta decay. The half-life for each decay is indicated next to the arrow. The series ends with the stable isotope lead-206.

short half-lives, the daughter products in this series are present wherever uranium is found. The relative abundance of each is established by a balance between its own decay and its formation in the preceding decay. Note that one of the daughters in the uranium series is the chemically inert gas radon, which, as we saw in Example 43-5, can be a serious health hazard in enclosed buildings.

In 1934 Marie Curie's daughter Irène and her husband Frédéric Joliot-Curie (Fig. 43-22) were the first to induce artificial radioactivity, by bombarding stable isotopes with alpha particles. Enrico Fermi soon demonstrated that bombardment with slow neutrons could also produce new radioactive nuclei. Today we obtain radioisotopes in a variety of ways, including bombardment with particle beams, irradiation with neutrons from nuclear reactors, or by extracting them from the by-products of nuclear fission. As one example, oxygen-15, used in positron emission tomography (PET), is produced by bombarding ordinary nitrogen-14 with cyclotron-accelerated deuterons ($^2_1$H):

$$^{14}_{7}\text{N} + ^2_1\text{H} \longrightarrow ^{15}_{8}\text{O} + ^1_0 n,$$

where $^1_0 n$ is a free neutron (0 charge, 1 mass unit) that also results from the reaction. Because ^{15}O has a two-minute half-life, the cyclotron must be located right in the hospital doing the PET diagnosis (see Fig. 29-14).

FIGURE 43-22 Irène and Frédéric Joliot-Curie, discoverers of artificial radioactivity. The couple shared the 1935 Nobel Prize in Chemistry.

APPLICATION	*Uses of Radioactivity*

Here we survey just a few of the myriad uses for radioactive materials in our technological civilization, some of which are shown in Fig. 43-23.

Radioactive Tracers The use of small quantities of radioactive materials allows scientists, engineers, and physicians to trace material flows in a wide range of systems. Medical techniques such as brain, thyroid, or lung scans use isotopes of materials incorporated into these organs to study organ function. The more general technique of positron-emission tomography sends short-lived positron emitters through the body to provide internal images. Biologists use radioactive tracers routinely to trace the uptake and distribution of chemical compounds in biological organisms; "labeling" carbon dioxide with ^{14}C, for example, helps scientists unravel the carbon cycle and learn, perhaps, how plants can help alleviate global warming caused by carbon emissions from fossil fuels. Engineers use radioisotope tracers to study wear in mechanical systems. Incorporating radioisotopes in an engine bearing, for example, and then measuring the radioactivity of the

| (a) | (b) | (c) |

FIGURE 43-23 Some uses of radiation. (*a*) Treating cancer. (*b*) Food preservation. Both rolls are two weeks old, but the one on the right was irradiated to retard spoilage. (*c*) Bomb detection. This unit at New York's Kennedy Airport uses neutrons to produce short-lived radioisotopes in airline luggage; analysis of the decay radiation permits detection of explosives.

lubricating oil provides a measure of the rate of wear of the bearing surface.

Cancer Treatment Radiation destroys living cells, especially those that, like cancer cells, divide rapidly. By bombarding cancer tissue with radiation, the cancer cells can be selectively destroyed. Early efforts used gamma radiation from cobalt-60, but today's treatments rely on particle beams whose energy and therefore penetration can be precisely tailored to minimize damage to nearby tissue (Fig. 43-23a). Alternately, small amounts of radioisotopes can be embedded in the tumor region itself. Gold-98, a 65-hour beta emitter, is often used for this purpose, while injection of phosphorus-32 into the bloodstream helps fight blood cancers such as leukemia.

Food Preservation High radiation doses destroy bacteria and enzymes that cause food spoilage (Fig. 43-23b). Food does not become radioactive, although some argue that it may undergo chemical changes that alter nutritional value or introduce harmful compounds. In the United States, food irradiation with ^{60}Co gamma rays has been practiced from many years with spices, poultry, and some other foods. In the wake of the 1998 National Food Safety Initiative, though, food irradiation will likely become more widespread—although not without continuing controversy.

Insect Control Since radiation preferably damages rapidly dividing cells, high doses to the reproductive organs cause sterility. If large numbers of harmful insects are sterilized in this way, then released into the wild, they mate with normal insects but no offspring result. The population then dwindles. The medfly, a serious pest of citrus crops, has been partially controlled in this way.

Activation Analysis Bombarding materials with neutrons or other particles results in excitation of nuclei or the production of unstable isotopes. Analyzing the resulting radiation helps identify unknown materials. Art historians use this technique to detect art forgeries; chemists and environmental scientists identify unknown substances and pollutants; and airport security systems increasingly include activation analysis to search for explosives in luggage (Fig. 43-23c). Activation analysis has the advantage of being nondestructive, in that no material need be removed for the analysis. Half-lives of the activated nuclei are usually so short as to make any radiation hazard negligible.

Biological Effects of Radiation

In the early years of the twentieth century, radiation was believed beneficial. People actually paid for the privilege of spending time in old uranium mines, breathing the radioactive vapors. But we now know that radiation, despite its many beneficial uses in medicine, is harmful to living tissue.

Radiation works its damage because alpha, beta, and gamma particles have sufficient energy to ionize and otherwise disrupt biological molecules. Results include cell death, loss of necessary biological functions, mutations that lead to cancer, and mutations in reproductive cells that cause genetic changes. Numerous studies confirm that these effects occur. In the 1920s, watch hands were painted with radium-containing paint so they would glow in the dark. Factory workers painting the hands licked their brushes to keep the bristles together; nearly all of them died of bone cancer or radiation-induced anemia. Uranium miners, exposed to radon gas trapped in the mines, have much higher than average lung cancer rates. Marie Curie and her daughter Irène, along with other early nuclear scientists, succumbed to leukemia and other cancers that undoubtedly resulted from radiation exposure. (Pierre Curie was spared this fate; he was killed in the street when a horse bolted from its carriage.)

The energy absorbed in a dose of radiation is a crude measure of its biological danger. The SI unit of absorbed dose is the **gray** (Gy), defined as 1 J of energy absorbed per kg of absorbing material. (An older unit, the rad, equal to 0.01 Gy, is still widely used.) A more appropriate measure of biological radiation dose is the **sievert** (Sv), which is the absorbed energy per unit mass weighted by factors that compensate for differences in biological effectiveness among different types and energies of radiation. (An older unit, the rem, is equal to 0.01 Sv.) Alpha particles, for example, cause much more damage per unit energy delivered than do gamma rays, so 1 Gy of alpha radiation is more harmful than 1 Gy of gamma radiation. But 1 Sv of alphas and 1 Sv of gammas cause essentially the same biological damage.

The biological effects of high radiation doses are well known; exposure to 4 Sv, for example, causes death in about half of its human victims. But doses in the 0.1-Sv range and below are far more controversial. There are only a few cases of exposures to populations large enough that small radiation effects can be determined with statistical accuracy. One of the most important such populations are the survivors of the Hiroshima bombing, from whom many of our radiation-exposure standards derive. But today, more than half a century after the bombing, controversy continues over the doses delivered at Hiroshima. Our radiation safety standards remain, therefore, uncertain.

Even less certain are the effects of very low doses, such as the 10^{-5}-Sv average dose to people living near the 1979 Three Mile Island nuclear accident. Today these effects are estimated by linearly extrapolating from the effects of higher doses. But many radiation biologists believe that natural repair mechanisms in cells may act to diminish the effects of low doses, resulting in an overestimate of radiation dangers. On the other hand, varying susceptibility of different individuals, especially the young and unborn, may mean the danger is greater than current standards suggest.

The public has a deeply held fear of things nuclear, and of radiation in particular. Yet we're exposed to a great deal more natural radiation than we are to radiation associated with nuclear technologies. The average U.S. citizen, for example, receives about 3.6 mSv yearly; Fig. 43-24 shows the breakdown among natural and artificial sources. That number varies dramatically with geographical location—high altitudes and uranium-rich rocks lead to higher exposures—and with occupation. Airline crews, for example, often receive cosmic-ray radiation doses in excess of the doses allowed for nuclear power plant workers. And the average risk to your health from radiation—with the possible exception of indoor radon exposure in high-radon areas—is very small compared with many other risks you knowingly accept.

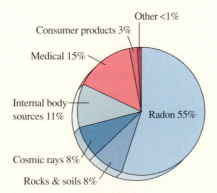

FIGURE 43-24 Natural (bluish) and artificial (reddish) sources of radiation, as percentages of the average yearly dose of 3.6 mSv. "Other" includes nuclear power plants, radioactive waste, weapons tests, and other sources.

(a)

(a)   *(b)*

FIGURE 43-25 *(a)* When a high-energy particle or gamma-ray photon enters the gas-filled tube of a Geiger counter, it knocks electrons from some gas atoms. Accelerated toward the positive central electrode, the electrons ionize additional atoms to produce a measurable pulse of current in the resistor. The output can drive an amplifier and loudspeaker, a meter, a digital counter, or sophisticated electronics for determination of particle energies. *(b)* Worker using a Geiger counter (arrow) to check for radiation during cleanup after the Three Mile Island nuclear accident.

Detecting Radiation

Our human senses can't detect nuclear radiation, but a number of technological devices make detection and measurement of radiation a routine process. Becquerel's 1896 discovery of radioactivity by the accidental fogging of photographic film showed that film can detect radiation. A more widely used detector is the **Geiger counter**, described in Fig. 43-25. Geiger counters and related detectors use radiation-induced ionization of gas atoms to produce pulses of electric current. Another class of detectors uses **scintillation**—the production of light flashes as atoms excited by incident radiation drop to lower energy states. A number of common plastics, as well as some crystalline materials and liquids, make efficient scintillators. The light flashes are usually detected by a photomultiplier (see Application: The Photomultiplier in Chapter 39) that converts them to electrical signals. Semiconductors provide another approach to radiation detection. As we discussed in the preceding chapter, electric current in semiconductors is carried by negative electrons, positive holes, or both. Radiation passing through a semiconductor can excite electrons to the conduction band, creating electron-hole pairs and producing a brief pulse of current that marks the passage of radiation. Detectors using these various mechanisms not only indicate the presence and quantity of radiation, but many can also provide accurate measures of alpha, beta, or gamma energies (recall Fig. 43-20).

43-6 MODELS OF NUCLEAR STRUCTURE

We've seen how the interplay of nuclear and electric forces, along with the phenomenon of weak-force beta decay, determines the general mix of neutrons and protons in the stable nuclei shown in Fig. 43-4. But look more closely at that figure: There are many more stable nuclei for even values of Z, and some—those with the so-called **magic numbers** 2, 8, 20, 28, 50, 82, or 126 protons or neutrons—have many more stable nuclei than do nearby values. Why? And why do unstable nuclei decay with the particular particle energies and half-lives they exhibit? Any detailed model of the nucleus must answer these questions.

Unfortunately there is no complete nuclear theory, analogous to the atomic theory presented in Chapter 41, that explains all aspects of all nuclei. The reasons for this have to do with our still imperfect knowledge of the nuclear force, and with the fact that nucleons are so tightly packed that each interacts with many others; we therefore cannot solve for nuclear structure with a simple two-particle model like we used for the single electron and nucleus in hydrogen-like atoms. Instead, nuclear physicists resort to several models to explain different aspects of nuclear structure.

The Liquid-Drop Model

We've seen that the density of nuclear matter is essentially independent of nuclear size—just as the density of a liquid drop doesn't depend on the drop size. Treating the nucleus as a liquid drop is a reasonable approximation for larger nuclei, whose many nucleons are like the molecules of a liquid. In this **liquid-drop model** the nuclear binding energy is analogous to the heat of vaporization, and surface-tension effects show why nucleons near the surface of a nucleus are less tightly bound. A semi-empirical formula derived from the liquid-drop model gives good agreement with observed nuclear binding energies, and the model helps explain the process of nuclear fission. A liquid-drop nucleus can rotate, vibrate, and change shape as long as its volume stays the same, and the resulting quantized energy levels lead to predictions of nuclear spectral lines—analogous to those we discussed for molecules in the preceding chapter—that are in good agreement with observation (Fig. 43-26).

Because a liquid is a continuum, however, the liquid-drop model cannot explain effects resulting from small changes in nucleon number. Thus, the model sheds no

FIGURE 43-26 Regularly spaced gamma rays from dysprosium-152. The dysprosium nuclei are set spinning rapidly by collisions, and the spectral lines result as they decay through the different rotational states.

FIGURE 43-27 Maria Goeppert Mayer with her daughter. Mayer and J. Hans Jensen shared the Nobel Prize for their work on the nuclear shell model.

light on why even atomic numbers are more stable or on why there are magic numbers with exceptional stability.

The Shell Model

The **nuclear shell model**, advanced in the late 1940s by physicists Maria Goeppert Mayer (Fig. 43-27) and J. Hans Jensen, treats the nucleus as having a shell structure similar to that of atoms. The shells occur because neutrons and protons obey the exclusion principle as they move in a square-well-like average potential. By including spin-orbit interaction in their model, Goeppert Mayer and Jensen explained the magic numbers as resulting from closed-shell structures similar to the electronic structures of the inert gases. The nucleons of a closed shell are tightly bound, giving the nucleus a high degree of stability. Additional nucleons are not bound into the closed-shell structure and stay primarily on the outskirts of the nucleus where they are more readily excited to higher energy levels.

Neutrons and protons in the shell model act independently, and each has its own set of quantum numbers. Closed-shell structure, and thus enhanced stability, occurs at magic numbers of either protons or neutrons, and thus, there are several stable nuclei for each magic value of atomic number Z or neutron number $A - Z$. Some nuclei, such as $_{20}^{40}\mathrm{Ca}$ ($Z = 20, A - Z = 20$) or $_{82}^{208}\mathrm{Pb}$ ($Z = 82$, $A - Z = 126$) are "doubly magic" and show exceptional stability.

In addition to explaining the magic numbers, the shell model also predicts the angular momenta of most nuclei. It is generally most successful for lighter nuclei and for those with proton or neutron numbers near the magic values.

The Collective Model

Neither the liquid-drop nor the shell model is a complete theory of the nucleus. Each is good at explaining certain aspects of nuclear structure but neither gives the whole picture. The **collective model**, advanced in part by Niels Bohr's son Aage, combines aspects of the liquid-drop and shell models. As in the shell model, magic numbers of nucleons group into filled "cores," surrounded by "extra" nucleons orbiting much like atomic electrons. But as in the liquid-drop model, the core shape, and with it the nuclear potential, can change as the nucleons undergo collective motions that give the model its name. These motions include wave-like circulations that help account for nuclear angular momenta and magnetic moments. One remarkable prediction of the collective model is that nuclei need not be spherical. The reason is that the values of the magic numbers change if the nuclear potential changes due to distortion from spherical shape. That means nuclei with nonmagic numbers may actually be more stable if they take on nonspherical shapes. In 1975 Aage Bohr shared the Nobel Prize with James Rainwater of the United States and Ben Mottelson of Denmark for their work on nonspherical nuclei in the collective model.

CHAPTER SYNOPSIS

Summary

1. Nearly all the mass of an atom is in the **nucleons**—protons and neutrons—that make up the **atomic nucleus**. Although massive, the nucleus occupies only a tiny fraction of the atomic volume. The number of protons is the **atomic number**, Z, and determines the chemical element. The total number of nucleons is the **mass number**, A. Nuclei with the same atomic number but different mass numbers are **isotopes** of the same element.

2. Nuclei are held together by the strong but short-range **nuclear force** that acts between nucleons. The nuclear force must overcome the long-range repulsive electric force between protons, which explains why nuclei must include neutrons, why there are more neutrons than protons in the more massive nuclei, and why stable nuclei do not exist for atomic number $Z > 83$.

3. The mass of a nucleus differs from the sum of the masses of its constituent nucleons by E_b/c^2, where E_b is the **binding energy**. The binding energy per nucleon is greatest for nuclei with mass numbers around 60, showing that these are the most stable nuclei. All nuclei have essentially the same density, and their radii are given approximately by $R = R_0 A^{1/3}$.

4. Coupling of nucleon spins gives many nuclei a net angular momentum and thus a net magnetic moment. Flipping the magnetic moment results in slight energy differences when the nucleus is in a magnetic field, either external or arising from atomic electrons. This energy shift causes **hyperfine splitting** of the hydrogen ground-state energy and is the basis of **nuclear magnetic resonance**.

5. Unstable nuclei are **radioactive**, and decay by emitting particles. In **alpha decay** the nucleus emits a helium-4 nucleus (alpha particle), and its mass number and atomic number drop by four and two, respectively. In most **beta decays**, a neutron decays into a proton, an electron (beta particle), and an antineutrino. The electron and antineutrino escape, leaving a nucleus with its atomic number increased by one but with no change in mass number. Some beta decays emit a positron and a neutrino, converting a proton to a neutron and dropping the atomic number by one. In **gamma decay**, an excited nucleus drops to a lower energy state, emitting a high-energy photon (gamma ray) with no change in either its atomic or mass number.

6. The decay of radioactive nuclei is characterized by the **decay constant** λ or, equivalently, the **half-life** $t_{1/2}$. A sample containing initially N_0 nuclei decays exponentially with time:

$$N = N_0 e^{-\lambda t} = N_0 2^{-t/t_{1/2}}.$$

The **activity** of a radioactive sample is the number of decays per unit time. The SI activity unit is the **becquerel** (Bq), equal to one decay per second. Activities decay with the same exponential factors given in the equation above. Short-lived isotopes occur naturally as a result of **decay series** of longer-lived species. Particle accelerators and nuclear reactors help create a variety of radioactive isotopes, with numerous uses in science, engineering, medicine, and industry.

7. Radiation is harmful to living tissues, in very rough proportion to the energy absorbed. The **gray** (Gy), equal to 1 J of energy absorbed per kg, is the SI measure of absorbed dose. The dose unit weighted for biological effectiveness is the **sievert** (Sv). Radiation doses of a few Sv are lethal to humans, but much controversy surrounds the effect of lower doses. A typical human receives several mSv of radiation dose yearly, most of it from natural sources.

8. There is not yet a complete theory of nuclear structure, but several models can explain different aspects of the nucleus. The **liquid-drop model** treats the nucleus as a deformable drop of constant-density liquid and is reasonably successful in predicting nuclear binding energies. The **shell model** explains the **magic numbers** of high nuclear stability in terms of closed nucleon shells analogous to the closed electron shells of inert gases. The **collective model** combines aspects of the two others and helps explain nonspherical nuclei.

Terms You Should Understand

(Pairs are closely related terms whose distinction is important; number in parentheses is chapter section where term first appears.)

nucleon (43-1)
nuclear force (43-2)
atomic number, mass number (43-2)
element, isotope (43-2)
hyperfine splitting (43-3)
nuclear magnetic resonance (NMR) (43-3)
binding energy (43-4)
radioactivity (43-5)
decay constant, half-life (43-5)
alpha, beta, gamma radiation (43-5)
neutrino (43-5)
decay series (43-5)
magic numbers (43-6)
liquid-drop, shell, and collective models (43-6)

Symbols You Should Recognize

A, Z (43-2)
^4_2He and similar nuclear symbols (43-2, 43-5)
λ, τ, $t_{1/2}$ (43-5)
ν, $\bar{\nu}$ (43-5)

Problems You Should be Able to Solve

formulating nuclear symbols (43-2)
calculating nuclear radii (43-3)
evaluating energies associated with nuclear magnetic moments (43-3)
relating binding energy and nuclear mass (43-4)

using half-lives or mean lifetimes to analyze radioactive decay (43-5)
writing equations for nuclear decay reactions (43-5)

Limitations to Keep in Mind

We do not yet have a complete theory of the atomic nucleus.

QUESTIONS

1. How did alpha-particle scattering experiments reveal the existence of the nucleus?
2. Why do nuclei contain neutrons?
3. Why do different isotopes have the same chemical behavior?
4. Why are there no stable nuclei for sufficiently high atomic numbers?
5. Lighter nuclei have roughly equal numbers of protons and neutrons, but heavier nuclei have more neutrons. Why?
6. Give an order-of-magnitude comparison between the sizes of atoms and nuclei.
7. Why is the MeV/c^2 a useful unit for the masses of nuclei and elementary particles?
8. How is the existence of a peak in the curve of binding energy significant for the evolution of stars?
9. Could a nucleus have more mass than that of its separate constituents?
10. Why might future archaeologists have problems dating samples from the second half of the twentieth century?
11. What types of radiation are affected by magnetic fields? Why?
12. Beta decay by positron emission is soon followed by a pair of 511-keV gamma rays. Why?
13. Cobalt-60 undergoes beta decay. Why, then, do gamma rays result?
14. What is the difference between a gray and a sievert?
15. How does the shell model explain magic numbers?
16. Why would it have been easier to make bombs fueled with uranium-235 a few billion years ago?
17. Why doesn't the level of carbon-14 in a living organism continue to build up until death?
18. Why are the nuclear magnetic resonance frequencies of the hydrogen nuclei in water (H_2O) and methane (CH_4) different? After all, the nuclei in both cases are just protons.
19. Why are iodine-131 and strontium-90 particularly dangerous radioisotopes?
20. List some beneficial and some harmful aspects of radiation.
21. Which model, liquid-drop or shell, would do a better job explaining (a) nuclear fission and (b) nuclear gamma-ray spectra?
22. When the Sun is very active it can lower the flux of cosmic rays entering the solar system. If this happened at some earlier time, what effect would it have on radiocarbon dating of samples from that time?

PROBLEMS

Section 43-1 Discovery of the Nucleus

1. In a head-on collision of a 9.0-MeV α particle and a nucleus in a gold foil, what is the minimum distance before electrical repulsion reverses the α particle's direction? Assume the gold nucleus remains at rest. Your answer shows how Rutherford set upper limits on the size of nuclei.
2. In a Rutherford-type experiment performed with 6.0-MeV α particles incident on iron, what is the closest approach to the center of an iron nucleus by an α particle on a head-on course?

Section 43-2 Building Nuclei: Elements, Isotopes, and Stability

3. Three of the isotopes of radon ($Z = 86$) have 125, 134, and 136 neutrons, respectively. Write symbols for each.
4. Write the symbol for the germanium isotope with 44 neutrons.

5. How do (a) the number of nucleons and (b) the nuclear charge compare in the two nuclei $^{35}_{17}Cl$ and $^{35}_{19}K$?
6. Examine Fig. 43-4 to find the element that has the most stable isotopes.
7. Use Fig. 43-4 to find atomic numbers $Z < 83$ for which there are no stable isotopes. What elements are these?

Section 43-3 Properties of the Nucleus

8. Compare the radius of the proton (the $A = 1$ nucleus) with the Bohr radius of the hydrogen atom.
9. Write the symbol for a boron nucleus with twice the radius of ordinary hydrogen (^1_1H).
10. How does the minimum distance calculated in Problem 1 compare with the radius of the gold nucleus?
11. A uranium-235 nucleus splits into two roughly equal-size pieces. What is their common radius?
12. To what size would Earth have to collapse to be at nuclear density?

13. Find the energy needed to flip the spin state of a proton in Earth's magnetic field, whose magnitude is about 30 μT.

14. An NMR spectrometer is described as a "300-MHz instrument," meaning 300 MHz is the frequency supplied to its transmitter coil to flip the spin states of bare protons. What is the strength of its unperturbed magnetic field?

15. The permanent magnet from an old NMR spectrometer has a field strength of 1.4 T. What frequency of electromagnetic radiation was used in the instrument?

Section 43-4 Binding Energy

16. What is the total binding energy of oxygen-16, given that its nuclear mass is 15.9905 u?

17. Determine the atomic mass of nickel-60, given that its binding energy is very nearly 8.8 MeV/nucleon.

18. Find the nuclear mass of plutonium-239, given its atomic mass 239.052157 u.

19. Find the total binding energy of sodium-23, given its atomic mass of 22.989767 u.

20. Iron-56 is among the most tightly bound nuclei. Its mass is 55.9206 u. Find the binding energy per nucleon, and check against Fig. 43-11.

21. The mass of a lithium-7 nucleus is 7.01435 u. Find the binding energy per nucleon.

22. Calculate the binding energy per nucleon for gold-197, given its atomic mass of 196.96654 u.

23. By what percentage is the binding energy of $^{64}_{30}$Zn different from that of $^{64}_{29}$Cu, another isotope with the same mass number? Use atomic masses of 63.92915 for ^{64}Zn and 63.92977 for ^{64}Cu.

Section 43-5 Radioactivity

24. Use the half-life given in Table 43-2 to calculate the decay constant for uranium-235.

25. The decay constant for argon-46 is 0.0835 s^{-1}. What is its half-life?

26. Copper-64 can decay by any of the three beta-decay processes. Write equations for each decay.

27. Referring to Fig. 43-21, write equations describing the decays of (a) radon-222 and (b) lead-214.

28. Carbon-10 is a positron emitter with a 19.3-s half-life. (a) Write an equation to describe its decay. (b) A ^{10}C sample has initial activity 48 MBq. What will be its activity after 1 minute, in Bq?

29. A milk sample shows an iodine-131 activity level of 450 pCi/L. What is its activity in Bq/L?

30. Nuclear bomb tests of the 1950s deposited a layer of strontium-90 over Earth's surface. How long will it take—from the time of the bomb tests for (a) 99% and (b) 99.9% of this radioactive contaminant to decay?

31. In the 4.5-billion-year lifetime of the Earth, how many half-lives have passed of (a) carbon-14, (b) uranium-238, and (c) potassium-40? For each isotope, give the number of atoms remaining today from a sample of 10^6 atoms at Earth's formation.

32. You measure the activity of a radioactive sample at 2.4 MBq. Thirty minutes later, the activity level is 1.9 MBq. What is the decay constant for the radioactive material?

33. How many atoms in a 1-g sample of U-238 decay in 1 minute? *Hint:* This time is so short compared with the half-life that you can consider the activity essentially constant.

34. Plot the fraction of the initial atoms remaining in a carbon-14 sample for times from $t = 0$ to $t = 40,000$ years.

35. The Swedish standard for I-131 in milk is 2000 Bq/liter. How long would Swedes have waited following the Chernobyl accident for their milk to be considered safe, assuming the same initial contamination level as in Example 43-6?

36. If radon infiltration into a house were stopped, what fraction of the initial radon would be left after (a) 1 day, (b) 1 week, (c) 1 month?

37. Nitrogen-13 has a 10-min half life. A sample of ^{13}N contains initially 10^5 atoms. Plot the number of atoms as a function of time from $t = 0$ to $t = 1$ hour. Make your horizontal axis (time) linear, but your vertical axis logarithmic. Why is the curve a straight line? What is the significance of its slope?

38. Thorium-232 is an α emitter with a 14-billion-year half life. Radium-228 is a β^- emitter with a 5.75-year half-life. Actinium-228 is a β^- emitter with a 6.13-hour half-life. (a) What is the third daughter in the thorium-232 decay series? (b) Make a chart similar to Fig. 43-21 showing the first three decays in the thorium series.

39. How much cobalt-60 ($t_{1/2} = 5.24$ years) must be used to make a laboratory source whose activity will exceed 1 GBq for a period of 2 years?

40. Archaeologists unearth a bone (Fig. 43-28) and find its carbon-14 content is 34% of that in a living bone. How old is the archaeological find?

FIGURE 43-28 How old is the bone? (Problem 40)

41. Marie Curie and Pierre Curie won the 1903 Nobel Prize for isolating 0.1 g of radium-226 chloride (^{226}RaCl$_2$). What was the activity of their sample, in Bq and in curies?

42. Below are reported levels of iodine-131 contamination in milk for several countries affected by the 1986 Chernobyl accident, along with each country's safety guideline.

Given that I-131's half-life is 8.04 days, how long did each country have to wait for I-131 levels to decline to the level it considered safe?

LEVEL (Bq/L)

COUNTRY	REPORTED	SAFETY GUIDELINE
Poland	2000	1000
Sweden	2900	2000
Austria	1500	370
Germany	1184	500

43. A mixture initially contains twice as many atoms of sodium-24 ($t_{1/2} = 15$ h) as it does of potassium-43 ($t_{1/2} = 22.3$ h). Plot the decay curves for the two isotopes, and use your graph to determine when the numbers of atoms are equal.

44. How many atoms are in a radioactive sample with activity 12 Bq and half-life 15 days?

45. Analysis of a moon rock shows that 82% of its initial K-40 has decayed to Ar-40, a process with a half-life of 1.2×10^9 years. How old is the rock?

46. Neutron activation of the stable nitrogen isotope $^{15}_{7}$N in a bomb-detection system turns it into unstable $^{16}_{7}$N, which decays by beta emission with a 7.13-s half-life. How long after activation will the N-16 activity have dropped by a factor of one million?

47. An excited state of the isotope technetium-99, designated Tc-99*, has a 6.01-hour half-life and is widely used in nuclear medicine. Tc-99* decays to Tc-99, an unstable isotope with a 2.13×10^5-year half-life. A 0.10-μg sample initially contains pure Tc-99*. Find its activity level at (a) $t = 0$, (b) $t = 1$ day, (c) $t = 1$ week, (d) $t = 1$ year, and (e) $t = 10$ years. The atomic weight of Tc-99 is 98.9 u.

48. In the preceding problem, what fraction of the activity after 1 week is due to Tc-99* decay?

49. Today, uranium-235 comprises only 0.72% of natural uranium; essentially all the rest is U-238. Use the half-lives given in Table 43-2 to determine the percentage of uranium-235 in natural uranium when Earth formed, about 4.5 billion years ago.

50. Figure 43-29 shows the activity as a function of time for a mixture of two radioisotopes. Find (a) the half-life and (b) the initial activity for each. *Hint:* Note that the activity is plotted logarithmically, and that the graph consists approximately of two straight lines connected by a curved stretch.

Section 43-6 Models of Nuclear Structure

51. For each of the magic numbers, give an isotope that has a magic number of protons and one with a magic number of neutrons.

FIGURE 43-29 Problem 50.

52. List all the doubly magic stable isotopes—those with magic numbers of both protons and neutrons.

Paired Problems

(Both problems in a pair involve the same principles and techniques. If you can get the first problem, you should be able to solve the second one.)

53. The nuclear mass of $^{48}_{22}$Ti is 47.9359 u. Find the binding energy per nucleon.

54. Find the atomic mass of iridium-193, for which the binding energy is 7.94 MeV/nucleon.

55. Geologists are looking for underground sites that could store nuclear wastes securely for 250,000 years. What fraction of plutonium-239 initially in such waste would remain after that time? See Table 43-2.

56. One of the longer-lived isotopes produced in nuclear explosions is cesium-137, with a 30-year half-life. Atmospheric testing by the major nuclear powers ended in 1963. What fraction of the Cs-137 in the environment at that time remains today?

57. A sample of oxygen-15 ($t_{1/2} = 2.0$ min) is produced in a hospital's cyclotron. What should be the initial activity concentration if it takes 3.5 min to get the O-15 to a patient undergoing a PET scan for which an activity of 0.5 mCi/L is necessary?

58. Technetium-99*, an excited state with a 6.01-hour half-life, is widely used in brain scans. How much Tc-99* must a manufacturer produce if it is to arrive at a hospital 3.5 hours later with 10 mg of Tc-99* remaining?

Supplementary Problems

59. If a solid sphere with the 1.9-cm radius of a ping-pong ball were made entirely of nuclear matter, what would be the acceleration due to gravity at its surface? Compare with that at Earth's surface.

60. Show that the gravitational acceleration at the surface of a sphere of nuclear matter is $g = 6.4 \times 10^7 R$, where g is in m/s^2 and R is the sphere's radius in meters. (General relativity limits this result to cases where the surface escape speed is much less than the speed of light.)

61. How cool would you have to get a material before the thermal energy kT was insufficient to excite protons in a 35-T magnetic field from their lower to upper spin state?

62. Alpha particles emitted by polonium-209 describe a circle of radius r in a magnetic field B, where the product Br has the value 0.332 T·m. What are (a) the speed and (b) the energy of the alpha particles?

63. The atomic masses of uranium-238 and thorium-234 are 238.050784 u and 234.043593 u, respectively. Find the energy released in the alpha decay of ^{238}U.

64. In the alpha decay of the preceding problem, what percentage of the original U-238 mass gets converted to energy?

65. Some human lung cancers in smokers may be caused by polonium-210, which arises in the decay series of uranium-238 that occurs naturally in fertilizers used on tobacco plants. Write equations for (a) the production and (b) the decay of ^{210}Po. (c) How might the health effects of Po-210 differ if its half-life were 1 day or 10 years instead of the actual 138 days?

66. When lithium-7 absorbs a neutron, the resulting nucleus first beta decays by electron emission, then splits into two identical pieces. Identify the pieces.

67. In the preparation of a transuranic nucleus, $^{209}_{83}$Bi and $^{54}_{24}$Cr interact to form a heavy nucleus plus a neutron. Identify the heavy nucleus.

68. It is possible but difficult to realize the alchemists' dream of synthesizing gold. One reaction involves bombarding mercury-198 with neutrons, producing, for each neutron captured, a gold-197 nucleus and another particle. Write the equation for this reaction.

69. Nickel-65 beta decays by electron emission with decay constant $\lambda = 0.275$ h^{-1}. (a) What is the daughter nucleus? (b) In a sample that is initially pure Ni-65, how long will it be before there are twice as many daughter nuclei as parent nuclei?

70. Free neutrons have a half-life of about 10 min. A nuclear fission reaction produces neutrons with kinetic energies around 1 MeV. (a) How far, on average, can such a neutron get from the reaction site before it decays? (b) Repeat for a neutron with thermal energy kT typical of room temperature.

71. The **mean lifetime**, τ, of a radioactive nucleus is the average of the times from $t = 0$ to $t = \infty$, weighted by the number dN of nuclei that decay in each time interval dt, that is,

$$\tau = \frac{\int_{t=0}^{\infty} t \, dN}{\int_{t=0}^{\infty} dN}.$$

Evaluate this quantity to show that $\tau = 1/\lambda$. *Hint:* Write dN as $\left| dN/dt \right| dt$.

44 Nuclear Energy: Fission and Fusion

Nuclear energy is 10 million times more concentrated than chemical energy. One consequence is the awesome destructive power of nuclear weapons.

Nuclear energy plays a major role in the natural universe, powering the Sun and other stars. On Earth, applications of nuclear energy became a serious possibility in the 1930s, a reality in the 1940s, and a fact of life in the 1950s. Since then, the use of nuclear energy for both peaceful and military purposes has increased substantially. Proliferation of nuclear weapons throughout the world presents a serious threat to human survival. On a more positive note, nuclear fission supplies more than 15% of the world's electrical energy, and nuclear fusion raises the prospect of a cleaner, safer nuclear power source with nearly limitless fuel resources.

FIGURE 44-1 The curve of binding energy, showing regimes where fusion and fission can result in the release of nuclear energy.

44-1 ENERGY FROM THE NUCLEUS

In the preceding chapter we saw how the curve of binding energy leads to the possibility of nuclear energy release through **fusion**—the joining of lighter nuclei to form heavier ones nearer the peak of the binding energy curve (Fig. 44-1). Fusion in the interiors of long-dead stars created the nuclei of which we're made, while fusion in the Sun provides the energy that sustains life on Earth. Fusion energy released in our thermonuclear weapons, on the other hand, has the potential to destroy that life.

For nuclei heavier than iron (Fe), energy release can occur if a nucleus splits, or **fissions**, into two lighter nuclei. In contrast to fusion, nuclear fission seems to be of little consequence in the cosmic scheme of things. Technologically, though, nuclear fission is important: it is the one major energy-releasing nuclear reaction that we can now sustain in a controlled manner to produce electricity, while uncontrolled fission reactions play an important role in nuclear weapons.

44-2 NUCLEAR FISSION

The 1932 discovery of the neutron by James Chadwick gave physicists an excellent tool for probing the atomic nucleus. Unlike protons or alpha particles, the neutron carries no electric charge and therefore can penetrate the nucleus without having to overcome the coulomb repulsion. Following earlier work by the Italian physicist Enrico Fermi, the German chemists Otto Hahn and Fritz Strassmann in 1938 bombarded uranium (atomic number $Z = 92$) with neutrons. They were puzzled to find among the reaction products radioactive versions of the much lighter elements barium ($Z = 56$) and lanthanum ($Z = 57$). Physicist Lise Meitner (Fig. 44-2) and her nephew Otto Frisch soon interpreted

FIGURE 44-2 Lise Meitner and Otto Hahn. Meitner and her nephew Otto Frisch interpreted Hahn and Strassmann's experiments as evidence of neutron-induced fission of uranium. Meitner, an Austrian physicist, had fled to Sweden to escape Hitler. Earlier, as a woman in a male-dominated field, she had braved sexist policies that denied her access to laboratories when men were present. By the 1930s she had become one of the world's most respected nuclear physicists. Element 109 (meitnerium) now bears her name.

FIGURE 44-3 Painting of the first nuclear reactor, built under the stands of the University of Chicago stadium during World War II. No photographs were taken because of wartime secrecy. Note the man in the pit, manually adjusting a control rod to increase the nuclear reaction rate.

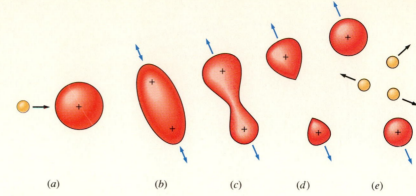

(a)   (b)   (c)   (d)   (e)

FIGURE 44-4 Neutron-induced fission of ^{235}U.

these unusual findings, concluding that neutron bombardment had caused the uranium nuclei to fission into two parts. Word of the discovery spread throughout the world's physics community, and with it the realization that fission represented an energy source many orders of magnitude more potent than chemical reactions. It was the eve of World War II, and the military implications were obvious and ominous. The United States initiated a program to develop fission explosives, hoping to produce nuclear weapons before the Germans did. With the help of the international physics community, many of whom had fled Europe to escape Fascism, the U.S. effort succeeded. In 1945, only seven years after the discovery of fission, the world's first nuclear explosion was detonated in the New Mexico desert. A few weeks later, fission bombs devastated the Japanese cities of Hiroshima and Nagasaki, bringing World War II to an end. In the course of development work on the bomb, scientists led by Enrico Fermi had also constructed the world's first nuclear reactor. Built under the stands of the University of Chicago stadium, the reactor became operational in 1942 (Fig. 44-3).

Nuclear fission occurs when a massive nucleus splits into two lighter parts. Although the process can occur spontaneously in a variety of heavy nuclei, such **spontaneous fission** is extremely rare. More common is **induced fission**, occurring typically when a heavy nucleus absorbs a neutron. (High-energy protons and gamma rays can also induce fission.) In a common fission reaction, a ^{235}U nucleus absorbs a neutron to form a highly excited ^{236}U nucleus. The ^{236}U undergoes vigorous oscillations that deform it into a dumbbell shape (Fig. 44-4). The two ends of the dumbbell are positively charged, and they still experience the long-range electrical repulsion. But the short-range nuclear force is less effective in holding the distant ends together. If the oscillation amplitude is great enough, the nucleus reaches a point of no return at which the coulomb repulsion dominates and the two ends of the dumbbell fly apart. This fissioning occurs very rapidly; the intermediate ^{236}U nucleus lasts only about 10^{-12} s.

The two lighter nuclei formed in the fission process are **fission fragments**. The composition of the fission fragments varies, depending on exactly how a given nucleus splits. The most likely outcome is a pair of unequal-mass fragments, with mass numbers in the ranges 90 to 100 and 135 to 145 more probable; other distributions occur with lower probability (Fig. 44-5). As we saw in the preceding chapter, the ratio of neutrons to protons for stable nuclei increases with increasing mass number. Since the fission fragments share the protons and neutrons of the parent uranium, these lighter nuclei have an excess of neutrons

FIGURE 44-5 Distribution of fission fragments for fission of ^{235}U induced by slow neutrons.

for their mass, and are therefore highly unstable (Fig. 44-6). This instability manifests itself in an almost immediate "boiling off" of neutrons from the fission fragments. Typically two to three neutrons are emitted for each fission event; the average number is about 2.47 neutrons per ^{235}U fission. The fragments remaining are still highly unstable and therefore very radioactive; in particular, their radioactivity greatly exceeds that of the original uranium prior to fissioning.

To summarize, induced fission of ^{235}U involves these steps:

1. A neutron strikes a ^{235}U nucleus and is absorbed, forming highly excited ^{236}U.
2. The ^{236}U undergoes violent oscillations that distort it into a dumbbell shape. In the dumbbell configuration, the electrical repulsion overcomes the strong nuclear force, and in about 10^{-12} s the ^{236}U nucleus fissions into two unequal-mass fragments.
3. The fission fragments almost immediately "boil off" several excess neutrons, leaving two nuclei that are still highly radioactive.

We can describe the fission event using the standard notation for nuclear reactions. In writing such a reaction, we generally skip the short-lived intermediate stages of ^{236}U and the fission fragments before neutron emission; the general neutron-induced fission of ^{235}U can then be written

$$ {}_0^1 n + {}_{92}^{235}\text{U} \longrightarrow X + Y + b\,{}_0^1 n, \qquad (44\text{-}1) $$

where X and Y stand for the fission fragments and b is the number of neutrons emitted. The reaction conserves electric charge and the number of nucleons, so the atomic numbers Z that subscript the element symbols, and separately the mass number superscripts, must sum to the same values on both sides of the reaction. A specific example is a ^{235}U fission that produces isotopes of molybdenum (Mo) and tin (Sn):

$$ {}_0^1 n + {}_{92}^{235}\text{U} \longrightarrow {}_{42}^{102}\text{Mo} + {}_{50}^{131}\text{Sn} + 3\,{}_0^1 n $$

(This particular reaction is significant because ^{131}Sn quickly undergoes a series of beta decays leading to ^{131}I, an especially dangerous environmental contaminant that lodges in the thyroid gland; see Example 43-6.)

FIGURE 44-6 Chart of the stable nuclei, simplified from Fig. 43-4. Thick curve is the region of stability, extended to include the long-lived unstable isotopes through uranium. Thin straight line has the neutron-to-proton ratio of uranium and thus of fission products. The latter lie well above the stable region, and are therefore radioactive.

EXAMPLE 44-1	*A Fission Reaction*

Neutron-induced fission of ^{235}U results in the formation of ^{141}Ba, three neutrons, and a second fission fragment. What is that fragment?

Solution

From Appendix D we find that barium, Ba, has atomic number $Z = 56$. So the reaction is

$$ {}_0^1 n + {}_{92}^{235}\text{U} \longrightarrow {}_{56}^{141}\text{Ba} + X + 3\,{}_0^1 n, $$

where X is the unknown fragment. The atomic numbers on the left sum to 92, reflecting the presence of 92 protons. So there must be 92 protons on the right; with $Z = 56$ for barium, element X must have $Z = 92 - 56 = 36$. Consulting Appendix D, we see that element 36 is krypton, Kr.

On the left side of the equation, the neutron contributes one unit to the total mass number and the uranium contributes 235, reflecting a total of 236 nucleons participating in the reaction. With mass number $A = 141$ for the barium fragment, and with the three neutrons each contributing one mass unit, the mass number of the krypton fragment must be $236 - (141 + 3) = 92$. So the second fragment is krypton-92, or $_{36}^{92}\text{Kr}$.

EXERCISE Neutron-induced fission of ^{235}U produces rubidium-93 and two neutrons. What is the other fission product?

Answer: cesium-141

Some problems similar to Example 44-1: 4–6

Something besides fission fragments and neutrons appears in the fission. That something is energy—a substantial amount of energy. Energy is released because the total binding energy of the original nucleus is less than the sum of the binding energies of the widely separated fission fragments, as you can see in Fig. 44-1. A slight potential barrier, associated with unbalanced forces on the nucleons at the edge of the nucleus, keeps the unexcited uranium nucleus from fissioning. When a neutron is absorbed, the binding energy of the neutron appears as vibrational energy of the nucleus, and the fission process shown in Fig. 44-4 ensues. As the fission fragments fly apart, they gain energy from the electrical repulsion.

A total energy of about 200 MeV is released in a typical uranium fission event. Most of this—about 165 MeV—appears as kinetic energy of the fission fragments. Neutron kinetic energy accounts for about 5 MeV, while approximately 7 MeV are released almost instantaneously as gamma rays emitted by the excited fission fragments. After a slight delay, the fission fragments emit somewhat over 20 MeV additional energy in the form of high-speed beta particles (electrons or positrons), gamma rays, and neutrinos.

EXAMPLE 44-2	*Fission Energy*

Assuming 200 MeV per fission event, how much pure ^{235}U should be fissioned to provide the same energy as the burning of 1 metric ton (1000 kg) of coal?

Solution
Consulting the "Energy Content of Fuels" table in Appendix C, we find that coal burning releases 2.9×10^7 J/kg, so 1000 kg of coal releases an energy $E = 2.9 \times 10^{10}$ J. With 1.6×10^{-19} J/eV, uranium fission releases about $(200 \times 10^6 \text{ eV/fission}) \times (1.6 \times 10^{-19} \text{ J/eV}) = 3.20 \times 10^{-11}$ J/fission.

So we need a total of

$$\frac{2.9 \times 10^{10} \text{ J}}{3.2 \times 10^{-11} \text{ J/fission}} = 9.06 \times 10^{20} \text{ fission events}.$$

Each fission event involves one ^{235}U nucleus, so we need this many nuclei. The mass of a ^{235}U nucleus is approximately 235 u, so the mass of ^{235}U required is

$(9.06 \times 10^{20} \text{ nuclei})(235 \text{ u})(1.66 \times 10^{-27} \text{ kg/u}) = 3.5 \times 10^{-4}$ kg.

Thus, 0.35 g of pure ^{235}U—about one one-hundredth of an ounce—can supply the same energy as a ton of coal! The ratio of coal to uranium mass is about 3×10^6, showing that uranium packs over a million times as much energy per unit mass as does coal. This ratio is typical of the difference between nuclear and chemical fuels, and shows why a nuclear power plant may be refueled once a year, while a comparable coal-burning plant needs many hundred-car trainloads of coal each week.

EXERCISE Estimate the explosive energy yield in joules and in megatons (see Appendix C) for a fission bomb made from 18 kg of U-235, under the unrealistic assumption that all the U-235 fissions.

Answer: 1.48×10^{15} J $= 0.35$ Mt

Some problems similar to Example 44-2: 7, 8, 41, 42

Besides ^{235}U, a number of other nuclides can undergo neutron-induced fission; these are said to be **fissionable**. Fissionable nuclides that will fission with neutrons of any energy—especially relatively low thermal energy—are called **fissile**. In a fissile nucleus, the barrier is very low and the required neutron energy is therefore negligible. Only three fissile nuclides are known: they include the uranium isotopes ^{233}U, ^{235}U, and the plutonium isotope ^{239}Pu. By far the most important of these are ^{235}U and ^{239}Pu.

Although ^{235}U occurs naturally, it presently constitutes only about 0.72% of natural uranium (nearly all the rest—99.27%—is ^{238}U). For most uses, uranium must be enriched in ^{235}U, to several per cent for commercial power reactors and often over 90% for weapons. **Uranium enrichment** is a difficult and expensive

FIGURE 44-7 Uranium-enrichment technologies. (*a*) In gaseous diffusion, uranium is first combined chemically with fluorine to form uranium hexafluoride gas (UF_6). The gas is pumped into a chamber divided by a membrane. The lighter, faster moving $^{235}UF_6$ (light color) is more likely to get through the membrane. After many such cells the desired concentration is reached. (*b*) United Nations worker destroying an Iraqi centrifuge following the 1991 Persian Gulf war. The high-speed centrifuge had been built to separate uranium isotopes as part of Iraq's well-advanced nuclear weapons program.

process; since the isotopes ^{235}U and ^{238}U are chemically similar, enrichment schemes make use of the very slight mass difference between the two. Among the techniques used are centrifuging, gaseous diffusion (Fig. 44-7), and selective ionization of ^{235}U by lasers. Enrichment technology is highly sensitive because a nation possessing it can readily produce weapons-grade uranium. Pakistan, the newest member of the nuclear weapons "club," used centrifuges built in part with smuggled American components to enrich uranium for its bombs.

Plutonium-239, with a half-life of 24,110 years, does not occur in nature. It is produced artificially by neutron bombardment of ^{238}U. This reaction first produces the highly unstable isotope ^{239}U, which undergoes beta decay with a half-life of 23.5 minutes to form ^{239}Np. The ^{239}Np again decays by beta emission, with a half-life of 2.35 days, leaving ^{239}Pu. This sequence of reactions leading to plutonium can be written

$$\begin{align} {}^{1}_{0}n + {}^{238}_{92}U &\longrightarrow {}^{239}_{92}U \\ {}^{239}_{92}U &\longrightarrow {}^{239}_{93}Np + e^- + \overline{\nu} \\ {}^{239}_{93}Np &\longrightarrow {}^{239}_{94}Pu + e^- + \overline{\nu}. \end{align}$$

Although ^{239}Pu is produced in copious amounts in nuclear reactors (see Problem 55), **reprocessing** spent reactor fuel to extract plutonium is difficult and dangerous. Contamination with other plutonium isotopes further complicates the separation of fissile plutonium. Like uranium enrichment, plutonium separation is a sensitive technology; nevertheless, all the nations except China and Pakistan that are known to have developed nuclear weapons chose plutonium over uranium for their first nuclear explosions. And in the 1990s, several European countries and Japan had embarked on ambitious plutonium reprocessing programs for commercial power reactors, including intercontinental shipments of plutonium.

A number of other isotopes, most importantly ^{238}U, are fissionable with fast neutrons that bring in enough kinetic energy to overcome the potential barrier. But ^{238}U fission does not result in significant neutron emission, so, for reasons we will discuss in the next section, a self-sustaining ^{238}U fission reaction is not possible. However, fast-neutron fission of ^{238}U plays a significant role in thermonuclear weapons, as we will see later.

Chain Reactions

Neutrons induce fission, and fission itself may release neutrons—the very things needed to induce fission. This fact makes possible a **chain reaction**, in which each fission event supplies neutrons that give rise to more such events.

What's required to sustain a chain reaction? At a minimum, neutrons from each fission event must, on the average, cause one more nucleus to fission. Otherwise the reaction will fizzle to a halt, in which case the configuration of fissile material is called a **subcritical mass**. If each fission event causes, on the average, exactly one additional fission event, then the reaction continues with a constant rate of energy release; the fissile material then comprises a **critical mass**. Recall that the average number of neutrons released in ^{235}U fission is 2.47; thus in a critical reaction most of the neutrons do not cause additional fission. If, on the other hand, an average of more than one neutron from each fission causes additional fission, then the reaction rate grows exponentially with time (Fig. 44-8). In this case the configuration is **supercritical**.

Quantitatively, the criticality of a fissile mass is described by the **multiplication factor** k, which is just the average number of neutrons from a fission event that cause additional fission. A subcritical mass has $k < 1$, a critical mass has

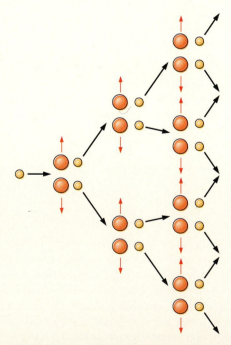

FIGURE 44-8 A supercritical chain reaction. In this case the multiplication factor $k = 2$, since two neutrons from each fission cause additional fission. In reality some neutrons are lost, and some fission events produce more than two neutrons.

$k = 1$, and a supercritical mass has $k > 1$. The value of k is determined by several factors, including the proportion of fissile isotope, the concentration of neutron-absorbing substances, and the size and configuration of the mass. With $k = 2$, for example, a single fission event results in two more fissions; they cause a total of four more fissions, then eight, and so on. In general, the number of fissions increases by a factor of k with each successive generation of fission events; therefore, the number of fission events occurring after n such generations is k^n. The *total* number, N, of fission events that occur by the nth generation is the sum $N = 1 + k + k^2 + k^3 + \cdots + k^n$. You may recognize this sum as a geometric series; in any event, you can show by mathematical induction that it has the value

$$N = \frac{k^{n+1} - 1}{k - 1}. \qquad (44\text{-}2)$$

The average time between successive generations of fission events is called the **generation time**. As the example below indicates, short generation times can lead to explosive release of nuclear energy.

EXAMPLE 44-3	*Explosion!*

A 14-kg sample of pure ^{235}U is assembled into a supercritical mass with multiplication factor $k = 1.8$. If the generation time between fission events is $\tau = 10$ ns, how long does it take the entire mass to fission? How much energy is released in the process?

Solution

With an approximate atomic weight of 235 u, our 14-kg mass contains a number N of nuclei given by

$$N = \frac{14 \text{ kg}}{(235 \text{ u/nucleus})(1.66 \times 10^{-27} \text{ kg/u})} = 3.59 \times 10^{25} \text{ nuclei}.$$

We need to find the number of fission generations that must occur before this number of nuclei have fissioned. Solving Equation 44-2 for k^{n+1} gives

$$k^{n+1} = N(k - 1) + 1.$$

With $N \simeq 10^{25}$, the $+1$ on the right-hand side is totally negligible. Dropping it, and taking logarithms of both sides, we have

$$\ln(k^{n+1}) = \ln[N(k - 1)].$$

But $\ln(x^y) = y \ln x$, so

$$(n + 1) \ln k = \ln[N(k - 1)].$$

Solving for n then gives

$$n = \frac{\ln[N(k - 1)]}{\ln k} - 1 = \frac{\ln[(3.25 \times 10^{25})(1.8 - 1)]}{\ln(1.8)} - 1$$

$$= 99,$$

where we have rounded to the nearest whole number. Thus, the time for the entire mass to fission is 99τ—that is, 990 ns or just under one microsecond.

We can get the energy released from the "Energy Content of Fuels" table in Appendix C or by using the approximate figure of 200 MeV per fission event. Choosing the latter approach, we have

$$E = (200 \times 10^6 \text{ eV/fission})(1.6 \times 10^{-19} \text{ J/eV})$$

$$\times (3.59 \times 10^{25} \text{ fissions})$$

$$= 1.1 \times 10^{15} \text{ J}.$$

To get a feel for this number, note that it is about 3.2×10^8 kWh, or nearly 2 weeks' energy output of a large electric power plant, released instead in a microsecond! Or, comparing our 14-kg supercritical mass with the 0.35 g of ^{235}U that we found in Example 44-2 to be equivalent to a ton of coal, we find that the energy released in the 1-μs supercritical chain reaction is equivalent to burning $(14 \text{ kg})/(0.35 \times 10^{-3} \text{ kg } ^{235}\text{U/ton coal}) = 40{,}000$ tons of coal in one microsecond. The reaction is truly explosive, and our supercritical mass is, of course, a nuclear bomb.

EXERCISE A terrorist group designs a crude nuclear weapon made from 25 kg of ^{235}U with generation time $\tau = 2.0$ μs and multiplication factor $k = 1.08$. How long would it take this weapon to consume its fuel, assuming—as is unlikely—that it could remain together during the process.

Answer: 1.5 ms

Some problems similar to Example 44-3: 14–16, 43, 44, 48

44-3 APPLICATIONS OF NUCLEAR FISSION

Rarely has a basic discovery led as quickly to significant technological applications as did nuclear fission. Following the 1938 discovery of fission, the pressures of wartime brought the first nuclear reactor to criticality in 1942 and the first nuclear weapon to detonation in 1945. Both nuclear reactors and nuclear weapons have developed considerably since that time. Today, reactors are used for power generation, propulsion of ships, research, and the production of plutonium for weapons. Fission in weapons now serves to trigger and to complement fusion in the most destructive nuclear devices. Some milestones in fission technology include the first full-scale commercial nuclear power plant at Calder Hall in England going on line in 1956 and the launching of the first nuclear-powered submarine, the *U.S. Nautilus,* also in 1956. Other nuclear projects have fared less well; early enthusiasm for nuclear-powered rockets and aircraft was quickly dampened, although proposals for flights to Mars have revived interest in nuclear rockets.

Fission Weapons

As Example 44-3 shows, assembling a critical mass of fissile material can result in a colossal explosion. And the mass required is not large—about 15 kg for uranium-235 and less than 5 kg for plutonium-239. Weapons based on these simple facts of nuclear fission have been a frightening reality for half a century. Here and again later in this chapter we discuss the physics of nuclear weapons not out of any enthusiasm for these horrific devices, but because, for better or worse, they play a central and potentially devastating role in the modern world. The threat of war between the nuclear superpowers has waned with the end of the Cold War and the nuclear arms reduction treaties of the 1990s—although the United States and Russia still possess many thousands of nuclear weapons each, and with them comes a nonzero probability of a devastating nuclear exchange. More threatening is the proliferation of nuclear weapons to other countries, a threat made all the more real with the 1998 detonation of nuclear devices by India and Pakistan, triggering a potential nuclear arms race on the South Asian subcontinent. The danger of nuclear proliferation arises directly from physics of the fission chain reaction and from the small quantity of fissile material needed for a critical mass.

The major technological difficulty in actually producing a fission weapon is to assemble a supercritical mass rapidly enough that the explosive chain reaction consumes most of the fissile material before it blows apart. The occurrence of spontaneous fission means that a supercritical mass, once assembled, will almost instantly explode. Fission weapons therefore start with fissile material in a subcritical configuration and, at detonation time, rapidly reconfigure the material into a supercritical arrangement. The simplest fission weapon is the gun-type device shown in Fig. 44-9. The weapon that destroyed Hiroshima was a gun-type uranium bomb whose critical mass was about the size of a large grapefruit (Fig. 44-10). The explosive yield was equivalent to about 12,500 tons

Chemical explosive
Gun barrel
U-235 target
U-235 bullet
Steel tamper

FIGURE 44-9 A gun-type fission weapon. The U-235 "bullet" and target are both subcritical masses. Igniting the chemical explosive propels the bullet into the target, forming a critical mass. The heavy steel tamper holds the fissioning mass together briefly, ensuring more complete fission.

FIGURE 44-10 The gun-type nuclear bomb that destroyed Hiroshima. The device had a mass of just over 3000 kg, of which only about 50 kg was uranium. Of that, less than 1 kg was consumed in the fission reaction—yet it yielded the explosive equivalent of 12,500 tons of chemical explosive.

FIGURE 44-11 The Hiroshima bomb, technologically primitive and low in explosive yield, nevertheless devastated the city and caused 100,000 immediate deaths.

(12.5 kt) of the chemical explosive trinitrotoluene (TNT). [Explosive yields of nuclear weapons are commonly expressed by giving the weight in tons of TNT that would release the same explosive energy. One kiloton (kt) is equivalent to 4.18×10^{12} J; for larger weapons the megaton (MT), equivalent to 4.18×10^{15} J, is used.] Despite its awesome destructive power (Fig. 44-11), the Hiroshima weapon was quite inefficient. As the following example shows, fission actually occurred in less than 1 kg of its approximately 50 kg of enriched uranium.

EXAMPLE 44-4	*Destruction of Hiroshima*

What mass of ^{235}U actually fissioned to produce the 12.5-kt explosive yield of the Hiroshima bomb?

Solution
Converting kilotons to joules and then to MeV, we have for the explosive energy

$$E = \frac{(12.5 \text{ kt})(4.18 \times 10^{12} \text{ J/kt})}{(1.6 \times 10^{-19} \text{ J/eV})(10^6 \text{ eV/MeV})} = 3.27 \times 10^{26} \text{ MeV} .$$

At 200 MeV per fission event, this energy requires

$$\frac{3.27 \times 10^{26} \text{ MeV}}{200 \text{ MeV/fission}} = 1.63 \times 10^{24} \text{ fissions} .$$

Each fission event involves one ^{235}U nucleus, with an approximate mass of 235 u; therefore, the total mass M of ^{235}U undergoing fission is

$$M = (1.63 \times 10^{24} \text{ nuclei})(235 \text{ u/nucleus})(1.66 \times 10^{-27} \text{ kg/u})$$
$$= 0.64 \text{ kg} .$$

This is only about 1% of the total uranium mass in the bomb. Modern fission explosives are far more efficient, consuming about 30% of their fissile fuel.

EXERCISE How much uranium-235 would be needed to produce a 150-kt fission weapon with 30% efficiency?

Answer: 25 kg

Some problems similar to Example 44-4: 9–13

A more sophisticated design is required for a plutonium fission weapon, where neutrons from spontaneous fission make it more likely that the weapon will "preignite" and blow apart before complete fission takes place. Consequently plutonium weapons are generally of the implosion type shown in Fig. 44-12. A barely subcritical sphere of plutonium is surrounded symmetrically by chemical explosives. When the explosives detonate, they compress the sphere to a supercritical density, and the nuclear explosion follows. The world's first nuclear explosive, detonated at the Trinity Site in New Mexico, was a 20-kt implosion-type plutonium device, as was the bomb dropped on Nagasaki.

FIGURE 44-12 An implosion-type fission weapon. Chemical explosives compress a barely subcritical sphere of ^{239}Pu to critical density, initiating an explosive chain reaction.

Practical weapons have some efficiency-boosting features not shown in our simple diagrams. For example, neutron-emitting material is often embedded in the fissile mass to ensure more complete fission. For the same reason, neutron-reflecting material surrounds the uranium or plutonium to reduce neutron losses and therefore increase the multiplication factor k.

Although the construction of a fission weapon is straightforward, acquisition of weapons-grade fissile material is not. As we noted earlier, enrichment of uranium with the rare isotope ^{235}U is difficult and costly. Removal and purification of ^{239}Pu from spent reactor fuel poses similar problems. For these reasons, efforts to halt the spread of nuclear weapons have concentrated on keeping enriched uranium and plutonium under strict control, and on curbing the spread of enrichment and reprocessing technologies. Nevertheless, the spread continues. Seven nations have openly tested nuclear explosives, and several more almost certainly have them. Investigation of Iraq's nuclear-weapons program following the Persian Gulf War shows that a nation seeking nuclear weapons capability can approach that goal undetected. The breakup of the Soviet Union in 1991 poses a further proliferation peril, as fissile materials and perhaps even weapons themselves become items of illicit international commerce. These developments, coupled with the immense destructive power in the hands of any nation or terrorist group possessing nuclear weapons, makes our nuclear age a dangerous time.

Nuclear Reactors

In a **nuclear reactor**, the fission chain reaction is controlled rather than explosive; that is, the reaction is held just critical, with $k = 1$ and with a constant rate of energy release. Since the maximum possible k is 2.47 (the average number of neutrons per fission event in ^{235}U), a reactor must be designed so most neutrons don't induce additional fission events; otherwise it would quickly run out of control. In commercial power reactors k is limited in part by keeping the concentration of fissile ^{235}U low—typically a few per cent—so that many neutrons are absorbed by ^{238}U instead of causing fission. Since uranium enrichment is expensive, this is also an economical approach, and in addition ensures that reactor fuel cannot readily be converted to weapons use.

The high-energy neutrons that emerge in fission turn out not to be very effective at causing additional fission events, and in most reactor designs the concentration of ^{235}U is too low for these so-called fast neutrons to sustain the chain reaction. In contrast, slow neutrons—those moving at about the mean thermal speed—are much more effective in causing fission. Consequently most reactors include a substance called the **moderator** that slows neutrons to thermal speeds. A reactor using these **thermal neutrons** is called a **thermal reactor**. In a thermal reactor, neutrons collide elastically with moderator nuclei, transferring energy to them. In Chapter 11 we found that the maximum energy transfer occurs when colliding objects have equal mass; therefore, the best moderators are low-mass nuclei (see Problem 18). Water, with the protons of its hydrogen atoms serving as moderator nuclei, is the most widely used moderator in U.S. power reactors. Its main disadvantage is that it can absorb neutrons, forming the hydrogen isotope deuterium (^2H); the absorbed neutrons are then lost to the chain reaction. Canadian reactors use heavy water (deuterium oxide, ^2H_2O) as the moderator. The probability of neutron absorption by deuterium is much less than for ordinary hydrogen, so fewer neutrons are lost. With more neutrons available, Canadian reactors can use natural, unenriched uranium as fuel.

FIGURE 44-14 A pressurized-water reactor, the most common type of commercial power reactor in the United States.

Canadian-style heavy-water reactors may have been used in India's development of nuclear weapons.)

Whatever the reactor details, the ultimate purpose of a nuclear power plant is to produce steam that drives a turbine that, in turn, is connected to a generator. Electromagnetic induction in the spinning generator drives electric current that is the power plant's useful product. The second law of thermodynamics (see Chapter 22) limits the efficiency with which heat energy can be converted to mechanical and electrical energy; as a result, about two-thirds of the energy generated in the core of a typical power reactor is dumped to the environment as waste heat. This heat is extracted in the power plant's condenser, where spent steam from the turbine is condensed to water.

In the simplest water-cooled reactor designs, the core cooling water is itself boiled to produce steam. A U.S. light-water reactor of this type is called a **boiling-water reactor**. Although simple and economical, the BWR has the disadvantage that water circulating through the turbine and condenser is highly radioactive. A more common U.S. design is the **pressurized-water reactor** (PWR), in which the core cooling water is kept under pressure to prevent boiling. Heat is transferred from this primary coolant to a secondary loop where water boils and drives the turbine. Figure 44-14 is a diagram of a typical PWR, while Fig. 44-15 shows a power plant using the PWR design. CANDU and gas-cooled reactors, like the PWR, also transfer heat from the primary coolant to a secondary boiling-water loop.

FIGURE 44-15 The San Onofre nuclear power plant hugs the pacific coast near San Diego. Its two pressurized-water reactors are cooled with ocean water; hence the absence of large cooling towers.

FIGURE 44-16 New fuel rod bundles being lowered into the core of a nuclear reactor. The blue glow, called Cerenkov radiation, is from high energy beta particles (electrons) interacting with the reactor's cooling water.

As fission proceeds in a reactor core, the concentration of fission fragments in the fuel increases. Before the ^{235}U is exhausted, these fission products absorb enough neutrons to interfere with the chain reaction. Fuel rods must therefore be replaced at regular intervals. In U.S. light-water reactors, about one-third of the fuel is replaced each year, so a given fuel rod remains in place for 3 years (Fig. 44-16). In Canadian CANDU and Russian RBMK reactors, coolant circulates through pipes rather than in a pressure vessel surrounding the entire reactor. In these reactors, fuel rods can be reached while the reactor is in operation, and refueling takes place on a nearly continuous basis. Another reaction that occurs in the nuclear fuel is the conversion of ^{238}U to plutonium, via the neutron-capture reaction we discussed earlier. As plutonium builds up, it begins to fission in significant quantities. Near the end of a fuel rod's 3-year residency in the reactor core, in fact, only 30% of the energy production is from ^{235}U fission. Most of the rest—54%—is from fission of ^{239}Pu, with the remainder from other plutonium isotopes. Figure 44-17 outlines the evolution of fuel in the reactor.

As Fig. 44-17 indicates, spent reactor fuel is contaminated with highly radioactive fission products. It also contains significant amounts of fissile ^{235}U and ^{239}Pu. Through the expensive, technologically complex, and dangerous technique of nuclear fuel reprocessing, these fissile isotopes can be extracted and used to make new reactor fuel or nuclear weapons. The high cost and threat of

FIGURE 44-17 Evolution of 1000 kg of 3.3% enriched uranium over its 3-year stay in a reactor core. Of the 33 kg of ^{235}U initially present, 25 kg are consumed in fission. In addition, 24 kg of ^{238}U are converted to plutonium and other transuranic isotopes, some of which also undergo fission. The fuel mass remains largely ^{238}U, contaminated with 35 kg of radioactive fission products and lesser amounts of transuranic elements formed by neutron capture.

FIGURE 44-18 Containers of radioactive waste being stored at an underground facility in France, which gets 70% of its electricity from nuclear power. The United States, which has just over 20% nuclear-generated electricity, has yet to complete a long-term disposal site for spent fuel waste.

weapons proliferation in a plutonium-reprocessing economy have led some nations—notably the United States—to forgo reprocessing. Others have eagerly developed reprocessing technology, and international shipments of plutonium began in the 1990s. Whether or not nuclear waste is reprocessed, the spent fuel ultimately requires disposal. The presence of long-lived radioactive isotopes means that the material remains dangerous for thousands of years. To date, no entirely satisfactory method of disposal has been developed, although it is generally agreed that underground storage will prove the safest alternative (Fig. 44-18).

The thermal reactors we have been discussing use uranium fuel only slightly enriched in ^{235}U. Although some of the ^{238}U comprising the bulk of the fuel is converted to plutonium, most remains energetically useless. In contrast, a **breeder reactor** is designed to convert large amounts of ^{238}U to plutonium, "breeding" more fissile fuel. Use of breeder reactors would greatly extend our supplies of fissile fuels, by making much of the 99.27% of natural uranium that is nonfissile ^{238}U into fissile ^{239}Pu. For this reason, breeder technology has been pursued in a number of countries, particularly in France and Japan. Breeders operate at higher temperatures and use fast instead of slow neutrons to induce fission. They have no moderator and their coolant is liquid sodium. High temperature and the use of fast neutrons make these reactors less stable and technologically more sensitive than nonbreeding designs (see Problem 23). Furthermore, a breeder-based power system inherently involves reprocessing and the movement of large amounts of plutonium—of which only a few kg suffice to make a fission weapon. Because of these technological, safety, and proliferation concerns, the future of breeder reactors is unclear.

APPLICATION	*The Chernobyl Accident*

In April 1986, the Chernobyl nuclear power plant in the then-Soviet Ukraine exploded, spewing more than 10^{16} Bq of radioactive material into the atmosphere (Fig. 44-19). Two people were killed in the explosion, and hundreds more died of radiation sickness. Fallout affected much of eastern Europe and Scandinavia, and tens of thousands of people are expected to develop fatal cancers as a result. Details of the Chernobyl accident illustrate many aspects of basic nuclear physics and of reactor engineering and control.

Ironically, the accident occurred during a test of the power supply for the **emergency core-cooling system** (ECCS), designed to dump water on the reactor core in the event of a loss-of-coolant accident. Mismanagement of the test, serious operator errors, and reactor design all contributed to the

FIGURE 44-19 Technicians checking for radiation inside the damaged Chernobyl nuclear power plant.

FIGURE 44-20 Sequence of events leading to the Chernobyl reactor explosion, shown as a plot of reactor power as a function of time. Time scale is not linear.

Chernobyl disaster. Figure 44-20 outlines the sequence of events leading to the April 26 explosion.

Preparation for the test began at 1:00 A.M. on April 25 as operators slowly decreased the reactor output from its normal 3200 MW thermal power to 1600 MW, a process that took 12 hours. Then, following the test plan, they disconnected the turbine-generator and disabled the emergency core-cooling system. Although the plan called for shutting off the ECCS to prevent its coming on and disturbing the test, this move violated the reactor's operating procedures. When the operators resumed lowering the reactor power, one of them failed to set an automatic control that would have maintained a thermal power at 700 to 1000 MW. The power level plunged to a mere 30 MW.

We noted earlier that fission products act as "poisons," absorbing neutrons and thereby inhibiting fission. A particularly virulent reactor poison is xenon-135, whose cross section for neutron absorption is 4400 times that of ^{235}U. Xenon-135 forms in the 6.7-h-half-life decay of ^{135}I. In normal operation, the ^{135}Xe concentration reaches a steady level in which neutron capture destroys the isotope as quickly as ^{135}I decay creates it. But when reactor power decreases, neutron production drops and with it the destruction of ^{135}Xe. But ^{135}I continues to decay into ^{135}Xe, so xenon concentration increases. The "poisoning" effect then makes it difficult to raise the reactor power until several of the xenon's 9.2-h half-lives have passed.

At Chernobyl, an operator's error had resulted in a rapid power drop, leading to high ^{135}Xe concentration. Impatient to complete the test, operators committed another safety violation: To compensate for neutron absorption in the xenon, they withdrew too many control rods. By 1:19 A.M. on April 26, they had managed to raise the power to 200 MW, still well below the 700 MW minimum needed for the test. About the same time they turned on two additional cooling water pumps, as called for in the test procedures.

In a U.S. light-water reactor, the cooling water is also the moderator. But in a graphite-moderated reactor such as Chernobyl, the dominant effect of water is to absorb neutrons. So the additional water required withdrawal of still more control rods. Now the reactor was in a dangerous situation: An increase in power would boil water, decreasing neutron absorption and thus increasing the fission rate. That would make the water boil even faster, increasing the power even more, and a runaway reaction could result. Worse, with so few control rods in place, the reaction might be sustained by prompt neutrons alone, resulting in a power increase too rapid to halt with mechanical control rods.

The Chernobyl operators realized they had too much water, and at 1:22 A.M. they reduced the flow. But they did not immediately reinsert control rods. Thirty seconds later a computer warned that the reactor should be shut down. Ignoring the warning, operators continued the test by diverting steam from the turbine-generator. The decreased load caused more water to boil, again reducing neutron absorption. The reactor went supercritical from prompt neutrons alone, and the power level soared by a factor of 4000 in 5 seconds. The power surge ruptured water pipes, causing a steam explosion that lifted the heavy concrete reactor cover. A second explosion followed, caused perhaps by hydrogen generated from steam reacting with the zirconium cladding on the fuel rods. The graphite moderator caught fire, and heavy smoke carried highly radioactive fission products into the atmosphere. Substantial radiation release continued for 10 days, and fallout dropped on much of Europe (see Example 43-6).

Could a Chernobyl accident happen in the United States? For commercial light-water reactors the answer is decidedly no. Loss of the water coolant/moderator in a light-water reactor immediately halts the chain reaction, making a runaway reaction impossible. But that doesn't mean light-water reactors are entirely safe. Even after the chain reaction stops, the immense heat generated by radioactive decay is enough to melt the core. A partial **meltdown** occurred in the 1979 Three Mile Island accident in Pennsylvania, but fortunately the reactor's containment structure held in nearly all the radioactivity. The threat of a hydrogen explosion during that accident—a possibility not previously considered in reactor accident scenarios—showed that we may not yet realize all the potentially dangerous situations possible in a system as complex as a nuclear power plant.

Is Nuclear Power Safe?

Opinions are strong on both sides of this question. The small but nonzero probability of a catastrophic accident makes nuclear power seem especially dangerous in the public mind. Radiation—invisible and poorly understood by most people—makes the environmental and health effects of nuclear power seem particularly insidious. And its relation to nuclear weapons makes nuclear power seem far from benign.

The important question is not whether nuclear power is safe but how it compares with other energy sources—especially its main competitor, coal. Coal mining accidents regularly kill as many people as did Chernobyl, and the latter's excess cancer fatalities are no match for the 10,000 deaths that occur *each year* in the United States alone as a result of air pollution from coal-burning power plants. And the continued use of fossil fuels, especially carbon-rich coal, has started us on the path of irreversible climatic change that could bring environmental disaster.

Statistically, the risks from nuclear power seem low compared with those of its fossil-fuel alternatives. And those risks are orders of magnitude lower than others many people willingly endure, like smoking, failing to wear seat belts, or persisting in high-fat diets. Nevertheless, unanswered questions about the long-term effects of low-level radiation, about nuclear waste disposal, and most significantly about the proliferation of nuclear technology in a politically un-stable world raise doubts even for those who agree that nuclear power is statis-tically quite safe. Many take the compromise position that views nuclear fission as a bridge to a time when safer, more economical, and more sustainable power sources become available. In Chapter 42 we argued that solar photovoltaic cells may soon be one such source. Nuclear fusion, which we now discuss, may be another.

44-4 NUCLEAR FUSION

The curve of binding energy (Fig. 44-1) shows that nuclear energy can be released either by fission of heavy nuclei or by fusion of light nuclei. The binding-energy curve is steepest near its left end, showing that the most energy per nucleon is released by fusing the very lightest element—hydrogen. Indeed, the fusion reactions powering the Sun and many other stars begin with the fusion of hydrogen to form deuterium:

$$^1_1H + \,^1_1H \longrightarrow \,^2_1H + e^+ + \nu \quad (0.42 \text{ MeV}); \quad (44\text{-}3a)$$

a positron (e^+) and neutrino (ν) are also released, and the total energy liberated in the reaction is 0.42 MeV. Deuterium then fuses with hydrogen to form the helium isotope ^3_2He:

$$^1_1H + \,^2_1H \longrightarrow \,^3_2He + \gamma \quad (5.49 \text{ MeV}). \quad (44\text{-}3b)$$

19.3

Fusion

Here γ represents a gamma ray, and the quantity in parentheses is the total energy released. Helium-3 nuclei from two such reactions then react to give a single ^{4}He nucleus and a pair of protons; this event liberates 12.86 MeV:

$$^3_2He + \,^3_2He \longrightarrow \,^4_2He + 2\,^1_1H \quad (12.86 \text{ MeV}). \quad (44\text{-}3c)$$

There is one additional energy-producing reaction associated with these events: the positron from reaction 44-3a annihilates with an electron, forming two gamma rays with a total energy of $2mc^2$ or 1.022 MeV:

$$e^+ + e^- \longrightarrow 2\gamma \quad (1.022 \text{ MeV}). \quad (44\text{-}3d)$$

The reactions 44-3a–c constitute the **proton-proton cycle**. In the full cycle, reactions 44-3a and b occur twice for each occurrence of reaction 44-3c. The net

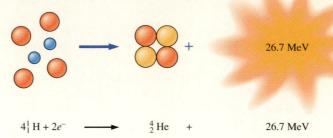

$$4\,{}^{1}_{1}H + 2e^{-} \longrightarrow {}^{4}_{2}He + 26.7\ \mathrm{MeV}$$

FIGURE 44-21 The net result of the proton-proton cycle in Equations 44-3a–d is the conversion of 4 protons and 2 electrons into a helium-4 nucleus, in the process releasing 26.7 MeV of energy.

effect, including two occurrences of the annihilation reaction 44-3d, is to convert four protons and two electrons into a single helium-4 nucleus; a total of 26.7 MeV is released in the process (Fig. 44-21; see also Problem 30). In massive stars, ${}^{4}_{2}He$ is then the building block for the formation of still heavier nuclei by additional fusion reactions, as we discussed in the preceding chapter.

Although ordinary hydrogen (${}^{1}_{1}H$) is abundant, the reaction 44-3a does not occur readily. Terrestrial fusion research has therefore focused on reactions involving the heavier hydrogen isotopes. Of most immediate interest are deuterium-deuterium (D–D) and deuterium-tritium (D–T) reactions, listed below along with the energy released in each:

$${}^{2}_{1}H + {}^{2}_{1}H \longrightarrow {}^{3}_{2}He + {}^{1}_{0}n \qquad (3.27\ \mathrm{MeV}) \qquad (44\text{-}4a)$$

$${}^{2}_{1}H + {}^{2}_{1}H \longrightarrow {}^{3}_{1}H + {}^{1}_{1}H \qquad (4.03\ \mathrm{MeV}) \qquad (44\text{-}4b)$$

$${}^{2}_{1}H + {}^{3}_{1}H \longrightarrow {}^{4}_{2}He + {}^{1}_{0}n \qquad (17.6\ \mathrm{MeV}). \qquad (44\text{-}4c)$$

The two possible outcomes of the D–D reaction have nearly equal probability.

The electrical repulsion between nuclei makes it difficult to bring them close enough to fuse. In potential-energy terms, nuclei must overcome the potential barrier associated with the electrostatic force before they can drop into the deep potential well of the stronger but shorter-range nuclear force (Fig. 44-22). As

FIGURE 44-22 Potential-energy diagram for two nuclei, showing electrostatic potential barrier and deep well associated with the attractive nuclear force.

FIGURE 44-23 Power loss by radiation and power produced by D–D and D–T fusion reactions, as functions of temperature on a log-log plot. Ignition temperature (dashed lines) is the temperature at which fusion power equals radiation loss.

we discussed in Chapter 40, nuclei can quantum-mechanically tunnel through the potential barrier when they lack sufficient energy to overcome it. Although the possibility of tunneling lowers the energy needed to initiate fusion, that energy still remains high. In the Sun's core, for example, fusing nuclei approach one another with energies on the order of 1 keV, corresponding to a temperature of 15 MK.

In terrestrial applications, simply achieving fusion is not sufficient for net energy production. At fusion temperatures, atoms are stripped of their electrons and the material forms a hot plasma, or ionized gas. As electrons and nuclei interact in the plasma, their speeds and directions of motion change. Since the particles are charged, those accelerations result in electromagnetic radiation that represents a loss of energy from the plasma. Radiation losses increase with increasing temperature, but power generated by fusion increases even faster (Fig. 44-23). The temperature at which fusion-generated power exceeds power loss by radiation is the **critical ignition temperature**. For the D–D reactions of Equations 44-4a and b, Fig. 44-23 shows that ignition temperature is about 6×10^8 K; for the D–T reaction of Equation 44-4c it is roughly an order of magnitude lower, or about 5×10^7 K.

The high temperatures required for fusion pose two problems: first, how to achieve those temperatures and, second, how to contain the fusing material. The stars solve both problems with their immense gravitational fields. A star is born as interstellar material—mostly hydrogen and helium—collapses under its own gravity; the resulting compression heats the material to fusion temperatures. Once fusion has begun, the star settles into an equilibrium in which the high pressure of the fusing material is balanced by the gravitational force. In stars more massive than the Sun, fusion proceeds through stages in which the products of one fusion stage become fuel for the next. In that way, the stars produce all the elements from lithium through iron—elements that are ultimately recycled into space in the stellar explosions called supernovae. Those elements condense to form new stars, planets, and sometimes living things. With the exception of hydrogen, essentially all the atoms in our bodies were formed by this process of fusion in ancient stars.

Gravitational confinement is not possible in terrestrial fusion applications, and we must find another means of confining the fusing material. For net fusion energy production, confinement must last long enough for the energy produced by fusion to exceed the energy needed to heat the material. The heat input required is proportional to the number of nuclei being heated, or, on a per-unit-volume basis, to the density, n. (Fusion researchers generally use the particle density, n—the number of particles per unit volume—rather than the mass density.) However, the rate of fusion energy production per unit volume is proportional to the *square* of the density. You can see this by considering half the nuclei as targets to be struck by the other half. If you double the number of target nuclei alone, you double the fusion rate. Doubling the density doubles the number of targets and the number of projectiles hitting them, and therefore quadruples the fusion rate. The total energy produced by fusion is, in turn, given by the fusion rate multiplied by the time τ during which fusion is occurring. Requiring that the fusion energy, proportional to $n^2\tau$, exceed the heating energy, proportional to n, gives a minimum value for the quantity $n\tau$—the product of density and confinement time—that must be met in an energy-producing fusion device. This condition on $n\tau$ is called the **Lawson criterion**. For the D–D and D–T reactions of Equation 44-4, the Lawson criteria are approximately

$$n\tau > 10^{22} \text{ s/m}^3 \qquad \text{(D–D)}$$
$$n\tau > 10^{20} \text{ s/m}^3 \qquad \text{(D–T)}. \tag{44-5}$$

The factor-of-100 difference between these two Lawson criteria shows that D–T fusion is much more readily achieved.

The Lawson criterion offers a choice in the design of a fusion device: Strive for a high plasma density with a short confinement time or a lower density with a longer time. Two distinct approaches emphasize these two possibilities. **Inertial confinement** relies on the inertia of the reacting particles—that is, on their inability to be accelerated instantaneously away from the reaction site—to provide confinement; very short confinement times are required in this scheme. Inertial confinement occurs in fusion weapons and in particle-beam and laser-fusion devices. In **magnetic confinement**, the more traditional approach to controlled fusion, complicated magnetic field configurations confine the fusion plasma at lower densities but for longer times. However the Lawson criterion is met, it is of course also necessary to surpass the critical ignition temperature.

Fusion Weapons

Although peaceful fusion devices still elude scientists and engineers, fusion weapons have been with us since the 1950s. Explosives powered substantially by nuclear fusion are often called "hydrogen bombs" to distinguish them from nuclear fission explosives (incorrectly called "atomic bombs"). Fusion-based devices are more correctly called **thermonuclear weapons**, a term reflecting the high temperature needed to ignite fusion. As we will see, there is at present no such thing as a pure fusion weapon, and the distinction between fission and fusion weapons is blurred by a variety of fission-fusion hybrids.

The earliest use of fusion in nuclear weapons was to boost the yield of fission explosives. The D–D and D–T reactions of Equations 44-4a and c release neutrons. If a small quantity of deuterium-tritium mixture is placed at the center of an implosion-type fission weapon (see Fig. 44-12), the material will be compressed to ignition temperature by the fission explosion, and the neutrons released in the subsequent fusion will ensure more complete fissioning of the uranium or plutonium. Such fusion-boosted fission weapons were first tested in 1951, and today tritium is an essential element in weapons of the U.S. nuclear arsenal.

The effort to produce fusion weapons was fraught with controversy; many scientists who had worked on fission bombs during World War II were horrified at what they had produced and felt that the world did not need the 1000-fold increase in destructive power that fusion promised. For a while it also seemed technologically impossible to produce a fusion weapon compact enough to be delivered by aircraft; early designs involved enormous refrigerators to maintain tanks of liquid deuterium and tritium. But in the early 1950s, Edward Teller and Stanislaw Ulam devised a clever configuration wherein the energy of a fission explosion is focused on a mixture of solid lithium deuteride (LiD) and plutonium-239 (Fig. 44-24). The mixture is compressed to fusion ignition temperature, while fission-produced neutrons react with lithium to produce tritium and helium:

$$^{6}_{3}\text{Li} + ^{1}_{0}n \longrightarrow ^{4}_{2}\text{He} + ^{3}_{1}\text{H}.$$

The deuterium and tritium undergo fusion that provides typically half the weapon's explosive yield. The remainder comes from fission in a layer of natural uranium (predominantly ^{238}U) surrounding the fusion fuel. Although ^{238}U cannot sustain a fission chain reaction, it will fission when bombarded by fast neutrons. In a thermonuclear weapon, the D–T fusion reaction is the source of those neutrons. The high-yield thermonuclear weapons in today's nuclear

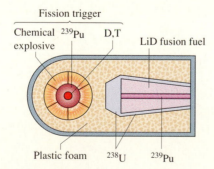

Fission trigger

Chemical ^{239}Pu D,T   LiD fusion fuel
explosive

Plastic foam ^{238}U ^{239}Pu

FIGURE 44-24 A fission-fusion-fission weapon. Although details remain classified, the device is known to involve a sequence in which energy from the fission trigger is focused by plastic foam onto the lithium deuteride and plutonium-239. Neutrons from plutonium fission convert lithium to tritium. Deuterium-tritium fusion ensues, releasing energy and neutrons. The neutrons induce fission in the ^{238}U surrounding the fusion fuel.

FIGURE 44-25 A 10-megaton thermonuclear test explosion in the Pacific Ocean during the 1950s.

arsenals generally involve the fission-fusion-fission sequence outlined here (Fig. 44-25).

In a fission weapon, explosive yield is limited by the amount of material that can participate in the chain reaction before the weapon blows apart. But neither the fusion explosion in a thermonuclear weapon nor the fissioning of its ^{238}U requires a chain reaction. For this reason, there is no upper limit to the explosive yield of a thermonuclear weapon. Devices as large as 58 MT have been exploded, yielding almost 5000 times the energy of the fission bomb that destroyed Hiroshima. Following development of thermonuclear weapons in the 1950s, arsenals of the major nuclear powers contained many weapons in the 10- to 20-MT range. The development of ballistic missiles then led to an emphasis on smaller devices with higher yield-to-weight ratios (Fig. 44-26). The yields of strategic warheads today are typically in the 40 kT to 1 MT range, and the missiles that deliver them may carry ten or more such warheads (Fig. 44-26).

The impact of any nuclear weapon is horrifying, but thermonuclear weapons are particularly destructive. A 1-MT explosion causes third-degree burns in

(a)

(b)

FIGURE 44-26 Nuclear weapons have become smaller and more efficient. (a) A model of the 21-kt Nagasaki bomb beside a 335-kt MK-12A warhead. (b) The Minuteman II missile carries three MK-12A's, each capable of destroying a city. When fully implemented in 2007, the Strategic Arms Reduction Treaty (START II) will eliminate such multiple-warhead missiles based on land.

FIGURE 44-27 Destruction of a wood-frame house during a nuclear test in the 1950s. At left, house is illuminated with visible radiation from the blast. At right, 2.3 s later, shock wave arrives and the house disintegrates from the 5 psi overpressure. In a 1-Mt nuclear explosion, this level of destruction covers an area of 80 km²—somewhat more than the area of Manhattan.

people as far as 11 km from the blast; shock-wave over-pressures in excess of the 5 psi needed to collapse most buildings extend out to 5 km (Fig. 44-27; see also Problem 54). Fission-fusion-fission weapons also produce a great deal of radioactive fallout from uranium fission. Depending on wind patterns, this fallout can deliver lethal radiation doses hundreds of kilometers from the blast. With tens of thousands of nuclear weapons still in the world's arsenals, and with many nations striving to join the nuclear "club," this application of nuclear technology is truly a threat to human civilization.

Controlled Fusion: Inertial Confinement

In a fusion weapon, confinement occurs because the reactions are over before the explosion blows the reacting material apart. **Inertial confinement** fusion (ICF) schemes seek to apply the same strategy, with miniature fusion explosions taking place under controlled conditions (Fig. 44-28). Experimental ICF devices use laser or particle beams to compress pellets of deuterium-tritium fuel to enormous densities for confinement times on the order of 10^{-11} to 10^{-9} s. If the pellet is struck symmetrically from many directions, rapid heating of its outer layers drives an inward-propagating shock wave that compresses the material about 10,000-fold, resulting in densities higher than at the center of the Sun! The Nova laser fusion experiment at Lawrence Livermore National Laboratory in California achieves a peak power of about 10^{14} W—well above the output of all the world's electric power generating plants. The Nova device delivers its energy in 0.1 to 1 ns through 10 converging beams

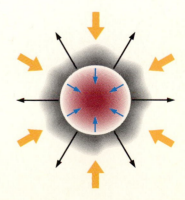

FIGURE 44-28 Implosion of a fusion pellet. Laser beams (yellow arrows) vaporize the outer surface of the pellet, which expands outward. By Newton's third law, the core of the pellet experiences an inward force that compresses it to fusion conditions.

(a)

(b)

FIGURE 44-29 (a) Lasers converge on the target chamber of the NOVA laser-fusion facility at Lawrence Livermore National Laboratory. (b) For a brief instant the fusing pellet—only a few millimeters in diameter—shines like a miniature star.

(Fig. 44-29). Inertial-confinement fusion has made impressive strides in recent years, but the technological problems of delivering huge energies symmetrically to the target in a short time remain formidable.

Controlled Fusion: Magnetic Confinement

Controlled fusion research began in the 1950s with **magnetic confinement** of the fusion plasma. In Chapter 29 we showed how charged particles in a highly conducting plasma are essentially "frozen" to magnetic field lines, able to move easily along the field lines but not across them. There, and again in Chapter 31, we discussed some aspects of magnetic confinement in fusion reactors. Magnetic confinement is appealing because it offers the possibility of relatively long confinement times in "magnetic bottles" that keep the 100-MK fusion plasma from contact with any material surface. Because of the long confinement times—several seconds in some devices—densities needed to achieve the Lawson criterion are much lower than atmospheric density.

The first job of any magnetic confinement scheme is to create a magnetic configuration that keeps plasma away from the relatively cool walls of the device. Plasma particles can reach the walls in three general ways: (1) If magnetic field lines penetrate the walls, particles spiraling along those field lines may hit the walls (Fig. 44-30a). This mechanism is known as end loss. (2) Collisions

(a)   (b)   (c)

FIGURE 44-30 Plasma loss in magnetic confinement. (a) End losses occur when field lines intersect the device walls. (b) Curvature of the field lines results in cross-field drifts. (c) Large-scale instabilities distort the plasma and magnetic field. Here the so-called sausage instability causes alternate narrowing and bulging of the plasma column.

FIGURE 44-31 A toroidal fusion device has no end losses, since its magnetic field lines don't penetrate the device walls. The toroidal shape of the Tokamak Fusion Test Reactor at Princeton University shows clearly in this photo taken inside the device during its assembly.

among particles, and inhomogeneities in the magnetic field, result in particles drifting across the field lines toward the walls of the fusion device (Fig. 44-30b). (3) Plasmas are notoriously unstable. A wide variety of waves can propagate in plasma, and some of these waves can grow exponentially at the expense of particle energy, resulting in gross distortion of the plasma and field configuration that lets plasma hit the walls (Fig. 44-30c).

The most promising magnetic confinement devices eliminate end loss altogether, using a toroidal magnetic field whose field lines do not penetrate the device walls (Fig. 44-31). But bending the field lines into circles makes the field nonuniform, giving rise to cross-field drifts (Fig. 44-30b). Making the toroid larger decreases the inhomogeneities in the magnetic field and increases the time necessary for drifting particles to reach the walls. Both these trends are promising, as they suggest that appropriate confinement times should be achievable simply by building bigger devices. In practical toroidal devices, additional field components enhance confinement and suppress instabilities. A particularly promising toroidal fusion device is the **tokamak**, a Russian invention now under study throughout the world. We described the tokamak at the end of Section 31-4; see Fig. 31-33.

A successful fusion device must not only confine plasma but also heat it to ignition temperature. A variety of heating schemes have been tried. In tokamaks and related devices, the fusion plasma carries considerable electric current, and the I^2R joule heating increases the plasma temperature. Unfortunately, the resistance of a plasma drops with increasing temperature, so joule heating alone is unable to bring the plasma to ignition temperature. A second approach is to inject energy in the form of electromagnetic waves at the cyclotron frequency of the ions. Recall from Chapter 29 that the cyclotron frequency is the frequency at which charged particles execute circular motion about magnetic field lines; when electromagnetic waves at the same frequency are present, a resonance condition exists and the particles absorb energy from the waves. Finally, one of the most successful heating methods is the injection of high-energy beams of neutral particles into the plasma. The particles must be neutral in order to cross the magnetic field lines providing plasma confinement; once in the plasma they are ionized by collisions and thus help maintain not only the plasma temperature but also the density of charged particles. Figure 44-32 describes neutral-beam injection.

FIGURE 44-32 Neutral-beam heating of a fusion plasma. A high-energy beam of neutral atoms crosses the magnetic field that confines the ionized plasma, dumping energy that heats the plasma.

Prospects for Fusion Energy

When work on fusion power began in the 1950s, researchers confidently predicted that limitless energy sources would be available in a few decades. But plasma confinement and heating have proved more complex and subtle than expected, and the newer inertial confinement scheme has revealed its own technical problems. Nevertheless, the promise of nearly unlimited energy—a gallon of seawater equivalent to more than 300 gallons of gasoline (see Problem 39)—remains, and progress toward controlled fusion continues. An important milestone was reached in the early 1990s, when experimental fusion devices in Europe, Japan, and the United States produced megawatt levels of fusion power for times of several seconds. By the mid-1990s the Tokamak Fusion Test

b. Other fissile isotopes include ^{233}U and ^{239}Pu; the latter is produced by neutron absorption in the common uranium isotope ^{238}U.

c. A fission **chain reaction** occurs when the neutrons from a fission event go on to induce more fission. The average number of neutrons causing additional fission is the **multiplication factor**, k. If $k < 1$ the reaction is **subcritical** and quickly comes to a halt, if $k = 1$ the reaction is **critical** and proceeds at a steady rate, and if $k > 1$ the reaction is **supercritical** and the reaction rate grows exponentially. A single fission event results, after n generations of fission have passed; in a total of N such events:

$$N = \frac{k^{n+1} - 1}{k - 1}.$$

The actual rate at which the reaction grows is determined by the multiplication factor k and the **generation time** between fissions.

d. **Fission weapons** are detonated by rapidly assembling a supercritical configuration of fissile ^{235}U or ^{239}Pu. In explosive yield, several kilograms of fissile material are equivalent to tens of thousands of tons of chemical explosive.

e. In a **nuclear reactor**, the chain reaction is controlled to maintain a steady reaction rate. In addition to the fissile fuel itself, key components of a reactor include
   • **Control rods**, made of neutron-absorbing material and moved in and out of the fuel mass to adjust the reaction rate.
   • A **moderator**, such as water or graphite, that slows neutrons and thereby enhances the likelihood of fission.
   • A **coolant** that carries off heat from the nuclear reaction. In a power reactor, heat produces steam that drives a turbine connected to an electric generator.

f. Safety of nuclear power is a matter of public and scientific debate; issues range from uranium mining to reactor accidents to disposal of radioactive waste. A related concern is the diversion of reactor fuel or reactor-produced plutonium for nuclear weapons. Statistically, however, nuclear fission may well be safer than some other means of large-scale electric power generation, especially the burning of coal.

2. **Nuclear fusion** is the joining of light nuclei to form heavier nuclei. Fusion is an important natural process, powering the stars and thereby creating elements from helium through iron. Fusion requires temperatures high enough that average particle energies are sufficient to overcome electrical repulsion. The most promising reactions for terrestrial fusion applications are the deuterium-tritium (D–T) and deuterium-deuterium (D–D) reactions:

$$^{2}_{1}\text{H} + {}^{2}_{1}\text{H} \longrightarrow {}^{3}_{2}\text{He} + {}^{1}_{0}n + 3.27 \text{ MeV}$$

$$^{2}_{1}\text{H} + {}^{2}_{1}\text{H} \longrightarrow {}^{3}_{1}\text{H} + {}^{1}_{1}\text{H} + 4.0 \text{ MeV}$$

$$^{2}_{1}\text{H} + {}^{3}_{1}\text{H} \longrightarrow {}^{4}_{2}\text{He} + {}^{1}_{0}n + 17.6 \text{ MeV}.$$

a. **Fusion weapons** actually involve a combination of fission, fusion, and more fission.

b. Practical energy generation from **controlled fusion** has yet to be achieved. Any controlled fusion scheme must satisfy the **Lawson criterion**, stating that the product of density and confinement time must exceed a certain value.
   • **Inertial confinement** schemes use powerful lasers or particle beams to compress pellets of fusion fuel to high density and temperature. Fusion occurs before the reacting material has time to leave the reaction site.
   • **Magnetic confinement** schemes use magnetic fields to confine fusion plasma at low densities but for relatively long times.

Terms You Should Understand

(Pairs are closely related terms whose distinction is important; number in parentheses is chapter section where term first appears.)

fission, fusion (44-1)
fission fragment (44-2)
fissionable, fissile (44-2)
subcritical, critical, supercritical mass (44-2)
chain reaction (44-2)
multiplication factor (44-2)
generation time (44-2)
moderator (44-3)
control rods (44-3)
BWR, PWR (44-3)
thermal reactor, breeder reactor (44-3)
prompt, delayed neutrons (44-3)
proton-proton cycle (44-4)
D–T fusion, D–D fusion (44-4)
inertial confinement, magnetic confinement (44-4)
thermonuclear weapon (44-4)
kiloton, megaton (44-4)

Symbols You Should Recognize

τ (44-2)
k (44-2)
kt, Mt (44-4)

Problems You Should Be Able to Solve

calculating energy release in fission and fusion reactions (44-2, 44-4)
determining fuel requirements and energy yields for fission and fusion devices (44-3, 44-4)
analyzing super- and subcritical chain reactions (44-2)

Limitations to Keep in Mind

Descriptions of the nuclear technologies given here are highly simplified; much more subtle considerations of physics and engineering are required to make working devices.

QUESTIONS

1. If the nuclear binding-energy curve increased monotonically with atomic number, would energy release by nuclear fission be possible? By nuclear fusion? Explain.
2. On a per-nucleon basis, does fission or fusion release more energy? Explain by invoking Fig. 44-1.
3. On an energy-release-per-unit-mass basis, by approximately what factor do nuclear reactions exceed chemical reactions?
4. A chain reaction in natural uranium is very difficult to achieve today, because the fissile isotope ^{235}U comprises only 0.72% of natural uranium. Yet 3 billion years ago, a natural chain reaction occurred in what is now a uranium mine in Africa. Speculate on why this might have been possible, using the fact that the half-lives of ^{235}U and ^{238}U are about 700 million years and 4.5 billion years, respectively.
5. Explain and distinguish the roles of the control rods and moderator in a nuclear reactor.
6. Explain how delayed neutrons make possible mechanical control of a nuclear chain reaction.
7. Compare the multiplication factors in a thermal fission reactor and in a fission bomb.
8. Why are kilotons and megatons—actually units of weight—used as units of nuclear explosive yield?
9. Why is a water-moderated reactor intrinsically safer in a loss-of-coolant accident than a graphite-moderated reactor?
10. What is the essential distinction between a boiling-water reactor and a pressurized-water reactor?
11. Name an advantage of the Canadian CANDU and Russian RBMK reactor designs over U.S. light-water reactors.
12. What is prompt criticality? Why is it a dangerous situation in a nuclear reactor?
13. Name some advantages and disadvantages of nuclear fuel reprocessing.
14. Is ^{238}U fissionable? Is it fissile? Explain the distinction.
15. Uranium-238 cannot sustain a fission chain reaction. Is it therefore useless in nuclear power reactors? Explain.
16. Uranium-238 cannot sustain a fission chain reaction, yet ^{238}U fission provides about half the explosive yield of a thermonuclear weapon. How does this happen?
17. Why are the names "atomic bomb" and "hydrogen bomb" inappropriate?
18. Explain the role of ^{135}Xe in the Chernobyl reactor accident.
19. Why are fission fragments radioactive? Invoke Fig. 44-6 in giving your answer.
20. Why are high temperatures required for nuclear fusion?
21. Nuclear fusion powers the stars. Besides energy, what other important product results from stellar fusion?
22. The D–T reaction has the disadvantage of involving radioactive tritium fuel. Why is this reaction nevertheless the most likely to see first commercial use?
23. Why is there an upper limit to the yield of an effective fission weapon but not of a fusion weapon?
24. Why don't we talk of chain reactions and multiplication factors in discussing fusion?
25. Contrast the inertial and magnetic approaches to fusion confinement.
26. Why are many magnetic fusion reactors toroidal in shape?
27. What are some advantages of D–D over D–T fusion for a controlled fusion reactor?

PROBLEMS

Section 44-1 Energy from the Nucleus

1. The masses of the neutron, the deuterium nucleus, and the ^{3}He nucleus are 1.008665 u, 2.013553 u, and 3.014932 u, respectively. Use the Einstein mass-energy relation to verify the 3.27-MeV energy release in the D–D fusion reaction of Equation 44-4a.
2. In the fission reaction of Example 44-1, the masses are 235.043915 u, 140.9139 u, and 91.8973 u for ^{235}U, ^{141}Ba, and ^{92}Kr, respectively. Use these values, along with the neutron mass from the preceding problem, to find the energy released in this reaction.
3. Using the values 200 MeV for ^{235}U fission and 17.6 MeV for D–T fusion, calculate and compare the energy release per kilogram for each reaction.

Sections 44-2 and 44-3 Nuclear Fission and Its Applications

4. A ^{235}U nucleus undergoes neutron-induced fission, resulting in a ^{141}Cs nucleus, three neutrons, and another nucleus. What is that nucleus?

5. Neutron-induced fission of ^{235}U results in the fission fragments iodine-139 and yttrium-95. How many neutrons are released?
6. Write a complete equation for the neutron-induced fission of plutonium-239 that results in barium-143, 2 neutrons, and another nucleus.
7. Assuming 200 MeV per fission, determine the number of fission events occurring each second in a reactor whose thermal power output is 3200 MW.
8. How much ^{235}U would be needed to fuel the reactor of the preceding problem for one year? (Your answer is an overestimate because fission of ^{239}Pu also contributes to the power output.)
9. Find the explosive yield in equivalent tonnage of TNT for the chain reaction analyzed in Example 44-3.
10. Arguments over the feasibility of a nuclear test ban have been focused on the ability to distinguish nuclear explosions in the 1-kt range from chemical explosions and seismic events. How much U-235 would have to fission to produce an explosive yield of 1.0 kt?

11. How much uranium-235 would be consumed in a fission bomb with 20-kt explosive yield?

12. What is the efficiency of a nuclear weapon that uses 28 kg of ^{235}U to produce a 44-kt explosive yield?

13. The 1974 Threshold Test Ban Treaty limited underground nuclear tests to a maximum yield of 150 kt. (a) How much ^{235}U would be needed for a device with this yield, assuming 30% efficiency? (b) What would be the diameter of this mass, assembled into a sphere? The density of uranium is 18.7 g/cm^3.

14. Suppose that all of the approximately 2.5 neutrons per fission event caused additional fission in a U-235 mass with generation time $\tau = 10$ ns. (a) How long would it take to fission the first 10^{24} nuclei? (b) How long would it take for ten times this number to fission?

15. The effective multiplication factor in a typical nuclear weapon is about 1.5, and the generation time is about 10 ns. Under these conditions, (a) how many generations would it take to fission a 10-kg mass of ^{235}U? (b) How long would the process take?

16. What is the generation time in a fission reaction that consumes 2.0×10^{20} uranium-235 nuclei in its first 10 s, if the multiplication factor is 1.0012?

17. The temperature in a typical reactor core is 600 K. What is the thermal speed of a neutron at this temperature? *Hint:* See Section 20-1.

18. A neutron is emitted from a fission reaction with kinetic energy 2.0 MeV. If it loses 88.9% of its energy in a collision with a moderator nucleus (typical for head-on collisions in heavy water), how many such collisions must it undergo before it reaches 0.05 eV, the thermal energy at typical reactor temperatures?

19. A neutron collides elastically and head-on with a stationary deuteron in a reactor moderated by heavy water. How much of its kinetic energy is transferred to the deuteron? *Hint:* Consult Chapter 11.

20. A buildup of fission products "poisons" a reactor, dropping the multiplication factor to 0.992. How long will it take the reactor power to drop in half, assuming a generation time of 0.10 s?

21. If a reactor is shut down abruptly, neutron absorption by ^{135}Xe prevents rapid start-up. The xenon has a 9.2-h half-life. How long must reactor operators wait until the ^{135}Xe level drops to one-tenth of its peak value? (Your answer neglects formation of additional ^{135}Xe from the decay of ^{135}I.)

22. It takes n generations to fission the first N nuclei in a chain reaction with multiplication factor k, where $N \gg 1$. Find an expression for the number of generations that will pass before a total of $10N$ nuclei have been fissioned.

23. One reason breeder reactors are considered potentially dangerous is that the reaction in a breeder is sustained by fast neutrons alone; there is no moderator. The generation time for prompt neutrons is then 100 ns, down from 100 ms in a thermal reactor. For the conditions of Example 44-5, compute the power-doubling time in a breeder with $\tau = 100$ ns.

24. It is generally considered impossible for a slow-neutron reactor to blow up like a bomb, even in the event of a runaway reaction. To help confirm this, compare the time for the reaction rate to double in (a) an out-of-control reactor with prompt-neutron generation time $\tau = 100$ μs and multiplication factor $k = 1.01$ (unrealistically high) and (b) in a bomb with $\tau = 10$ ns and $k = 1.5$.

25. In the dangerous situation of prompt criticality in a fission reactor, the generation time drops to 100 μs as prompt neutrons alone sustain the chain reaction. If a reactor goes prompt critical with $k = 1.001$, how long does it take for a 100-fold increase in reactor power?

26. During an accident, a reactor's fission rate increases by a factor of 100 in 20 s. If the generation time is 3.0×10^{-4} s, what is the multiplication factor k?

27. Operators seek to double the power output of a nuclear reactor over a period of 1 hour. If the generation time is 120 ms, by how much should the multiplication factor k be increased from 1 during this period?

28. Find the generation time in a reactor whose power increases by a factor of 4 in 5 minutes, given that the multiplication factor is $k = 1.00022$.

Section 44-4 Nuclear Fusion

29. Fusion researchers often express temperature in energy units, giving the value of kT rather than T. What is the temperature in kelvins of a 2-keV plasma?

30. Verify from Equations 44-3 that the proton-proton cycle yields a net energy of 26.7 MeV.

31. In a magnetic-confinement fusion device with confinement time 0.5 s, what density would be required to meet the Lawson criterion for D–T fusion?

32. An early tokamak, the T-4 device at Moscow's Kurchatov Institute, achieved a particle density of 3×10^{19} m^{-3} for 10 ms. By what factor would its confinement time need to be increased to meet the D–T Lawson criterion?

33. How much heavy water (deuterium oxide, ^{2}H$_2$O or D$_2$O) would be needed to power a 1000-MW D–D fusion power plant for 1 year?

34. A D–D fusion-powered car (Fig. 44-34) gets 30 miles per gallon of seawater. How many times could it circle the Earth on a gallon of pure heavy water (^{2}H$_2$O)?

FIGURE 44-34 Fusion-powered car from the film *Back to the Future* (Problem 34).

35. Inertial-confinement schemes generally involve confinement times on the order of 0.1 ns. What is the corresponding density needed to meet the Lawson criterion for D–T fusion?

36. (a) Estimate the D–D fusion energy content of Lake Erie, whose volume is 480 km^3. (b) How long would this resource supply humanity's energy needs at the current rate of 10^{13} W?

37. The proton-proton cycle consumes four protons while producing about 27 MeV of energy. (a) At what rate must the Sun consume protons to produce its power output of about 4×10^{26} W? (b) The present phase of the Sun's life will end when it has consumed about 10% of its original protons. Estimate how long this phase will last, assuming the Sun's 2×10^{30} kg mass was initially 71% hydrogen.

38. Consider deuterium undergoing fusion via the reactions of Equation 44-4a and 44-4b. Since reaction 44-4b also produces one tritium nucleus, reaction 44-4c also occurs. Furthermore, ^{3}He nuclei from reaction 44-4a fuse with deuterium, yielding additional energy:

$$\textstyle^3_2\text{He} + {}^2_1\text{H} \longrightarrow {}^4_2\text{He} + {}^1_1\text{H} + 18.3 \text{ MeV}.$$

(a) How many deuterons are needed for one occurrence of each of these four reactions? (b) Assuming that reactions 44-4a and b occur with equal probability, and that all the $^3_2$He and $^3_1$H formed undergo the remaining two reactions, use the result of (a) and the energies listed for the various reactions to determine the average energy release per deuteron.

39. About 0.015% of hydrogen nuclei are actually deuterium. (a) How much energy would be released if all the deuterium in a gallon of water underwent fusion? (Use an average of 7.2 MeV per deuteron; see preceding problem.) (b) In terms of energy content, to how much gasoline does a gallon of water correspond? Gasoline's energy content is 36 kWh/gal, as listed in Appendix C.

40. The Princeton Large Torus (PLT) fusion device has a plasma column 90 cm in diameter, with particle density about 4×10^{19} m^{-3}. Heating by neutral-beam injection results in a plasma temperature about 6×10^7 K, and the confinement time is 25 ms. (a) What is the Lawson parameter for this device? (b) Keeping other factors the same, what confinement time would be needed to meet the D–T Lawson criterion?

Paired Problems

(Both problems in a pair involve the same principles and techniques. If you can get the first problem, you should be able to solve the second one.)

41. The total power generated in a nuclear power reactor is 1500 MW. How much ^{235}U does it consume in a year?

42. New Hampshire's Seabrook nuclear power plant produces electrical energy at the rate of 1.2 GW, and consumes 1311 kg of ^{235}U each year. Find (a) its thermal power output and (b) its efficiency, assuming the plant operates continuously.

43. Two fission chain reactions have $\tau = 20$ ns, but reaction A has $k = 1.3$ and reaction B has $k = 1.4$. Compare the total number of nuclei that fission in the first μs of each reaction.

44. A fission chain reaction has $\tau = 10$ ns and $k = 1.35$. Find the total number of nuclei that fission (a) in the first 1.0 μs and (b) in the interval from 1.0 μs to 1.1 μs.

45. This problem and the next one explore the differences between magnetic and inertial-confinement fusion. (a) What is the mean thermal speed of a deuteron in the PLT device of Problem 40? (b) At this speed, how long would it take a deuteron to cross the plasma column? Using this value as a "confinement time" in the absence of magnetic confinement and the density given in Problem 40, calculate the Lawson parameter and show that it falls far short of the value needed for D–T fusion.

46. A laser fusion device compresses a fusion pellet to a diameter of 0.1 mm and a density of 250 g/cm^3. (a) If the temperature is comparable to that of the PLT plasma (see preceding problem), how long does it take a deuteron with thermal speed to cross the pellet? (b) Estimate the number density of deuterons in the pellet, approximating it as all deuterium. (c) Use the results of (a) and (b) to calculate the Lawson parameter for the pellet, and show that it exceeds the required value for D–T fusion.

Supplementary Problems

47. A laser-fusion fuel pellet has mass 5.0 mg and is composed of equal parts (by mass) of deuterium and tritium. (a) If half the deuterium, and an equal number of tritium nuclei, participate in D–T fusion, how much energy is released? (b) At what rate must pellets be fused in a power plant with 3000 MW thermal power output? (c) What mass of fuel would be needed to run the plant for 1 year? Compare with the 3.6×10^6 tons of coal needed to fuel a comparable coal-burning power plant.

48. A fission chain reaction has $\tau = 10$ ns. In the first microsecond of the reaction, a total of 2.4×10^{23} fission events occur. In the next 100 ns an additional 5.8×10^{25} fissions occur. Find the multiplication factor k.

49. Use a graphical or numerical method to find the multiplication factor necessary to fission 10^{22} uranium-235 nuclei in 2.0 μs, if the generation time is 10 ns.

50. The half-lives of ^{235}U and ^{238}U are 7.04×10^8 years and 4.46×10^9 years, respectively. Uranium-235 presently constitutes 0.72% of natural uranium, with essentially all the rest ^{238}U. Three billion years ago, a natural fission chain reaction occurred in what is now a uranium mine in Africa. What percentage of natural uranium was ^{235}U at that time? Compare with the 3% ^{235}U typical of fuel for today's power reactors.

51. Roughly half the yield of a thermonuclear weapon comes from D–T fusion and the rest from ^{238}U fission induced by fast neutrons from the D–T fusion. Each D–T reaction releases about 18 MeV and each fission about 200 MeV. Estimate the masses of fusion fuel and ^{238}U in a 1-Mt

thermonuclear bomb, assuming both types of fuel react completely.

52. An important cooling mechanism in fusion plasmas is the loss of electromagnetic radiation by plasma electrons. Analysis of Maxwell's equations shows that the power radiated by an electron is proportional to the *square* of its acceleration. (a) Write an expression for the acceleration of an electron a distance r from a fully ionized nucleus of atomic number Z. (b) Show from your result that, for a given r, the radiated power is proportional to Z^2. Your result shows why contamination with high-Z materials, such as the metals of the reactor walls, can seriously degrade the plasma temperature.

53. The volume of the fireball produced in a nuclear explosion is roughly proportional to the weapon's explosive energy yield. For weapons exploded at ground level, how would the land area subject to a given level of damage scale with the weapon's yield? Your result shows one reason military strategists favor multiple smaller warheads.

54. The peak overpressure generated in a nuclear explosion is approximately

$$\Delta P = 22.4\left(\frac{Y}{r^3}\right) + 15.8\left(\frac{Y}{r^3}\right)^{1/2},$$

where ΔP is the overpressure in pounds per square inch (psi), Y is the weapon's yield in megatons, and r is the distance in miles from the explosion. An overpressure of 5 psi is sufficient to destroy a typical wood-frame house (see Fig. 44-27). Determine the radius of the 5-psi zone for (a) the 12.5-kt Hiroshima bomb, (b) a 300-kt MX missile warhead, and (c) the 9-Mt bombs that were once deployed on U.S. bombers.

55. Of the neutrons emitted in eaach fission event in a light-water reactor, an average of 0.6 neutrons are absorbed by ^{238}U, leading to the formation of ^{239}Pu. (a) Assuming 200 MeV per fission, how much ^{239}Pu forms each year in a 30% efficient nuclear power plant whose electric power output is 1000 MW? (b) With careful design, a fission explosive can be made from 5 kg of ^{239}Pu. How many potential bombs are produced each year in the power plant of part (a)? (Extracting the ^{239}Pu is not an easy job.)

56. In the D–D reaction of Equation 44-4b, assume the two deuterons approach head on with equal speeds. Use conservation of momentum to determine how the 4-MeV reaction energy is shared between the tritium nucleus and the proton that result. Assume the initial kinetic energy of the deuterons is negligible.

From Quarks to the Cosmos 45

Jets of particles emerge in this image obtained with the DELPHI detector at the Large Electron-Positron Collider of the European Laboratory for Particle Physics (CERN). Production of new particles in high-energy collisions such as this one gives our best experimental evidence for the structure and organization of the fundamental particles.

The past six chapters have extended the realm of physics to the atomic and molecular scale, then down to the even smaller scale of the atomic nucleus. Here we go further still, probing the structure of nucleons themselves and trying to make sense of the host of subatomic particles nature reveals. We'll be asking questions about the ultimate nature of matter at the tiniest scales, but in the process we will find a remarkable connection with questions of the largest scale—questions about the origin and ultimate fate of the entire universe.

45-1 PARTICLES AND FORCES

By 1932 there were four "elementary" particles of matter known: the electron, the proton, the neutron, and the neutrino. In addition, there were the positron, antiparticle to the electron, and the photon of electromagnetic radiation. There were also the seemingly fundamental forces—gravity, the electromagnetic force, the nuclear or strong force, and the weak force that manifests itself in beta decay.

In Chapter 39 we saw how the photoelectric effect shows that the interaction of electromagnetic waves with matter ultimately involves individual photons—the quanta of the electromagnetic field. In the quantum-mechanical view of electromagnetism, the force between two charged particles also involves photons, now exchanged between the interacting particles. Imagine two astronauts floating in space, tossing a ball back and forth (Fig. 45-1a). Catching or throwing the ball, one astronaut gains momentum in a direction away from the other, so the exchange results in a net average repulsive force. If the two astronauts struggle for possession of the ball, each trying to wrest it from the other, then the ball mediates what appears as an attractive interaction (Fig. 45-1b). Figure 45-1 gives classical analogs for the attractive and repulsive electrical interactions involving photon exchange.

So far we've found photons emitted when a particle jumps into a lower energy state, with the photon carrying off energy equal to the energy difference between the two states. The process obviously conserves energy. But now we're saying that a single, free electron emits photons that it exchanges with another particle to produce what we call the electromagnetic force. How can that process conserve energy? The energy-time uncertainty relation (Equation 39-6) says that an energy measured in a time Δt is necessarily uncertain by an amount $\Delta E \geq \hbar/\Delta t$. The photon exchanged by two particles lasts only a short time, and therefore, its energy is uncertain. So we can't really say that energy conservation is violated. A photon created in this way and lasting only for the short time it takes to exchange with another particle is called a **virtual photon**. We never "see" the virtual photon since it's emitted by one particle and absorbed by the other.

(a)   (b)

FIGURE 45-1 (a) Two astronauts toss a ball back and forth. The resulting forces cause an apparent repulsion of the two. This is analogous to the exchange of virtual photons between particles of like charge. (b) The astronauts struggle for possession of the ball, resulting in an attractive force. This is analogous to the exchange of virtual photons between particles of opposite charge.

The quantum theory of the electromagnetic interaction is called **quantum electrodynamics** (QED). Although begun by Dirac, it was finally brought to consistent form in 1948 by Richard Feynman, Sin-Itiro Tomonaga, and Julian Schwinger. The fundamental event in QED is the interaction of a photon with an electrically charged particle. Two such events joined by a common virtual photon give the quantum electrodynamical description of the electromagnetic force (Fig. 45-2). The predictions of quantum electrodynamics have been confirmed experimentally to a remarkably high precision, and today QED is our best-verified theory of physical reality.

Mesons

In 1935 the Japanese physicist Hideki Yukawa (Fig. 45-3) proposed that the nuclear force should, like the electromagnetic force, be mediated by exchange of a particle. Yukawa called his hypothetical particle a **meson**. Because the range of the strong force is limited, Yukawa argued, the meson should have finite mass. The reason for this connection between mass and range again lies in the energy-time uncertainty relation.

Consider first the electromagnetic force: It falls off as $1/r^2$ and thus has an essentially infinite range. Two particles can be very far apart and still interact electromagnetically. Since photons travel at the finite speed of light, the time Δt for a photon interaction can be arbitrarily long. The energy-time uncertainty principle $\Delta E \Delta t \geq \hbar$ thus shows that the energy uncertainty ΔE can be arbitrarily small. Thus, the possible energies for virtual photons must extend downward toward zero—and that can happen only if photons have zero rest energy.

The nuclear force, in contrast, has a finite range of about 1.5 fm. Moving at close to the speed of light, the longest a particle mediating this force would need to exist is for a time Δt given by

$$\Delta t = \frac{\Delta x}{c} = \frac{1.5 \times 10^{-15} \text{ m}}{3.0 \times 10^8 \text{ m/s}} = 5.0 \times 10^{-24} \text{ s}.$$

Then energy-time uncertainty gives

$$\Delta E \geq \frac{\hbar}{\Delta t} = \frac{1.05 \times 10^{-34} \text{ J·s}}{5.0 \times 10^{-24} \text{ s}} = 2.1 \times 10^{-11} \text{ J} = 130 \text{ MeV}.$$

Since the meson energy need not go below this value, this could be its rest energy. Yukawa therefore proposed the existence of a new particle with mass $130 \text{ MeV}/c^2$, about 250 times that of the electron. Yukawa's prediction was eventually confirmed—but not before nature introduced yet another particle.

45-2 PARTICLES AND MORE PARTICLES

In the 1930s, the most available source of high-energy particles was cosmic radiation—streams of high-energy protons and other particles of extraterrestrial origin that, on striking Earth's atmosphere, induce nuclear and elementary-particle interactions. In 1937 the American physicist Carl Anderson (who had earlier discovered the positron) and his colleagues identified in cosmic rays a particle with a mass 207 times that of the electron. Now called the **muon**, this particle had the same charge and spin as the electron and seemed to behave much

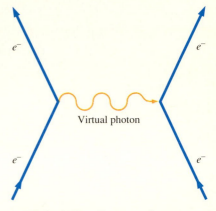

FIGURE 45-2 A Feynman diagram, showing the interaction of two electrons through the exchange of a virtual photon. Time proceeds upward in the diagram, so the two electrons are initially on converging paths. After exchanging the photon their paths diverge, so the interaction represents a repulsive force.

FIGURE 45-3 Richard Feynman and Hideki Yukawa.

like a heavier version of the latter. Two muons were found: the negatively charged μ^- and its antiparticle μ^+. Although the muon mass was close to Yukawa's prediction, the muon interacted only weakly with nuclei and, therefore, could not be the mediator of the nuclear force. "Who ordered that?" quipped Columbia University physicist I. I. Rabi, meaning that there seemed to be no role for the new particle in the then-understood scheme of things.

The real Yukawa particle was discovered 10 years later in 1947, again in cosmic rays, and turned out to have a mass about 270 times that of the electron. This time there were three related particles, now called **pions**: the positive π^+, negative π^-, and neutral π^0.

The new particles are all unstable, undergoing decays that ultimately result in the well-known stable particles. The negative pion, for example, decays with a mean lifetime of 26 ns into a negative muon and an antineutrino:

$$\pi^- \rightarrow \mu^- + \overline{\nu}.$$

The muon then decays with a 2.2-μs lifetime into an electron and a neutrino-antineutrino pair:

$$\mu^- \rightarrow e^- + \nu + \overline{\nu}.$$

APPLICATION	*Detecting Particles*

Despite their minute size, we can, remarkably, follow the trajectories of individual subatomic particles. Early particle detectors used the fact that high-energy particles create ions as they tear through matter. One of the first detectors was the **cloud chamber**, built by C. T. R. Wilson in 1910. The chamber contains a supersaturated vapor that condenses on the ions to produce a visible track. A more recent version is the **bubble chamber**, developed in 1952. Here rapid expansion brings a liquid beyond its boiling point, and bubbles form on the ions left by high-energy particles (Fig. 45-4). The **multiwire proportional chamber** works somewhat like the Geiger counter

we described in Chapter 43. In the chamber, criss-crossed grids of fine wire record current pulses from electrons liberated as charged particles passing through the gas-filled chamber; analyzing the pulse distribution reveals the particle trajectory (Fig. 45-5). Development of the multiwire proportional

FIGURE 45-5 A multiwire detector under construction at the Stanford Linear Accelerator Center. High-energy particles induce electrical signals in nearby wires; processing signals from the thousands of wires then allows a computer to reconstruct the particles' trajectories.

(a) (b)

FIGURE 45-4 (a) Installation of the Big European Bubble Chamber (BEBC) at CERN, the European Laboratory for Particle Physics. (b) Subatomic particle tracks in a BEBC photo.

chamber won Georges Charpak the 1992 Nobel Prize. Magnetic fields cause the trajectories to curve, allowing determination of their charge-to-momentum ratios. **Scintillation detectors** give off flashes of light as particles pass through them; photomultiplier tubes detect the flashes, whose intensity gives a measure of the particle energy. **Calorimeters**, consisting of composite layers of scintillators and energy-absorbing material, analyze the showers of secondary particles produced by a single high-energy particle to provide accurate measurement of the original particle's energy. Modern detectors are often huge agglomerations of several basic detector types, arranged to extract as much information as possible from particle interactions (Fig. 45-6). Detector outputs are analyzed by computer, allowing identification of events so rare that they may occur only once in a million interaction events.

These detection schemes work for charged particles, but what about neutrals such as the neutron, the neutrino, or the π^0? Although their trajectories do not show in the detectors' particle-trajectory images, they often lead to interactions involving charged particles whose detection then infers the properties of the neutral particle.

FIGURE 45-6 Technician working inside the OPAL particle detector at CERN. OPAL consists of a variety of detector types arranged concentrically around the point where particles collide.

Particle Classification

Soon more particles were discovered, first in cosmic rays, then with the increasingly energetic particle accelerators that were becoming available. By 1980 there were well over 100 "elementary" particles. Physicists sought to explain this "particle zoo," and a first step was to classify similar particles into general categories. Early classification schemes separated particles by mass, but a more enlightening approach is based on the fundamental forces.

Leptons are particles that do not experience the strong force. They include the familiar electron, the muon, a more massive particle called the tau, and three types of neutrinos, one associated with each of the charged leptons. The neutrinos were long thought to be massless, but experimental results announced in 1998 suggest that neutrinos have small nonzero mass, and that they "oscillate," changing among the three types. Each of the leptons has an antiparticle as well. There are thus a total of six lepton-antilepton pairs, and experimental evidence strongly suggests that no others can exist. The leptons all have spin $\frac{1}{2}$, and are therefore **fermions**—a term we introduced in Chapter 41 to describe particles that have half-integer spin and therefore obey the exclusion principle. Leptons are believed to be true elementary point particles with zero size and no internal structure.

Hadrons are particles that do experience the strong force. They fall into two subclasses: **mesons** have integer spin and are therefore **bosons** that do not obey the exclusion principle. Mesons include Yukawa's pions and a host of others; all are unstable. **Baryons** have half integer spins and are therefore fermions. They include the familiar proton and neutron and similar but more massive particles. The baryons are grouped into pairs, triplets, and higher multiple groupings of closely related particles. The neutron and proton, for example, form a pair, differing in charge and very slightly in mass. Each baryon has an antiparticle, as do most mesons; but some neutral mesons are their own antiparticles.

The third class of particles comprises the **field particles** or **gauge bosons**, quanta of the different force fields and "carriers" of those forces. These include the familiar photon for the electromagnetic force, three particles called the W^+, W^-, and Z for the weak force, a particle called the gluon, and a hypothetical graviton that would carry the gravitational force in an as yet incomplete theory of quantum gravity. All the field particles are bosons, carrying spin 1 or, for the graviton, spin 2. You might think Yukawa's meson should be in this category in its role as carrier of the nuclear force. That it doesn't appear here is a hint that the nuclear force is not really fundamental; as we'll soon see, it is the gluon that plays the more fundamental role.

Table 45-1 lists just a few of the particles known even before full confirmation of today's elementary particle theories.

Particle Properties and Conservation Laws

Many new particles can be characterized by known properties such as mass, spin, and electric charge. Of these, spin and charge are associated with important conservation laws—the conservation of angular momentum and the conservation of electric charge. Allowed interactions among particles *must* conserve

TABLE 45-1 *Selected Particles*[1]

CATEGORY/ PARTICLE	SYMBOL (PARTICLE/ ANTIPARTICLE)	SPIN	MASS (MeV/c^2)	BARYON NUMBER, LEPTON NUMBERS, STRANGENESS (ANTIPARTICLE HAS OPPOSITE SIGN)					LIFETIME(s)
				B	L_e	L_μ	L_τ	S	
Field particles									
Photon	γ, γ	1	0	0	0	0	0	0	Stable
Z^0	Z^0, Z^0	1	91177	0	0	0	0	0	$\sim10^{-25}$
Leptons									
Electron	e^-, e^+	1/2	0.511	0	+1	0	0	0	Stable
Muon	μ^-, μ^+	1/2	105.7	0	0	+1	0	0	2.2×10^{-6}
Tau	τ^-, τ^+	1/2	1784	0	0	0	+1	0	3.0×10^{-13}
Electron neutrino	$\nu_e, \overline{\nu}_e$	1/2	small[2]	0	+1	0	0	0	Stable
Muon neutrino	$\nu_\mu, \overline{\nu}_\mu$	1/2	small[2]	0	0	+1	0	0	Stable
Tau neutrino	$\nu_\tau, \overline{\nu}_\tau$	1/2	small[2]	0	0	0	+1	0	Stable
Hadrons									
Mesons									
Pion	π^+, π^-	0	139.6	0	0	0	0	0	2.6×10^{-8}
Pion	π^0, π^0	0	135.0	0	0	0	0	0	0.83×10^{-16}
Eta	η^0, η^0	0	549	0	0	0	0	0	$\sim5\times10^{-19}$
Rho	ρ^0, ρ^0	1	768	0	0	0	0	0	$\sim4\times10^{-24}$
Kaon	K^+, K^-	0	494	0	0	0	0	1	1.2×10^{-8}
Kaon	$K^0, \overline{K}^0$	0	498	0	0	0	0	1	0.89×10^{-10} 5.2×10^{-8} [3]
Baryons									
Proton	$p, \overline{p}$	1/2	938.3	1	0	0	0	0	Stable[4]
Neutron	$n, \overline{n}$	1/2	939.6	1	0	0	0	0	889
Lambda	$\Lambda^0, \overline{\Lambda}^0$	1/2	1115.6	1	0	0	0	−1	2.6×10^{-10}
Sigma	$\Sigma^+, \overline{\Sigma}^-$	1/2	1189.4	1	0	0	0	−1	0.80×10^{-10}
Sigma	$\Sigma^0, \overline{\Sigma}^0$	1/2	1192.6	1	0	0	0	−1	7.4×10^{-20}
Sigma	$\Sigma^-, \overline{\Sigma}^+$	1/2	1197.4	1	0	0	0	−1	1.5×10^{-10}
Omega	$\Omega^-, \overline{\Omega}^+$	3/2	2285.2	1	0	0	0	−3	0.82×10^{-10}

[1] Does not include any hadrons with c, b, or t quarks.
[2] Neutrino masses are probably on the order of 10 eV, and differ by about 0.07 eV.
[3] The neutral kaon exists as a quantum-mechanical superposition of states with two different lifetimes.
[4] If unstable, proton lifetime exceeds 10^{33} years.

these quantities. The annihilation of an electron-positron pair, for example, is allowed because the initial particles have no net charge and neither do the resulting photons. Similarly, beta decay of the neutron produces an electron, a proton, and a neutral antineutrino and thus conserves charge:

$$n \rightarrow p + e^- + \overline{\nu}_e.$$

Here we subscript the antineutrino to indicate that it's an electron antineutrino, as opposed to the muon or tau variety.

Other particle properties appear to be conserved as well. Associated with each baryon or antibaryon is its **baryon number**, assigned the value $+1$ for a baryon and -1 for an antibaryon. All experimental evidence to date points to conservation of baryon number: In all observed particle reactions, the sums of the baryon numbers before and after the reaction are equal. An example is, again, beta decay of the neutron: The process starts with a neutron of baryon number $B = 1$ and ends up with a proton ($B = 1$), an electron, and an antineutrino. The last two are leptons, so their baryon numbers are zero, and thus, baryon number is conserved. Some theories, which we describe shortly, suggest that baryon number is only approximately conserved. If that is so, then the proton itself is an unstable particle with a mean lifetime in excess of 10^{33} years.

Lepton number seems also to be conserved; specifically, there are three separate lepton numbers associated with the electron, the muon, the tau, and their corresponding neutrinos. Again beta decay provides an example: The neutron and proton, being baryons, have electron-lepton number zero, while the resulting electron and *anti*neutrino have electron-lepton numbers of $+1$ and -1, respectively.

In the late 1950s particles called K, Λ, Σ, and Ξ were discovered. Strange characteristics of these particles' decays could be explained by introducing a new fundamental property, called **strangeness**. Previously known particles had zero strangeness, but the **strange particles** had strangeness ±1 or ±2. Strangeness is conserved in strong and electromagnetic interactions, but in weak interactions its value can change. We have listed strangeness along with other particle properties in Table 45-1. We'll soon see that several other new properties are needed to characterize matter fully.

19.5
Particle Physics

EXAMPLE 45-1	*Conservation Laws*

A pion collides with a proton, producing a neutral kaon and a lambda particle:

$$\pi^- + p \rightarrow K^0 + \Lambda^0.$$

(a) Which of electric charge, baryon number, lepton numbers, and strangeness are conserved? (b) The kaon and lambda are unstable and undergo further decays. Could the end result be an electron and a proton?

Solution

(a) Table 45-1 shows that all the particles are hadrons, so lepton numbers are zero on both sides of the equation. On the left, with a negative and a positive particle, the net charge is zero. On the right, both particles are neutral, so charge is conserved. On the left, the pion is a meson and the proton a baryon, so the total baryon number is $+1$. On the right, Table 45-1 shows that the kaon is a meson and the Λ^0 a baryon,

so baryon number, too, is conserved. Finally, neither the pion nor the proton is strange, so the total strangeness on the left is zero. Table 45-1 shows that the kaon has strangeness $+1$ and the lambda particle has strangeness -1. Thus strangeness, too, is conserved.

(b) Having an electron and a proton as the final state would conserve charge (0), strangeness (0), and baryon number (1). But it would not conserve electron-lepton number, which is zero for the original particles and would become $+1$. Adding an electron antineutrino would, however, satisfy all conservation laws.

EXERCISE Is the interaction $p + n \rightarrow p + p + n + \overline{p}$ permitted by conservation of baryon number?

Answer: Yes

Some problems similar to Example 45-1: 5–9

(a)

(b)

FIGURE 45-7 *(a)* Theorists C. N. Yang (left) and T. D. Lee suggested that parity need not be conserved. *(b)* Experimentalist Chien-Shiung Wu and her collaborators verified the Lee-Yang suggestion.

Symmetries

Watch a physical process in a mirror and you would expect the image ought also to be a possible physical process. If so, then the laws of physics should exhibit **symmetry** with respect to mirror reflection. At the subatomic level, the statement that a process and its mirror image are equally likely is called **conservation of parity**. Mathematically, a system has parity $+1$ if its wave function is unchanged on reflection through the origin—that is, on a change of coordinates $x \rightarrow -x$, $y \rightarrow -y$, $z \rightarrow -z$. If the wave function changes sign then the parity is -1. Parity is conserved if its value before a system undergoes some interaction is the same, either $+1$ or -1, as it is afterward.

In 1956 there were two particles, then called τ and θ, that seemed identical in all respects except that their decay products had different total parity. In that year theoretical physicists T. D. Lee and C. N. Yang (Fig. 45-7) noted that conservation of parity, known to hold for electromagnetic and strong interactions, had not been tested for interactions involving the weak force. They made the revolutionary suggestion that parity need not be conserved—a statement tantamount to suggesting that nature can, at a fundamental level, distinguish right- from left-handed systems that are otherwise identical.

Experimentalists quickly set out to test the Lee-Yang suggestion. A group led by Chien-Shiung Wu (Fig. 45-7) studied the beta decay of cobalt-60 by aligning ^{60}Co nuclei with a magnetic field at the very low temperature of 0.01 K. If the beta-decay electrons were emitted with equal probability along and opposite the field direction, Wu and her co-workers reasoned, then the original process and its mirror image would be equally likely (Fig. 45-8*a*). But what they found was a preferential beta emission opposite the field direction—a clear indication that the mirror world behaved differently (Fig. 45-8*b*).

Although parity might be violated, theorists held that a combination of parity reversal (P) and charge conjugation (C)—changing particles into their antiparticles—would result in physical systems whose behavior was indistinguishable. But in 1964 even this so-called CP conservation was found to be violated, albeit only in the rare decay of the neutral kaon into a pion-antipion pair. The Russian physicist Andrei Sakharov suggested that this asymmetric decay into matter and

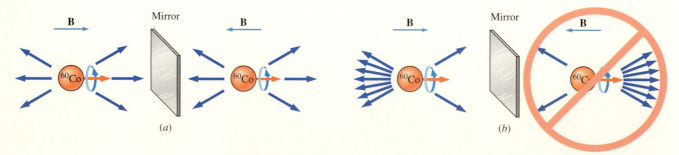

FIGURE 45-8 Experimental evidence for nonconservation of parity. At left of mirror, a ^{60}Co nucleus has its spin aligned with a magnetic field; in a classical picture, this would correspond to the nucleus spinning with its top part emerging from the page. Reflected in a mirror, this sense of spin does not change, so the spin vector in the mirror image still points to the right, even though the magnetic field is reversed. *(a)* In order that this mirror image be an equally likely physical situation, beta emission (blue arrows) must occur with equal probability along and against the spin direction. *(b)* Experiment shows that beta emission occurs preferentially opposite the spin direction, so the mirror-image situation at right in *(b)* does not occur.

antimatter might have led to a slight imbalance between the two in the early universe and thus could be responsible for the present situation in which we find large quantities of matter but essentially no antimatter.

It still appears that *CPT* conservation holds—that is, a combination of mirror reflection, charge conjugation, and reversal of the time coordinate $t \rightarrow -t$ makes a new physical process indistinguishable from the original. For most processes *C*, *P*, and *T* hold individually; for weak decays *C* and *P* are separately violated but *CP* and *T* still hold, while for the rare kaon events even time reversal does not give an equally probable physical process. There may be a deep philosophical connection here involving the direction of time, but the full implications of *CPT* symmetry and the failure of its individual components are not fully understood.

45-3 QUARKS AND THE STANDARD MODEL

The proliferation of particles distressed physicists used to finding an ultimate simplicity underlying the phenomena of nature. What kind of a universe is it that requires hundreds of "elementary" particles? Were they all really "elementary," or was there some still more fundamental level exhibiting the sought-after simplicity?

The Eightfold Way

In 1961, physicists Murray Gell-Mann and Yu val Ne'eman independently noted patterns in the groupings of the then-known particles. They plotted strangeness versus electric charge and found that the particles fell into groups that formed simple geometric patterns (Fig. 45-9). The patterns were called collectively the **Eightfold Way**, after a set of Buddhist principles for right living.

Clearly the Eightfold-Way patterns revealed some underlying organization or structure, in much the way that the periodic table of the elements hints at the building of atoms from more basic particles in an organizational scheme that produces the chemical elements. A triumph came when Gell-Mann arranged the spin-$\frac{3}{2}$ baryons into an Eightfold-Way pattern and found an empty spot where a particle with strangeness $S = -3$ should be. Experimentalists soon found the particle—the Ω^-—in bubble-chamber photographs from previous experiments.

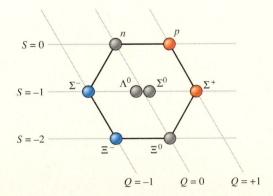

FIGURE 45-9 There are eight spin-$\frac{1}{2}$ baryons. On a plot of strangeness versus charge, they form a distinct hexagonal pattern. Similar patterns exist for other particle groups.

FIGURE 45-10 Protons and neutrons consist of the quark combinations *uud* and *udd*, respectively.

FIGURE 45-11 Mesons consist of a quark and an antiquark. The meson shown here is the π^+, made from an up quark and a down antiquark.

FIGURE 45-12 The color force binds quarks through the interchange of gluons. Shown here is a proton, consisting of the quark combination *uud*. The three quarks are continually exchanging gluons, changing their colors but not flavors. The composite proton remains colorless throughout.

Quarks

Success of the Eightfold Way convinced physicists that many "elementary" particles were not really elementary. In 1964 Gell-Mann and his Cal Tech colleague George Zweig independently proposed a set of three particles that could be combined to form the then-known hadrons. Gell-Mann called these particles **quarks**, a word used in James Joyce's novel *Finnegan's Wake*. Gell-Mann's quarks became known as the **up quark** (*u*), the **down quark** (*d*), and the **strange quark** (*s*); the latter was needed to give strange particles their "strangeness." For each quark there was also an antiquark.

One surprising thing about quarks is that they carry fractional electric charges. The two least massive quarks, the up and the down, carry $+\frac{2}{3}e$ and $-\frac{1}{3}e$, respectively; their antiparticles have the opposite charges. The quarks combine in pairs or triplets to make the two classes of hadrons. Baryons, like the proton and the neutron, consist of three quarks (Fig. 45-10). Mesons, in contrast, contain a quark and its antiquark (Fig. 45-11). The quarks all have spin $\frac{1}{2}$, which explains why the three-quark baryons all have odd half integer spin, and why the two-quark mesons have integer spin.

The Pauli exclusion principle precludes three particles having the same quantum numbers, and therefore, there must be an additional property distinguishing quarks. Called **color**, this property is a kind of "charge"—not to be confused with electric charge—that can take on any of three values called, whimsically, red, green, and blue. The force binding quarks of different colors is the **color force**, and the quark theory is therefore known as **quantum chromodynamics (QCD)**. In QCD, **gluons** are the field particles that play the role of photons in quantum electrodynamics, binding particles subject to the color force. Particles formed from quarks—the mesons and the baryons—are always **colorless**. This is true of mesons because they contain a quark of one color and another of its anticolor. It is true of baryons because they contain three quarks of different colors which combine to give the baryon zero net color charge. The nuclear force, once thought to be fundamental, is actually a residual manifestation of the color force, acting between the quarks in colorless particles—in much the same way that the van der Waals force between neutral gas molecules is a "residue" of the stronger electric force among the particles comprising the molecules.

Photons mediate the electromagnetic force between charged particles, but are themselves uncharged. In contrast, gluons, like the quarks they bind, carry color charge. There are eight different gluons; six carry combinations like redantiblue ($R\overline{B}$), green-antired ($G\overline{R}$), etc.; the other two are colorless. Exchange of a colored gluon, unlike photon exchange in quantum electrodynamics, thus changes the colors of the particles involved (Fig. 45-12).

Another surprising aspect of quarks is that the color force binding them into composite particles does not decrease with separation. For that reason it appears impossible to isolate a single quark (Fig. 45-13). As a result we never see individual free particles with fractional electric charge.

Today the quark model is universally accepted. Many of the previously "elementary" particles—including protons and neutrons—are now considered composites of quarks. Electrons and other leptons seem, on the other hand, to be truly elementary.

The up, down, and strange quarks soon proved insufficient to account for all the observed particles. Theorist Sheldon Glashow argued for a fourth quark, which he called the **charmed quark**. Ten years later, following intensive

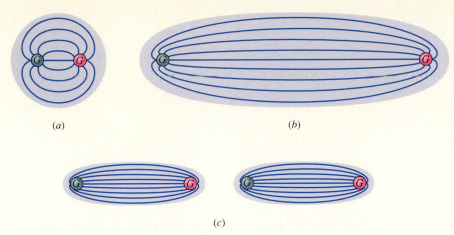

(a)

(b)

(c)

FIGURE 45-13 Quark confinement in a meson consisting of a green quark and its antiquark. (a) Field lines represent the "color field" that joins the two. (b) The field remains confined as the quarks are moved apart, so the field strength stays essentially constant. Infinite energy would therefore be required to separate the quarks. (c) Pulling the quarks far apart builds up enough energy to create a quark-antiquark pair, with the result that the energy goes into creating new composite particles rather than isolating individual quarks.

searches, experimental teams at Brookhaven National Laboratory and the Stanford Linear Accelerator Center announced the discovery of a particle that implies the existence of the charmed quark. The Brookhaven group called the new particle J and the Stanford group called it ψ; to this day it is called the J/ψ. The charmed and strange quarks form a related pair, similar to the up/down quark pair.

There's still one more quark pair. A 1977 experiment confirmed the existence of the **bottom quark**, and in 1994 experiments at Fermilab gave evidence for the **top quark**. The more exotic quarks are more massive, and therefore, through mass-energy equivalence, it takes more energy to produce particles containing them. This need for higher energy is what drives the push for ever more powerful and expensive particle accelerators (see Application: Particle Accelerators, later in this chapter.) Table 45-2 lists some properties of the six quarks.

TABLE 45-2 *Matter Particles of the Standard Model*

QUARK NAME	SYMBOL	MASS* (MeV/c²)	CHARGE	CORRESPONDING LEPTONS (SYMBOL, MASS IN MeV/c²)
Down	d	5–15	$-\frac{1}{3}e$	Electron (e, 0.511), electron neutrino (ν_e, $t^\dagger$)
Up	u	2–8	$+\frac{2}{3}e$	
Strange	s	100–300	$-\frac{1}{3}e$	Muon (μ, 106), muon neutrino (ν_μ, $t^\dagger$)
Charmed	c	1000–1600	$+\frac{2}{3}e$	
Bottom	b	4100–4500	$-\frac{1}{3}e$	Tau (τ, 1777), tau neutrino (ν_τ, $t^\dagger$)
Top	t	1.8×10^5	$+\frac{2}{3}e$	

* Quark masses are estimates; specifying quark masses is difficult because individual quarks are not found in isolation and because the mass equivalent of hadron binding energy is very large.

† Neutrino upper mass limits are $<$7.3 eV, $<$270 keV, and $<$35 MeV, for ν_e, ν_μ, ν_τ, respectively.

| **EXAMPLE 45-2** | *Quark Composition* |

Evaluate the charge and strangeness of the Λ^0 particle, which has quark composition *uds*.

Solution

From Table 45-2 we see that the net charge is $(\frac{2}{3}e) + (-\frac{1}{3}e) + (-\frac{1}{3}e) = 0$. Since the *s* quark has strangeness -1. while the up and down are not strange, the total strangeness is -1.

EXERCISE Find the charge and strangeness of the Ξ^-, whose quark composition is *dss*.

Answer: $-e, -2$

Some problems similar to Example 45-2: 14–17

The Standard Model

We now have six **flavors** of quarks—up, down, strange, charmed, top, bottom—that seem to be truly elementary constituents of matter. As we've seen, quarks join to form all the hadrons—the baryons and mesons. But other particles, namely the leptons and the field particles, are not made from quarks. They, like the quarks, are believed to be truly indivisible, elementary particles.

In this "zoo" of elementary particles, physicists recognize three distinct "families." The up and the down quarks make up the neutron and proton; together with the electron and its related neutrino, they account for the properties of ordinary matter. A second family consists of the strange and charmed quarks, the electron-like muon, and the muon neutrino. The quarks of this family are more massive than the up and down quarks, and the muon is more massive than the electron. More massive still are the particles of the third family, consisting of the top and bottom quarks, the electron-like tau particle, and the tau neutrino. Table 45-2 shows these three families from which all known matter is constructed.

You may be expecting that future editions of this book will tell of additional quarks, and thus of additional families of matter. Whether such additional families exist was an open question until 1991, when physicists at the Large Electron Positron Collider in Geneva, Switzerland, examined over half a million particle decay events to conclude that the number of different types of neutrinos that can exist is 2.99 ± 0.06. Since there is presumably a neutrino type for each family, this result seems to preclude the existence of additional families.

The theory that currently describes elementary particles and their interactions is called the **standard model**. In addition to the constituent particles of matter shown in Table 45-2, the standard model includes the photons that mediate the electromagnetic interaction, the gluons of the color force, the *W* and *Z* particles that mediate the weak force, and an as yet undetected particle called the Higgs boson, believed responsible for other particles' masses (Table 45-3). The standard model is very successful in explaining the phenomena of particle physics, but it leaves many fundamental questions unanswered. Why, for example, do the quarks and leptons have the particular masses they do? Why are there only three families of elementary particles? Why are leptons and quarks distinct? Are these particles really elementary, or are there even smaller structures at hitherto unexplored scales? Continuing theoretical work and experiments at ever higher energies may someday answer these questions.

TABLE 45-3 *Field Particles and Forces*

PARTICLE	MASS (GeV/c²)	ELECTRIC AND COLOR CHARGE	FORCE MEDIATED	RANGE	APPROXIMATE STRENGTH AT 1 fm (RELATIVE TO COLOR FORCE)
Graviton	0	0, 0	Gravity	Infinite	10^{-38}
$W^{\pm}$	80.2	$\pm 1, 0$	Weak	$< 2.4 \times 10^{-18}$ m	10^{-13}
Z^0	91.2	0, 0			
Photon, γ	0	0, 0	Electromagnetic	Infinite	10^{-2}
Gluon, g (8 varieties)	0	0, 6 color/anticolor combinations, 2 colorless	Color	Infinite*	1
Higgs boson, H^0	40–1000?	0, 0	The Higgs boson is an as yet undetected particle needed to account for the nonzero masses of the $W^{\pm}$ and Z^0, as well as of other particles.		

* The nuclear force is the residual color force between colorless particles and has a range of about 1 fm (10^{-15} m).

45-4 UNIFICATION

We first introduced the fundamental forces of nature in Chapter 5, indicating that today physicists recognize just three such forces: gravity, the electroweak force, and the color force. It was not always that simple, though. In Chapters 23 through 34 we studied the electric and magnetic forces, first separately but then with realization that they fell under the single umbrella of electromagnetism. The unification of electricity and magnetism was a major step forward in our understanding of physical reality. Physicists continue to strive for further unification, with the ultimate hope that someday all the forces will be understood as a manifestation of a single common interaction.

Electroweak Unification

In the 1960s and early 1970s, a century after Maxwell formalized the unification of electromagnetism, physicists Steven Weinberg, Abdus Salam, and Sheldon Glashow proposed that the electromagnetic force and the weak force are really aspects of the same thing. Their theory predicted the existence of the particles W^+, Z^0, and W^-, the "carriers" of the unified electroweak force. In 1983 a huge international consortium headed by Carlo Rubbia discovered the W and Z particles, using advances in accelerator technology developed by Simon van der Meer (Fig. 45-14). That discovery confirmed the electroweak unification, and

(a)

(b)

FIGURE 45-14 (*a*) Carlo Rubbia and Simon van der Meer at CERN, the European Laboratory for Particle Physics. (*b*) Computer image showing particle tracks resulting from decay of a W particle.

Rubbia and van der Meer joined a long list of physicists who had won the Nobel Prize for contributions to our understanding of the structure of matter (Table 45-4).

TABLE 45-4 *Nobel Prizes for Interpreting the Structure of Matter*

PRIZE YEAR	PRIZE RECIPIENT	RESEARCH CONTRIBUTION[+]
1903	Antoine H. Becquerel	Discovery of radioactivity (1896)
	Marie and Pierre Curie	Studies of radioactivity
1906	J. J. Thomson	Discovery of the electron and other studies (1897)
1908	Ernest Rutherford*	Disintegration of elements using alpha particles (1902)
1935	James Chadwick	Discovery of the neutron (1932)
	Irène and Frédéric Joliot-Curie	Synthesizing new radioactive elements (1934)
1936	Victor F. Hess	Discovery of cosmic rays (1911)
	Carl D. Anderson	Discovery of the positron (1932)
1938	Enrico Fermi	Production of new radioactive elements (1934–1937)
1949	Hideki Yukawa	Theoretical prediction of mesons (1934)
1951	John D. Cockcroft and Ernest T. S. Walton	Transmutation of nuclei (1932)
1957	C. N. Wang and T. D. Lee	Theoretical prediction of parity violation (1956)
1959	Emilio G. Segrè and Owen Chamberlain	Production of the antiproton (1955)
1963	Maria Goeppert Mayer and J. Hans Jensen	Nuclear shell model (1949)
	Eugene P. Wigner	Application of symmetry to elementary particle theories
1968	Luis Alvarez	Contributions to particle accelerators (1946)
1969	Murray Gell-Mann	Theory classifying elementary particles (1964)
1975	Aage Bohr, Ben Mottelson, and James Rainwater	Studies of nuclear structure and dynamics
1976	Burton Richter and Samuel C. C. Ting	Discovery of the J/ψ particle, the first known particle made from charmed quarks (1973–1974)
1979	Sheldon L. Glashow, Abdus Salam, and Steven Weinberg	Unification of the electromagnetic and weak forces (1961–1972)
1980	Val L. Fitch and James W. Cronin	Experimental proof of violation of conservation of charge-parity (1964)
1984	Carlo Rubbia and Simon van der Meer	Discovery of W particles, predicted by the electroweak unification (1983); stochastic cooling for focusing particle beams (1968)
1988	Leon M. Lederman, Melvin Schwartz, and Jack Steinberger	Production of neutrino beams and discovery of the μ neutrino (1960–1962)
1990	Jerome Friedman, Henry Kendall, and Richard Taylor	Using scattering of electrons from protons and neutrons to explore the quark model
1992	Georges Charpak	Developing electronic detectors for detecting subatomic particles in accelerators
1995	Martin L. Perl	Discovery of the tau particle (1974–1977)
	Frederick Reines	Pioneering work with neutrinos, including demonstrating the existence of the electron antineutrino (1950s)

[+]Number in parentheses is year discovery was made.

* Rutherford and Joliot-Curie prizes in chemistry; all others in physics.

FIGURE 45-15 The Super Kamiokande experiment in Japan consists of 50,000 tons of pure water in a deep underground chamber, surrounded by 11,200 photomultiplier tubes that detect light flashes from rare nuclear reactions. The device functions as a neutrino detector, and can also search for evidence of proton decay. Experiments with Super Kamiokande show that the proton has a lifetime of at least 10^{33} years, and that neutrinos probably have nonzero mass.

Further Unification

Electroweak unification led to the present situation in physics, which has gravity, electroweak, and color forces as the fundamental interactions of matter. A further step, the **grand unification theories** (GUTs), attempt to merge the electroweak and color forces. Some versions of the GUT predict that the proton is not stable, but should decay on the very long timescale of some 10^{31} years. We can't wait that long, but we can put 10^{34} protons together in the form of tens of thousands of tons of water and watch for proton decay (Fig. 45-15). Such experiments have not found the predicted decays, but another GUT prediction—that neutrinos have mass—was tentatively verified in 1998, also in the device of Fig. 45-15. Many physicists believe that some form of grand unification will soon be achieved.

Even grand unification would still leave two forces, one of them gravity. Attempts to reconcile our current theory of gravity—Einstein's general theory of relativity—with quantum mechanics have so far made little progress. Yet such a reconciliation is a necessary prerequisite for a final unification of all known forces. In the 1980s a flurry of interest developed in so-called **superstring theories**, which picture elementary particles as fundamental vibration modes on stringlike structures that may be as short as 10^{-35} m (Fig. 45-16). This model

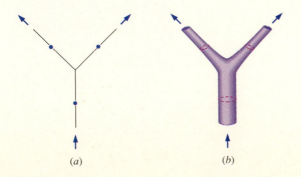

(a)　　　　　(b)

FIGURE 45-16 (*a*) A particle decay involving point particles in the standard model. (*b*) A similar decay in string theory. At a given instant in time, the particle is represented by a stringlike loop. Its trajectory in spacetime traces out the cylindrical structures shown. Note that there is no single point in spacetime at which the decay takes place.

is set not in the four-dimensional spacetime to which we have become accustomed, but in a ten-dimensional spacetime with six of its dimensions "compactified" in a way that makes them undetectable in normal interactions. To some physicists, string theories hold the promise of a "theory of everything," explaining all our observations about the behavior of the universe. To others they are another in a long list of unsuccessful attempts at a comprehensive explanation of physical phenomena. Only further research will tell.

Symmetry Breaking

Unification theories predict that phenomena that appear distinct under normal conditions will be clearly seen as one at sufficiently high energies. The observed unification represents a kind of symmetry that is "broken" as the energy level drops. A crude mechanical analog for such **symmetry breaking** is a ball rolling around the rim of a bowl. As long as the ball keeps rolling, there is no preferred angular direction. But if the ball stops, it drops into the bowl in a particular direction. The symmetry of the original situation has been broken. Analogously, at energies above 100 GeV, what we call the electromagnetic and weak forces are one and the same. But at lower energies the symmetry is broken, and we see two distinct forces. Particle accelerators now being planned will exceed the energy of electroweak symmetry breaking, allowing us to explore that interaction in its fundamental simplicity. But the energy at which symmetry breaking occurs increases to some 10^{15} GeV as we move from electroweak to grand unification, making it unlikely that we will achieve that energy in the foreseeable future. And the energy at which gravity, too, would join a single unified force is an even more remote 10^{19} GeV.

APPLICATION	*Particle Accelerators*

Most particles are more massive than the proton and some, like the weak-force mediators $W^{\pm}$ and Z, are extremely massive. Since the more massive particles are all unstable, discovering them involves first creating them—and that requires energy of at least mc^2, with m the particle mass. That energy requirement, along with the hint of new phenomena such as force unification, drives particle physicists' seemingly insatiable desire for particle accelerators of ever higher energies.

The earliest accelerators were electrostatic devices that established large potential differences between conducting electrodes and used the associated electric field to accelerate charged particles (Fig. 45-17). But such accelerators are limited to maximum energies around 20 MeV because of the difficulties of handling high voltages. We saw in Chapter 29 how this problem is cleverly circumvented in the cyclotron, an accelerator that uses a magnetic field to keep particles in circular orbits so they can gain energy on each orbit from a modest electric field. But cyclotrons work only for nonrelativistic particles, for which the cyclotron frequency is independent of particle energy. Today's high-energy experiments call for

FIGURE 45-17 This van de Graaf generator, now at the Boston Museum of Science, develops a potential difference of 1.5 MV. It was originally used as a particle accelerator.

(a)

(b)

FIGURE 45-18 (*a*) Aerial view of Fermilab, the national accelerator laboratory outside Chicago. The main ring is the 2-km-diameter Tevatron that accelerates protons and antiprotons to 1 TeV. Beams of the two particle species then collide to produce scores of new particles. (*b*) Inside the main ring. The red structures comprise a ring of conventional electromagnets that steer the particles until their energy is large enough for injection into the superconducting magnet ring (yellow, below).

ultrarelativistic particles, whose speeds differ only minutely from the speed of light. As a result, today's accelerators are primarily variations on the **synchrotron**, a device in which the magnetic field increases with the particle energy to maintain particles in a circular orbit of fixed radius (Fig. 45-18). An alternative to the synchrotron is the **linear accelerator**, the largest of which is the Stanford Linear Accelerator shown in Fig. 38-15.

A head-on collision between two cars is much more damaging than a collision of a moving car with a stationary one since in the former case all the energy goes into damaging the cars while in the latter a great deal of it goes into accelerating the

initially stationary "target" car. For the same reason head-on collisions of high-energy particles make much more energy available for creation of new particles. As a result, most of the highest energy accelerators today are colliders, with particle beams accelerated in opposite directions and brought to collide inside elaborate detectors (see Application: Detecting Particles earlier in this chapter). The largest accelerator currently planned is the Large Hadron Collider (LHC) at CERN, the European Laboratory for Particle Physics, at Geneva, Switzerland. When it becomes operational around 2005, LHC will collide proton beams at energies up to 14 TeV. This high energy will permit a whole new realm of particle physics experiments, including searches for the hypothetical Higgs boson that is believed responsible for symmetry breaking and for other particles' masses. The LHC will also create conditions that existed when the universe was only 10^{-12} s old; more on this in the next section. Figure 45-19 compares some of the world's most energetic accelerators.

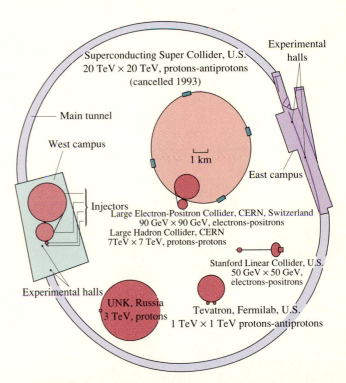

FIGURE 45-19 Drawings of the world's largest accelerators, shown on the same scale. The Superconducting Super Collider, a U.S. project whose construction was halted in 1993, would have measured 87 km in circumference—about the same size as the beltway surrounding Washington, DC. Note that the larger machines have substantial injector systems, consisting of one or more smaller accelerators to bring the particles to high energy before injection into the main ring. Energies of the separate particle beams are listed for collider machines. CERN's Large Electron-Positron Collider and the LHC will share the same 27-km-circumference tunnel.

45-5 THE EVOLVING UNIVERSE

We come at the end to the broadest possible questions about physical reality: How did the universe come to be? What is its overall structure? What will its future bring? Remarkably, these cosmological questions are closely linked with the questions of particle physics.

Expansion of the Universe

Early in the twentieth century, astronomers argued about the nature of certain fuzzy patches visible in telescope photographs. Many thought they were gas clouds scattered among the visible stars, but others made a more radical proposal: that some of these "nebulae" were gravitationally bound systems containing billions of stars and that they were almost inconceivably distant.

In the 1920s, the opening of the 2.5-m telescope at California's Mt. Wilson Observatory finally resolved the issue. There, astronomer Edwin Hubble (Fig. 45-20) proved that some nebulae were indeed distant galaxies like our own Milky Way, each containing some 10^9 to 10^{13} stars. Today cosmologists think of galaxies as individual "point particles" whose distribution traces the overall structure of the universe.

Hubble continued to study the galaxies through the 1920s, and by analyzing their spectra he made a remarkable discovery: spectral lines from distant galaxies are shifted toward the red, with the amount of shift dependent on the distance to the galaxies. The most reasonable and widely held explanation is that the redshift is caused by the Doppler effect (see Section 17-7). Then the implication of Hubble's work is that the distant galaxies are receding from us at speeds proportional to their distances. This result is known as the **Hubble law**:

$$v = H_0 d, \tag{45-1}$$

where v is the recession speed, d the distance, and H_0 the **Hubble constant**. Observations of distant galaxies give a value for H_0 of approximately 20 kilometers per second per million light-years of distance. Astronomers now use the Hubble relation to find the distances to remote galaxies, measuring their redshifts and using Equation 45-1 to infer their distances.

It may sound like Hubble's law puts us right in the center of things, in grotesque violation of modern science's view that Earth and its inhabitants do not occupy a favored position in the universe. But actually the inhabitants of any other galaxy would observe the same thing: all the distant galaxies would be receding from them at speeds proportional to their distances. As long as the universe is infinite in extent, none can claim to be at the center. And if it's not infinite, then Einstein's general theory of relativity gives it a closed-curve shape which still has no center (Fig. 45-21).

FIGURE 45-20 Edwin Hubble in about 1960, at the prime focus of the 5-m telescope at Palomar Observatory in California.

FIGURE 45-21 In Einstein's general theory of relativity, a finite universe would be like the 4-dimensional analog of the surface of a sphere. If the sphere expands, all the galaxies move apart at speeds proportional to their distances from each other. Yet there is no "center" to this expanding universe—i.e., to the sphere's surface.

FIGURE 45-22 The Hubble deep field is a long exposure photo taken with the Hubble Space Telescope of a region devoid of nearby stars. Most objects that appear in the HDF are distant galaxies that yield a wealth of information about the origin of the universe.

The Hubble relation implies that the universe is expanding, as all the galaxies move apart. Extrapolating the expansion backward in time suggests there was a time when all the galaxies were together. Thus Hubble's discovery is evidence that the universe may have had a definite beginning, in the form of a colossal explosion that flung the matter of the universe into an expansion that continues today. Based on additional evidence that we will describe shortly, most scientists now accept the theory that the universe began with such an explosive **big bang**. Hubble's pioneering work on the universe's origin is honored in the naming of the Hubble Space Telescope (Fig. 45-22).

EXAMPLE 45-3	*How Old Is the Universe?*

Using the value $H_0 = 20$ km/s/Mly and assuming the expansion rate has been constant, find how long the universe has been expanding.

Solution

Today, a galaxy a distance d from our Milky Way is moving with speed $v = H_0 d$ (Equation 45-1). If the galaxy has been moving at constant speed, then $d = vt$, and the time since the expansion started must be

$$t = \frac{d}{v} = \frac{d}{H_0 d} = \frac{1}{H_0}.$$

This quantity is called the **Hubble time**. To find its value, we need to convert from the mixed units used for the Hubble constant:

$$t = \frac{1}{H_0} = \frac{1}{(20 \text{ km/s/Mly})/[3.0 \times 10^5 \text{ (km/s)}/(\text{ly/y})]} = 15 \text{ Gy},$$

where we used the speed of light—3.0×10^5 km/s—to convert km/s to ly/y, thus giving the answer in years. So the universe is about 15 billion years old under the assumptions of this example.

EXERCISE What is the distance to a galaxy whose redshift shows it to be moving away from Earth at 1% of the speed of light?

Answer: 200 Mly

Some problems similar to Example 45-3: 25–29

The Cosmic Background Radiation

In 1965 Arno Penzias and Robert Wilson at Bell Laboratories found a faint "noise" of microwave radiation in a satellite communications antenna they were testing. They could not eliminate the noise, which seemed to come from all directions. Theorists at Princeton pointed out that Penzias and Wilson had detected radiation dating to a much earlier era in the universe. This **cosmic microwave background radiation** is the strongest evidence for the big bang theory.

The big bang theory suggests that the universe started very hot and then cooled as it expanded, doing work against its own gravitation. At first it was so hot that any atoms that formed would be dashed apart by collisions at the high thermal energy prevailing. Thus, in its early times the universe was populated by individual charged particles. These interacted readily with electromagnetic radiation, making the universe opaque. But by about 10^5 years the temperature had dropped to several thousand kelvins, and at that point atoms of hydrogen and

FIGURE 45-23 Spectrum of the cosmic microwave background, taken with the Cosmic Background Explorer satellite (COBE). The observational plot so perfectly matches that of a blackbody at 2.726 K that the discrepancies are less than the width of the curve drawn here. COBE was built with the specific goal of making detailed studies of the cosmic background radiation.

FIGURE 45-24 The positions of 1065 galaxies in a wedge-shaped region extending more than one-third of the way around the sky. The radial distance from Earth extends outward from the apex to a maximum of about 1 billion light-years. The galaxies are distributed on the edges of large bubblelike voids.

helium could form (there were almost no heavier nuclei present; see Chapter 43). Since neutral atoms interact much more weakly with electromagnetic radiation, the universe became transparent, and photons emitted as the atoms formed could travel throughout the universe with little chance of being subsequently absorbed. Those photons became the cosmic background radiation, permeating the entire universe.

Expansion of the universe would cool not only matter but also radiation; the big bang scenario suggests that radiation from the time the atoms formed should now be cooled to that of a blackbody at 2.7 K—a radiation distribution that peaks in the microwave region of the electromagnetic spectrum. Analysis of the microwave background shows that it has indeed precisely that blackbody character and temperature (Fig. 45-23). Thus, the observed cosmic microwave background is a direct reflection of conditions 10^5 years after the universe began.

An outstanding problem in cosmology is the origin of large-scale structure in the universe. Why should an initially homogeneous "soup" of individual particles eventually organize itself into large "clumps" like galaxies? Increasingly, astronomical observations show that the galaxies themselves are organized into much larger structures, concentrated on the edges of giant bubblelike voids hundreds of millions of light-years across (Fig. 45-24). How could this structure arise from an initially structureless big bang? A partial answer came in 1992, with the discovery that the cosmic microwave background itself contains very faint "ripples" that may be the "seeds" of larger structures that developed later as the universe expanded.

The Earliest Times

With the microwave background, we can "see" the universe as it was 10^5 years after its beginning. We also have a direct tie to an earlier time, from about 1 second to tens of minutes, when the lightest nuclei were forming. The first composite nuclei were the simplest: deuterium, consisting of a proton and a neutron. The rate of deuterium formation is critically sensitive to the expansion rate of the universe. If the expansion was rapid, then the density dropped fast enough that a deuteron, once formed, was unlikely to be disrupted in a subsequent collision. But a slower expansion rate would result in less deuterium, as deuterons and protons interacted to form tritium and helium-3. Very sensitive measurements of deuterium abundance, based on spectral lines from interstellar deuterium, therefore provide direct evidence for conditions early in the big bang.

We may soon be able to look further back, by studying a "neutrino background," believed to date from about 1 second after the big bang began, a time when the density became low enough that neutrinos ceased to have much interaction with other matter. Evidence for still earlier times comes indirectly from applying elementary particle physics to the behavior of matter at extreme temperatures and seeing if it predicts an evolution to conditions we do observe. Today's accelerators achieve conditions that existed only 10^{-9} s from the beginning, and the Large Hadron Collider will push that back to 10^{-12} s. Thus, these early times provide a test of elementary particle physics, and that physics in turn helps us to understand how the universe began and how it evolved to its present state. Figure 45-25 summarizes our understanding of the evolution of the universe, showing that the phenomena of particle physics and unification are inextricably tied with cosmic questions.

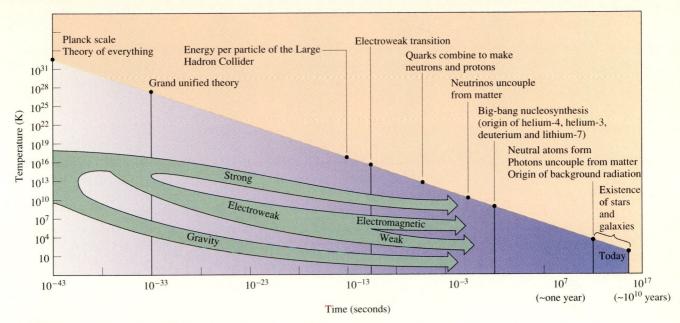

FIGURE 45-25 The evolution of the universe from the earliest times to the present. Note the highly logarithmic scale. As the universe has expanded, its average temperature and energy have fallen, creating epochs in which different physical processes dominated. At first the fundamental forces were all one, but symmetry breaking caused them to separate as the energy dropped. Present-day physics cannot address times before 10^{-43} s, which requires a merging of quantum theory and general relativity.

The Inflationary Universe

The original big bang theory has problems with several observed features of the universe. Why, for example, do we find only matter but not antimatter? Why does the whole universe seem essentially homogeneous and in thermodynamic equilibrium on the largest scales, when the most distant regions are so far apart that light could not have traveled between them in the time since the beginning? And why—as we elaborate in the next section—do the kinetic and potential energy of the universe seem exactly equal?

The current solution to these conundrums is the **inflationary universe** theory, first advanced by Alan Guth of MIT. This theory incorporates the ideas of grand unification and thus links the evolution of the universe directly with particle physics. At the earliest times, the temperature and thermal energy were so high that all the forces were unified. The inflationary theory proposes that, at 10^{-35} s, the universe expanded at a phenomenal rate, growing 10^{50} times more than in the original big bang theory (Fig. 45-26). This rapid expansion lasted a mere 10^{-32} s, but that short period was sufficient to establish many of the characteristics we now observe.

Because of the tremendous expansion, now-distant locations would have once been much, much closer—so close that they could have reached the thermodynamic equilibrium now evident at large scales. As the universe cooled, it acted like a very pure liquid that can cool well below its normal freezing point without freezing. Analogously, the thermal energy of the universe dropped below the value at which the forces all appear unified, but unification still held. In this

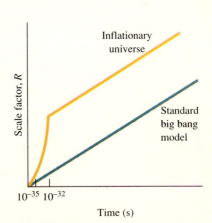

FIGURE 45-26 Expansion in the standard and inflationary big-bang theories. The scale factor R measures the amount of expansion.

supercooled state, the universe had excess energy, and this energy led to the inflation in a way that can be predicted from theory. When the symmetry was finally broken and the individual forces appeared distinct, the excess energy was transformed into the matter and energy we observe today. Thus, in this theory we ourselves and all that we deal with are consequences of the delay in this "phase change" of the universe.

What of the Future?

Will the universe expand forever, or will the expansion eventually stop and then reverse? That question is like asking whether a spacecraft launched from Earth will eventually return or not, and the answer is the same as the one we found in Chapter 9: If the kinetic energy exceeds the magnitude of the (negative) potential energy, then the expansion will proceed forever; otherwise a collapse will eventually ensue. The ratio of potential to kinetic energy is designated Ω:

$$\Omega = \frac{|\text{ potential energy of the universe }|}{\text{kinetic energy of expansion}}.$$

This single parameter determines the universe's ultimate fate; $\Omega > 1$ implies eventual collapse, while $\Omega < 1$ means the expansion will continue forever. An alternative way of expressing the same dichotomy is in terms of the average density, as the example below illustrates.

| **EXAMPLE 45-4** | *Critical Density* |

Assuming a spherically symmetric universe obeying Newton's laws, find the **critical density** above which the universe is destined ultimately to collapse.

Solution

Let R be a typical cosmological distance for which the Hubble relation applies, and m the mass of a test particle (i.e., a galaxy). Since the universe is homogeneous on the large scale, it suffices to determine the potential to kinetic energy ratio for this single test particle. If M is the total mass within a sphere of radius R, then

$$\Omega = \frac{GMm/R}{\frac{1}{2}mv^2} = \frac{GM/R}{\frac{1}{2}v^2}.$$

We can substitute for the expansion speed v from Hubble's law to get

$$\Omega = \frac{GM/R}{\frac{1}{2}H_0^2R^2} = \frac{2GM}{H_0^2R^3}.$$

But the density $\rho = M/\frac{4}{3}\pi R^3$, so we can solve for M/R^3 and substitute to get

$$\Omega = \frac{8\pi G\rho}{3H_0^2}.$$

The critical density ρ_c occurs when $\Omega = 1$, so

$$\rho_c = \frac{3H_0^2}{8\pi G}.$$

Although we derived this result on the assumption of a homogeneous, spherically symmetric, Newtonian universe, it is in fact correct even for a universe governed by Einstein's general relativity.

EXERCISE Find the approximate value of ρ_c, using the Hubble constant from Example 45-3.

Answer: 5×10^{-27} kg/m^3

Some problems similar to Example 45-4: 30, 31

Observationally, cosmologists have several ways to determine the density ρ of the universe or, equivalently, the expansion parameter Ω. By studying the most distant objects, we look back into the past to a time when the expansion rate is expected to have been faster; measuring how much faster then gives information on the parameter Ω. Until very recently, data from the most distant reaches of the universe weren't good enough to draw firm conclusions about the behavior of cosmic expansion over time. Then, in 1998, came a startlingly unexpected discovery: measurements by two different groups suggested that cosmic expansion is actually accelerating! If this result is confirmed, it means a kind of "antigravity" force is operating on the large scale. Albert Einstein actually proposed such a force as part of his general relativity, but rescinded the idea and later called it a "great blunder." Time will tell whether he was right.

Another approach to cosmic expansion is to measure the amount of matter and then calculate the density. Using the density of visible matter alone gives considerably less than the critical density. But the motions of galaxies in large galaxy clusters imply the presence of much more matter than we can see—enough to bring the universe closer to the critical density. Some of that **dark matter** may consist of burned-out stars, or perhaps neutrinos—but most of it must be something entirely new, consisting of as-yet undetected, exotic particles. It is a sobering thought that most of the matter in the universe is probably in a form about which we know essentially nothing.

One of the most remarkable features of the universe is that its density appears even close to the critical density, since it seems the density could have *any* value. The inflationary theory provides an explanation, suggesting that the entire universe we now observe was, before inflation, smaller than a proton. The tremendous expansion flattened out the curves of spacetime which, in general relativity, are the manifestation of gravity. The result is a universe that may be right on the borderline between expansion and gravitational collapse, a universe that will expand forever but just barely so.

Understanding the Universe

With this chapter's brief survey, we have reached the limits of present understanding of the universe. In the process we have seen that particle physics and cosmology are inextricably linked. To understand our universe, we need to understand all its aspects from the largest to the smallest—and that means understanding all the forces, from gravity to the weak force; all the physical laws, from Newton's and Maxwell's on to the laws of quantum mechanics; and the nature of the particles that comprise all matter. We have touched, in many cases only briefly, on all these topics. We hope this text has given you a foundation for further understanding and appreciation of the richness and diversity of the physical universe.

CHAPTER SYNOPSIS

Summary

1. What appear in classical physics as forces are actually interactions mediated by the exchanges of particles called **field particles** or **gauge bosons**; an example is the photon of the electromagnetic force. The mass of a field particle is zero for forces with infinite range, but nonzero for short-range forces.

2. There are a large number of subnuclear particles. Major groups include the **leptons**, which do not experience the nuclear or strong force, and the **hadrons**, which do experience the strong force. Leptons include the familiar electron and two similar but more massive and unstable particles, the **muon** and the **tau**, along with three types of neutrinos. Each also has an antiparticle. **Hadrons** include the integer-spin **mesons** and the half-integer-spin **baryons**, of which the proton and neutron are familiar examples. A third group are the field particles that mediate the fundamental forces.

3. Particles exhibit a number of properties like electric charge and spin. Some, like electric charge, angular momentum, and baryon number, are conserved in all interactions. Others are conserved in only some interactions; an example is **strangeness**, which is conserved in strong and electromagnetic interactions but which may change in weak interactions. Particle systems may also exhibit **symmetries**, including invariance under parity inversion, charge conjugation, and time reversal. However, each of these symmetries is violated individually in some interactions.

4. Leptons and field particles appear to be truly elementary, but hadrons are composed of **quarks**. A quark and an

antiquark together make a meson, while three quarks make a baryon. Quarks come in the six **flavors** up, down, strange, charmed, top, and bottom, and in three **colors**. Particles made from quarks are **colorless**, and the nuclear force between such particles is actually a residue of the **color force** between their quark constituents. **Gluons** are the field particles that mediate the color force. The color force does not decrease with increasing quark separation, making it impossible to isolate individual quarks.

5. The **standard model** incorporates the quarks and leptons into three families of particles, with ordinary matter involving only the first family: the up and down quark and the electron and its associated neutrino.

6. **Unification** of the fundamental forces is a goal of physics. In the 1970s, physicists Weinberg, Salam, and Glashow showed that the weak force and electromagnetism are manifestations of the same underlying force. Physicists today are striving to unify this electroweak force with the color force, to produce a **grand unification theory**. Unification with gravity is a more remote goal. At high enough energies, forces now distinct should seem as one. The appearance of distinct forces as energy drops is called **symmetry breaking**.

7. Questions of **cosmology** are closely linked with those of particle physics. The **Hubble law** shows that the universe is expanding, providing evidence that the universe had a definite beginning in a cosmic explosion called the **big bang**. Further evidence for the big bang comes from the **cosmic microwave background radiation**, left over from the time the atoms formed, about 10^5 years after the beginning. Evidence of earlier times comes from primordial nucleosynthesis, especially of deuterium, and from applying particle physics to conditions believed to exist within the first second. The **inflationary universe** theory helps explain several features of the present universe, including its density being very nearly the **critical density** that separates a forever-expanding universe from one that will eventually collapse under its own gravity.

Terms You Should Understand

(Pairs are closely related terms whose distinction is important; number in parentheses is chapter section where term first appears.)

virtual photon (45-1)	flavor, color (45-3)
muon, pion (45-2)	standard model (45-3)
lepton, hadron, field particle (45-2)	unification (45-4)
meson, baryon (45-2)	Hubble's law (45-5)
fermion, boson (45-2)	cosmic microwave background radiation (45-5)
strangeness (45-2)	big bang (45-5)
parity (45-2)	inflationary universe (45-5)
quark (45-3)	critical density (45-5)

Symbols You Should Recognize

uud, $u\bar{d}$, and similar quark symbols (45-3)

H_0 (45-5) Ω, ρ_c (45-5)

Problems You Should Be Able to Solve

evaluating conserved quantities in particle interactions (45-2)

determining quark composition of baryons and mesons (45-3)

applying Hubble's law (45-5)

Limitations to Keep in Mind

This chapter deals with the frontiers of physics, and some of the material presented is likely to undergo change.

QUESTIONS

1. Why did Yukawa conclude that the particle mediating the strong force should have nonzero mass?
2. Researchers looking for Yukawa's meson found instead the muon, a particle with approximately the predicted mass. How did they know they had not found the right particle?
3. How can we detect tracks of individual particles?
4. Which interactions conserve parity? Which conserve *CPT*?
5. How are baryons fundamentally different from leptons?
6. Why was discovery of the J/ψ particle significant?
7. What coordinates are changed under the inversion processes *P* and *T*?
8. Why are we unlikely to observe an isolated quark?
9. Describe the relation between the color force and the nuclear force.
10. What is the role of gluons?
11. Classify (a) mesons and (b) baryons as fermions or bosons, and relate your classification to the particles' quark compositions.
12. A hadron decays into leptons. Is the hadron a baryon or a meson? Explain.
13. What is the difference between color and flavor?
14. Why are most particle accelerators circular?
15. Photons and neutrinos are both massless particles. So how are they different?
16. Why is the Fermilab accelerator called a "tevatron?"

17. Name the fundamental force involved in (a) binding of a proton and neutron to make a deuterium nucleus, (b) decay of a neutron into a proton, an electron, and a neutrino, (c) binding of an electron and proton to make a hydrogen atom.
18. What forces are unified in the electroweak theory?
19. What forces would be unified by GUTs?
20. Why do we need higher-energy particle accelerators to explore fully the standard model?
21. How can Hubble's law hold without the universe having a center?
22. Is it possible for a charged particle to be its own antiparticle?
23. Is it possible for a neutral particle *not* to be its own antiparticle? Explain.
24. Describe the origin of the cosmic microwave background.
25. Why is deuterium an especially good probe of conditions in the first few minutes after the big bang?
26. Why are particle physics and cosmology linked?
27. The Large Hadron Collider will create conditions like those that existed 10^{-12} s after the big bang. Explain.

PROBLEMS

Section 45-1 Particles and Forces

1. How long could a virtual photon of 633-nm red laser light exist without violating conservation of energy?
2. Some scientists have speculated on a possible "fifth force," with a range of about 100 m. Following Yukawa's reasoning, what would be the mass of the field particle mediating such a force?
3. The mass of the photon is assumed to be zero, but experiments put only an upper limit of 5×10^{-63} kg for the photon mass. What would be the range of the electromagnetic force if the photon mass were actually at this upper limit?

Section 45-2 Particles and More Particles

4. Write the equation for the decay of the positive pion into a muon and neutrino, being sure to label the type of neutrino. *Hint:* The positive muon is an antiparticle.
5. Use Table 45-1 to find the total strangeness before and after the decay $\Lambda^0 \rightarrow \pi^- + p$, and use your answer to determine the force involved in this reaction.
6. The η^0 particle is a neutral nonstrange meson that can decay into a positive, a negative, and a neutral pion. Write the reaction for this decay, and verify that it conserves charge, baryon number, and strangeness.
7. Are either or both of the following decay schemes possible for the tau particle? (a) $\tau^- \rightarrow e^- + \overline{\nu}_e + \nu_\tau$, (b) $\tau^- \rightarrow \pi^- + \pi^0 + \nu_\tau$
8. Is the interaction $p + p \rightarrow p + \pi^+$ allowed? If not, what conservation law does it violate?
9. Which of the following reactions (a) $\Lambda^0 \rightarrow \pi^+ + \pi^-$ and (b) $K^0 \rightarrow \pi^+ + \pi^-$ is not possible, and why?
10. Both the neutral kaon and the neutral ρ meson can decay into a pion-antipion pair. Which of these decays is mediated by the weak force, and why?
11. Grand unification theories suggest that the decay $p \rightarrow \pi^0 + e^+$ may be possible, in which case all matter may eventually become radiation. Are (a) baryon number and (b) electric charge conserved in this hypothetical proton decay?
12. Consider systems described by wave functions that are proportional to the terms (a) xy^2z, (b) x^2yz, and (c) xyz, where x, y, and z are the spatial coordinates. Which pairs of these systems could be transformed into each other under an interaction obeying conservation of parity?
13. What happens to the sign of the wave functions proportional to the terms (a) xy^2t and (b) xy^2t^2 under the operation PT?

Section 45-3 Quarks and the Standard Model

14. What is the quark composition of the π^-?
15. The Eightfold Way led Gell-Mann to predict a baryon with strangeness -3. What must be its quark composition?
16. The Σ^+ and Σ^- have quark compositions uus and dds, respectively. Are the Σ^+ and Σ^- each other's antiparticles? If not, give the quark compositions of their antiparticles.
17. The J/ψ particle is an uncharmed meson that nevertheless includes charmed quarks. What is its quark composition?
18. List all the possible quark triplets formed from any combination of the up, down, and charmed quarks, along with the charge of each.

Section 45-4 Unification

19. Estimate the volume of the 4000 tons of water used in the proton-decay experiment of Fig. 45-15.
20. Estimate the temperature in a gas of particles such that the thermal energy kT is enough to make electromagnetism and the weak force appear as a single phenomenon.
21. Repeat the preceding problem for the 10^{15} GeV energy of grand unification.
22. The Tevatron at Fermilab accelerates protons to an energy of 1 TeV. (a) How much is this in joules? (b) How far would a 1-g mass have to fall in Earth's gravitational field to gain this much energy?

23. (a) What would be the relativistic factor γ for a 7-TeV proton in the Large Hadron Collider? (b) Find an accurate value for the proton's speed.
24. How long, as measured in the lab frame, would it take a 7-TeV proton to circle the 27-km-circumference of the Large Hadron Collider?

Section 45-5 The Evolving Universe

Note: In working these problems, take $H_0 = 15$ km/s/Mly unless otherwise indicated.

25. Express the Hubble constant in SI units.
26. What is the distance to a galaxy receding from us at 2×10^4 km/s?
27. What is the recession speed of a galaxy 300 Mly from Earth?
28. What would be the age of the universe, assuming constant expansion rate, if the Hubble constant were 25 km/s/Mly?
29. A widely used value for H_0 is 17 km/s/Mly. What age does this value imply for the universe, under the simple assumptions of Example 45-3?
30. Estimate the diameter to which the Sun would have to be expanded for its average density to be the critical density of Example 45-4.
31. Find the critical density if the Hubble constant is 34 km/s/Mly.

Supplementary Problems

32. A baryon called Λ^0 has mass 1116 MeV/c^2. Find the minimum speed necessary for the particles in a proton-antiproton collider to produce lambda-antilambda pairs.
33. A so-called muonic atom is a hydrogen atom with the electron replaced by a muon; the muon's mass is 207 times the electron's. Find (a) the size, and (b) the ground-state energy of a muonic atom.

34. (a) By what factor must the magnetic field in a proton synchrotron be increased as the proton energy increases by a factor of 10? Assume the protons are highly relativistic, so $\gamma \gg 1$. (b) By what factor must the diameter of the accelerator be increased to raise the energy by a factor of 10 without changing the magnetic field?
35. How long would it take the proton in Problem 24 to circle the LHC, as measured in the proton's frame? Assume you can apply special relativity even though the path is curved.
36. A galaxy's hydrogen-β spectral line, normally at 486.1 nm, appears at 495.4 nm. (a) Use the Doppler shift of Chapter 17 to find the galaxy's recession speed, and (b) infer the distance to the galaxy. Are you justified in using Chapter 17's nonrelativistic Doppler formulas? Use $H_0 = 15$ km/s/Mly.
37. Use Wien's law (Equation 39-2) to find the wavelength at which the 2.7-K cosmic microwave background has its peak intensity.
38. One of the earliest measurements of the cosmic microwave background was made at a wavelength of 7 cm. By what factor would the intensity have been greater had the measurement been made at the wavelength of peak intensity? See preceding problem.
39. Many particles are far too short-lived for their lifetimes to be measured directly. Instead, tables of particle properties often list "width," measured in energy units, and indicating the width of distribution of measured rest energies for the particle. For example, the Z^0 has a mass of 91.18 GeV and a width of 2.5 GeV. Use the energy-time uncertainty relation to estimate the corresponding lifetime.
40. A mix of particles starts with equal numbers of the three types of sigma particles listed in Table 45-1. What is the relative portion of each after (a) 5×10^{-20} s and (b) 5×10^{-10} s? Give your answer in a reference frame in which the particles are at rest.

APPENDIX A
Mathematics

A-1 ALGEBRA AND TRIGONOMETRY

Quadratic Formula

If $ax^2 + bx + c = 0$ then $x = \dfrac{-b \pm \sqrt{b^2 - 4ac}}{2a}$

Circumference, Area, Volume

Where $\pi \simeq 3.14159\ldots$

circumference of circle	$2\pi r$
area of circle	πr^2
surface area of sphere	$4\pi r^2$
volume of sphere	$\frac{4}{3}\pi r^3$
area of triangle	$\frac{1}{2}bh$
volume of cylinder	$\pi r^2 \ell$

Trigonometry

definition of angle (in radians): $\theta = \dfrac{s}{r}$

2π radians in complete circle
1 radian $\simeq 57.3°$

Trigonometric Functions

$$\sin \theta = \frac{y}{r}$$

$$\cos \theta = \frac{x}{r}$$

$$\tan \theta = \frac{\sin \theta}{\cos \theta} = \frac{y}{x}$$

Values at Selected Angles

$\theta \rightarrow$	0	$\dfrac{\pi}{6}$ (30°)	$\dfrac{\pi}{4}$ (45°)	$\dfrac{\pi}{3}$ (60°)	$\dfrac{\pi}{2}$ (90°)
$\sin\theta$	0	$\dfrac{1}{2}$	$\dfrac{\sqrt{2}}{2}$	$\dfrac{\sqrt{3}}{2}$	1
$\cos\theta$	1	$\dfrac{\sqrt{3}}{2}$	$\dfrac{\sqrt{2}}{2}$	$\dfrac{1}{2}$	0
$\tan\theta$	0	$\dfrac{\sqrt{3}}{3}$	1	$\sqrt{3}$	∞

Graphs of Trigonometric Functions

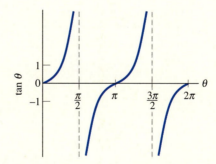

Trigonometric Identities

$$\sin(-\theta) = -\sin\theta$$

$$\cos(-\theta) = \cos\theta$$

$$\sin\left(\theta \pm \frac{\pi}{2}\right) = \pm\cos\theta$$

$$\cos\left(\theta \pm \frac{\pi}{2}\right) = \mp\sin\theta$$

$$\sin^2\theta + \cos^2\theta = 1$$

$$\sin 2\theta = 2\sin\theta\cos\theta$$

$$\cos 2\theta = \cos^2 \theta - \sin^2 \theta = 1 - 2\sin^2 \theta = 2\cos^2 \theta - 1$$

$$\sin(\alpha \pm \beta) = \sin\alpha\cos\beta \pm \cos\alpha\sin\beta$$

$$\cos(\alpha \pm \beta) = \cos\alpha\cos\beta \mp \sin\alpha\sin\beta$$

$$\sin\alpha \pm \sin\beta = 2\sin[\tfrac{1}{2}(\alpha \pm \beta)]\cos[\tfrac{1}{2}(\alpha \mp \beta)]$$

$$\cos\alpha + \cos\beta = 2\cos[\tfrac{1}{2}(\alpha + \beta)]\cos[\tfrac{1}{2}(\alpha - \beta)]$$

$$\cos\alpha - \cos\beta = -2\sin[\tfrac{1}{2}(\alpha + \beta)]\sin[\tfrac{1}{2}(\alpha - \beta)]$$

Laws of Cosines and Sines

Where A, B, C are the sides of an arbitrary triangle and α, β, γ the angles opposite those sides:

Law of cosines

$$C^2 = A^2 + B^2 - 2AB\cos\gamma$$

Law of sines

$$\frac{\sin\alpha}{A} = \frac{\sin\beta}{B} = \frac{\sin\gamma}{C}$$

Exponentials and Logarithms

Graphs

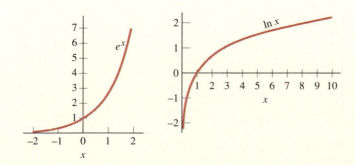

Exponential and Natural Logarithms Are Inverse Functions

$$e^{\ln x} = x, \quad \ln e^x = x \quad e = 2.71828\ldots.$$

Exponential and Logarithmic Identities

$$a^x = e^{x \ln a} \qquad \ln(xy) = \ln x + \ln y$$

$$a^x a^y = a^{x+y} \qquad \ln\left(\frac{x}{y}\right) = \ln x - \ln y$$

$$(a^x)^y = a^{xy} \qquad \ln\left(\frac{1}{x}\right) = -\ln x$$

$$\log x \equiv \log_{10} x = \ln(10)\ln x \simeq 2.3\ln x$$

Expansions and Approximations

Series Expansions of Functions

Note: $n! = n(n-1)(n-2)(n-3) \cdots (3)(2)(1)$

$$e^x = 1 + x + \frac{x^2}{2!} + \frac{x^3}{3!} + \cdots \quad \text{(exponential)}$$

$$\sin x = x - \frac{x^3}{3!} + \frac{x^5}{5!} - \cdots \quad \text{(sine)} \left.\vphantom{\begin{array}{c} x \\ x \\ x \end{array}}\right\}$$

$$\cos x = 1 - \frac{x^2}{2!} + \frac{x^4}{4!} - \cdots \quad \text{(cosine)} \qquad (x \text{ in radians})$$

$$\ln(1 + x) = x - \frac{x^2}{2} + \frac{x^3}{3} - \cdots \quad \text{(natural logarithm)}$$

$$(1 + x)^p = 1 + px + \frac{p(p-1)}{2!}x^2 + \frac{p(p-1)(p-2)}{3!}x^3 + \cdots$$

(binomial, valid for $|x| < 1$)

Approximations

For $|x| \ll 1$, the first few terms in the series provide a good approximation; that is,

$$
\left.
\begin{array}{l}
e^x \simeq 1 + x \\[4pt]
\sin x \simeq x \\[4pt]
\cos x \simeq 1 - \frac{1}{2}x^2 \\[4pt]
\ln(1 + x) \simeq x \\[4pt]
(1 + x)^p \simeq 1 + px
\end{array}
\right\} \quad \text{for } |x| \ll 1
$$

Expressions that do not have the forms shown may often be put in the appropriate form. For example:

$$\frac{1}{\sqrt{a^2 + y^2}} = \frac{1}{a\sqrt{1 + \frac{y^2}{a^2}}} = \frac{1}{a}\left(1 + \frac{y^2}{a^2}\right)^{-1/2}.$$

For $y^2 \ll a^2$, this may be approximated using the binomial expansion $(1 + x)^p \simeq 1 + px$, with $p = -\frac{1}{2}$ and $x = y^2/a^2$:

$$\frac{1}{a}\left(1 + \frac{y^2}{a^2}\right)^{-1/2} \simeq \frac{1}{a}\left(1 - \frac{1}{2}\frac{y^2}{a^2}\right).$$

Vector Algebra

Vector Products

$\mathbf{A} \cdot \mathbf{B} = AB\cos\theta$

$|\mathbf{A} \times \mathbf{B}| = AB\sin\theta$, with direction of $\mathbf{A} \times \mathbf{B}$ given by right-hand rule:

Unit Vector Notation

An arbitrary vector $\mathbf{A}$ may be written in terms of its components A_x, A_y, A_z and the unit vectors $\hat{\mathbf{i}}, \hat{\mathbf{j}}, \hat{\mathbf{k}}$ that have length 1 and lie along the x, y, z axes:

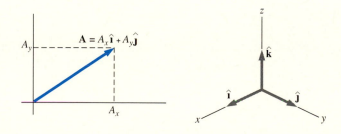

In unit vector notation, vector products become

$\mathbf{A} \cdot \mathbf{B} = A_x B_x + A_y B_y + A_z B_z$

$\mathbf{A} \times \mathbf{B} = (A_y B_z - A_z B_y)\hat{\mathbf{i}} + (A_z B_x - A_x B_z)\hat{\mathbf{j}} + (A_x B_y - A_y B_x)\hat{\mathbf{k}}$

Vector Identities

$\mathbf{A} \cdot \mathbf{B} = \mathbf{B} \cdot \mathbf{A}$

$\mathbf{A} \times \mathbf{B} = -\mathbf{B} \times \mathbf{A}$

$\mathbf{A} \cdot (\mathbf{B} \times \mathbf{C}) = \mathbf{B} \cdot (\mathbf{C} \times \mathbf{A}) = \mathbf{C} \cdot (\mathbf{A} \times \mathbf{B})$

$\mathbf{A} \times (\mathbf{B} \times \mathbf{C}) = (\mathbf{A} \cdot \mathbf{C})\mathbf{B} - (\mathbf{A} \cdot \mathbf{B})\mathbf{C}$

A-2 CALCULUS

Derivatives

Definition of the Derivative

If y is a function of x $[y = f(x)]$, then the **derivative of y with respect to x** is the ratio of the change Δy in y to the corresponding change Δx in x, in the limit of arbitrarily small Δx:

$$\frac{dy}{dx} = \lim_{\Delta x \to 0} \frac{\Delta y}{\Delta x}.$$

Algebraically, the derivative is the rate of change of y with respect to x; geometrically, it is the slope of the y versus x graph—that is, of the tangent line to the graph at a given point:

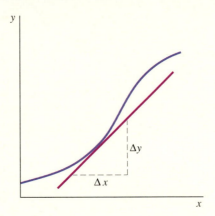

Derivatives of Common Functions

Although the derivative of a function can be evaluated directly using the limiting process that defines the derivative, standard formulas are available for common functions:

$$\frac{da}{dx} = 0 \quad (a \text{ is a constant})$$

$$\frac{dx^n}{dx} = nx^{n-1} \quad (n \text{ need not be an integer})$$

$$\frac{d}{dx}\sin x = \cos x$$

$$\frac{d}{dx}\cos x = -\sin x$$

$$\frac{d}{dx}\tan x = \frac{1}{\cos^2 x}$$

$$\frac{de^x}{dx} = e^x$$

$$\frac{d}{dx}\ln x = \frac{1}{x}$$

Derivatives of Sums, Products, and Functions of Functions

1. Derivative of a constant times a function

$$\frac{d}{dx}[af(x)] = a\frac{df}{dx} \quad (a \text{ is a constant})$$

2. Derivative of a sum

$$\frac{d}{dx}[f(x) + g(x)] = \frac{df}{dx} + \frac{dg}{dx}$$

3. Derivative of a product

$$\frac{d}{dx}[f(x)g(x)] = g\frac{df}{dx} + f\frac{dg}{dx}$$

Examples

$$\frac{d}{dx}(x^2\cos x) = \cos x\frac{dx^2}{dx} + x^2\frac{d}{dx}\cos x = 2x\cos x - x^2\sin x$$

$$\frac{d}{dx}(x\ln x) = \ln x\frac{dx}{dx} + x\frac{d}{dx}\ln x = (\ln x)(1) + x\left(\frac{1}{x}\right) = \ln x + 1$$

4. Derivative of a quotient

$$\frac{d}{dx}\left[\frac{f(x)}{g(x)}\right] = \frac{1}{g^2}\left(g\frac{df}{dx} - f\frac{dg}{dx}\right)$$

Example

$$\frac{d}{dx}\left(\frac{\sin x}{x^2}\right) = \frac{1}{x^4}\left(x^2\frac{d}{dx}\sin x - \sin x\frac{dx^2}{dx}\right) = \frac{\cos x}{x^2} - \frac{2\sin x}{x^3}$$

5. Chain rule for derivatives

If f is a function of u and u is a function of x, then

$$\frac{df}{dx} = \frac{df}{du}\frac{du}{dx}.$$

Examples

a. Evaluate $\dfrac{d}{dx}\sin(x^2)$. Here $u = x^2$ and $f(u) = \sin u$, so

$$\frac{d}{dx}\sin(x^2) = \frac{d}{du}\sin u\frac{du}{dx} = (\cos u)\frac{dx^2}{dx} = 2x\cos(x^2).$$

b. $\dfrac{d}{dt}\sin\omega t = \dfrac{d}{d\omega t}\sin\omega t\dfrac{d}{dt}\omega t = \omega\cos\omega t.$ (ω a constant)

c. Evaluate $\dfrac{d}{dx}\sin^2 5x$. Here $u = \sin 5x$ and $f(u) = u^2$, so

$$\frac{d}{dx}\sin^2 5x = \frac{d}{du}u^2\frac{du}{dx} = 2u\frac{du}{dx} = 2\sin 5x\frac{d}{dx}\sin 5x$$

$$= (2)(\sin 5x)(5)(\cos 5x) = 10\sin 5x\cos 5x = 5\sin 2x.$$

Second Derivative

The second derivative of y with respect to x is defined as the derivative of the derivative:

$$\frac{d^2y}{dx^2} = \frac{d}{dx}\left(\frac{dy}{dx}\right).$$

Example

If $y = ax^3$, then $dy/dx = 3ax^2$, so

$$\frac{d^2y}{dx^2} = \frac{d}{dx}3ax^2 = 6ax.$$

Partial Derivatives

When a function depends on more than one variable, then the partial derivatives of that function are the derivatives with respect to each variable, taken with all other variables held constant. If f is a function of x and y, then the partial derivatives are written

$$\frac{\partial f}{\partial x} \quad \text{and} \quad \frac{\partial f}{\partial y}.$$

Example

If $f(x, y) = x^3\sin y$, then

$$\frac{\partial f}{\partial x} = 3x^2\sin y \quad \text{and} \quad \frac{\partial f}{\partial y} = x^3\cos y.$$

Integrals

Indefinite Integrals

Integration is the inverse of differentiation. The **indefinite integral**, $\int f(x)\,dx$, is defined as a function whose derivative is $f(x)$:

$$\frac{d}{dx}\left[\int f(x)\,dx\right] = f(x).$$

If $A(x)$ is an indefinite integral of $f(x)$, then because the derivative of a constant is zero, the function $A(x) + C$ is also an indefinite integral of $f(x)$, where C is any constant. Inverting the derivatives of common functions listed in the preceding section gives the integrals on the next page (a more extensive table appears at the end of this appendix).

$$\int a\,dx = ax + C$$

$$\int x^n\,dx = \frac{x^{n+1}}{n+1} + C, \quad n \neq -1$$

$$\int \sin x\,dx = -\cos x + C$$

$$\int \cos x\,dx = \sin x + C$$

$$\int e^x\,dx = e^x + C$$

$$\int x^{-1}\,dx = \ln x + C$$

Definite Integrals

In physics we are most often interested in the **definite integral**, defined as the sum of a large number of very small quantities, in the limit as the number of quantities grows arbitrarily large and the size of each arbitrarily small:

$$\int_{x_1}^{x_2} f(x)\,dx \equiv \lim_{\substack{\Delta x \to 0 \\ N \to \infty}} \sum_{i=1}^{N} f(x_i)\,\Delta x,$$

where the terms in the sum are evaluated at values x_i between the limits of integration x_1 and x_2; in the limit $\Delta x \to 0$, the sum is over all values of x in the interval.

The definite integral is used whenever we need to sum over a quantity that is changing—for example, to calculate the work done by a variable force (Chapter 7), the entropy change in a system whose temperature varies (Chapter 22), or the flux of an electric field that varies with position (Chapter 24).

The key to evaluating the definite integral is provided by the **fundamental theorem of calculus**. The theorem states that, if $A(x)$ is an *indefinite* integral of $f(x)$, then the *definite integral* is given by

$$\int_{x_1}^{x_2} f(x)\,dx = A(x_2) - A(x_1) \equiv A(x) \Big|_{x_1}^{x_2}.$$

Geometrically, the definite integral is the area under the graph of $f(x)$ between the limits x_1 and x_2:

Evaluating Integrals

The first step in evaluating an integral is to express all varying quantities within the integral in terms of a single variable. For example, in evaluating $\int E\,dr$ to calculate an electric potential (Chapter 25), it is necessary first to express E as a function of r. This procedure is illustrated in many examples throughout this text; Example 23-7 provides a typical case.

Once an integral is written in terms of a single variable, it is necessary to manipulate the integrand—the function being integrated—into a form whose integral you know or can look up in tables of integrals. Two common techniques are especially useful:

1. **Change of variables**

   An unfamiliar integral can often be put into familiar form by defining a new variable. For example, it is not obvious how to integrate the expression

   $$\int \frac{x\,dx}{\sqrt{a^2 + x^2}}.$$

   where a is a constant. But let $z = a^2 + x^2$. Then

   $$\frac{dz}{dx} = \frac{da^2}{dx} + \frac{dx^2}{dx} = 0 + 2x = 2x,$$

   so $dz = 2x\,dx$. Then the quantity $x\,dx$ in our unfamiliar integral is just $\frac{1}{2}dz$, while the quantity $\sqrt{a^2 + x^2}$ is just $z^{1/2}$. So the integral becomes

   $$\int \tfrac{1}{2}z^{-1/2}\,dz = \frac{\frac{1}{2}z^{1/2}}{(1/2)} = \sqrt{z},$$

   where we have used the standard form for the integral of a power of the independent variable. Substituting back $z = a^2 + x^2$ gives

   $$\int \frac{x\,dx}{\sqrt{a^2 + x^2}} = \sqrt{a^2 + x^2}.$$

2. **Integration by parts**

   The quantity $\int u\,dv$ is the area under the curve of u as a function of v between specified limits. In the figure below, that area can also be expressed as the area of the rectangle shown minus the area under the curve of v as a function of u. Mathematically, this relation among areas may be expressed as a relation among integrals:

   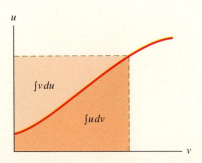

   $$\int u\,dv = uv - \int v\,du. \qquad \text{(integration by parts)}$$

This expression may often be used to transform complicated integrals into simpler ones.

Example

Evaluate $\int x \cos x \, dx$. Here let $u = x$, so $du = dx$. Then $dv = \cos x \, dx$, so $v = \int dv = \int \cos x \, dx = \sin x$. Integrating by parts then gives

$$\int x \cos x \, dx = (x)(\sin x) - \int \sin x \, dx = x \sin x + \cos x,$$

where the $+$ sign arises because $\int \sin x \, dx = -\cos x$.

Table of Integrals

[More extensive tables are available in many mathematical and scientific handbooks; see, for example, *Handbook of Chemistry and Physics* (Chemical Rubber Co.) or Dwight, *Tables of Integrals and Other Mathematical Data* (Macmillan).] Increasingly, sophisticated computer software is used instead of tables for the symbolic evaluation of integrals; among the most widely used are *Maple*, *Mathematica*, and *Derive*.

In the expressions below, a and b are constants. An arbitrary constant of integration may be added to the right-hand side.

$$\int e^{ax} \, dx = \frac{e^{ax}}{a}$$

$$\int \sin ax \, dx = -\frac{\cos ax}{a}$$

$$\int \cos ax \, dx = \frac{\sin ax}{a}$$

$$\int \tan ax \, dx = -\frac{1}{a}\ln(\cos ax)$$

$$\int \sin^2 ax \, dx = \frac{x}{2} - \frac{\sin 2ax}{4a}$$

$$\int \cos^2 ax \, dx = \frac{x}{2} + \frac{\sin 2ax}{4a}$$

$$\int x \sin ax \, dx = \frac{1}{a^2}\sin ax - \frac{1}{a}x\cos ax$$

$$\int x \cos ax \, dx = \frac{1}{a^2}\cos ax + \frac{1}{a}x\sin ax$$

$$\int \frac{dx}{\sqrt{a^2 - x^2}} = \sin^{-1}\left(\frac{x}{a}\right)$$

$$\int \frac{dx}{\sqrt{x^2 \pm a^2}} = \ln(x + \sqrt{x^2 \pm a^2})$$

$$\int \frac{dx}{x^2 + a^2} = \frac{1}{a}\tan^{-1}\left(\frac{x}{a}\right)$$

$$\int \frac{x \, dx}{\sqrt{a^2 - x^2}} = -\sqrt{a^2 - x^2}$$

$$\int \frac{x \, dx}{\sqrt{x^2 \pm a^2}} = \sqrt{x^2 \pm a^2}$$

$$\int \frac{dx}{(x^2 \pm a^2)^{3/2}} = \frac{\pm x}{a^2\sqrt{x^2 \pm a^2}}$$

$$\int x e^{ax} \, dx = \frac{e^{ax}}{a^2}(ax - 1)$$

$$\int x^2 e^{ax} \, dx = \frac{x^2 e^{ax}}{a} - \frac{2}{a}\left[\frac{e^{ax}}{a^2}(ax - 1)\right]$$

$$\int \frac{dx}{a + bx} = \frac{1}{b}\ln(a + bx)$$

$$\int \frac{dx}{(a + bx)^2} = -\frac{1}{b(a + bx)}$$

$$\int \ln ax \, dx = x \ln ax - x$$

The International System of Units (SI)

This material is from the United States edition of the English translation of the sixth edition of "Le Système International d'Unités (SI)," the definitive publication in the French language issued in 1991 by the International Bureau of Weights and Measures (BIPM). The year the definition was adopted is given in parentheses.

unit of length (meter): The meter is the length of the path traveled by light in vacuum during a time interval of 1/299 792 458 of a second. (1983)

unit of mass (kilogram): The kilogram is the unit of mass; it is equal to the mass of the international prototype of the kilogram. (1889)

unit of time (second): The second is the duration of 9 192 631 770 periods of the radiation corresponding to the transition between the two hyperfine levels of the ground state of the cesium-133 atom. (1967)

unit of electric current (ampere): The ampere is that constant current which, if maintained in two straight parallel conductors of infinite length, of negligible circular cross section, and placed 1 meter apart in vacuum, would produce between these conductors a force equal to 2×10^{-7} newton per meter of length. (1948)

unit of thermodynamic temperature (kelvin): The kelvin, unit of thermodynamic temperature, is the fraction 1/273.16 of the thermodynamic temperature of the triple point of water. (1957) Also, the unit kelvin and its symbol K should be used to express an interval or a difference of temperature.

unit of amount of substance (mole): (1) The mole is the amount of substance of a system that contains as many elementary entities as there are atoms in 0.012 kilogram of carbon 12. (1971) (2) When the mole is used, the elementary entities must be specified and may be atoms, molecules, ions, electrons, other particles, or specified groups of such particles.

unit of luminous intensity (candela): The candela is the luminous intensity, in a given direction, of a source that emits monochromatic radiation of frequency 540×10^{12} hertz and that has a radiant intensity in that direction of (1/683) watt per steradian. (1979)

▲ *SI Base and Supplementary Units*

QUANTITY	SI UNIT	
	NAME	SYMBOL
Base Unit		
Length	meter	m
Mass	kilogram	kg
Time	second	s
Electric current	ampere	A
Thermodynamic temperature	kelvin	K
Amount of substance	mole	mol
Luminous intensity	candela	cd
Supplementary Units		
Plane angle	radian	rad
Solid angle	steradian	sr

▲ *SI Prefixes*

FACTOR	PREFIX	SYMBOL
10^{24}	yotta	Y
10^{21}	zetta	Z
10^{18}	exa	E
10^{15}	peta	P
10^{12}	tera	T
10^{9}	giga	G
10^{6}	mega	M
10^{3}	kilo	k
10^{2}	hecto	h
10^{1}	deka	da
10^{0}	—	—
10^{-1}	deci	d
10^{-2}	centi	c
10^{-3}	milli	m
10^{-6}	micro	μ
10^{-9}	nano	n
10^{-12}	pico	p
10^{-15}	femto	f
10^{-18}	atto	a
10^{-21}	zepto	z
10^{-24}	yocto	y

▲ *Some SI Derived Units with Special Names*

QUANTITY	SI UNIT			
	NAME	SYMBOL	EXPRESSION IN TERMS OF OTHER UNITS	EXPRESSION IN TERMS OF SI BASE UNITS
Frequency	hertz	Hz		s^{-1}
Force	newton	N		$m \cdot kg \cdot s^{-2}$
Pressure, stress	pascal	Pa	N/m^2	$m^{-1} \cdot kg \cdot s^{-2}$
Energy, work, heat	joule	J	$N \cdot m$	$m^2 \cdot kg \cdot s^{-2}$
Power	watt	W	J/s	$m^2 \cdot kg \cdot s^{-3}$
Electric charge	coulomb	C		$s \cdot A$
Electric potential, potential difference, electromotive force	volt	V	J/C	$m^2 \cdot kg \cdot s^{-3} \cdot A^{-1}$
Capacitance	farad	F	C/V	$m^{-2} \cdot kg^{-1} \cdot s^4 \cdot A^2$
Electric resistance	ohm	Ω	V/A	$m^2 \cdot kg \cdot s^{-3} \cdot A^{-2}$
Magnetic flux	weber	Wb	$V \cdot s$	$m^2 \cdot kg \cdot s^{-2} \cdot A^{-1}$
Magnetic field	tesla	T	Wb/m^2	$kg \cdot s^{-2} \cdot A^{-1}$
Inductance	henry	H	Wb/A	$m^2 \cdot kg \cdot s^{-2} \cdot A^{-2}$
Radioactivity	becquerel	Bq	1 decay/s	s^{-1}
Absorbed radiation dose	gray	Gy	J/kg, 100 rad	$m^2 \cdot s^{-2}$
Radiation dose equivalent	sievert	Sv	J/kg, 100 rem	$m^2 \cdot s^{-2}$

Conversion Factors

The listings below give the SI equivalents of non-SI units. To convert from the units shown to SI, multiply by the factor given; to convert the other way, divide. For conversions within the SI system see table of SI prefixes in Appendix B, Chapter 1, or inside front cover. Conversions that are not exact by definition are given to, at most, 4 significant figures.

Length

1 inch (in.) = 0.0254 m
1 foot (ft) = 0.3048 m
1 yard (yd) = 0.9144 m
1 mile (mi) = 1609 m
1 nautical mile = 1852 m

1 angstrom (Å) = 10^{-10} m
1 light-year (ly) = 9.46×10^{15} m
1 astronomical unit (AU) = 1.5×10^{11} m
1 parsec = 3.09×10^{16} m
1 fermi = 10^{-15} m = 1 fm

Mass

1 slug = 14.59 kg
1 metric ton (tonne; T) = 1000 kg

1 unified mass unit (u) = 1.660×10^{-27} kg

Force units in the English system are sometimes used (incorrectly) for mass. The units given below are actually equal to the number of kilograms multiplied by *g*, the acceleration of gravity.

1 pound (lb) = weight of 0.454 kg
1 ton = 2000 lb = weight of 908 kg

1 ounce (oz) = weight of 0.02835 kg

Time

1 minute (min) = 60 s
1 hour (h) = 60 min = 3600 s

1 day (d) = 24 h = 86 400 s
1 year (y) = 365.2422 d* = 3.156×10^7 s

---

* The length of the year changes very slowly with changes in Earth's orbital period.

Area

1 hectare (ha) $= 10^4$ m^2	1 acre $= 4047$ m^2
1 square inch (in.^{2}) $= 6.452 \times 10^{-4}$ m^2	1 barn $= 10^{-28}$ m^2
1 square foot (ft^2) $= 9.290 \times 10^{-2}$ m^2	1 shed $= 10^{-30}$ m^2

Volume

1 liter (L) $= 1000$ cm$^3 = 10^{-3}$ m^3 1 gallon (U.S.; gal) $= 3.785 \times 10^{-3}$ m^3
1 cubic foot (ft^3) $= 2.832 \times 10^{-2}$ m^3 1 gallon (British) $= 4.546 \times 10^{-3}$ m^3
1 cubic inch (in.^{3}) $= 1.639 \times 10^{-5}$ m^3
1 fluid ounce $= 1/128$ gal $= 2.957 \times 10^{-5}$ m^3
1 barrel $= 42$ gal $= 0.1590$ m^3

Angle, Phase

1 degree (°) $= \pi/180$ rad $= 1.745 \times 10^{-2}$ rad
1 revolution (rev) $= 360° = 2\pi$ rad
1 cycle $= 360° = 2\pi$ rad

Speed, Velocity

1 km/h $= (1/3.6)$ m/s $= 0.2778$ m/s	1 ft/s $= 0.3048$ m/s
1 mi/h (mph) $= 0.4470$ m/s	1 ly/y $= 3.00 \times 10^8$ m/s

Angular Speed, Angular Velocity, Frequency, and Angular Frequency

1 rev/s $= 2\pi$ rad/s $= 6.283$ rad/s (s^{-1})   1 rev/min (rpm) $= 0.1047$ rad/s (s^{-1})
1 Hz $= 1$ cycle/s $= 2\pi$ s^{-1}

Force

1 dyne $= 10^{-5}$ N	1 pound (lb) $= 4.448$ N

Pressure

1 dyne/cm^2 $= 0.10$ Pa	1 lb/in.^{2} (psi) $= 6.895 \times 10^3$ Pa
1 atmosphere (atm) $= 1.013 \times 10^5$ Pa	1 in. H_2O (60°F) $= 248.8$ Pa
1 torr $= 1$ mm Hg at 0°C $= 133.3$ Pa	1 in. Hg (60°F) $= 3.377 \times 10^3$ Pa
1 bar $= 10^5$ Pa $= 0.987$ atm	

Energy, Work, Heat

1 erg $= 10^{-7}$ J	1 Btu* $= 1.054 \times 10^3$ J
1 calorie* (cal) $= 4.184$ J	1 kWh $= 3.6 \times 10^6$ J
1 electron-volt (eV) $= 1.602 \times 10^{-19}$ J	1 megaton (explosive yield; Mt)
1 foot-pound (ft·lb) $= 1.356$ J	$= 4.18 \times 10^{15}$ J

* Values based on the thermochemical calorie; other definitions vary slightly.

Power

$1 \text{ erg/s} = 10^{-7} \text{ W}$	$1 \text{ Btu/h (Btuh)} = 0.293 \text{ W}$
$1 \text{ horsepower (hp)} = 746 \text{ W}$	$1 \text{ ft·lb/s} = 1.356 \text{ W}$

Magnetic Field

$1 \text{ gauss (G)} = 10^{-4} \text{ T}$	$1 \text{ gamma } (\gamma) = 10^{-9} \text{ T}$

Radiation

$1 \text{ curie (ci)} = 3.7 \times 10^{10} \text{ Bq}$	$1 \text{ rad} = 10^{-2} \text{ Gy}$
	$1 \text{ rem} = 10^{-2} \text{ Sv}$

▲   *Energy Content of Fuels*

ENERGY SOURCE	ENERGY CONTENT
Coal	2.9×10^7 J/kg $= 7300$ kWh/ton $= 25 \times 10^6$ Btu/ton
Oil	43×10^6 J/kg $= 39$ kWh/gal $= 1.3 \times 10^5$ Btu/gal
Gasoline	44×10^6 J/kg $= 36$ kWh/gal $= 1.2 \times 10^5$ Btu/gal
Natural gas	55×10^6 J/kg $= 30$ kWh/100 ft^3 $= 1000$ Btu/ft^3
Uranium (fission)	
Normal abundance	5.8×10^{11} J/kg $= 1.6 \times 10^5$ kWh/kg
Pure U-235	8.2×10^{13} J/kg $= 2.3 \times 10^7$ kWh/kg
Hydrogen (fusion)	
Normal abundance	7×10^{11} J/kg $= 3.0 \times 10^4$ kWh/kg
Pure deuterium	3.3×10^{14} J/kg $= 9.2 \times 10^7$ kWh/kg
Water	1.2×10^{10} J/kg $= 1.3 \times 10^4$ kWh/gal $= 340$ gal gasoline/gal
H_2O	
100% conversion, matter to energy	9.0×10^{16} J/kg $= 931$ MeV/u $= 2.5 \times 10^{10}$ kWh/kg

APPENDIX D
The Elements

The atomic weights of stable elements reflect the abundances of different isotopes; values given here apply to elements as they exist naturally on Earth. For stable elements, parentheses express uncertainties in the last decimal place given. For elements with no stable isotopes (indicated in boldface), sets of most important isotopes are given. (Exceptions are the unstable elements thorium, protactinium, and uranium, for which atomic weights reflect natural abundances of long-lived isotopes.) See also periodic table inside back cover.

ATOMIC NUMBER	NAMES	SYMBOL	ATOMIC WEIGHT
1	Hydrogen	H	1.00794 (7)
2	Helium	He	4.002602 (2)
3	Lithium	Li	6.941 (2)
4	Beryllium	Be	9.012182 (3)
5	Boron	B	10.811 (5)
6	Carbon	C	12.011 (1)
7	Nitrogen	N	14.00674 (7)
8	Oxygen	O	15.9994 (3)
9	Fluorine	F	18.9984032 (9)
10	Neon	Ne	20.1797 (6)
11	Sodium (Natrium)	Na	22.989768 (6)
12	Magnesium	Mg	24.3050 (6)
13	Aluminum	Al	26.981539 (5)
14	Silicon	Si	28.0855 (3)
15	Phosphorus	P	30.973762 (4)
16	Sulfur	S	32.066 (6)
17	Chlorine	Cl	35.4527 (9)
18	Argon	Ar	39.948 (1)
19	Potassium (Kalium)	K	39.0983 (1)
20	Calcium	Ca	40.078 (4)
21	Scandium	Sc	44.955910 (9)
22	Titanium	Ti	47.88 (3)
23	Vanadium	V	50.9415 (1)
24	Chromium	Cr	51.9961 (6)
25	Manganese	Mn	54.93805 (1)
26	Iron	Fe	55.847 (3)
27	Cobalt	Co	58.93320 (1)

ATOMIC NUMBER	NAMES	SYMBOL	ATOMIC WEIGHT
28	Nickel	Ni	58.69 (1)
29	Copper	Cu	63.546 (3)
30	Zinc	Zn	65.39 (2)
31	Gallium	Ga	69.723 (1)
32	Germanium	Ge	72.61 (2)
33	Arsenic	As	74.92159 (2)
34	Selenium	Se	78.96 (3)
35	Bromine	Br	79.904 (1)
36	Krypton	Kr	83.80 (1)
37	Rubidium	Rb	85.4678 (3)
38	Strontium	Sr	87.62 (1)
39	Yttrium	Y	88.90585 (2)
40	Zirconium	Zr	91.224 (2)
41	Niobium	Nb	92.90638 (2)
42	Molybdenum	Mo	95.94 (1)
43	**Technetium**	**Tc**	**97, 98, 99**
44	Ruthenium	Ru	101.07 (2)
45	Rhodium	Rh	102.90550 (3)
46	Palladium	Pd	106.42 (1)
47	Silver	Ag	107.8682 (2)
48	Cadmium	Cd	112.411 (8)
49	Indium	In	114.82 (1)
50	Tin	Sn	118.710 (7)
51	Antimony (Stibium)	Sb	121.75 (3)
52	Tellurium	Te	127.60 (3)
53	Iodine	I	126.90447 (3)
54	Xenon	Xe	131.29 (2)
55	Cesium	Cs	132.90543 (5)
56	Barium	Ba	137.327 (7)
57	Lanthanum	La	138.9055 (2)
58	Cerium	Ce	140.115 (4)
59	Praseodymium	Pr	140.90765 (3)
60	Neodymium	Nd	144.24 (3)
61	**Promethium**	**Pm**	**145, 147**
62	Samarium	Sm	150.36 (3)
63	Europium	Eu	151.965 (9)
64	Gadolinium	Gd	157.25 (3)
65	Terbium	Tb	158.92534 (3)
66	Dysprosium	Dy	162.50 (3)
67	Holmium	Ho	164.93032 (3)
68	Erbium	Er	167.26 (3)
69	Thulium	Tm	168.93421 (3)
70	Ytterbium	Yb	173.04 (3)
71	Lutetium	Lu	174.967 (1)
72	Hafnium	Hf	178.49 (2)
73	Tantalum	Ta	180.9479 (1)
74	Tungsten (Wolfram)	W	183.85 (3)
75	Rhenium	Re	186.207 (1)
76	Osmium	Os	190.2 (1)
77	Iridium	Ir	192.22 (3)
78	Platinum	Pt	195.08 (3)
79	Gold	Au	196.96654 (3)
80	Mercury	Hg	200.59 (3)
81	Thallium	Tl	204.3833 (2)
82	Lead	Pb	207.2 (1)
83	Bismuth	Bi	208.98037 (3)

ATOMIC NUMBER	NAMES	SYMBOL	ATOMIC WEIGHT
84	Polonium	Po	209, 210
85	Astatine	At	210, 211
86	Radon	Rn	211, 220, 222
87	Francium	Fr	223
88	Radium	Ra	223, 224, 226, 228
89	Actinium	Ac	227
90	Thorium	Th	232.0381 (1)
91	Protactinium	Pa	231.03588 (2)
92	Uranium	U	238.0289 (1)
93	Neptunium	Np	237, 239
94	Plutonium	Pu	238, 239, 240, 241, 242, 244
95	Americium	Am	241, 243
96	Curium	Cm	243, 244, 245, 246, 247, 248
97	Berkelium	Bk	247, 249
98	Californium	Cf	249, 250, 251, 252
99	Einsteinium	Es	252
100	Fermium	Fm	257
101	Mendelevium	Md	255, 256, 258, 260
102	Nobelium	No	253, 254, 255, 259
103	Lawrencium	Lr	256, 258, 259, 261
104	Rutherfordium	Rf	257, 259, 260, 261
105	Dubnium	Db	260, 261, 262
106	Seaborgium	Sg	259, 260, 261, 263
107	Bohrium	Bh	261, 262
108	Hassium	Hs	264, 265
109	Meitnerium	Mt	266
110	—	—	269
111	—	—	272
112	—	—	277

Astrophysical Data

SUN, PLANETS, PRINCIPAL SATELLITES

BODY	MASS (10²⁴ kg)	MEAN RADIUS (10⁶ m EXCEPT AS NOTED)	SURFACE GRAVITY (m/s²)	ESCAPE SPEED (km/s)	SIDEREAL ROTATION PERIOD* (days)	MEAN DISTANCE FROM CENTRAL BODY† (10⁶ km)	ORBITAL PERIOD	ORBITAL SPEED (km/s)
Sun	1.99×10^6	696	274	618	36 at poles 27 at equator	2.6×10^{11}	200 My	250
Mercury	0.330	2.44	3.70	4.25	58.6	57.6	88.0 d	48
Venus	4.87	6.05	8.87	10.4	−243	108	225 d	35
Earth	5.97	6.37	9.81	11.2	0.997	150	365.3 d	30
Moon	0.0735	1.74	1.62	2.38	27.3	0.385	27.3 d	1.0
Mars	0.642	3.38	3.74	5.03	1.03	228	1.88 y	24.1
Phobos	9.6×10^{-9}	9-13 km	0.001	0.008	0.32	9.4×10^{-3}	0.32 d	2.1
Deimos	2×10^{-9}	5-8 km	0.001	0.005	1.3	23×10^{-3}	1.3 d	1.3
Jupiter	1.90×10^3	69.1	26.5	60.6	0.414	778	11.9 y	13.0
Io	0.0889	1.82	1.8	2.6	1.77	0.422	1.77 d	17
Europa	0.478	1.57	1.3	2.0	3.55	0.671	3.55 d	14
Ganymede	0.148	2.63	1.4	2.7	7.15	1.07	7.15 d	11
Callisto	0.107	2.40	1.2	2.4	16.7	1.88	16.7 d	8.2
and 13 smaller satellites								
Saturn	569	56.8	11.8	36.6	0.438	1.43×10^3	29.5 y	9.65
Tethys	0.0007	0.53	0.2	0.4	1.89	0.294	1.89 d	11.3
Dione	0.00015	0.56	0.3	0.6	2.74	0.377	2.74 d	10.0
Rhea	0.0025	0.77	0.3	0.5	4.52	0.527	4.52 d	8.5
Titan	0.135	2.58	1.4	2.6	15.9	1.22	15.9 d	5.6
Iapetus	0.0019	0.73	0.2	0.6	79.3	3.56	79.3 d	3.3
and 12 smaller satellites								
Uranus	86.6	25.0	9.23	21.5	−0.65	2.87×10^3	84.1 y	6.79
Ariel	0.0013	0.58	0.3	0.4	2.52	0.19	2.52 d	5.5
Umbriel	0.0013	0.59	0.3	0.4	4.14	0.27	4.14 d	4.7
Titania	0.0018	0.81	0.2	0.5	8.70	0.44	8.70 d	3.7
Oberon	0.0017	0.78	0.2	0.5	13.5	0.58	13.5 d	3.1
and 11 smaller satellites								
Neptune	103	24.0	11.9	23.9	0.768	4.50×10^3	165 y	5.43
Triton	0.134	1.9	2.5	3.1	5.88	0.354	5.88 d	4.4
and 7 smaller satellites								
Pluto	0.015	1.2	0.4	1.2	−6.39	5.91×10^3	249 y	4.7
Charon	0.001	0.6			−6.39	0.02	6.39 d	0.2

*Negative rotation period indicates retrograde motion, in opposite sense from orbital motion. Periods are sidereal, meaning the time for the body to return to the same orientation relative to the distant stars rather than the Sun.

†Central body is galactic center for Sun, Sun for planets, and planet for satellites.

Got It! Answers

Chapter 2
p. 24   Average speed and average velocity are the same for direct and Kansas City stop flights; average speed is greater for Minneapolis stop flight
p. 27   (b) has constant speed; (a) reverses direction; (d) speeds up

Chapter 3
p. 52   (c)

Chapter 4
p. 70   None are true

Chapter 5
p. 95   The straight line
p. 99 $T > F_g$ for upward a; $T < F_g$ for downward a; $T = F_g$ for constant speed in either direction
p. 102   No; for example, a car slowing down is moving in the direction opposite the force that's slowing it
p. 107   The statement is nonsense. If there were no gravity in space, spacecraft, astronauts, and the moon would all move in straight lines and we would never see them again!
p. 109   Upward acceleration does not imply upward velocity; cable tension exceeds weight in (a) and (d)
p. 111   No, the force on the larger block is 1.5 N

Chapter 6
p. 134   Both 1N
p. 139   Because the gravitational force is just enough to keep them accelerating in a circular path
p. 144   Because the objects were on sloping surfaces, and therefore the normal force was not equal to the weight
p. 153   The larger one; it has a greater ratio of mass to surface area

Chapter 7
p. 172   No, it's only 5 J, as the spring hasn't exerted 2 N the whole way
p. 177   Both do the same work; the faster hiker produces more power

Chapter 8
p. 200   Because an *increase* in potential energy (positive dU/dx) means one is doing work moving *against* a force

Chapter 10
p. 244   Second one from the top
p. 247 $30\hat{\jmath}$ kg·m/s
p. 249   Both skaters' momenta increase; the basketball's momentum reverses sign; system momentum is unchanged

Chapter 11
p. 264   The first two
p. 271   The object initially at rest

Chapter 12
p. 290   12-6*b*
p. 297   Because its mass is farther from the rotation axis
p. 304   The solid ball

Chapter 13
p. 323   The boy brings no additional angular momentum to the system, while the girl does
p. 327   Upward

Chapter 14
p. 339   The diagonal pair

Chapter 15
p. 369   Frequency and period are the same; amplitude and phase constants are different
p. 375   (a) no change; (b) doubles; (c) doubles

Chapter 17
p. 422   No, it drops by 20 dB

Chapter 18
p. 450   No, the water level stays the same
p. 451   Because you'll put the center of gravity over the center of buoyancy, making the canoe unstable as in Fig. 18-17*b*
p. 458   Lower

Chapter 19
p. 480   The rock
p. 486   Radiation; (a) conduction; (b) convection

Chapter 20
p. 504   100°C

Chapter 21
p. 526   In an isothermal process T is constant so pressure and volume are inversely proportional. But temperature changes in an adiabatic process, so P and V are not inversely proportional. Thus there's no contradiction between the two equations

Chapter 23
p. 576   (a) $-\hat{\imath}$; (b) $\sqrt{2}/2(-\hat{\imath} + \hat{\jmath})$
p. 583   (a) 200 N/C; (b) 100 N/C
p. 585   Because P isn't the same distance from **all** parts of the rod

Chapter 24
p. 606   (a) $\phi = 0$ for A, B; $\phi = Es^2$ for C; (b) $\phi = 0$ for A, $\phi = s^2 E\cos 45° = \sqrt{2}s^2 E/2$ for B, C
p. 608 ϕ doesn't change; $\mathbf{E}$ generally does
p. 611   (a) remains zero; (b) doubles
p. 617   (a) $E = \dfrac{Q}{-(2)(1\ m)^2(\varepsilon_0)}$ (b) $E = \dfrac{kQ}{(1000\ m)^2}$
p. 619   No

Chapter 25
p. 633   (a) doubles; (b) doubles; (c) becomes zero; (d) changes sign
p. 633 $W_\alpha = 2W_p$
p. 635   Proton: toward negative ΔV; electron: toward positive ΔV
p. 636   b, a, c
p. 637   More
p. 644   (a)
p. 645   Yes to both

Chapter 26
p. 666   They're the same
p. 670   Twice the working voltage
p. 673   (a) parallel; (b) series; (c) the series combination 1/2 C

Chapter 27
p. 686   (a) current to left; (b) upward current; (c) current to left; (d) no current; (e) no current
p. 700   (a) (2); (b) (1)

Chapter 28
p. 713   (a) 3 V; (b) 0 V; (c) 6 V
p. 716   (a) 2R; (b) 1/2 R
p. 718   A will be brightest; with C disconnected both A and B will have equal brightness
p. 732   (a) 6 mA; (b) 2 mA

Chapter 29
p. 746   Greatest force in (a); least (0) in (c); direction in (a) and (b) is into page
p. 749   Counterclockwise
p. 755   Upward

Chapter 30
p. 777   (a) into the page at P, out of the page at Q; (b) stronger at P
p. 778   Tighten, independent of the current direction
p. 780   Into the page
p. 782   (b) only

Chapter 31
p. 811   Same direction as in Fig. 31-11; you still do work
p. 812   Counterclockwise
p. 814   Harder
p. 817   (a) clockwise; (b) counterclockwise

Chapter 32
p. 838   Increasing
p. 843   Only when it's opened

Chapter 33
p. 857   The capacitor
p. 864   Smaller, by a factor of 10^{-4}
p. 866   Above

Chapter 34
p. 895   Both have the same speed; light has a much smaller wavelength
p. 905   (a) $B_1 = 2B_2$; (b) $S_1 = 4S_2$; (c) $\lambda_1 = \lambda_2$

Chapter 35
p. 923 $n_3 > n_1 > n_2$

Chapter 36
p. 943   At the mirror's center of curvature
p. 950   (a) The intensity diminishes, but the full image remains (b) there would be no image (c) there would be no image, or it would be very out of focus (d) yes, but not obviously visible except when looking straight toward lens
p. 951   Virtual, convex
p. 960   +2.25 diopter

Chapter 37
p. 976   Closer together

Chapter 38
p. 1020   When the twins move at the same speed they are the same age; the twin making the longer trip is younger
p. 1029   Less, although insignificantly so at these low relative speeds
p. 1033   Very nearly c

Chapter 39
p. 1050 $P_A = 16P_B$; $\lambda_{\max A} = 1/2\lambda_{\max B}$
p. 1052   Slope remains h; intercept changes
p. 1056   Much less, by m_e/m_p

Chapter 40
p. 1082   B's is 10^6 times A's

Chapter 41
p. 1106 $5a_0$
p. 1111   6
p. 1122 $5p \rightarrow 3s$

Chapter 43
p. 1177   About 12,000 years

Answers to Odd-Numbered Problems

Chapter 1
3. 100,000 times bigger
   5.0.108 782 775 7 ns
7. 10^8
9. 0.79 rad
11. 28 g
13. 10^6
15. 7%
17. Less by about 3 mi/h
19. 30 AU
21. 143 m^2/kg
23. Approximately 0.0175 rad
25. L/T^2
27. (c) Actually, the speed is given by
   $v = \sqrt{\lambda g/2\pi}$
29. 2.5×10^6 m
31. 7.4×10^6 m/s^2
33. (a) 2.5×10^{-11} m^2; (b) 5.0 μm
35. 280 K
37. 41 m
39. 7
43. $\sim 1.3 \times 10^6$
45. About 0.1%
47. (a) $\sim 1.4 \times 10^{18}$ m^3; (b) $\sim 1.3 \times 10^{12}$
49. ~ 10 times as much in stars
51. $\sim 10^4$
53. $d_{Sun}/d_{moon} = 380$;
   $R_{Sun} \sim 7 \times 10^5$ km
   (a) 4 μm; (b) 7500
55. (a) $\sim 10^{28}$; (b) $\sim 10^{14}$
59. $\sqrt{F_0/\mu}$; this is in fact the equation
   for the wave speed

Chapter 2
1. 10.16 m/s
3. 5.30 m/s
5. (a) 24 km north; (b) 9.6 km/h;
   (c) 16 km/h; (d) 0; (e) 0
7. 28.5 km/h
9. 48 mi/h
11. 1 m/s = 2.24 mi/h
13. 51 ft/s = 35 mi/h
15. (a) 84 km/h; (b) 55 h
17. 2.6 h later; 1800 km from New
   York
21. (a) 2.0 m/s; (b) 0; (c) -5.0 m/s; (d)
   1.2 m/s; (e) 0.17 m/s
23. (a) $b - 2ct$; (b) 8.4 s after launch
25. (a) $t = 0$ s, 0.13 s, 2.5 s;
   (b) $v = 3bt^2 - 2ct + d$;
   (c) $v_0 = 1.0$ m/s;
   (d) $t = 0.065$ s, 1.7 s
27. 125 m/s^2
29. 15 m/s^2
31. 31 s
33. (a) 126 m/s; (b) 0.46 m/s^2
35. $a = 6bt - 2c$
37. 100 m by both methods
41. (a) 46 m/s^2; (b) 61 s
43. (a) 2.0 m/s^2; (b) 150 m
45. 5×10^{-12} m
47. 12 km/s
49. (a) 0.014 s; (b) 0.51 m
51. 886 m
53. Yes, $a = 370$ m/s^2
55. No collision; 10 m apart
57. 4.6×10^{-3} m/s^2
59. Collide at 12 km/h
61. 11.3 m/s
63. (a) 27 m; (b) 4.7 s
65. Mars
67. 273 m
69. 2.0 m/s
71. 2.4 s
73. (a) 25 km/h; (b) 13 km/h
75. (a) 4.6 s; (b) 8.7 m/s
77. 0.196 s
79. 5.0 s; 17 km/h
81. (a) 7.0 m/s; (b) in 2.3 s
83. 83.113 m

85. (b) 3.8 s; (c) 19 m; (d) 100 m
87. (a) $v = \omega x_0 \cos \omega t$,
   $a = -\omega^2 x_0 \sin \omega t$;
   (b) $v_{max} = \omega x_0$, $a_{max} = \omega^2 x_0$
89. (a) 20 cm
91. 1.19 m

Chapter 3
1. 266 m, 34° N of W
3. 702 km, 21.3° E of N
5. $C = 5.0$ m; 127° clockwise from $\mathbf{A}$
7. (a) 0.68A vertically upward;
   (b) 1.88A, to the right
9. (a) L, $-x$ direction; (b) $\sqrt{3}L$,
   $+ y$ direction; (c) 0; (d) 2L,
   $- x$ direction
11. $\mathbf{V} = 105\hat{\imath} + 58\hat{\jmath}$ km
13. $\mathbf{A} = 8.19\hat{\imath} + 5.74\hat{\jmath}$;
   $\mathbf{B} = -3.44\hat{\imath} - 4.91\hat{\jmath}$;
   $\mathbf{C} = -3.38\hat{\imath} + 7.25\hat{\jmath}$
15. $\mathbf{A} + \mathbf{B} = 4.8\hat{\imath} + 0.82\hat{\jmath}$;
   $\mathbf{A} - \mathbf{B} = 12\hat{\imath} + 11\hat{\jmath}$;
   $\mathbf{A} + \mathbf{C} = 4.8\hat{\imath} + 13\hat{\jmath}$;
   $\mathbf{A} + \mathbf{B} + \mathbf{C} = 1.4\hat{\imath} + 8.1\hat{\jmath}$
17. $\mathbf{C} = -15\hat{\imath} + 9\hat{\jmath} - 18\hat{k}$
19. (a) $1.37\hat{\imath} + 8.07\hat{\jmath}$;
   (b) $8.25\hat{\imath} + 17.9\hat{\jmath}$; (c) $10.5\hat{\imath} - 17.6\hat{\jmath}$
21. $0.19\hat{\imath} + 0.88\hat{\jmath}$ km;
   (b) 1.0 km at 79° N of E
23. (a) $A_x = 5.9$, $A_y = 8.1$; $A_x' = 5.4$,
   $A_y' = 8.4$ units
25. $\hat{\imath} + \hat{\jmath} + \hat{k}$; magnitude $= \sqrt{3}$
27. (a) in x-y system $\Delta\mathbf{r} = 5.0\hat{\imath} +$
   $3.5\hat{\jmath}$ km; in x'-y' system
   $\Delta\mathbf{r} = 6.1\hat{\imath} + 0.50\hat{\jmath}$ km;
   (b) $\Delta r = 6.1$ km
29. (a) 5.4 mi at 32° E of N;
   (b) 15 mi/h at 32° E of N
31. (a) 264 km/h, 29° west of north;
   (b) $-128\hat{\imath} + 231\hat{\jmath}$ km/h

33. $19\hat{\imath} + 4.5\hat{\jmath} + 0.26\hat{k}$ km/h
35. (a) $24\hat{\imath} + 10\hat{\jmath}$ m;
    (b) $12\hat{\imath} + 5.0\hat{\jmath}$ m/s;
    (c) $12\hat{\imath} - 5.0\hat{\jmath}$ m/s
37. 5.12 m/s^2, 41° S of W (221° from
    x axis)
39. (a) $\bar{\mathbf{v}} = -0.8\hat{\jmath}$ cm/h;
    (b) $\bar{\mathbf{a}} = -0.42\hat{\imath}$ cm/h^2
41. (a) $10\hat{\imath} - 21\hat{\jmath}$ m/s; (b) 13.0 m/s;
    (c) 89.0°
43. $6.7\hat{\imath}$ m/s; (b) 1.66 s; (c) 8.24 m/s
45. (a) 18 km/h, 14 km/h, 10 km/h,
    14 km/h; (b) 21 km/h, 17 km/h,
    13 km/h, 17 km/h
47. $30\hat{\imath} + 64\hat{\jmath}$ km/s
49. (a) 1.6 m, 2.8 m/s, both vertically
    downward; (b) 2.03 m, 3.57 m/s,
    both at 38.1° to the vertical;
    (c) 9.8 m/s^2, vertically downward in
    both frames of reference
51. 16.2 m
53. (a) 1.8 km/h/s = 0.50 m/s^2;
    (b) 142°
55. (a) 0.32 cm/s; (b) 0.034 cm/s^2;
    (c) 90°
57. 25° upstream
59. (a) 1.30×10^6 m; (b) 7.78 km/s;
    (c) 9.19 m/s^2, almost g
61. $-b\hat{\imath} + a\hat{\jmath}$, $b\hat{\imath} - a\hat{\jmath}$
63. $\frac{\sqrt{2}}{2}\hat{\imath} + \frac{\sqrt{2}}{2}\hat{\jmath}$
67. 53.8°, 13.9 km/h

Chapter 4
1. 0°
3. $1.3\hat{\imath} + 2.3\hat{\jmath}$ m/s
5. 6.0 m/s at 53° below x axis
7. (a) 23°; (b) 5.4×10^3 km
9. 1.83 m
11. (a) $t = 18$ s; (b) 300 m; (c) 22 m/s,
    at 120° to x axis
13. (a) 2.6×10^{17} cm/s^2, upward;
    (b) parabolic
15. (a) 1.34 s; (b) 14.7 m
17. 5.7 m/s
19. 34 nm
21. 8.3 m/s at 61°
23. Yes
25. −14.6 m/s
27. (a) 6.64 km/s; (b) 16.0 min;
    (c) 8.28 km/s
29. 1.1 s
31. 1090 m

35. (a) 8.8 m; (b) 0.53 m
37. 11.2 m/s
41. 31.2° or 65.7°
43. 2.8×10^{-3} m/s^2
45. 54 min
47. 2.27 km
59. 5.07×10^3 s
51. $a_r = 0.947$ m/s^2;
    $a_t = 6.28\times10^{-2}$ m/s^2
53. $t = \sqrt{r/a_t}$
55. 89 m/s
57. 32 m
59. 19 m
61. 300 m, 119 m
63. $\mathbf{v}_0 = 6.36\hat{\imath} + 10.3\hat{\jmath}$ m/s,
    or 12.1 m/s at 58.3°
65. 83°
71. 7.2 m/s at 77° to horizontal
75. 38°
77. $2v_0 \sin\theta_0/g$
79. 892 m/s^2
81. 364 m

83. (a) $\tan^{-1}\left(\tan\theta_0 - \dfrac{gt}{v_0\cos\theta_0}\right)$;

    (b) $\tan^{-1}\left(\tan\theta_0 - \dfrac{gx}{v_0^2\cos^2\theta_0}\right)$

Chapter 5
1. 3.8 MN
3. 1.5×10^3 kN
5. 10^{-4} N
7. (a) $m_B = 3.0m_A$; (b) 2.0 m/s^2
9. Quadruples
11. (a) 11 m; (b) 24 m; (c) 43 m;
    (d) 53 m
13. $F_{\text{driver}} = 5.7$ kN; $F_{\text{passenger}} = $
    125 kN, 22 times greater
15. 5.77 N at 72.2° to the x axis
17. Venus
19. (a) 3.3 N; (b) 12 oz
21. 9.1×10^3 kg
23. 7.7×10^{12} m/s^2
25. 4.26 m/s^2
27. 385 N
29. 2.9 m/s^2, downward
31. (a) 3.1×10^7 N; (b) 9.4×10^2 N
33. 0.53 s
35. 55 kN
37. 2.0 N
39. 1.3×10^{-21} cm

41. (a) 5.26 kN; (b) 1.08 kN;
    (c) 494 N; (d) 589 N
43. 130 N
45. 33 cm
47. 1.9 m/s^2
49. 830 g
51. (a) 132 cm; (b) 127 cm;
    (c) 120 cm; (d) 40 m/s^2
53. 30.7 kN
55. 4.1 cm
57. Apparent weight is 55% of actual
    weight
59. 240 N
61. 4.3 cm
63. 7.2 m
65. (a) $0.40mg$; (b) $2.40mg$; (c) $1.40mg$
67. (a) 16 kN; (b) 1.5 kN

69. (a) $a = \dfrac{m_f - m_s}{m_s}g$;

    (b) $y = \dfrac{m_f a_s h}{(m_f - m_s)(g + a_s)}$

71. $\ell + nm(a + g)/k$, where n is the
    spring number measured from the
    bottom
75. 900 N

Chapter 6
1. $2.7\hat{\imath} + 5.5\hat{\jmath}$ N, or 6.1 N at 64° to
    the x axis
3. (a) 2.0 kN; (b) 1.4 kN
7. 43 cm
9. 530 N; 3.6 times the weight
11. 98 N in horizontal string; $98\sqrt{2}$ N
    in vertical string
13. 230 N in short rope; 84 N in long
    rope
15. (a) 6.3 m/s^2; (b) 0.44 s
17. Right-hand mass 2.5 times left-hand
    mass
19. (a) 7.1 kg; (b) 3.9 kg
21. Left to right, 56.9 N, 34.4 N,
    89.2 N
23. 8.18×10^{-8} N
27. (a) 13 m/s; (b) 4.4°
29. 132 m
31. (a) 310 N; (b) 0; (c) nothing
33. (a) 35°; (b) 22°
35. 24 m/s = 86 km/h
37. 17 m/s
39. 579 N
41. 5.43 kN
43. 0.21

45. 25.3 s
47. 11.4 m/s^2
49. 133 m
51. 2.7 cm
53. (a) 1.6 m/s^2; (b) 3.3 N
55. 4.2 m/s^2
57. 95 km/h
61. 0.72
63. 2.6 N
65. 4.7 m/s
67. 0.70 m/s^2, upward
69. 7.75 m/s
71. 1.40 s
73. (a) 0.12; (b) toward the inside of the turn
75. 2.0 rpm
79. (a) lower at equator, where a net downward force is needed to keep you accelerating with Earth's rotation; (b) 0.34%
85. 28 cm

Chapter 7
1. 900 J
3. 0.15 NJ
5. 9.6 MJ
7. (a) 400 J; (b) 31 kg
9. 4.4 MN
13. (a) 86.9 units^2; (b) 20.3 units^2
15. (a) 0; (b) 90°
19. 169 J
21. 622 J
23. (a) 60 kJ; (b) 20 kJ
25. (a) 360 J; (b) 350 J; (c) 357.5 J; (d) 359.375 J
27. $k_B = 8k_A$
29. 190 J
31. (a) 33 J; (b) 60 J; (c) 78 J
33. $F_0\left(x + \dfrac{x^2}{2\ell_0} + \dfrac{\ell_0^2}{\ell_0 + x} - \ell_0\right)$
35. (a) 0; (b) $2FR$; (c) πFR
37. 135 J
41. 2.56×10^{-15} J = 16.0 keV
43. 116 km/h
45. 1 : 2
47. 113 m/s
49. 2.82 kN = 635 lbs
51. (a) 83.6 MJ; (b) 256 km/h; (c) 2.14 s; (d) 33.3 m/s^2
53. (a) 60 kW; (b) 1 kW; (c) 40 W
55. 9.4 MJ
57. (a) 36 MW; (b) 1.1 MW

61. 2.7 h
63. 430 million gal/day
65. 2.1 MJ
67. 7.7°
69. 1.9 m
71. $\frac{1}{2}F_0 x_0$
73. 70.5°
75. (a) 28 kJ; (b) 18 kJ
77. (a) 450 W; (b) 8.0 kJ
79. (a) $W = \frac{1}{2}bt^2$; (b) $a = \sqrt{b/m}$
81. (a) 71 kW·y; (b) 93 kW·y
83. About 450 J
85. (a) 33 J; (b) 167 J

Chapter 8
1. (a) $-2\mu mg\ell$; (b) $-\sqrt{2}\mu mg\ell$
3. (a) 0; (b) $F_0 a$
5. (a) 1.3 MJ; (b) 59 kJ
9. (a) 1.07 J; (b) 1.12 J
11. 5.2 J
13. 55 cm
15. 2.9 J
17. $U = -\frac{1}{3}ax^3 - bx$
19. $U = F_0\left(x + \dfrac{x^2}{2\ell} + \dfrac{\ell^2}{\ell + x} - \ell\right)$
21. 50 m/s (180 km/h)
23. 92 m
27. (a) 16 m/s; (b) 29%
29. 26 MN/m
31. 3.6 m/s = 13 km/h
35. 2.6 m/s
39. ±2.0 m
43. (a) $U = -\dfrac{a}{2}x^2 + \dfrac{b}{4}x^4$; (b) $x = 0.66$ m, $x = 2.1$ m
45. (a) −6.7 N; (b) 0; (c) 4.5 N
49. (a) 30 cm; (b) 10.4 N/m
51. 35%
53. 19%
55. 0.36
57. 2.6 m/s
59. 0.036
61. 62 cm from left end of frictional zone
63. 1000 MW, twice that of the coal plant
65. $x = 2(h_1 - h_2)$
69. (a) $v = \sqrt{\sqrt{2}g\ell}$; (b) $T = \dfrac{3}{\sqrt{2}}mg$
71. ±25 cm

73. 14 m
75. (a) 2.53×10^5 m/s; (b) 2.91×10^5 m/s; (c) 2.93×10^5 m/s
77. 54.6 mJ
79. 75 cm
81. (a) 1.74 cm; (b) 0.78 cm; (c) 7.4×10^7 m/s
83. (a) $v = \left[\dfrac{x^2}{m}(a - \frac{1}{2}bx^2)\right]^{1/2}$; (c) $v_{\max} = a/\sqrt{2mb}$
85. $v = [2ax^{|b+1|}/m|b + 1|]^{1/2}$
87. 14.5 m

Chapter 9
1. $R_E/\sqrt{2}$
3. 58%
5. 8.6 kg
7. 46 nN
9. $(\sqrt{2} - 1)R_{\text{planet}} \simeq 0.41 R_{\text{planet}}$
11. 1.2×10^{-7}
13. 3.1 km/s
15. 1.8 days
17. 1.0 hour
19. 2.6×10^{41} kg, about 10^{11} solar masses
21. 6.3×10^{10} m
23. 2.64×10^{10} m
25. 2.47 times Earth's orbital radius
27. No
29. 3.2×10^9 J
31. 530 km
33. 58 MJ
35. $R_E/99 = 64$ km; underestimate
37. 3%
39. $\sqrt{2}$
41. 7.7 Mm
43. (a) 11.2 km/s; (b) 9.74 km/s; (c) no
47. $v = \sqrt{2GM\left(\dfrac{3}{R} + \dfrac{1}{r}\right)}$
49. 8.1×10^{11} m, just beyond Jupiter
53. 109 min
55. 15 km/s, 23 km/s
57. 7.95 km/s
59. 8.85×10^5 m
61. (a) 9.0×10^{10} m; (b) 5.3×10^{11} J; (c) 38 km/s
63. 4.60×10^{10} m
65. (b) 8.8 mm; (c) 2.9 km
67. 1400 km lower

Chapter 10

1. 0.75 m from the center
3. 14 m to the right, 4.5 m upward in Fig. 10-23
5. $\ell/2\sqrt{3}$ along the perpendicular bisector of any side
7. $X = 44$ cm, $Y = 55$ cm, with origin at lower left
9. $0.115a$ above the vertex of the missing triangle
11. 6.5 pm from the oxygen
13. $\mathbf{R} = (t^2 + \frac{10}{3}t + \frac{7}{3})\hat{\mathbf{i}} + (\frac{2}{3}t + \frac{8}{3})\hat{\mathbf{j}}$; $\mathbf{V} = (2t + \frac{10}{3})\hat{\mathbf{i}} + \frac{2}{3}\hat{\mathbf{j}}$; $\mathbf{A} = 2\hat{\mathbf{i}}$
15. $\sim 10^{-10}$ m, the diameter of a hydrogen atom
17. $m_{\text{mouse}} = \frac{1}{4}m_{\text{bowl}}$
19. Moves in the opposite direction at 67 cm/s
21. 21 g, $-x$ direction
23. 498 m/s
25. 8.4 km/h
27. (a) 0.14 N/m²; (b) 0.014 mm
29. (a) 0.99 m; (b) 3.9 m/s
31. 3.9 km/h
33. $26\hat{\mathbf{i}} + 16\hat{\mathbf{j}}$ m/s
35. 1100 kg
37. (a) 3.3×10^6 N (b) 3.4×10^5 kg
39. 0.22
41. $K_{\text{cm}} = 1.1\times10^{-14}$ J before and after; $K_{\text{int}} = 0$ before, $K_{\text{int}} = 1.3\times10^{-14}$ J after
43. before: $K_{\text{cm}} = 1.6$ MJ, $K_{\text{int}} = 21$ kJ; after: $K_{\text{cm}} = 1.6$ MJ, $K_{\text{int}} = 0$
45. (a) 1.7 cm above the bottom; (b) 2.7 cm above the bottom
47. 20 m
49. $-47\hat{\mathbf{i}} - 68\hat{\mathbf{j}}$ m/s
51. (a) 3.5 m; (b) 1.3 m/s; (c) 0
53. (a) 0.096 m/s²; (b) 6.2 m/s
55. 9.3 m/s
57. (a) $m = \dfrac{\pi\rho h^2}{2a}$; (b) $Z = \frac{2}{3}h$
59. 3.0 mN
61. (a) 37.7°; (b) 0.657 m/s
63. 5.8 s after explosion
65. (a) thrust $= m\dfrac{dv}{dt} = [(1 + f)V_{\text{ex}} - V]\dfrac{dM_{\text{in}}}{dt}$; (b) 1504 lb

67. $v_1 = \left(\dfrac{m_2 kx^2}{m_1^2 + m_1 m_2}\right)^{1/2}$; $v_2 = \left(\dfrac{m_1 kx^2}{m_2^2 + m_1 m_2}\right)^{1/2}$

Chapter 11

1. 52 N·s
3. 4.3×10^3 N, $7.1mg$
5. (a) $-2.6\hat{\mathbf{i}} + 0.74\hat{\mathbf{j}}$ N·s; (b) $-51\hat{\mathbf{i}} + 14\hat{\mathbf{j}}$ N
7. 0.46 s
9. (a) 7.3 MN·s; (b) 5.6 MN
11. (a) 6.8×10^{-3} N·s; (b) 2.3 N
13. $\Delta P/P = 2\%$
15. 12 ms
17. (a) 5.4 mi/h; (b) 13%
19. 19 kg
23. 10^{21} kg
25. $4.0\hat{\mathbf{i}} + 21.5\hat{\mathbf{j}}$ Mm/s
29. 1.3 µJ
31. 120°
33. 46 m/s
35. $v_{1f} = -11$ Mm/s; $v_{2f} = 6.9$ Mm/s
37. $3 + 2\sqrt{2} \simeq 5.8$; it doesn't matter which is more massive
39. $v_A = -\frac{1}{3}v$; $v_B = \frac{2}{9}v$; $v_C = \frac{8}{9}v$
45. 22°
47. ($v_{1i} = 0.833$ m/s, $v_{2i} = 1.22$ m/s, $\theta_{2i} = 28.3°$); ($v_{1i} = 1.20$ m/s, $v_{2i} = 1.12$ m/s, $\theta_{2i} = 31.2°$)
49. 13 m/s at 27° to horizontal
51. $v_A = -\frac{1}{5}v_0\hat{\mathbf{i}}$; $v_B = \frac{3}{5}v_0\hat{\mathbf{i}} + \frac{1}{5}\sqrt{3}v_0\hat{\mathbf{j}}$; $v_C = \frac{3}{5}v_0\hat{\mathbf{i}} - \frac{1}{5}\sqrt{3}v_0\hat{\mathbf{j}}$
53. 350 N
55. 52 km/h at 33° north of east
57. $m_{\text{truck}} = 7.6m_{\text{car}}$
59. (a) $m_1 = 3m_2$; (b) $v_{2f} = 2v$
61. $v_{1f} = 1.66$ m/s; $v_{2f} = 0.703$ m/s; $\theta_{2f} = 67°$ clockwise from initial velocity of the first ball
63. (a) 12.0 m; (b) 15.4 m/s
65. 0.88
69. $v_{1\text{kg}} = 4.0$ m/s; $v_{4\text{kg}} = 1.0$ m/s at 50° clockwise from the x axis
71. $v_{1200} = 2.2$ km/h, $v_{1800} = 18$ km/h
73. (a) $v_1 = 0.28v_0$, $v_2 = 0.48v_0$; (b) 3, $0.26v_0$, $0.31v_0$

Chapter 12

1. (a) 7.27×10^{-5} rad/s; (b) 1.75×10^{-3} rad/s; (c) 1.45×10^{-4} rad/s; (d) 31.4 rad/s

3. (a) 75.4 rad/s; (b) 0.24 mrad/s; (c) 6283 rad/s; (d) 2.0×10^{-7} rad/s
5. (a) 66 rpm; (b) 3.7 s
7. (a) 21.7 rad/s, 207 rpm; (b) 34.7 rad/s, 331 rpm
9. (a) 0.068 rpm/s; (b) 7.1×10^{-3} rad/s²
11. (a) 12 min; (b) 2.2×10^4
13. 1.3 rad/s²
15. (a) 2.0 s; (b) 1.0 rev
17. 1.2 m
19. 0.079 N·m
21. 0.15 N·m
23. (a) 0.70 N·m, counterclockwise; (b) $\mathbf{F}_1$ and $\mathbf{F}_5$
25. 614 kg·m²
27. (a) $\frac{2}{3}m\ell^2$; (b) $\frac{2}{3}m\ell^2$; (c) $\frac{4}{3}m\ell^2$
29. 45 kg·m²
33. (a) 1×10^{38} kg·m²; (b) 2.6×10^{19} N·m
35. (a) 1.29×10^{38} kg·m²; (b) 6.45×10^{33} N·m
39. (a) 430 min; (b) 1900 rev
41. 1900 N·m
43. 170 rpm
45. $m_{\text{pulley}} = 0.49$ kg; $m_1 = 0.41$ kg; $m_2 = 0.58$ kg
47. (a) 450 J; (b) 140 W
49. 0.089%
51. 12.2 rad/s
53. 7.0 m/s
55. $v = \sqrt{\dfrac{2gh}{\alpha + 1}}$
57. 17%
59. hollow
61. (a) 0.156; (b) 0.070 rad/s
63. $\dfrac{253}{512}MR^2 = 0.494MR^2$
65. (a) 310 N; (b) 165 kg
67. (a) 3.5 m/s; (b) 24%
69. $\frac{27}{10}R$
71. $\omega = \sqrt{2A/I}$
73. $I = \frac{1}{2}Mb^2$
75. $2\sqrt{2g/3R}$
77. $\frac{1}{2}Mg\ell\cos\phi$

Chapter 13

1. 63 rad/s, west
3. 0.52 rad/s², $-37°$
5. 16.6 rad/s

7. (a) $-z$; (b) z; (c) in the x-y plane, 45° clockwise from the x axis

9. (a) $-12\hat{\mathbf{k}}$ N·m; (b) $36\hat{\mathbf{k}}$ N·m; (c) $12\hat{\mathbf{i}} - 36\hat{\mathbf{j}}$ N·m

11. $-9.0\hat{\mathbf{k}}$ N·m

13. Parallel to the x axis or 120° clockwise from x axis

17. $F_x = 1.33F_y - 3.13$ N

19. 414 kg·m²/s

21. 7.6 rad/s

23. 0.017 kg·m²/s

25. 37 kg·m²/s

27. 2.7×10^5 kg·m²/s, out of page in Fig. 13-30

29. 0.21 kg·m²

31. (a) 0.17 rev/s; (b) 386 J

33. (a) 142 rpm; (b) 21%

35. 2.5 days

37. (a) 23.7 rpm; (b) 3.49 mJ

39. (a) $\dfrac{2M\omega_0}{2M + 3m}$; (b) $\dfrac{M\omega_0}{M + 6m}$; (c) same as (b)

41. (a) 0.537 rad/s; (b) 6.44 m/s; (c) 207 N

43. 6.0 cm

45. $I = \dfrac{mgd}{2\omega\Omega}$

47. $\tan^{-1}(\tfrac{1}{2}) = 26.6°$

49. 0.37 rev/s

51. 22 g

53. $\dfrac{I}{I + mR^2}$

55. (a) 1.61; (b) 2.22

59. Sun's rotation 2.8%; Jupiter's orbital motion 60%

61. $v = \left[\dfrac{8(m + M)g\ell}{m^2}(\tfrac{1}{4}m + \tfrac{1}{3}M)\right]^{1/2}$

63. (a) $\dfrac{m\sqrt{2gh}}{(\tfrac{1}{2}M + m)R}$;

(b) $\dfrac{\sqrt{2gm[hm - (\tfrac{1}{2}M + m)R]}}{(\tfrac{1}{2}M + m)R}$;

(c) $h = \dfrac{(\tfrac{1}{2}M + m)R}{m}$

65. (a) $\tfrac{2}{7}\omega_0$; (b) $2\omega_0R/7\mu_kg$

Chapter 14

1. (a) $\tfrac{1}{3}\ell F_3 - \tfrac{2}{3}\ell F_2 + \ell F_1 = 0$; (b) $\tfrac{1}{2}\ell F_1 - \tfrac{1}{6}\ell F_2 - \tfrac{1}{6}\ell F_3 + \tfrac{1}{2}\ell F_4 - \tfrac{1}{2}\ell F_5 = 0$

3. (b) $\boldsymbol{\tau}_{\text{origin}} = -7\hat{\mathbf{k}}$ N·m

5. (a) A vector of length $\sqrt{2}\,F$, oriented 45° clockwise from the negative y axis, applied at the point $x = 0$ m, $y = +1$ m or anywhere on the line $y = (1 + x)$ m. (b) Not possible; the first two vectors sum to zero but produce a nonzero torque, so any other vector applied to balance torques will upset force balance.

7. Both sets have $-F_1 + F_2 \sin\phi + F_3 = 0$, $-F_2\cos\phi + F_4 = 0$; torque equations are (a) $\ell_2 F_4 \cos\phi - \ell_2 F_3 \sin\phi - \ell_1 F_2 = 0$; (b) $(\ell_2 - \ell_1)F_2 - \ell_2 F_1 \sin\phi = 0$.

9. (a) $\tau_A = \tfrac{1}{2}\ell mg$; (b) $\tau_B = 0$; (c) $\tau_C = \tfrac{1}{2}\ell mg$

11. $\tfrac{1}{6}L$

13. (a) 61 cm from left end; (b) 1.4 m from left end

15. 120 N

17. 11.7 kN

19. (a) 40 N·m; (b) 1300 N

21. Vertical forces both 73.5 N, downward, horizontal forces both 33.6 N, away from door jamb at top, toward jamb at bottom

23. 5.0 kN; tension

25. 0.87

27. 500 N

29. 7.3°

31. 50 kN

33. 6.05 kN

35. $\tfrac{1}{8}$

37. Maximum height of CM is at sphere center; lower for clown (b)

39. Two equilibria for $|a| > 2\sqrt{3}$; one metastable, other unstable

41. 1.2 m

43. 170 N

45. 74 kg

47. $\tfrac{1}{2}(\sqrt{3} - 1)mgs$

49. $\mu = \dfrac{\sin 2\theta}{3 + \cos 2\theta}$

51. $mg/2k$

53. 28°

55. Left scale 16.3 N; right scale 22.9 N

57. Tip

59. Slide

61. (a) $0.44mg$, at 12° to Earth's polar axis; (b) $0.036mgR_E$, out of the plane of Fig. 14-57

Part 1 Cumulative Problems

1. 16.5 m from the post

3. $a = \dfrac{2g[(m_1 + m_2)\sin\theta - \mu m_1\cos\theta]}{2m_1 + 3m_2}$

5. (a) $v = \tfrac{2}{7}\omega R$; (b) $\Delta x = \dfrac{2\omega^2 R^2}{49\mu g}$

Chapter 15

1. $T = 0.780$ s; $f = 1.28$ Hz

3. 11.5 fs

5. $A = 20$ cm, $\omega = \pi/2$ s⁻¹, $\phi = 0$; $A = 30$ cm, $\omega = 2.0$ s⁻¹, $\phi = -\pi/2$; $A = 40$ cm, $\omega = \pi/2$ s⁻¹, $\phi = \pi/4$

7. 63.3 kg

9. 1.7 kN/m

11. 0.69 s

13. (a) $\pi\sqrt{m/k}$; (b) $v_0\sqrt{m/k}$

17. (a) 1.0 cm; (b) 6.2 m; (c) 3.6 km

19. 0.11 N·m/rad

21. (a) $2\pi\sqrt{\ell/g}$; (b) $2\pi\sqrt{2\ell/3g}$; (c) $2\pi\sqrt{2\ell/g}$; (d) infinite

23. 0.34 s

25. $R = \sqrt{2\kappa/k}$

27. Within 1 μm

33. 5.0 g

35. $\omega^2 = \dfrac{k_1 k_2}{m(k_1 + k_2)}$

37. a and b are the amplitudes in the x and y directions, respectively

39. 400 J, or 1.4×10^{-3} of the total KE

41. $t = (0.14 + n)$ s, $t = (0.53 + n)$ s, n an integer; $x = \pm37$ cm

45. $\omega = \sqrt{2k/3M}$

47. 34

49. 77% at $0.90\,\omega$; 66% at 1.1ω

53. (a) 19 s⁻¹; (b) 0.33 s; (c) 92 m/s²

55. (a) 6.5 cm; (b) 0.51 s

57. 1.64 s

59. 300 g

61. (a) $E_2 = \tfrac{1}{4}E_1$; (b) $a_{2\,\text{max}} = \tfrac{1}{4}a_{1\,\text{max}}$

63. 2.1 m/s²

65. 0.44

67. $2\pi\sqrt{7/10ga}$

69. $f = 0.54$ Hz; $A = 22$ cm; $\phi = -0.11$ rad

71. $T = 2\pi\sqrt{R/g}$

73. $T = 2\pi\sqrt{mL/2F_0}$

Chapter 16

1. 3.4 s
3. 3.38 m
5. 1.81×10^8 m/s $= 0.604c$
7. (a) 400 nm; (b) 0.3 mm
9. 11 m
11. (a) 0.58 m^{-1}; (b) 1.53 s^{-1}
13. (a) 13.7 s^{-1}; (b) 0.393 cm^{-1};
    (c) $y = 2.5\cos(0.393x + 13.7t)$
15. (a) 25 cm; (b) 0.37 Hz; (c) 12 m;
    (d) 4.4 m/s
17. $y = \dfrac{2}{(x - 3t)^4 + 1}$
19. (a) 3.0 m; (b) 1.5 s; (c) 2.0 m/s;
    (d) $+x$
21. 250 m/s
23. (a) 7.6 N; (b) 1.7 m/s
25. 364 m/s
27. 94 N
29. 7.64 g/cm^3
31. 585 m/s
33. 9.9 W
35. 35 cm
37. $4\pi^2 A^2 F/\lambda$
39. 12 mW/m^2
41. (a) 9.1 kW/m^2; (b) 0.88 W/m^2
43. (a) 6.4 kW/m^2; (b) 4.9 W/m^2
45. 5.1 m
47. (a) 2 cm; (b) pulse 1 at $x = 0$,
    direction $+x$, pulse 2 at $x = 5$,
    direction $-x$; (c) $t = 2.5$ s
49. Every 6 s
51. 5.34 m
53. $\sqrt{gh}$
55. (a) 1.5 cm; (b) 63 cm; (c) 11 ms;
    (d) 56 m/s;
    (e) 18 W
57. $v = \sqrt{k\ell(\ell - \ell_0)/m}$
59. 10 m
61. Every 30 s
63. $u < 0.063v$
67. 5.2 km
69. 67 m

Chapter 17

3. $\lambda = 34$ cm; $T = 1.0$ ms;
   $\omega = 6.3 \times 10^3$ s^{-1};
   $k = 0.18$ cm^{-1}
5. 0.29 s
7. 0.14 kg/m^3
9. monatomic
11. 739 m/s
13. 190 m/s

15. 4.4 nm
17. (a) 3.8 mW/m^2; (b) 96 dB;
    (c) 1.8 N/m^2; (d) 1.6 μm
19. 1 kHz to 6.5 kHz
21. (a) 3.2 μW/m^2, 0.051 N/m^2;
    (b) 3.2×10^{-13} W/m^2;
    1.6×10^{-5} N/m^2
23. (a) 20 dB; (b) approximately
    250 Hz
27. 6.3 m
29. 1
31. 3.2 km/s
33. 0.75 mm
35. 39 μs
37. no; $F = 31$ kN
39. (a) 280 Hz; (b) 70 Hz; (c) 210 Hz
45. (a) 16.6 cm; (b) 424–457 Hz
47. 0.33 Hz
49. 91 Hz
51. 253 m/s
53. 400 Hz
55. 43 m/s $= 154$ km/h
57. (a) 2800 Hz; (b) 933 Hz
59. 25 m/s
63. $u/v = 1/\sin 45° = 1.4$
65. (a) 5.5×10^{-4} W/m^2; (b) 87 dB
67. (a) 112 m/s; (b) 4, 5
69. 960 m/s
71. 0.445 s
73. (a) $\lambda = 5.0$ m, $f = 0.56$ Hz;
    (b) $\lambda = 2.5$ m; $f = 0.79$ Hz
75. 1.36
77. $\Delta f = 2uf/(v - u)$
79. 16 kHz

Chapter 18

1. 1.2 kg
3. 10^{-14}
5. (a) 81 N; (b) 65 N
7. 1 in. H$_2$O $= 249$ Pa
9. 2×10^7 N, or 2000 tons
11. 21 N
13. 0.25 m^2
15. No; $F = 2.3 \times 10^4$ N, or 2.5 tons
17. 1700 kg/m^3
19. ~90 m
21. 890 Pa gauge
23. 8.1×10^{10} N
25. 3.6 mm
27. 93 cm higher in the eye
29. 8.11×10^3 kg
31. 44 kg
33. 0.75 %
35. 59 g

37. 27 m
39. (a) 49 kg; (b) 2500 kg
43. (a) 1.8×10^4 m^3/s; (b) 1.5 m/s
45. 1.75 m/s
47. (a) $h_2 = h_1$; (b) $h_2 = h_1 - \dfrac{3v^2}{2g}$
49. 14.3 m
51. 7.2 cm^3/s
53. (a) no; (b) yes
55. 13
57. 70%
59. $A\sqrt{2gh}$
61. (a) 14 m/s; (b) 2.2 m
65. (a) 25 L/s; (b) 55 m/s; (c) 1.8 kPa
67. $t = \dfrac{A_0}{A_1}\sqrt{\dfrac{2h}{g}}$
69. $P = P_a + \rho g h_0 + \frac{1}{2}\rho\omega^2 r^2$;
    (b) $h = h_0 + \dfrac{\omega^2 r^2}{2g}$
71. (a) $\rho(h) = \dfrac{P_0}{h_0 g} e^{-h/h_0}$; (b) 5.7 km

Part 2   Cumulative Problems

1. (a) $\ell = \dfrac{4M}{\pi d^2 \rho}$; (b) $T = 4\sqrt{\dfrac{\pi M}{d^2 \rho g}}$
5. 17.2 cm

Chapter 19

1. 720
3. 20°C
5. -40
7. -196°C, -321°F
9. (a) 138 kPa; (b) 33.4 kPa;
   (c) 233 kPa
11. 1.37 L
13. 586 mm
15. 240 kcal
17. 0.36 kg
19. 24 days
21. (a) 23 kJ; (b) 337 kJ; (c) 65 kJ
23. 7.5 kW
25. 2.4 kg
27. (a) 560 g; (b) 0.27 K/s
29. 1.8 kg
31. 0.70 K
33. 1.6 K/s
35. 197 g
37. 56.2°C
39. 55 kW
41. 0.293 W
43. 25 ft^2·°F·h/Btu
45. 200 W

47. (a) 12.3 ft²·°F·h/Btu;
    (b) 715 Btu/h
49. Will save 10 gallons/month
51. 23°C
53. 80%
55. −25°C
57. Drop by 5.9 K
59. 24°C
61. 480 W
63. 1151 K
65. −2.5°C
67. (a) $87; (b) $10
69. 4.65%
71. $\Delta T_{copper} = 0.16$ K; $\Delta T_{iron} = 2.1$ K
73. 2.9 J
77. 10 hours

Chapter 20

1. 2.6 m³
3. 1.8 MPa
5. (a) 27 L; (b) 330 K
7. 2.7×10^7
9. 11 L
11. 515 kPa
13. (a) 1.27 atm; (b) 0.0268 mol;
    (c) 0.786 atm
15. H_2
17. 268 K, compared with ideal gas
    292 K
19. 1.76 MPa
21. (a) 9.1×10^{20}; (b) 2.0×10^{20}
23. 22 kJ
25. 3.9 kg
27. 5.96 MJ
29. 1.3×10^{10} kg
31. 564 W
33. 44 minutes
35. 48 min
37. (a) 117 s
39. 3.55 MJ
41. 64°C
43. 177 g
45. 135 g solid in 865 g liquid, at 234 K
47. 5.0 kg
49. 4.9°C
51. 1.00021 cm
53. 3.9 km
55. 43.6 mm³
57. 307 K
59. $d = \frac{1}{2}L_0\sqrt{2\alpha\,\Delta T + (\alpha\,\Delta T)^2}$
61. 120 mol/m³
63. 19 kW
65. 79 g
67. 1.2 kg ice, 0.80 kg water, all at 0°C

69. 50 min
71. (a) 61 h; (b) 52 h
77. 34 km

Chapter 21

1. 29 kJ
3. Increases by 250 J
5. (a) 310 MW; (b) 54%
7. 0.02°C
9. $2P_1V_1$
11. 1.2 kJ
13. 4.3 kJ
15. 190 K
17. 1.99 kJ
19. (a) 399 J; (b) 264 kPa
21. (a) 571 kPa; (b) 438 J
23. 440°C
25. $V = 0.18V_0$
27. (b) Gas does 13 J of work;
    (c) 22 J heat lost from gas
29. (a) 300 K, 1.5 kJ; (b) 336 K, 0 J;
    (c) 326 K, 429 J
31. (a) 39.9 kPa; (b) 83.3 kPa;
    (c) 80.2 kJ
33. 928 J
35. (a) 211 J; (b) 12.9 L
37. 75°C
39. 128 J
41. (a) −9.0°C; (b) stable
43. 57.7%
45. 79%
47. 20 mol
49. Drops 23.1 K
51. 343 K
53. 28 kPa
55. (a) 598 J; (b) 2500 J flows in
57. 25 m
59. $\frac{4}{3}P_1V_1$
63. (a) 2.5 kJ; (b) 447 K
69. (a) $M = \dfrac{P_0A}{g}\left[\dfrac{h_1}{h_2} - 1\right]$;

    (b) $M = \dfrac{P_0A}{g}\left[\left(\dfrac{h_1}{h_2}\right)^{\gamma} - 1\right]$

Chapter 22

1. (a) 12!/6!, or 6.7×10^5 states;
   (b) about 1 in 1000
3. 5×10^{24} J, assuming oceans cover
   75% of Earth to an average depth
   of 3 km; this is about 20,000 times
   annual use
5. (a) 27%; (b) 7.0%; (c) 77%
7. 0.95 K
9. 52% winter, 48% summer
11. (a) 1.75 GW; (b) 43%; (c) 505 K
13. 2×10^7 kg/s, slightly more than the
    Mississippi's flow
15. (a) 39%; (b) 550 J; (c) 190°C
17. 53.3 kJ
19. (a) 4.3; (b) maximum COP = 11
23. (a) $COP_{summer} = 13$, $COP_{winter} = 3.5$;
    (b) 0.076 J; (c) 0.22 J
25. (a) 561.7 J; (b) 464.1 J;
    (c) 97.66 J; (d) 17.4%;
    (e) $T_c = 403$ K, $T_h = 487$ K
29. 718 K
31. 1.22 kJ/K
33. 8.9°C
35. 1.36×10^8 J/K
39. (a) 53 J/K; (b) 74 J/K; (c) 0
41. (a) −109 J/K; (b) 122 J/K;
    (c) 13 J/K
43. $\Delta S_{AB} = 68.5$ J/K, $\Delta S_{BC} = 45.7$ J/K,
    $\Delta S_{CA} = -114.2$ J/K
45. 470 kPa
47. (a) 69%; (b) 967 K
49. Decrease in $T_{minimum}$
51. $58
53. (a) $W = Q = 345$ J; (b) $e = 24\%$
55. 598 J/K
57. About 30 km for 200 kg of water
    and a 60 kg bather
59. 166 MW
61. (a) $1 - 5^{1-\gamma}$; (b) $3T_{min}(5^{\gamma-1})$;
    (c) $e_{Carnot} = 1 - \frac{1}{3}(5^{1-\gamma})$
63. (a) 7.94; (b) 5.26; (c) $P_n = 2.96P_s$
65. (a) $T_h = T_{h0}e^{-P_0t/mc(T_{h0}-T_c)}$;
    (b) $P = 0$ at
    $t = \dfrac{mc(T_{h0} - T_c)}{P_0}\ln\left(\dfrac{T_{h0}}{T_c}\right)$
67.

	P	V	T	$U - U_A$	$S - S_A$
A					
B			$3.4\,T_0$	$6.0\,P_0V_0$	$3.1\,P_0V_0/T_0$
C	$1.5\,P_0$	$2.2\,V_0$	$3.4\,T_0$	$6.0\,P_0V_0$	$3.8\,P_0V_0/T_0$
D			$3.0\,T_0$	$5.0\,P_0V_0$	$3.8\,P_0V_0/T_0$

Part 3 Cumulative Problems

1. $e = 1 - r^{1-\gamma}\left[\dfrac{r_c^{\gamma} - 1}{\gamma(r_c - 1)}\right]$

3. $W = an^2\left(\dfrac{1}{V_2} - \dfrac{1}{V_1}\right) +$

$nRT \ln\left(\dfrac{V_2 - bn}{V_1 - bn}\right)$

5. (a) $t_1 = \dfrac{L_f M T_h}{P_h T_0}$;

(b) $P = P_h\left[1 - \dfrac{T_0}{T_h}e^{P_h(t-t_1)/McT_h}\right]$;

(c) $t_2 = t_1 + \dfrac{McT_h}{P_h}\ln\dfrac{T_h}{T_0}$, with L_f the heat of fusion of ice and c the specific heat of water.

Chapter 23

1. Several coulombs
3. (a) uud; (b) udd
5. About 10^9 N; about 10^6 times typical human weight
7. 8.2×10^{-8} N
9. $21.8\ \mu$C
11. $-0.18\hat{\imath} + 0.64\hat{\jmath}$ nN
13. (a) $\hat{\jmath}$; (b) $-\hat{\imath}$; (c) $(\hat{\imath} + 3\hat{\jmath})/$ $\sqrt{10} \approx 0.032\hat{\imath} + 0.95\hat{\jmath}$
15. $14\hat{\imath} - 7.4\hat{\jmath}$ N
17. $15\ \mu$C
19. $1.6\hat{\imath} - 0.33\hat{\jmath}$ N
21. $\dfrac{kq^2}{a^2}(\sqrt{2} + \tfrac{1}{2})$
23. $q_2 = 143\ \mu$C; $q_3 = 116\ \mu$C
25. 3.8×10^9 N/C
27. (a) 2.2 MN/C; (b) 77 N
29. 5.15×10^{11} N/C
31. (a) $26\hat{\jmath}$ MN/C; (b) $-5.15\hat{\jmath}$ MN/C; (c) $-58\hat{\jmath}$ MN/C
33. $-4e$
35. (a) $\mathbf{E} = \dfrac{2kqy}{(a^2 + y^2)^{3/2}}\hat{\jmath}$; (b) $y = \pm a/\sqrt{2}$
37. (a) $8.0\hat{\jmath}$ GN/C; (b) $190\hat{\jmath}$ MN/C; (c) $216\hat{\jmath}$ kN/C
39. 39 pm
41. (a) $\mathbf{E} = 2\sqrt{3}\,kqa/x^3\hat{\imath}$; (b) $p = \sqrt{3}\,qa$
43. 2.1 MN/C
45. $-\dfrac{2k\lambda_0}{\pi\ell}\hat{\imath}$

47. -137 nC
51. 1.1 kN/C
53. (a) $2.5\ \mu$C/m; (b) 3.0×10^5 N/C; (c) 1.8 N/C
55. 3.3×10^{-12} kg
57. (a) 1.35 cm; (b) reverses direction, accelerates and exits field region at 3.8×10^5 m/s
59. $v > \ell\sqrt{qE/md}$
61. 2.8 Mm/s
63. $-14\ \mu$C/m
65. (a) 3.0 mN·m; (b) 11 mJ
67. (b) Attractive
69. $x = -8.09$ nm
71. $\dfrac{k\lambda_0}{\ell}\hat{\imath}$
73. $2\sqrt{2}kQ/\pi a^2$
75. 1.4 cm
77. 33.3 μC
79. $-4q$, a distance $3a$ to the right of $-q$
81. 50.7 kN/C, downward
83. 7.0 cm; 0.54 μC
85. (a) $(2kep/r^3)\hat{\imath}$; (b) $(-2kep/r^3)\hat{\imath}$, with the $+x$ direction to the right

Chapter 24

1. $+3\ \mu$C
5. (a) 1.7 kN·m²/C; (b) 1.2 kN·m²/C; (c) 0
7. 490 N·m²/C
9. $\pi R^2 E$
11. (a) $-q/\varepsilon_0$; (b) $-2q/\varepsilon_0$; (c) 0; (d) 0
13. 4.9×10^4 N·m²/C
15. (a) 0.69 MN·m²/C (b) -0.69 MN·m²/C (c) 0
17. 10 kN/C
19. 1.8×10^{12} N/C
21. (a) 2.2×10^5 N/C, outward; (b) 2.5×10^4 N/C, outward; (c) 4.0×10^3 N/C, inward
23. (a) $8kQ/R^2$, inward; (b) $kQ/4R^2$, inward; (c) (a) would not change, (b) would become 0
25. (a) 3.6 MN/C; (b) 3.8 MN/C; (c) (a) would not change, (b) would nearly double
27. as $1/r$
29. 6.3 μC/m³
33. 3.6 mC/m³
35. 58 nC/m²
37. $E_1 = \sigma/2\varepsilon_0$, left; $E_2 = \sigma/2\varepsilon_0$, right; $E_3 = 3\sigma/2\varepsilon_0$, right; $E_4 = \sigma/2\varepsilon_0$, right

39. 18 N/C
41. (a) $x < 1.83$ cm; (b) $x > 54.5$ cm
43. 1.6×10^5 N/C
45. (a) $\rho = 0$; (b) $\sigma = 4.0$ mC/m²; (c) other charges would destroy the symmetry, making σ nonuniform
47. (a) 0.50 μC/m²; (b) 56 kN/C
49. (b) $-Q$
51. 1.8 MN/C
53. (a) 0; (b) 180 kN/C; (c) 0; (d) 20 kN/C
57. (a) $4kq/R^2$; (b) $3kq/4R^2$
59. (a) 0; (b) 1.3 MN/C; (c) 0
61. (a) 1.9×10^{11} N/C; (b) 3.6×10^{10} N/C
63. $\dfrac{\rho r}{3\varepsilon_0} - \dfrac{\rho a^3}{3\varepsilon_0 r^2}$
65. $\tfrac{1}{3}E_0 a^2$
67. 0.39 μs
69. $E_{\text{in}} = \rho_0 x^2/2\varepsilon_0 d$; $E_{\text{out}} = \rho_0 d/8\varepsilon_0$
73. (a) 0; (b) $(ac/\varepsilon_0 r^2)(e^{-1} - e^{-r/a})$; (c) $(ac/\varepsilon_0 r^2)(e^{-1} - e^{-b/a})$
77. $+10.6\ \mu$C/m² on both outer faces, $\pm36.9\ \mu$C/m² on inner faces

Chapter 25

1. 600 μJ
3. 3.0 kV
5. 910 V
7. 5.6 kV/m
9. Proton and He$^+$ both gain 100 eV $= 1.6\times10^{-17}$ J; α gains 200 eV $= 3.2\times10^{-17}$ J
11. 4.5 V
13. 0.23 MC
17. 6.1 μC
19. 27.2 V
21. $Q = 5.4$ nC, $r = 17$ cm
23. (a) 442 kV; (b) 9.2 Mm/s
25. kQ/R.
27. $V(x) = -\tfrac{1}{2}ax^2$
29. 52 nC/m
31. $x = -a/2$, $x = a/4$
33. (a) 2.6 kV; (b) 1.8 kV; (c) 0
35. $kQ\left(\dfrac{2}{c} + \dfrac{1}{b} - \dfrac{1}{a}\right)$
37. $2kQ/R$
39. $2\pi k\sigma(\sqrt{x^2 + b^2} - \sqrt{x^2 + a^2})$
41. (a) $V(x, y) = -E_0(x + y) = -150(x + y)$ V/m; (b) 150V
45. (a) $\mathbf{E} = -ay\hat{\imath} - ax\hat{\jmath}$
47. (a) 4 V; (b) $E_x = 1$ V/m, $E_y = -12$ V/m, $E_z = 3$ V/m

49. $\mathbf{E} = \dfrac{kQx}{(x^2 + a^2)^{3/2}}\hat{\mathbf{i}}$

51. $E = V_0/R$, radially outward

53. 3 kV

55. (a) 34 kV, -9.0 kV; (b) 12.6 kV on each; (c) 24 nC

57. (a) 43 kV; (b) 1.7 MV/m; (c) 540 V; (d) 0

59. 1.55 keV $= 2.47 \times 10^{-16}$ J

61. (a) $\dfrac{2kqx}{x^2 - a^2}$; (b) $\dfrac{2kq}{x}$

63. (a) 27 kV; (b) no change

65. -7.5 V

67. (a) $x = -3$ m, 0 m, 1 m; (b) $\mathbf{E} = (3x^2 + 4x - 3)\hat{\mathbf{i}}$; (c) $x = -1.87$ m, 0.535 m

69. (a) 7.2 kV; (b) 14.4 kV

71. 14 cm, 1.7 nC

73. 23 nC/m

75. (a) $V(x) = \dfrac{k\lambda_0 x}{\ell^2}\left[x \ln\left(\dfrac{2x + \ell}{2x - \ell}\right) - \ell\right]$; (b) $\frac{1}{12}\lambda_0\ell$; (c) $\dfrac{k\lambda_0\ell}{12x}$

77. $y^2 + (x - \frac{5}{3}a)^2 = (\frac{4}{3}a)^2$; i.e., a circle

79. $\dfrac{kq}{2a}\ln\left(\dfrac{\sqrt{2} + 1}{\sqrt{2} - 1}\right) \simeq \dfrac{0.881\, kq}{a}$

Chapter 26

1. $3kq^2/\ell$

3. 4.88 kJ

5. $v = 2q\sqrt{\dfrac{k}{m\ell}}$

7. $6kq^2/a$

9. (a) 2.0 MV/m; (b) 9.9 kV; (c) 5.5 mJ

11. (a) 0.74 μC; (b) 40 kV

13. $\frac{1}{2}kQ^2\left(\dfrac{1}{a} - \dfrac{1}{b}\right)$

15. (b) $dW = 2kq\, dq/a$; (c) $W = kQ^2/a$

17. 1 km^3

19. 9×10^{30} J/m^3

21. 24 μJ

25. $(kQ^2/R)(2^{2/3} - 1) = 0.60$ mJ

27. $U/\ell = \pi a^4 \rho^2/16\varepsilon_0$

29. ± 14 C

31. 6.5 mF

33. 0.74 nF

---

35. 55 pF

39. 70 nF

41. 1 μF stores 15 times as much energy

43. (a) 30 μF; (b) 0.1 μF; (c) 0.01 μF

45. (a) 5.0 kW; (b) 250 μF; (c) 0.50 W

47. (a) Increases by factor of 2.5; (b) drops to 40% of its original value

51. Equal

53. (a) 6.0 μF; (b) 0.55 μF; (c) 0.83 μF, 1.3 μF, 1.5 μF, 2.2 μF, 2.8 μF, 3.7 μF

55. (a) 2 series pairs in parallel or two parallel pairs in series; (b) 4 in series

57. 0.86 μF

61. $\pm 1\%$

63. (a) 64 V; (b) drops from 16 mJ to 14 mJ

65. (a) 3.5 mm; (b) 87 kV

67. 126 pF

69. (a) 50 nC, 170 nC; (b) 23 μJ, 77 μJ

71. 1.4 mm

73. (a) 0.90 J; (b) 1.8 J; energy comes from work done by agent separating the plates

75. (a) 1.2 μF; (b) 24 μC

77. 13 min

79. $\dfrac{kq^2}{a}(2\sqrt{3} - \frac{15}{2}) \simeq -4.04\dfrac{kq^2}{a}$

81. (a) $C = \frac{1}{2}(\kappa + 1)C_0$; (b) $U = \dfrac{C_0 V_0^2}{\kappa + 1}$; (c) $F = \dfrac{2C_0 V_0^2(\kappa - 1)}{L(\kappa + 1)^2}$, into capacitor

83. (a) 75 mF; (b) 1.4×10^{12} J

85. (a) $Q^2/2\varepsilon_0 a$; (b) $Q^2/\varepsilon_0 a$; (a) is right since (b) includes the fields of *both* plates, and a plate doesn't experience a force from its own field

87. $31d^2kp^2/1280\ell^5$

89. 7.2 pF/m

Chapter 27

1. 9.4×10^{18}

3. 2.9×10^5 C

5. 1.9×10^{11}

7. 7.65 MA/m^2

---

9. 1.23 cm/s

11. 0.31 mA

13. (a) 37 A/m^2; (b) 86%

15. 9.71 μC

17. 0.17 V/m

19. 2.2×10^{-6} $\Omega \cdot$m

21. (a) 5.95×10^7 $(\Omega \cdot$m$)^{-1}$; (b) 4.5 $(\Omega \cdot$m$)^{-1}$

23. 6.4×10^{-15} s

25. 253°C

27. 25 Ω

29. 2.34 mA

31. 25 mA

33. 1.5 kA

35. (a) 17 mΩ; (b) 86 A; (c) 18 MA/m^2; (d) 1.7 V/m

37. $d_{\mathrm{Al}} = 1.26\, d_{\mathrm{Cu}}$

39. (a) 2.07 cm; (b) 2.60 cm; (c) aluminum

41. 34 mΩ

43. 1.38 kW

45. 160 μA

47. 240 Ω

49. 48 W

51. 960 W·h $= 3.5$ MJ

53. Resistor with more power has $\sqrt{2}$ times greater diameter

55. (a) 150 A; (b) 3.4 km

57. 0.54 mA

59. 0.94 Ω

61. (a) 8.7 kA; (b) 15%

63. 2.9 A

65. 2.5 A

67. 120 A

69. 17 J/K

Chapter 28

5. 1.4 hours

7. 6.0 V

9. 229 kΩ

11. 1.5 A

13. 0.02 Ω

15. 45 Ω

17. 30 A

19. R_1 for each

21. 24

23. (a) 162 Ω; (b) 125 mW

25. (a) $\dfrac{R_1 \mathcal{E}}{R + 2R_1}$; (c) $\frac{1}{2}\mathcal{E}$

27. 2.45 W

29. $(\mathcal{E}_2 R_1 - \mathcal{E}_1 R_2)/(R_1 + R_2)$

31. $\frac{7}{5}R$

33. $I_1 = 2.79$ A, $I_2 = 2.36$ A, $I_3 = 0.429$ A
35. $\mathcal{E}_2 > 5.49$ V
37. 8.5 nA, downward
39. 1.6% low
41. 1.2 kW41.24.99 kΩ
43. (a) 20 V; (b) 3.0 mA
49. (a) 9.0 V; (b) 1.5 ms; (c) 0.32 μF
51. (a) 0.35 s; (b) 0.17 s
53. 3.4 μF
55. (a) $I_1 = 25$ mA, $I_2 = 0$, $V_C = 0$; (b) $I_1 = I_2 = 10$ mA, $V_C = 60$ V; (c) $I_1 = 0$, $I_2 = 10$ mA, $V_C = 60$ V; (d) $I_1 = I_2 = 0$, $V_C = 0$
57. (a) $3\mathcal{E}/4R$; (b) $2\mathcal{E}/3R$
59. 3.4 kΩ
61. (a) 4.51 V; (b) 35.2 Ω
63. 1.07 A, left to right
65. (a) 13.0 V; (b) 2.23 mA
67. 15.9 ms
69. 83 μs
71. (a) 0; (b) 1.0 A; (c) 0.75 A; (d) 0; (e) 1.0 A; (f) 3.0 A; (g) 3.0 A; (h) 1.0 A
73. (b) 22 hours

Chapter 29
1. (a) 1.6 mT; (b) 2.3 mT
3. (a) 2.0×10^{-14} N; (b) 1.0×10^{-14} N; (c) 0
5. (a) $-1.1\hat{\imath} + 1.5\hat{\jmath} + 1.7\hat{k}$ mN
7. $\mathbf{B} = 0.13\hat{k}$ T; $\mathbf{v}_2 = -14\hat{\imath}$ km/s
9. $-24.9\hat{\imath} - 40.1\hat{\jmath} - 87.3\hat{k}$ fN
11. 40.1° or 140°
13. (a) $1.2\hat{\imath} + 0.45\hat{\jmath}$ fN; (b) $-1.2\hat{\imath} + 35\hat{\jmath} - 15\hat{k}$ fN
15. 3.9 mm
17. 1.5 mT
19. $r_{proton} = 43\, r_{electron}$
21. (a) 86 mT; (b) 1.01 keV
23. 1.3 μs
25. (a) 15 MHz; (b) 19 MeV; (c) 6467
29. 0.43%
31. 1.1 mm, 2.6 mm
33. 0.38 N
35. (a) 49 mT; (b) 0.73 N/m
37. 43 kN
39. 0.12 T
41. 21 mN, diagonally toward the upper right
45. 76 mT
47. (a) 1.1 mA·m²; (b) 1.0 mN·m

49. 0.15 A·m²
51. (a) 0.35 A·m²; (b) 4.2×10^{-2} N·m
53. 1.97×10^{-25} J $= 1.23$ μe V
55. $42\hat{\imath} + 88\hat{\jmath} - 25\hat{k}$ fN
57. 30 km
59. 0.27 N, to right
61. 0.010 T
63. 6.8 mm
65. (a) 12.8 N/m; (b) 24.9 cm
67. $I = Mg/2Bd$
69. 77 mT
73. $\mu = 9.25 \times 10^{-24}$ A·m²

Chapter 30
1. 12 cm
3. 1.23 mT
5. 3.8 GA
7. 0.875 cm²
9. 732
11. Between the wires, 2.0 cm from center of 5.0-A wire
13. 23° west of magnetic north
15. $\mu_0 I/4a$, into page
17. $\dfrac{\mu_0 I}{4ab}(b - a)$, out of the page
19. 5 μN
21. 3.8 mm
23. 13.2 μN/m, at 71.6° below right-pointing horizontal
25. 23.5 mN
27. 7.0 A
29. 24 A
31. 123 mA
33. (a) 0; (b) 0.36 mT; (c) 1.9 mT; (d) 0.50 mT
35. (a) $\mu_0 J_0 r^2/3R$; (b) $\mu_0 J_0 R^2/3r$
37. (a) 5.3 mm; (b) maximum; (c) 130 A
41. (a) 0; (b) $\dfrac{\mu_0 I(r^2 - a^2)}{2\pi r(b^2 - a^2)}$; (c) $\dfrac{\mu_0 I}{2\pi r}$
43. 17 T
45. (a) 38 mT; (b) 5.9 μT
47. 3.3 m
49. $1/2\pi n R$
51. $\chi_M = -1.6 \times 10^{-3}$; diamagnetic
53. 7.2×10^3
55. $I_{outer} = 2I_{inner}$, in opposite direction
57. Out of page
59. 1 km
61. (a) 8.0 μT; (b) 4.0 μT; (c) 0
63. (a) 0.40 mT; (b) 1.0 nT
65. $B = \mu_0 I/\ell$ inside, $B = 0$ outside
67. 55A

Chapter 31
3. 1.4×10^{-4} T·m²
5. 160 T/s
7. (a) 0.30 A; (b) 0.20 A
9. $B^2 \ell^3 v/R$
11. 39 V
13. 0.32 V
15. 9.0 μA, counterclockwise in Fig. 31-5
17. (a) $I = -25$ mA for $0 < t < 2$ s; $I = 0$ for 2 s $< t < 3$ s; $I = +25$ mA for 3 s $< t < 5$ s; (b) $P = 3.1$ mW for $0 < t < 2$ s and 3 s $< t < 5$ s; otherwise 0
19. (a) Left to right; (b) 140 μA; (c) 28 μA
21. (a) $I_R = -I_{peak} \cos \omega t$, where (b) $I_{peak} = 110$ mA; (c) 0
23. 0.16 s
25. 7.1 mV
27. (a) Downward; (b) $(B\ell v)^2/R$
29. $v_{final} = \mathcal{E}/B\ell$; R affects time to reach final speed
31. (a), (b) Both 6.7 mA, counterclockwise; (c) 0.44 mW in both cases
33. (a) 25 mA; (b) 1.25 mN; (c) 2.5 mW; (d) 2.5 mW
35. (a) 77 mV; (b) 94 mV
37. 1.57 μN
39. $dB/dt = 10$ T/s
41. $E = \frac{1}{2}bh$, to the left above the field region and to the right below
43. $\frac{16}{3} B_0 x_0^2$
45. 42 mA, clockwise
47. $mgR/B^2 \ell^2$
49. $\frac{1}{2} B R^2 \omega$
51. 58 T/ms
53. $I = at^2$, with $a = 0.81$ A/s²
57. (a) 2.4 km; (b) 2.89 kW; (c) \$6.94; (d) their generators get harder to turn, so they use more fuel
59. (a) $J = rb/2\rho$; (b) $P = \pi b^2 a^4 h/8\rho$

Chapter 32
1. 120 V
3. 12 V
5. (a) $-2\pi f M I_p \cos 2\pi ft$; (b) 133 mH
7. $\mu_0 n^2 \pi R^2 \ell$
9. $\dfrac{\mu_0 \ell}{2\pi} \ln\left(\dfrac{a + w}{a}\right)$
11. 3.2 mH

13. 40 kV
15. 1350
17. (a) 2.1 A; (b) −1.5 A (minus indicates direction reversal)
19. 26 A/ms
21. 11 A
25. 130 Ω
27. 15 s
29. (a) 0.11 H; (b) 20 mA
31. (a) 76 mA; (b) 4.4 V (c) 7.6 V; (d) 2.2 A/s; (e) 0.58 W
33. 400 Ω
35. 50 Ω
37. (a) 2.0 A; (b) 6.0 A; (c) 60 V
39. 100 mA
41. (a) 52 J; (b) 1.8 s
43. (a) 2.5 kW; (b) no
45. 15 W
47. (a) 5.7 MJ; (b) 31 mΩ; (c) 39 s
49. 9.9×10^8 J/m^3
51. 10^{11} times that of gasoline; 2600 times that of U-235
53. Smaller by factor $1/4n^2R^2$
55. 10^{18} J
57. $|\mathcal{E}| = 2Mbt$
59. 48 H
61. (a) 4.4 A; (b) 0; (c) 70 V
63. $(\mu_0 I^2 \ln 100)/4\pi$
65. (a) $L = \dfrac{\mu_0 N^2 \ell}{2\pi} \ln\left(\dfrac{R + \ell}{R}\right)$
67. (a) $P = I_0^2 R e^{-2Rt/L}$
69. $E/B = 1/\sqrt{\mu_0 \varepsilon_0} = 3.0 \times 10^8$ m/s $= c$

Chapter 33

1. $V = 325 \sin(314t)$, with V in volts and t in seconds.
3. 9.9 V
5. (a) 350 mA; (b) 1.50 kHz
9. $V_{\text{rms}} = V_p/\sqrt{3}$
11. $I_1 = I_p \cos\phi$, $I_2 = I_p \sin\phi$
13. 45 mA
15. (a) 804 Ω; (b) 48.2 Ω; (c) 2.41 Ω
17. 1.47 μF
19. 15.9 kHz
21. (a) 10 mA; (b) 14 V; (c) 7.7 mA
23. 2.7 mC
25. 0.32 A
27. 8.23 kHz or 5.17×10^4 s^{-1}
29. 4.9 pF to 42 pF
31. 3.65 F
33. (a) 1.63 A; (b) 4.58 μs

35. (a) 63 V; (b) 89 V; (c) 764 mA; (d) 99 μJ
37. (a) $1/\sqrt{2}$; (b) 1/2; (c) $-1/\sqrt{2}$; (d) 1/2
39. Close switch B for 702 ms, then simultaneously close A and open B. After 351 ms, open A; (b) 400 V
41. 0.22 μF
43. 16.2 V
45. (a) 0.43 μH; (b) $R > 16$ Ω
47. (a) $V_{Rp} = 23.86$ V, $V_{Cp} = 28.19$ V, $V_{Lp} = -28.05$ V; (b) $V_{R\,\text{rms}} = 16.9$ V, $V_{C\,\text{rms}} = 256$ V, $V_{L\,\text{rms}} = 255$ V
49. (a) above; (b) current lags by approximately 49°
51. (a) I leads V by 63.6°; (b) I lags V by 81.7°
53. 500 W
55. (a) 80 Ω; (b) 97 Hz
57. (a) 5.5%; (b) 9.1%; (c) large
59. 5.0 mA
61. (a) 8.91 V; (b) 1.38 kΩ
63. $4f_1$
65. 13.2 V
67. 0.64 Ω
69. 8.25 V
71. 36.2 nF, 11.5 nF
73. $C = \dfrac{L[\ln(U_0/U_1)]^2}{4\pi^2 N^2 R^2}$
77. $R = 400$ Ω, $L = 67$ mH, $C = 0.094$ μF
79. 1.3 kN/m

Chapter 34

1. 1.3 nA
5. (a) 0; (b) 8.33×10^{-13} T; (c) 9.26×10^{-13} T
7. $-x$
9. (a) $\sqrt{2}E$; (b) $(\hat{\mathbf{j}} - \hat{\mathbf{i}})/\sqrt{2}$
13. 0.24 s
15. 1 ns
17. 600 m
19. 2.32
21. 5000 km
23. 27.3 MHz
25. 15 kV/m
27. 1.1 pT
29. 63°
31. 30°
33. $0.34S_0$
35. $0.304S_0$

37. $S = S_0 \cos^2 \omega t$, with $\omega = 20\pi$ s^{-1}
39. 12 GW/m^2
41. 2.7×10^{-10} W/m^2
43. (a) 157 W/m^2; (b) 344 V/m; (c) 1.15 μT
45. 1.1%
47. 3.9×10^{26} W
49. (a) 8.3×10^{-26} W/m^2; (b) 7.9×10^{-12} V/m
51. (a) 4.6 kW; (b) 52.5 mV/m
53. (a) as $1/r$; (b) as $1/r^2$
55. 6.0 mPa
57. 5.0 μN
59. 6.18×10^3 years
61. $E_p = 40$ V/m, $B_p = 0.13$ μT
63. 1/4
65. 4.2 μN
67. 17.7 μPa
71. (a) 1.0 MV/m; (b) 4.3 mm; (c) 64 mJ; (d) 2.1×10^{-10} kg·m/s; (e) 64 W
73. Long, with cylindrical symmetry
75. 0.3 μm

Part 4   Cumulative Problems

1. 6.0 MV
3. 15.7 mA, top to bottom
5. (b) $c/\sqrt{\kappa}$

Chapter 35

1. 15°
3. ±0.5°
5. (a) 2; (b) 210°
9. 126 nm
11. 1.57
13. Ethyl alcohol
15. 1.83
17. 5.1 m
19. (a) 3.20×10^{14} Hz; (b) 937 nm
21. (a) 49.8°; (b) 42.2°; (c) 22.4°
23. (a) 61.3°; (b) 80.9°; (c) there is none
27. $\sqrt{2}$
29. 2.62×10^8 m/s
31. 53.5°
35. 36.5° to 38.1°
37. About 3.4 cm
39. 67.5°
41. 36.9°
43. 1.96×10^8 m/s
45. 2.32 m
47. 28.8°

49. (a) base; (b) 63.2°
51. 35.2°
53. $n = 1.17$
59. (b) 42.1°

Chapter 36
1. Mirror surface 55° to horizontal
3. (a) 30 cm; (b) virtual, 60 cm behind mirror
5. (a) Image height is 1/4 object height; (b) inverted
7. (a) 12.4 cm in front of mirror; (b) 3.22 times
9. (a) 24.3 cm behind the mirror; (b) 29 mm; (c) virtual
11. 18 cm in front of mirror
13. 7.42 cm
15. 40 cm
17. (a) 20 cm; (b) reduced to 55% actual size
19. 12 cm
21. 47.7 mm
23. 2.3 mm
27. (a) Real, inverted, 7.7 cm high; (b) virtual, upright, 7.7 cm high
29. 29 cm or 41 cm from the candle
31. 40 cm
35. $\ell' = -67.9$ cm
37. 2
39. 7.56 mm
41. Virtual image, 81 cm from lens on same side as object
43. (a) 40 cm; (b) 160 cm; (c) −170 cm (the lens becomes a diverging lens)
45. (a) 15.3 cm; (b) 1.63
47. Real image, magnified 2.74 times
49. (a) 5.0 m; (b) 37 cm
51. (a) Enlarged 2.9 times at 110 mm
55. 1.85 cm
57. −17.7 cm
59. (a) 2.0 m; (b) 3.0 m in front of the mirror
61. 25 cm
63. 96 cm
65. ($\ell = 85.4$ cm, $m = -1.81$); ($\ell = 155$ cm, $m = -0.552$)
67. Real image, 109 cm on other side of lens
69. (a) 8.31 mm; (b) 92 cm from my eye
71. No, one is virtual, the other real

75. (a) 1.1 m from primary; (b) elliptical

Chapter 37
1. 484 nm
3. (a) 95 cm; (b) 3.8 mm
5. 1.25 mm
7. 17.7 cm, 44.7 cm
9. 0.034°
11. (a) 17.1 cm; (b) 20.0 cm
13. 4
15. (a) 4.8°, 9.7°; (b) 2.9°, 6.8°
17. (a) 8.97°; (b) 51.3°
19. (a) 2nd; (b) 1st
21. 6th order
23. 3200
25. 415 nm, 581 nm
27. Echelle grating has 60% greater resolving power
29. 2000
31. 103 nm
33. 424 nm, 594 nm
35. (a) 75.2 nm; (b) violet
37. 368 nm
39. 5
41. 236
43. 375 nm (i.e., $\frac{3}{4}\lambda$)
45. 545 nm
47. 1022
49. $a/\lambda = 1$
51. 26.6°
53. 0.0162
55. 1.3 m
57. 2.0 μm
59. 6.9 km
61. For diameters greater than 14 cm
63. 7645
65. 0.01 nm
67. 77.2 nm
69. 46 m
71. 484 nm
75. 5542
77. Every 2.8 s
79. 6.0 km/s
81. 3.0 μm

Part 5 Cumulative Problems
1. $m_0 = (n - 1)d(\sin\alpha)/\lambda_0$
3. 3.2 min
5. (b) $x = \dfrac{\sqrt{3}}{2a}\ln\left[2(1 - ay - \sqrt{\tfrac{1}{4} - 2ay + a^2y^2})\right]$

Chapter 38
1. (a) 4.50 h; (b) 4.56 h; (c) 4.62 h
3. (a) $v/c = \tan\theta$, with θ the amplitude of the sine curve. Here $\theta = 20''$ (20/3600 degree), giving $v = 29$ km/s; (b) the orbit is nearly circular
5. 41 cm
7. (a) 8.26 h; (b) 6.28 h
9. (a) 0.857c; (b) 9.69 min
11. 0.14c
13. $c/\sqrt{2}$
15. Twin A: 83 years; twin B: 40 years
17. 0.96c
19. 0.996c
21. (a) B first, by 5.74 min; (b) A first, by 13.1 min; (c) essentially simultaneous
23. A, by 6.4 My
27. (a) 21.7 m; (b) 12.4 m
29. 0.918c
31. 3.4
33. (a) 2.1 MeV; (b) 1.6 MeV
35. 0.36c
37. 5.3×10^{-8} kg·m/s, about the same as the insect
39. 5.9×10^{-30} kg
41. (a) 0.020c; (b) 0.55c; (c) 0.94c; (d) essentially c
47. (a) -0.75×10^{10} ly^2; (b) 99×10^{10} ly^2; events can be causally related only if $(\Delta s)^2 > 0$
49. 0.274c
51. (a) 16 ns; (b) 4.8 m
53. (a) 4.2 ly; (b) −2.4 years (i.e., B occurs earlier)
55. 0.395c
57. 0.9999953c
59. 598 m
61. 0.95c

Chapter 39
1. 16
3. 10 μm
5. 1.4 eV
7. 500 nm
9. 4.1 kK–7.2 kK
11. $R_{200\text{ nm}} = 0.056R_{500\text{ nm}}$
15. 191 nm; ultraviolet
17. $P_{\text{blue}} = 1.44P_{\text{red}}$
19. (a) 1.7×10^{28} s^{-1}; (b) 3.2×10^{15} s^{-1}; (c) 1.3×10^{18} s^{-1}

21. 223 nm
23. (a) 1.12×10^{15} Hz; (b) 2.80 eV
25. Copper, nickel, silicon
27. 17.9%
29. (a) 243 nm; (b) 2.3 eV
31. 295 eV
33. 255 keV
35. β ($n_1 = 5$)
37. 91.2 nm; ultraviolet
39. 253 nm
41. $n = 7$
43. 2.95 eV
45. (a) 26 pm; (b) 40.8 eV
47. (a) 3.7×10^{-63} m; (b) 73 nm
49. $v_p = 5.46 \times 10^{-4} v_e$
51. 151 V
53. 110 eV
55. 6.3×10^7 m/s
57. 126 nm
59. 23 keV
61. 21.1 pm
63. 1 ps
65. (a) 8.2 eV; (b) 6.1×10^{-8}
67. (a) 1.32 μm; (b) $R_{690\,nm} = 0.27 R_{max}$
69. No
71. 9.72 pm
73. (a) 10.2 eV; (b) $n = 2, n = 1$
75. The proton
77. 6×10^5 m/s
79. 18 pm
85. $\frac{1}{2} m_e c^2 [(\gamma - 1) + \sqrt{(\gamma - 1)(\gamma + 3)}]$

Chapter 40
1. $m^{-1/2}$
3. $1/\sqrt{a}$
5. $A = \sqrt{\dfrac{\sqrt{2}}{a\sqrt{\pi}}}$
7. $U = \infty$ for $|x| > b$;
   $U = E - \hbar^2/m(b^2 - x^2)$
   for $-b \le x \le b$
9. 1.43 pm
11. Electron
13. (a) 3.48×10^{-19} J $= 2.18$ eV;
    (b) 571 nm
15. 21.3 μm
17. $E_1 = 200$ MeV
19. 2.38 nm
21. At $x = \frac{1}{4} L$ or $\frac{3}{4} L$
23. 0.196

25. 33.0 eV
27. 493 N/m
29. $n > 99$
33. (a) 0.168; (b) $10^{-7.9 \times 10^{36}}$
35. 9.87×10^{-11} J $= 616$ MeV
39. (a) 1×10^{10} K; (b) 2×10^{13} K
41. 0.23 MeV
43. 0.987
45. 2.66×10^{14} Hz
47. 0.93 nm
49. $n = 3$
51. (a) 4; (b) 2
57. (a) No bound states; (b) one bound
    state, at $\varepsilon = 6.44$; (c) two bound
    states, at $\varepsilon = 7.53$, $\varepsilon = 29.3$

Chapter 41
3. (a) $1/\pi a_0^3 e = 7.91 \times 10^{29}$ m^{-3};
   (b) $1/a_0 e = 6.95 \times 10^9$ m^{-1}
5. $\sqrt{42}\hbar$
7. -1.51 eV
9. $n = 4, \ell = 3$
11. 1.7×10^{40}
13. $3d$
15. (a) Possible with $n = 5, \ell = 3$;
    (d) possible with $n = 2, \ell = 1$
17. (a) 3.66×10^{-34} J·s;
    (b) 3.17×10^{-34} J·s
23. (a) $\frac{\sqrt{15}}{2}\hbar$; (b) 4
25. $\frac{3}{2}, \frac{5}{2}$
29. (a) 2 at $n = 1$, 2 at $n = 2$, 1 at
    $n = 3$; (b) $19h^2/8mL^2$
31. (a) 5; (b) $\frac{5}{2}\hbar\omega$
33. $1s^2 2s^2 2p^6 3s^2 3p^6 3d^1 4s^2$
35. $1s^2 2s^2 2p^6 3s^2 3p^6 3d^{10} 4s^1$
37. $n_2 = 3$
39. 2.19 eV
41. 3.193 eV
43. 0.34 nm
45. 980 V
47. (a) 18.7 eV; (b) 9.5%
49. (a) 645 nm ($4 \to 3$),
    502 nm ($5 \to 4$),
    410 nm ($6 \to 5$); (b) 2
51. 0.11
53. $\sqrt{2}\hbar$
55. (a) $3D_{5/2}$; (b) 6
57. 1.09 keV
63. (b) He: 54.4 eV; O: 870 eV;
    Fe: 9.19 keV; Pb: 91.4 keV;
    U: 115 keV

Chapter 42
1. 0, 2.63 meV, 7.89 meV, 15.8 meV
3. 1.25 meV
5. 9.40×10^{-46} kg·m^2
7. $\hbar^2 j/I$
9. 0.121 nm
11. (a) 0.179 eV; (b) 0.359 eV
13. 7.07×10^{13} Hz
15. (a) 4.73×10^{13} Hz;
    (b) 1.95×10^{-46} kg·m^2
17. 181 kcal/mol
19. 10.2
21. (a) $F = \dfrac{\alpha k e^2}{r_0^2} \left[\left(\dfrac{r_0}{r} \right)^{n+1} - \left(\dfrac{r_0}{r} \right)^2 \right]$;
    (b) 3.0×10^{-6} N, about 260 times
    the electrostatic attraction
23. 4.7 eV
25. 1 meV
27. 6.35×10^4 K, over 200 times room
    temperature
29. 345 nm; ZnS
31. 1.91 eV
33. (a) 345 nm; (b) no; not much of the
    solar energy will be absorbed
35. 2.78 eV
37. 1.79 kA
39. 15.2 meV
41. 35.8 μm
43. -8.40 eV
45. 9.41 eV
47. $E_{min} = 15$ meV; $kT = 8.6$ meV
49. 4.1 K; no, this is below the boiling
    point of O_2
51. (a) 2.87×10^{-20} J $= 179$ meV;
    (b) 8 pm; (c) about 6% of bond
    length

Chapter 43
1. 25 fm
3. $^{211}_{86}$Ra, $^{220}_{86}$Ra, $^{222}_{86}$Ra
5. (a) same; (b) $q_K = q_{Cl} + 2e$
7. $Z = 43$, technetium and $Z = 61$,
   promethium
9. $^8_5$B
11. 5.9 fm
13. 5.3×10^{-12} eV
15. 60 MHz
17. 59.93 u
19. 187 MeV
21. 5.61 MeV/nucleon
23. 0.037%
25. 8.3 s

27. (a) $^{222}_{86}\text{Rn} \rightarrow {}^{218}_{84}\text{Po} + {}^{4}_{2}\text{He}$;
    (b) $^{214}_{82}\text{Pb} \rightarrow {}^{214}_{83}\text{Bi} + e^- + \bar{\nu}$
29. 16.7 Bq/L
31. (a) 7.85×10^5, essentially 0;
    (b) 1.009, 5.0×10^5;
    (c) 3.6, 8.2×10^4
33. 7.5×10^5
35. 4.3 days
37. Slope $= -\log_{10}e/\tau$, with τ the mean lifetime
39. 3.1×10^{17} atoms = 31 μg
41. 2.78 GBq = 75.2 mCi
43. At $t = 45.8$ h
45. 3.0×10^9 y
47. (a) 19.5 GBq; (b) 1.22 GBq;
    (c) 138 Bq; (d) 62.8 Bq;
    (e) 62.8 Bq
49. 23%
51. 2: $^{3}_{2}\text{He}$, $^{4}_{2}\text{He}$; 8: $^{16}_{8}\text{O}$, $^{15}_{7}\text{N}$; 20: $^{40}_{20}\text{Ca}$, $^{39}_{19}\text{K}$; 28: $^{58}_{28}\text{Ni}$, $^{52}_{24}\text{Cr}$; 50: $^{119}_{50}\text{Sn}$, $^{89}_{39}\text{Y}$; 82: $^{208}_{82}\text{Pb}$, $^{138}_{56}\text{Ba}$; 126: no magic Z, $^{208}_{82}\text{Pb}$
53. 8.7 MeV/nucleon
55. 0.076%
57. 1.68 mCi/L
59. 1.2×10^6 m/s^2, more than 10^5 times g_{Earth}
61. 0.072 K
63. 4.3 MeV
65. (a) $^{210}_{83}\text{Bi} \rightarrow {}^{210}_{84}\text{Po} + e^- + \bar{\nu}$;
    (b) $^{210}_{84}\text{Po} \rightarrow {}^{206}_{82}\text{Pb} + {}^{4}_{2}\text{He}$
67. $^{262}_{107}\text{Ns}$
69. (a) copper-65; (b) 4.0 h

Chapter 44

3. 8.2×10^{13} J/kg for fission; 3.4×10^{14} J/kg for fusion; fusion is about 4 times greater
5. 2
7. 1.0×10^{20} fissions/s
9. 263 kt
11. 1.02 kg
13. (a) 25.6 kg; (b) 13.8 cm
15. (a) 142; (b) 1.42 μs
17. 3.86 km/s
19. 88.9%
21. 31 h
23. 69 μs
25. 0.46 s
27. $\Delta k = 2.3 \times 10^{-5}$
29. 23 MK
31. 2×10^{20} m^{-3}
33. 454 kg
35. 10^{30} m^{-3}
37. (a) 3.7×10^{38} proton/s;
    (b) 7.3×10^9 years
39. 4.4×10^{10} J; (b) 340 gal
41. 576 kg
43. $N_B = 33N_A$
45. (a) 8.6×10^5 m/s; (b) 4×10^{13} s/m^3, less than 1 millionth of the Lawson criterion

47. (a) 1.1 GJ; (b) 2.8 pellets/s;
    (c) 450 kg
49. 1.28
51. 6.0 kg fusion fuel; 26 kg ^{238}U
53. Area $\propto$ yield$^{2/3}$
55. (a) 783 kg; (b) 157

Chapter 45

1. 3×10^{-16} s
3. 7×10^{19} m
5. $S = -1$ for Λ^0; $S = 0$ for decay products; weak force
7. Both
9. (a), violates conservation of baryon number
11. (a) No; (b) yes
13. (a) Stays same; (b) changes sign
15. sss
17. $c\bar{c}$
19. 3.6×10^3 m^3, about 1 million gallons
21. 10^{28} K
23. (a) 7463; (b) 0.9999999910 c
25. 1.6×10^{-18} s^{-1}
27. 4.5×10^3 km/s
29. 18 Gy
31. 2×10^{-26} kg/m^3
33. (a) 256 fm; (b) -2.81 keV
35. 12 ns
37. 1.07 mm
39. 2.6×10^{-25} s

Credits

Page abbreviations are as follows: (T) top, (C) center, (B) bottom, (L) left, (R) right.

429BR Courtesy of T.D. Rossing/Northern Illinois University

430 Courtesy of Fox Products, South Whitley, IN

431 From PSSC Physcis, 2nd ed., 1965 D.C. Heath and Co. with Education Development Center

432 ©Camerique/H. Armstrong Roberts

435L Courtesy of Dr. Gary Settles

435C ©The Harold E. Edgerton 1992 Trust/Palm Press

435R ©H. Armstrong Roberts

437 ©David R. Frazier Photolibrary

439L ©Greg Cranna/Stock Boston

439R ©Philippe Plailly/SPL/Photo Researchers, Inc.

441 ©Peter Rauter/Tony Stone Images

447 ©Bill Gallery/Stock Boston

448CL ©J.Cochin//Explorer/Photo Researchers, Inc.

448CR ©Fred McConnaughey/Photo Researchers, Inc.

448B Courtesy of Lawrence Livermore National Laboratory

449L ©Fredrik D. Bodin/Stock Boston

449R ©W.Eastep/The Stock Market

450 ©Wolfgang Kaehler

452TR Courtesy of Mercedes Benz

452TL ©Diane Schiumo/Fundamental Photographs

452C Courtesy of NASA

454 ©Dan McCoy/Rainbow

458T ©Tod Planck/Tony Stone Images

458B ©Larry Mulvehill/Photo Researchers, Inc.

459T ©J. Toers/George Hall Photos

459B ©Globus Brothers

460 ©Peter Menzel/Stock Boston

461 ONERA photo, H. Werle 1974, from Milton van Dyke, An Atlas of Fluid Motion

464B ©Bruce Thomas/The Stock Market

471 ©Ray Ellis/Photo Researchers, Inc.

472 ©NASA/SS/Photo Researchers, Inc.

474C ©Jay M. Pasachoff

474B ©1998 PhotoDisc, Inc.

477 ©G.R. Roberts

478 ©Robert Fried/Stock Boston

483 ©John Wilson White/Addison Wesley Longman

484 ©Manuel G. Velarde

485T ©Keith Wood/Tony Stone Images

485B ©Camerique/H. Armstrong Roberts

496 ©Paolo Koch/Photo Researchers, Inc.

505 ©Alexander Tsiaras/Photo Researchers, Inc.

507C ©Paul Silverman/Fundamental Photographs

507B ©Richard Megna/Fundamental Photographs

511T Courtesy of the Hartford Steam Boiler and Insurance Company

511C ©AP/Wide World

516 ©Bruce M. Wellman/Stock Boston

517 ©Jim Richardson/H. Armstrong Roberts

519 ©Jan Halaska/Photo Researchers, Inc.

523 ©David R. Austen/Stock Boston

527 Courtesy of Price Weber

529T ©Geoff Juckes/The Stock Market

529B ©Thomas Ives/The Stock Market

543 ©Sinclair Stammers/SPL/Photo Researchers, Inc.

546L ©David R. Frazier Photolibrary/Photo Researchers, Inc.

546R Courtesy of the Environmental Protection Agency

550 ©Rick Browne/Stock Boston

551 ©Martin Bond/SPL/Photo Researchers, Inc.

552T Courtesy of Lockheed

552B ©Hank Morgan/Photo Researchers, Inc.

569 ©Astrid and Hans Frieder Michler/SPL/Photo Researchers, Inc.

570 ©Gary Ladd

571L ©Richard Megna/Fundamental Photographs

571R Courtesy of Intel

573T ©Lawrence Berkeley Laboratory/SPL/Photo Researchers, Inc.

573C ©Richard Megna/Fundamental Photographs

928 ©Simon Fraser/SPL/Photo Researchers, Inc.

930 ©Bruno J. Zehnder/Peter Arnold, Inc.

931 ©Leonard Lessin/Peter Arnold, Inc.

933L Courtesy of the Library of Congress

933R Courtesy of James E. Kettler, Ohio University, Belmont

938 ©Dan McCoy/Rainbow

941CR ©Ken Kay/Fundamental Photographs

941BL ©Tony Craddock/SPL/Photo Researchers, Inc.

941BR ©Richard Megna/Fundamental Photographs

942C Courtesy of NASA

942B Courtesy of the Space Telescope Institute/NASA

943 ©Erik Borg/Addison Wesley Longman

944 Courtesy of The Exploratorium

946 ©Perkin Elmer

947 Courtesy of Roland Winston/Argonne National Laboratory

948C ©David Parker/SPL/Photo Researchers, Inc.

948B ©David Parker/SPL/Photo Researchers, Inc.

950 ©Erik Borg/Addison Wesley Longman

952 ©Erik Borg/Addison Wesley Longman

957T Courtesy of The Exploratorium

957L ©Christopher Johnson/The Stock Market

957R Courtesy of 3M

958 ©OMIKRON/Photo Researchers, Inc.

961 ©Chris Rogers/Rainbow

964 ©Roger Ressmeyer/Starlight

966 Courtesy of Jay M. Pasachoff

967L Courtesy of Jay M. Pasachoff

967R Courtesy of The Exploratorium

968 Courtesy of Yerkes Observatory

971 ©1998 PhotoDisc, Inc.

974 Courtesy of the US Navy

975 © Erik Borg/Addison Wesley Longman

976 ©Erik Borg/Addison Wesley Longman

979T ©Chris Jones

979B ©J. Amos/H. Armstrong Roberts

982 ©CSHL/Peter Arnold, Inc.

985 ©Richard Megna/Fundamental Photographs

987B ©Erik Borg/Addison Wesley Longman

987C Courtesy of Jay M. Pasachoff

994 Courtesy of Cagnet Atlas

995 Courtesy of Cagnet Atlas

997 ©Chris Jones

998 Courtesy of the US Navy

999 ©NRAO/AUI

1001 ©C.C. Jones/Union

1004L Courtesy of the Association of Universities for Resesarch in Astronomy

1004R Courtesy of NASA

1007 Courtesy of Almaden Research Center/Courtesy of IBM.

1008 ©UPI/Corbis Bettmann

1013 ©UPI/Corbis Bettmann

1021 Courtesy of the Stanford Linear Accelerator

1032 ©Brookhaven National Laboratory/SPL/Photo Researchers, Inc.

1037 Courtesy of the Space Telescope Science Institute, operated by the Association of Universities for Research in Astronomy, Inc.

1041 Courtesy of David Malin/Anglo-Australian Telescope Board

1042L ©C. Powell and D. Perkins/SPL/Photo Researchers, Inc.

1042R ©John R. McGregor/Peter Arnold, Inc. Standard

1044 Courtesy of CERN

1046 Courtesy of The Exploratorium

1049T Courtesy of R. Giovanelli and H.R. Gillett/ CSIRO National Measurement Laboratory, Australia

1049B Courtesy of Alfred Leitner/Rensselaer Polytechnic Institute

1050 Courtesy of Margrethe Bohr Collection, AIP

1052T ©Skelton Photography

1052B ©Paul Shambroom/Photo Researchers, Inc.

1053L Courtesy of AT&T Bell Laboratories

1054 Courtesy of the Meggers Gallery, AIP

1057T Courtesy of the Institute of Theoretical Physics, Lund, Sweden

1057B Courtesy of Carl D. Anderson/California Institute of Technology

1058 ©AP/Wide World

1060T ©U.S. Navy/SPL/Photo Researchers, Inc.

1060B Courtesy of Ullstein Bilderdienst

1061 Courtesy of Ullstein Bilderdienst

1062 Courtesy of the Stein Collection, AIP

1064L Courtesy of Fermilab Visual Media Services

1064R Courtesy of CERN

1065T Courtesy of Ralph Crane

1065B ©Patrice Loiez/CERN/SPL/Photo Researchers, Inc.

1067 ©Joe Stancampiano and Karl Luttrell/National Geographic Society

1071 ©Dan McCoy/Rainbow

1072R ©Mark Marten/Los Alamos National Laboratory/Photo Researchers, Inc.

Extended

1044 ©Wexler/Burgess/SPL/Photo Researchers, Inc.

1046 Courtesy of The Exploratorium

1057T Courtesy of R. Giovanelli and H.R. Gillett/CSIRO National Measurement Laboratory, Australia

1057B Courtesy of Alfred Leitner/Rensselaer Polytechnic Institute

1058 Courtesy of the AIP Niels Bohr Library, photo by Paul Ehrenfest

1063L ©Prof. P.M. Motta/SPL/Photo Researchers, Inc.

1063R ©CNRI/SPL/Photo Researchers, Inc.

1064 Courtesy of AT&T Bell Laboratories

1068 Courtesy of the AIP Niels Bohr Library

1074 Courtesy of the Lawrence Livermore National Laboratory

1076 Courtesy of Elisha Huggins, Dartmouth

1077T From C. Jonsson, Zeitscrift fer Physik, Springer-Verlag, Berlin, Heidelberg, New York

1077B Courtesy of the AIP, Meggers Gallery

1079 Courtesy of Paul Ehrenfest

1088 Courtesy of Don Eigler, IBM Almaden

1089L Courtesy of IBM

1089R Courtesy of IBM

1090L Courtesy of Don Eigler, IBM Almaden

1090R Courtesy of R.M. Penner, University of California, Irvine

1094 Courtesy of Fermilab Visual Media Services

1100 ©Ken Eward/Photo Researchers, Inc.

1113 Courtesy of the National Radio Astronomy Observatory

1115T Courtesy of Mike Matthews/JILA

1115B ©Richard Megna/Fundamental Photographs

1121T Courtesy of the Astronomical Data Center (ADC) catalog A6016, by Reader J., Corliss Ch.H.:1981, 'Line Spectra of the Elements', CRC Handbook of Chemistry and Physics; NSRDS-NBS 68

1121B Courtesy of Osram/Sylvania, Inc.

1124 ©SPL/Photo Researchers, Inc.

1126 Courtesy of D.A. Calvert/Royal Greenwich Observatory

1127TL ©Skelton Photography

1127TC ©Dan McCoy/Rainbow

1127TR ©Paul Shambroom/Photo Researchers, Inc.

1127C ©Dagmar Schilling/Peter Arnold, Inc.

1127B ©Dan McCoy/Rainbow

1133 ©Science Source/Photo Researchers, Inc.

1136 ©Geoff Tomplinson/SPL/Photo Researchers, Inc.

1150T ©1998 PhotoDisc, Inc.

1150B ©John Mead/SPL/Photo Researchers, Inc.

1152L ©IBM Research/Peter Arnold, Inc.

1152C Courtesy of Brookhaven National Laboratory and NYU Medi-Cal Center

1153T Courtesy of Conductus

1153B Courtesy of the American Superconductor Corp.

Index